Electronic Communication Techniques

Third Edition

Paul H. Young, P.E.
Tacan Corporation

Merrill, an imprint of
Macmillan Publishing Company
New York

Maxwell Macmillan Canada
Toronto

Maxwell Macmillan International
New York   Oxford   Singapore   Sydney

Cover art/photo: NASA (Media Services Corp.)
Editor: Dave Garza
Developmental Editor: Monica Ohlinger
Production Editor: Colleen Brosnan
Art Coordinator: Ruth A. Kimpel
Text Designer: Susan E. Frankenberry
Cover Designer: Thomas Mack
Production Buyer: Pamela D. Bennett
Illustrations: Precision Graphics

This book was set in Times Roman by Bi-Comp, Inc. and was printed and bound by R. R. Donnelley & Sons Company. The cover was printed by Lehigh Press, Inc.

Macmillan Publishing Company
866 Third Avenue
New York, NY 10022

Macmillan Publishing Company is part of the Maxwell Communication Group of Companies.

Maxwell Macmillan Canada, Inc.
1200 Eglinton Avenue East, Suite 200
Don Mills, Ontario M3C 3N1

The Publisher offers discounts on this book when ordered in bulk quantities. For more information, write to:
  Special Sales Department
  Macmillan Publishing Company
  445 Hutchinson Avenue
  Columbus, OH 43235
or call
  1-800-228-7854

Photo credits: p. 59, courtesy of Western Union Corporation; p. 142, courtesy of AT&T Bell Laboratories; pp. 307–8, courtesy of Cubic Communications, Inc.; p. 606, courtesy of Hewlett-Packard Company; p. 617, provided by the author; pp. 655–56, courtesy of Comtech Telecommunications Corp.

Library of Congress Cataloging-in-Publication Data
Young, Paul H., P.E.
    Electronic communication techniques / Paul H. Young. — 3rd ed.
        p.   cm.
    Includes bibliographical references and index
    ISBN 0-02-431201-0
    1. Telecommunication. 2. Electronics. I. Title.
  TK5105.Y68 1994
  621.382—dc20   93-1028
                                          CIP

Printing: 1 2 3 4 5 6 7 8 9   Year: 4 5 6 7

MERRILL'S INTERNATIONAL SERIES IN ENGINEERING TECHNOLOGY

Zanger & Zanger, *Fiber Optics: Communication and Other Applications*, 0-675-20944-7

Microcomputer Servicing

Adamson, *Microcomputer Repair*, 0-02-300825-3

Asser, Stigliano, & Bahrenburg, *Microcomputer Servicing: Practical Systems and Troubleshooting, 2nd Edition*, 0-02-304241-9

Asser, Stigliano, & Bahrenburg, *Microcomputer Theory and Servicing, 2nd Edition*, 0-02-304231-1

Programming

Adamson, *Applied Pascal for Technology*, 0-675-20771-1

Adamson, *Structured BASIC Applied to Technology, 2nd Edition*, 0-02-300827-X

Adamson, *Structured C for Technology*, 0-675-20993-5

Adamson, *Structured C for Technology (with disk)*, 0-675-21289-8

Nashelsky & Boylestad, *BASIC Applied to Circuit Analysis*, 0-675-20161-6

Instrumentation and Measurement

Berlin & Getz, *Principles of Electronic Instrumentation and Measurement*, 0-675-20449-6

Buchla & McLachlan, *Applied Electronic Instrumentation and Measurement*, 0-675-21162-X

Gillies, *Instrumentation and Measurements for Electronic Technicians, 2nd Edition*, 0-02-343051-6

Transform Analysis

Kulathinal, *Transform Analysis and Electronic Networks with Applications*, 0-675-20765-7

Biomedical Equipment Technology

Aston, *Principles of Biomedical Instrumentation and Measurement*, 0-675-20943-9

Mathematics

Monaco, *Essential Mathematics for Electronics Technicians*, 0-675-21172-7

Davis, *Technical Mathematics*, 0-675-20338-4

Davis, *Technical Mathematics with Calculus*, 0-675-20965-X

INDUSTRIAL ELECTRONICS/INDUSTRIAL TECHNOLOGY

Bateson, *Introduction to Control System Technology, 4th Edition*, 0-02-306463-3

Fuller, *Robotics: Introduction, Programming, and Projects*, 0-675-21078-X

Goetsch, *Industrial Safety and Health: In the Age of High Technology*, 0-02-344207-7

Goetsch, *Industrial Supervision: In the Age of High Technology*, 0-675-22137-4

Geotsch, *Introduction to Total Quality: Quality, Productivity, and Competitiveness*, 0-02-344221-2

Horath, *Computer Numerical Control Programming of Machines*, 0-02-357201-9

Hubert, *Electric Machines: Theory, Operation, Applications, Adjustment, and Control*, 0-675-20765-7

Humphries, *Motors and Controls*, 0-675-20235-3

Hutchins, *Introduction to Quality: Management, Assurance, and Control*, 0-675-20896-3

Laviana, *Basic Computer Numerical Control Programming*, 0-675-21298-7

Pond, *Fundamentals of Statistical Quality Control*

Reis, *Electronic Project Design and Fabrication, 2nd Edition*, 0-02-399230-1

Rosenblatt & Friedman, *Direct and Alternating Current Machinery, 2nd Edition*, 0-675-20160-8

Smith, *Statistical Process Control and Quality Improvement*, 0-675-21160-3

Webb, *Programmable Logic Controllers: Principles and Applications, 2nd Edition*, 0-02-424970-X

Webb & Greshock, *Industrial Control Electronics, 2nd Edition*, 0-02-424864-9

MECHANICAL/CIVIL TECHNOLOGY

Dalton, *The Technology of Metallurgy*, 0-02-326900-6

Keyser, *Materials Science in Engineering, 4th Edition*, 0-675-20401-1

Kokernak, *Fluid Power Technology*, 0-02-305705-X

Kraut, *Fluid Mechanics for Technicians*, 0-675-21330-4

Mott, *Applied Fluid Mechanics, 4th Edition*, 0-02-384231-8

Mott, *Machine Elements in Mechanical Design, 2nd Edition*, 0-675-22289-3

Rolle, *Thermodynamics and Heat Power, 4th Edition*, 0-02-403201-8

Spiegel & Limbrunner, *Applied Statics and Strength of Materials, 2nd Edition*, 0-02-414961-6

Spiegel & Limbrunner, *Applied Strength of Materials*, 0-02-414970-5

Wolansky & Akers, *Modern Hydraulics: The Basics at Work*, 0-675-20987-0

Wolf, *Statics and Strength of Materials: A Parallel Approach to Understanding Structures*, 0-675-20622-7

DRAFTING TECHNOLOGY

Cooper, *Introduction to VersaCAD*, 0-675-21164-6

Ethier, *AutoCAD in 3 Dimensions*, 0-02-334232-3

Goetsch & Rickman, *Computer-Aided Drafting with AutoCAD*, 0-675-20915-3

Kirkpatrick & Kirkpatrick, *AutoCAD for Interior Design and Space Planning*, 0-02-364455-9

Kirkpatrick, *The AutoCAD Book: Drawing, Modeling, and Applications, 2nd Edition*, 0-675-22288-5

Kirkpatrick, *The AutoCAD Book: Drawing, Modeling, and Applications, Including Release 12, 3rd Edition*, 0-02-364440-0

Lamit & Lloyd, *Drafting for Electronics, 2nd Edition*, 0-02-367342-7

Lamit & Paige, *Computer-Aided Design and Drafting*, 0-675-20475-5

Maruggi, *Technical Graphics: Electronics Worktext, 2nd Edition*, 0-675-21378-9

Maruggi, *The Technology of Drafting*, 0-675-20762-2

Sell, *Basic Technical Drawing*, 0-675-21001-1

TECHNICAL WRITING

Croft, *Getting a Job: Resume Writing, Job Application Letters, and Interview Strategies*, 0-675-20917-X

Panares, *A Handbook of English for Technical Students*, 0-675-20650-2

Pfeiffer, *Proposal Writing: The Art of Friendly Persuausion*, 0-675-20988-9

Pfeiffer, *Technical Writing: A Practical Approach, 2nd Edition*, 0-02-395111-7

Roze, *Technical Communications: The Practical Craft, 2nd Edition*, 0-02-404171-8

Weisman, *Basic Technical Writing, 6th Edition*, 0-675-21256-1

Preface

Electronic Communication Techniques is intended to bridge the gap between circuit design and the system concepts that predetermine circuit requirements in particular applications. The results of theoretical research are combined with engineering principles, design equations, charts, and tables for those of us who will design and produce hardware and software. The mathematics level is typical of that used by engineering practitioners, with calculus and transform techniques occasionally employed.

Enough circuit detail and topical coverage have been included to provide material for two or even three courses in analog and digital communications technology, depending upon the depth and pace desired. Use in a one-term course is also feasible, as is shown shortly. Any sequence of topics needed for a particular program may be selected.

The second edition included changes in every chapter. The most pervasive change was introduction of computer-aided engineering (CAE) software for analysis, design, and problem solving. Other major second-edition changes included the introduction of Laplace and Fourier analysis, correlation and noise, and design and simulation of mixers, filters, and microstrip amplifiers. New sections were added on such system applications as packet switching with some queuing theory and basic network engineering analysis, OSI and the Integrated Services Digital Network (ISDN), satellite multiple-access techniques, spread spectrum, and local, national, and international fiber-optic networks.

The main thrust of the third edition is to improve readability and consistency of nomenclature, especially in the early chapters, and then concentrate in the last few chapters on providing updates of the latest technology. Thus, Chapter 16 on television has a new section on high-definition television (HDTV) and video compression with motion compensation. Chapter 17 has an improved description of quadrature amplitude modulation (QAM), which is a very important digital modulation technique used in HDTV. Chapter 18 has been entirely rewritten due to the rapid improvements and deployment of lasers, fiber, fiber-optic amplifiers, and fiber-optic-based networking.

Chapter 1 provides a review of tuned circuits and amplifiers for continuity with previous courses. There is more material on low-noise amplifiers and saturation characteristics in the receiver circuits chapter (Chapter 7). RF oscillators in Chapter 2 are considered as stable tuned amplifiers with well-defined feedback arrangements. Oscillators also provide the carrier signal for subsequent modulation in transmitters.

Development of the frequency domain and the signal spectra of common periodic waveforms in Chapter 3 includes Fourier transforms and Fourier series analysis. Noise fundamentals in Chapter 4 include a typical treatment enhanced with circuit noise power analysis using shot and thermal noise spectral density techniques. System noise figure and temperature are included, as is the concept of partial correlation of uncorrelated random noise due to finite bandwidth circuits. A computer simulation reinforces the theory.

A detailed derivation of the equation for the system noise figure is presented in Chapter 5, which also introduces communication systems and amplitude modulation. Other aspects of noise and signal-to-noise ratio are introduced in the context of their effects on analog receivers in Chapter 5 and in their statistical nature as it affects transmission error in pulsed and digital data systems in Chapter 13. The continuous information signals discussion in Chapter 5 along with the digital information signal discussion of Chapter 12 constitute a broad overview of information theory from a nonstatistical perspective.

Chapters 6, 7, and 9, as well as Chapters 1 and 2, provide most of the analog circuit analysis and design component of the textbook. The circuit details may be skipped by those who have time only for system-level concepts. Chapters 5 and 8 and the first five sections of Chapter 7 provide coverage for receiver systems and AM/AM-sideband communication. Mixer circuit design is presented in Chapter 7, along with a section on Butterworth and Chebyshev bandpass filter circuit design. As is the case with all the chapters that include circuit analysis and design, computer circuit simulations and design program examples are included.

Chapter 8 completes the system-level discussion of amplitude modulation begun in Chapter 5. In addition to various sideband systems, Chapter 8 presents frequency-division multiplex and quadrature-multiplex concepts, including Costas loop analysis for the demodulation of quadrature-multiplexed signals. Chapter 9 covers frequency and phase modulation circuits and systems. The voltage-controlled oscillator, including integrated-circuit VCOs, and the phase detector of Chapter 9 are combined to produce the phase-locked loop of Chapter 10. Phase-locked loop applications in communication and instrumentation are so widespread that a full chapter is devoted to them. The most successful approach is to present PLL basics (through Section 10–2) in an introductory communication course and then discuss loop dynamics with FSK and synthesizers in a later course.

AM, FM, and PM basics are the same today as they have always been, but they will be implemented in digital form as ASK, OOK, FSK, PSK, and compound modulations such as QAM. Chapters 11, 12, 13, 17, and 18 address these topics, with Chapters 11 and 12 forming the conceptual bridge for the applications in Chapters 13, 17, and 18. Sampling theory and intersymbol interference concepts are presented in Chapter 11, as are information theory concepts and digital signal power and quantization noise equations of Chapter 12. Chapter 13 presents data communication in a noise environment with techniques for detecting and correcting data errors, a table of some standard data modems, a short introduction to queuing and basic digital network engineering relationships, OSI, ISDN, and a probability-of-error (bit error rate) calculation program.

Chapters 14 and 15 cover transmission lines and waveguides, antennas, and propagation. Stub-tuner design, complex-impedance S-parameter 2-port problems with simulations on the Touchstone microwave and RF circuit-simulator software providing a good lead-in to a section on microstrip amplifier design. Another Touchstone simulation confirms the microstrip design results.

Chapter 16 on television has been broadened beyond North American/Japan's NTSC with the inclusion of PAL and SECAM format standards. 1993 saw the launching of digital high-definition television (HDTV) in the United States. The third edition includes a full section on the digital video compression and motion-compensation techniques that made HDTV possible. A new section also discusses all the digital techniques suggested by HDTV, including transmission formats and techniques.

Chapter 17 presents circuit and systems technology for digital radio, space transmission systems with applications for QPSK, and multilevel QAM. The increasingly important areas of the satellite multiple-access techniques of FDM TDMA, and CDMA as well as spread-spectrum system concepts are also in in Chapter 17.

Fiber-optic communication-link applications are included in Ch chapter has been rewritten to include ultra linear DFB lasers wi and thermal electric coolers, Mach-Zehnder external modulators and dispersion-shifted fiber technology. New sections present t ters of passive optical couplers and analog fiber-optic link a OMI, laser, and carrier-to-noise performance. An expansion o especially FDDI with dual counter-rotating rings for self-he networking, augments SONET networks.

A one-semester survey of communication transmission the following outline (by chapter numbers): 4, "Elements of tion and AM Systems"; 6, "Transmitter Circuits," Section circuits; 7, through Section 7–3 on receiver circuits; 8, "Frequency and Phase Modulation"; 10, through Se loops; 11, "Pulse and Digital Modulation"; 12, th communications; 13, "Data Communication T Lines and Waveguides"; 15, "Antennas and through Section 17–1 on modems and digital modu PSK and QAM); 18, "Fiber Optic Communication."

A one-year, two-course sequence could be structured as follows:

- Course 1, *Electronic Communications* (4 hours): Chapters 1, 2, 3, 4, 5, 6, 7, 8, 9, 10
- Course 2, *Digital, Satellite, and Optical Communications* (4 hours): Chapters 11, 12, 13, 14, 15, 16, 17, 18

If there is one course devoted to systems and a second course devoted to circuits, you might need to skip around a little more (moving Chapters 6, 7, and 9 to the second course, for example). This book was written to allow just such separation, if desired, by keeping system and circuit coverage in balance.

Acknowledgments

Knowledge is cumulative. It is not based on each person's reinventing everything that is known, but rather on accumulating what has been learned in the past, synthesizing new ideas from old formulas and principles, and creating completely new insights. I hope elements of all three can be found in these pages, and I want to acknowledge the contributions of all those whose work has added to my understanding of the various topics in this book.

I want to thank those who reviewed the third edition of this manuscript: Thomas K. Grady, Western Washington University; A. Kisha, DeVry Institute of Technology—Columbus; Jim Stewart, DeVry Institute of Technology—Woodbridge, and Melvin C. Vye, University of Akron.

I also want to acknowledge those who reviewed the second edition manuscript: Wayne Brown, DeKalb Technical College; Leonard Bundra, Lincoln Technical Institute; Charles Corkhill, ITT Technical Institute; Joseph Ennesser, DeVry Institute of Technology; Gary Granneman, California Polytechnical State University–San Luis Obispo; F. David Harris, Purdue University–Calumet; Alex Kisha, DeVry Institute of Technology—Columbus; Jack Leischman, Northwestern Michigan College; and Dr. Lee Rosenthal, Fairleigh Dickinson University.

A special thanks goes to two friends and former students: Mark Tucker, who initiated me to SPICE and provided some of the specialized circuit analyses; and Kenneth Field, without whose computer wizardry the computer material would not have been completed.

I would like to thank these companies for providing diagrams and specification sheets for this book: National Semiconductor; Motorola Semiconductor Products; RCA, Solid State Division; J. W. Miller Company; Cubic Communications (my former associates); Hewlett-Packard; AT&T; Western Electric; ITT; IBM; DEC; and AMI.

Finally, I am grateful to my former associates at Cubic Corporation in San Diego, where for ten years I gained the insight that real understanding comes from experience rather than from a textbook.

But the most deeply felt appreciation is for the patience, love, and understanding of my wife Beryl and my children James, John, and Sara (listed alphabetically!). The importance of this support is immeasurable. In addition to all that, Beryl typed the original manuscript with ''all those awful equations,'' wrote the solutions manuals, and has most recently endured my office spreading to the living room.

Contents

3   Signal Spectra   85

4   Elements of Noise   114

14   Transmission Lines and Waveguides   585

15   Antennas and Radiowave Propagation   636

16

Basic Television   669

1

Radio Frequency Amplifiers

■ INTRODUCTION

Electronic communication involves the study of how information is transferred from one place to another over electronic channels. The emphasis in this text will be on the techniques for transferring the electrical signals.

Information in the form of sound or visual images is processed into electrical signals by transducers. A microphone is the transducer for converting energy in the form of sound pressure waves into electrical energy; a television camera is the transducer for conversion of visual images into electrical video signals. Further electronic processing can convert these signals into a digital format.

Once the information is processed into suitable electrical signals, we must determine how these signals can be transmitted over long distances and then made

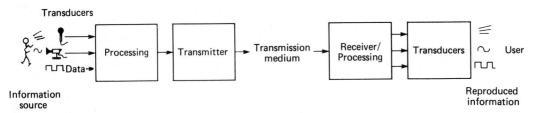

FIGURE 1-1   Information transfer system block diagram.

available for one or millions of users. There are numerous approaches and technologies for accomplishing this feat, but the most general breakdown of the basic elements involved in a single channel of information transfer is shown in Figure 1-1.

This figure illustrates, in a block-diagram form, information gathered, transformed into electrical signals, and processed for transmission. Typical transmission media are transmission lines or optical (light) cable, and space—from antenna to antenna. The distances and broadcast coverage involved usually result in vast energy loss, thus requiring very sensitive receiving and processing systems before the signals can be reproduced into the desired form for the user. Our objective will be to break down each block of Figure 1-1 and study the requirements and techniques used to realize typical communication systems.

Most of the information being communicated over electronic networks falls in a frequency range below 4 MHz. The information is found in discrete and continuous form. However, the transmission facilities and circuits are primarily analog (that is, continuous) and can usually be characterized by radio frequency (RF) sinusoids.

Below microwave frequencies (about 1000 MHz), passive circuit components come in typical packages—resistors, capacitors, and inductors. These are called *lumped-element* components. This text emphasizes circuit design and analysis with lumped-element components. Because these components become physically too small at microwave frequencies, "distributed-element" design characterized by transmission-line theory must be used at microwave frequencies. Transmission-line techniques are deferred until Chapter 14.

Since communication systems and circuits can be characterized by their response to sinusoidal signals, the study of electronic communication techniques begins with a review of impedance concepts and complex impedance algebra.   ■

1-1 ■ IMPEDANCE REVIEW

The voltage drop across a resistance R due to any current i is directly proportional to the *amount* of current: $v_R = iR$. As simple as this is, it will serve you well to remember that any current waveform in a resistor will produce the exact same voltage waveform across the resistor. The same does not hold true for inductors and capacitors. For an inductor, the voltage drop is directly proportional to how fast the current is *changing* with time: $v_L \propto di/dt$. A constant of proportionality L is used to equate voltage and the rate of change of current. Thus,

$$v_L = L\frac{di}{dt} \tag{1-1}$$

where the inductance L is entirely determined by the physical properties and materials of the inductor.

A sinusoidal current $i = I \sin 2\pi ft$ will produce a voltage drop across the inductor of

$$v_L = L\frac{di}{dt} = L\frac{d(I \sin 2\pi ft)}{dt}$$

$$= L(2\pi fI \cos 2\pi ft) = 2\pi fL(Ij \sin 2\pi ft)$$

where j indicates the 90° phase relationship between cosine and sine. The result can be written as

$$v_L = j2\pi fLi$$

or

$$v_L/i_L = jX_L$$

where

$$X_L = 2\pi fL \tag{1-2}$$

is called the *reactance* and has the same units as resistance (ohms).

A similar derivation can be performed for a capacitor in which the current-voltage relationship is

$$i_C = C\frac{dv}{dt} \tag{1-3}$$

The constant of proportionality C, the capacitance, is entirely determined by the physical properties and materials of the capacitor. The result of the derivation for a sinusoidal current or voltage yields

$$\frac{v_C}{i_C} = \frac{1}{j2\pi fC} = -jX_C$$

where

$$X_C = \frac{1}{2\pi fC} \tag{1-4}$$

is the reactance, in ohms, for a capacitance of C farads to a sinusoid of frequency f (hertz).

Much computation time may be saved by recognizing that a sinusoid has a continuous change of phase of 2π radians for every cycle, so that $2\pi f = \omega$ describes the angular frequency in radians/second and Equations 1-2 and 1-4 can be written compactly as

$$X_L = \omega L \tag{1-5}$$

FIGURE 1–2   Series circuit.

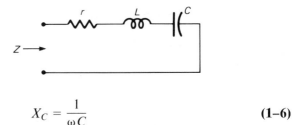

and $$X_C = \frac{1}{\omega C} \tag{1-6}$$

for sinusoidal signals.

Series Circuit Impedance

Series circuits are most easily handled by summing the complex impedances to determine the total impedance. As an example, the circuit of Figure 1–2 is a series combination of r, X_L, and X_C, with total impedance given by

$$Z = r + jX_L - jX_C$$

where lowercase r is used for resistance because small resistance values are the rule in series RLC circuits. This can also be written as $Z = r + jX$, where X is the net reactance as determined from

$$\begin{aligned} Z &= r + j(\omega L - 1/\omega C) \\ &= r + j(2\pi f L - 1/2\pi f C) \end{aligned}$$

A lot of algebra is saved and familiarization with impedance concepts gained by computing the individual reactance values in ohms once the frequency of interest is known.

■ ■ ■ ■

EXAMPLE 1–1

The series circuit of Figure 1–2 has the following component values: $r = 10\ \Omega$, $L = 10\ \mu H$, and $C = 100$ pF. Determine the impedance Z, the current for $V_z = 10$ V rms,* the resulting voltage dropped across the capacitor, and the power dissipated by the circuit—all at $f = 5.5$ MHz.

Solution:

1. At $f = 5.5$ MHz, $\omega = 2\pi(5.5\ \text{MHz}) = 34.6$ M rad/s, $X_L = \omega L = 345.6\ \Omega$, $X_C = 1/\omega C = 289.4\ \Omega$. Hence, $Z = 10 + j345.6 - j289.4 = 10 + j56.2 = \textbf{57.1}\ \Omega\ \underline{/80°}$, where $|Z| = \sqrt{r^2 + X^2}$ and $\theta = \tan^{-1}(X/r)$, as diagramed in Figure 1–3.

2. $i = v/Z = 10\ \text{V}/57.1\ \underline{/80°} = \textbf{175 mA}\ \underline{/-80°}$. The phase angle $\theta = -80°$ indicates that the current lags the applied voltage (or the voltage leads the current) for this essentially inductive circuit.

3. $v_C = iX_C = (175\ \text{mA}\ \underline{/-80°})(289.4\ \Omega\ \underline{/-90°}) = \textbf{50.6 V}\ \underline{/-170°}$.

4. The power dissipated in the circuit is that lost in the resistance, $P_r = i^2 r = (175\ \text{mA})^2\ (10\ \Omega) = \textbf{306 mW}$.

---

* All ac voltage and current values are rms unless specified as peak (pk) or peak-to-peak (pk-pk).

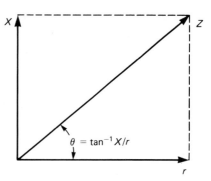

FIGURE 1-3 Impedance diagram.

$\theta = \tan^{-1} X/r$

1-2 ■ SERIES RESONANCE

The magnitude of the impedance for the circuit of Figure 1–2, $|Z| = |r + j(X_L - X_C)|$, has a minimum value at the frequency that causes the reactive components to cancel each other. This condition, called *resonance*, leaves a purely resistive impedance, $Z = r$. To determine the resonant frequency, set $X_L = X_C$ or $2\pi f_0 L = 1/(2\pi f_0 C)$. Solving for f_0, we get

$$f_0 = \frac{1}{2\pi\sqrt{LC}} \tag{1-7a}$$

and

$$\omega_0 = \frac{1}{\sqrt{LC}} \tag{1-7b}$$

To show what a dramatic difference resonance can produce, repeat Example 1–1 at a frequency just 10% below the 5.5 MHz used before.

■ ■ ■ ■

EXAMPLE 1-2

Series Resonance

The series LRC circuit of Example 1–1 has component values of $r = 10\ \Omega$, $L = 10\ \mu H$, and $C = 100$ pF. Determine the frequency for which resonance occurs, and repeat the calculations of Example 1–1 at resonance.

Solution:

Resonance occurs at $\omega_0 = 1/\sqrt{(10 \times 10^{-6})(100 \times 10^{-12})} = 31.6$ M rad/s, $f_0 = 5.03$ MHz.

1. $X_L = \omega L = 316\ \Omega$ and $X_C = 316\ \Omega$; $Z = 10 + j316 - j316\ \Omega = \mathbf{10\ \Omega}$

2. $i = v/Z = 10$ V/10 Ω = 1 A = I_{max}

3. $v_C = iX_C = (1\text{ A})(-j316\ \Omega) = \mathbf{-j316\ V}$ (316 V $\underline{/-90°}$)

4. $P_r = i^2 r = (1\text{ A})^2(10\ \Omega) = \mathbf{10\ W}$

For a series LRC circuit, the current reaches a maximum at resonance—it is limited only by the circuit resistance. At frequencies above and below the resonant frequency, the current is less than the maximum, as illustrated in Figure 1–4.

FIGURE 1-4 Current versus frequency in a series-resonant circuit.

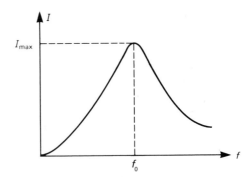

Another remarkable result of resonance can be seen from the voltage across the capacitor in Example 1–2. The voltage dropped across the capacitor is 316 V (and lags the current by 90°), but the voltage applied to the circuit is only 10 V! This is an increase of 31.6 times. The ratio of 31.6 is called the Q of the circuit and can also be calculated from 316 Ω/10 Ω, which formally is

$$Q = X_L/r \tag{1-8}$$

or

$$Q = X_C/r \tag{1-9}$$

Q is the ratio of maximum energy stored in the capacitor to the energy lost in the resistance per radian of RF current.

1–3 ■ CIRCUIT Q AND BANDWIDTH

Inductors store energy in the magnetic field surrounding the device; capacitors store energy in the space between conductors. The energy is stored during one-half of the ac cycle and returned during the other half. Any energy lost during the cycle is associated with a dissipative resistance and gives rise to a quality factor, Q. Circuit Q is defined as the ratio of maximum energy stored to the amount lost per ac cycle. The Q of a circuit is very important in electronic communications because it determines the 3-dB bandwidth of resonant circuits. The bandwidth of the overall system will limit the amount of information that can be transmitted through the system and the amount of noise that can enter the system. Bandwidth is calculated from Q and the resonant frequency of the circuit by

$$BW = f_0/Q \tag{1-10}$$

As an example, for the series circuit of Example 1–2, the 3-dB bandwidth is

$$BW = \frac{5.03 \text{ MHz}}{31.6} = 159.2 \text{ kHz}$$

If different values of r are substituted in the series LRC circuit, the Q will change in accordance with Equations 1–8 and 1–9, but the center frequency f_0 will

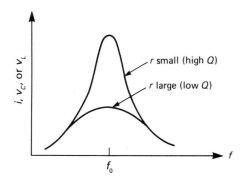

FIGURE 1–5   Result of varying r in a series-resonant circuit.

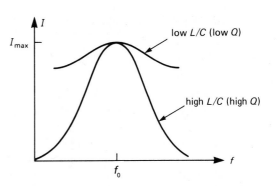

FIGURE 1–6   Current in an LRC circuit with r and LC constant but L/C is varied to change the Q.

not. This is illustrated in Figure 1–5. If r remains unchanged but L is increased and C decreased proportionately, the center frequency, $f_0 = 1/(2\pi\sqrt{LC})$, will remain unchanged but the Q will increase ($Q = X/r$). The curves for v_C and v_L versus frequency will be the same as Figure 1–5, and the maximum current, $I_{max} = V_i/r$, will also remain unchanged. Figure 1–6 illustrates the resonance curves for current in a constant-r series LRC circuit in which the LC product remains constant but L/C changes in order to increase Q.

1–4 ■ PARALLEL RESONANCE

Parallel resonant circuits are used when a high-impedance, tuned circuit is required. This high-impedance property is derived for Figure 1–7 with an ideal inductor and capacitor as follows:

$$Z = \frac{1}{Y}$$

$$Y = \frac{1}{R} + \frac{1}{jX_L} + \frac{1}{-jX_C}$$

$$= \frac{-j^2 X_L X_C + jX_L R - jX_C R}{-j^2 X_L X_C R}$$

FIGURE 1–7   Parallel RLC circuit.

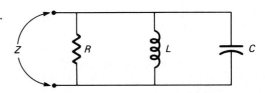

Since $-j^2 = +1$ and $Z_p = 1/Y$, then

$$Z_p = \frac{X_L X_C R}{X_L X_C + jR(X_L - X_C)} \tag{1-11}$$

At resonance, $X_L = X_C$, and therefore $Z_p = R$. The currents in X_L (v_i/jX_L) and X_C are always equal and opposite in phase. With no power losses in these two (ideal) components, the parallel LC combination requires no additional input current and they present an infinite impedance in parallel with R.

Equation 1–11 can be manipulated into a convenient expression for impedance as a function of frequency by the substitution of Equations 1–5 and 1–6 as follows:

$$Z = \frac{(\omega L/\omega C)R}{(\omega L/\omega C) + jR(\omega L - 1/\omega C)}$$

$$= \frac{R}{1 + jR(\omega C/\omega L)(\omega L - 1/\omega C)}$$

$$= \frac{R}{1 + j(R/L)(\omega LC - 1/\omega)}$$

Now substitute Equation 1–7, $\omega_0^2 = 1/LC$, cross multiply, and bring the common ω_0 out of the brackets to get Z as a function of frequency:

$$Z(\omega) = \frac{R}{1 + j(R/\omega_0 L)(\omega/\omega_0 - \omega_0/\omega)}$$

For a parallel resonant circuit,

$$Q = R/X_L \tag{1-12}$$

With this substitution and $\omega_0 = 2\pi f_0$, we have the parallel impedance at any frequency f:

$$Z(f) = \frac{R}{1 + jQ(f/f_0 - f_0/f)} \tag{1-13}$$

A plot of the complex impedance versus frequency is shown in Figure 1–8, including the magnitude $|Z|$ and phase θ.

Note that for frequencies well above resonance, $f \gg f_0$,

$$Z = \frac{R}{1 + jQf/f_0} \approx \frac{R}{jQf/f_0}$$

$$= -j\left(\frac{Rf_0}{Q}\right)\left(\frac{1}{f}\right)$$

The impedance gets more and more capacitive (the capacitor becomes more like a short circuit) and decreases with $1/f$, so that for a doubling of frequency, the

FIGURE 1-8   Parallel RLC circuit impedance. Magnitude $|Z|$ and phase θ versus frequency.

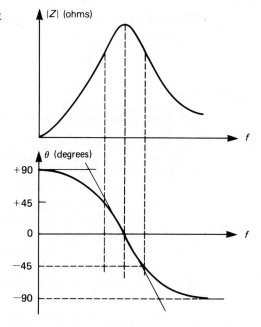

impedance is cut to one-half. Also, for $f \ll f_0$,

$$Z = \frac{R}{1 + jQ(-f_0/f)} \approx \frac{Rf}{-jQf_0}$$

$$= j\left(\frac{R}{Qf_0}\right)f$$

which is inductive.

■ ■ ■ ■

EXAMPLE 1-3

A circuit has $R = 10$ kΩ, $L = 100$ µH, and $C = 400$ pF, in parallel. Determine

1. The resonant frequency and Q.
2. The impedance at 16 kHz below resonance.

Solution:

1. $f_0 = 1/2\pi\sqrt{(100 \times 10^{-6})(400 \times 10^{-12})} = \textbf{795.8 kHz}$

$$Q = \frac{10 \text{ k}\Omega}{2\pi(796 \times 10^3)(100 \times 10^{-6})} = \textbf{20}$$

2. $Z(780 \text{ kHz}) = \dfrac{10 \text{ k}\Omega}{1 + j20[(780/796) - (796/780)]} = \dfrac{10 \text{ k}\Omega}{1 - j0.812}$

$$= 7.76 \text{ k}\Omega \; \underline{/39.1°} = \textbf{(6 + }j\textbf{4.9) k}\Omega$$

FIGURE 1-9   Tuning coils.
(a) Solenoid. (b) Toroid.

(a)   (b)

Nonideal Parallel Resonant Circuits

Inductors used as tuning coils for radio frequency (RF) applications are usually wound in the form of a solenoid or on a toroid (doughnut-shaped) form, as shown in Figure 1–9. The toroid coil is shown wound on a permanent form called the *core*. Materials used for coil cores can have the same permeability as air (μ_0) for VHF and UHF applications. But for HF, and certainly below, ferrite cores with $\mu \gg \mu_0$ are used in order to minimize the number of turns of wire necessary. Also these coil forms give structural support and, more importantly, confine the magnetic fields in order to prevent inductive coupling between separate circuits.

Air core inductors have fairly high Q values; that is, the ratio of energy stored in the magnetic field to the $i^2 r$ power lost is high, on the order of 200–300 for 22-gauge wire. The resistive losses occur primarily through wire (usually copper) resistance, including skin effects at higher frequencies. *Skin effect* is the tendency for current to concentrate near the conductor surface as the frequency is increased. Since resistance is inversely proportional to the cross-sectional area through which the current flows, the effective resistance of a wire will increase with frequency. The low-frequency resistance of a wire with radius r, length ℓ, and resistivity ρ is $R = \rho\ell/(\pi r^2)$ and becomes

$$ R = \frac{\ell\sqrt{\mu\pi f\rho}}{2\pi r} \tag{1-14} $$

at RF frequencies due to the skin effect. To reduce the skin effect, RF coils are often wound with stranded wire called *Litz* (Litzendraht) wire and even hollow tubing for large current-carrying capacity.

Litz wire is used to give high Q at the lower frequencies (below 2 MHz). Each strand in Litz wire is insulated from the others, generally with enamel, and the strands make contact only at the ends. The strands are braided so that any one conductor is encircled by as much flux as any other, ensuring that all strands will carry equal current.

Inductors made with cores of ferrite or other materials have eddy current and hysteresis losses that decrease their Q. Ferrite-core coils have Qs in the range of 50–250 if the core is properly chosen for the appropriate frequency from manufacturer's data sheets. If available, a Q-meter is used when winding RF inductors and transformers. Such an instrument allows the measurement of the Q, inductance L, and the coefficient of coupling between primary and secondary windings for transformers.*

---

* See Equation 1–61b in the section ''Measuring the Coupling Coefficient.''

When coils of finite Q are used in a tuned circuit such as that illustrated in Figure 1–10, then the frequency response is analyzed as follows:
The series circuit admittance is written as

$$Y(j\omega) = j\omega C + \frac{1}{r + j\omega L} \tag{1-15}$$

To make the denominator real, multiply and divide the right-hand term by the complex conjugate of its denominator. Then

$$Y(j\omega) = j\omega C + \frac{r - j\omega L}{r^2 + \omega^2 L^2} = j\omega C + \frac{r}{r^2 + \omega^2 L^2} - \frac{j\omega L}{r^2 + \omega^2 L^2} \tag{1-16a}$$

Therefore, the real and imaginary parts are separated as

$$Y(j\omega) = \frac{r}{r^2 + \omega^2 L^2} + j\left[\omega C - \frac{\omega L}{r^2 + \omega^2 L^2}\right] \tag{1-16b}$$

At resonance, $\omega = \omega_0$ and the imaginary part goes to zero, which requires that

$$\omega_0 C = \frac{\omega_0 L}{r^2 + \omega_0^2 L^2} \tag{1-17}$$

When solved for ω_0, this yields

$$\omega_0 = \sqrt{\frac{1}{LC} - \left(\frac{r}{L}\right)^2} \tag{1-18}$$

Now, with the nonideal finite-Q coil modeled as a series r-L combination, as illustrated in Figure 1–10, the quality factor of the circuit with no external resistive loading is referred to as the *circuit unloaded Q* and is computed from

$$Q_u \triangleq \frac{\omega_0 L}{r} \tag{1-19}$$

Then using Equation 1–18, we have

$$\frac{1}{LC} = \omega_0^2 + \left(\frac{r}{L}\right)^2 \tag{1-20}$$

FIGURE 1–10 Series and parallel ac equivalents for a finite-Q coil in an RLC circuit.

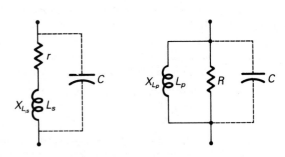

Replace $r/L = \omega_0/Q_u$ from Equation 1–19 to arrive at

$$\omega_0^2 = \frac{1}{LC}\left(\frac{Q_u^2}{Q_u^2 + 1}\right) \tag{1–21}$$

Equation 1–21 shows that a finite-Q coil has the effect of reducing the undamped, natural frequency of an ideal LC circuit by the factor $Q_u^2/(Q_u^2 + 1)$. With this in mind, we can replace the series rL_s branch of Figure 1–10 with an R, L_p branch pair to have the equivalent of the parallel RLC circuit of Figure 1–7 if the values of the parallel R and parallel L (denoted L_p in contrast to L_s of the series equivalent) are

$$R = r(Q_u^2 + 1) \tag{1–22a}$$

and

$$L_p = L_s\left(\frac{Q_u^2 + 1}{Q_u^2}\right) \tag{1–22b}$$

These two equations are determined by replacing ω and L_s with ω_0 and L_p in the first and last terms of the right-hand side of Equation 1–16b and equating these terms to $1/R$ and $1/X_{L_p}$, respectively. Equation 1–22a gives the important result that a small resistor in series with an inductor will have the same ac result as a large resistor in parallel. As an example, for a Q_u of 10, the power loss of a 1-Ω resistor in series with an ideal inductor will be equivalent to the loss of a 101-Ω parallel resistor.

Finally, with the series-to-parallel conversions made, the parallel-tuned circuit impedance as a function of frequency can be written from Equation 1–13 as

$$Z(f) = \frac{R}{1 + jQ_u(f/f_0 - f_0/f)} \tag{1–23}$$

A simple approximate analysis is as follows: For $X_{L_s} \gg r$, the series and parallel circuits are essentially inductive. If the circuits are equivalent, then $Q_s = Q_p$ so that $X_{L_s}/r = R/X_{L_p}$. For essentially inductive circuits, $X_{L_p} \approx X_{L_s}$ and $R = X_{L_s}^2/r = X_{L_s}Q_u = rQ_u^2$.

1–5 ■ SMALL-SIGNAL RF AMPLIFIER ANALYSIS

Among the myriad uses for tuned networks in communication systems, series and parallel resonant circuits are used at radio frequencies to reduce the voltage and current losses that result from series inductances and shunt capacitances. Series circuits are used where low impedances are encountered, such as in solid-state UHF power amplifiers and for microwave circuits. See Table 1–1 for frequency-range names used in electronics. Parallel resonance produces high impedance. Below UHF, solid-state and vacuum-tube impedances are high enough that parallel-tuned circuits are used. Shunt capacitance in untuned amplifiers will drastically reduce the gain at high frequencies. Parallel resonance can restore the gain. As an example, the common-emitter amplifier is used at RF frequencies by parallel-

TABLE 1–1 Frequency-Range Names Used in Electronics

Name	Frequency Range	Military Designations
DC	0 Hz	
Power frequencies	10–1000 Hz	
Audio	20–20,000 Hz	
Video	50 Hz–4.5 MHz	
Supersonic or ultrasonic	25 kHz–2 MHz	
Very low radio frequencies (VLF)	10–30 kHz	
Low radio frequencies (LF)	30–300 kHz	
Medium radio frequencies (MF)	300 kHz–3 MHz	
High radio frequencies (HF)	3–30 MHz	
Very high radio frequencies (VHF)	30–300 MHz	
Ultra high radio frequencies (UHF)	300 MHz–3 GHz	$\begin{cases} \text{L-Band: 1–2 GHz} \\ \text{S-Band: 2–3.7 GHz} \end{cases}$
Super high radio frequencies (SHF)	3–30 GHz	$\begin{cases} \text{C: 3.7–6.5 GHz} \\ \text{X: 7–11 GHz} \\ \text{Ku: 11–18 GHz} \\ \text{Kc: 18–21 GHz} \\ \text{Ka: 26–40 GHz} \end{cases}$
Extremely high radio frequencies (EHF)	30–300 GHz millimeter band	$\begin{cases} \text{W: 50–70 GHz} \\ \text{V: 75–110 GHz} \end{cases}$
Heat or infrared	1×10^{12}–4.3×10^{14} Hz*	
Visible light—red to violet	4.3×10^{14}–1×10^{15} Hz	
Ultraviolet	1×10^{15}–6×10^{16} Hz	
X rays	6×10^{16}–3×10^{19} Hz	
Gamma rays	3×10^{19}–5×10^{20} Hz	
Cosmic rays	5×10^{20}–8×10^{21} Hz	

* 10^{12} Hz = 1000 GHz, or 1 THz (terahertz).

resonating the collector circuit and then transformer-coupling to the next stage. The gain and frequency response is then determined in the same way as for the RC-coupled, low-frequency prototype except that the collector load is a parallel resonant circuit. Broadband amplifiers use microwave integrated circuits (MIC) and transmission lines; see Chapter 14.

An analysis of the tuned RF amplifier of Figure 1–11 is as follows: The input signal voltage v_{be} produces a base current $i_b = v_{be}/R_{in}$, where $R_{in} = (\beta + 1)r_e$.

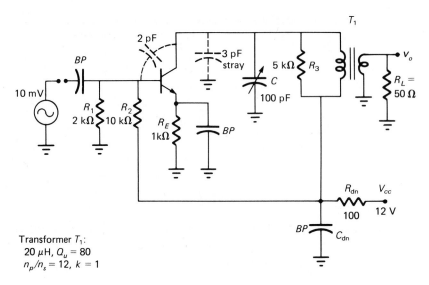

FIGURE 1–11   Small-signal tuned RF amplifier for analysis.

Resistance r_e is the effective dynamic resistance of the emitter junction given by $r_e = V_T/I_E$, where I_E is the emitter dc bias current, and the junction potential V_T varies between 25 and 40 mV, depending on the base-emitter junction geometry. Using 26 mV, $r_e = (26 \text{ mV})/I_E$ and is referred to as $1/g_m$ in the hybrid-π model. Collector current $i_c = \beta i_b$ adds to the base current to provide emitter current $i_e = i_c + i_b = (\beta + 1)i_b$. For small-signal applications, $\beta \geq 100$ so that $i_c \approx i_e$ with an error of less than 1%. In any case, i_c flowing through a collector circuit impedance of Z_c produces a voltage $v_c = i_c Z_c$, so the magnitude of the voltage gain A_v is given by

$$|A_v| = \frac{v_c}{v_b} = \frac{i_c Z_c}{i_b R_{in}} = \beta \frac{Z_c}{R_{in}} = \left(\frac{\beta}{\beta + 1}\right)\frac{Z_c}{r_e}$$

$$= \alpha \frac{Z_c}{r_e} \approx \frac{Z_c}{r_e}$$

with less than 1% of error when $\beta \geq 100$. Including a notation for the phase inversion of a common-emitter amplifier, we write

$$A_v = -Z_c/r_e \qquad\qquad\qquad \textbf{(1–24)}$$

where Z_c is the effective ac impedance from collector to ground and $r_e = 26 \text{ mV}/I_E$ is the transistor emitter resistance.

Gain in Decibels

The gain of a system is computed as the product of individual stage gains or losses. However, for convenience, the gain of individual stages is converted to decibels (dB) to allow simple addition of the individual stage gains for determining the complete system gain. Converting a numerical gain value to a decibel value requires the use of logarithms.

The replacement of gain multiples by decibel-gain sums is based on the well-established principle that the addition of the logarithms (logs) of two numbers is equivalent to the log of the product of the original two numbers. The base of the logarithm system that we use is 10 and the expression for the base 10 logarithm of N is

$$\log_{10} N = x \qquad\qquad \text{(1-25a)}$$

This equation is equivalent to

$$10^x = N \qquad\qquad \text{(1-25b)}$$

Equations 1-25a and 1-25b *together* form the definition of logarithms. Equation 1-25b can also be read as "the antilog of x equals N." For example, form the product $N \cdot M$ as follows: Let

$$10^x = N \quad \text{and} \quad 10^y = M \qquad\qquad \text{(1-26)}$$

where x and y are the logarithms of N and M, respectively. The sum of the logarithms is $x + y$, for which the antilog is computed as

$$10^{x+y} \qquad\qquad \text{(1-27)}$$

However, $\qquad\qquad 10^{x+y} = 10^x \cdot 10^y = N \cdot M \qquad\qquad$ (1-28)

Hence, summing logarithms of numbers (and taking the antilog) is the equivalent to multiplying the numbers.

In electronics, the *decibel* is defined for the ratio of power or, for example, power gain p_o/p_i. The logarithm base is 10 and, to minimize the occurrence of magnitudes less than unity, the logarithm is multiplied by 10. The definition of gain in decibels (dB, where capital B is used in honor of Alexander Graham Bell) is

$$\text{gain in dB} \equiv 10 \log_{10}(p_o/p_i) = A_p \text{ (dB)} \qquad\qquad \text{(1-29)}$$

where A_p rather than G, which denotes conductance throughout this text, is used for amplification of power. Also, since $p = v^2/R$, then

$$A_p \text{ (dB)} = 10 \log_{10}\left(\frac{v_o^2/R_o}{v_i^2/R_i}\right) = 10 \log_{10}\left(\frac{v_o}{v_i}\right)^2 + 10 \log_{10}\left(\frac{R_i}{R_o}\right) \qquad \text{(1-30a)}$$

Of course, since $\log(1) = 0$, if $R_i = R_o$, then

$$A_v \text{ (dB)} = 10 \log(v_o/v_i)^2 = 20 \log(v_o/v_i) \qquad\qquad \text{(1-30b)}$$

for voltage gain, and

$$A_i \text{ (dB)} = 20 \log(i_o/i_i) \qquad\qquad \text{(1-30c)}$$

for current gain, where input and output impedances are equal. Observe in Equation 1-30a that when signal *voltages* are measured, the amplifier gain is inversely proportional to circuit impedance levels.

Determine (**1**) the output voltage v_o for 40-mV input, (**2**) the maximum voltage gain, and (**3**) the bandwidth for the class A (small-signal) tuned amplifier of Figure 1–11.

Solution:

1. The maximum voltage gain will occur at the resonant frequency of the "tank" circuit. *Tank* in this context is an often-used term referring to the overall parallel-tuned circuit in which the RF signal energy is stored. With very little error (due to the power losses in the transformer), the resonant frequency will be $f_0 = 1/(2\pi\sqrt{LC_T})$, where C_T is the total capacitance from collector to ground. $C_T = 2 \text{ pF} + 3 \text{ pF} + 100 \text{ pF} = 105 \text{ pF}$, so

$$f_0 = \frac{1}{2\pi\sqrt{(20 \times 10^{-6})(105 \times 10^{-12})}} = \textbf{3.47 MHz}$$

   The reactance of L is $X_L = 2\pi(3.47 \times 10^6)(20 \times 10^{-6}) = 436 \ \Omega$. $R_{\text{coil}} = Q_u X_L = \textbf{34.9 k}\boldsymbol{\Omega}$. The load resistor R_L is reflected into the tank circuit by the turns ratio squared, so that $R'_L = (n_p/n_s)^2 R_L = (12)^2 50 \ \Omega = \textbf{7.2 k}\boldsymbol{\Omega}$. Since the transistor collector dynamic impedance r_c is not given, we determine the entire tank loading as $R'_c = R_3 \parallel R'_L \parallel R_{\text{coil}} = 5 \text{ k}\Omega \parallel 7.2 \text{ k}\Omega \parallel 34.9 \text{ k}\Omega = \textbf{2.72 k}\boldsymbol{\Omega}$. (The 5-k$\Omega$ tank resistor, R_3, is often included in order to increase bandwidth and can have the effect of reducing circuit parasitics. Of course, it does reduce the amplifier gain.)

   The emitter dynamic impedance $r_e = 26/I_E(\text{mA})$ must be determined in order to calculate the transistor voltage gain, A_{vt}. The voltage drop across R_{dn} should be less than 0.1 V, so, ignoring this,

$$V_B = [2 \text{ k}\Omega/(2 \text{ k}\Omega + 10 \text{ k}\Omega)]12 \text{ V} = 2 \text{ V}$$
$$V_E = V_B - 0.7 \text{ V} = 2 - 0.7 = 1.3 \text{ V dc}$$

   $I_E = V_E/R_E = 1.3 \text{ V}/1 \text{ k}\Omega = 1.3 \text{ mA}$; consequently, $r_e = 26/1.3 = 20 \ \Omega$. The input signal generator is ideal ($Z_0 = 0$), so the voltage gain from base to collector for this common-emitter amplifier is $A_{vt} = -R'_c/r_e = 2.72 \text{ k}/20 = 136$ with a 180° phase inversion. The input 40 mV will be inverted and amplified to $v_c = v_b A_{Vt} = (40 \text{ mV})(136) = 5.44 \text{ V rms}$. This voltage is stepped down by the turns ratio of T_1 to $v_o = v_c(n_s/n_p) = (5.44 \text{ V}) (\text{1/12}) = \textbf{453 mV.}$

2. If the transformer inverts the phase of the primary voltage, then the maximum voltage gain of the amplifier is seen to be

$$A_v = (A_{vt})(-n_s/n_p)$$
$$(-136)(-\text{1/12}) = 11.3 \qquad \text{or, in dB}$$
$$A_v(\text{dB}) = 20 \log A_v$$
$$= 20 \log 11.3 = \textbf{21.1 dB}$$

3. Finally, the bandwidth is given by BW $= f_0/Q_L$, where Q_L is the loaded or effective Q of the circuit. $Q_L = R'_C/X_L = 2.72$ kΩ/436 Ω $= 6.2$ and BW $= f_0/Q_L$ $= 3.47$ MHz/6.2 $=$ **556 kHz.**

See the PSpice results at the end of the chapter.

1-6 ■ SMALL-SIGNAL RF AMPLIFIER DESIGN

Circuit design always involves choice and good judgment. However, logically developed rules of thumb (ROT) can make the design process less mystifying.

For example, a transistor bias scheme with excellent temperature stability is shown in Figure 1–12. R_E gives a stabilizing effect to base-emitter junction voltage variations and collector leakage current variations by providing negative feedback for these very slow changes. A typical rule of thumb is to set the voltage drop across R_E at about 1 V, or approximately 10% of the supply voltage in typical low-voltage systems. With $V_E = 1$ V, V_B will be 1.6–1.7 V, depending on the amount of emitter current. Then R_1 and R_2 are chosen to divide V_{cc} to establish this base bias voltage. $R_1 = V_B/I_1$, where $I_1 \approx 10I_B$ is a good rule of thumb. This leaves R_2 to drop the remainder; $R_2 = (V_{cc} - V_B)/(1.1I_1)$.

Appropriate bias current is based on the maximum output power P_o and power supply voltage V_{cc}. A transformer-coupled class A amplifier can achieve as much as 50% collector efficiency at full output voltage swing (V_{cc} peak collector ac voltage swing). If 10% of V_{cc} is used up for V_E and the collector saturation voltage, then the power supply is supplying

$$P_{dc} = 0.9V_{cc}I_c \qquad\qquad (1\text{--}31)$$

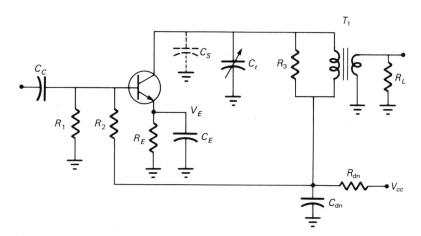

FIGURE 1–12   Small-signal tuned RF amplifier for design.

to the transistor collector. With 50% efficiency,

$$P_{dc} = P_o/\eta = 2P_0$$

so that $I_c = 2P_o/(0.9V_{cc})$ **(1–32)**

As an example, for $P_o = 10$ mW and $V_{cc} = 12$ V, $I_c = 20$ mW/$(0.9 \times 12$ V$) = 1.85$ mA. Then, $R_E = 0.1(12$ V$)/1.85$ mA $= 649$ Ω. With the dc design thus completed, we turn to the ac design.

R_{dn} and C_{dn} form a low-pass filter called a *decoupling network*. The decoupling network provides an ac low-impedance point between the collector circuit and the base (input), as well as circuit isolation from the power supply line and possible feedback to other amplifiers. A choke is *not* recommended as a replacement for R_{dn} because of resonances with circuit capacitance. For low-power receiver circuits, a typical 100-Ω resistor for R_{dn} will cost you less than a couple tenths of a volt of supply voltage to the amplifier. C_{dn} is a bypass (*BP*) capacitor that is chosen for a reactance of one *order of magnitude* (10:1) less than the resistance it is to bypass. For example, if the amplifier is used at 10 MHz and $R_{dn} = 100$ Ω, $X_c \leq 10$ Ω, so that $C_{dn} = 0.002$ μF will do. However, notice that the collector ac current path includes C_{dn} in series with C_E. Therefore, with C_E having a reactance of typically a couple of ohms, it is appropriate to make $C_{dn} = C_E$.

C_E is also a BP capacitor that bypasses R_E and provides a low impedance for the transistor emitter. For typically low-impedance receiver circuits with 50–75-Ω impedance levels, it is sufficient to make X_{CE} an order of magnitude less than the emitter's r_e. Since $r_e = 26$ mV/I_E, then

$$X_{CE} = 26 \text{ mV}/10I_E$$

C_c is a coupling capacitor or dc-block. Its value is determined in the same way as a bypass except that the reactance should be an order of magnitude less than the amplifier input impedance, Z_i. This will minimize the voltage division between X_{c_c} and Z_i. Hence, $X_{c_c} = Z_i/10$, where Z_i is the amplifier input impedance. In computing Z_i for RF amplifiers, the input capacitance must not be ignored.

The next steps in the design involve the collector circuit component values. The design procedure for a small-signal amplifier is quite different than for a large-signal power amplifier (see Chapter 6).

For large-signal power amplifiers, the design is based on determining the impedance levels that provide the required output power and bandwidth with maximum efficiency for a given supply voltage. For small-signal amplifiers, the output power level is important only in so far as setting a bias point capable of maintaining linearity for the maximum expected signal. This was considered in the dc part of the design.

The collector circuit design can be approached in at least three ways. One approach, based on a circuit gain requirement, would be to determine the required tank impedance at resonance to achieve the required gain, and then determine the reactances necessary to provide the required Q and bandwidth. A second approach is to determine the output transformer turns ratio to impedance-match the load to the transistor collector impedance for maximum power transfer and then calculate the values of L and C to tune the circuit with the required Q and bandwidth.

Another very straightforward approach for low-impedance solid-state amplifiers is to design for a reactance value of 200–300 Ω and then determine the transformer turns ratio, which provides the required Q and bandwidth. The gain is set by the resulting circuit impedance and transformer efficiency.

The design equations, based on an ideal transformer (coupling coefficient $k = 1$, and efficiency $\eta_T = 100\%$) are $X_L = 2\pi f L$, $X_C = 1/(2\pi f C)$, $Q = f_0/\text{BW}$, $R'_c = QX_L$, and transformer turns ratio,

$$n_p/n_s = \sqrt{R'_L/R_L} \qquad (1\text{–}33)$$

■ ■ ■ ■

EXAMPLE 1-5

Based on an ideal transformer and the schematic of Figure 1–12, design an amplifier for maximum power transfer at 10 MHz with a bandwidth of 500 kHz if the transistor has a collector dynamic impedance of 10 kΩ. $C_{BE} = 15$ pF, $C_{BC} = 1$ pF, and $\beta_{dc} = 50$. Maximum output signal power will be 5.4 mW in 50 Ω, and the power supply voltage is 12 V. Collector stray capacitance is 3 pF.

Solution:

1. $I_c = 2P_o/(0.9\ V_{cc}) = 2(5.4\ \text{mW})/(0.9 \times 12\ \text{V}) = \textbf{1 mA}$
2. Using the rule of thumb that $V_E = 1$ V, we get $R_E = V_E/I_E \approx 1/1\ \text{mA} = \textbf{1 k}\boldsymbol{\Omega}$. $R_1 = 1.7\ \text{V}/(10 \times 1\ \text{mA}/50) = 8.5$ kΩ. $R_2 = 10.3\ \text{V}/0.22\ \text{mA} = 47$ kΩ.
3. In order to determine an appropriate input coupling capacitor, we need to calculate the input impedance level. To determine Z_i we must know the approximate voltage gain from base to collector (to get C_i). With the collector impedance matched, $R'_c = (\frac{1}{2})(r_c) = 5$ kΩ. Then

$$A_v = R'_c/r_e = 5\ \text{k}\Omega/26\Omega = 192$$

$$C_i = C_{BE} + (1 + |A_v|)C_{BC} = 15 + (1 + 192)1\ \text{pF} = 208\ \text{pF}$$

$$X_{C_i} = [2\pi 10^7(208 \times 10^{-12})]^{-1} = 76.5\ \Omega$$

$$R_i = (\beta + 1)r_e \approx (51)(26\ \Omega) = 1.33\ \text{k}\Omega$$

where the ac beta* has been assumed equal to 50. $Z_i \approx -j76\ \Omega$. The coupling capacitor should be about $C = [2\pi 10^7(7.6\ \Omega)]^{-1} = \textbf{2000 pF.}$

4. If the output is to be designed for maximum power transfer, then the transformer should transform the 50-Ω load up to match r_c. Hence $n_p/n_s = \sqrt{r_c/R_L} = \sqrt{10\ \text{k}/50} = \textbf{14:1.}$
5. With the collector impedance thus matched, $R'_c = 5$ kΩ. The required bandwidth is 500 kHz, so $Q_L = 10\ \text{MHz}/0.5\ \text{MHz} = 20$ is the tank-loaded Q. $X_L =$

---

* Beta decreases at 6 dB/octave above a cutoff frequency given by $f_\beta = 1/[2\pi\beta_0 r_e(C_{BE} + C_{BC})] = 1/[2\pi(50)(26)(16\ \text{pF})] = 7.7$ MHz. At f_β the current gain is $\beta = \beta_0/\sqrt{2} = 35.4$ and continues down to unity at f_τ, the current gain bandwidth product. $f_\tau = \beta_0 f_\beta = 50 \times 7.7\ \text{MHz} = 385$ MHz. In any case, at 10 MHz, $\beta \approx 35$ for our transistor.

$X_C = R_c'/Q = (5 \text{ k}\Omega)/20 = \textbf{250 } \boldsymbol{\Omega}.$ The transformer primary inductance should be $L = X_L/2\pi f = 250/2\pi \times 10^7 = \textbf{4 } \boldsymbol{\mu}\textbf{H}.$ This will be tuned with $C = [2\pi 250 \times 10^7]^{-1} = \textbf{64 pF}.$ Since the collector circuit already has 1 pF + 3 pF = 4 pF to ground, the tuning capacitance must be about 60 pF. Typically, a 0.3–10-pF variable capacitor would be used in parallel with a 56-pF silver-mica capacitor.

6. If desired the power gain can be calculated as follows: The voltage gain from base to collector has been determined as 192. Power gain is voltage gain times current gain, $A_p = A_v A_I$. The current gain is complicated by the amplifier input circuitry. The equivalent circuit is shown in Figure 1–13. The base current that really produces transistor action ($i_c = \beta i_B$) is shown as i_B in Figure 1–13. For our amplifier, $R_{BB} = 8.5 \text{ k}\Omega \parallel 47 \text{ k}\Omega = 7.2 \text{ k}\Omega$, $R_i = 1.3 \text{ k}\Omega$ (actually closer to about $31 \times 26 = 806$ Ω) and $X_c = 76.5$ Ω. Clearly, most of the input current to the amplifier, i_i, will be shunted away by C_i. The amplifier input current is

$$i_i = v_i \, [1/7.2 \text{ k}\Omega + 1/1.3 \text{ k}\Omega + 1/-j76.5 \ \Omega] = 0.0131 v_i \ \underline{/86°}$$

The current in R_i is $i_B = (v_i)(1/1.3 \text{ k}\Omega) = 0.00077 v_i$. Hence the proportion of current to the amplification process is $i_B/i_i = 0.00077/0.0131 = 0.059$.

The current gain to the collector will be $A_I = 0.059\beta = 0.059 \times 50 = 3$, and the power gain will be

$$A_p = A_v A_I = 192 \times 3 = \textbf{576}$$

or $\qquad\qquad A_p \text{ (dB)} = 10 \log A_p = 27.6 \text{ dB}$

For the case of an ideal transformer, the power at the collector will equal the power delivered to the load, so the power gain for the amplifier at 10 MHz is **27.6 dB.**

The power gain of Example 1–5 (part 6) points out that the amplifier of Figure 1–12 has a problem due to lost current primarily in the Miller-effect capacitance at the input. It would be tempting to tune out C_i with a parallel-tuned circuit at the base of the transistor; don't yield to this temptation. The result is almost always circuit oscillation that makes the amplifier useless. This is explained in Chapter 2 under tuned-input, tuned-output oscillators.

An input transformer is more appropriate, but parasitic oscillations can still occur because of inductance in the base circuit. There are other solutions. One is neutralization in which a current is fed back from the collector, described more

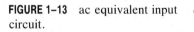
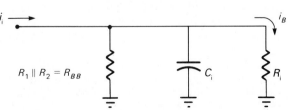

FIGURE 1–13 ac equivalent input circuit.

FIGURE 1–14 Tuned cascode amplifier.

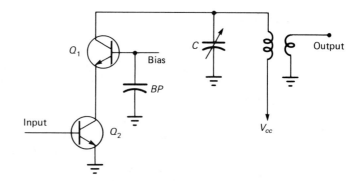

fully in Chapter 6. Another solution at frequencies below UHF is to eliminate the Miller capacitance by using the *cascode amplifier*, as seen in Figure 1–14.

The cascode consists of a unity gain common-emitter amplifier driving a tuned common-base amplifier. This circuit has the advantages of having the moderately high input impedance of a common-emitter (Q_2) but with no Miller-effect capacitance because of its low gain. All the gain is in the common-base (Q_1) section, which inherently has no Miller-effect capacitance because it is a noninverting amplifier. The cascode is popular in high-quality wideband audio amplifiers because of its low input capacitance (wide bandwidth), fairly high input impedance, and high gain characteristics.

A popular *integrated circuit* (IC) used in communications equipment up to frequencies of 120 MHz is the CA 3028A of Figure 7–21. The popularity of this differential amplifier/cascode (Q_3–Q_2) circuit derives from its stability and diversity of applications. It may be used for RF and IF amplifiers, oscillators, mixers/ frequency converters, limiters, coherent demodulators, and audio amplifiers. The most common RF and microwave IC (MIC) is the monolithic microwave IC (MMIC). It usually has four thin strip leads, of which two are grounds. The input and output impedances are 50 Ω and must be impedance-matched for proper microwave circuit behavior. Microwave-circuit matching is covered in Chapter 14.

1–7 ■ COUPLING TUNED CIRCUITS

Thus far we have studied RC-coupled circuits and ideal transformer-coupling of tuned circuits. RF circuits are also capacitively coupled and coupled with autotransformers.

Capacitance coupling is illustrated in Figure 1–15. C_1 and C_2 form an equivalent capacitance of

$$C_{eq} = \frac{C_1 C_2}{C_1 + C_2}$$

(1–34)

FIGURE 1-15 Capacitance coupling.

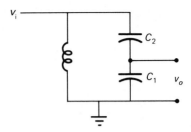

that will resonate with L at $f_0 = 1/2\pi\sqrt{LC_{eq}}$. The capacitors also form a voltage
divider and an impedance divider very much like an autotransformer (a tapped
inductor). As long as the impedance, connected at the output (across C_1), is an
order of magnitude or more larger than the reactance of C_1, the equations are
simply

$$\frac{v_o}{v_i} = \frac{C_{eq}}{C_1} = \frac{C_2}{C_1 + C_2} \tag{1-35}$$

This expression follows from the fact that the current through C_1 and C_2 is equal to
i and $v_i = iX_{C_{eq}}$, whereas $v_o = iX_{C_1}$. Hence

$$v_o/v_i = iX_{C_1}/iX_{C_{eq}} = \omega C_{eq}/\omega C_1$$

$$= C_1 C_2/[C_1(C_1 + C_2)] = C_2/(C_1 + C_2)$$

A similar analysis for R_i with R_o connected across C_1 (see Figure 1–15) yields

$$R_i = \left[\frac{C_1 + C_2}{C_2}\right]^2 R_o \tag{1-36}$$

for $R_o > 10X_{C_1}$.

  If R_o is not an order of magnitude or more greater than X_{C_1}, then the analysis
requires converting the parallel impedance formed by R_o and X_{C_1} to an equivalent
series impedance. The series equivalents are

$$R_{se} = \frac{R_o}{1 + Q^2} \tag{1-37}$$

and

$$C_{se} = C_1 \left(\frac{Q^2 + 1}{Q^2}\right) \tag{1-38}$$

where $Q = R_o/X_{C_1}$. The resulting series circuit with C_2, C_{se}, and R_{se} can easily be
converted to a single capacitor and series resistor R_{se}. The series RC values can be
further converted to equivalent parallel RC values if desired by using Equations
1–36 and 1–37 with R_o replaced with R_{pe} and C_1 replaced with C_{pe}. (C_{se} will have
to be replaced by the series equivalent of C_2 and C_{se}. Also, the new Q is for the
series circuit calculated from the single capacitor and series resistor values.)
Notice the similarities between Equations 1–37/1–38 and 1–22a/1–22b for making
RL series-parallel conversions.

■ ■ ■ ■

**EXAMPLE
1–6**

Determine R_i if $C_1 = 1000$ pF, $C_2 = 200$ pF, and $R_o = 200\ \Omega$ for the circuit of Figure 1–16. The circuit is resonant and operating at $\omega_0 = 10^7$ rad/s.

Solution:

1.
$$X_{C_1} = 1/\omega_0 C_1 = 1/(10^7 \times 10^{-9}) = 100\ \Omega$$

The Q of the $R_o C_1$ combination is

$$Q = R_o/X_{C_1} = 200/100 = 2$$

R_o loads down X_{C_1} too much to use the simple formula—Q must be at least 5. (For $Q_p > 10$, the error is less than 1%.) Using Equations 1–37 and 1–38, we get

$$R_{se} = 200/(1 + 2^2) = 40\ \Omega$$

and
$$C_{se} = 1000\text{ pF} \left(\frac{2^2 + 1}{2^2}\right) = 1250\text{ pF}$$

Figure 1–16 is now equivalent to Figure 1–17.

2.
$$C_{eq} = \frac{C_2 C_{se}}{C_2 + C_{se}} = \frac{(200\text{ pF})(1250\text{ pF})}{1450\text{ pF}} = 172.4\text{ pF}$$

To determine R_i use Equation 1–37 as follows: $R_i = R_{se}(1 + Q^2)$, where

$$Q = Q_{series} = X_{eq}/R_{se} = 1/[(10^7 \times 172.4 \times 10^{-12})(40\ \Omega)] = 14.5$$

With Q this high we could use a simple conversion, namely,

$$X_{pe} \approx X_{series}$$

and
$$R_{pe} \approx Q^2 R_{series}$$

This is because $Q^2 \approx Q^2 + 1$ and $Q^2/(Q^2 + 1) \approx 1$ when $Q > 10$. In any case,

$$R_i = [1 + (14.5)^2]40\ \Omega = \textbf{8.45 k}\boldsymbol{\Omega}$$

3. Compare 8.45 kΩ with that calculated from Equation 1–36:

$$R_i = \left(\frac{100\text{ pF} + 200\text{ pF}}{200\text{ pF}}\right)^2 200\ \Omega = 7.2\text{ k}\Omega$$

The error using Equation 1–36 is 14.8% because R_o is only $2 \times X_{C_1}$.

FIGURE 1–16 Capacitance coupling; impedance transformation.

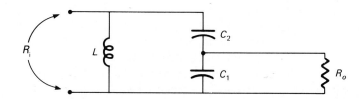

FIGURE 1-17 **FIGURE 1-17** Series equivalent
of Figure 1–16.

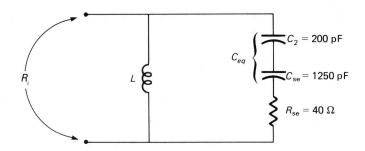

Capacitive coupling is not very common in RF work for two reasons. For one, transformers are simple to build, and the turns ratio can be changed with relative ease for maximum power transfer. Second, solid-state circuits involving oscillators and medium- to high-power amplifiers have high-order distortion products. The high-frequency distortion currents couple more readily through capacitors than inductors because as frequency increases, X_C decreases.

Inductive coupling may be accomplished using two inductors in a way analogous to capacitive coupling or with transformers, which involves mutual coupling. LC impedance-matching networks are covered in Section 6–4 and microwave-matching techniques are covered in Chapter 14.

The autotransformer of Figure 1–18 can be used when the dc isolation afforded by separate primary and secondary windings is not a consideration. Assuming that all the windings are mutually coupled like the familiar two-winding transformer, the equations are also the same; that is,

$$v_1 = v_2(n_1 + n_2)/n_1 \qquad \textbf{(1–39a)}$$

and
$$R_1 = [(n_1 + n_2)/n_1]^2 R_2 \qquad \textbf{(1–39b)}$$

These simple relationships are not valid when all the windings are not mutually coupled—that is, when the flux produced by current in some of the windings of n_2 does not cut n_1 turns. RF transformers rarely perform ideally because iron cores sometimes are not used or the cores do not have very high permeability. Such nonideal transformers are considered in Section 1–8.

FIGURE 1-18 Autotransformer; tapped inductor.

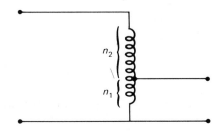

1-8 ■ NONIDEAL TRANSFORMER COUPLING

Transformer coupling calculations have typically assumed the ideal case where $k = 1$; that is, all the primary flux cuts secondary turns. This is approximately true for audio and power transformers containing a large amount of soft iron to concentrate and direct the flux from primary to secondary. When $k = 1$, the relationships between voltage and impedance transformation follow the familiar expressions of Equation 1–39.

The inductance and coupling for a toroidal-core transformer is derived as follows: If ψ_m is the amount of magnetic flux per turn produced by current $i(t)$, then the total flux in the toroidal inductor illustrated in Figure 1–19a is $N\psi_m$. The value $N\psi_m$ is also referred to as the total *flux linkage*. If a toroidal core of outer diameter R and material of permeability μ is uniformly wound, the magnetic flux density B is approximately uniform across the cross-sectional area $A = \pi r^2$ and is given by*

$$B = \frac{\mu N i(t)}{2\pi R} \tag{1–40}$$

where it is assumed that $R \gg r$. The core parameters are denoted in Figure 1–19b. The total flux per turn is BA, so that the total flux linkage of the toroidal-core inductor will be

$$N\psi_m = NBA = N\left(\frac{\mu N i}{2\pi R}\right)(\pi r^2) \tag{1–41}$$

By definition, the *inductance L* is the ratio of the total magnetic flux linkage to the current producing the flux; that is,

$$L = \frac{N\psi_m}{i} \tag{1–42}$$

* John D. Kraus, *Electromagnetics*, 3rd ed. (New York: McGraw-Hill, 1984), p. 166.

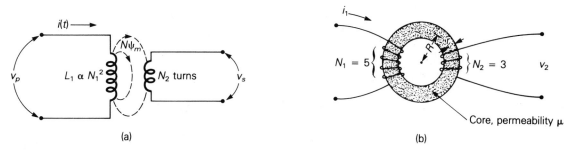

FIGURE 1-19 Nonideal transformer coupling. (a) Circuit model showing magnetic flux, ψ. (b) Physical illustration.

which, for the toroid using Equation 1–41, becomes

$$L = \frac{\mu N^2 r^2}{2R} \qquad (1\text{--}43)$$

Notice in Equation 1–43 that the inductance is determined entirely by the geometry and the magnetic property μ for the toroidal inductor; this is true for most other types of inductors as well.

To determine the voltage induced in a secondary winding due to current i_1 in the primary of a toroidal-core transformer, start with Faraday's law for an inductor, expressed as

$$v = -\frac{d\psi_m}{dt} \qquad (1\text{--}44a)$$

which, for an N-turn coil, is

$$v = -N \frac{d\psi_m}{dt} \qquad (1\text{--}44b)$$

The total flux due to i_1 in the primary winding of N_1 turns is given by

$$\psi_m = BA = \left(\frac{\mu N_1 i_1}{2\pi R}\right)(\pi r^2) = \frac{\mu N_1 r^2 i_1}{2R} \qquad (1\text{--}45)$$

Also, the voltage v_2 induced in the transformer secondary will be, from Equation 1–44b,

$$v_2 = -N_2 \left(\frac{d\psi_{m_2}}{dt}\right) \qquad (1\text{--}46)$$

Now we introduce the *coupling coefficient*,

$$k = \frac{\psi_{m_2}}{\psi_{m_1}} \qquad (1\text{--}47)$$

which gives the fraction of primary flux that cuts secondary turns, thereby inducing a secondary voltage. Thus, Equations 1–46 and 1–47 yield the secondary-winding induced voltage due to primary-winding flux as

$$v_2 = -N_2 k \left(\frac{d\psi_{m_1}}{dt}\right) \qquad (1\text{--}48)$$

where k is assumed constant.

Substituting Equation 1–45 into Equation 1–48 yields

$$v_2 = -N_2 k \frac{d}{dt}\left(\frac{\mu N_1 r^2 i_1}{2R}\right) \qquad (1\text{--}49a)$$

$$= -N_1 N_2 k \frac{\mu r^2}{2R} \frac{di_1}{dt} \qquad (1\text{--}49b)$$

Recalling from basic circuit theory that the voltage drop across an inductor is $v = L(di/dt)$ (Equation 1–1), we see that Equation 1–49b represents a voltage drop across some "mutually coupled" inductance. Indeed, replacing N_1 and N_2 in Equation 1–49b using Equation 1–43 yields

$$v_2 = -k\sqrt{\frac{L_1 2R}{\mu r^2}}\ \sqrt{\frac{L_2 2R}{\mu r^2}}\ \frac{\mu r^2}{2R}\frac{di_1}{dt} \qquad\qquad \textbf{(1–50a)}$$

$$= -k\sqrt{L_1 L_2}\ \frac{di_1}{dt} \qquad\qquad \textbf{(1–50b)}$$

From Equation 1–50 we may write

$$v_2 = -M\frac{di_1}{dt} \qquad\qquad \textbf{(1–51)}$$

where M, called the *mutual inductance*, is given by

$$M = k\sqrt{L_1 L_2} \qquad\qquad \textbf{(1–52)}$$

To compute the sinusoidal voltage induced in the secondary winding, represent the primary sinusoidal current as

$$i_1 = I_o e^{j\omega t} \qquad\qquad \textbf{(1–53)}$$

Then

$$\frac{di_1}{dt} = I_o(j\omega)e^{j\omega t} \qquad\qquad \textbf{(1–54)}$$

Substituting Equation 1–54 in Equation 1–51 yields

$$v_2 = -j\omega M I_o e^{j\omega t} \qquad\qquad \textbf{(1–55a)}$$

$$v_2 = -j\omega M i_1 \qquad\qquad \textbf{(1–55b)}$$

The product ωM may be viewed as the mutual reactance across which voltage is induced in the secondary winding due to the primary-winding current i_1.

■ ■ ■ ■

EXAMPLE 1-7

A sinusoidal current of 1 mA at 100 kHz flows in the primary of a transformer with a primary inductance of 1 mH, secondary inductance of 250 µH, and a coefficient of coupling 0.20. Determine the secondary induced voltage.

Solution:

$$M = k\sqrt{L_1 L_2} = 0.2\sqrt{(1 \times 10^{-3})(250 \times 10^{-6})} = 100\ \mu\text{H}$$

$$|v_s| = \omega M i_p = (2\pi \times 100\ \text{kHz})(100 \times 10^{-6}\ \text{H})(1 \times 10^{-3}\ \text{A}) = \textbf{62.8 mV}$$

A circuit model that is helpful in visualizing the effect of mutual inductance and the secondary voltage, v_s, is shown in Figure 1–20. Observe that the total primary inductance is L_1 and that the magnitude of the voltage seen looking back into the output leads is i_p times the mutual reactance, $2\pi f M$.

FIGURE 1-20 Equivalent transformer circuit showing mutual
inductance and secondary-
induced voltage.

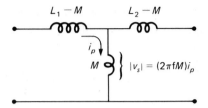

Transformer Loading

In Figure 1–21 a generator is connected across the primary of a high-frequency
transformer and v_L is measured with a high-impedance RF voltmeter. If the voltmeter impedance is high enough, very little current will flow in the secondary
winding and there will be essentially no interaction of the secondary back on the
primary.

However, when the secondary load impedance Z_L is low, secondary current
will flow due to the induced voltage and this current will induce a counter-EMF in
the primary. The same equations developed for secondary voltage due to primary
current hold for a "bucking" voltage induced into the primary by the current i_s
that flows in the secondary winding. Hence, the generator will experience a loading effect when Z_L is connected, and the amount of loading will be directly proportional to i_s, where

$$i_s = v_s/Z_{ss}$$

Z_{ss} is the *total* series impedance of the secondary circuit through which i_s flows;
that is,

$$Z_{ss} = Z_L + (r_2 + j\omega L_2)$$

where r_2 takes into account losses in the transformer secondary winding.

Figure 1–22 shows a model of the equivalent circuit we will use for transformer
coupling in the general (nonideal) case of $k < 1$. As illustrated at the end of this
section, k is easily determined from measurements with an inductance meter and,
if a Q-meter or impedance meter is available, all the elements of the model can be
determined.

The voltage induced in the secondary is, from Equation 1–55b, $v_s = -j\omega M i_p$.
This produces a current $i_s = v_s/Z_{ss}$, where

$$Z_{ss} = (j\omega L_2 + r_2) + Z_L \qquad\qquad (1\text{--}56)$$

FIGURE 1-21 Generator-driven
transformer with load.

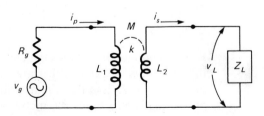

FIGURE 1-22 Equivalent circuit
model showing secondary-
induced voltage v_s and reflected
impedance Z_r due to the loading
effect of the secondary.

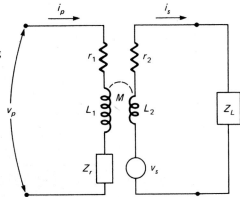

A current flowing in the transformer secondary will induce a counter-EMF in the
primary winding. This could be modeled with a $-v_p$ bucking voltage, but instead
this effect is modeled as a reflected impedance, Z_r, in series with the primary
windings. The bucking voltage would be $-v_p = +j\omega M i_s$, and since i_p flows
through this counter-EMF,

$$Z_p = j\omega M i_s / i_p = j\omega M(-j\omega M i_p / Z_{ss})/i_p \qquad (1\text{-}57)$$

Hence $\qquad\qquad Z_r = (\omega M)^2 / Z_{ss} \qquad\qquad\qquad\qquad\quad (1\text{-}58)$

The following example illustrates how an inductive secondary circuit is reflected as a capacitance and a capacitive secondary looks inductive in the primary.

■ ■ ■ ■

**EXAMPLE
1-8**

An air-coupled transformer yields the following Q-meter measurements at 2.5
MHz: $L_1 = 100$ μH, $Q_1 = 157$, $L_2 = 10$ μH, $Q_2 = 50$, and $k = 0.05$. Determine
the following:

1. Mutual inductance.

2. Reflected impedance for

    a. $Z_L = \infty$ (open-circuit);

    b. $Z_L = -j200$ Ω ($C_s = 318$ pF);

    c. $Z_L = 0$.

3. Primary impedance for the loads in part 2.

Solution:

1. $M = k\sqrt{L_1 L_2} = 0.05\sqrt{(100 \times 10^{-6})(10 \times 10^{-6})} = \mathbf{1.58\ \mu H}$

2. $\omega = 2\pi f = 2\pi(2.5\ \text{MHz}) = 15.7 \times 10^6$ rad/s

    $\omega M = (15.7 \times 10^6)(1.58 \times 10^{-6}) = 24.8$ Ω (mutual reactance)

    $X_{L_2} = \omega L_2 = (15.7 \times 10^6)(10 \times 10^{-6}) = 157$ Ω

    $r_2 = X_{L_2}/Q_2 = 157$ Ω$/50 = 3$ Ω

a. $Z_L = \infty$, $Z_{ss} = \infty$, $Z_r = (\omega M)^2/Z_{ss} = 0$

b. $Z_L = -j200\ \Omega$, $Z_{ss} = r_2 + j\omega L_2 + Z_L = 3 + j157 - j200 = 3 - j43\ \Omega$,

$$Z_r = \frac{(\omega M)^2}{Z_{ss}} = \frac{(24.8)^2}{43.1\ \Omega\ \underline{/-86^\circ}} = 14.3\ \underline{/86^\circ} = 1 + j14.2\ \Omega$$

Notice that the impedance reflected into the primary is essentially inductive for the capacitive secondary circuit.*

c. $Z_L = 0$, $Z_{ss} = 3 + j157 + 0 = 3 + j157\ \Omega \approx 157\ \underline{/90^\circ}$. The secondary is highly inductive and should reflect into the primary as a capacitive effect. $Z_r = (24.8\ \Omega)^2/j157\ \Omega = -j3.92\ \Omega$.

3. The impedance of the loaded primary for the secondary loads of part 2 are determined from $Z_p = r_1 + j\omega L_1 + Z_r$. $X_{L_1} = \omega L_1 = (15.7 \times 10^6) \times (100 \times 10^{-6}) = 1.57\ \text{k}\Omega$. $r_1 = X_{L_1}/Q_1 = 1.57\ \text{k}\Omega/157 = 10\ \Omega$.

a. For $Z_L = \infty$, $Z_r = 0$, and $Z_p = 10\ \Omega + j1.57\ \text{k}\Omega$

b. For $Z_L = -j200\ \Omega$, $Z_r = 1 + j14.2\ \Omega$, so $Z_p = (10 + 1) + j(1570 + 14.2) = 11 + j1584.2\ \Omega$. The capacitive load has resulted in an effective primary circuit inductance of $L_p' = 1584.2/(15.7 \times 10^6) = 100.9\ \mu\text{H}$.

c. For $Z_L = 0$, $Z_r = -j3.92$, and $Z_p = 10 + j(1570 - 3.92) = 10 + j1566\ \Omega$. The effective primary inductance is $L_p' = 1566/(15.7 \times 10^6) = 99.75\ \mu\text{H}$. Part (c) of the example suggests a method for determining the coefficient of coupling k for a transformer.

Measuring the Coupling Coefficient

To determine k for a transformer, measure L_p with the secondary open-circuited. Then short-circuit the secondary leads and measure L_p', the effective primary inductance for $Z_L = 0$. The coupling coefficient is calculated from these measurements using

$$L_p' = L_p(1 - k^2) \tag{1-61a}$$

$$k = \sqrt{1 - L_p'/L_p} \tag{1-61b}$$

The proof of this is as follows: With $Z_L = 0$,

$$Z_{ss} = r_2 + j\omega L_2 + 0 \approx j\omega L_2 \tag{1-62}$$

---

* It is often convenient to compute reflected resistance and reactances separately. However, in so doing you must remember that the reflected reactance is of opposite type (phase) to the net reactance of the secondary. The equations for magnitudes are

$$R_r = \frac{(\omega M)^2}{|Z_{ss}|^2} R_{ss} \tag{1-59}$$

and

$$X_r = \frac{(\omega M)^2}{|Z_{ss}|^2} X_{ss} \tag{1-60}$$

for a high Q secondary. Using Equation 1–57 with Equation 1–62, the primary impedance becomes

$$Z_p = r_1 + j\omega L_1 - j\omega M^2/L_2 \approx j\omega(L_1 - M^2/L_2) \qquad \textbf{(1–63)}$$

for a high Q primary; that is, $X_{L_1} \gg r_1$. The primary is inductive with an effective inductance of

$$L'_p = L_1 - M^2/L_2 \qquad \textbf{(1–64)}$$

Since $M^2 = k^2 L_1 L_2$, then

$$L'_p = L_1 - k^2 L_1 L_2/L_2 = L_1(1 - k^2) \qquad \textbf{(1–65)}$$

That is, $$L'_p = L_p(1 - k^2)$$

1–9 ■ DOUBLE-TUNED CIRCUITS

Coupling circuits with nonideal transformers present more mathematical complexity than ideal transformers, as demonstrated in Example 1–8. However, the ability to vary the mutual coupling between circuits leads to very useful applications such as multiresonator filters and the detectors used in FM (frequency modulation) receivers. (FM circuits are covered in Chapter 9.) An example of a two-resonator bandpass filter (a two-pole filter) is illustrated in Figure 1–23. The secondary is tuned to resonance by C_2 and has a Q of $Q_s = \omega_0 L_2/r_2$ if all the losses are represented by r_2. The primary circuit, independent of the secondary, has a Q of $Q_p = \omega_0 L_1/r_1$ where all the losses are represented by r_1. C_1 tunes the primary to resonance ($\omega_0 = 2\pi f_0$) and the entire circuit is referred to as a *double-tuned circuit*.

The output voltage-versus-frequency response of the double-tuned circuit is highly dependent on the amount of coupling between primary and secondary. As illustrated in Figure 1–24, the critically coupled response rises to a maximum and

FIGURE 1–23   Double-tuned circuit.

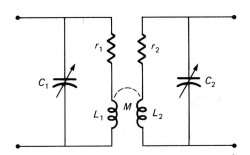

FIGURE 1–24 Effect of the coupling coefficient in a double-tuned circuit. Actual results from iterations of PSpice analysis of Figure 1–34a.

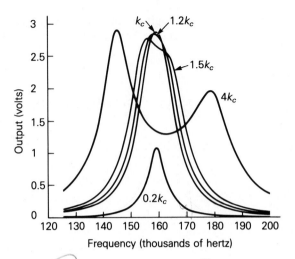

has a rounded peak. Critical coupling occurs when r_2 couples into the primary with a value equal to r_1. The critical coupling coefficient can be calculated as

$$k_c = 1/\sqrt{Q_p Q_s} \tag{1–66}$$

If the actual circuit coupling is less than critical $(k < k_c)$, then the response will have a very sharp peak that, as seen in Figure 1–24, does not reach the maximum achievable.

With actual circuit coupling 50% greater than k_c, the optimum coupling occurs:

$$k_{\text{opt}} = 1.5 k_c \tag{1–67}$$

This gives maximum output with a nearly flat response in the passband (see SPICE results in Figure 1–24). The analysis of the circuit of Figure 1–25 with the computer simulation program PSpice* was done using iterations of the input circuit file of Figure 1–34a in the "Computer Analysis" section at the end of this chapter. PSpice is described in Appendix E. The node numbers used in the circuit file correspond to the circled numbers of Figure 1–25 with a 100-kΩ resistor, R_1, to provide a finite current source impedance.

Coupling in excess of k_{opt} will result in the overcoupled, double-hump response shown in Figure 1–24. For the very overcoupled double-tuned circuit, the maximum gain will not occur for input signals at or near f_0 but will occur when the input signal frequency is

$$\frac{f_0}{\sqrt{1 + k}} \tag{1–68a}$$

and also at

$$\frac{f_0}{\sqrt{1 - k}} \tag{1–68b}$$

---

* PSpice is a registered trademark of MicroSim Corporation, Laguna Hills, Calif.

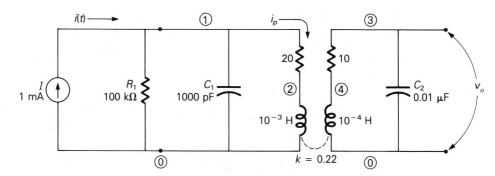

FIGURE 1–25 Double-tuned circuit example.

This is because the impedance coupled into the primary circuit at f_0 is a fairly large resistance that reduces the primary-winding Q and current. The result is that less voltage is induced into the secondary. However, above and below the secondary resonant frequency, the net secondary reactance offsets some of the reflected resistance, and the primary Q and increased current results in more induced secondary voltage. Beyond the peaks, the reflected secondary reactance detunes the primary to the point where the primary circuit impedance and voltage decrease, and the output falls off. An example will illustrate this point.

■■■■

EXAMPLE 1–9

Double-Tuned Circuit with Current Source

1. At what frequency ω_0 will the primary and secondary circuits of Figure 1–25 be resonant simultaneously?

2. Determine the primary- and secondary-*winding* Qs.

3. Will the circuit be critically coupled, undercoupled, optimally coupled, or overcoupled?

4. Find v_o with $i(t) = 1$ mA at ω_0.

5. Find v_o at the lowest frequency for which a peak response occurs.

Solution:

1. For the secondary, $\omega_0 = 1/\sqrt{L_2 C_2} = 1/\sqrt{10^{-4}10^{-8}} = \mathbf{10^6}$ **rad/s.** The resonant secondary will have $Z_{ss} = 10\ \Omega$ pure resistance, so that Z_r is also purely resistive. Assuming that the primary circuit has $Q_p \geq 10$, then the primary circuit is resonant at approximately $\omega_0 = 1/\sqrt{L_1 C_1} = 1/\sqrt{10^{-3}10^{-9}} = \mathbf{10^6}$ **rad/s** $= \omega_0$ of the secondary also.

2. For the primary *winding*, $Q_{u_1} = \omega_0 L_1/r_1 = 10^6 10^{-3}/20 = \mathbf{50}$, and for the secondary *winding*, $Q_{u_2} = \omega_0 L_2/r_2 = 10^6 10^{-4}/10 = \mathbf{10}$.

3. To compute Q_p and Q_s we must consider the entire primary *circuit* and secondary *circuit* independently of the other. Q_s is the same as $Q_{u_2} = \mathbf{10}$, but Q_p

includes the 100 kΩ as well as r_1. For convenience, take R and C together and perform a parallel-to-series conversion (Equation 1–22):

$$r = \frac{R}{1 + Q_c^2} = \frac{100 \text{ k}\Omega}{1 + (R\omega_0 C_1)^2} = \frac{100 \text{ k}\Omega}{1 + (100)^2} = 9.999 \approx 10 \ \Omega$$

Since $Q_{u_1} = 50$ and $Q_c = 100$, $Q_p = [(1/Q_{u_1}) + (1/Q_c)]^{-1} = $ **33.3**. This result is confirmed by looking at the series circuit r, C_1, r_1, L_1 and writing $Q_p = X/(r + r_1) = 1 \text{ k}\Omega/30 \ \Omega = $ **33.3**.

Finally, $k_c = 1/\sqrt{Q_p Q_s} = 1/\sqrt{33.3 \times 10} = $ **0.055**, and $k_{opt} = 1.5k_c = $ **0.082**. Since the actual circuit is coupled with $k = 0.22$, which is 4 times greater than critical coupling, the circuit is *overcoupled*. Some of the effects of the overcoupled response will be shown in parts 4 and 5.

4. $M = k\sqrt{L_1 L_2} = 0.22\sqrt{10^{-3}10^{-4}} = 69.6 \ \mu\text{H}$. $\omega M = 10^6 \times 69.6 \ \mu\text{H} = 69.6 \ \Omega$. $Z_{ss}(\omega_0) = 10 \ \Omega$ (resonance). Hence, $Z_r = (69.6)^2/10 = 484 \ \Omega$. $X_{L_1} = 1 \text{ k}\Omega$ from part 2, so the effective primary winding has $Z_p = (20 + j1 \text{ k}\Omega) + 484 \ \Omega = 504 + j1000 \ \Omega$. Now, use the current-divider rule, ignoring the less than 20 μA in the 100 kΩ, to compute i_p:

$$i_p = \frac{(r - jX_{c_1})i}{Z_p + (r - jX_{c_1})} = \frac{(10 - j1000)(1 \text{ mA})}{514 \ \Omega} = -j1.95 \text{ mA}$$

Now, $v_s = -j\omega M i_p = -j(69.6 \ \Omega)(-j1.95 \text{ mA}) = -135.4 \text{ mV}$. So, $i_s = v_s/Z_{ss} = (-135.4 \text{ mV})/10 \ \Omega = 13.5 \text{ mA}\underline{/180°}$, and $v_o = i_s X_{c_2} = (13.5 \text{ mA} \underline{/180°})(100 \underline{/-90°}) = $ **1.35** $\underline{/90°}$ with phase referenced to the input current source. (Note that the secondary circuit current shows a phase reversal compared to the input current source.)

5. For this overcoupled circuit, we would anticipate peaks in the frequency response, one below ω_0 and one above. The lower-frequency peak is computed from Equation 1–68a as $\omega = \omega_0/\sqrt{1 + k} = \omega_0/\sqrt{1 + 0.22} = 0.905\omega_0$ (144.1 kHz). In recalculating circuit reactances, we may use $\omega = 0.905$ M rad/s or, more simply, scale all reactances by the factor 0.905. Hence the 100-Ω (at ω_0) secondary inductor becomes 90.5 Ω and the capacitor becomes 110.5 Ω. $Z_{ss} = 10 + j(90.5 - 110.5) = 10 - j20 \ \Omega = 22.4 \ \Omega \underline{/-63.4°}$. $Z_r = (0.905 \times 69.6 \ \Omega)^2/(22.4 \ \Omega \underline{/-63.4°}) = 177.3 \underline{/63.4°} = 79.4 + j158.5 \ \Omega$. (Note that this represents a substantial change from the purely resistive 484 Ω, at ω_0.) Now, $Z_p = (20 + j905.5) + (79.4 - j158.5) = 99.4 + j1064 \ \Omega$ and the current-divider rule yields

$$i_p = \frac{(10 - j1105)(1 \text{ mA})}{99.4 + j1064 + 10 - j1105} = \frac{-j1105 \text{ mA}}{109.4 - j41} = 9.46 \text{ mA} \underline{/-69.5°}$$

Secondary voltage is

$$v_s = -j\omega M i_p = -j(63 \ \Omega)(9.46 \text{ mA} \underline{/-69.5°}) = 596 \text{ V} \underline{/-159.5°}$$

and

$$v_o = i_s Z_L = [(0.596 \text{ V} \underline{/-159.5°})/(22.4 \ \Omega \underline{/-63.4°})](-j110.5 \ \Omega)$$
$$= \textbf{2.94 V} \underline{/\mathbf{-186°}}$$

This is an increase of 6.76 dB above the output at resonance. The reason for the increase is seen to be the almost fivefold rise in primary winding current with only a slight detuning of the primary, while the secondary Z_{ss} only doubled (X_c increased only 10%).

Figure 1–24 shows the excellent agreement between our results and the $k = 4k_c$ (0.219) curve of the set of curves generated with a computer simulation using PSpice. The PSpice input and output files are presented in Figure 1–34. If the secondary circuit capacitor and resistor are interchanged, then the higher-frequency peak is only 1.5 dB below the lower-frequency peak as compared with the 3.6 dB for the capacitor load of Figure 1–25 and the plot of Figure 1–24.

1–10 ■ LAPLACE TRANSFORMS

Introduction

Electronic circuit and system behavior are most commonly characterized in the time domain by differential equations. A simple example of this is the response of an inductor to a current. That is, the voltage drop $v(t)$ across an inductor of L henrys due to a current $i(t)$ is described by $v(t) = L\,di/dt$. Complex circuit equations and system transfer functions can be written as sums of these simple differential equations.

One of the most powerful and easy techniques for solving a differential equation is the Laplace transform technique. The technique consists of mathematically transforming the time-domain differential equation into an algebraic equation in s and then manipulating the algebraic equation by the simple rules of algebra to obtain a solution in the s domain. The variable s is most generally complex and is written as

$$s = \alpha + j\omega \qquad\qquad (1\text{–}69)$$

where the real part α is associated with the time constant of exponentially rising or decaying responses and the imaginary part is associated with sinusoidal time variations and is the frequency ω in radians per second, or simply s^{-1}.

The Laplace transform, denoted $F(s)$, is the *frequency-domain* solution of the differential equation and often provides enough information about the system, in particular the system *frequency response*, that there may be no need to determine the time-domain solution $f(t)$. If the time-domain behavior is desired, the mathematical procedure of inversely transforming $F(s)$ is performed to find $f(t)$. Thus, the Laplace transform pair, which may be written as

$$f(t) \leftrightarrow F(s) \qquad\qquad (1\text{–}70)$$

indicates that the time-to-frequency-domain transformation can be reversed (inverted) by an appropriate mathematical process symbolized by the arrow.

Definition.   If a function of time $f(t)$ is multiplied by e^{-st} and integrated over the present and all future time, the result is the function $F(s)$ called the *one-sided Laplace transform* of $f(t)$. This is written as

$$\mathcal{L}[f(t)] = F(s) \triangleq \int_0^\infty f(t)e^{-st}\, dt \qquad (1\text{--}71)$$

A more general definition will include past time, but we will merely include the value of the function at $t = 0$ to represent the result of past behavior.

Some examples will aid in demonstrating how Laplace transforms are derived using Equation 1–71. Then Table 1–2 of Laplace transforms for common functions will be given, followed by a discussion and listing of important theorems and other operational fundamentals in Table 1–3.

TABLE 1–2 Laplace Transform Pairs (page 1 of 2)

Time Domain to Complex Frequency Domain		
$f(t)$	$\leftrightarrow$	$F(s)$
1. $\delta(t)$, unit impulse (defined in Equation 1–72)		1
2. 1, or unit step, $u(t)$		$\dfrac{1}{s}$
3. t		$\dfrac{1}{s^2}$
4. $\dfrac{1}{(n-1)!}\, t^{n-1}$		$\dfrac{1}{s^n}$
5. e^{-at}		$\dfrac{1}{s+a}$
6. $\dfrac{1}{(n-1)!}\, t^{n-1}e^{-at}$		$\dfrac{1}{(s+a)^n}$
7. $\dfrac{1}{a}(1 - e^{-at})$		$\dfrac{1}{s(s+a)}$
8. $\dfrac{1}{ab}\left[\dfrac{1 - be^{-at}}{b-a} + \dfrac{ae^{-bt}}{b-a}\right]$		$\begin{cases}\dfrac{1}{s(s+a)(s+b)} \\[2ex] = \dfrac{1}{s(s^2 + s(a+b) + ab)}\end{cases}$
9. $\dfrac{1}{ab}\left[a - \dfrac{b(\alpha - a)e^{-at}}{b-a} + \dfrac{a(\alpha - b)e^{-bt}}{b-a}\right]$		$\dfrac{s+\alpha}{s(s+a)(s+b)}$
10. $\dfrac{1}{b-a}(e^{-at} - e^{-bt})$		$\dfrac{1}{(s+a)(s+b)}$
11. $\dfrac{1}{a-b}(ae^{-at} - be^{-bt})$		$\dfrac{s}{(s+a)(s+b)}$
12. $\dfrac{1}{b-a}[(\alpha - a)e^{-at} - (\alpha - b)e^{-bt}]$		$\dfrac{s+\alpha}{(s+a)(s+b)}$

TABLE 1–2 Laplace Transform Pairs (page 2 of 2)

Time Domain to Complex Frequency Domain		
$f(t)$	$\leftrightarrow$	$F(s)$

13. $\sin(\omega t)$ $\qquad\qquad$ $\dfrac{\omega}{s^2 + \omega^2}$

14. $\cos(\omega t)$ $\qquad\qquad$ $\dfrac{s}{s^2 + \omega^2}$

14a. $(\sqrt{A^2 + (B/\omega)^2})(\cos(\omega t + \theta))$ $\qquad\qquad$ $\dfrac{As + B}{s^2 + \omega^2}$

$\qquad$ where $\theta = \arctan(-B/A)$

15. $\dfrac{\sqrt{a^2 + \omega^2}}{\omega} \sin(\omega t + \theta)$ $\qquad\qquad$ $\dfrac{s + a}{s^2 + \omega^2}$

$\qquad$ where $\theta = \arctan(\omega/a)$

16. $\dfrac{1}{\omega^2}(1 - \cos(\omega t))$ $\qquad\qquad$ $\dfrac{1}{s(s^2 + \omega^2)}$

17. $\dfrac{a}{\omega^2} - \dfrac{\sqrt{a^2 - \omega^2}}{\omega^2} \cos(\omega t + \theta)$ $\qquad\qquad$ $\dfrac{s + a}{s(s^2 + \omega^2)}$

$\qquad$ where $\theta = \arctan(\omega/a)$

18. $\dfrac{1}{b} e^{-at} \sin(bt)$ $\qquad\qquad$ $\begin{cases} \dfrac{1}{(s + a)^2 + b^2} \\ = \dfrac{1}{s^2 + 2as + (a^2 + b^2)} \end{cases}$

19. $e^{-at} \cos(bt)$ $\qquad\qquad$ $\dfrac{s + a}{(s + a)^2 + b^2}$

20. $\dfrac{1}{\omega_n\sqrt{1 - \delta^2}} e^{-\delta\omega_n t} \sin(\omega_n\sqrt{1 - \delta^2}t)$ $\qquad\qquad$ $\dfrac{1}{s^2 + 2\delta\omega_n s + \omega_n^2}$

21. $\dfrac{1}{a^2}(at - 1 + e^{-at})$ $\qquad\qquad$ $\dfrac{1}{s^2(s + a)}$

22. $\dfrac{1}{a^2}(1 - e^{-at} - ate^{-at})$ $\qquad\qquad$ $\dfrac{1}{s(s + a)^2}$

23. $\dfrac{\sqrt{(\alpha - a)^2 + b^2}}{b} e^{-at} \sin(bt + \theta)$ $\qquad\qquad$ $\begin{cases} \dfrac{s + \alpha}{(s + a)^2 + b^2} \\ = \dfrac{s + \alpha}{s^2 + 2as + (a^2 + b^2)} \end{cases}$

24. te^{-at} $\qquad\qquad$ $\dfrac{1}{(s + a)^2}$

25. $te^{-at} \sin(\beta t)$ $\qquad\qquad$ $\dfrac{2\beta(s + a)}{[(s + a)^2 + \beta^2]^2}$

26. $te^{-at} \cos(\beta t)$ $\qquad\qquad$ $\dfrac{(s + a)^2 - \beta^2}{[(s + a)^2 + \beta^2]^2}$

TABLE 1-3 Laplace Theorems and Properties

---

1. $f(t) \leftrightarrow F(s)$
                  where $F(s) = \int_{-\infty}^{\infty} f(t)e^{-st}\, dt$

2. $Af(t) \leftrightarrow AF(s)$
                  where $A = $ constant

3. $Af_1(t) + Bf_2(t) \leftrightarrow AF_1(s) + BF_2(s)$
                  linearity theorem

4. $\dfrac{d[f(t)]}{dt} \leftrightarrow sF(s) - f(0^+)$
                  first-order derivative*

4a. Initial-value theorem: $f(0^+) = \lim sF(s)$, $s \to \infty$.

5. $\dfrac{d^2 f(t)}{dt^2} \leftrightarrow s^2 F(s) - sf(0^+) - f'(0^+)$
                  $f'(t)$ is first derivative

5a. Initial value of derivative: $f'(0) = \lim [s^2 F(s) - sf(0^+)]$, $s \to \infty$

6. $f^{(n)}(t) \leftrightarrow s^n F(s) - s^{n-1}f(0^+) - s^{n-2}f'(0^+) \cdots - f^{(n-1)}(0^+)$

7. $\int f(t)\, dt \leftrightarrow \dfrac{1}{s} F(s) + \dfrac{1}{s} (\int f(t)\, dt)|_{t=0}$

8. $t^n f(t) \leftrightarrow (-1)^n F^{(n)}(s)$
                  (n) is nth derivative

9. $e^{-at}f(t) \leftrightarrow F(s + a)$
                  frequency shift

10. $2f(t) \cos(\omega_0 t) \leftrightarrow F(s + j\omega_0) + F(s - j\omega_0)$
                  DSB-SC modulation

11. $f(t - a)[u(t - a)] \leftrightarrow e^{-as}F(s)$
                  phase shift (time delay) only (Note the left-side truncation of $f(t)$ by $u(t - a)$.)

12. $f(at) \leftrightarrow \dfrac{1}{|a|} F\left(\dfrac{s}{a}\right)$
                  time scaling

13. $f_1(t) * f_2(t)$
                  convolution

    $= \int_{-\infty}^{\infty} f_1(\tau)f_2(t - \tau)\, d\tau \leftrightarrow F_1(s) F_2(s)$

---

* The plus superscript is used to avoid the confusion that often exists as to the value of $f(t)$ immediately after $t = 0$ when $f(t = 0)$ has a discontinuity (typical circuit initial conditions).

■ ■ ■ ■

EXAMPLE 1-10

Derive $F(s)$ for $f(t) = e^{-at}$ using the defining integral.

Solution:

$$F(s) = \int_0^{\infty} e^{-at}e^{-st}\, dt = \int_0^{\infty} e^{-(s+a)t}\, dt = \frac{1}{s + a}$$

Note that, since the Laplace integral starts at $t = 0$, $f(t) = e^{-at}$ is, *for Laplace*, the same as $e^{-at}u(t)$, where $u(t)$ is the unit step function defined by

$$u(t) = \begin{cases} 0, & t < 0 \\ 1, & t \geq 0 \end{cases}$$

FIGURE 1-26 (a) Unit impulse or Dirac delta function. (b) Unit step function.

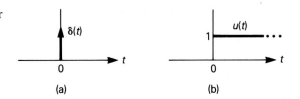

It should be noted that Table 1–2 does not list *every* transform pair that has been worked out by investigators. In fact, entire books containing nothing but transform pairs have been published.

The *unit impulse* or *Dirac delta* function of Equation 1 in Table 1–2 is a special function and is used for mathematical convenience in representing an ideal, zero-width pulse that is unrealizable in practice. It is commonly defined as

$$\int_{-\infty}^{\infty} \delta(t)\, dt = 1 \tag{1–72}$$

where $\delta(t) = 0$ for $t \neq 0$ (see Figure 1–26), and has the important property that

$$\delta(t) = \frac{du(t)}{dt} \tag{1–73}$$

Because of Equation 1–73, including the property that $\delta(t) = 0$ for $t \neq 0$, the value at $t = 0$ of any function $f(t)$ that is continuous at $t = 0$ can be expressed mathematically as

$$\int_{-\infty}^{\infty} f(t)\delta(t)\, dt = f(0) \tag{1–74}$$

Equation 1–74 is often used as the definition of the Dirac delta function, $\delta(t)$. See Appendix A.

■ ■ ■ ■

EXAMPLE 1–11

Derive the Laplace transform $F(s)$ for $\cos(\omega_0 t)$.

Solution:

From Euler's equation, $\cos(\omega_0 t)$ may be written as

$$\cos(\omega_0 t) = \frac{1}{2} e^{j\omega_0 t} + \frac{1}{2} e^{-j\omega_0 t} \tag{1–75}$$

and from Equation 5 of Table 1–2, $e^{-at} \leftrightarrow 1/(s + a)$, we may immediately write the Laplace transform pair

$$\cos(\omega_0 t) \leftrightarrow \frac{1}{2}\frac{1}{s - j\omega_0} + \frac{1}{2}\frac{1}{s + j\omega_0} = \frac{s}{s^2 + \omega_0^2}$$

The result is Equation 14 of Table 1–2.

Use of the Table of Transform Pairs

The process of finding a time-domain function that corresponds to a given transform is called *inverse transformation*. The symbol $\mathcal{L}^{-1}[F(s)]$ is called the inverse transform of $F(s)$. Finding the inverse transform is analogous to finding an antilogarithm. Thus, for example, since

$$\mathcal{L}[e^{-at}] = \frac{1}{s + a}$$

it is also true that

$$\mathcal{L}^{-1}\left[\frac{1}{s + a}\right] = e^{-at}$$

Also, $\mathcal{L}^{-1}(5) = 5\delta(t)$, $\mathcal{L}^{-1}[1/(s^2 + \omega^2)] = \sin(\omega t)$, and so on. It is often necessary to manipulate a transform algebraically in order to make it fit a form in the table.

■ ■ ■ ■

**EXAMPLE
1–12**

Find the inverse Laplace transforms of $F(s) = 40/(s^2 + 25)$.

Solution:

First, use simple algebra to put $F(s)$ in a form recognizable in the table.

$$F(s) = \frac{40}{s^2 + 25} = 8\left[\frac{5}{s^2 + 5^2}\right]$$

and, from pair 13 in Table 1–2,

$$f(t) = 8\sin(5t)$$

Circuit Impedance and Transfer Function

Determining input and output impedances of an electrical circuit is most conveniently accomplished by first expressing each component reactance in the complex frequency domain, $s = \sigma + j\omega$. Then the usual rules of Kirchhoff, Thevenin, and Norton may be applied.

The series RLC circuit of Figure 1–27a illustrates the resistance R, inductive reactance $X_L(s) = sL$, and capacitive reactance $X_C(s) = 1/\omega C$. Also, mutual inductive reactance would be $X_M(s) = sM$. Then, the circuit input impedance $Z_i(s)$ is

$$Z_i(s) = sL + 1/sC + R \qquad \text{(1–76a)}$$

$$= \frac{s^2LC + sRC + 1}{sC} \qquad \text{(1–76b)}$$

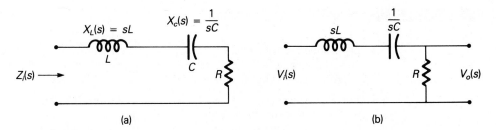

FIGURE 1-27

where Equation 1-76b is in *standard polynomial form* expressed generally by

$$F(s) = \frac{a_m s^m + \cdots + a_1 s + a_0}{b_n s^n + \cdots + b_1 s + b_0} \qquad \text{(1-77)}$$

The circuit impedance to a sinusoidal excitation is determined by letting $s = j\omega$ in Equation 1-76b as

$$Z(j\omega) = \frac{(j\omega)^2 LC + j\omega RC + 1}{j\omega C}$$

$$= \frac{-\omega^2 LC + j\omega RC + 1}{j\omega C} \qquad \text{(1-78a)}$$

$$= R + j(\omega L - 1/\omega C) \qquad \text{(1-78b)}$$

where the (expected) result in Equation 1-78b expresses the complex impedance in real and imaginary parts.

The *transfer function* of a network is a mathematical expression of the ratio of an output parameter to an input parameter. Thus, in Figure 1-27b the output parameter is voltage (as a function of complex frequency) and the input parameter indicated is $V_i(s)$. As shown, then, the transfer function $H(s)$ is written as

$$H(s) = \frac{V_o(s)}{V_i(s)} \qquad \text{(1-79)}$$

We can determine $H(s)$ in terms of circuit component values by using the voltage-divider rule and writing

$$H(s) = \frac{R}{X_L(s) + X_C(s) + R}$$

$$= \frac{R}{sL + 1/sC + R} \qquad \text{(1-80a)}$$

$$= \frac{sR/L}{s^2 + sR/L + 1/LC} \qquad \text{(1-80b)}$$

If the transient response of the circuit is not of interest but the steady-state sinusoidal response is, then make the substitution $s = j\omega$ in Equation 1-80 to get $H(j\omega)$.

■ PROBLEMS

1. Determine the reactance of a 300-pF capacitor and for a 1-μH inductor at 5 MHz, 10 MHz, and 15 MHz. Plot the reactances versus frequency on the same graph. At approximately what frequency do the "curves" cross?

2. For Figure 1–2, $r = 5\ \Omega$, $C = 300$ pF, and $L = 0.844\ \mu$H.
   a. Find the resonant frequency in rad/s and Hz.
   b. Calculate both reactances at resonance. Are they equal?
   c. What is the series impedance Z at resonance?

3. For the circuit of Problem 2, calculate the reactances and the impedance Z at 11 MHz (10% above the resonant frequency).

4. a. Determine the circuit Q and 3-dB bandwidth for the circuit of Problem 2.
   b. Determine the result of f_0, Q, and bandwidth for
      (1) r doubled to 10 Ω.
      (2) $r = 5\ \Omega$ but C doubled to 600 pF.
      (3) $r = 5\ \Omega$, $C = 300$ pF but L doubled.

5. $R = 500\ \Omega$, $C = 300$ pF, and $L = 0.844\ \mu$H are connected in parallel.
   a. Determine Z, Q, and bandwidth at resonance.
   b. If 10 V rms at f_0 is applied to the circuit, determine the current in R, C, and L.
   c. What is the ratio of i_C to i_R?

6. A small-signal tuned amplifier has a voltage gain of 100 at the resonant frequency 1 MHz.
   a. If the parallel-tuned circuit Q is 10, determine the gain and phase shift at 0.95 MHz.
   b. Repeat at 1.1 MHz.

7. An RF coil has a Q of 50 and reactance of $j1$ kΩ at 200 kHz. Determine the equivalent series and parallel RL circuit values r, L_s, R, and L_p.

8. Repeat Problem 7 if $Q = 2$.

9. A common-emitter amplifier employs a transistor with an emitter-bias current of 0.5 mA dc and a dynamic collector impedance of 20 kΩ in parallel with 1 pF. The collector circuit consists of a 1-kΩ load in parallel with $C = 49$ pF and $L = 2\ \mu$H ($Q_u = 50$).
   a. Determine the resonant frequency of the amplifier.
   b. If a signal voltage of 50 mV at the resonant frequency is measured on the transistor base, determine the output (collector) voltage.

10. a. Determine the bandwidth and write $A_v(f)$ in terms of Q and f_0 for the amplifier of Problem 9.
    b. Sketch $A_v(f)$, both magnitude and phase.

11. The circuit of Figure 1–11 has the following circuit values: $V_{cc} = 10$ V, $C_{bc} = 2$ pF, C (stray) = 2 pF, $C = 91$ pF, $R_1 = 2$ kΩ, $R_2 = 10$ kΩ, $R_E = 2.8$ kΩ, R_3 infinite, R_L is removed. The transformer has $k = 1$, $L = 4\ \mu$H with $Q_u = 104$, $n_p/n_s = 5$. Determine the following:
    a. Resonant frequency
    b. Base-to-collector voltage gain at the resonant frequency
    c. Base-to-v_o voltage gain at resonance
    d. Bandwidth.

e. How much would the gain decrease (in dB) and bandwidth increase if a 50-Ω resistor is connected to the output at v_o?

12. A 50% efficient, transformer-coupled grounded-emitter amplifier with $V_{cc} = 12$ V should have how much bias current if 100 mW is to be delivered to the load?

13. Design a 5-mW ideal-transformer small-signal RF amplifier for 20-MHz operation with 2 MHz of bandwidth and a 75-Ω load and generator impedance. Use Figure 1–12 with C (stray) = 5 pF, $C_{CB} = 2$ pF, $\beta = 100$, $r_c = 10$ kΩ, $R_3 = 10$ kΩ, and $V_{cc} = 9$ V. (Assume 50% collector efficiency.)

14. Design an RF amplifier for P_o (max) = 7 mW into $R_L = 100$ Ω at $\omega_0 = 10^7$ rad/s and bandwidth of 10^6 rad/s. T_1 consists of $X_{L_1} = 100$ Ω with $Q_u = 100$ at ω_0, and $k = 1$. The transistor has $V_{BE} = 0.7$, $\beta = 100$, $C_{BC} = 3$ pF, $C_{BE} = 27$ pF. Determine
a. R_1, R_2, and R_E.
b. Turns ratio of T_1 given the bandwidth requirement.
c. Amplifier ac input impedance, R_i and X_i (input resistance and parallel capacitive reactance).

15. The circuit of Figure 1–15 has $C_1 = 500$ pF, $C_2 = 75$ pF, and $L = 220$ nH.
a. Determine the resonant frequency ω_0 and f_0.
b. If a resistor of 580.8 Ω is in parallel with the inductor, what is the circuit Q?
c. What is the equivalent value of resistance as seen from the output for **b**?
d. If the voltage across the inductor is 10 V, what voltage will be measured across C_1?

16. Inductor $L_1 = 100$ mH and $L_2 = 4$ mH are close enough to each other that for every 10 lines of flux created by RF current in L_1, 1 line of flux couples to L_2. What is the mutual inductance between the inductors?

17. Two inductors are wound to form a transformer. Measurements at 7.9 MHz yield the following results: $L_p = 20$ μH and $Q = 220$ is the primary inductance. $L_s = 4$ μH and $Q = 100$ is the secondary.
a. What value is M if $k = 0.2$?
b. If the secondary is open, how much power loss will result due to 10 mA of primary current?
c. Determine the secondary voltage magnitude and phase for 10 mA of primary current.

18. A transformer was wound in the laboratory and measured at 7.9 MHz with the following results: $L_p = 4.25$ μH with $Q = 220$ and $L_s = 1.25$ μH with $Q = 100$.
a. What value of k (coefficient of coupling) is needed to produce a mutual inductance, M, of 0.3 μH?
b. If $k = 0.5$ and $f = 7.9$ MHz, what value of primary current is needed to produce 200 mV across a 50-Ω secondary load resistor?
c. Repeat **b** for a frequency of 500,000 rad/s.
d. Determine the primary- and secondary-winding resistances.
e. What value of C would be required in order to resonate the secondary circuit at 7.9 MHz?

19. For the air-coupled transformer of Figure 1–22, assume $\omega M = 10$ Ω, $r_1 = 5$ Ω, $r_2 = 3$ Ω, $X_{L_1} = 40$ Ω, $X_{L_2} = 20$ Ω, and $i_p = 10$ mA.
a. Computer v_L for (i) $Z_L = 0$ and (ii) $Z_L = j5$ Ω.
b. Repeat **a** for $v_p = 2$ V rms (i_p not initially known).

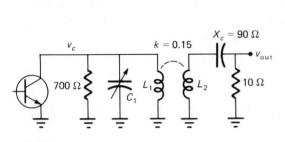

FIGURE 1-28   FIGURE 1-29

20. In Figure 1-22 $L_1 = 100$ μH, $L_2 = 20$ μH, $r_1 = 2$ Ω, $r_2 = 1$ Ω, and $k = 0.8$.
    a. Determine v_L for $Z_L = 50 - j50$ Ω and $i_p = 0.1$ A, 0.5 MHz.
    b. Repeat a for $i_p = 0.1$ A, 397887 Hz. Compute the Q_u of the transformer windings, primary and secondary.

21. The input signal frequency is $\omega = 10^7$ rad/s for Figure 1-28. $L_1 = L_2 = 10$ μH with Q_u infinity. Calculate each of the following:
    a. X_{L_2}
    b. Z_{ss}
    c. M
    d. Z_r (reflected into primary winding)
    e. Value of C_1 to tune the circuit to resonance

22. Figure 1-29 is a double-tuned circuit operating at resonance.
    a. Determine k_c, the coefficient of coupling for a critically coupled circuit.
    b. If $k = 0.09$, will the circuit be overcoupled, undercoupled, or optimally coupled? (Show calculations.)

23. A double-tuned circuit is shown in Figure 1-30. If $f = 159$ kHz ($\omega = 10^6$ rad/s), determine
    a. C_s to resonate the secondary.
    b. Q_s (Q of secondary circuit).
    c. Value of C_p for resonance.
    d. Q_p, primary circuit Q with secondary open. (*Hint:* Think "series resonance.")
    e. The value of k that would make the complete circuit critically coupled.
    f. Is the circuit shown undercoupled, overcoupled, critically or optimally coupled? Why?

FIGURE 1-30

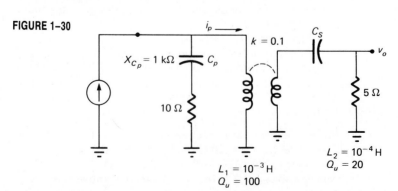

g. Sketch an *approximate* frequency response for this circuit. Include the value of v_o for a circuit input of 0.12 mA rms, and show the approximate 3-dB bandwidth.

24. Use the circuit of Figure 1–30 with the following circuit value changes: $L_1 = 4 \; \mu H$, $L_2 = 1 \; \mu H$, $C_s = 100 \; pF$, $R_L = 50 \; \Omega$, $k = 0.5$, and $i(t) = 0.01 \; A$ rms. Change the 10-Ω resistor in Figure 1–30 to 0 Ω and find

a. C_p for a double-tuned circuit at resonance.

b. Is the overall double-tuned circuit overcoupled, undercoupled, critically or optimally coupled, or what?

c. Find v_o at ω_0.

d. Repeat **c** at $\omega = \omega_0/\sqrt{1-k}$.

25. a. Interchange the secondary circuit capacitor and resistor in Figure 1–25 and plot the resulting frequency response for $k = 0.22$ using a circuit simulator software package such as SPICE or Micro-CAP II.

b. Use circuit simulator software to plot the frequency response of Problem 23.

26. Show that $Z(s)$ for Figure 1–7 can be written as $s/[C(s^2 + s/RC + 1/LC)]$.

27. Show that the transfer function, $H(s)$, for the tuned circuit with capacitive voltage divider of Figure 1–16 is

$$H(s) = \frac{V_o(s)}{V_i(s)} = \frac{C_2}{C_1 + C_2} \frac{s}{s + (1/[R_o(C_1 + C_2)])}$$

28. Show that

$$Z_i(s) = \frac{sL[sR_o(C_1 + C_2) + 1]}{s^3 LC_1 C_2 R_o + s^2 LC_2 + sR_o(C_1 + C_2) + 1}$$

for Figure 1–16.

29. a. Derive $Z_i(s) = [V_i(s)/I_i(s)]$ for the circuit of Figure 1–31. (You do not need to work with $I(s)$ and $V(s)$.) Write the results in a standard polynomial form.

b. Derive the Laplace transfer function, $H(s) = V_o(s)/V_i(s)$, for the circuit. Write the result in standard polynomial form.

30. a. Write $Z_i(j\omega)$ in standard polynomial form for the circuit of Figure 1–31.

b. Repeat **a** for $H(j\omega)$.

31. Place a capacitor C_2 from the input pin of Figure 1–31 to ground and repeat Problem 29 using $H(s) = V_o(s)/I_i(s)$. Show that the results are

$$Z_i(s) = \frac{s^2 RLC_1 + sL + R}{s^3 RLC_1 C_2 + s^2 LC_2 + sR(C_1 + C_2) + 1}$$

and

$$H(s) = \frac{V_o(s)}{I_i(s)} = \frac{R}{s^3 RLC_1 C_2 + s^2 LC_2 + sR(C_1 + C_2) + 1}$$

FIGURE 1–31

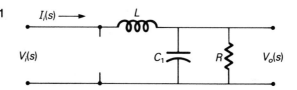

1-11 ■ COMPUTER ANALYSIS

Using the computer for communications system and circuit analysis has become a virtual necessity in education and industry. What follows are examples of programs and/or readily available simulation software for solving problems of the type covered in this chapter. Communications analysis software is discussed in Appendix E.

The BASIC program is interactive in the sense that an introductory sentence or short paragraph describes the purpose of the program and the user is then prompted for input data.

The programs for Chapter 1 are as follows:

- **PSpice tuned amplifier analysis:** A small-signal RF amplifier was analyzed using the student version of PSpice. The circuit file is Figure 1–32a, and circuit diagram with node numbers is Figure 1–32b; note that a 50-Ω generator is used. Figure 1–32c shows the output voltage and phase versus frequency for a 40-mV input. Figure 1–32d shows the transistor collector signal voltage versus frequency.

- **Transformer coupling characteristics program:** This program, written in BASIC, will solve double-tuned circuit problems based on the circuit of Figure 1–25. Thus, you can verify the results of Example 1–9. The listing is in Figure 1–33. The user is prompted for circuit component values, transformer coupling coefficient, and input ac current magnitude.

- **Coupling transformer simulation:** A PSpice simulation of the double-tuned circuit of Figure 1–25 with $k = 0.082155$ is provided. The circuit file is in Figure 1–34a and the output magnitude and phase are plotted versus frequency in Figure 1–34b. Note that the transformer is very nearly critically coupled.

- **RF amplifier analysis:** The tuned amplifier of Example 1–5 was analyzed on PSpice. The results are presented in Figure 1–35.

```
*** 07/11/89 *** Evaluation PSpice (Jan. 1988) *** 16:45:45 ***
    SPICE SMALL SIGNAL RF AMPLIFIER
****        CIRCUIT DESCRIPTION
****************************************************************
    R1 1 2 50
    R2 3 0 2K
    R3 3 6 10K
    R4 4 0 1K
    R5 5 6 5K
    R6 7 0 50
    R7 6 0 10MEG
    R8 6 8 100
    C1 2 3 .01UF
    C2 4 0 .1UF
    C3 3 5 2PF
    C4 5 0 3PF
    C5 5 0 100PF
    C6 6 0 .01UF
    Q1 5 3 4 QNL
    .MODEL QNL NPN BF=100
    L1 5 6 20UH
    L2 7 0 .138UH
    K1 L1 L2 0.9999
    VCC 8 0 DC 12
    VIN 1 0 AC .04
    .AC DEC 100 2.2MEG 5.0MEG
    .PLOT AC VM(7) VP(7)
    .PLOT AC VM(5)
    .PLOT AC VM(3)
    .PROBE
    .END
****** SMALL SIGNAL BIAS SOLUTION     TEMPERATURE = 27.000 DEG C
    NODE  VOLTAGE   NODE  VOLTAGE   NODE  VOLTAGE   NODE  VOLTAGE
    (1)   0.0000    (2)   0.0000    (3)   1.9451    (4)   1.1671
    (5)  11.7860    (6)  11.7860    (7)   0.0000    (8)  12.0000
```

FIGURE 1–32a   PSpice tuned amplifier analysis.

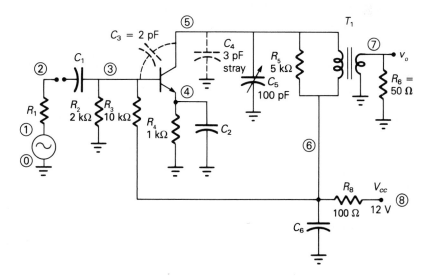

FIGURE 1–32b   Circuit with nodes for PSpice analysis.

```
******** AC ANALYSIS                      TEMPERATURE =   27.000 DEG C
*: VM(7)
+: VP(7)
  FREQ          VM(7)
(*)---------- 1.00E-2   1.00E-1    1.00E+0    1.00E+1    1.00E+2
(+)----------2.00E+2   -1.00E+2    0.00E+0    1.00E+2    2.00E+2
  2.4E+6  8.2E-2.         *+          .          .          .
  2.47E+6 8.8E-2.         X           .          .          .
  2.5E+6  9.5E-2.         X           .          .          .
  2.58E+6 1.0E-1.         +*          .          .          .
  2.65E+6 1.1E-1.        +.*          .          .          .
  2.7E+6  1.2E-1.        +. *         .          .          .
  2.8E+6  1.3E-1.        +. *         .          .          .
  2.83E+6 1.4E-1.        +.  *        .          .          .
  2.9E+6  1.6E-1.       + .  *        .          .          .
  2.97E+6 1.9E-1.       + .    *      .          .          .
  3.04E+6 2.1E-1.      +  .      *    .          .          .
  3.1E+6  2.5E-1.      +  .        *  .          .          .
  3.2E+6  2.9E-1.     +   .        *  .          .          .
  3.25E+6 3.4E-1.   +     .       *   .          .          .
  3.33E+6 3.8E-1.  +      .      *     .          .          .
  3.4E+6  4.1E-1. +       .      *     .          .          .
  3.49+6  3.9E-1.         .      *     .          .        +    .
  3.57E+6 3.5E-1.         .      *     .          .      +      .
  3.65E+6 3.1E-1.         .    *       .          .   +         .
  3.74E+6 2.6E-1.         .  *         .          . +           .
  3.8E+6  2.3E-1.         . *          .          .+            .
  3.9E+6  1.9E-1.        . *           .          .+            .
  4.0E+6  1.7E-1.        . *           .          . +           .
  4.1E+6  1.5E-1.        . *           .          . +           .
  4.2E+6  1.4E-1.       . *            .          .+            .
  4.3E+6  1.3E-1.       . *            .          .+            .
  4.39E+6 1.2E-1.        .*            .          .+            .
  4.49E+6 1.1E-1.        .*            .          .+            .
  4.6E+6  9.7E-2.        *             .          .+            .
  4.7E+6  9.1E-2.        *             .          +             .
  4.8E+6  8.4E-2.        *             .          +             .
  4.9E+6  7.9E-2.       *.             .          +             .
  5.04E+6 7.4E-2.       *.             .          +             .
         - - - - - - - - - - - - - - - - - - - - - - - - - - -
```

FIGURE 1–32c   PSpice phase and magnitude plot for circuit of Figure 1–32a.

```
SPICE SMALL SIGNAL RF AMPLIFIER
**** AC ANALYSIS                        TEMPERATURE = 27.000 DEG C
*: VM(5)


   FREQ        VM(5)
(*)----------    1.0000E-01    1.0000E+00    1.0000E+01
                         - - - - - - - - - - - - - -
 2.200E+06   7.604E-01 .              *.            .
 2.251E+06   8.046E-01 .              *.            .
 2.304E+06   8.534E-01 .              *             .
 2.357E+06   9.078E-01 .              *             .
 2.412E+06   9.688E-01 .              *             .
 2.468E+06   1.037E+00 .              .*            .
 2.526E+06   1.116E+00 .              .*            .
 2.585E+06   1.205E+00 .              . *           .
 2.645E+06   1.309E+00 .              . *           .
 2.707E+06   1.430E+00 .              .  *          .
 2.770E+06   1.573E+00 .              .  *          .
 2.834E+06   1.744E+00 .              .   *         .
 2.900E+06   1.953E+00 .              .   *         .
 2.968E+06   2.209E+00 .              .    *        .
 3.037E+06   2.528E+00 .              .     *       .
 3.108E+06   2.927E+00 .              .      *      .
 3.180E+06   3.418E+00 .              .      *      .
 3.254E+06   3.985E+00 .              .       *     .
 3.330E+06   4.530E+00 .              .        *    .
 3.407E+06   4.830E+00 .              .        *    .
 3.487E+06   4.699E+00 .              .        *    .
 3.568E+06   4.225E+00 .              .        *    .
 3.651E+06   3.650E+00 .              .       *     .
 3.736E+06   3.125E+00 .              .      *      .
 3.823E+06   2.691E+00 .              .     *       .
 3.912E+06   2.343E+00 .              .    *        .
 4.003E+06   2.064E+00 .              .    *        .
 4.097E+06   1.838E+00 .              .   *         .
 4.192E+06   1.653E+00 .              .  *          .
 4.290E+06   1.500E+00 .              .  *          .
 4.390E+06   1.371E+00 .              . *           .
 4.492E+06   1.261E+00 .              . *           .
 4.596E+06   1.166E+00 .              .*            .
 4.704E+06   1.084E+00 .              .*            .
 4.813E+06   1.012E+00 .              .*            .
 4.925E+06   9.480E-01 .              *             .
 5.040E+06   8.912E-01 .              *             .
                         - - - - - - - - - - - - - -
```

FIGURE 1–32d PSpice plot of signal magnitude at node 5 (to ground).

```
10 REM **   COUPLING CHARACTERISTICS OF A DOUBLE-TUNED CIRCUIT **
20 REM **          EXAMPLE OF STUDENT COMPUTER ASSIGNMENT        **
30 REM ***            FOR PROFESSOR P. YOUNG, AT ASU            ***
40 CLS
50 PRINT"THIS PROGRAM WILL CALCULATE THE COUPLING CHARACTERISTICS
60 PRINT"OF DOUBLE-TUNED CIRCUITS AT RESONANCE. INPUTS CONSIST OF
70 PRINT"COMPONENT VALUES, COUPLING COEFFICIENT AND INPUT CURRENT.
75 PRINT "See Figure 1-25 of ELECTRONIC COMMUNICATION TECHNIQUES"
80 PRINT:PRINT:INPUT "Enter coefficient of coupling (k<1), k=";K
90 PRINT:INPUT "Enter current source value in amperes, I=";IP
100 PRINT:INPUT"Enter parallel primary resistance in ohms, R1=";R1
110 INPUT"Enter primary parallel capacitor value in farads,C1=";C1
120 INPUT "Enter series primary resistor value in ohms, R2=";R2
130 PRINT:INPUT "Enter primary coil inductance in henrys, L1=";L1
140 PRINT:INPUT"Enter secondary coil inductance in henrys,L2=";L2
150 PRINT:INPUT "Enter secondary resistor value in ohms, R3=";R3
160 PRINT:INPUT"Enter secondary capacitor value in farads,C2=";C2
170 WO = 1/SQR(L2*C2) :REM Wo = RESONANT FREQUENCY
180 XL1 = WO*L1 :REM REACTANCE OF L1
190 QU1 = XL1/R2 :REM UNLOADED Q OF PRIMARY
200 XL2 = WO*L2:QS = XL2/R3
210 XC2 = XL2 :REM AT RESONANCE THE REACTANCES ARE EQUAL
220 XC1 = XL1
230 QUPAR = R1/XC1 :REM Q OF PARALLEL CIRCUIT R1-C1 IN PRIMARY
240 RSER = R1/(1+(QUPAR^2)) : REM SERIES EQUIVALENT OF R
250 XCSER = (XC1*(QUPAR^2))/(QUPAR^2 + 1) : REM SERIES EQUIV CAP.
260 RTOT = RSER + R2 :REM TOTAL RESISTANCE OF PRIMARY BEFORE ZR
270 QP = XL1/RTOT : REM PRIMARY EFFECTIVE Q
280 KC = 1/SQR(QP*QS):KOPT = 1.5*KC
290 M = K*SQR(L1*L2) :REM MUTUAL INDUCTANCE
300 WOM = WO*M :REM MUTUAL REACTANCE
310 ZSS = R3 :REM ZSS IS PURELY RESISTIVE
320 ZR = WOM^2/ZSS :REM REFLECTED RESISTANCE, PURELY RESISTIVE
330 XBR1MAG = SQR(RSER^2 + XCSER^2) :REM Z OF CAPACITIVE BRANCH
340 IPMAG = (IP*XBR1MAG)/(RTOT + ZR) :REM MAGNITUDE OF PRI. CURR
350 VSMAG = WOM*IPMAG :REM VSMAG=MODELED SECONDARY SOURCE VOLTAGE
360 ISMAG = VSMAG/ZSS :REM SECONDARY CURRENT
370 VOMAG = ISMAG * XC2 :REM OUTPUT VOLTAGE IN SECONDARY
380 PHARAD = -(ATN(XCSER/RSER)) : REM PHASE ANGLE FOR PRI-I, RAD
390 IPPHA = PHARAD*(180/3.14) :REM CONVERTS RADS TO DEG.
400 VSPHA = IPPHA - 90 : REM SEC. VOLTAGE IS -90  OUT OF PHASE
410 VOPHA = VSPHA - 90 :REM FOR OUTPUT CAP., SHIFT -90 DEGREES
420 IF K > KOPT THEN 450 :REM OVERCOUPLING CONDITION
430 IF K < KC THEN 460 : REM UNDER COUPLING CONDITION
440 IF K = KOPT THEN 470:IF K = KC THEN 480:GOTO 490
450 PRINT "THE CIRCUIT IS OVERCOUPLED":GOTO 490
460 PRINT "THE CIRCUIT IS UNDERCOUPLED":GOTO 490
470 PRINT "k = KOPT" :GOTO 490
480 PRINT "k = KC"
490 PRINT "The output voltage is "VOMAG" V"
500 PRINT "The output phase is "VOPHA" degrees"
510 PRINT "The resonant frequency in radians per second is "WO
520 PRINT "The critical coupling coefficient is "KC
530 PRINT "The optimum coupling coefficient is "KOPT
540 PRINT "Mutual Inductance = " M" henrys
550 PRINT "Reflected Impedance = " ZR" ohms" :PRINT:PRINT
560 PRINT:INPUT "WOULD YOU LIKE TO DO ANOTHER? (YES/NO) ";QUES$
570 IF QUES$ = "YES" THEN 40:IF QUES$ = "NO" THEN 580:GOTO 560
580 END
```

FIGURE 1-33 Transformer coupling characteristics program.

```
******** 5/19/88******** Demo PSpice (May 1986) *******11:24:07********
COUPLING TRANSFORMER SIMULATION

IAC 1 0 AC .001
C1 1 0 1000PF
C2 3 0 10000PF
R1 1 0 100K
R2 1 2 20
R3 3 4 10.0
L1 2 0 .001
L2 4 0 .0001
* COUPLING COEFFICIENT K=1.5Kc
K1 L1 L2 .082155
.AC LIN 40 100K 210K
.PLOT AC VM(3) VP(3)
.END
```

FIGURE 1-34a   Coupling transformer simulation. The circuit is Figure 1-25.

```
****      AC ANALYSIS                    TEMPERATURE =    27.000 DEG C

LEGEND:
*: VM(3)
+: VP(3)
(*)------------  3.162D-02    1.000D-01    3.162D-01    1.000D+00    3.162D+00

            - - - - - - - - - - - - - - - - - - - - - - - - - -

(+)------------ -2.000D+02   -1.000D+02    .000D+00    1.000D+02    2.000D+02
    FREQ      VM(3)    - - - - - - - - - - - - - - - - - - - - - - - -
  1.197D+05  1.032D-01 .              X           .           .           .
  1.226D+05  1.201D-01 .           +. *           .           .           .
  1.254D+05  1.413D-01 .           +.   *         .           .           .
  1.282D+05  1.686D-01 .           +.     *       .           .           .
  1.310D+05  2.044D-01 .           +.       *     .           .           .
  1.338D+05  2.528D-01 .          + .         *   .           .           .
  1.367D+05  3.203D-01 .          + .           * .           .           .
  1.395D+05  4.185D-01 .          + .             .*          .           .
  1.423D+05  5.684D-01 .         + .              .       *   .           .
  1.451D+05  8.107D-01 .        + .               .           *           .
  1.479D+05  1.223D+00 .     +    .                .           .*          .
  1.508D+05  1.902D+00 .  +       .                .           .        *   .
  1.536D+05  2.628D+00 .          .                .           .    +    *  .
  1.564D+05  2.784D+00 .          .                .           .      +    *.
  1.592D+05  2.662D+00 .          .                .           .   +.       *
  1.621D+05  2.592D+00 .          .                .       +    .        *   .
  1.649D+05  2.387D+00 .          .                . +         .        *
  1.677D+05  1.825D+00 .          .                +           .        *
  1.705D+05  1.233D+00 .          .            +   .           .*
  1.733D+05  8.334D-01 .          .          +     .       *   .
  1.762D+05  5.877D-01 .          .        +       .    *      .
  1.790D+05  4.331D-01 .          .      +     .   *          .
  1.818D+05  3.312D-01 .          .    +       .*           .           .
  1.846D+05  2.611D-01 .          .   +      * .            .           .
  1.874D+05  2.109D-01 .          . +      *   .            .           .
  1.903D+05  1.737D-01 .          . + *       .            .           .
  1.931D+05  1.456D-01 .          . X          .            .           .
            - - - - - - - - - - - - - - - - - - - - - - - - - -
```

FIGURE 1–34b   PSpice magnitude and phase plot for the double-tuned circuit of Figure 1–25.

```
        ** 4/11/89*** Demo PSpice (May 1986) *** 0:16:26***

RF AMPLIFIER ANALYSIS

****       CIRCUIT DESCRIPTION
*****************************************************
 * Analysis of EXAMPLE 1-5   Electronic Communication Techniques

 *Resistors
 RG 1 2 1
 R1 3 0 8.5K
 R2 3 7 47K
 RE 4 0 1K
 R4 5 7 10K
 RDN 6 7 100
 RL 8 0 50
 *CAPACITORS
 Cc 2 3 2000PF
 CE 4 0 .0075UF
 CS 5 0 3PF
 Ct 5 0 60PF
 CDN 7 0 .002UF
 *TRANSFORMER
 L1 5 7 4UF
 L2 8 0 20.4NF
 K1 L1 L2 1
 *VOLTAGE SOURCES
 VCC 6 0 DC 12V
 VG 1 0 AC 10MV
 *TRANSISTOR
 Q1 5 3 4 GNL
 .MODEL GNL NPN PE=.6 VA=17.5 BF=50 CJE=15PF CJC=1PF
 *OUTPUT SPECIFICATIONS
 .WIDTH OUT=70
 .AC LIN 25 8.5MEG 11.5MEG
 .PRINT AC VDB(8) VM(5)
 .PLOT AC VM(8) VP(8)
 .END
```

FIGURE 1–35a

```
 ** 4/11/89*** Demo PSpice (May 1986) *** 0:16:26***
 ****       BJT MODEL PARAMETERS

             GNL
 TYPE        NPN
 IS          1.00D-16
 BF          50.000
 NF          1.000
 VAF         1.75D+01
 BR          1.000
 NR          1.000
 CJE         1.50D-11
 VJE          .600
 CJC         1.00D-12
```

FIGURE 1–35b

```
        ** SMALL SIGNAL  BIAS SOLUTION      TEMPERATURE = 27.0   DEG C ***

        NODE  VOLTAGE     NODE  VOLTAGE     NODE  VOLTAGE     NODE  VOLTAGE

        (1)   .0000       (2)   .0000       (3)   1.7325      (4)   .9712
        (5)  11.8825      (6)  12.0000      (7)  11.8825      (8)   .0000
```

FIGURE 1–35c

```
*** 4/11/89****** Demo PSpice (May 1986) ******** 0:16:26********
RF AMPLIFIER ANALYSIS
*****  AC ANALYSIS           TEMPERATURE = 27 DEG C
**************************************************************

LEGEND:

*: VM(8)
+: VP(8)

     FREQ       VM(8)

(*)----------- 1.585D-02    2.512D-02    3.981D-02    6.31D-02    1.0D-01
                          - - - - - - - - - - - - - - - - - - - - - - -

(+)---------- -2.0D+02      -1.0D+02      .0D+00      1.0D+02     2.0D+02
                          . - - - - - - - - - - - - - - - - - - - - - -
8.500D+06/1.770D-02.  *           .+          .           .          .
8.625D+06/1.925D-02.    *         .+          .           .          .
8.750D+06/2.107D-02.        *     +           .           .          .
8.875D+06/2.320D-02.          *   +           .           .          .
9.000D+06/2.576D-02.            +*            .           .          .
9.125D+06/2.887D-02.            +    *        .           .          .
9.250D+06/3.271D-02.           +.       *     .           .          .
9.375D+06/3.756D-02.           +.          *  .           .          .
9.500D+06/4.381D-02.           +.            *.           .          .
9.625D+06/5.201D-02.          + .             .       *   .          .
9.750D+06/6.282D-02.         + .              .           *          .
9.875D+06/7.657D-02.      +    .              .           . *        .
1.000D+07/9.133D-02.    +      .              .           .      *   .
1.013D+07/9.989D-02.   +       .              .           .          *
1.025D+07/9.520D-02.           .              .           .      +  *.
1.038D+07/8.187D-02.           .              .           .    +*    .
1.050D+07/6.808D-02.           .              .         . *   +      .
1.063D+07/5.689D-02.           .              .       * . +          .
1.075D+07/4.833D-02.           .              .   *     . +          .
1.088D+07/4.180D-02.           .            .*         . +           .
1.100D+07/3.675D-02.           .          * .          . +           .
1.113D+07/3.275D-02.           .      *     .          . +           .
1.125D+07/2.953D-02.           .    *       .          .+            .
1.138D+07/2.689D-02.        . *             .          .+            .
1.150D+07/2.469D-02.           *            .          .+            .
                          - - - - - - - - - - - - - - - - - - - - - - -

        TOTAL JOB TIME           67.72
```

FIGURE 1–35d

2

Oscillators

■ INTRODUCTION

An oscillator is used as a signal source and is usually the first circuit needed in developing a communications system. In digital/data systems, the timing clock is derived from an oscillator and the information carrier for high-frequency transmissions is produced by an oscillator. There are numerous oscillator circuits, most of which are still referred to by the name of the originator, including Armstrong, Hartley, Colpitts, Clapp, Pierce, and Wien bridge.

   More important than naming oscillators is understanding how they work. Also, circuits that are not supposed to oscillate sometimes do, and knowing the fundamentals of oscillators can help to solve such knotty problems. The criteria for producing self-sustaining oscillations are

- Input power source
- Gain
- Frequency determining scheme
- Feedback must be positive (in-phase feedback; system loop gain = 1)

Input power is required because the oscillator must deliver power to be useful. In addition, the oscillator circuit itself consumes power and requires device biasing.

Some device providing signal *gain* is required to overcome internal signal losses.

Some circuit or inherent device resonance is required to control the frequency of an oscillator. All sorts of circuits and natural phenomena, from metal cavities and tuning forks to the atomic structures of gases and solids, are used to control the frequency of an oscillator.

Finally, an oscillator is a feedback system in which the feedback conforms to two criteria, called the *Barkhausen criteria*, namely:

- The feedback signal must be exactly in phase with the original input signal at the loop closure point.
- The overall steady-state gain around the feedback loop must be exactly equal to unity. Stated mathematically, using the notation of the classical feedback circuit, $A_vB = 1$.

The classical electronic system with feedback is block diagramed in Figure 2–1; the system gain is derived by noting that

$$v_{\text{fb}} = Bv_o \qquad \text{(2–1a)}$$

and

$$v_o = (v_i + v_{\text{fb}})A_v \qquad \text{(2–1b)}$$

Substituting Equation 2–1a in Equation 2–1b to eliminate v_{fb} and then collecting terms with v_o and v_i yields the following classical equation:

$$\frac{v_o}{v_i} = \frac{A_v}{1 - A_vB} = A_{\text{fb}} \qquad \text{(2–1c)}$$

FIGURE 2–1 Classical feedback system.

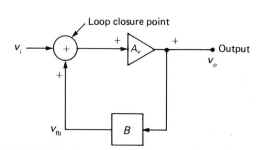

In Figure 2–1 and Equation 2–1c, A_v is the (forward) system gain (assumed to be noninverting), and B is the fraction of the output fed back to the input, also called the *feedback factor*. A_{fb} is the overall *system* gain with feedback.

Notice that, for a negative feedback system, the loop gain $A_v B$ must have a phase inversion to achieve the stability that is desired in such systems. Hence, for systems with negative feedback, $A_{fb} = A_v/(1 + A_v B)$.

In Equation 2–1 for positive (regenerative) feedback systems we see that when the Barkhausen criteria are met and $A_v B = 1$, the overall system gain goes to infinity; that is,

$$A_{fb} = A_v/(1 - 1) \rightarrow \infty$$

This is definitely undesirable in negative feedback systems but is an absolute necessity for oscillators.

Consider the behavior of an actual circuit for the usual condition of loop gain, $A_v B > 1$. The circuit of Figure 2–2 shows a tuned amplifier with a transformer-coupled output. R represents the total loading of the amplifier including any impedance from output to ground.

Suppose that the base-to-collector voltage gain is $A_v = -100 = +100 \; \underline{/180°}$. Also, assume that the transformer is inverting and ideal ($k = 1$ and no power losses). If the transformer turns ratio is $n_p : n_s = 90 : 1$, then $B = -\frac{1}{90} = 0.0111$ and $A_v B = +1.11$, which is more than enough loop gain. With $v_B = 10$ mV and $v_C = 1000$ mV, the output voltage will be $v_o = 1000 \, \text{mV}/90 = 11.1$ mV. Now if S_1 is switched to pin 1, the 11.1 mV is amplified to 1110 mV and stepped down to 12.3 mV, which will be fed back to the input and reamplified to 1.23 V, and the signal is building. Where will it stop? It will stop when this linear system saturates, and this will occur for a collector voltage of 24 V pk-pk. The 24 V pk-pk is based on the premise of a high-Q tuned circuit, a 12-V power supply, and $V_{CE}(\text{sat}) = 0$. Recall that, with $+12$ V dc to ground on the collector, the collector can swing from $+12$ V to 0 on the negative swing. Also, because of energy stored in the transformer during one-half of the cycle, the collector can swing 12 V above the bias point—to $+24$ V pk—on the other half-cycle. The total swing is 24 V pk-pk.

With 24 V pk-pk on the collector, the output voltage will be 24 V/90 = 266.7 mV. This, of course, is the voltage on the base because of the feedback. This voltage will be a little distorted, but that isn't the question here. The question is: What is the loop gain? We know that $B = -\frac{1}{90}$ because of the transformer. The

FIGURE 2–2

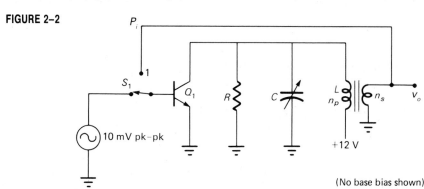

(No base bias shown)

gain of the saturated amplifier is $v_C/v_B = $ (24 V pk-pk)/(266.7 mV pk-pk) $= -90$. The loop gain is $A_v B = 90 \times \frac{1}{90} = +1$. The Barkhausen criteria are met exactly for this oscillator.

How does this oscillator get started? The same way that a (resonant) tuning fork gets started—by hitting it with an impulse. The signal generator in Figure 2–2 isn't actually needed. Suppose that the 12-V power supply has been off for a while and there is no current in the circuit. When the power supply is switched on, a surge of bias current rushes through the transformer primary winding, causing the resonant circuit to ring, to oscillate. This oscillating voltage is coupled across the transformer to the output, where it is fed back to the input and amplified to increase the collector signal. The signal continues to build until a steady state is reached with 24 V pk-pk on the collector and 266.7 mV pk-pk at the output.

The circuit will oscillate at the frequency for which the phase around the loop is exactly 360°. This is set by the tank resonant frequency

$$f_{\text{res}} = 1/(2\pi\sqrt{LC})$$

as derived in Equation 1–7a, and the quality factor Q of the transformer with circuit loading. The higher the transformer-loaded Q is, the faster the phase varies with frequency. The actual frequency of oscillation is given by

$$f_{\text{osc}} = f_{\text{res}}(Q^2 + 1)/Q^2 \qquad \text{(2–2)}$$

Hence, for $Q = 10$, f_{osc} is 0.5% above f_{res}. Q here is the effective Q of the tank (Q_{eff}) and is determined for the Hartley oscillator circuit of Figure 2–3 by analyzing all the loading effects on the tank circuit. This is done, along with a determination of loop gain to check that oscillations will be self-sustained, in the following analysis. ■

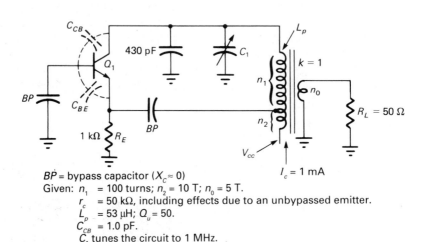

BP = bypass capacitor ($X_c \approx 0$)
Given: n_1 = 100 turns; n_2 = 10 T; n_0 = 5 T.
   r_c = 50 kΩ, including effects due to an unbypassed emitter.
   L_p = 53 µH; Q_u = 50.
   C_{CB} = 1.0 pF.
   C_1 tunes the circuit to 1 MHz.

FIGURE 2-3 Hartley oscillator.

2–1 ■ HARTLEY OSCILLATOR CIRCUIT ANALYSIS

The Hartley circuit of Figure 2–3 is first checked for positive feedback: assume that the emitter voltage is driven positive by an oscillation or noise. Q_1 will tend toward cutoff, and the collector will rise toward V_{cc}. This rising signal is voltage-divided by $n_2/(n_1 + n_2)$ but will have the same phase as the collector, except for a small amount due to loading by R_L. Hence, the signal fed back to the emitter through C_1 will be in phase with the initializing signal at f_{osc}.

Next, determine the loading on the tank circuit by the procedures developed in Chapter 1. The 50-Ω load, when reflected up to the collector, has a loading effect R_L', where

$$R_L' = R_L\left(\frac{n_1 + n_2}{n_0}\right)^2$$

$$= 50\left(\frac{110}{5}\right)^2 = 24.2 \text{ k}\Omega$$

(2–3)

Loading due to finite Q_u of the primary coil is computed as follows: at 1 MHz, $X_L = 300 \ \Omega$ and

$$R_{coil} = Q_u X_L$$
$$= (50)(333 \ \Omega) = 16.7 \text{ k}\Omega$$

(2–4)

A microwave antenna over-looking San Francisco Bay.

From the $n_1 - n_2$ tap looking to the left, we see $r_e \parallel R_E = 26\text{mV}/I_c \parallel 1 \text{ k}\Omega \approx 25 \text{ }\Omega$. Therefore, as seen from the "tank top,"

$$R'_e = \left(\frac{n_1 + n_2}{n_2}\right)^2 (r_e \parallel R_E) \tag{2-5}$$

$$= \left(\frac{110}{10}\right)^2 25 \text{ }\Omega = 3 \text{ k}\Omega$$

So the total collector loading is

$$R'_c = r_c \parallel R'_L \parallel R_{\text{coil}} \parallel R'_e \tag{2-6}$$
$$= 50 \text{ k}\Omega \parallel 24.2 \text{ k}\Omega \parallel 16.7 \text{ k}\Omega \parallel 3 \text{ k}\Omega = \textbf{2.2 k}\boldsymbol{\Omega}$$

or $G = 0.02 + 0.041 + 0.06 + 0.32 = 0.45 \text{ mS}$

so $R = 1/G = \textbf{2.2 k}\boldsymbol{\Omega}$

The amplification in the loop at resonance will be, for a common-base amplifier,

$$A_v = \frac{R'_c}{r_e} \tag{2-7}$$

$$= \frac{2.2 \text{ k}\Omega}{26\Omega} = 84.6 \text{ (38.5 dB)}$$

To determine if the circuit has enough gain, calculate the loop gain and hope that it is greater than unity.

But first, this is important: If you actually break the loop as shown in Figure 2–4 to get an idea of what the feedback voltage v_{fb} will be for a given v_i, you must include the *loading effects* ($r_e \parallel 1 \text{ k}\Omega$) that will pertain when the loop is closed and operating. This is illustrated in Figure 2–4. The gain, A_v, can be a lot different with and without the impedance loading effects.

$A_v = 84.6$ as calculated before, but if you ignore the $r_e \parallel 1 \text{ k}\Omega$ loading effects, the gain would appear to be $A_v = (50 \text{ k}\Omega \parallel 24.2 \text{ k}\Omega \parallel 15 \text{ k}\Omega)/26\Omega = 7.8 \text{ k}\Omega/26\Omega = 300$, which is 11 dB more than actually exists when the loop is closed. An error like this can result in an oscillator that doesn't oscillate. If $v_i = +100 \text{ mV}$, $v_c =$

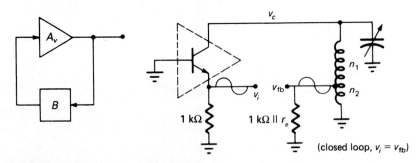

FIGURE 2-4   Circuit for open-loop gain analysis.

$A_v v_i = 8,460$ mV (no phase inversion for common-base). Also,

$$v_{\text{fb}} = \frac{n_2}{n_1 + n_2} v_c \tag{2–8}$$

$$= \left(\frac{10}{110}\right) (8460 \text{ mV}) = 769 \text{ mV}$$

which is much greater than the input 100 mV. This yields a loop gain of 7.69, which is calculated as $A_v B = 84.6 \times (^{10}/_{110}) = 7.69$ (17.7 dB). Since $7.69 \gg 1$, there is more than enough loop gain to ensure sustained oscillations.

To get an idea of the variable capacitor to use, we need to consider the capacitance already from collector to ground. Any capacitance associated with R_L will reflect up to the collector circuit as C/N^2, where $N = {}^{110}/_5$. Also, if we look to the left from the tap point, we see C_{BE} to ground (actually, a thorough analysis will show that you will see an *inductive* impedance looking into a moderately biased emitter). In any case, these capacitances are small and their effect at the collector is even smaller because they would reflect up by C/N^2.

Now look at C_{CB} from collector to ground. It would be tempting to apply the Miller effect increase to this capacitance, but in this case it would not be correct because we are dealing with a common-base amplifier. We are not driving the base (the base is grounded); the emitter is being driven. Hence, the ac current that has to be supplied to the emitter is not affected by C_{CB}. $C_{CB} = 1$ pF is seen only from collector to ground—across the tank. When added to the fixed capacitance of 430 pF, the tank has 431 pF to ground.

To tune 53 μH at 1 MHz requires

$$C = \frac{1}{(2\pi f)^2 L}$$

$$= 1/[(2\pi \times 10^6)^2 (53 \times 10^{-6})] = 478 \text{ pF}$$

So, in our case the variable should tune at around 47 pF. A 100-pF (max) variable can be used. The variable capacitor will easily make up for stray capacitances that are impossible to predict exactly.

2–2 ■ COLPITTS OSCILLATOR

The Colpitts oscillator illustrated in Figure 2–5 is like the Hartley except that the feedback is taken from a capacitive voltage divider instead of the tapped inductor or auto transformer. Notice that the feedback is to the emitter and not the base of Q_1. You should confirm that the phase is correct for oscillation. To do this, break the loop at the x near the emitter and follow the phase around the loop. The result is the same as for the Hartley oscillator.

Another thing to notice about Figure 2–5 is that a variable inductor is shown for tuning. This component is a transformer wound around a high-permeability fer-

FIGURE 2-5 (a) Colpitts oscillator with variable inductor for
tuning. (b) Equivalent circuit.

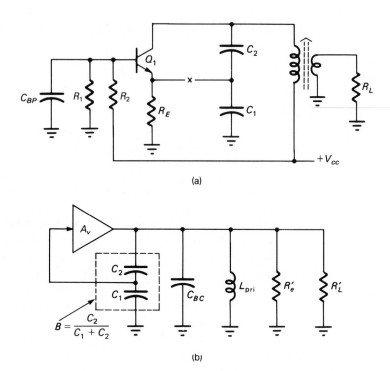

(a)

(b)

rous core. The core is threaded like a screw and can be raised or lowered in the
windings, thereby changing the inductance.

In analyzing the loop gain for a Colpitts oscillator it was shown in Chapter 1
that if the resistance to ground across C_1 is an order of magnitude (10 times) or
more greater than X_{C_1}, then the feedback factor is

$$B = \frac{v_e}{v_c} = \frac{C_2}{C_1 + C_2} = \frac{C_{eq}}{C_1} \qquad (2\text{--}9a)$$

where

$$C_{eq} = \frac{C_1 C_2}{C_1 + C_2} \qquad (2\text{--}9b)$$

Also, the resistive loading across the tank circuit is the parallel combination of R'_L
and R'_e, where

$$R'_L = \left(\frac{n_p}{n_s}\right)^2 R_L \qquad (2\text{--}10)$$

and

$$R'_e = \left(\frac{C_1}{C_{eq}}\right)^2 (r_e \parallel R_E) \qquad (2\text{--}11a)$$

$$\approx \left(\frac{1}{B}\right)^2 r_e \qquad (2\text{--}11b)$$

Let's work through another oscillator example using the Colpitts circuit. This example also requires more sophistication in determining the loop feedback factor.

■ ■ ■ ■

EXAMPLE 2–1

Colpitts Oscillator Analysis

Analyze the oscillator circuit of Figure 2–6.

1. Ignore resistance loading effects and determine the tuned circuit resonant frequency. (C_{OB} of the transistor is incorporated with stray capacitance in the 30 pF.)

2. Calculate the voltage gain of the common-base amplifier A_v (emitter-to-collector, as usual) at the resonant frequency.

3. Calculate the required base bias resistors (Figure 2–5a) assuming $V_{cc} = 12$ V, $V_{BE} = 0.7$ V, and $\beta = 50$.

4. Calculate an appropriate value for the base bypass capacitor, C_{BP} (Figure 2–5a).

5. Calculate loop gain, $A_v B$.

6. Will oscillations be sustained? Why or why not?

Solution:

1. This will be justified later, but the reactance of the 1500 pF is less than the 100 Ω in parallel with it. This, and the fact that $C_{eq} \approx 150$ pF means that we can find the effective C for tuning by

$$C_{eq} = \frac{(150 \text{ pF})(1500 \text{ pF})}{150 \text{ pF} + 1500 \text{ pF}} + 30 \text{ pF}$$

$$= 136.4 \text{ pF} + 30 \text{ pF} = 166.4 \text{ pF}$$

Then $f_0 = 1/2\pi\sqrt{LC_{eq}} = \textbf{10.8 MHz.}$

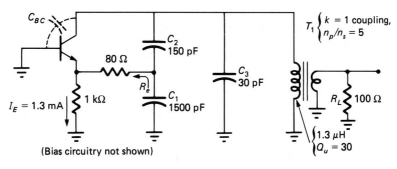

(Bias circuitry not shown)

FIGURE 2–6 Colpitts oscillator for analysis.

2. There are three loading effects on the collector: transformer primary (coil) losses, R_L', and R_e'.

$$R_{\text{coil}} = Q_u X_L = 30 \times 2\pi(10.8 \times 10^6)(1.3 \times 10^{-6}) = 2646\ \Omega$$

$$r_e = \frac{26\ \text{mV}}{I_E} = 20\ \Omega$$

$$80\ \Omega + (r_e \parallel 1\ \text{k}\Omega) = 99.6\ \Omega = R_e$$

$$R_e' = \left(\frac{C_1 + C_2}{C_2}\right)^2 R_e \approx \left(\frac{C_1}{C_2}\right)^2 R_e$$

$$= \left(\frac{1500 + 150}{150}\right)^2 99.6\ \Omega = 12\ \text{k}\Omega$$

$$R_L' = \left(\frac{n_p}{n_s}\right)^2 R_L = (5)^2 100\ \Omega = 2.5\ \text{k}\Omega$$

$$R_c' = R_{\text{coil}} \parallel R_e' \parallel R_L' = 2.65\ \text{k}\Omega \parallel 12\ \text{k}\Omega \parallel 2.5\ \text{k}\Omega = 1.16\ \text{k}\Omega$$

$$A_v = \frac{R_c'}{r_e} = \frac{1.16\ \text{k}\Omega}{20\Omega} = \textbf{58}\ (35.27\ \text{dB})$$

(Part 1 justification: $X_{1500\ \text{pF}} = 1/2\pi(1500 \times 10^{-12})(10.8 \times 10^6) = 9.8\ \Omega$. Since $100/9.8 = 10.2 = Q > 10$, the current through the C_1-C_2 string is 99% capacitive, and our approximation in part 1 is $\leq 1\%$ error.)

3. Since $V_E = I_E R_E = 1.3\ \text{mA} \times 1\ \text{k}\Omega = 1.3\ \text{V}$, then $V_B = 1.3\ \text{V} + V_{BE} = 2.0\ \text{V}$. Base bias current will be about 26 μA ($\beta = 50$) or less, so a bias resistor current of 1 mA is appropriate and convenient. Then $R_1 = 2\ \text{V}/1\ \text{mA} = 2\ \text{k}\Omega$, and $R_2 = (12\ \text{V} - 2\ \text{V})/1\ \text{mA} = 10\ \text{k}\Omega$.

4. $X_c(BP)$ should be less than one-tenth the base-to-ground impedance for adequate bypassing. Using 2 kΩ ∥ 10 kΩ ∥ 5 kΩ = 1.25 kΩ given that the 1500 pF (C_1) is less than 10 Ω, then $C_{BP} = 10/(2\pi 10.8\ \text{MHz} \times 1.25\ \text{k}\Omega) = 118\ \text{pF}$. The

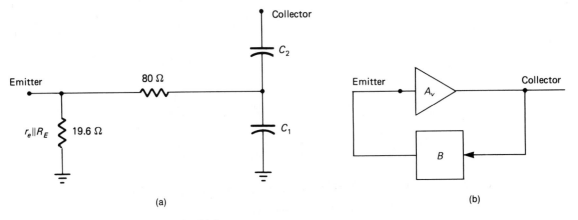

(a)   (b)

FIGURE 2–7 (a) Feedback network. (b) Loop parameters.

typical choice will be the standard 1000-pF ceramic capacitor (CK05-type with its low-inductance construction), providing plenty of margin. The 5 kΩ used in this calculation accounts for emitter-to-ground resistance reflected back to the base circuit. That is

$$\{r_e + [1 \text{ k}\Omega \parallel (80 \ \Omega + X_{c_1})]\}(\beta + 1) \approx [20 \ \Omega + 80 \ \Omega](101)$$

5. Determine the feedback factor B as illustrated in Figure 2–7. The ac collector voltage is stepped down to the input (emitter) in two steps, the capacitor voltage divider and the resistor divider. (The 80-Ω resistor might represent the resistance effect of a series resonant crystal.)

$$B = \left(\frac{C_2}{C_1 + C_2}\right) \left(\frac{19.6 \ \Omega}{80 + 19.6}\right) = (0.091)(0.197) = 0.0179 \ (-34.95 \text{ dB})$$

The loop gain is $A_v B = 58 \times 0.0179 = 1.038 \ (+0.32 \text{ dB})$, or $35.27 \text{ dB} + (-34.95 \text{ dB}) = +0.32 \text{ dB}.$

6. Yes, $A_v B > + 1.$

2–3 ■ CLAPP OSCILLATOR

The Clapp oscillator of Figure 2–8 is like the Colpitts except for a small capacitor, C_3, in series with the inductor L. Capacitor C_3 is made smaller than C_1 or C_2 so that its reactance is large, allowing it to have the most effect in determining the resonant frequency.

$$C_{\text{eff}} = \left\{\left[\left(\frac{1}{C_1} + \frac{1}{C_2}\right)^{-1} + C_{BC}\right]^{-1} + \frac{1}{C_3}\right\}^{-1} \tag{2–12}$$

for this circuit. In this scheme, C_1 and C_2 can be changed to get the optimum amount of feedback, and C_3 can be a variable capacitor for setting the frequency of oscillation. Also, C_3 could incorporate a negative temperature coefficient to

FIGURE 2–8 Clapp oscillator circuit. Circuit values shown are for 1.89 MHz.

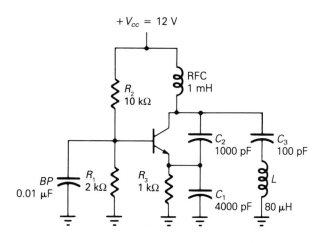

improve the oscillator frequency stability for applications in which the circuit temperature is expected to vary.

2–4 ■ TUNED-INPUT/TUNED-OUTPUT OSCILLATOR

The tuned-input/tuned-output oscillator was popular for vacuum tube oscillators because vacuum tubes have such a high input (control grid) impedance that a parallel resonant circuit at the input made sense. Although FETs have high gate impedance, this oscillator scheme is not reliable or frequency-stable because of its sensitivity to supply voltage and temperature variations.

However, this is a perfect circuit for demonstrating why too many multistage tuned amplifiers oscillate.

The feedback path for this oscillator, shown in Figure 2–9, is through C_{gd}. At the frequency for which the output tank Z_1 and the input tank Z_2 are resonant, the phase of the capacitive current through $C_{gd}(Z_3)$ will not be correct for oscillation. However, detailed analysis shows that the phase will be correct for oscillations at the frequency for which the net reactances of Z_1 and Z_2 are inductive, thus forming a series-resonant circuit with C_{gd}. Since a parallel LC circuit is inductive for frequencies below resonance (see Chapter 1), the circuit can oscillate at a frequency below the resonant frequencies of Z_1 and Z_2. Figure 2–10a illustrates the conditions when oscillations occur. Note that $L_o \neq L_1$ and $L_i \neq L_2$ of Figure 2–9. L_i and L_o are the net inductances for the input and output LC circuits for $f < f_{res}$.

Figure 2–10b shows a generalized oscillator configuration from which all the previously discussed oscillators can be modeled. Z_1, Z_2, and Z_3 should be high-Q

FIGURE 2–9   Tuned-input/tuned-output oscillator using a JFET.

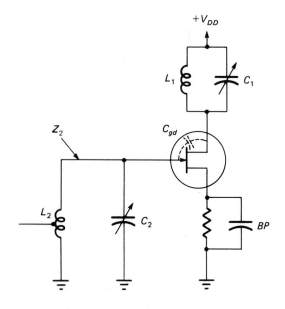

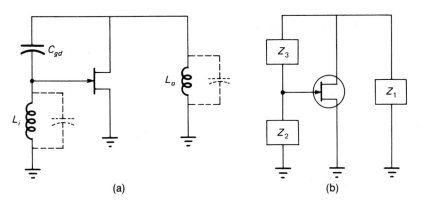

FIGURE 2–10   (a) Tuned-input/tuned-output oscillator equivalent circuit at resonance.
(b) Generalized oscillator impedance model.

reactances, and the reactance type (inductive or capacitive) must be the same for
Z_1 and Z_2. The reactance type for Z_3 must be opposite to that of Z_1 and Z_2.

2–5 ■ UNTUNED OSCILLATORS

Among the commonly used low-frequency oscillators that do not depend on a
resonant tuned circuit are the RC phase-shift oscillator, the Wien bridge oscillator, and the astable multivibrator. While these oscillators are implemented in
various circuit configurations, only the simple op-amp circuits are discussed here.

RC Phase-Shift Oscillator

As shown in Figure 2–11, an inverting amplifier is used to achieve signal gain, so
the RC network in the feedback path is designed to provide another 180° phase
inversion in order to achieve the required in-phase feedback.

The phase shift will be correct at a frequency given by

$$f = \frac{1}{2\pi \sqrt{6}RC} \tag{2–13}$$

At this frequency, the signal voltage loss through the RC network will be $\frac{1}{29}$, so
that the inverting amplifier must provide a voltage gain of 29. Since $|A_v| = R_f/R$,
then

$$R_f = 29R \tag{2–14}$$

As is the case for the Wien bridge oscillator, it is desirable to maintain a constant
loop gain so that the output signal stays sinusoidal in spite of changes due to aging
effects and temperature variations (temperature stability). This is usually accomplished by including in the feedback resistance a temperature-sensitive component that corrects for gain if the circuit temperature changes.

FIGURE 2-11   RC phase-shift
oscillator.

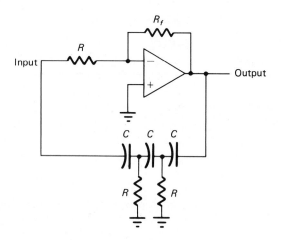

Wien Bridge Oscillator

The op-amp circuit of Figure 2–12 has both negative and positive feedback. R_f and
R_1 provide negative feedback and set the forward gain for the positive feedback
system. The positive feedback is from output to pin 2 through a series RC/parallel
RC, lead-lag network. The series RC provides a leading phase or "zero" in the
frequency response, and the parallel RC provides a lag or "pole" in the frequency
response. Another way to view this RC network is as a bandpass filter in which
the series RC stops the low frequencies, and the parallel RC stops the high
frequencies. The lead and lag phases cancel at a frequency given by $f_0 =
1/(2\pi RC)$, providing the necessary 0° net phase shift required for oscillation. The
RC network signal loss at f_0 is $\frac{1}{3}$, so that the noninverting amplifier (pin 2 to
output) must provide a gain of 3.

FIGURE 2-12   Wien bridge oscillator.

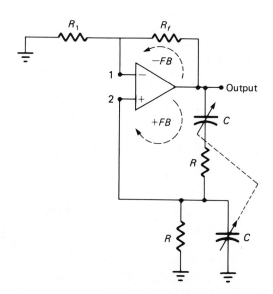

Astable Multivibrator

This oscillator is based on a very different phenomenon from what we have studied so far. The circuit has positive feedback and very high loop gain, but the waveforms are not sinusoidal. In fact, the output assumes one of two voltage states, except for short switching transitions.

The IC shown in Figure 2–13 is a voltage comparator so that, as long as v_3 is higher than v_2, the output is ideally a constant, $v_o = +V_{cc}$. Otherwise, the output is $v_o = -V_{EE}$. R_1 and R_2 divide v_o by the ratio $R_1/(R_1 + R_2)$, so that

$$v_3 = \frac{v_o R_1}{R_1 + R_2} \tag{2-15}$$

For the values shown in Figure 2–13, v_3 is either +6 V or −6 V. At the same time, capacitor C is charging from the previous switching point value of v_3 toward v_o through resistor R.

Figure 2–14 shows the voltage waveforms at the inputs and output of the IC. At time t_0 the output has just switched to $v_o \approx V_{cc} = +12$ V. The voltage division of R_1 and R_2 puts v_3 at +6 V. $v_2 = -6$ V at t_0 is the voltage across capacitor C. This voltage cannot change instantaneously so that at t_0+ the voltage drop across R is $v_o - v_3 = +12$ V − (−6 V) = 18 V, and the capacitor charging current is $I_c(t_0+) = I_R = (18 \text{ V})/47 \text{ k}\Omega = 0.38$ mA. The voltage across C, v_2, rises exponentially toward +12 V with a time constant of $RC = 47 \text{ k}\Omega \times 0.01 \text{ }\mu\text{F} = 0.47$ ms. v_2 never reaches 12 V because at t_1, the instant that v_2 exceeds $v_3 = +6$ V, the comparator switches to the next astable state. The time required for the voltage on pin 2 to rise from −6 V to +6 V is found from $v_2 = v_c = -6$ V + (18 V)(1 − $e^{-t/RC}$). Solving for t with $v_2 = +6$ V is as follows: 12 V/18 V = 1 − $e^{-t/RC}$; therefore, $e^{-t/RC}) = 1 − 2/3 = 1/3$. The natural logarithm of both sides yields = $-t/RC = \ln 1/3 = -1.1$. Hence, the time required is $t = 1.1RC = 0.517$ ms. Since

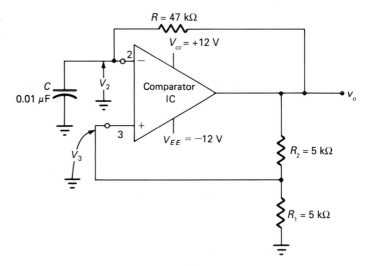

FIGURE 2-13 Astable multivibrator using an op-amp.

FIGURE 2–14   Waveforms for an
astable multivibrator.

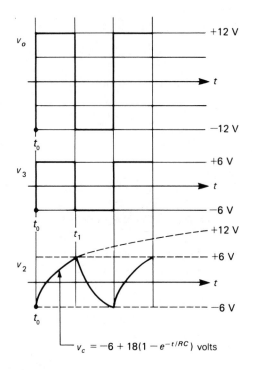

$$v_c = -6 + 18(1 - e^{-t/RC}) \text{ volts}$$

this is one-half of the output squarewave period, $f_0 = 1/T = 1/(2t) = 1/(2.2RC) = 967$ Hz.

In summary, the voltage $v_o = \pm V_{cc}$ when $|V_{EE}| = V_{cc}$,

$$v_3 = \pm v_o R_1/(R_1 + R_2) \qquad\qquad (2\text{–}16)$$

$$t = -RC \ln[R_2/(2R_1 + R_2)] \qquad\qquad (2\text{–}17)$$

and

$$f_0 = \frac{1}{-2RC \ln[R_2/(2R_1 + R_2)]} \qquad\qquad (2\text{–}18)$$

2–6 ■ OSCILLATOR STABILITY AND SPECTRAL PURITY

Oscillators are fairly simple circuits, but their performance is always critically important in communications systems. Among the oscillator parameters that affect the performance of communications transmitters and receivers are frequency stability, including short-term noise variations and long-term drift, and signal purity, including noise variations of signal amplitude and harmonic distortion content. Signal spectral purity can be controlled with filters and automatic gain control (AGC) circuitry. AGC can also improve the short-term noise variations, but special attention to shielding and power supply line filtering is required to minimize noise.

The most critical performance parameter for an oscillator, however, is frequency drift. The long-term frequency stability of an oscillator is affected by aging in the components that determine the frequency of oscillation. Also, temperature changes affect the frequency-controlling components and cause the oscillator to drift. These frequency changes are characterized in terms of the *temperature coefficient* of the components and how these affect the overall temperature stability of the oscillator.

Temperature Coefficient

The temperature coefficient (TC) of a system parameter is the fractional change in the parameter per degree change in temperature; that is,

$$\text{TC} = \Delta f_0/f_0 \tag{2–19}$$

The fractional change is usually given in percent or in parts per million parts (ppm), and the temperature change is given in degrees Celsius (°C). As an example, the frequency TC of an oscillator might be given as $+100$ ppm/°C. This means $\text{TC} = \Delta f_0/f_0 = (+100$ Hz/MHz) per °C change in temperature. If the temperature increases by 20°C, then a 5-MHz oscillator will have a frequency increase determined as follows:

$$\frac{\Delta f_0}{f_0} = \text{TC} \times \Delta T \tag{2–20}$$
$$= (+100 \text{ Hz/MHz/°C}) \times (20°\text{C})$$
$$= +2 \text{ kHz/MHz}$$

and
$$\Delta f_0 = \text{TC} \times \Delta T \times f_0$$
$$= (+2 \text{ kHz/MHz}) \times 5 \text{ MHz}$$
$$= +10 \text{ kHz}$$

That is, the oscillator frequency rises 10 kHz for a 20°C rise in temperature.

Temperature Stability of Oscillators

The frequency of an oscillator changes when the circuit temperature changes. This is because the circuit components that control the oscillator frequency have nonzero temperature coefficients. The principal offenders in this regard are capacitors.

Capacitors are typically made by sandwiching a dielectric (nonconductor) material between two conductors. The capacitance is determined by

$$C = \frac{\epsilon A}{d} \tag{2–21}$$

where A is the area of the conductors and d is the distance between the conductors; the conductors are separated by the dielectric material with dielectric constant ϵ. A change in ϵ or any of the physical dimensions of the capacitor results in a change in C.

An excellent capacitor for tuning an oscillator is a silver-mica capacitor, often called a *chocolate drop* because of its characteristic oval shape and glossy brown finish. The temperature coefficient is typically less than a few hundred ppm/°C of temperature change. If such a capacitor is used in a resonant circuit, the frequency change for a given change in temperature can be predicted from the capacitor TC and the resonant frequency relationship $f_0 = 1/(2\pi\sqrt{LC})$. The fractional change in f_0 as a function of capacitance change is found by differentiating as follows:

$$f_0 = \frac{1}{2\pi}(LC)^{-1/2} \tag{2-22}$$

$$\frac{df_0}{dC} = \frac{1}{2\pi}[-\tfrac{1}{2}(LC)^{-1/2-1}L] = \frac{-\tfrac{1}{2}L}{2\pi\sqrt{LC}(LC)}$$

Thus $\qquad df_0 = \dfrac{1}{2\pi\sqrt{LC}}\left(\dfrac{-1}{2C}\right)dC$

Now divide the left side by f_0 and the right side by the equivalent of f_0, namely, $1/(2\pi\sqrt{LC})$. The result is

$$\frac{df_0}{f_0} = -\frac{1}{2}\frac{dC}{C} \tag{2-23}$$

which, if the change in C is less than 10%, can be written approximately as

$$\frac{\Delta f_0}{f_0} = -\frac{1}{2}\frac{\Delta C}{C} \tag{2-24}$$

The result indicates that an increase of, say, 4% in capacitance ($\Delta C/C = 0.04$) causes a 2% decrease in oscillator frequency.

Improving the Frequency Stability of Oscillators

Experience with oscillators has shown that careful attention to circuit design and some tricks of the trade can help improve the frequency stability of oscillators. The following is a listing of important design considerations:

1. Use circuit components with known temperature coefficients. This is especially important with respect to capacitors.

2. Neutralize, or otherwise swamp-out with resistors, the effects of active device variations due to temperature, power supply, and circuit load changes.

3. Operate the oscillator at low power.

4. Use negative-TC capacitors to compensate for typically positive-TC tuned circuits.

5. Reduce noise; use shielding, AGC, and bias-line filtering.

6. Follow the oscillator with a buffer amplifier to reduce the effects of load changes.

7. Thermally isolate the oscillator; use an oven or other temperature-compensating circuitry such as thermistor- (sensistor-) capacitor networks.

8. Use a crystal to control the frequency.

Whereas typical LC tuned-circuit oscillators have temperature coefficients of approximately 500 ppm/°C (perhaps 100 ppm/°C with compensation), crystal-controlled oscillators (XOs) have better than 10 ppm/°C temperature coefficients. Temperature compensation circuits can improve XOs to a little less then 1 ppm/°C. These oscillators are called TCXOs. Finally, oven-controlled crystal oscillators can achieve stabilities of 10^{-10}, but typical ovens require in excess of 4 W to operate.

2-7 ■ CRYSTALS AND CRYSTAL OSCILLATORS

As mentioned in the last paragraph, crystal-controlled oscillators have excellent frequency stability. A crystal is a device that is usually made by cutting a pure quartz crystal in a very thin slice and then plating the faces with a conductor in order to make an electrical connection. This is illustrated in Figure 2–15. The property that makes the quartz crystal useful in electronics is the *piezoelectric effect*.

The piezoelectric (pressure/electric) effect exhibited by quartz is a phenomenon whereby the application of an electrical voltage along certain axes or planes, called *cuts*, causes a deformation of the quartz material. The converse is also true; that is, if the crystal is deformed by pressure, an electrical charge develops along certain axes or faces of the cut crystal. What makes this property useful in oscillators and high-Q filters is that the thickness, stiffness, and type of cut determines to a very high degree the frequency of vibration of the quartz. In addition, the physical properties of quartz are extremely stable with temperature.

FIGURE 2–15 Packaging for a crystal.

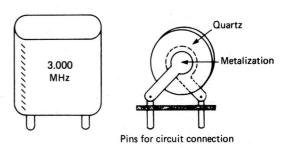

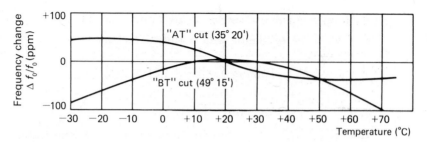

FIGURE 2-16 Temperature characteristics for two Y-cut quartz crystals (angle of cut given in degrees and minutes of arc).

Crystal Cuts (Y, AT, and BT)

Quartz has a hexagonal (6-sided) crystal structure, and a vertical cut parallel to any one of the six flat faces is called a cut. If the Y-cut is not made vertically but instead is cut at 35° 20′ from the vertical axis, an AT cut is obtained. The temperature stability characteristic of quartz crystals depends on the type of cut (Figure 2–16).

As an example of how knowledge about the crystal cut can be useful, consider the use of either an AT or BT cut crystal in a temperature-compensated crystal oscillator, TCXO. The simplest temperature-compensation circuits are linear and have the effect of rotating the $\Delta f_0/f_0$ curve of Figure 2–16 about the $T = 20°C$ axis. Clearly, the AT-cut would be preferable to the BT-cut for this application.

Equivalent Circuit, Impedance, and Q

The equivalent circuit for a crystal is shown in Figure 2–17a. Capacitance C_p includes package capacitance and can be on the order of 7–10 pF for small packages; C_s is on the order of 0.05 pF. On the other hand, L is huge for quartz

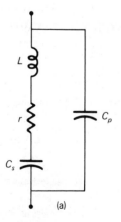

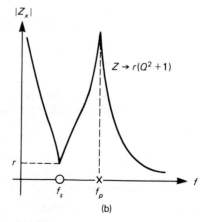

FIGURE 2-17 (a) Equivalent circuit of crystal. (b) Impedance plot versus frequency.

crystals, on the order of tens of henrys; and r is small because r represents the internal losses and relates to Q. The Q of crystals is commonly in excess of 10^5, so r is less than a few ohms. Notice from this equivalent circuit that the crystal has a parallel and a series resonance, and with $C_p \approx 140C_s$, the separation between f_s and f_p would be on the order of 0.36%.

Series resonance occurs at a lower frequency than f_p because

$$f_s = \frac{1}{2\pi\sqrt{LC_s}}$$

whereas

$$f_p = \frac{1}{2\pi\sqrt{L\dfrac{C_sC_p}{C_s + C_p}}} \tag{2-25}$$

With these two expressions, it can be shown that

$$f_p \approx \left(1 + \frac{C_s}{2C_p}\right) f_s \tag{2-26}$$

The magnitude of the impedance for a crystal is shown as a function of frequency in Figure 2–17b, where the distance between f_s and f_p is greatly exaggerated for ease of illustration.

Crystal Oscillators

The circuit symbol for a crystal is shown in the feedback path of the Colpitts crystal-controlled oscillator of Figure 2–18.

Notice that this symbol suggests a capacitor structure in which the dielectric is special. In this case, of course, the dielectric is a special cut of quartz with piezoelectric properties. This circuit is analyzed or designed in the same manner as the LC Colpitts (Figure 2–5). The LC tuned circuit provides a narrowband (high-Q) collector impedance for the grounded-base amplifier. Positive feedback is

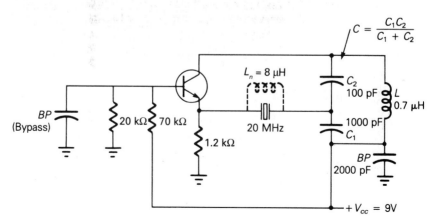

FIGURE 2–18   Colpitts crystal-controlled oscillator. Component values are for 20 MHz.

FIGURE 2–19 Colpitts crystal-
controlled oscillator.

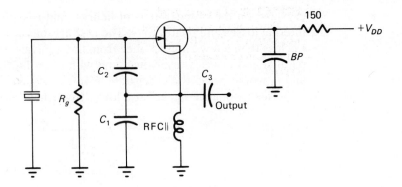

provided through the crystal, which is operating near its series-resonant mode frequency, f_s. Of course, the LC tuned circuit should also be tuned at or near f_s.

Inductor L_n is included in oscillators operating above about 20 MHz. The purpose of L_n is to neutralize the 7- to 10-pF package capacitance, C_p, which at low frequencies would not be a problem. However, the impedance of a 10-pF capacitance above 20 MHz is less than 1000 Ω, so that high-frequency feedback signals can sneak by and never excite the quartz. You will know this has happened by the dumbfounded look on the designer's face when he or she tunes the LC tank circuit, and the oscillator output frequency changes a lot—it is not crystal-controlled. To avoid this embarrassment, neutralize C_p by parallel-resonating it with an inductance of approximately

$$L_n = \frac{1}{(2\pi f_0)^2 C_p} \qquad\qquad (2\text{–}27)$$

Another practical crystal-controlled Colpitts oscillator circuit is shown in Figure 2–19. The crystal is operating near its parallel resonant frequency, with C_1 and C_2 providing feedback and source-to-gate impedance transformation. The high-impedance RF choke (RFC) provides a dc-bias current path while C_1, chosen for about 500 Ω reactance, essentially becomes the source-circuit impedance. If a very low-impedance load is used at the output, C_3 will have to be adjusted downward in value so that the impedance at the source does not swamp X_{C_1}.

Harmonic Operation

To increase the frequency at which the quartz crystal vibrates, the quartz blank is ground thinner and made smaller. Obviously, there is a physical limit to this process. The limit for sturdy, reliable fundamental-mode crystals is about 25 MHz. Like all mechanical resonances, there are higher-order modes so that you can purchase specially finished crystals for operation at the third overtone (to about 75 MHz) and fifth or even seventh overtones (to about 125 MHz).

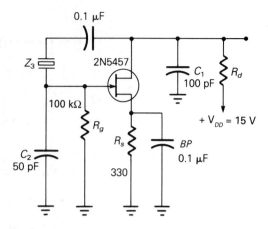

FIGURE 2–20 Pierce crystal oscillator. Parts values are for use with a 3.5789-MHz standard TV chroma subcarrier oscillator crystal.

2–8 ■ PIERCE CRYSTAL OSCILLATOR

The Pierce circuit shown in Figure 2–20 is an oscillator that depends on the inductive component of the crystal equivalent circuit to provide a feedback signal of correct phase. Recall from the generalized impedance model of Figure 2–10b that Z_3, which is represented by the crystal in the Pierce circuit, must have an opposite reactance to Z_1 and Z_2 (which are capacitors in this case). Consequently, if R_d is large enough to provide the common-source amplifier with enough gain, the circuit will oscillate at a frequency between f_s and f_p where the crystal is inductive. This provides optimum stability. Bias is provided to the FET from V_{DD} through R_d; R_d may be replaced by a choke. R_g is on the order of 100 kΩ and provides a dc connection between the FET gate and ground. R_s, bypassed for high gain by the BP capacitor, provides self-bias ($V_s = I_d R_s$) to the FET source lead. R_s and its bypass may be removed, if desired, and the source lead connected to ground.

The Pierce oscillator is often used in applications where different fundamental-mode crystals are switched in and out to provide various frequencies of crystal-controlled stability. Only first-overtone crystals are used because, despite special finishing and mounting, higher-overtone crystals will vibrate most strongly in the fundamental mode.

■ PROBLEMS

1. A three-stage amplifier is shown in Figure 2–21. Points A and B are both possible inputs to feedback the output. To which pin, A or B, would you connect the output in order to have the correct phase for an oscillator? Why?

FIGURE 2-21

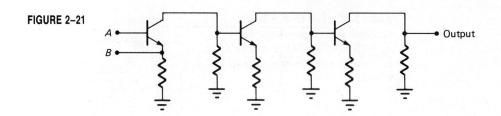

A •

B •

Output

2. An integrated circuit with high input and output impedances provides a voltage gain of +50 in the circuit of Figure 2–22.
   a. Name the oscillator circuit configuration.
   b. Calculate the loop gain.
   c. Will oscillations be sustained? Comment on both Barkhausen criteria.

3. a. Determine the inductance value required for oscillations at 2 MHz for the circuit of Figure 2–22.
   b. Show that the Barkhausen criteria are met, with a sketch (phase) and calculation.
   c. What will be the circuit Q?
   d. If $A_v = 100$, determine the maximum value for the 100-pF capacitor.

4. Make the following changes in the integrated circuit oscillator of Figure 2–22: Change the 1 kΩ to 2 kΩ, the 9 kΩ to 6 kΩ, and the 400 pF to 900 pF, and make the inductor 10 μH with a tap two turns aboveground connected to a 50-Ω load; the total autotransformer turns are 20.
   a. Should the internal output impedance Z_0 of the IC be high or low? Why?
   b. Determine the minimum voltage gain, A_{v0}.
   c. What would $|A_v|$ drop to if the IC output impedance Z_0 was as low as 1 kΩ?
   d. Compute the frequency of oscillation.

5. Determine the frequency of oscillation for the Clapp oscillator of Figure 2–8 with $C_1 = 1000$ pF, $C_2 = 4000$ pF, $C_3 = 100$ pF, and $L = 80$ μH. Ignore resistive loading effects.

6. For the circuit of Figure 2–23, assume the following: $n_1 = 20$ turns (T), $n_2 = 10$ T, $n_3 = 2$ T, and n_2 is independent of n_3 and $k = 1$. $L_1 = 50$ μH, $Q_u = $ infinity. For the transistor β = 20 and $I_E = 1$ mA.
   a. What value of C will tune the circuit to 1 MHz? (Ignore transformer secondary reactances.)

FIGURE 2-22

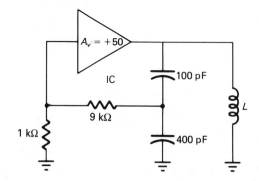

$A_v = +50$

IC

100 pF

9 kΩ

1 kΩ

400 pF

L

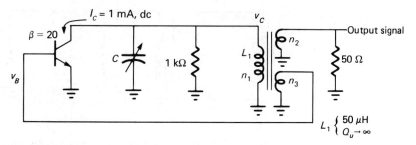

FIGURE 2-23

    b. Calculate base-collector voltage gain.
    c. Calculate loop gain.
    d. Should the $n_1 : n_3$ transformer invert the phase in order to make an oscillator? Why/why not?
    e. Is there enough gain for oscillations to be sustained?

7. The circuit of Figure 2–24 is to be an oscillator. (The value of $r_e = 1\ k\Omega$ is not practical, but it is convenient for this exercise.) Determine the following:
    a. Minimum voltage gain of the IC, in decibels.
    b. Should the connection for the switch, S_1, be to pin 1 or pin 2? Why? (Show this on the sketch.)
    c. Value of L to tune the output to 100 MHz.

8. For the circuit of Figure 2–25:
    a. Calculate the loop gain of this oscillator circuit. (Ignore mutual inductance.)
    b. Will there be sustained oscillations? Why/why not?

9. Colpitts oscillator analysis (Figure 2–26):

    Given: $V_{GSQ} = -1.9V$, $I_{DQ} = 3.2$ mA
                 $V_{DSQ} = 7.36V$, $V_p = -4$ V (pinch-off)
                   $I_{DSS} = 12$ mA, $C_1 = 100$ pF, $C_2 = 2000$ pF
    Assume: $r_d = 100\ k\Omega$, Z(gate) $\approx \infty$, C_3 is an ac short circuit.

    a. Determine resonant frequency of drain circuitry.
    b. Determine tank impedance at resonance, Z_d. (Tank impedance includes all loading effects (power losses) as seen from drain to ground.)

FIGURE 2-24

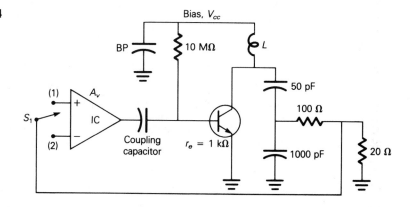

FIGURE 2–25

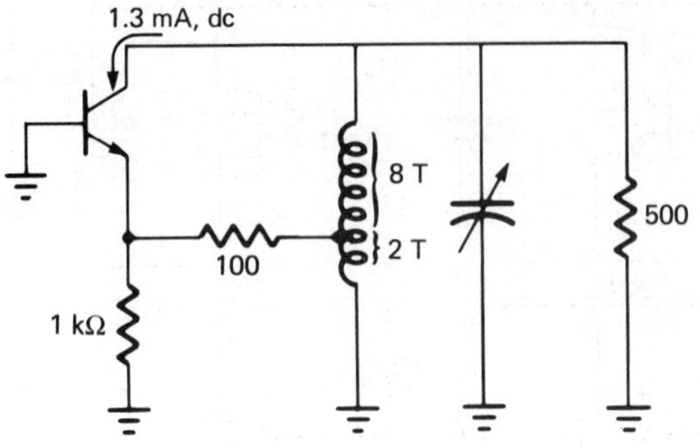

FIGURE 2–26

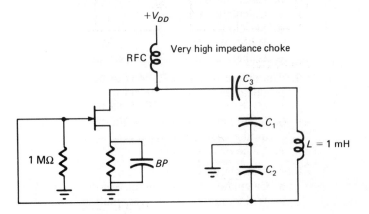

c. Calculate gate-to-drain voltage gain, A_v. (The data given will yield g_m.)

d. Will oscillations be sustained? Why or why not?

e. What is the function of C_3?

f. If L changes by 5%, how much will the resonant frequency change?

10. Which of the circuits in Figure 2–27 is suitable as a crystal oscillator if the crystal must act inductively?

11. Determine the frequency that the crystal should be cut for in Figure 2–28. Should it be series mode or parallel mode? If the temperature stability of the crystal frequency is 5.0 ppm/°C, determine the oscillator frequency after a rise in temperature of 50°C. Why L_2?

12. For Figure 2–28, let $I_c = 0.26$ mA, dc, $R = 1$ kΩ, crystal resistance 400 Ω, and the capacitor on ground, $C_1 = 1650$ pF. The overall equivalent C remains 150 pF. Compute the loop gain in dB.

13. Design a Colpitts crystal oscillator similar to the circuit of Figure 2–18, except that the inductor shown is replaced with a transformer ($L_p = 10$ μH, $n_p/n_s = 4$, $k = 1$) with a 50-Ω load connected across the secondary winding. Also, $V_{cc} = 10$ V, $C_1 = 900$ pF, $C_2 = 100$ pF, $V_{BE} = 0.7$ V, β = 100, and 11.7 mW must be delivered to the 50-Ω load.

a. Determine values of R_1 (base-to-ground), R_2, and R_E.

b. Find the frequency to which the crystal should be cut.

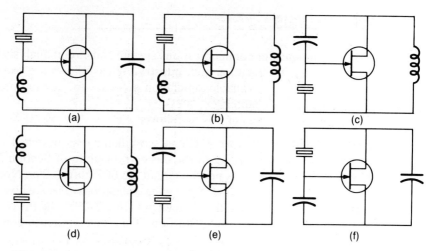

(a)   (b)   (c)

(d)   (e)   (f)

FIGURE 2–27

FIGURE 2–28

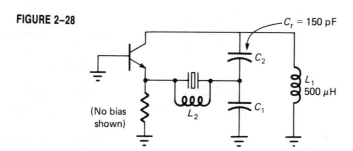

c. Determine loop gain and phase if the crystal is considered to have 10-Ω resistance.

d. Should the crystal be series or parallel resonant? *Why?*

e. What is the purpose of L_n?

f. Compute L_n if the crystal package capacitance is 10 pF.

14. Use a circuit simulator software package (Micro-CAP II, SPICE, or other) to perform the following for the circuit of Problem 13:

a. Determine bias voltage values.

b. Carry out transient analysis to see the output signal waveform.

c. Carry out Fourier analysis to examine the harmonics in the output signal.

15. An astable multivibrator has V_{cc} = 15 V, V_{EE} = −15 V, R_1 = 4.7 kΩ, R_2 = 5.6 kΩ, C = 1000 pF, and an ideal IC.

a. Determine the value of feedback resistor R to produce a 50-kHz oscillator.

b. Draw the waveforms at each input and output pin of the IC (with respect to ground). Include the peak voltage and an accurate time scale.

16. Design an astable multivibrator oscillator with R_1 = R_2 and f_0 = 100 kHz. Sketch the complete schematic with your component values.

2-9 ■ COMPUTER ANALYSIS

Using the computer for communications system and circuit analysis has become a virtual necessity in education and industry. What follows are examples of programs and/or readily available simulation software for solving problems of the type covered in this chapter.

The program for Chapter 2 is as follows:

- **SPICE Colpitts oscillator:** A Colpitts oscillator was designed for a 10-MHz center frequency according to the design procedure of Section 2–2 and modeled for a SPICE analysis per the circuit file of Figure 2–29a. Note that a pulse was input for the purpose of starting the oscillations. The steady-state transient response is clearly sinusoidal, as seen in Figure 2–29b, and the oscillator output signal appears to have low distortion. This observation is confirmed by the results of the Fourier analysis (using the .FOUR statement in the circuit file), which shows that the total harmonic distortion was 0.775% (Figure 2–29c).

```
****** 5/26/88******** Demo PSpice (May 1986) *******14:35:19********
SPICE COLPITTS OSCILLATOR
R1 1 3 12K
R2 3 0 3.7K
RE1 4 5 44
RE2 5 0 700
L1 1 2 1.2U
C1 2 6 218PF
C2 6 0 2842PF
CBY 3 0 .03U
VDC 1 0 DC 17
VIN 5 6 PULSE(0 100 1NS 1NS 10NS)
.OPTIONS LIMTIM=6 LIMPTS=400 ITL5=10000
.TRAN .01US .35US
.PLOT TRAN V(2)
.FOUR 10MEG V(2)
Q1 2 3 4 BJT
.MODEL BJT NPN PE=.6 VA=30
.END
```

```
****     INITIAL TRANSIENT SOLUTION       TEMPERATURE =    27.000 DEG C

  NODE    VOLTAGE     NODE    VOLTAGE     NODE    VOLTAGE     NODE    VOLTAGE

(  1)    17.0000    (  2)    17.0000    (  3)     3.9243    (  4)     3.1225

(  5)     2.9378    (  6)     2.9378
```

FIGURE 2-29a

```
****      TRANSIENT ANALYSIS                 TEMPERATURE =   27.000 DEG C

                     1.600D+01    1.650D+01    1.700D+01    1.750D+01    1.800D+01
          TIME     V(2)   - - - - - - - - - - - - - - - - - - - - - - - - - - -
       .000D+00   1.700D+01 .          .            *           .            .
      1.000D-08   1.676D+01 .          .       *               .            .
      2.000D-08   1.656D+01 .        *         .               .            .
      3.000D-08   1.654D+01 .        *         .               .            .
      4.000D-08   1.669D+01 .           *      .               .            .
      5.000D-08   1.697D+01 .          .           *           .            .
      6.000D-08   1.726D+01 .          .           .       *               .
      7.000D-08   1.745D+01 .          .           .            *           .
      8.000D-08   1.747D+01 .          .           .            *           .
      9.000D-08   1.731D+01 .          .           .        *              .
      1.000D-07   1.704D+01 .          .           *           .            .
      1.100D-07   1.675D+01 .          .      *                .            .
      1.200D-07   1.657D+01 .         *        .               .            .
      1.300D-07   1.654D+01 .        *         .               .            .
      1.400D-07   1.670D+01 .           *      .               .            .
      1.500D-07   1.698D+01 .          .           *           .            .
      1.600D-07   1.727D+01 .          .           .       *               .
      1.700D-07   1.745D+01 .          .           .             *          .
      1.800D-07   1.747D+01 .          .           .             *          .
      1.900D-07   1.730D+01 .          .           .        *              .
      2.000D-07   1.702D+01 .          .           *           .            .
      2.100D-07   1.674D+01 .          .      *                .            .
      2.200D-07   1.656D+01 .         *        .               .            .
      2.300D-07   1.655D+01 .        *         .               .            .
      2.400D-07   1.672D+01 .           *      .               .            .
      2.500D-07   1.699D+01 .          .           *           .            .
      2.600D-07   1.727D+01 .          .           .        *              .
      2.700D-07   1.746D+01 .          .           .            *           .
      2.800D-07   1.746D+01 .          .           .            *           .
      2.900D-07   1.729D+01 .          .           .        *              .
      3.000D-07   1.701D+01 .          .           *           .            .
      3.100D-07   1.673D+01 .          .     *                 .            .
      3.200D-07   1.655D+01 .        *         .               .            .
      3.300D-07   1.656D+01 .        *         .               .            .
      3.400D-07   1.673D+01 .          .    *                  .            .
      3.500D-07   1.701D+01 .          .           *           .            .
                     - - - - - - - - - - - - - - - - - - - - - - - - - - - - -
```

FIGURE 2–29b

```
****      FOURIER ANALYSIS                  TEMPERATURE =   27.000 DEG C

FOURIER COMPONENTS OF TRANSIENT RESPONSE V(2)

DC COMPONENT =   1.701D+01
```

HARMONIC NO	FREQUENCY (HZ)	FOURIER COMPONENT	NORMALIZED COMPONENT	PHASE (DEG)	NORMALIZED PHASE (DEG)
1	1.000D+07	4.744D-01	1.000000	-.664	.000
2	2.000D+07	2.734D-03	.005764	-176.525	-175.862
3	3.000D+07	1.561D-03	.003290	-174.540	-173.876
4	4.000D+07	1.122D-03	.002366	-172.363	-171.699
5	5.000D+07	8.858D-04	.001867	-170.124	-169.461
6	6.000D+07	7.378D-04	.001555	-167.854	-167.191
7	7.000D+07	6.378D-04	.001345	-165.652	-164.988
8	8.000D+07	5.705D-04	.001203	-163.544	-162.880
9	9.000D+07	5.304D-04	.001118	-161.237	-160.574

```
          TOTAL HARMONIC DISTORTION =       .774917  PERCENT
```

FIGURE 2-29c

3

Signal Spectra

■ INTRODUCTION

The oscillators of Chapter 2 differ in the purity of their output signals. A tuned-circuit oscillator with a carefully controlled feedback level to eliminate distortion can approximate a pure signal source. Such a sinusoidal signal source output can be written as a function of time: $v(t) = A \sin 2\pi f_0 t$, where A is the peak amplitude, f_0 is the frequency (assumed constant), and t is the variable, time.

When distortion exists, as is always the case, higher harmonics of the fundamental frequency f_0 are present. In addition, dc, a voltage at zero frequency, usually exists so that the total output signal at any instant in time can be determined by adding the instantaneous values of each component. This is expressed mathematically as

$$v(t) = V_{dc} + V_1 \sin 2\pi f_0 t + V_2 \sin 2\pi (2f_0) t$$
$$+ \cdots + V_n \sin 2\pi n f_0 t + \cdots \qquad (3\text{--}1)$$

where n is the harmonic number and nf_0 is the nth harmonic of the fundamental frequency.

This can be written more compactly using the summation symbol, Σ, with the N harmonics. Hence,

$$v(t) = V_0 + \sum_{n=1}^{N} V_n \sin 2\pi n f_0 t \qquad (3\text{–}2)$$

where V_n is the peak amplitude of the nth harmonic, and $n = 1, 2, 3, 4, \ldots, N$.

The term V_0 is the average (dc) value of the signal, and the sinusoidal terms, $V_n \sin 2\pi(n f_0)t$, show the periodic variation of the signal from the average value. A *periodic signal,* such as $V_1 \sin 2\pi f_0 t$, is one that repeats itself every T seconds, where T is the period and is related to the repetition frequency by $f_0 = 1/T$. ■

3–1 ■ FOURIER SERIES AND SIGNAL ANALYSIS

Signal analysis is very important in communication theory and system and circuit design. In order to predict and understand electronic system and circuit behavior, we use the results of mathematical analysis. In particular, we need to know the frequency, frequency range (bandwidth), and power level of signals.

Most signals used in electronics are periodic in practice and can be analyzed in terms of the power, voltage, or current at the various frequencies in the signal. As an example, the waveform seen on an oscilloscope (time-domain picture) at the output of a multivibrator oscillator studied in Chapter 2 is shown in Figure 3–1.

If this signal, usually referred to as a *squarewave* (notice the symmetry along the time axis) is analyzed for its frequency content, some remarkable discoveries are made. Suppose the squarewave oscillator signal is connected through a very narrow, variable-frequency, bandpass filter to a voltmeter, as shown in Figure 3–1. Starting at $f = 0$ (dc), we get a voltage reading equal to $A/2$. This is the average value of the input squarewave with peak amplitude A. So there is a dc component present. This is not very remarkable because our eye and brain integrate this waveform to conclude that, if the signal is on for one-half the period and off for the other half, there is a positive average value—the dc component.

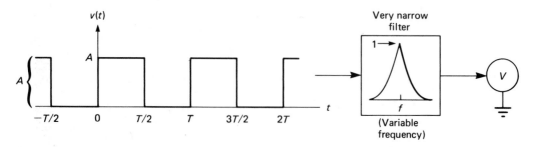

FIGURE 3–1 Wave analyzer set up to analyze the frequency components of a square-wave signal.

What is remarkable, however, is that if the filter center frequency is slowly increased, the voltmeter reads zero until a frequency f_0, which is the fundamental frequency of the signal. The frequency f_0 is the repetition rate of the squarewave and is equal to $f_0 = 1/T$. A mathematical analysis predicts that the peak value of the voltage at this "first harmonic" f_0 will be $2A/\pi$ volts.

In addition, there will be absolutely no reading on the voltmeter except at frequencies that are exact harmonics of the first harmonic f_0 and, for these squarewaves, only odd harmonics are present. This has been analyzed mathematically using a procedure known as Fourier series* analysis.

To write a Fourier series for approximating a time-varying function $v(t)$, the time function must be defined over one full period T. In general, a complex periodic signal includes a phase shift at $t = 0$ so that orthogonal (or quadrature) components must be included. Thus, the most general expression for the periodic time function is

$$v(t) \approx V_0 + \sum_{n=1}^{N} (a_n \cos 2\pi n f_0 t + b_n \sin 2\pi n f_0 t) \tag{3-3}$$

where the approximation of $v(t)$ is more accurate as the number of components included, N, is increased. V_0 is the average value (dc), and the peak amplitudes a_n and b_n are determined as follows:

$$V_0 = \frac{1}{T} \int_0^T v(t) \, dt \tag{3-4}$$

$$a_n = \frac{2}{T} \int_0^T v(t) \cos(2\pi n f_0 t) \, dt \tag{3-5}$$

$$b_n = \frac{2}{T} \int_0^T v(t) \sin(2\pi n f_0 t) \, dt \tag{3-6}$$

■ ■ ■ ■

EXAMPLE 3-1

The function $v(t)$ is defined over one period as shown in Figure 3–1; that is,

$$v(t) = \begin{cases} A & 0 \le t \le T/2 \\ 0 & T/2 < t < T \end{cases}$$

Then

$$V_0 = \frac{1}{T} \int_0^{T/2} A \, dt = \frac{A}{T} t \Big|_0^{T/2} = \frac{A}{T}\left(\frac{T}{2}\right)$$

$$= \frac{A}{2}, \quad \text{the average (dc) value}$$

$$a_n = \frac{2}{T} \int_0^{T/2} A \cos 2\pi n f_0 t \, dt = \frac{2A}{2\pi n f_0 T} \sin 2\pi n f_0 \frac{T}{2}$$

* The French mathematician Baron Jean Baptiste Joseph Fourier, 1768–1830, developed the Fourier series while studying heat-conduction problems.

Since
$$f_0 = \frac{1}{T}, \qquad a_n = \frac{A}{\pi n} \sin \pi n = 0$$

because $\sin n\pi = 0$ for all integers n. But,

$$b_n = \frac{2}{T} \int_0^{T/2} A \sin 2\pi n f_0 t \; dt$$

$$= \frac{-A}{\pi n f_0 T} \left(\cos 2\pi n f_0 \frac{T}{2} - \cos 0 \right)$$

$$= \frac{A}{\pi n} (1 - \cos \pi n)$$

$$= \frac{A}{\pi n} (1 + 1) \qquad \text{for } n \text{ odd}$$

and
$$= \frac{A}{\pi n} (1 - 1) \qquad \text{for } n \text{ even}$$

Consequently,
$$b_n = \frac{2A}{n\pi} \qquad (n \text{ odd only})$$

If an infinite number of harmonics are included, the function $v(t)$ can be written exactly. Hence,

$$v(t) = \frac{A}{2} + \sum_{n,\text{odd only}}^{\infty} \left(\frac{2A}{\pi n} \right) \sin 2\pi n f_0 t$$

Recognizing *even* and *odd* functions will be very helpful in saving computation time. The function of Figure 3–1 is an *odd function* and, as in Example 3–1, $a_n = 0$. This is true for all odd periodic functions.

Even and Odd Functions

A function $y = f(t)$ is *even* if it has perfect symmetry about the y-axis; that is,

$$f(+t) = f(-t) \tag{3–7}$$

for all t. This notion is illustrated in Table 3–1d for the "full-wave rectified" waveform. The waveform of Table 3–1d has the same even symmetry as $y = \cos(\omega_0 t)$ and, when b_n is computed for the Fourier series, it is always found that $b_n = 0$ for $f(t)$ even.

The reason that $b_n = 0$ in Equation 3–3 for the cosine series is that sine and cosine are *orthogonal* (at right angles) to each other and, when their product is integrated over one cycle as in Equation 3–6, the result is zero.

Likewise, an *odd* function, such as $\sin \omega_0 t$, has $a_n = 0$ in the Fourier series and need not be computed. An *odd* function is one for which

$$f(+t) = -f(-t) \tag{3–8}$$

TABLE 3-1 Some Periodic Waveforms and Their Fourier Series
Mathematical Expressions

a.
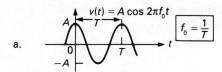

$$f_0 = \frac{1}{T}$$

b.
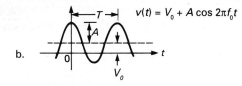

c.
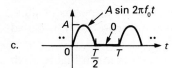

$$v(t) = \frac{A}{\pi} + \frac{A}{2} \sin 2\pi f_0 t - \frac{2A}{3\pi} \cos 2\pi(2f_0)t - \frac{2A}{15\pi} \cos 2\pi(4f_0)t + \cdots$$

$$= \frac{A}{\pi} + \frac{A}{2} \sin 2\pi f_0 t + \sum_{n=2}^{\infty} \frac{A[1+(-1)^n]}{\pi(1-n^2)} \cos 2\pi(nf_0)t \ (nf_0)t$$

d.
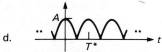

* (the rectifier input signal
will have a period of 2 *T*)

$$v(t) = \frac{2A}{\pi} + \frac{4A}{3\pi} \cos 2\pi f_0 t - \frac{4A}{15\pi} \cos 2\pi(2f_0)t + \cdots$$

$$= \frac{2A}{\pi} + \sum_{n=1}^{\infty} \frac{4A(-1)^n}{\pi[1-(2n)^2]} \cos 2\pi(nf_0)t$$

e.
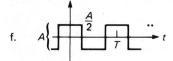

$$v(t) = \frac{2A}{\pi} \sin 2\pi f_0 t + \frac{2A}{3\pi} \sin 2\pi(3f_0)t + \cdots$$

$$= \sum_{n,\ \text{odd only}}^{\infty} \frac{2A}{n\pi} \sin 2\pi(nf_0)t$$

f.
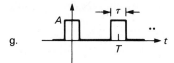

$$v(t) = \frac{2A}{\pi} \cos 2\pi f_0 t - \frac{2A}{3\pi} \cos 2\pi(3f_0)t + \frac{2A}{5\pi} \cos 2\pi(5f_0)t + \cdots$$

$$= \sum_{n=1}^{\infty} \left(A \frac{\sin n\pi/2}{n\pi/2} \right) \cos 2\pi(nf_0)t$$

g.
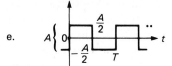

$$v(t) = \frac{A\tau}{T} + \sum_{n=1}^{\infty} \left(2A \frac{\tau}{T} \right) \left(\frac{\sin n\pi\tau/T}{n\pi\tau/T} \right) \cos 2\pi(nf_0)t$$

h.
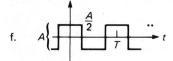

$$v(t) = \sum_{n,\ \text{odd only}}^{\infty} \left(A \frac{\sin n\pi/4}{n\pi/4} \right) \cos 2\pi(nf_0)t$$

(special case of 50% "alternate inversion")

i.
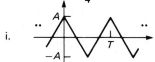

$$v(t) = \frac{8A}{\pi^2} \cos 2\pi f_0 t + \frac{8A}{9\pi^2} \cos 2\pi(3f_0)t + \frac{8A}{25\pi^2} \cos 2\pi(5f_0)t + \cdots$$

$$= \sum_{n\ \text{odd}}^{\infty} \frac{8A}{(n\pi)^2} \cos 2\pi(nf_0)t$$

j.
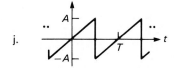

$$v(t) = \frac{2A}{\pi} \left[\sin 2\pi f_0 t - \tfrac{1}{2} \sin 2\pi(2f_0)t + \tfrac{1}{3} \sin 2\pi(3f_0)t + \cdots \right]$$

$$= \sum_{n=1}^{\infty} [(-1)^{n+1}] \left(\frac{2A}{n\pi} \right) \sin 2\pi(nf_0)t$$

That is, the function for negatively increasing time is the inverse of that for $t > 0$. The visual test is to perform a double mirror image—first about the y-axis ($t = 0$) and then about the t-axis ($y = 0$). $f(t)$ should exactly overlap $f(-t)$ after the double flip (try Table 3–1e).

The results of Example 3–1 may be plotted in the frequency domain by letting time stand still and recalling the results of the wave analyzer experiment illustrated in Figure 3–1. The frequency-domain plot, shown in Figure 3–2b, is called the *signal spectrum*.

Figure 3–3a shows a time-domain plot of the signal components in Figure 3–2b when $\theta = 2\pi f_0 t$. Also shown is the sum at each instant of time. Here we see the approximation to the squarewave with only the dc and first three (odd) harmonics present. Figure 3–3b illustrates a computer-generated 5-V squarewave limited to (odd) harmonics 1 through 9.

It can be concluded from these figures that a squarewave can be produced with nothing but sinewave generators. All that must be done is to gather together and phase-synchronize many generators, set their frequencies to harmonics of the fundamental rate f_0, set the amplitudes of each harmonic to the value calculated from the Fourier series equation, and connect the outputs together into an oscilloscope to see the result. What would be observed is that the more higher harmonics included, the sharper the squarewave. Put in another context, to transmit good sharp squarewaves, a very wide bandwidth is required. Observe in Figure 3–3a that if a squarewave is input to a linear circuit with absolute minimum bandwidth so that only the fundamental is output, the rise time of the circuit is effectively reduced. The effective rise time can be visualized conveniently here as the slope of the waveform in the first 15 electrical degrees. If, however, the bandwidth is increased sevenfold to allow the seventh harmonic, then the effective rise time increases nearly four times.

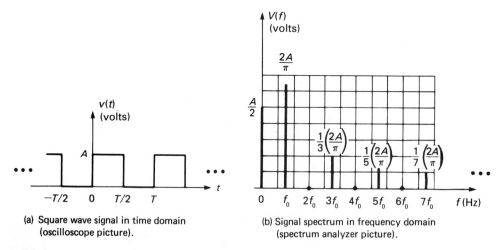

(a) Square wave signal in time domain (oscilloscope picture).

(b) Signal spectrum in frequency domain (spectrum analyzer picture).

FIGURE 3–2   (a) Squarewave signal in time domain (oscilloscope picture). (b) Signal spectrum in frequency domain (spectrum analyzer picture).

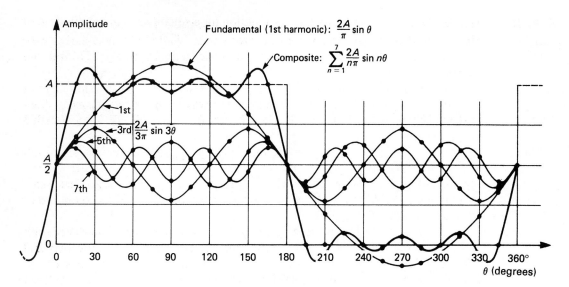

FIGURE 3–3a   Time plot of individual components of a squarewave. The first seven sine components are added to the dc value.

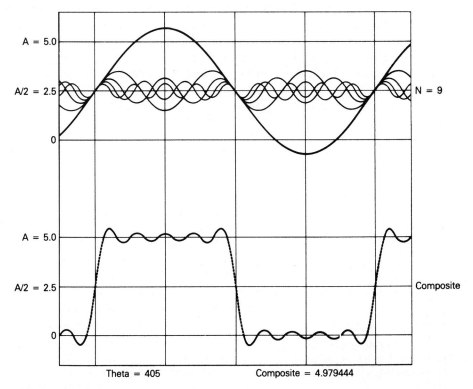

FIGURE 3–3b   Computer results illustrating the sum of harmonics producing a square-wave approximation (odd harmonics, 1–9).

■ ■ ■ ■

**EXAMPLE
3–2**

**Fourier
Series
Problem
(Even
Function)**

Derive the sine-cosine Fourier series function for a squarewave of amplitude A as shown in Figure 3–4.

Solution:

$$DC: \quad a_0 = \frac{1}{T_0} \int_{-T_0/2}^{T_0/2} v(t)\, dt = \frac{1}{T_0} \int_{-T_0/4}^{T_0/4} A\, dt = \frac{At}{T_0} \Big|_{-T_0/4}^{T_0/4} = \frac{A}{T_0}\left(\frac{T_0}{4} - \frac{-T_0}{4}\right) = \frac{A}{2}$$

$$AC: \quad a_n = \frac{2}{T_0} \int_{-T_0/2}^{T_0/2} v(t)\cos\left(\frac{n2\pi t}{T_0}\right) dt \qquad \left(\text{where } \omega_0 = \frac{2\pi}{T_0}\right)$$

$$= \frac{2}{T_0} \int_{-T_0/2}^{T_0/2} A\,\cos\left(\frac{n2\pi t}{T_0}\right) dt = \frac{2A}{T_0}\frac{T_0}{n2\pi}\sin\left(\frac{n2\pi t}{T_0}\right)\Big|_{-T_0/4}^{T_0/4}$$

$$= \frac{A}{n\pi}\sin\left(\frac{n\pi}{2}\right) - \sin\left(\frac{-n\pi}{2}\right)$$

but $\sin(-\theta) = -\sin(\theta)$ (it is an odd function); hence

$$a_n = \frac{A}{n\pi}\left[2\,\sin\left(\frac{n\pi}{2}\right)\right] = \frac{A\,\sin(n\pi/2)}{n\pi/2}$$

Also, $b_n = 0$. (The given squarewave is an even function.) The proof of $b_n = 0$ is as follows:

$$b_n = \frac{2A}{T_0} \int_{-T_0/4}^{T_0/4} \sin\left(\frac{n2\pi t}{T_0}\right) dt = \frac{2A}{T_0(2\pi n/T_0)}\left[\frac{-\cos(n2\pi t)}{T_0}\right]\Big|_{-T_0/4}^{T_0/4}$$

$$= -\frac{A}{n\pi}\left[\cos\left(\frac{n\pi}{2}\right) - \cos\left(\frac{-n\pi}{2}\right)\right]$$

But, $\cos(-\theta) = +\cos(\theta)$; therefore, $b_n = 0$. The complete Fourier series is then

$$v(t) = \frac{A}{2} + \sum_{n=1}^{\infty} \frac{A\,\sin(n\pi/2)}{n\pi/2}\cos(n\omega_0 t)$$

$$v(t) = \frac{A}{2} + \frac{2A}{\pi}\cos(\omega_0 t) + \frac{2A}{3\pi}\cos(3\omega_0 t) + \cdots \qquad \text{odd only}$$

Fourier Series: Complex Exponential Form

A very compact form of the Fourier series is written by substituting the complex exponential $e^{j\theta(t)}$ for the individual quadrature sinusoidal components. Since, by

FIGURE 3–4 Symmetrical squarewave with period T_0 and amplitude A.

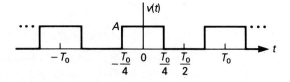

Euler's equations (see Equation 1–75 for the cosine equivalent), the exponential representation has a component on both the positive and negative frequency axes, the resulting Fourier series will be an infinite summation from minus to plus infinity.

The *complex exponential* Fourier series is defined by

$$f(t) = \sum_{n=-\infty}^{\infty} c_n e^{j2\pi n f_0 t} \qquad (3\text{--}9)$$

where the peak amplitude *coefficients* are derived from

$$c_n = \frac{1}{T}\int_{-T/2}^{T/2} f(t)e^{-j2\pi n f_0 t}\, dt \qquad (3\text{--}10)$$

■■■■

EXAMPLE 3–3

Complex Exponential Fourier Series

Derive the complex exponential Fourier series for the even-function rectangular pulse waveform of Table 3–1g.

Solution:

We first derive the coefficients from Equation 3–10 and insert them in Equation 3–9. The function, expressed over the single period from $-\tau/2$ to $T - \tau/2$, is

$$f(t) = \begin{cases} A & -\dfrac{\tau}{2} \le t \le \dfrac{\tau}{2} \\[2ex] 0 & \dfrac{\tau}{2} < t < \left(T - \dfrac{\tau}{2}\right) \end{cases}$$

From Equation 3–10,

$$c_n = \frac{1}{T}\int_{-\tau/2}^{\tau/2} Ae^{-j2\pi n f_0 t}\, dt = \frac{A}{T}\left. \frac{e^{-j2\pi n f_0 t}}{-j2\pi n f_0}\right|_{-\tau/2}^{\tau/2}$$

$$= \frac{A(e^{j2\pi n f_0(\tau/2)} - e^{-j2\pi n f_0(\tau/2)})}{j2\pi n f_0 T} = \frac{A\tau}{T}\frac{\sin(\pi n f_0 \tau)}{\pi n f_0 \tau}$$

Since the fundamental frequency of the waveform is $f_0 = 1/T$, we may also write

$$c_n = \frac{A\tau}{T}\frac{\sin(n\pi\tau/T)}{n\pi\tau/T}$$

Hence,

$$f(t) = \sum_{n=-\infty}^{\infty}\left[\frac{A\tau}{T}\frac{\sin(n\pi\tau/T)}{(n\pi\tau/T)}\right]e^{j2\pi n f_0 t}$$

Fourier Series of Other Waveforms

Other common signals have been analyzed by the Fourier series procedure and the results are given in Table 3–1. Notice, as illustrated for the squarewaves of (e) and (f), that the presence of sines or cosines in the series depends on where the signal crosses the $t = 0$ axis. The reason is that the cosines peak at $\theta = 0°$ like the squarewave in (f), whereas sines are zero at $\theta = 0°$ like the squarewave in (e). This is also seen in Figure 3–3, where if the $t = 0$ axis is taken at $\theta = 90°$, the Fourier series will consist of cosines with alternating $+$ and $-$ signs. This series of cosines is written in Table 3–1f (with no dc component).

Another signal important in pulsed communication systems deserves comment. The rectangular pulses of Table 3–1g have a signal spectrum that includes odd and even harmonics of the fundamental f_0. Also, the amplitude of each harmonic is proportional to $(\sin n\pi d)/n\pi d$, where $d = \tau/T$ is called the *duty cycle* of the pulses. As illustrated in Figure 3–5, for a 25% duty cycle rectangular pulse train, the spectrum goes through zero at integer multiples of $f = 1/\tau$. The spectrum has nulls at $f = n/\tau$ because sine $n\pi\tau/T \rightarrow 0$ whenever $n\tau/T$ is a whole number. This is because sine $n\pi = 0$ (sketch a sinewave and prove this for yourself). For the particular case of $T = 4\tau$, $n/4$ is a whole number for every fourth harmonic of f_0. The spectrum of harmonics up to the first null, $f = 1/\tau$, is called the first *lobe* or the *main lobe*.

Since the first lobe includes frequency components up to $f = 1/\tau$, it is clear that as the pulses get narrower, more bandwidth is required to transmit the first lobe of signal power. This is illustrated in Figure 3–6. The limit of this is shown in part (c), in which $\tau \rightarrow 0$ and the pulses approach impulses. Since $\tau \rightarrow 0$, $f = (1/\tau) \rightarrow \infty$ and the spectral components would have constant amplitudes for frequencies extending out to infinity. This implies that, in order to produce a train of perfect impulses, infinite bandwidth and infinite power are required. Of course, neither of these conditions can be realized in practice, so true impulses are a mathematical construct.

Figure 3–7 shows the effect of keeping the pulse width constant while decreasing the pulse repetition frequency; that is, $f_0 = 1/T$, where T is increasing. As seen in part (b), more and more components are included in the first lobe because the

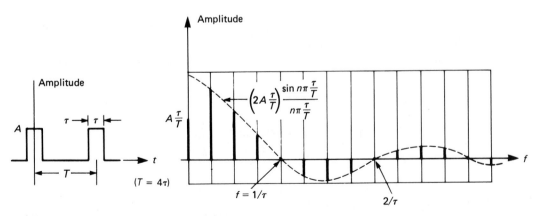

FIGURE 3–5   Time waveform and frequency spectrum of 25% duty-cycle pulses.

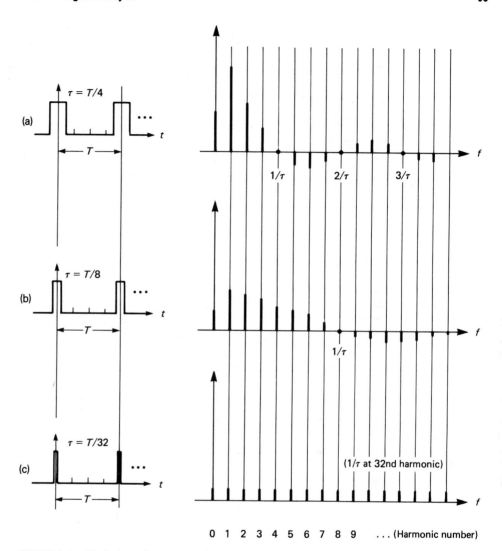

$\tau = T/4$

(a)

T

$1/\tau$ $2/\tau$ $3/\tau$

$\tau = T/8$

(b)

T

$1/\tau$

$\tau = T/32$

(c)

T

(1/τ at 32nd harmonic)

0 1 2 3 4 5 6 7 8 9   . . . (Harmonic number)

FIGURE 3–6   Variation of spectrum for narrowing pulses (smaller τ).

various harmonics, nf_0, are getting closer together as f_0 decreases. In the limit, as the period increases without bound, $T \rightarrow \infty$, the spectral components become indistinguishable from one another and the amplitudes blend into a continuous curve described by

$$V(f) = A\tau \frac{\sin \pi\tau f}{\pi\tau f} \qquad\qquad (3\text{–}11)$$

This result for a single rectangular pulse is derived by using the Fourier transformation of $v(t)$ to the frequency domain as $V(f)$.

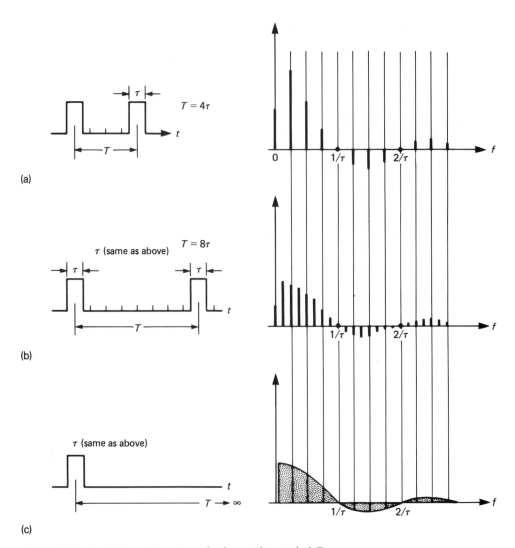

FIGURE 3–7   Variation of spectrum for increasing period T.

3–2 ■ FOURIER TRANSFORMS

As seen in the previous section, the Fourier series is a mathematical technique for writing functions so as to give both time and frequency behavior. Usually the function is periodic, but this is not a requirement.

The Fourier transform is used for transforming a time-function signal into the frequency domain (spectrum) and the inverse transformation transforms a spectrum function into an equivalent time function. Our primary interest here is in amplitude frequency spectrum analysis of nonperiodic functions.

The *Fourier transform* of the time function $v(t)$ is defined by

$$\mathscr{F}\{v(t)\} = V(f) = \int_{t=-\infty}^{\infty} v(t)e^{-j2\pi ft}\, dt \qquad (3\text{-}12)$$

TABLE 3-2 Fourier Transform Pairs

Function	Time Waveform $v(t)$	Spectrum $V(f)$
1. Rectangular pulse	$\Pi\left(\dfrac{t}{T}\right)$	$T\dfrac{\sin(\pi fT)}{\pi fT}$ (T is the time pulsewidth)
2. Triangular pulse	$\Lambda\left(\dfrac{t}{T}\right)$	$T\left(\dfrac{\sin(\pi fT)}{\pi fT}\right)^2$ (T is the time pulse base width)
3. Constant	1	$\delta(f)$
4. Impulse at $t = t_0$	$\delta(t - t_0)$	$e^{-j2\pi ft_0}$
5. Sinc, Sa($2\pi Bt$) (known as the sampling function)	$\dfrac{\sin(2\pi Bt)}{2\pi Bt}$	$\dfrac{1}{2B}\Pi\left(\dfrac{f}{2B}\right)$ ($2B$ is the frequency pulsewidth)
6. Phasor	$e^{j(\omega_0 t + \Phi)}$	$e^{j\Phi}\delta(f - f_0)$
7. Sinusoid	$\cos(\omega_0 t + \Phi)$	$\dfrac{1}{2}e^{j\Phi}\delta(f - f_0)$ $+\dfrac{1}{2}e^{-j\Phi}\delta(f + f_0)$
8. Gaussian	$e^{-\pi(t/t_0)^2}$	$t_0 e^{-\pi(ft_0)^2}$
9. Exponential, one-sided	$\begin{cases} e^{-t/T}, & t > 0 \\ 0, & t < 0 \end{cases}$	$\dfrac{T}{1 + j2\pi fT}$
10. Exponential, two-sided	$e^{-|t|/T}$	$\dfrac{2T}{1 + (2\pi fT)^2}$
11. Impulse train	$\displaystyle\sum_{k=-\infty}^{\infty} \delta(t - kT)$	$\displaystyle f_0 \sum_{n=-\infty}^{\infty} \delta(f - nf_0),$ where $f_0 = \dfrac{1}{T}$
12. Unit step	$u(t) = \begin{cases} 1, & t \geq 0 \\ 0, & t < 0 \end{cases}$	$\dfrac{1}{2}\delta(f) + \dfrac{1}{j2\pi f}$
13. Signum	$\text{sgn}(t) = \begin{cases} +1, & t > 0 \\ -1, & t < 0 \end{cases}$	$\dfrac{1}{j\pi f}$

Equation 3–12 is to be compared to the Laplace transform (*one-sided*), which is

$$\mathcal{L}[f(t)] = F(s) = \int_0^\infty f(t)e^{-st}\, dt, \quad \text{where } s = \sigma + j2\pi f$$

Like the Laplace transform, the output signal pair $v_o(t) \Leftrightarrow V_o(f)$ for a system of transfer function $H(f)$, with input signal $v_i(t) \Leftrightarrow V_i(f)$, is denoted by

$$v_o(t) = \mathcal{F}^{-1}\{V_o(f)\} \tag{3–13}$$

TABLE 3-3 Some Fourier Transform Properties and Theorems

Operation	Function	Fourier Transform
1. Linearity	$a_1v_1(t) + a_2v_2(t)$	$a_1V_1(f) + a_2V_2(f)$
2. Time delay	$v(t - T_d)$	$V(f)e^{-j2\pi fT_d}$
3. Scale change	$v(at)$	$\dfrac{1}{\lvert a \rvert} V\!\left(\dfrac{f}{a}\right)$
4. Conjugation	$v^*(t)$	$V^*(-f)$
5. Duality	$V(t)$	$v(-f)$
6. Real signal frequency translation [$v(t)$ is real]	$v(t)\cos(\omega_0 t + \theta)$	$\dfrac{1}{2}[e^{j\theta}V(f - f_0) + e^{-j\theta}V(f + f_0)]$
7. Complex signal frequency translation	$v(t)e^{j2\pi f_0 t}$	$V(f - f_0)$
8. Bandpass signal	$\text{Re}\{g(t)e^{j2\pi f_0 t}\}$	$\dfrac{1}{2}[G(f - f_0) + G^*(-f - f_0)]$
9. Differentiation	$\dfrac{d^n v(t)}{dt^n}$	$(j2\pi f)^n V(f)$
10. Integration	$\displaystyle\int_{-\infty}^t v(\lambda)d\lambda$	$(j2\pi f)^{-1}V(f) + \dfrac{1}{2}V(0)\delta(f)$
11. Time-domain convolution	$v_1(t) * v_2(t)$ $= \displaystyle\int_{-\infty}^\infty v_1(\lambda) \cdot v_2(t - \lambda)d\lambda$	$V_1(f)V_2(f)$
12. Time-domain multiplication	$v_1(t)v_2(t)$ $= \displaystyle\int_{-\infty}^\infty V_1(\lambda)V_2(f - \lambda)d\lambda$	$V_1(f) * V_2(f)$
13. Time-domain multiplication by t^n	$t^n v(t)$	$(-j2\pi)^{-1}\dfrac{d^n V(f)}{df^n}$

Superscript * means *complex conjugate*; in 11 and 12, * is a multiplication operator, *convolution*. Observe that there are notable similarities between the properties and theorems of Tables 3–3 for Fourier transformation and 1–3 for Laplace transformation.

where
$$V_o(f) = H(f)V_i(f) \qquad (3\text{-}14)$$

and the inverse Fourier transform denoted in Equation 3–13 is defined by

$$v(t) = \mathcal{F}^{-1}\{V(f)\} = \int_{-\infty}^{\infty} V(f)e^{j2\pi ft}\, df \qquad (3\text{-}15)$$

Tables 3–2 and 3–3, respectively, provide a convenient summary of some of the more commonly encountered Fourier transform pairs and important properties and theorems.

■ ■ ■ ■

EXAMPLE 3-4

Fourier Transform

Derive the Fourier transform for the single pulse of amplitude A and width τ (Figure 3–7c).

Solution:

The function is $v(t) = A$, $-\tau/2 \le t \le \tau/2$. From Equation 3–12,

$$V(f) = \int_{-\infty}^{\infty} v(t)e^{-j2\pi ft}\, dt = \int_{-\tau/2}^{\tau/2} Ae^{-j2\pi ft}\, dt$$

$$V(f) = \frac{Ae^{-j2\pi ft}}{-j2\pi f}\Big]_{-\tau/2}^{\tau/2} = \frac{A(e^{-j2\pi f\tau/2} - e^{j2\pi f\tau/2})}{-j2\pi f}$$

$$= \frac{A\,\sin(\pi\tau f)}{\pi f} = A\tau\,\frac{\sin(\pi\tau f)}{\pi\tau f} = A\tau\,\mathrm{Sa}(\pi\tau f)$$

where $\mathrm{Sa}(x)$ is known as the sampling function. Note that the result is Equation 3–11. It is also the same as Table 3–2, Equation 1. That is,

$$v(t) = A\Pi\!\left(\frac{t}{T}\right) \Leftrightarrow V(f) = AT\,\mathrm{Sa}(\pi fT) \qquad (3\text{-}16)$$

is the Fourier transform pair for a rectangular time pulse of width T and amplitude A.

3-3 ■ EFFECT OF FILTERS ON SIGNALS

In this section we consider the effects on periodic signals due to ideal, linear-phase filters. The objective is to introduce an important concept in an idealized fashion.

The frequency response of a filter is shown in Figure 3–8. This figure shows graphically the ratio of output to input amplitude at any frequency and is also referred to as a graphical transfer characteristic. The curve shows that an input sinusoid at 1 kHz passes through the filter unattenuated—$V_o/V_i = 1$. However, the amplitude of a 2-kHz sinusoid will be decreased (attenuated) to one-half of its input amplitude; $V_o/V_i = 0.5$. This is an attenuation of 6 dB ($20\log_{10}0.5 = -6$

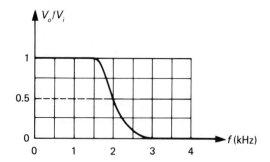

FIGURE 3–8   Low-pass filter frequency response (graphical transfer characteristic). Vertical axis shows gain (loss) versus frequency.

dB). Sinusoids of 3 kHz or greater are reduced to zero; that is, they are completely filtered out. Also notice that any dc ($f = 0$) component of the input signal will pass unattenuated through the filter. The frequency response is that of a low-pass filter that we will assume to have *no time delays or phase shifts*.

Now, suppose that a squarewave signal with 1000 pulses/s is applied to this idealized filter. What will an oscilloscope show at the output? Consider Figure 3–9 which illustrates the system. The oscilloscope picture (time domain) of the input 4-V squarewave is shown in Figure 3–9a as an input to the low-pass filter (LPF). The transfer characteristic of the LPF is also shown. Since the filter is characterized in terms of its effect on the various input signal frequency components, we

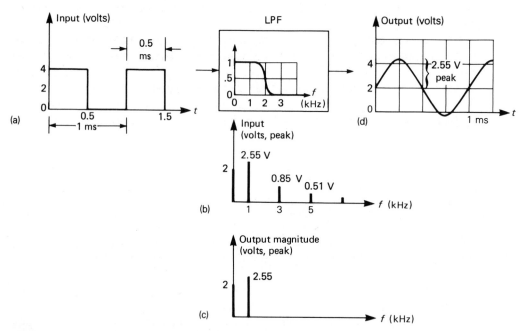

FIGURE 3–9   Input/output waveforms by Fourier analysis (low-pass filtering of squarewave). (a) Input squarewave. (b) Input signal spectrum. (c) Filter output spectrum. (d) Output signal after low-pass filtering.

must determine what these components are. Fortunately, the mathematical work has been done for us, and the results for various signals are listed in Table 3–1; for this particular signal, Equation 3–1 is used because it includes the dc component. The frequency components for a 4-V, 1-kHz squarewave are determined to be

$$v(t) = 2 + \frac{8}{\pi} \sin 2\pi(1 \text{ kHz})t + \frac{8}{3\pi} \sin 2\pi(3 \text{ kHz})t$$

$$+ \frac{8}{5\pi} \sin 2\pi(5 \text{ kHz})t + \cdots \text{ volts}$$

The spectrum is plotted in Figure 3–9b and is aligned with the filter transfer characteristic in order to see what the effect will be on each input frequency component. The filter response curve gives the gain (actually a loss) at each input frequency. So, just as an amplifier input voltage is multiplied by the gain of the amplifier to get the output voltage, the amplitude of each input frequency component is multiplied by the filter gain to determine the output amplitude of that component. The dc component gets multiplied by 1 in the filter. The amplitude of the 1-kHz sinusoid is also multiplied by 1, as seen on the filter response curve. Now notice that any frequency component at 3 kHz or more will be multiplied by zero so that all of the higher harmonics are completely filtered out in the low-pass filter. The result is a frequency spectrum shown in Figure 3–9c, where we see that the output signal will consist of a 2.55-V pk, 1-kHz sinusoid, superimposed on a 2-V dc voltage. This is easily sketched (see Table 3–1b) on a time axis and is shown in Figure 3–9d. Thus, an oscilloscope will show a 1-kHz sinusoid varying between +4.55 V (2 V + 2.55 V pk) and −0.55 V (2 V − 2.55 V pk).

Please note that if other harmonics are present in the output, you must sketch each component carefully on a time axis. The resultant signal, as seen by an oscilloscope, is obtained by adding the voltage values together *for a particular instant of time* and plotting this point for that time. As the resultant (sum) voltage for each instant of time is plotted, you will quickly see the overall shape of the resultant waveform. An example is now given that also illustrates the effect that phase distortion will have on the signal.

■ ■ ■ ■

EXAMPLE 3–5

1. Sketch the harmonics and the resultant waveform for the first three nonzero components of the squarewave shown in Figure 3–10. This illustrates the result of band-limiting a squarewave in a low-pass filter with cutoff frequency above $f \gtrsim 5f_0$.

2. Now, show the results of phase distortion on this frequency-limited square-wave as observed on an oscilloscope. Let the first harmonic suffer a 30° leading phase shift in the input circuit (high-pass filter) of an oscilloscope.

Solution:

1. Table 3–1e gives the Fourier series for this 4.71-V squarewave as $v(t) = [4(4.71)/\pi] [\sin\theta + \frac{1}{3} \sin 3\theta + \frac{1}{5} \sin 5\theta + \cdots]$, where $2\pi f_0 t$ is changed to θ so

FIGURE 3–10 Waveform for
Example 3–5.

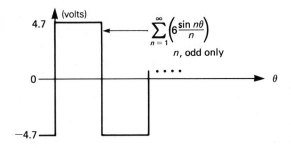

that we can plot $v(t)$ for different phase angles, θ. First, $6 \sin \theta$ is plotted for θ in steps of $10°$ to get good accuracy. This is repeated for $(6/3)\sin 3\theta = 2 \sin 3\theta$ and for $(6/5)\sin 5\theta = 1.2 \sin 5\theta$. These individual component plots are shown in Figure 3–11 along with the sum (resultant) of the components written mathematically as

$$v(t) = \sum_{n=1}^{5} (6/n)\sin n\theta$$

with n odd only.

2. Now, leave the third and fifth harmonic signals alone and shift the fundamental, $f = f_0$, by $30°$ to the left. Plot the resultant.

   The result, shown in Figure 3–12, illustrates the *tilt* or *sag* that you observe when a (band-limited) squareware is ac-coupled to an oscilloscope.

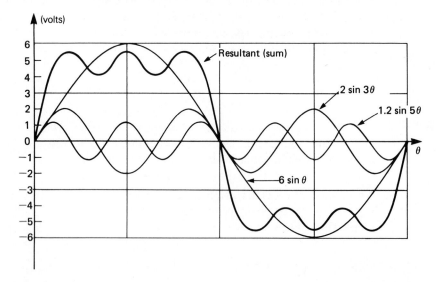

FIGURE 3–11 Squarewave after complete, ideal attenuation of all harmonics above the fifth ($f > 5f_0$).

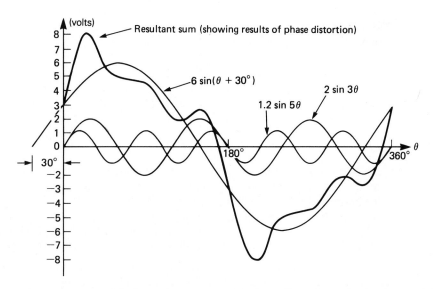

FIGURE 3-12 Band-limited squarewave with phase distortion. The fundamental is shifted 30° to the left (lead) due to phase shift in an actual high-pass filter, such as the input circuit of an oscilloscope, ac-coupled.

3-4 ■ HARMONIC AND PHASE DISTORTION

The results of Example 3–5 can be used to illustrate the meaning of harmonic and phase distortion. Suppose a 1-kHz sinusoidal generator is connected to a two-stage amplifier, as shown in Figure 3–13. If the generator output voltage is set too high, say, 2 V pk, as illustrated, then the voltage at the collector of Q_1 will be nearly a squarewave due to severe clipping. If the amplifier is intended for linear operations, the result is amplitude distortion.

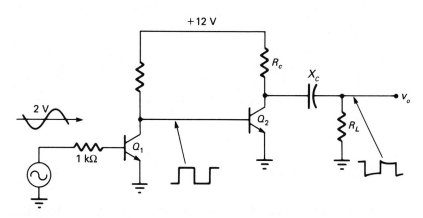

FIGURE 3–13

FIGURE 3-14   Linear-phase
response.

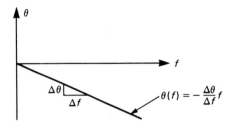

In addition, the input is a single sinusoid, and the output includes many harmonics of the 1-kHz input frequency. When harmonics are present at the output of a linear amplifier but not present at the input, the result is called *harmonic distortion*. Hence, amplitude distortion gives rise to harmonic distortion.

The input to the second stage of the system is a squarewave. However, if the reactance of the coupling capacitor at 1 kHz is $X_C \lesssim (R_C + R_L)$, then the 1-kHz fundamental frequency component will not lose much amplitude but will be advanced in phase more than normal for a linear-phase system, and the result will be the sag or tilt shown for v_0 in Figure 3–13. This is *phase distortion*.

A linear-phase system is one in which the phase through the system is directly proportional to frequency. This is illustrated for a low-pass filter in Figure 3–14. Stated mathematically, a linear-phase system is one for which $\Delta\theta/\Delta f$ is constant over the signal frequency range.

Total Harmonic Distortion

Total harmonic distortion (THD) is given in most linear amplifier manufacturers' data sheets. It is most easily measured with a wave analyzer (Figure 3–1) or a spectrum analyzer. A very low distortion sinusoid is applied at the amplifier input and turned up to produce a specified power output. The amplitudes of the various harmonics measured at the output are compared with the fundamental to determine the percentage of distortion for each harmonic. For instance, if the second-harmonic amplitude is 2 V and the fundamental is 10 V, then the second-harmonic distortion is $D_2 = 2/10 = 0.20$, or 20%.

Total harmonic distortion is calculated as the rms sum of the individual harmonic distortion ratios; that is,

$$\text{THD} = \sqrt{(D_2)^2 + (D_3)^2 + \cdots + (D_n)^2} \qquad (3\text{–}17)$$

where D_n is the distortion ratio of the nth significant harmonic *produced in the amplifier*.

3–5 ■ NONDETERMINISTIC SIGNALS

Thus far we have considered the frequency spectra for deterministic signals; that is, given the amplitude, waveshape, frequency, and relative phase of a periodic

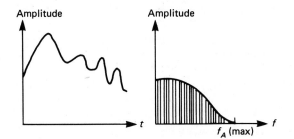

FIGURE 3–15 Generalized time signal and frequency spectra for audio.

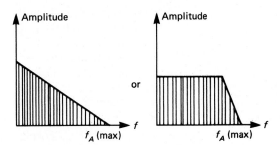

FIGURE 3–16 Idealized audio frequency spectra.

signal, these signals are entirely predictable and no more information is available (or necessary).

If information is to be processed using electrical systems, we will be dealing with nondeterministic signals—that is, signals that are unpredictable at any given instant of time. Such signals are treated statistically and as time averages.

A few examples of nondeterministic signals encountered in electronics are those derived from music, voice, video programming, digital data, and random noise phenomena. Since most interesting signals are nondeterministic, we need to learn how to treat such signals without getting overwhelmed by mathematics.

The electrical signals derived from audio, video, and data programming change continuously and nondeterministically with time; consequently they are usually characterized as time averages, and their relative amplitude and frequency content are specified. An as example, voice and music typically have strong low-frequency energy and progressively weaker high-frequency content. Hence, generalized time and frequency spectra plots might look like those in Figure 3–15.

The frequency components will vary as the music varies, but on the average the energy decreases to a negligible amount at $f_A(\text{max})$, as illustrated in the two idealized spectra of Figure 3–16. Other nondeterministic signals, especially noise-type, are discussed in Chapter 5.

■ PROBLEMS

1. The squarewave of Figure 3–2 has a period of 20 ms and a peak (and peak-to-peak) value of 10 V. Determine the pulse repetition frequency and dc (average) voltage.

2. The squarewave of Figure 3–2 has a peak-to-peak voltage of 5 V. Determine the dc value and the peak values of the first five sinusoidal components, f_0 through $5f_0$.

3. Sketch the frequency spectrum of Problem 2.

4. **a.** Sketch the time plot of the components of Problem 2 versus phase angle θ.
   b. Sketch the composite (sum of all components) to show what the squarewave would look like if limited to the first five harmonics.

5. Determine the (peak) voltages of the dc and first three nonzero sinusoidal components for the following signals:
   a. Half-rectified wave (Table 3–1c) with 6-V pk amplitude
   b. Full-rectified wave with 6-V pk amplitude
   c. Triangle signal (Table 3–1i) of 6-V pk-pk amplitude
   d. Sawtooth of 6-V pk-pk amplitude
   e. Symmetrical squarewave of Table 3–1f with 6-V pk-pk amplitude
   f. 50% alternate inversion rectangular pulses with 6-V pk-pk amplitude

6. A 6-V pk, 60-Hz sinusoid is half-wave rectified with no loss of peak amplitude. If no filter capacitor exists, determine
   a. The dc output voltage;
   b. Peak voltage of the output 60-Hz component;
   c. Peak voltage of the 120-Hz component;
   d. Power dissipated by a 500-Ω output load for **a–c** and the sum of these powers;
   e. Input power assuming a real 1-kΩ input impedance. Compare to **d.**

7. Sketch the frequency spectrum for each of the signals of Problem 5.

8. Write the Fourier series of each of the signals of Problem 5 assuming $T = 16\frac{2}{3}$ ms.

9. A 6-V pk squarewave is applied to the low-pass filter whose frequency response is shown in Figure 3–8. Determine the dc voltage and peak amplitudes of the output components.

10. If the period is 1 ms long,
   a. Sketch the input and output spectra for Problem 9;
   b. Sketch the output waveform as seen on an oscilloscope.

11. A 25% duty cycle, 3-V pk, rectangular pulse train centered at the origin with a period of 1 ms is an input to a low-pass filter; its frequency response is shown in Figure 3–17.
   a. Determine the dc voltage and peak amplitudes of the first five ac components (f_0 through $5f_0$) at the input.
   b. Sketch the filter output waveform as seen on an oscilloscope.

12. A half-wave rectifier with no output filter capacitor produces a 5-V pk output as shown in Figure 3–18.
   a. What is the frequency of the input sinusoid to the rectifier?
   b. Sketch the rectifier output spectrum up to and including the first four nonzero sinusoidal components.
   c. If the rectifier output signal is the input to a low-pass filter with the frequency response of Figure 3–18b, sketch the output spectrum and time-domain waveform.

FIGURE 3–17

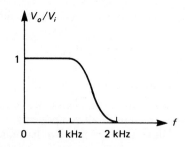

FIGURE 3–18

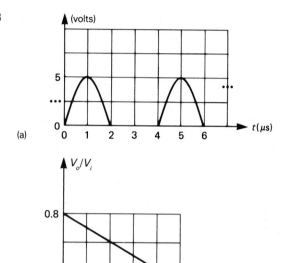

(a)

(b)

13. A linear amplifier with a gain of 40 dB is overdriven, resulting in the output shown in Figure 3–19a.
   a. Sketch the amplifier output frequency spectrum up to 5 kHz.
   b. The output signal is driving a low-pass filter whose exact frequency response (spectrum) is shown in Figure 3–19b. Sketch the filter output frequency spectrum, $V_0(f)$.
   c. Sketch accurately the filter output time waveform as seen on an oscilloscope.
   d. Write the Fourier series expression for the filter output waveform.
   e. Calculate the total harmonic distortion (THD).
   f. How could the distortion be eliminated? (A little thought and analysis will show that mere filtering will not work.)

FIGURE 3–19

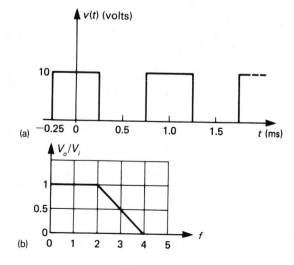

(a)

(b)

14. The following values are determined by waveform analysis measurements at the output of a linear amplifier driven by a pure sinusoid: $V_1 = 5V_{rms}$, $V_2 = 2V_{rms}$, $V_3 = 1V_{rms}$.
   a. Determine the percent of second-harmonic distortion.
   b. Determine the third-harmonic distortion.
   c. Determine the total harmonic distortion as a percent.
   d. Calculate the total output power into an 8-Ω load.

15. Show the results of phase distortion if the fundamental frequency component of Figure 3–11 is delayed (shifted to the right) by 30°.

16. Derive from first principles (the defining integral) the Fourier series for the waveform of Table 3–1f.

17. Derive from first principles (the defining integral) the Fourier transform for a rectangular pulse of amplitude A starting at $t = 0$ and ending at $t = T$.

18. Derive from first principles (the defining integral) the inverse Fourier transform for the ideal low-pass filter described as follows: $V(f)$ is a single rectangular pulse of unit amplitude starting at $f = -B$ and ending at $f = B$. Show that the result is given by

$$v(t) = \frac{2B \sin(2B\pi t)}{2B\pi t} = 2B \, Sa(2B\pi t)$$

3–6 ■ COMPUTER ANALYSIS

Using the computer for communications system and circuit analysis has become a virtual necessity in education and industry. What follows are examples of programs and/or readily available simulation software for solving problems of the type covered in this chapter.

All the programs are interactive in the sense that an introductory sentence or short paragraph describes the purpose of the program and the user is then prompted for input data.

The programs for Chapter 3 are as follows:

- **Squarewave approximation with multiple harmonics:** With this program the user can duplicate plots of the time-domain superposition of harmonic components of squarewaves. An example of the results is Figure 3–3 in the text. Just respond to prompts for the number of harmonics desired and the choices of 1°, 5°, or 10° resolution between plotted points. The listing is in Figure 3–20.

- **Graphing program for plotting (positive) points:** This BASIC program will plot graphs of X,Y data points for X and Y both positive. There are numerous applications possible such as plotting points computed for harmonic components of periodic, and nonperiodic, waveforms and plotting data from lab experiments. The listing is in Figure 3–21.

- **Fourier analysis of an oscillator circuit:** A distortion and harmonic content analysis of the output of a tuned-RF oscillator using SPICE is presented in Figure 2–29.

- **Filter effect on signal:** A three-pole Butterworth low-pass filter was designed and then simulated on Micro-CAP II. The frequency ("AC" analysis) response plot for the filter is shown in Figure 3–22a, clearly demonstrating the flat, zero-ripple passband, 50-kHz design cutoff frequency and the −18-dB/octave gain slope rolloff. The time response to an input 25-kHz squarewave is presented in Figure 3–22b, showing severe rounding indicative of the attenuation of the odd-only harmonics (75 kHz, 125 kHz, and so on). The overlaid $V(2)$ plot showing very little filtering is at the filter input where the observed effect is due to the high-frequency reactive input impedance of the filter; see Figure 3–22c for the schematic and node assignments.

```
10 REM *** Squarewave approximation with multiple harmonics ***
20 REM *****    Example of student computer assignment   *******
30 REM ***            For Professor Young at ASU            ***
40 CLS   :K=1  :PRINT
50 PRINT "This program will compute and list the points in the composite"
60 PRINT "waveform approximating a SQUAREWAVE (see Figure 3-3, ECTs)."
70 PRINT
80 PRINT "The squarewave is binary with voltage levels of zero or `A´ volts"
90 PRINT :INPUT "       Enter number of harmonics, N";N:PRINT
100 INPUT "        Enter the amplitude A";J:PRINT
110 INPUT "If you want a hard copy enter a 1; if not hit return.";QS:PRINT
120 PRINT "   Enter the resolution of the Composite to be printed.":PRINT
130 PRINT "1.)  Every 1 Deg.":PRINT:PRINT"2.)  Every 5 Deg.":PRINT:PRINT"3.)
    Every 10 Deg.":PRINT:PRINT"Make sure your printer is ready if you selected
    it!":
140 PRINT:INPUT KK:PRINT:PRINT
150 IF KK=1 THEN K=1 ELSE IF KK=2 THEN K=5 ELSE IF KK=3 THEN K=10 ELSE GOTO 120
160   CLS
170 PI = 3.141592654#
180 IF QS="1" THEN LPRINT"ANGLE","COMPOSITE":LPRINT
190 PP=1
200 NUM=0
210 PRINT"        ANGLE                        COMPOSITE"
220 FOR D = -45 TO 405
230 Q=0
240 QQ=0
250 FOR M=1 TO N STEP 2
260 V=(2/(M*PI))*(SIN(2*M*PI*D/360))
270 VV=(2*J/(M*PI)*SIN(2*M*PI*D/360))
280 W=V+Q
290 WW=VV+QQ
300 Q=W
310 QQ=WW
```

FIGURE 3–20   (page 1 of 2)

```
320 NEXT M
330 X=D+140
340 Z=W*100
350 REM PSET(X,300-Z),1
360 PP=PP-1:IF PP=0 THEN GOTO 370 ELSE GOTO 410
370 PP=K:IF Q$ = "1" THEN LPRINT D,(J/2)+WW
380 PRINT "THETA = ";:PRINT USING "###";D;:PRINT "          COMPOSITE = ";
    (J/2)+WW
390 IF NUM=20 THEN 430
400 NUM=NUM+1
410 NEXT D
420 GOTO 460
430 A=0:INPUT " ENTER A 1 TO QUIT OR HIT RETURN TO CONTINUE";A
440 IF A=1 THEN 460:NUM=0
450 NUM=0:CLS:PRINT "      ANGLE                  COMPOSITE":GOTO 390
460 END
```

FIGURE 3–20   (page 2 of 2)

```
10 REM *** GRAPHING PROGRAM FOR PLOTTING POSITIVE POINTS ***
20 REM ***     WRITTEN FOR PROFESSOR P. YOUNG AT ASU     ***
30 CLS:KEY OFF
40 PRINT "THIS PROGRAM WILL PLOT POSTIIVE GRAPHS ON THE SCREEN OF THE "
50 PRINT "COMPUTER.  THE X AND Y AXES CAN VARY FROM 0-99. THE COMMANDS"
60 PRINT "ARE PLOT POINT, ERASE POINT, NEW TITLE, CLEAR GRAPH, PRINT"
70 PRINT "GRAPH, AND END GRAPH. THE COMMAND STATUS IS IN THE BOTTOM"
80 PRINT "LEFT CORNER. ALL OF THE COMMANDS ARE SELF EXPLANATORY "
90 PRINT "EXCEPT THE PRINT. TO PRINT, GO TO THE PRINT MENU, EXECUTE"
100 PRINT "IT, THEN HIT `PRINT SCREEN´."
110 PRINT:Y=1:INPUT "HIT RETURN TO CONTINUE";X: IF Y=1 THEN 120
120 CLS:LOCATE 23,1:PRINT "MENU:"
130 FOR J=1 TO 21:LOCATE J,9:PRINT CHR$(222):NEXT J
140 FOR J=10 TO 78:LOCATE 21,J:PRINT CHR$(220):NEXT J
150 FOR J=10 TO 80 STEP 17:LOCATE 22,J:PRINT CHR$(254):NEXT J
160 FOR J=1 TO 21 STEP 5:LOCATE J,8:PRINT CHR$(223):NEXT J
170 LOCATE 24,1:INPUT:"INPUT X LIMITS LOW,HIGH";XLOW, XHIGH:
180 XL=XHIGH-XLOW:XM1=(XL/4)+XLOW:XM2=(XL/2)+XLOW:XM3=(XL*3/4)+XLOW
190 LOCATE 23,8:PRINT USING "##.##";XLOW;:LOCATE 23,25:PRINT USING "##.##";XM1;
200 LOCATE 23,42:PRINT USING "##.##";XM2;:LOCATE 23,59:PRINT USING "##.##";XM3;
210 LOCATE 23,76:PRINT USING "##.##";XHIGH;
220 GOSUB 390:LOCATE 24,1:INPUT:"INPUT Y LIMITS LOW,HIGH";YLOW, YHIGH:
230 YL=YHIGH-YLOW:YM=YLOW:YM1=(YL/4)+YLOW:YM2=(YL/2)+YLOW:YM3=(YL*3/4)+YLOW
240 LOCATE 1,1:PRINT USING "##.##";YHIGH;:LOCATE 6,1:PRINT USING "##.##";YM3;
250 LOCATE 11,1:PRINT USING "##.##";YM2;:LOCATE 16,1:PRINT USING "##.##";YM1;
```

FIGURE 3–21   (page 1 of 2)

```
260 LOCATE 21,1:PRINT USING "##.##";YLOW;
270 GOSUB 390:GOSUB 420:IF (Q1$="Y") OR (Q1$="y") THEN 550 ELSE 280 'PLOT PT
280 GOSUB 390:GOSUB 430:IF (Q2$="Y") OR (Q2$="y") THEN 500 ELSE 290 'ERASE PT
290 GOSUB 390:GOSUB 440:IF (Q3$="Y") OR (Q3$="y") THEN 520 ELSE 300 'NEW X TITLE
300 GOSUB 390:GOSUB 450:IF (Q7$="Y") OR (Q7$="y") THEN 530 ELSE 310'NEW Y1 TITLE
310 GOSUB 390:GOSUB 460:IF (Q8$="Y") OR (Q8$="y") THEN 540 ELSE 320'NEW Y2 TITLE
320 GOSUB 390:GOSUB 470:IF (Q4$="Y") OR (Q4$="y") THEN  120 ELSE 330 'CLR GRPH
330 GOSUB 390:GOSUB 480:IF (Q5$="Y") OR (Q5$="y") THEN 570 ELSE 340 'PRT GRAPH
340 GOSUB 390:GOSUB 490:IF (Q6$="Y") OR (Q6$="y") THEN 580 ELSE 350 'QUIT GRAPH
350 GOTO 270
360 LOCATE 24,1:INPUT;"X-AXIS LABEL";XASIS$:LOCATE 25,40:PRINT XASIS$;:RETURN
370 LOCATE 24,1:INPUT;"Y1-AXIS LABEL MAX 8-CHR'S";Y1ASIS$:LOCATE 3,1:PRINT Y1ASIS$;:RETURN
380 LOCATE 24,1:INPUT;"Y2-AXIS LABEL MAX 8-CHR'S";Y2ASIS$:LOCATE 4,1:PRINT Y2ASIS$;:RETURN
390 LOCATE 24,1:PRINT "                                ";:RETURN'CLEAR MENU
400 LOCATE 3,1:PRINT "         ";:RETURN 'CLEAR Y1XAIS
410 LOCATE 4,1:PRINT "         ";:RETURN 'CLEAR Y2XAIS
420 Q1$="N":LOCATE 24,1:INPUT;"PLOT POINT  ? Y=YES ";Q1$:RETURN 'PLOT A POINT
430 Q2$="N":LOCATE 24,1:INPUT;"ERASE POINT ? Y=YES ";Q2$:RETURN 'ERASE A POINT
440 Q3$="N":LOCATE 24,1:INPUT;"NEW X TITLE ? Y=YES ";Q3$:RETURN 'NEW X TITLE
450 Q7$="N":LOCATE 24,1:INPUT;"NEW Y1 TITLE  ? Y=YES ";Q7$:RETURN 'NEW Y2 TITLE
460 Q8$="N":LOCATE 24,1:INPUT;"NEW Y2 TITLE  ? Y=YES ";Q8$:RETURN 'NEW Y2 TITLE
470 Q4$="N":LOCATE 24,1:INPUT;"CLEAR GRAPH ? Y=YES ";Q4$:RETURN 'CLEAR GRAPH
480 Q5$="N":LOCATE 24,1:INPUT;"PRINT GRAPH ? Y=YES ";Q5$:RETURN 'PRINT GRAPH
490 Q6$="N":LOCATE 24,1:INPUT;"QUIT GRAPH  ? Y=YES ";Q6$:RETURN 'QUIT GRAPH
500 GOSUB 390:LOCATE 24,1:INPUT;"POINT X,Y ";XPOS,YPOS:XPT=INT(((XPOS-XLOW)/XL)*68+.5)+10
510 YPT=INT((1-((YPOS-YLOW)/YL))*20+.5)+1:LOCATE YPT,XPT:PRINT " ";:GOTO 270
520 GOSUB 390:LOCATE 25,40:PRINT "-                 ";:GOSUB 360:GOTO 270
530 GOSUB 400:LOCATE 3,1:PRINT "-      ";:GOSUB 390:GOSUB 370:GOTO 270
540 GOSUB 410:LOCATE 4,1:PRINT "-      ";:GOSUB 390:GOSUB 380:GOTO 270
550 GOSUB 390:LOCATE 24,1:INPUT;"POINT X,Y ";XPOS,YPOS:XPT=INT(((XPOS-XLOW)/XL)*68+.5)+10
560 YPT=INT((1-((YPOS-YLOW)/YL))*20+.5)+1:LOCATE YPT,XPT:PRINT "*";:GOTO 270
570 GOSUB 390:LOCATE 23,1:PRINT "     ";:LOCATE 25,1:INPUT;"   ",WAI$:GOTO 270
580 END
```

FIGURE 3-21   (page 2 of 2)

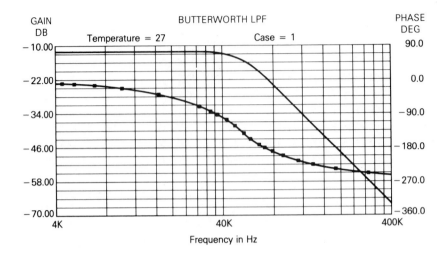

Frequency =   400.00000E+03 Hz   Gain = −65.675 Db
Phase angle = −255.248   Degrees   Group delay =   0.00000E+00
Gain slope = −180.57850E-01 Db/Oct   Peak gain = −12.050 Db/F= 400.00000E+01

FIGURE 3–22a

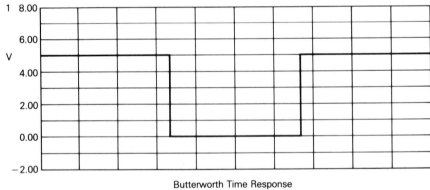

Butterworth Time Response

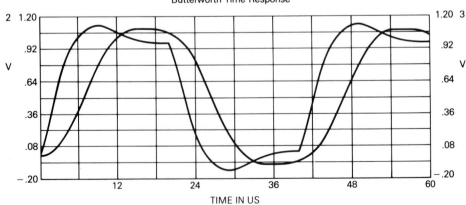

TIME IN US

FIGURE 3–22b   Time response of 3-pole low-pass filter to a 25-kHz squarewave input
V(1). V(3) is the rounded output signal.

BUTLPF Analysis Limits

Simulation time (Tmax/Tmin/Timestep)	60E-6
Display time (Dmax/Dmin/Printstep)	60E-6
Minimum accuracy (%)	5
Upper trace A waveform spec	1
Upper trace B waveform spec	1
Upper trace range (High/Low)	8/-2
Lower trace A waveform spec	2
Lower trace B waveform spec	3
Lower trace range (High/Low)	1.2/-.2
Zero, Read, Leave initial values (Z,R,L)	Z
Edit/review initial conditions (Y,N)	N
Dump node waveform to user file (Y,N)	N
Calculate D.C. operating point (Y,N)	N
Temperature (Low/High/Step)	27
Number of cases	1
Output to Disk, Printer, None (D,P,N)	N
Save, Retrieve, Normal run (S,R,N)	N

Micro-CAP II Student Version

BUTLPF NETLIST

No.	Type	A	B	C	D	Parameter
1	RESISTOR	–	1	–	2	200
2	CAPACITOR	–	2	–	0	.1016UF
4	INDUCTOR	–	2	–	3	230UH
5	CAPACITOR	–	3	–	0	34.5NF
8	RESISTOR	–	3	–	0	50
11	V (T)	–	0	–	1	SQR

Library Elements Used

SQR
P0= 0	P1= 5	P2= 0	P3= 1E-09
P4= .00002	P5= .2000E-04	P6= .00004	

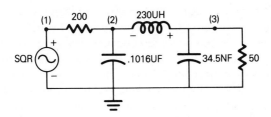

FIGURE 3–22c   Butterworth low-pass filter on Micro-CAP II.

4

Elements of Noise

■ INTRODUCTION

Noise can be considered anything that, when added to an information signal, makes it more difficult to extract the information. For instance, the low-frequency "hum" sometimes heard in the background when listening to amplified music or the random "speckles" sometimes seen on a television screen are examples of noise effects. Extraneous noise that occurs sporadically, such as the interference from electrical phenomena in storms, machinery, and lights, is specific to a given environment and must be dealt with by appropriate shielding, grounding, and filtering. However, other noise phenomena occur within the circuit components used to transmit and receive information. These must be understood and controlled within the system. Since the presence of noise puts a lower limit on the signals that can be detected and an upper limit on the amount of gain that can be used before noise overload or distortion occurs, we must identify the noise sources and their power level. ■

4–1 ■ SYSTEM NOISE SOURCES

Noise in electronic systems is basically due to the discrete character of the electrical charge. Electrical current is made up of individual packets of charge and is continuous only on a time-average basis. The noise generated by discrete charges in electronic systems has been identified and categorized as thermal noise, shot and partition noise, flicker ($1/f$) noise, and burst noise (also called *popcorn noise*).

Thermal Noise

A material is a conductor at room temperature if its electrons are free to move about within it. For example, the energy in the electrons of materials used to make resistors is greater at room temperature than the energy required to bind them to a particular molecule. As the temperature is raised, the electrons get more energetic and move around in the resistive material, interacting with the molecules and other electrons.

Since the random movement of electrical charge produces a current and a current in a resistor produces a voltage, we see that a resistor is a random noise generator. This random voltage or current is called *thermal noise* and is directly proportional to the temperature. This thermal noise is characterized by a waveform that never repeats itself exactly; that is, it is purely random and, as predicted by the kinetic theory of heat, the power spectrum is flat with frequency. Since all frequencies are present in this thermal random noise, it is, like the sun with all its colors, referred to as *white noise*.

The average noise power that can be delivered to a system at a temperature T is given by

$$N_{th} = kTB \tag{4–1}$$

where N_{th} = the average power in watts
 T = absolute temperature in kelvins, K
  = X°C + 273°
 k = 1.38×10^{-23} watt/K-Hz, Boltzmann's constant
 B = the bandwidth in which the measurement is made, in hertz

Notice that the power of thermal noise is independent of resistance. This is because any conductor, which by definition has electrons free to move around, is a source of noise power. On the other hand, if thermal noise is measured with a true rms voltmeter and if the noise meter *impedance is matched* to the impedance of the noise source, then the *meter will read $E_n = \sqrt{kTBR}$* in volts rms. However, where impedances are not matched and a resistor is considered *on its own*—that is, as an open-circuited equivalent noise generator with a series *resistor R*—then the *available noise voltage* would theoretically have to be

$$E_n = 2\sqrt{kTBR}$$
$$= \sqrt{4kTBR} \tag{4–2}$$

Figure 4–1 illustrates the 2-to-1 voltage drop when a 50-Ω noise generator is connected to a system for which the input impedance is 50 Ω.

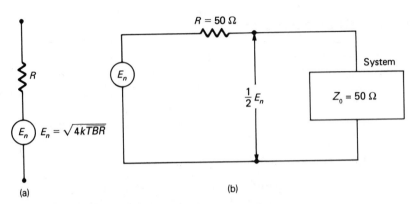

FIGURE 4–1   (a) No power delivered; open-circuited equivalent noise generator.
(b) Power delivered to the system is $N = kTB$, where B is the equivalent noise bandwidth, system is impedance-matched to noise source.

Thermal noise generally is not affected by the amount of bias current in a
resistor, although it is known that the ever-popular carbon resistor has an additional current-dependent noise that makes it unsuitable for critical applications.
Electric current does affect the noise produced in semiconductors and vacuum
tubes. The sources of noise that are bias-current-dependent are shot noise, flicker
noise, and popcorn noise.

Shot Noise

Historically, shot noise got its name in vacuum-tube days when someone imag-
ined that electrons crossing the tube and striking the metal anode would produce a
sound equivalent to pouring a bucket of shot (or BBs) over a metal plate.

The same kind of noise is generated in semiconductor junctions when the
electrons with charge q, comprising an average current I_{dc}, cross a potential
barrier. Indeed, a semiconductor diode anode, like its vacuum-tube counterpart,
will heat up considerably from this bombardment of electrons. In addition, mul-
tielectrode devices like transistors (triodes, pentodes, and so forth) produce a
noise from the random "selection" of electrons as some go into base (or perhaps
screen grid in vacuum tubes) current, while others go into collector (plate) current. This is sometimes referred to as *partition noise* and has a random shot noise
effect.

The power produced by shot noise is directly proportional to the bias (dc)
current. Like thermal noise, shot noise is purely random and its power spectrum is
flat with frequency. Measurements confirm that the mean-square value of shot
noise is given by

$$I_n^2 = 2qI_{dc}B$$

or

$$I_n = \sqrt{2qI_{dc}B} \qquad\qquad (4\text{--}3)$$

FIGURE 4–2   Shot noise model.

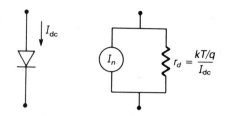

where I_n = rms-average noise current in amperes rms
   $q = 1.6 \times 10^{-19}$ coulombs (C), the charge/electron
   I_{dc} = dc bias current in the device, in amperes
   B = bandwidth in which the measurement is made, in hertz

Equation 4–3 is valid up to frequencies comparable to the inverse of the transit time of electrons across the device junction. This is typically in the thousands of megahertz range—that is, in the gigahertz range.

The equivalent noise generator concept can be used for a diode junction, just like the equivalent noise generator for a resistor with thermal noise. This is shown in Figure 4–2.

■ ■ ■ ■

EXAMPLE 4–1

Determine noise current and equivalent noise voltage for the diode in Figure 4–2 with $I_{dc} = 1$ mA. The noise is measured in a bandwidth of 10 MHz.

Solution:

1.
$$I_n = \sqrt{2qI_{dc}B}$$
$$= \sqrt{2(1.6 \times 10^{-19}\ \text{C})(1 \times 10^{-3}\ \text{A})(10 \times 10^{6}\ \text{Hz})}$$
$$= \mathbf{56.6\ nA}$$

2. The dynamic (not ohmic) junction resistance is $r_j = r_d = kT/qI_{dc} = 26\ \text{mV}/I_{dc} = 26\ \Omega$. Incidentally, kT/q can be between 25 and 40 mV at room temperature, depending on the junction doping characteristics.

$$E_n = I_n r_j = (56.6 \times 10^{-9}\ \text{A})(26\ \Omega) = \mathbf{1.47\ \mu V}$$

Flicker (1/f) Noise

Flicker noise is associated with crystal surface defects in semiconductors and is also found in vacuum tubes. The noise power is proportional to the bias current and, unlike thermal and shot noise, flicker noise decreases with frequency.

An exact mathematical model does not exist for flicker noise because it is so device-specific. However, the inverse proportionality with frequency is almost exactly $1/f$ for low frequencies, whereas for frequencies above a few kilohertz, the noise power is weak but essentially flat. Flicker noise is essentially random, but because its frequency spectrum is not flat, it is not a white noise. It is often referred to as *pink noise* because most of the power is concentrated at the lower end of the frequency spectrum.

The objection to carbon resistors mentioned earlier for critical low-noise applications is due to their tendency to produce flicker noise when carrying a direct current. In this connection, metal film resistors are a better choice for low-frequency, low-noise applications.

Burst Noise

Burst noise (or popcorn noise) is another low-frequency noise that seems to be associated with heavy-metal ion contamination. Measurements show a sudden shift in the bias current level that lasts for a short duration before suddenly returning to the initial state. Such a randomly occurring discrete-level burst would have a popping sound if amplified in an audio system.

Like flicker noise, popcorn noise is very device-specific, so that a mathematical model is not very useful. However, this noise increases with bias current level and is inversely proportional to the square of frequency ($1/f^2$).

4–2 ■ NOISE SPECTRAL DENSITY AND CIRCUIT CALCULATIONS

When making calculations of circuit noise, it is necessary to determine the contribution of noise power from each circuit component. Then the noise *power* from all sources is added to get the total circuit noise. Also, it is convenient to make the calculations independent of bandwidth and then, in the final mathematical step, to expand the answer to include the full noise bandwidth.

Let P_n be the total noise power measured in a bandwidth B. Then P_n/B is the noise power in 1 Hz of bandwidth and has units of watts per hertz of bandwidth. The quantity is called *noise spectral density* and, when multiplied* by the system noise bandwidth, yields the total noise power in watts. For instance, the thermal noise power delivered to a receiver, which is impedance-matched to its receiving antenna, can be calculated from $N_{th} = kTB$. The noise spectral density, N_0, is determined from

$$N_0 = N/B = kT \quad \text{(watts/Hz)} \qquad \textbf{(4–4)}$$

When considering the noise in circuits with semiconductors and resistors of various values, noise voltages and currents are used in the calculations. Also, the noise power spectral density is used and the circuit bandwidth is included in the final calculation. As an example, consider the silicon diode that is forward-biased by the resistor and 10-V power source in Figure 4–3. The total noise power, and subsequently the low-pass filter output noise voltage, is analyzed as follows.

---

* If either the noise power spectral density or the system bandpass frequency response is not constant, then the product must be integrated as in Equation 4–19.

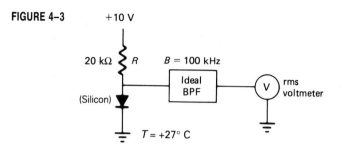

FIGURE 4–3

The thermal noise voltage, power spectral density magnitude, is—from Equation 4–2—defined by

$$\frac{E_n^2}{B} = 4kTR \tag{4–5}$$

with units of W/Hz. Henceforth, the quantity in Equation 4–5 will be called *voltage spectral density*. The equivalent thermal noise current spectral density, in units of A^2/Hz, is computed as

$$\left(\frac{I_n^2}{B}\right)_{\text{therm}} = \frac{E_n^2/R^2}{B} = \frac{E_n^2/B}{R^2} \tag{4–6}$$

where R is the resistance in which the thermal noise is created. The shot noise current (power) spectral density magnitude created in a diode is, from Equation 4–3,

$$I_n^2/B = 2qI_{\text{dc}} \tag{4–7}$$

If this current flows in a resistor r, then the shot noise voltage spectral density produced is given by

$$\left(\frac{E_n^2}{B}\right)_{\text{shot}} = \frac{I_n^2 r^2}{B} = r^2\left(\frac{I_n^2}{B}\right)_{\text{shot}} \tag{4–8}$$

Thermal and shot noise are produced by very wideband physical random processes and, over the bandwidths normally encountered in communication electronics, are considered to be uncorrelated and of infinite bandwidth—that is, white noise, containing energy at all frequencies with constant amplitude, on average. As such, summing of currents and voltages can only be done on a power basis. This means that the total noise voltage drop across the silicon diode in the series connection of diode, resistor R, and 10-V dc source can be found from

$$\left(\frac{E_n^2}{B}\right)_{\text{total}} = \left(\frac{E_n^2}{B}\right)_{\text{shot}} + \left(\frac{E_n^2}{B}\right)_{\text{therm},r_d} \tag{4–9a}$$

$$= \left(\frac{E_n^2}{B}\right)_{\text{shot}} + r_d^2\left(\frac{I_n^2}{B}\right)_{\text{therm}} \tag{4–9b}$$

where
$$\left(\frac{I_n^2}{B}\right)_{\text{therm}} = \frac{E_n^2/B}{R^2} \tag{4-10}$$

Equation 4–9b is consistent with an equivalent current spectral density summation by substitution of Equation 4–8 for the shot noise; that is,

$$\left(\frac{E_n^2}{B}\right)_{\text{total}} = r_d^2\left(\frac{I_n^2}{B}\right)_{\text{therm}} + r_d^2\left(\frac{I_n^2}{B}\right)_{\text{shot}}$$

$$= r_d^2\left(\frac{I_n^2}{B}\text{ therm} + \frac{I_n^2}{B}\text{ shot}\right) = r_d^2\left(\frac{I_n^2}{B}\right)_{\text{total}} \tag{4-11}$$

Equation 4–11 gives the noise voltage spectral density magnitude input to the bandpass filter in Figure 4–3. Given an ideal lossless LC filter of bandwidth B, the square root of the mean-square noise voltage at the output will measure

$$E_n = \sqrt{\left(\frac{E_n^2}{B}\text{ thermal} + \frac{E_n^2}{B}\text{ shot}\right)B} \tag{4-12}$$

which is the square root of the total of thermal and shot noise normalized power, in the bandwidth of measurement. Example calculations will help put this discussion into perspective.

■ ■ ■ ■

**EXAMPLE
4-2**

If the bandpass filter of Figure 4–3 has no insertion loss in the passband, what will the rms voltmeter read?

Solution:

1. Noise generated by R; this is thermal noise. The spectral density is, from Equation 4–5, $E_n^2/B = 4(1.38 \times 10^{-23} \text{ J/K})(273 \text{ K} + 27°\text{C})(20 \text{ k}\Omega) = \mathbf{3.32 \times 10^{-16} \ V^2/Hz}$. This is noise spectral density.

2. Noise generated by the diode: The thermal noise generated by the diode is very small compared to the 20-kΩ resistor because diode resistance is just a few ohms. Shot noise is far and away the greatest noise contribution of the diode. This is given as a spectral density by Equation 4–7, where the bias current is determined as

$$I_{\text{dc}} = \frac{10 \text{ V} - 0.7 \text{ V}}{20 \text{ k}\Omega} = 0.465 \text{ mA}$$

Thus
$$\frac{I_n^2}{B} = 2 \times (1.6 \times 10^{-19} \text{ C})(0.465 \times 10^{-3} \text{ C/s})$$

$$= \mathbf{1.488 \times 10^{-22} \ A^2/Hz}$$

The linear model of the circuit is shown in Figure 4–4, where $r_d = 0.026/I_{\text{dc}}$ is the dynamic junction resistance of the forward-biased diode. We could show the polarity on the noise generators, but this would have little meaning since the noise signals are random. We will measure the voltage from point A to

FIGURE 4-4 Noise model.

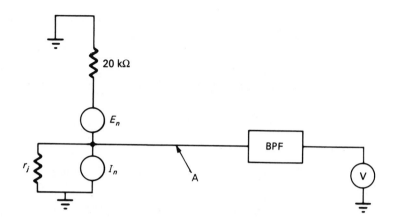

ground, after bandwidth limiting, so we need to determine the noise contribution of each source in volts. The thermal noise of the 20 kΩ will create a noise current (spectral density), from Equation 4–6, of

$$\frac{E_n^2/B}{R^2} = \frac{3.32 \times 10^{-16}}{(20 \text{ k}\Omega)^2} = 8.3 \times 10^{-25} \frac{A^2}{Hz}$$

This is I_n^2/B thermal, with units of A^2/Hz. This current will produce a voltage across the diode of $I_n r_d$, or, using Equation 4–8,

$$\left(\frac{E_n^2}{B}\right)_{therm} = r_d^2 \left(\frac{I_n^2}{B}\right)_{therm}$$

where

$$r_d = \frac{26 \text{ mV}}{0.465 \text{ mA}} = 55.9 \ \Omega$$

Thus

$$\left(\frac{E_n^2}{B}\right)_{therm} = (8.3 \times 10^{-25} \ A^2/Hz)(55.9 \ \Omega)^2$$

$$= 2.6 \times 10^{-21} \ V^2/Hz$$

The diode shot noise current will produce a voltage of I_n (shot) $\times r_d$, or, using noise spectral density,

$$\left(\frac{E_n^2}{B}\right)_{shot} = r_d^2 \left(\frac{I_n^2}{B}\right)_{shot} = 1.488 \times 10^{-22} \ A^2/Hz \times (55.9 \ \Omega)^2$$

$$= 4.65 \times 10^{-19} \ V^2/Hz$$

The total noise voltage spectral density magnitude across the diode is

$$\frac{E_n^2}{B} \text{ total} = \frac{E_n^2}{B} \text{ therm} + \frac{E_n^2}{B} \text{ shot}$$

$$= 0.026 \times 10^{-19} + 4.650 \times 10^{-19}$$

$$= 4.676 \times 10^{-19} \ V^2/Hz$$

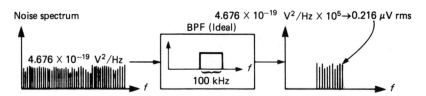

FIGURE 4–5   Filtering the noise.

Notice that, for this example (which is typical), the shot noise is a lot stronger than the thermal noise. Finally, the bandpass filter (BPF) takes in an infinite spectrum of noise (theoretically) and attenuates all except 100 kHz of it before the voltmeter makes the measurement. This is illustrated in Figure 4–5. To finish the problem, the voltmeter will measure a total noise voltage given by Equation 4–12 of

$$E_n = \sqrt{\left(\frac{E_n^2}{B} \text{ thermal} + \frac{E_n^2}{B} \text{ shot}\right)B}$$

$$= \sqrt{4.676 \times 10^{-14} \text{ V}^2} = 2.16 \times 10^{-7} \text{ V rms}$$

so $E_n = 0.216 \ \mu\text{V rms}$

This result for noise voltage may appear to show that not much noise is present; indeed, this is the case. However, we need to keep this in perspective relative to the amount of signal that arrives at the input of a high-gain system.

A final comment at this point is that the noise seen at the BPF output in Figure 4–5 is no longer white; it is *not uncorrelated* noise. More is said about this after equivalent noise bandwidth is discussed in Section 4–6.

4–3 ■ NOISE AND FEEDBACK

The calculations just completed demonstrate how noise from independent sources is combined. The same techniques are used for circuits with transistors and operational amplifiers. Negative feedback does not affect the amount of noise at the output. The noise from resistors and semiconductors is combined as if no feedback exists, although it should be obvious that the contribution of each component in the network must be included.

Remember that noise is the result of random processes. Consequently, the instantaneous phase and frequency is random and there is no possibility of canceling the input noise by feeding back noise from the output. If this well-established fact were not true, noise would no longer be a problem in communications systems because we would simply use a negative feedback amplifier to cancel it out.

Negative feedback can, however, help with certain interfering noise signals such as the 60-Hz "hum" often picked up in audio systems from ac power sources. Also, there are noise canceling circuits found in some systems such as television receivers, which can detect extraordinarily strong noise impulses and momentarily close a switch to short the impulses before they can affect the system.

4-4 ■ SYSTEM NOISE FIGURE

Resistors and semiconductor p-n junctions in a receiver produce noise that will cause a reduction in information due to an increase in noise that reduces the ratio of desired signal relative to noise. Because of internal noise, amplifiers and other circuits increase noise power by more than the circuit signal gain.

Receiver *noise ratio* (NR) is the ratio by which the input signal-to-noise ratio (S/N or SNR) deteriorates due to noisy receiver circuits. The *noise figure* (NF) is the decibel equivalent of the noise ratio: NF = 10 log NR. The additional noise degrades the input S/N so that, for a linear system, the noise ratio (NR) is expressed as

$$\text{NR} = \frac{(S/N)_i}{(S/N)_o} \qquad (4\text{--}13)$$

For example, suppose S/N = 25 dB at the input to an amplifier whose NF = 10 dB, then the S/N at the amplifier output will be 15 dB, as calculated from $(S/N)_o = (S/N)_i/\text{NR}$, or

$$(S/N)_o \text{ (dB)} = (S/N)_i \text{ (dB)} - \text{NF} \qquad (4\text{--}14)$$

These equations assume that the bandwidth used to compute the input S/N is the *net* bandwidth of the system.

Although it would appear that every amplifier will degrade the S/N by its NF, this is not the case. Each amplifier in a string of amplifiers produces a certain amount of noise over and above the thermal noise. As the input signal (and noise) is amplified, the signal becomes strong enough that the small amount of noise generated by succeeding stages adds relatively very little to the overall amount of noise. This concept leads to the formula derived in Chapter 5 for calculating the overall noise ratio of a system:

$$\text{NR} = \text{NR}_1 + \frac{\text{NR}_2 - 1}{A_{p_1}} + \frac{\text{NR}_3 - 1}{A_{p_1} \cdot A_{p_2}} + \cdots + \frac{\text{NR}_n - 1}{A_{p_1} \cdot A_{p_2} \cdot A_{p_3} \cdots A_{p_{n-1}}} \qquad (4\text{--}15)$$

where NR_1 = noise-power ratio of the first stage and A_{p_1} = power-gain ratio of the first stage.

Remember when using Equation 4–15, known as *Friis' formula*, that decibels are *not* used. Noise figures and gains are changed to power ratios. When the calculation is completed, convert back to system noise figure in decibels.

■ ■ ■ ■

**EXAMPLE
4-3**

$NF_1 = 2$ dB $= 1.58$, $NF_2 = 6$ dB $= 3.98$, $NF_3 = 10$ dB $= 10$, $A_{p_1} = 8$ dB $= 6.3$, and $A_{p_2} = 12$ dB $= 15.85$. Calculate the overall system noise figure.

Solution:

From Equation 4-15, $NR = 1.58 + (3.98 - 1)/6.3 + (10 - 1)/(6.3)(15.85) = 1.58 + 0.47 + 0.09 = $ **2.14**; therefore, **NF (dB) = 3.3 dB.**

Notice from Example 4–3 that, even though NR_2 is more than twice NR_1, the noise contribution of the second stage is much less than that of the first. So put your money in the first stage. Use a *low-noise* amplifier with *high gain* for the input amplifier. In fact, earth-station receiver systems ordinarily have a low-noise amplifier (LNA) placed right at the antenna output point to get the signal up out of the noise before further losses occur in the coaxial (or waveguide) feed-in cable to the receiver.

4-5 ■ NOISE TEMPERATURE

In addition to the noise figure of merit expressed in decibels, the amount of noise added by the noisy front end of a receiver is often expressed in terms of an *equivalent temperature*. The conversion between noise ratio NR and equivalent noise temperature T_{eq} is

$$T_{eq} = T_0(NR - 1) \qquad\qquad (4\text{-}16)$$

Here, $T_0 = 290$ K is a reference temperature in kelvins.

Equation 4–16 tells us that if the receiver adds no noise to that which comes in with the signal from the antenna—that is, $NF = 0$ dB, or $NR = 1$—then the equivalent receiver noise temperature is $T_{eq} = 290$ K$(1 - 1) = 0$ K. This is a "cool" receiver. On the other hand, if the receiver doubles the incoming noise, then $NR = 2$, or $NF = 3$ dB, and $T_{eq} = 290$ K$(2 - 1) = 290$ K.

The usefulness of this way of expressing the noisiness of a circuit or system is that, when two circuits are connected, the equivalent noise temperature at the connection is simply the sum of the separate equivalent noise temperatures. For example, a microwave antenna at an earth station pointed toward the darkness of space has a $T_{ant} \approx 25$ K, including the noise of space (≈ 3.5 K) and the antenna noise itself. If the receiver has $T_r = 290$ K (3 dB NF), then the resultant equivalent system noise temperature is

$$T_{eq} = T_{ant} + T_r = 25 + 290 = 315 \text{ K}$$

For convenience and for getting a feel for the numbers, Table 4–1 provides a short list of NF (dB) and equivalent noise temperatures. Incidentally, 0.5–2 dB tends to be the state-of-the-art range of noise figures for solid-state microwave receivers. Besides low-noise antennas and low-noise, high-gain amplifiers, an-

TABLE 4-1 Equivalence Between Noise Figure and Noise Temperature

NF (dB)	T_{eq} (K)
0.5	35.4
1	75.1
2	170
3	289
4	438
6	865
10	2610

Also, $N_0 = kT$ watts/hertz $= -174$ dBm/Hz $= -204$ dBW/Hz describes the thermal noise spectral density.

other way of improving the S/N of a received signal is by reducing the receiver bandwidth.

4-6 ■ NOISE BANDWIDTH

To this point in the discussion of circuit noise, the effects due to capacitors and inductors have not been mentioned. Also, an ideal rectangular (also called brick wall) system bandwidth has been assumed. Here we consider the noise spectrum in nonideal—that is, *real*—circuits.

Capacitors and inductors are present in all communications systems, but any noise power extracted from them is determined by the resistive component of their impedance. Low-Q devices can contribute noise due to the resistive components in the conductors and dielectric materials from which they are made. These effects are usually small compared to the noise from other resistors and semiconductors in the system.

The real effect on circuit noise due to the presence of capacitors and inductors is from their effect on the bandwidth. If a wideband random noise source is connected to a band-limiting filter and displayed on a spectrum analyzer, the noise spectrum is observed to have the same shape as the filter frequency response. This is illustrated in Figure 4-6 for a simple RC low-pass filter.

FIGURE 4-6 The output noise spectrum is the same as the filter frequency response.

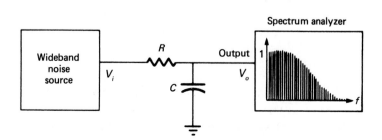

You will notice that, although the output noise spectrum follows the shape of the filter response, the noise never does go to zero as you might expect from the filter transfer function,

$$\frac{V_o}{V_i} = \frac{1}{1 + jf/f_c} \qquad (4\text{--}17)$$

which goes to zero as $f \to \infty$. The reason is that, even if an actual capacitor could become a perfect short at very high frequencies (which it cannot) or if an actual resistor had no "leak-by" internal capacitance, the flat noise observed on the spectrum analyzer is wideband noise generated in the analyzer itself. Thus the noise observed will never drop below the system "noise floor." In particular, the noise will not be less than $N_{th} = kTB$.

We now have to consider the value of B that should be used in calculating the total noise power at the output of the filter (disregarding the spectrum analyzer noise). It is clear from Figure 4–6 that there is more noise beyond f_c, the cutoff frequency of the filter. It turns out that there is a very simple solution for the case of a first-order RC (or RL) network such as that in Figure 4–6. The noise can be considered to have a rectangular shape of width equal to the *equivalent noise bandwidth* given by

$$B_{eq} = (\pi/2)f_c \qquad (4\text{--}18)$$

where f_c is the 3-dB cutoff frequency of the RC filter. This equation says that for this particular filter shape, the total noise power adds up to 57% more than that of a flat spectrum out to f_c. This is derived in the next section.

In general, the equivalent noise bandwidth for systems with more complex spectrum shapes must be determined mathematically if great accuracy is desired. The procedure is to integrate, mathematically or by graphical methods, the noise spectral density over the entire system frequency response shape. If the noise spectrum itself is not flat, such as for $1/f$ flicker noise, then its shape must be included, and the equivalent noise power bandwidth is determined from

$$B_{eq} = \int_0^\infty \left(\frac{V_o(f)}{V_i(f)}\right)^2 N(f)\, df \qquad (4\text{--}19)$$

where $N(f)$ is the noise power spectrum and the squared term is the system power spectrum transfer function.

Other aspects of noise in communication system design are considered in the chapters to follow. For instance, *noise figure, noise temperature,* and *signal-to-noise* ratio and their effects in communication receivers are introduced in Chapter 5.

4-7 ■ CORRELATION AND NOISE

In Chapter 13, the statistical nature of instantaneous noise voltage peaks is considered for its effects on errors in digital data communications. The peak voltage distribution of white noise is typically Gaussian and remains so when statistically independent random noise sources are combined. Gaussian distributions also remain Gaussian throughout linear systems. This fact tends to hold up reasonably well even in real systems that are not linear in a Gaussian sense, because statistically there must be occasional infinite peaks.

For random processes such as thermal and other noise phenomena, autocorrelation functions and power spectral density functions, respectively, are used for describing the correlation properties and the frequency content of the random process. For a linear system in which the noise process involved is wide-sense stationary, by definition the mean value of the noise function $x(t)$ is constant; that is, the process statistics do not change (drift) with time. Consequently, for wide-sense stationary noise, the random process ensemble averages are the same as the time averages, and the statistics (mathematics) are quite tractable, as will be demonstrated.

The *autocorrelation function* for a wide-sense stationary random process, $x(t)$, may be expressed as

$$R_x(\tau) = \overline{x(t_1)x(t_1 + \tau)} = \frac{1}{T}\int_{-T/2}^{T/2} x(t_1)x(t_1 + \tau)\, dt_1 \qquad (4\text{--}20)$$

where the overbar denotes the averaging operation, and $x(t_1)$ is a particular sample function (snapshot) of $x(t)$, as illustrated in Figure 4–7. In particular, the autocorrelation function for real processes is a maximum at, and symmetrical about, the origin, and the mean-square (rms squared) value, $\overline{x^2(t)}$, of the noise that gives the total noise power N is related to $R_x(\tau)$ by

$$\overline{x^2(t)} = R_x(0) = N \qquad (4\text{--}21)$$

Consider, now, the very practical case of thermal noise at the input to a filter. For all practical purposes, thermal noise can be considered as being a white noise process with a uniform power spectral density (PSD), constant over all frequencies; that is, as illustrated in Figure 4–8,

$$\mathscr{N}_x(f) = \frac{N_0}{2} \qquad (4\text{--}22)$$

FIGURE 4-7 White noise random process over a short time interval.

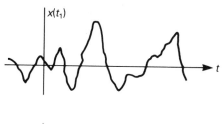

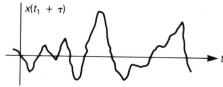

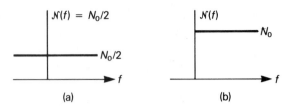

FIGURE 4-8 Power spectral density of noise process, $x(t)$, considered for (a) two-sided and (b) one-sided spectrum representations.

or, if only the positive half of the spectrum is considered, $\mathcal{N}_x(f) = N_0$. Furthermore, thermal noise has a Gaussian probability distribution, and the value of N_0 is $N_0 = kT$, from Equation 4–1.

Since the power spectral density $\mathcal{P}$ is the Fourier transform of the autocorrelation function

$$\mathcal{P}_x(f) = \mathcal{F}\{R_x(\tau)\} \tag{4–23}$$

it is particularly easy to find the autocorrelation function of the uniform PSD for white thermal noise. Hence, for a white, Gaussian thermal noise process, $x(t) = n(t)$,

$$R_x(\tau) = \mathcal{F}^{-1}\{\mathcal{P}(f)\} \tag{4–24a}$$

$$= \int_{-\infty}^{\infty} \left(\frac{N_0}{2}\right) e^{j2\pi f\tau}\, df \tag{4–24b}$$

$$= \frac{N_0}{2}\,\delta(\tau) \tag{4–24c}$$

where the Dirac delta function $\delta(\tau)$—illustrated in Figure 4–9 and defined in Section 1–10—is interpreted as meaning that white noise is completely uncorrelated (zero correlation) for any comparison of the waveform $x(t_1)$ with that same waveform delayed for any amount of time τ, where $\tau \neq 0$; review Figure 4–7, for example.

The underlying assumption in the preceding analysis is that the bandwidth is infinite. To return to reality, let us determine the consequences of passing the white noise through an RC low-pass filter (LPF) for which the transfer function, $H(f) = \mathcal{F}\{y(t)/x(t)\}$, is determined from Figure 4–10 as

$$y(t) = x(t)\left[\frac{1/j2\pi fC}{R + (1/j2\pi fC)}\right] = x(t)\left[\frac{1}{j2\pi fRC + 1}\right] \tag{4–25}$$

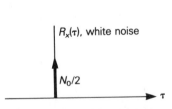

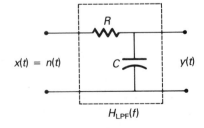

FIGURE 4-9 Autocorrelation function for white noise.

FIGURE 4-10 Low-pass filter (integrator).

Hence,

$$H_{\text{LPF}}(f) = \frac{1}{1 + j(f/f_c)} \qquad \textbf{(4-26)}$$

where $f_c = 1/(2\pi RC)$ is referred to as the cutoff frequency or half-power point.

Because of the stationarity of the white noise, the output PSD, $\mathcal{P}_y(f)$, is just the product of the input PSD with what is the equivalent to the *power* spectrum (for a voltage or current, for instance) transfer function, namely, $|H(f)|^2$; that is,

$$\mathcal{P}_y(f) = |H(f)|^2\,\mathcal{P}_x(f) \qquad \textbf{(4-27)}$$

In particular, substituting Equations 4–24a and 4–26 in Equation 4–27, we find that the power spectrum of the noise at the filter output is given by

$$\mathcal{P}_y(f) = \frac{N_0/2}{1 + (f/f_c)^2} \qquad \textbf{(4-28)}$$

which is illustrated in Figure 4–11.

Then, the correlation of noise at the output is determined by analyzing the autocorrelation function found from Equation 4–24a as the inverse Fourier transform of Equation 4–28 to be

$$R_o(\tau) = R_y(\tau) = \frac{N_0}{4RC}\,e^{-|\tau|/(RC)} \qquad \textbf{(4-29)}$$

which is illustrated in Figure 4–12.

In analyzing Equation 4–29 and Figure 4–12, we note first that the normalized low-pass filter output power is finite, as we would expect in practice, with value

$$P_o = R_y(0) = \frac{N_0}{4RC} \qquad \textbf{(4-30a)}$$

$$= \frac{N_0}{(2/\pi)2\pi RC} = \frac{\pi f_c}{2}\,N_0 \qquad \textbf{(4-30b)}$$

where the equivalent idealized (brick wall) bandwidth would be 1.57 times the cutoff frequency of the RC low-pass filter; note that $\pi f_c/2$ is Equation 4–18. But most important, we observe that the noise is *no longer uncorrelated* as it was at the filter input. This conclusion is based on a comparison of $R_x(\tau)$ for white noise (Figure 4–9) with $R_y(\tau)$ at the filter output. The correlation is strong for small lags, τ, but decreases exponentially as the $x(t_1 + \tau)$ of Figure 4–7 is shifted to the right.

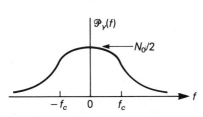

FIGURE 4-11 Output PSD of noise narrowing filter.

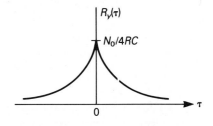

FIGURE 4-12 Autocorrelation of low-pass filtered noise.

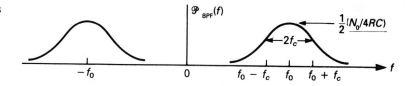

FIGURE 4–13 Narrow bandpass noise power spectral density showing energy in the frequency domain, near f_0.

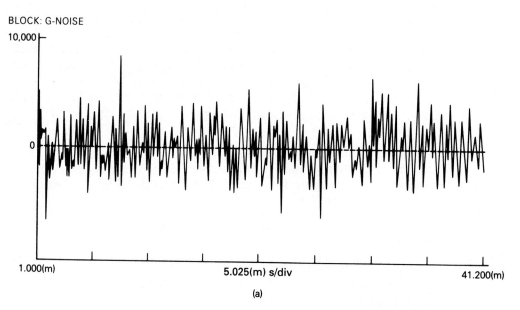

BLOCK: G-NOISE

10,000

0

1.000(m)   5.025(m) s/div   41.200(m)

(a)

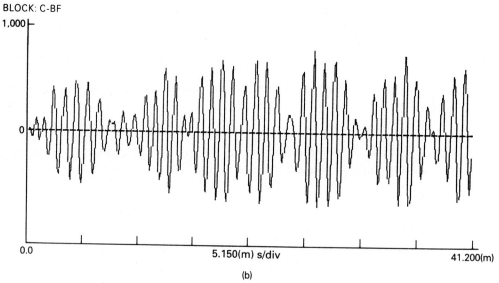

BLOCK: C-BF

1,000

0

0.0   5.150(m) s/div   41.200(m)

(b)

FIGURE 4–14 Computer simulation of (a) white Gaussian input noise and (b) narrowband filter output noise clearly showing strong correlation near frequency $f_0 = 1000$ Hz. Figure 4–14b is the time-domain equivalent of Figure 4–13 and gives visual evidence of noise correlation.

130

To bring this point home in a very practical, experiential way, consider the following: An equivalent bandpass filter centered at f_0 would have a PSD as illustrated in Figure 4–13, expressed simply as

$$\mathcal{P}_{BPF}(f) = \tfrac{1}{2}\mathcal{P}_y(f \pm f_0) \tag{4–31}$$

Observe in Figure 4–13 that frequency components near f_0 have much more power than those outside of the 3-dB bandwidth, $2f_c$. This is the reason that noise (only) at the input to a bandpass process (filter) gives rise to the observed *shaped noise* in the time domain at the output, as illustrated in the computer simulations of Figure 4–14. Keep in mind that there is *no* signal from any sinusoidal generator at the input to the bandpass filter, only the white Gaussian noise shown in Figure 4–14a. Therefore, correlated noise is present in a circuit without an input signal.

■ PROBLEMS

1. Calculate the thermal noise delivered to a system with a bandwidth of 1 Hz operating at 17°C. Distinguish between the type of noise produced by a 1-kΩ resistor and a semiconductor junction.

2. Determine the open-circuit noise voltage in 1 Hz of bandwidth at 290 K for the following:
   a. 1-kΩ resistor
   b. 26-Ω resistor
   c. Diode biased at 1 mA
   d. Diode biased at 16 mA
   e. Distinguish between the types of noise for a–d.

3. Determine the total noise voltage for the circuits of Figure 4–15. The equivalent noise bandwidth is 1 MHz and $T = 300$ K.

4. Determine the equivalent noise bandwidth for a single RC low-pass filter if $R = 20$ kΩ and $C = 0.1$ μF.

5. A Tektronix 475, 200-MHz oscilloscope has $Z_i = 1$ MΩ shunted by 20 pF.
   a. Ignoring the 20 pF, calculate the theoretical noise voltage at the input.
   b. Calculate the noise voltage if the bandwidth is determined by the 20 pF. (Assume a rectangular bandwidth.)

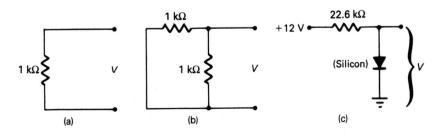

(a)   (b)   (c)

FIGURE 4–15

FIGURE 4–16

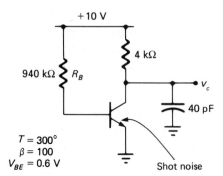

c. If the scope internal circuitry adds 26 dB of noise, determine the effective noise voltage at the input.

d. If the scope input sensitivity is 2 mV/division, how many divisions (or fractions of a division) should be observed on the scope screen? (Use the rms noise value.)

6. For the circuit of Figure 4–16:
   a. Calculate the thermal noise spectral density of R_B.
   b. Calculate the shot noise spectral density generated in the emitter-base junction of Q_1.
   c. Ignoring the thermal noise of the 4-kΩ resistor (insignificant compared to the rest), determine the total rms noise voltage at v_c. For convenience, assume that the frequency response due to the 40-pF capacitor is flat up to the cutoff frequency f_c and then breaks down sharply.

7. A television receiver has circuits with the following specifications: Two RF amplifiers with gain of 10 dB each and noise figures of 4 dB each, followed by a mixer circuit with a gain of –6 dB and noise figure of 8 dB. The mixer is followed by two amplifiers with 20 dB of gain each and noise figures of 15 dB. Determine
   a. The system noise figure.
   b. The noise temperature.

8. Compute the noise figure of a receiving system with a noise temperature of 200°C.

9. A receiving system is receiving a signal of 100 μV pk from a 50-Ω antenna that is matched to the receiver impedance. The receiver has a noise figure of 7 dB, gain of 20 dB, and an equivalent bandwidth of 200 kHz. Compute
   a. The receiver output noise power.
   b. The output signal-to-noise ratio.

4–8 ■ COMPUTER ANALYSIS

Using the computer for communications system and circuit analysis has become a virtual necessity in education and industry. What follows are examples of programs and/or readily available simulation software for solving problems of the type covered in this chapter.

All the programs are interactive in the sense that an introductory sentence or short paragraph describes the purpose of the program and the user is then prompted for input data.

The programs for Chapter 4 are as follows:

- **Communications systems noise figure:** The program of Figure 4–17 computes the noise figure and gain of a multistage heterodyne receiver system. The program even analyzes the effect on noise figure due to signal loss ahead of the first amplifier; an example would be a receiver preselect filter or long cables between the antenna and first amplifier. Any circuit function block can effectively be eliminated by responding to prompts for gain and noise figure by entering 0 dB. Also, gains and NFs may be entered as power ratios or decibels.

- **Discrete autocorrelation function calculation:** This program, listed in Figure 4–18, calculates the autocorrelation function of a discrete function such as random noise values or pseudorandom data code generators. Sequences up to and including 31 discrete values will be accepted.

```
10 REM *       COMMUNICATION SYSTEMS         *
20 REM *          NOISE FIGURE               *
30 REM *        BY P. YOUNG,  ASU            *
40 CLS: PRINT
50 PRINT "    This program computes the noise figure and gain of"
60 PRINT "      a multistage, heterodyne receiver system."
70 PRINT:PRINT:PRINT"The receiver block diagram is as follows:":PRINT
80 PRINT "   _____      _____      _____      _____   "
90 PRINT"Input: RF-BPF  :  : RF-AMP1 :  :  LOSS   :  : RF-AMP2 : "
100 PRINT"--->:   (L1)   :-->: (NR1,G1):-->:   (L2)   :-->: (NR2,G2):*"
110 PRINT"    :_____:  :_____:  :_____:  :_____:  "
120 PRINT"                                                        "
130 PRINT"   _____      _____      _____      _____   "
130 PRINT"   : MIXER   :  : IF-BPF  :  : IF-AMP1 :  : IF-AMP2 :  "
140 PRINT"*->:   (L3)   :-->:   (L4)   :-->: (NR4,G4):-->: (NR5,G5):->"
150 PRINT"   :_____:  :_____:  :_____:  :_____OUT"
160 PRINT
170 PRINT"G=Power gain ratio, G(dB)=10 log G, L=Loss (power ratio),"
180 PRINT"L(dB)=10logL, NR=Noise power ratio, NF=10log NR, noise fig"
190 PRINT :PRINT
200 PRINT "The loss for the preselect bandpass filter, RF-BPF, is"
210 INPUT "(either -dB, or power ratio)";L1: PRINT
220 INPUT "This loss is in  1) dB , or  2) power ratio. (WHICH?)",X
230 IF X=1 THEN 240 ELSE GOTO 250
240 L1= 10^(L1/10) :PRINT :REM  change dB to noise power ratio < 1
250   PRINT "RF-AMP1, or LNA at the antenna feed point, noise figure"
260   PRINT "and power gain:"  :PRINT
270    INPUT "The noise figure is (either dB, or power ratio)? ",NR1
280    INPUT "This value is in  1) dB , or  2) power ratio" ;NR1X
290    IF NR1X=1 THEN 300 ELSE GOTO 310
300    NR1= 10^(NR1/10)   :REM  change dB to noise power ratio
310    INPUT "The RF gain is (either dB, or power ratio)" ;G1
320    INPUT "This value is in  1) dB , or  2) power ratio" ;G1X
330    IF G1X=1 THEN 340 ELSE GOTO 350
340    G1= 10^(G1/10)  :PRINT : REM   change dB to power gain ratio
350 PRINT:PRINT"Any loss between amplifiers effectively reduces the"
360 PRINT "gain and increases the noise figure of the second amp."
```

FIGURE 4–17 (page 1 of 2)

```
370 PRINT "Enter the loss between RF-AMP1 and RF-AMP2. If there is
 no loss, enter 0 dB."
380 INPUT "The loss is (either -dB, or power ratio)" ;L2
390 INPUT "This loss is in  1) dB , or  2) power ratio" ;L2X
400 IF L2X=1 THEN 410 ELSE GOTO 420
410 L2= 10^(L2/10)  :REM    change dB to power ratio < 1
420    PRINT :PRINT "      RF-AMP2 noise figure and gain:" :PRINT
430    INPUT "The noise figure is (either dB, or power ratio) " ;NR2
440    INPUT "This value is in  1) dB , or  2) power ratio" ;NR2X
450    IF NR2X=1 THEN 460 ELSE GOTO 470
460    NR2= 10^(NR2/10)   :PRINT
470    INPUT "The gain is (either dB, or power ratio)";G2
480    INPUT "This value is in  1) dB , or  2) power ratio";G2X
490    IF G2X=1 THEN 500 ELSE GOTO 510
500    G2= 10^(G2/10) :PRINT  :PRINT
510 PRINT"A MIXER is usually modeled as a pure loss (-6dB gain,typ.)"
520 PRINT"This accounts for both decreased gain and increased noise."
530 PRINT"If more specific knowledge of gain and noise figure is"
540 PRINT"known, then enter 0 (dB) here and use the next amplifier
 block as a mixer."
550 PRINT  :INPUT "The mixer gain, in -dB, is ";L3
560 L3 = 10^(L3/10)   :PRINT  :PRINT
570    INPUT "Enter the IF filter loss, in -dB: ";L4  :PRINT
580    L4 = 10^(L4/10)   :PRINT
590 PRINT "The first stage IF amplifier noise figure usually affects"
600 PRINT "the overall system noise figure because of mixer and
 IF-BPF loss."
610 PRINT :PRINT "PLease enter the noise figure and gain." :PRINT
620    INPUT "The noise figure is (either dB, or power ratio)" ;NR4
630    INPUT "This value is in  1) dB , or  2) power ratio";G4X
640    IF NR4X=1 THEN 650 ELSE GOTO 660
650    NR4= 10^(NR4/10)   :REM  change dB to noise power ratio
660    INPUT "The IF gain is (either dB, or power ratio)";G4
670    INPUT "This value is in  1) dB , or  2) power ratio";G4X
680    IF G4X=1 THEN 690 ELSE GOTO 700
690    G4= 10^(G4/10)  :PRINT : REM   change dB to power gain ratio
700 PRINT "It is very rare that more than two IF stages affect"
710 PRINT "the system noise figure. For the second IF stage,"
720    INPUT "the noise figure is (either dB, or power ratio)" ;NR5
730    INPUT "This value is in  1) dB , or  2) power ratio" ;NR5X
740    IF NR5X=1 THEN 750 ELSE GOTO 760
750    NR5= 10^(NR5/10)    :PRINT
760    INPUT "The IF gain is (either dB, or power ratio)" ;G5
770    INPUT "This value is in  1) dB , or  2) power ratio" ;G5X
780    IF G5X=1 THEN 790 ELSE GOTO 800
790    G5= 10^(G5/10)  :PRINT : REM   change dB to power gain ratio
800 PRINT  : REM Compute system NF and gain
810 SG1 = L1*G1*L2*G2*L3*L4*G4
820 NR=NR1/L1+(NR2/L2-1)/L1/G1+(NR4/L3/L4+1)/L1/G1/L2/G2+(NR5-1)/SG1
830 PRINT "The system noise figure is, NF =";10*LOG(NR)/LOG(10)
840 GB = SG1*G5 :PRINT "The predetection gain assuming only two IF"
850 PRINT "stages is " ; 10*LOG(GB)/LOG(10) ;"dB."  :PRINT
860 PRINT "Enter the gain of all other amplifiers ahead of the"
870 INPUT "demodulator, in dB" ;G6
880 PRINT "The total predetection system gain is "
 ;G6+10*LOG(GB)/LOG(10); " dB."
890 END
```

FIGURE 4-17 (page 2 of 2)

```
10 REM *  DISCRETE AUTOCORRELATION FUNCTION CALCULATION  *
20 REM *      EXAMPLE OF STUDENT COMPUTER ASSIGNMENT       *
30 REM *           FOR PROFESSOR P. YOUNG, AT ASU          *
40 CLS:KEY OFF
50 LOCATE 10,20
60 DIM RES1(32),RES2(32),X(64)
70 GOSUB 610
80 CLS:PRINT"This program calculates the autocorrelation function"
83 PRINT"of a discrete function. The user must input the sequence"
86 PRINT "whose autocorrelation function is to be calculated. The"
88 PRINT "result is computed for 3, 7, 15 or 31 bit sequences."
90 LOCATE 20,15: INPUT "       PRESS RETURN TO CONTINUE",A$
100 CLS:LOCATE 10,15: PRINT "LENGTH OF THE SEQUENCE";:INPUT LENGTH
110 CLS: LOCATE 8,15
120 IF LENGTH=3 THEN GOTO 500
130 IF LENGTH=7 THEN GOTO 520
140 IF LENGTH=15 THEN GOTO 540
150 IF LENGTH=31 THEN GOTO 560
160 CLS: LOCATE 10,10
165 PRINT"SEQUENCE LENGTH MISMATCH. PRESS 3, 7, 15, OR 31."
168 GOSUB 610:GOTO 100
170 FOR I=1 TO LENGTH
180 IF X(I)=0 THEN X(I)=-1
190 NEXT I
200 '********************************************************
210 'TO CALCULATE THE AUTOCORRELATION FUNCTIONS ************
220 '********************************************************
230 FOR I=0 TO 31: RES1(I)=0:RES2(I)=0:NEXT I
235 J=0
240 FOR K=0 TO (LENGTH-1)
250 FOR I=1 TO LENGTH: J=I+K: RES1(K)=X(I)*X(J)+RES1(K)
260 IF J=LENGTH THEN GOTO 280
270 NEXT I
280 NEXT K
290 RES2(1)=0: J=0
300 FOR K=1 TO (LENGTH-1)
310 FOR I=1 TO LENGTH : J=I+K: RES2(K)=X(J)*X(I)+RES2(K)
320 IF J=LENGTH THEN 340
330 NEXT I
340 NEXT K
350 '********************************************************
360 ' TO PRINT THE RESULTS                     ************
370 '********************************************************
380 FOR I=-(LENGTH-1) TO 0
390 IF I=0 THEN GOTO 410
400 PRINT "R(";I;")=",RES1(-I): NEXT I
410 PRINT "R( 0 )=",RES1(0)
420 FOR I=1 TO LENGTH-1: PRINT "R(";I;")=",RES2(I): NEXT I
430 INPUT "DO TOU WANT TO DRAW A GRAPH(Y/N)",A$
440 IF A$="Y" THEN GOTO 670
450 IF A$="N" THEN 460 ELSE 430
460 CLS: LOCATE 10,10
461 INPUT "DO YOU WANT TRY ANOTHER SEQUENCE(Y/N): ",A$
462 IF A$="Y" THEN GOTO 100
465 IF A$="N" THEN END ELSE 460
470 '
480 ' ROUTINES FOR DIFFERENT SEQUENCE LENGTHS *****************
490 '
500 PRINT "ENTER THE SEQUENCE STARTING WITH"
```

FIGURE 4–18   (page 1 of 2)

```
505 INPUT "THE BIT AT t=0:",X(3),X(2),X(1)
510 GOTO 170
520 PRINT "ENTER THE SEQUENCE STARTING WITH"
525 INPUT "THE BIT AT t=0:",X(7),X(6),X(5),X(4),X(3),X(2),X(1)
530 GOTO 170
540 PRINT "ENTER THE SEQUENCE STARTING WITH"
545 INPUT "THE BIT AT t=0:",X(15),X(14),X(13),X(12),X(11),X(10),
X(9),X(8),X(7),X(6),X(5),X(4),X(3),X(2),X(1)
550 GOTO 170
560 PRINT "ENTER THE SEQUENCE STARTING WITH"
565 INPUT "THE BIT AT t=0:",X(31),X(30),X(29),X(28),X(27),X(26),
X(25),X(24),X(23),X(22),X(21),X(20),X(19),X(18),X(17),X(16),
X(15),X(14),X(13),X(12),X(11),X(10),X(9),X(8),X(7),X(6),X(5),
X(4),X(3),X(2),X(1)
570 '*********************************************
580 '   DELAY ROUTINE
590 '*********************************************
600 GOTO 170
610 R=0
620 FOR I=1 TO 75: R=R+1: NEXT I
630 RETURN
640 '*********************************************
650 '               GRAPHICS
660 '*********************************************
670 CLS:SCREEN 2:Q=LENGTH+1:WINDOW(-Q,-Q)-(Q,Q)
680 LINE (-Q,-Q)-(Q,Q),,B:LINE (-Q,0)-(Q,0)
690 FOR I=-(LENGTH-1) TO 0
700 IF I=0 THEN 720
710 LINE (I,0)-(I,RES1(-I)):NEXT I
720 LINE (0,0)-(0,RES1(0))
730 FOR I=1 TO (LENGTH-1):LINE (I,0)-(I,RES2(I)):NEXT I
740 SCREEN 2: LOCATE 2,3:PRINT "X AXIS SCALE IS: ";-Q;"TO";Q
750 PRINT "  Y AXIS SCALE IS: ";-Q;"TO";Q
755 PRINT :INPUT " PRESS RETURN TO CONTINUE",A$
760 GOTO 460
```

FIGURE 4-18   (page 2 of 2)

5

Modulation and Amplitude-Modulated Systems

■ INTRODUCTION

In previous chapters the groundwork was laid to begin the study of information transfer systems. Basic amplifier and oscillator circuits were reviewed, followed by signals, filters, and noise concepts. We now turn to some of the techniques for transferring electrical signals containing useful information.

Information in the form of sound or visual images is processed into electrical signals by transducers. A microphone is the transducer for converting energy in the form of sound pressure waves into electrical energy; a television camera is the transducer for conversion of visual images into electrical video signals. Further electronic processing can convert these signals into a digital format.

Once the information is processed into suitable electrical signals, we must determine how these signals can be transmitted over long distances and then made available for one or millions of users. There are numerous approaches and technologies for accomplishing this feat, but the most general breakdown of the basic elements involved in a single channel of information transfer is shown in Figure 5-1.

This figure illustrates, in a block-diagram form, information gathered, transformed into electrical signals, and processed for transmission. Typical transmission media are transmission lines or optical (light) cable, and space—from antenna to antenna. The distances and broadcast coverage involved usually result in vast energy loss, thus requiring very sensitive receiving and processing systems before the signals can be reproduced into the desired form for the user. Our objective is to break down each block of Figure 5-1 and study the requirements and techniques used to realize typical communications systems.   ■

5-1 ■ COMMUNICATIONS SYSTEMS

Communications system design requires knowledge of the physical form of the information and the distance over which it is to be transmitted. With this and the desired quantity and quality of the information transfer, an appropriate signal form and transmission scheme can be devised. The quantity aspect involves determining the amount, speed, and efficiency of information transfer through a finite bandwidth channel. The quality aspect involves the determination of how precisely the received information represents what was transmitted.

The information must be converted to electrical energy in an analog (continuous) or pulse (discrete) form for processing in electronic circuits. Analog information is usually characterized by the power/frequency spectrum content of the electrical signals, whereas information in a discrete or pulsed form is character-

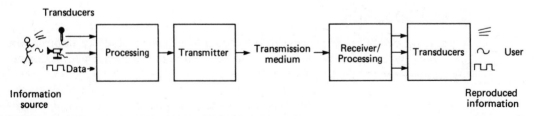

FIGURE 5-1 Information transfer.

ized with respect to the basic unit of information, the bit. Digital transmission quantity is given in bits of information per second, and the number of incorrectly received bits per unit time, the bit error rate, is a measure of the transmission system quality. Much more on this and other aspects of information theory is given in Chapter 12, which introduces digital and data transmission systems.

As a practical matter the transmission of information over long distances involves the use of analog (that is, continuous-wave) facilities such as telephone and cable TV transmission wires, coaxial cables, fiber-optic (glass) cables, and other dielectric-material media, including the earth's atmosphere and space.

Most of the information that people have traditionally communicated falls into the sound (audio) band of frequencies. The audio band also includes low-speed digital data pulses for printed text and limited computer terminal signals. Transmission of high-speed data from a computer requires in excess of 100 kHz of bandwidth, whereas fast-changing video signals such as those from a television broadcast require about 4 MHz of bandwidth.

The next few chapters cover basic analog communication techniques and, for simplicity, emphasize audio information signals. Later chapters are devoted to the study of video and high-speed data transmissions. You will notice as you study various transmission techniques that a few concepts keep repeating; these are the common threads of all communication systems. Each is defined in the text, but here is a list of the terms to look for:

- **Bandwidth** (usable frequency range)—Information signal bandwidth versus system/circuit bandwidth.
- **Information rate** (versus bandwidth)—Measures of the maximum signal rate of change that affects audio fidelity and brillance, video sharpness and resolution, and digital data bit rate.
- **Noise interference** (versus bandwidth)—*Signal-to-noise* ratio for analog and digital signals; timing jitter and bit error rate for digital systems.
- **Distortion**—Frequency and phase distortion (versus bandwidth) and amplitude distortions in analog systems. Digital pulse spreading and intersymbol interference (versus bandwidth). Dispersion, multipath, voice echoes, and video ghosts.
- **Multiplexing**—Simultaneous signaling separated by time, frequency, or phase.

5-2 ■ VOICE TRANSMISSION AND MULTIPLEXING

The audio (sound) frequency spectrum is limited to less than 20 kHz because the human ear cannot detect sound above this frequency. The important part of the frequency spectrum for speech decreases above 1 kHz, so voice transmission over typical analog telephone circuits requires less than 3.5 kHz of total bandwidth for reasonable intelligibility. However, high-fidelity music extends into the region of 15 kHz, so uncompensated telephone circuits are not appropriate for transmitting music.

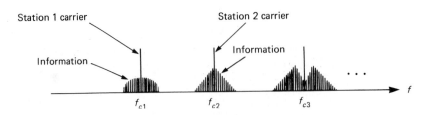

FIGURE 5–2   Frequency-division multiplexing (FDM) of three (or more) information sources.

Suppose you and 99 others want to transmit 100 different music programs to thousands of homes in your area. One solution might be to set up 100 large speakers and as many large audio power amplifiers, resulting in flooding the community with sound pressure waves.

Even if the sound pressure waves could be converted to electromagnetic waves, to which ears don't respond, you would still have at least two serious technological challenges in transmitting audio frequencies through space. First, antennas would be HUGE—measured in miles (see Chapter 15). Second, all signals would be received on the same antenna, and one station could not be separated from the next unless everyone agreed to transmit at separate times of the day. Such a transmission arrangement is a rough example of what is called *time-division multiplexing* (TDM).

The solution to both of these challenges is to send the audio (or video) signal on a high-frequency "carrier" and have each station use a different carrier frequency—this is called *frequency-division multiplexing* (FDM) and is illustrated in Figure 5–2.

5-3 ■ MODULATION

The process of transferring information signals to a high-frequency carrier is called *modulation*.

A high-frequency carrier signal has three different parameters that could be modulated (varied) in order to make it carry the information that we want to transmit: these are amplitude, frequency, and phase. When the amplitude of the carrier is varied in accordance with the information signal, *amplitude modulation* (AM) is produced. Likewise, *frequency modulation* (FM) or *phase modulation* (PM) results when the information signal varies the carrier frequency or phase.

Amplitude Modulation

An unmodulated sinusoidal carrier signal can be described mathematically as

$$e_c(t) = E_c \cos 2\pi f_c t \qquad (5\text{–}1)$$

where E_c is the peak continuous-wave (CW) amplitude and f_c is the carrier frequency in hertz. Figure 5–3 illustrates the result of amplitude modulation of the

FIGURE 5-3 Amplitude modulation.

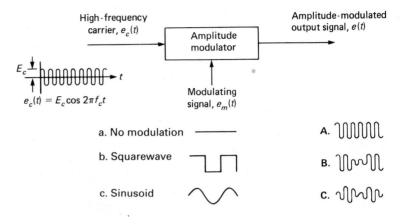

a. No modulation —————   **A.**

b. Squarewave   **B.**

c. Sinusoid   **C.**

carrier by a squarewave and a sinusoid. The sinusoidal modulating signal of Figure 5-3c can be described by $e_m(t) = E_m\cos 2\pi f_m t$, where E_m is the peak voltage of the modulation signal of frequency f_m.

Notice that the carrier amplitude in Figure 5-3C is varying at the same rate f_m as the modulating signal. Also, the amount by which the amplitude varies is proportional to the peak modulating voltage E_m. Hence, the modulated output signal can be described mathematically as

amplitude of carrier

$$e(t) = (E_c + E_m\cos 2\pi f_m t)\cos 2\pi f_c t \tag{5-2}$$

peak value of carrier   sinusoidal variation of
sinusoid (average value   carrier amplitude
of modulated signal peaks)   due to modulating signal

The sinusoidally modulated AM signal is shown in Figure 5-4. For an arbitrary information signal $m(t)$, the AM signal is described mathematically as

$$e(t) = [E_c + m(t)]\cos 2\pi f_c t \tag{5-3}$$

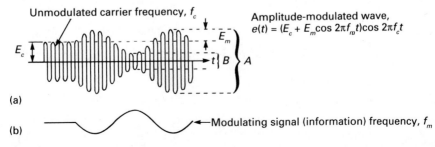

(a)

(b)

FIGURE 5-4 (a) Amplitude-modulated signal. (b) Information signal to be transmitted by AM.

AT&T microwave relay station near Boulder Junction, Colorado.

Modulation Index

By increasing the amplitude E_m of the modulating signal $E_m \sin 2\pi f_m t$, the amount of power in the information part of the transmitted signal is increased relative to the power used for the carrier. Since we would like the total power transmitted to contain as much information power as possible, a figure of merit called the *modulation index, m,* is used. The AM modulation index is defined by $m_a = E_m/E_c$. Hence, the AM signal can be written for sinusoidal modulation as $e(t) = E_c(1 + m_a \cos 2\pi f_m t) \cos 2\pi f_c t$. The most convenient way to measure the AM index is to use an oscilloscope: simply display the AM waveform as in Figure 5–4, and measure the maximum excursion A and the minimum excursion B of the amplitude "envelope" (the information is in the envelope). The AM index is computed from Figure 5–4 as

$$A = 2(E_c + E_m) \qquad\qquad \textbf{(5–4a)}$$

$$B = 2(E_c - E_m) \qquad\qquad \textbf{(5–4b)}$$

and solving for E_c and E_m in terms of A and B will then yield

$$m_a = \frac{E_m}{E_c} = \frac{A - B}{A + B} \qquad\qquad \textbf{(5–4c)}$$

It should be clear that peak measurements $A/2$ and $B/2$ will yield m_a also. The numerical value of m_a is always in the range of 0 (no modulation) to 1.0 (full modulation) and is usually expressed as a percentage of full modulation.

■ ■ ■ ■

**EXAMPLE
5-1**

If $A = 100$ V and $B = 20$ V, determine the percent of modulation, the peak carrier voltage, and the peak value of the information voltage.

Solution:

1. $m_a = \dfrac{100 \text{ V} - 20 \text{ V}}{100 \text{ V} + 20 \text{ V}} = \textbf{0.667,}$ which is expressed as **66.7% AM.**

2. The average of the two peak-to-peak measurements is the peak-to-peak amplitude of the unmodulated carrier $2E_c$. Hence, the peak carrier voltage can be computed for measurements of A and B as follows:

$$2E_c = \frac{100 \text{ V} + 20 \text{ V}}{2} = 60 \text{ V pk-pk,} \quad \text{and} \quad E_c = \textbf{30 V pk}$$

3. $E_m = m_a E_c = 0.667 \times 30$ V pk $= \textbf{20 V pk}$

If more than one sinusoid, such as a musical chord (that is, a triad, three tones), modulates the carrier, then we get the resultant AM index by RMS-averaging the indices that each sine wave would produce. Thus, in general,

$$m_a = \sqrt{m_1^2 + m_2^2 + m_3^2 + \cdots + m_n^2} \qquad \textbf{(5-5)}$$

5-4 ■ AM SPECTRUM AND BANDWIDTH

The amplitude-modulated signal of Figure 5–4 for sinusoidal modulation consists of three high-frequency components and no low-frequency components. Consequently, the AM signal can be transmitted from a high-frequency antenna of reasonable size. To show the three high-frequency components, called the *carrier, upper sideband,* and *lower sideband,* let us analyze the mathematical expression for the AM signal from Equation 5–2,

$$\begin{aligned} e(t) &= (E_c + E_m \cos 2\pi f_m t)\cos 2\pi f_c t \\ &= E_c \cos 2\pi f_c t + E_m \cos 2\pi f_m t \cos 2\pi f_c t \end{aligned}$$

The second term of this expression can be expanded by the trigonometric identity

$$\cos A \cos B = \tfrac{1}{2}[\cos(A - B) + \cos(A + B)]$$

so that

$$e(t) = E_c \cos 2\pi f_c t + \tfrac{1}{2} E_m \cos 2\pi (f_c - f_m)t + \tfrac{1}{2} E_m \cos 2\pi (f_c + f_m)t \qquad \textbf{(5-6)}$$

Since the carrier frequency is usually at least 100 times higher than the highest modulation frequency ($f_c > 2f_m$ is absolutely required), the AM is seen to be the sum of three high-frequency signals

$$e(t) = E_c \cos 2\pi f_c t \qquad\qquad \text{carrier}$$

$$+ \frac{E_m}{2} \cos 2\pi (f_c - f_m)t \qquad \text{lower sideband, LSB}$$

$$+ \frac{E_m}{2} \cos 2\pi (f_c + f_m)t \qquad \text{upper sideband, USB}$$

There is a lot of information to be gained from this expression, not the least of which are *frequency spectrum, bandwidth,* and *power relationships.* A plot of voltage versus frequency, the frequency spectrum, is shown in Figure 5–5. Figures 5–5 and 5–4 correspond to spectrum analyzer and oscilloscope displays of the AM signal in which the three high-frequency sinusoids beat together, sometimes adding and sometimes subtracting, to produce the sinusoidal amplitude variation of the carrier. In Figure 5–6, the AM signal is shown to be the instantaneous phasor sum of the carrier f_c, the lower-side frequency $f_c - f_m$, and the upper-side frequency $f_c + f_m$.

The phasor addition is shown for six different instants, illustrating how the instantaneous amplitude of the AM signal can be constructed by phasor addition. Notice how the USB ($f_c + f_m$), which is a higher frequency than f_c, is steadily gaining on the carrier, whereas the LSB ($f_c - f_m$), a lower frequency, is steadily falling behind.

The last phasor sketch in Figure 5–6 shows the phasor relationship of sidebands to carrier at the instant corresponding to the minimum amplitude of the AM signal. You can see in this sketch that if the amplitude of each sideband is equal to one-half of the carrier amplitude, then the AM envelope goes to zero. This corresponds to the maximum allowable value of E_m; that is, $E_m = E_c$ and $m = 1.0$, or 100% modulation.

As illustrated in Figure 5–7b and c, an excessive modulation voltage will result in peak clipping and harmonic distortion, which means that additional sidebands are generated. Not only does overmodulation distortion result in the reception of distorted information, but also the additional sidebands generated usually exceed the maximum bandwidth allowed. For commercial AM broadcast, the FCC (Federal Communications Commission) specifies 10-kHz bandwidth because carrier frequency assignments are allocated from 540 to 1600 kHz in 10-kHz increments. This means that the maximum modulating frequency must be an undistorted 5 kHz; otherwise the transmitted spectrum spills over into an adjacent channel (station), causing interference called *cross talk.*

FIGURE 5–5 Frequency spectrum of the AM signal shown in Figure 5–4.

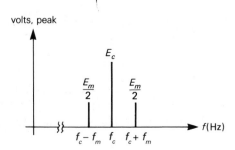

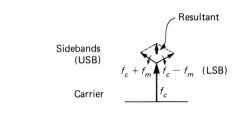

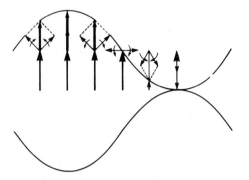

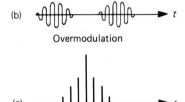

FIGURE 5–6   AM represented as the vector sum of sidebands and carrier.

FIGURE 5–7   (a) $m_a = 1.0$ (100% AM). (b) The result of overmodulation that corresponds to the spectrum in (c).

The bandwidth of the AM signal is very easy to determine from the frequency spectrum of Figure 5–5 because bandwidth is measured in terms of frequency. The entire AM signal is seen to exist in a bandwidth of $2f_m$ centered at the carrier frequency. This can also be calculated from $(f_c + f_m) - (f_c - f_m) = 2f_m = $ BW. For example, if a 1-MHz carrier is modulated by a 5-kHz tone, the bandwidth required to transmit and receive all the AM signal is 2×5 kHz = 10 kHz centered around 1 MHz. This *information bandwidth* is not necessarily the appropriate 3-dB bandwidth required of a single tuned circuit that might be expected to pass the AM signal without distortion. To pass this signal without distortion, the overall system bandwidth should be flat and the overall system phase response should be linear over the 10-kHz information bandwidth range.

5–5 ■ POWER IN AN AM SIGNAL

To determine the power in an AM signal, consider Equation 5–6. If this voltage signal is present on an antenna of effective real impedance R, then the power of each component will be determined from the peak voltages of each *sinusoid*.

$$P = \frac{(V_{\text{rms}})^2}{R} = \frac{(V_{\text{pk}}/\sqrt{2})^2}{R} = \frac{V_{\text{pk}}^2}{2R}$$

So for the carrier, $P_c = E_c^2/2R$, and for *each* of the two sideband components,

$$\frac{[(m_a/2)E_c/\sqrt{2}]^2}{R} = \frac{m_a^2}{4} \cdot \frac{E_c^2}{2R} = \frac{m_a^2}{4} P_c$$

Therefore, $$P_{1sb} = m_a^2 P_c/4 \qquad\qquad (5\text{--}7)$$

where P_{1sb} denotes the power in one sideband only. The total power in the AM signal will be the sum of these powers.

$$
\begin{aligned}
P_{total} &= P_c + P_{LSB} + P_{USB} \\
&= P_c + \frac{m^2}{4} P_c + \frac{m^2}{4} P_c \\
&= P_c\left(1 + \frac{m^2}{2}\right) = P_t \qquad\qquad (5\text{--}8)
\end{aligned}
$$

■ ■ ■ ■

EXAMPLE 5-2

Determine the power in each spectral component of the AM signal of Example 5–1 in which $E_c = 30$ V pk and m is 66.7%. Let the effective impedance be 50 Ω.

Solution:

$$P_c = \frac{E_c^2}{2R} = \frac{(30 \text{ V pk})^2}{2 \times 50\Omega} = \mathbf{9\ W}$$

$$P_{LSB} = \frac{m^2}{4} P_c = \frac{(0.667)^2}{4} 9 \text{ W} = \mathbf{1\ W}$$

$$P_{USB} = P_{LSB} = \mathbf{1\ W}$$

Also, the total power in the AM signal is

$$P_t = 9 \text{ W} + 1 \text{ W} + 1 \text{ W} = 9 \text{ W}\left[1 + \frac{(0.667)^2}{2}\right] = \mathbf{11\ W}$$

5-6 ■ NONSINUSOIDAL MODULATION SIGNALS

When complex nonsinusoidal signals amplitude-modulate a carrier, the frequency spectrum tells us more than the oscilloscope (time-domain) display. This is illustrated in Figure 5–8 for a generalized time-varying signal as might be produced in typical broadcast programming. The information signal is continuously changing but has a maximum range (band) of frequencies. Figure 5–8c shows that the information signal includes frequencies almost down to dc ($f = 0$) and as high as $f_m(\text{max})$, but most of the information is in the lower midband of frequencies. The information signal in modulated systems, as illustrated by Figure 5–8a, is often referred to as the *baseband signal*, and the spectrum of Figure 5–8c is the (one-sided) *baseband spectrum*, where only positive frequencies are shown.

FIGURE 5-8 (a) Information
signal (modulation). (b) AM
output (time domain). (c) One-
sided frequency spectrum of
$m(t)$. (d) AM output frequency
spectrum (one-sided).

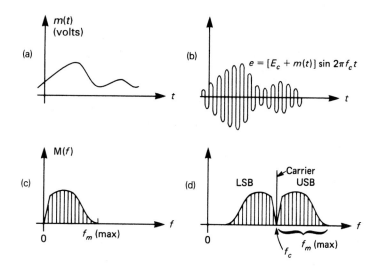

The modulated signal spectrum (one-sided) of Figure 5-8d consists of the upper
and lower sidebands on either side of the carrier. This figure clearly shows that the
information bandwidth of the AM signal is $2f_m(max)$.

The mathematically formal method of determining the frequency spectrum of a
time-varying signal is to employ the Fourier transform. From Equation 5-3,

$$v_{AM}(t) = E_c(t)\cos 2\pi f_c t + m(t)\cos 2\pi f_c t \qquad (5-9)$$

and from Table 3-2, Equation 7, and Table 3-3, Equation 6, with $\phi = 0$ and $\theta = 0$,
respectively, the Fourier transform yields

$$V_{AM}(f) = E_c\frac{1}{2}[\delta(f - f_c) + \delta(f + f_c)] + \frac{1}{2}[M(f - f_c) + M(f + f_c)] \qquad (5-10)$$

For the *two-sided* idealized audio frequency spectrum of Figure 5-9a, the plot of
the Fourier transform of $v_{AM}(t)$, $V_{AM}(f)$, is illustrated in Figure 5-9b.

If an AM transmitter is modulated by a digital-data signal, the carrier amplitude
shifts between the digital states, and the resultant transmitted signal is called
amplitude-shift key (ASK). In the special case wherein the transmitter is keyed
entirely *off* and *on* to transmit the information, the process is called *on-off keying*
(OOK). This is illustrated for a narrow-pulse train in Figure 5-10. If the carrier

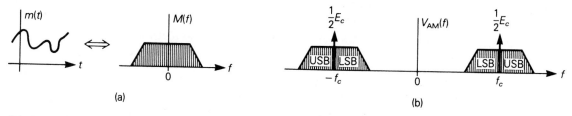

FIGURE 5-9 (a) Idealized audio time-domain signal and baseband Fourier transform
spectrum. (b) Fourier transform spectrum of $m(t)$ amplitude modulated on a (cosine)
carrier signal of frequency f_c.

FIGURE 5–10 25% duty cycle
on-off key (OOK) signal and
(one-sided) frequency spectrum.

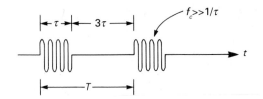

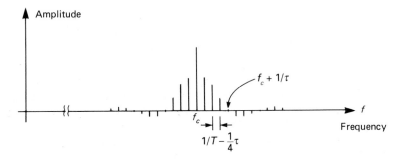

burst is defined as

$$v(t) = \begin{cases} A \cos 2\pi f_c t & \dfrac{-\tau}{2} < t < \dfrac{\tau}{2} \\[2mm] 0 & \text{elsewhere} \end{cases} \qquad (5\text{--}11)$$

then the Fourier transform may be expressed as

$$V_n(f) = \frac{Ad}{2}\left[\frac{\sin(nd - f_c\tau)\pi}{(nd - f_c\tau)\pi} + \frac{\sin(nd + f_c\tau)\pi}{(nd + f_c\tau)\pi}\right] \qquad (5\text{--}12)$$

where

$$d = \frac{\tau}{T} \qquad (5\text{--}13)$$

is the duty cycle of the bursts, and T is the pulse repetition period (inverse of the pulse repetition frequency, PRF). The 25% duty-cycle input baseband spectrum is from Figure 3–5. Notice how the spectral components, the amplitudes of which decrease in $(\sin x)/x$ fashion, continue above and below the carrier and cover a very wide bandwidth. Spectral nulls occur at

$$f_c \pm \frac{n}{\tau}, \qquad n \neq 0 \qquad (5\text{--}14)$$

which, for the 25% duty cycle illustrated in Figure 5–10, is every fourth spectral component above and below the carrier component. The spectrum between

$$f = f_c \pm \frac{1}{\tau} \qquad (5\text{--}15)$$

is referred to as the *first lobe* spectrum, or bandwidth. The keying circuit of most OOK transmitters is often equipped with a somewhat long RC time constant (slow rise time) that rounds off the leading and trailing edges of the carrier bursts. The technique reduces the spectrum "splatter" and subsequent out-of-band interference.

5–7 ■ AM DEMODULATION

When an amplitude modulated signal is received, it must be demodulated in order to recover the original information. The AM signal to be demodulated is shown in Figure 5–11 as a 35-kHz carrier modulated by a 5-kHz tone. (Notice that there are 7 RF cycles per audio amplitude variation of the AM envelope.)

So, how can the 5-kHz tone be recovered from the AM signal? It would be impossible with just a filter because, as Figure 5–11d shows, all the power in the signal is at 30, 35, and 40 kHz. But notice from Figure 5–11b and c that the *peaks* of the AM signal—that is, the envelope—follow the 5-kHz tone. Hence, the simplest answer is to use a peak-amplitude detector or rectifier. Such a circuit is shown in Figure 5–12; the diode can be connected as shown or reversed because the positive and the negative peaks follow the (5-kHz) information.

The peak detector is considered in two steps, without the capacitor and then with the capacitor included (see Figure 5–13).

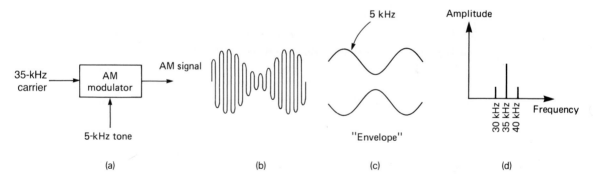

FIGURE 5–11   Amplitude-modulated signal. (a) Generating the AM signal. (b) The AM signal (time domain). (c) The "envelope." (d) One-sided frequency spectrum of the AM signal (frequency domain).

FIGURE 5–12   Peak amplitude detector.

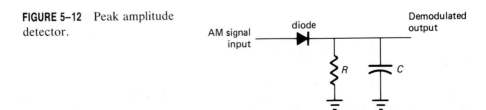

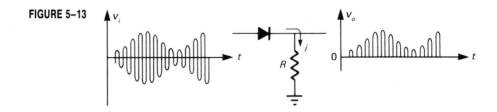

FIGURE 5–13

The diode shown in Figure 5–13 conducts when v_i exceeds v_o. This occurs whenever v_i exceeds the diode cut-in voltage of about 0.2 V for germanium. Hence, with no capacitor, the detector output is just the positive peaks of the input AM signal. The average value of v_o will rise and fall at the same rate as the information—5 kHz in this case. All that is required is some filtering to smooth out the recovered information.

If a capacitor is added to the circuit, as shown in Figure 5–14, not only is filtering provided but also the average value of the demodulated signal is increased, thereby increasing the efficiency of the demodulator. With this circuit the capacitor charges up to the positive peak value of the carrier pulses while the diode is conducting. When the input voltage goes below the peak, the capacitor is holding its charge so that v_o exceeds v_i and the diode stops conducting. The capacitor is allowed to discharge just slowly enough through the resistor that the very next carrier peak will exceed v_o, thereby allowing the diode to conduct and charge the capacitor up to the new peak value. The result is that the output voltage will follow the input AM peaks with a loss of only the voltage dropped across the diode. v_o still has some high-frequency ripple, as seen in Figure 5–14, but this is easily smoothed with additional filtering.

Notice that the demodulated information signal at the detector output v_o is a voltage with an average value of about 0.8 V. The average value of v_o is what the rectified peak value of the *unmodulated* carrier would be. As such, it is an indicator of the received signal strength and will be used in the AM receiver for *automatic gain control* (AGC).

Diagonal Clipping Distortion

The values of R and C at the output of the AM detector must be chosen to optimize the demodulation process. In particular, as illustrated in Figure 5–15, if

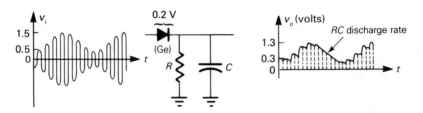

FIGURE 5–14   AM demodulation.

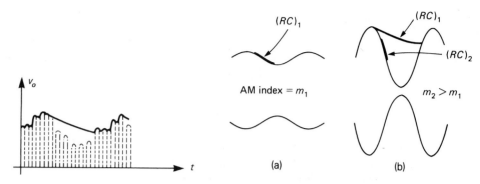

FIGURE 5–15   Diagonal clipping.

FIGURE 5–16   A higher-index AM requires a shorter RC time constant.

the capacitor is too large, it will not be able to discharge fast enough for v_o to follow the fast variations of the AM envelope. The result will be that much of the information will be lost during the discharge time. This effect is called *diagonal clipping* because of the diagonal appearance of the discharge curve. The distortion that results, however, is not just poor sound quality (fidelity) as for peak clipping; it can also result in a considerable loss of information.

The optimum time constant is determined by analyzing the diagonal clipping problem. Compare the RC discharge rate required for the low modulation index illustrated in Figure 5–16a with that required for the same modulating signal but higher index seen in b. Clearly, the modulation index is an important parameter, and the appropriate RC time constant depends not only on the highest modulating frequency $f_m(\max)$, but also on the depth or percentage of modulation m_a. In fact, the maximum value of C is determined from Equation 5–16.

$$C \le \frac{\sqrt{(1/m^2) - 1}}{2\pi R f_m(\max)} \tag{5–16}$$

Negative Peak Clipping

Typically the resistor R used in the detector output is a potentiometer for controlling the audio output power level, the volume control. As illustrated in Figure 5–17, the demodulated information signal is ac coupled by capacitor C_c to the audio amplifiers. Coupling capacitor C_c is made large enough to pass the lowest audio frequencies while blocking the dc bias of the audio amplifier and the average (dc) value of v_o. The audio amplifier input impedance Z_A should be much greater than the output impedance of the detector R to avoid peak clipping distortion that occurs when the peak ac current required by Z_A is greater than the average current available. This is the same problem as encountered when ac coupling any electronic circuits.

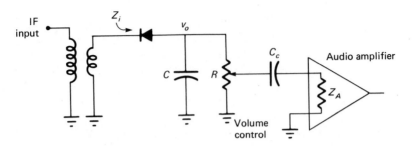

FIGURE 5-17   Complete AM detector and volume (loudness) control.

■ ■ ■ ■

EXAMPLE
5-3

For the AM demodulator of Figure 5–14, determine the following:

1. Total power delivered to the detector circuit if the input impedance is 1 kΩ (detector circuit $R = 2$ kΩ as derived in Section 7–6)
2. $v_o(\text{max})$, $v_o(\text{min})$, and $V_o(\text{dc})$
3. Average current if $R = 2$ kΩ
4. Appropriate value of C if $R = 2$ kΩ, $f_m(\text{max}) = 5$ kHz, and $m_a = 0.9(\text{max})$

Solution:

1. $v_i(average) = (1.5 \text{ V} + 0.5 \text{ V})/2 = 1$ V pk. Peak input voltage for an unmodulated carrier would be 1 V pk.

$$P_c = \frac{(V_c\text{pk})^2}{2R_i} = \frac{(1 \text{ V})^2}{2 \times 1 \text{ k}\Omega} = 0.5 \text{ mW}, \qquad m = \frac{1.5 - 0.5}{1.5 + 0.5} = 0.5$$

Thus $P_t = [1 + (0.5)^2/2](0.5 \text{ mW}) = \textbf{562.5 } \boldsymbol{\mu}\textbf{W}.$

2. $v_o(\text{max}) = 1.5 \text{ V} - 0.2 \text{ V} = \textbf{1.3 V pk}$
   $v_o(\text{min}) = 0.5 \text{ V} - 0.2 \text{ V} = \textbf{0.3 V pk}$
   $v_o(\text{dc}) = v_o(\text{average}) = (1.3 \text{ V} + 0.3 \text{ V})/2 = \textbf{0.8 V}$
3. $I_o(\text{dc}) = V_o(\text{dc})/R = 0.8 \text{ V}/2 \text{ k}\Omega = \textbf{400 } \boldsymbol{\mu}\textbf{A}$

4. $C = \dfrac{\sqrt{(1/m^2) - 1}}{2\pi R f_m(\text{max})} = \dfrac{\sqrt{(1/0.9)^2 - 1}}{2\pi(2 \times 10^3)(5 \times 10^3)} = \textbf{0.008 } \boldsymbol{\mu}\textbf{F}$

5–8 ■ AM RECEIVER SYSTEMS

A simple peak detector requires a couple of milliwatts of total input power if high-modulation index AM is to be demodulated. Since a signal received on an antenna far from the transmitter will be very weak, amplification ahead of the detector is required.

The simple receiver of Figure 5–18 is called a *tuned-radio-frequency* (TRF) receiver. Many RF amplifier stages would be required in order to provide a couple of milliwatts of signal to the RF detector. Although the TRF system is a straightforward concept, there are various reasons for not using this type of receiver: at higher RF and microwave frequencies, amplifiers are more difficult to build, are less efficient, have less gain than intermediate- and low-frequency amplifiers, and the bandwidth changes when the circuit is tuned over a broad range. For these and other reasons that will be understood later, the TRF technique became virtually extinct as soon as the superheterodyne effect of a mixer was implemented.

Superheterodyne Receiver Systems

Conversion of an RF signal from one part of the spectrum to another is accomplished by combining the RF signal with a locally generated periodic signal in a nonlinear device.

The locally generated signal is provided by an oscillator called the *local oscillator* (LO), and the nonlinear device is part of a circuit called the *mixer* or *frequency converter*. The nonlinearity results in the creation of sum and difference frequencies of the LO and RF signals. The process is called *heterodyning*, and the nonlinearity is necessary to provide the mathematical equivalent of time multiplication between the LO voltage and the RF signal voltage. The mathematical derivation of the heterodyne principle is given in Chapter 7 along with greater detail of mixer/frequency converter circuits.

If conversion to a low or intermediate frequency is desired, the local oscillator is usually tuned to a frequency above that of the received RF carrier, and then the difference frequency signal ($f_{LO} - f_{RF}$) is selected while the other signals (f_{LO}, f_{RF}, and $f_{LO} + f_{RF}$) are rejected by filtering. Virtually every high-frequency receiver uses the heterodyne technique to down-convert the received signal to an intermediate frequency (IF) signal, $f_{IF} = f_{LO} - f_{RF}$. Such a receiver is referred to

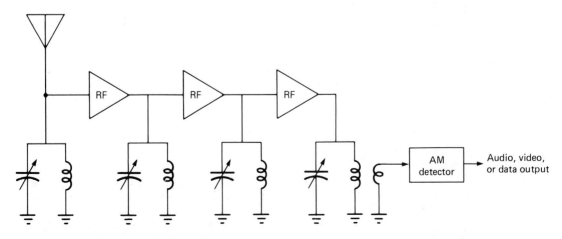

FIGURE 5–18   Tuned-RF (TRF) receiver.

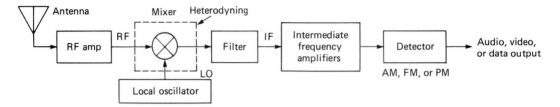

FIGURE 5–19 Superheterodyne receiver. Note that the detector could be for AM, FM, or PM because virtually every modern receiver uses the heterodyning (mixing) effect. Indeed, many use a second down-conversion and second IF amplifier system following the first IF. Such a superheterodyne receiver is called a double-conversion receiver.

as a *superheterodyne*, or *superhet*, receiver, and the circuit most used in modern radios for frequency conversion is called a *mixer*. The mixer allows the oscillator circuit to function separately from the nonlinear mixing process. The block diagram of Figure 5–19 illustrates a superheterodyne receiver with a mixer and separate local oscillator (LO). The filter shown is tuned to the difference (IF) frequency and is followed by IF amplification ahead of the demodulator. As indicated, the superheterodyne principle is used in AM, FM, and PM receivers. Details and design of the circuits which make up the blocks in Figure 5–19 are given in Chapter 7; system considerations including bandwidth, distortion, and noise performance are discussed in the sections to follow.

5–9 ■ SUPERHETERODYNE RECEIVERS FOR AM

The standard AM broadcast band in North America extends from 535 to 1605 kHz, with transmitted carrier frequencies every 10 kHz from 540 to 1600 kHz (±20 Hz tolerance). The 10 kHz of separation between AM stations allows for a maximum modulation frequency of 5 kHz, since the modulation process produces sidebands above and below the carrier. Another standard used for commercial broadcast AM radio is a receiver IF fixed at 455 kHz. The 455 kHz is chosen to be below the lowest transmitted signal frequency but high enough to avoid radio-frequency interference (RFI) from signals called the *image*. Interference caused by an image-frequency signal is discussed in the next section.

The most important reason for using the heterodyne (mixing) technique in a multichannel receiver is to allow for a fixed-tuned IF system. Thus most of the gain and sensitivity of the superhet receiver is provided by a factory-tuned IF section. Then, selecting different stations is a matter of retuning the RF input to the new station frequency and, simultaneously, retuning the local oscillator to a frequency such that the difference $f_{LO} - f_{RF}$ is equal to the intermediate frequency f_{IF}. The LO frequency is almost always higher than the RF carrier frequency, a characteristic referred to as *high-side injection* to the mixer. For example, to receive the AM station whose RF carrier is $f_{RF} = 560$ kHz, the LO must be tuned to

$$
\begin{aligned}
f_{LO} &= f_{RF} + f_{IF} \\
&= 560 \text{ kHz} + 455 \text{ kHz} \\
&= 1015 \text{ kHz}
\end{aligned}
\tag{5–17}
$$

Physically, tuning to a new station is conveniently accomplished by "ganging" the RF and LO tuning capacitors on a single control—"your radio dial"—and the best part is that the user does not have to retune three or more narrowband, high-gain IF amplifiers. Incidentally, the single-control ganging could be implemented by a single mechanical shaft simultaneously rotating movable plates of the RF and LO variable tuning capacitors, or it could be implemented with a potentiometer varying the bias voltages of varicap (or varactor) tuning diodes as in *electronic tuning*.

Choice of IF Frequency and Image Response

When designing a receiver for standard broadcast or for applications such as microwave radio links, numerous considerations go into an appropriate choice for the frequency of the IF system. Most of the decision rests on minimizing receiver *interference* and optimizing *selectivity*.

Receiver *selectivity*, tuning in one station while rejecting interference from all others, is determined by filtering at the receiver RF input and in the IF. One of the first steps in the receiver system design is to make a thorough search and analysis of the frequency spectrum at the site or sites where the receiver is to be located. The spectrum analysis will expose potential sources of interference both external and internal to the receiver hardware. Internal interference comes from oscillators, power supplies, and even the demodulator circuit. External interference comes from artificial and naturally occurring electrical phenomena as well as from other transmitting sources. RFI from other transmitting sources includes that from adjacent channels and image-frequency sources.

Adjacent channels, those immediately above and below the desired channel or station, are rejected primarily by IF filtering. (See Figure 5–20 for an illustration of adjacent-channel rejection by multipole IF filtering.) It would seem natural to filter out adjacent channel transmissions right at the RF input, but two considerations make this difficult. The first is that the RF input circuit may be required to

FIGURE 5–20 IF filter at mixer output.

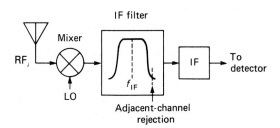

tune over a relatively wide frequency range. Standard broadcast AM receivers, for example, tune over a 3 : 1 frequency range (540–1600 kHz). Maintaining a high Q and constant bandwidth in such a circuit is very difficult, especially with solid-state circuits, which are characteristically of low impedance. Good filtering requires highly selective, high-Q circuits.

The second consideration is that multipole filter networks are employed where a high degree of attenuation to adjacent channel interference is required. However, tuning a multipole filter from station to station over even a moderate tuning range is not practical.

Another consideration in the IF choice is the need to reject interference from an *image* signal. The *image frequency* is that frequency which is exactly one IF frequency above the LO when *high-side injection* is used; that is,

$$f_{\text{image}} = f_{\text{LO}} + f_{\text{IF}} \tag{5–18}$$

From this we could also write $f_{\text{image}} = f_{\text{RF}} + 2f_{\text{IF}}$ for high-side injection. Image response rejection is achieved by filtering *before* the mixer. If a signal at the image frequency gets into the mixer, it is too late to do anything about it. Suppose, for instance, a signal from some unknown source transmitting at 1.910 MHz gets into the mixer when the AM receiver (IF = 455 kHz) is tuned to receive a station whose carrier frequency is $f_{\text{RF}} = 1$ MHz. The LO is 1.455 MHz and the interfering signal is at 1.910 MHz; consequently the difference frequency is 1.910 MHz − 1.455 MHz = 455 kHz—exactly our IF center frequency! There is absolutely no way for the IF filter to prevent this signal from interfering with our enjoyment of the station to which we tuned. RF filtering for a specific amount of image rejection is discussed in the section on RF bandwidth design considerations of Chapter 7. Meanwhile, more needs to be said about the receiver system before the circuit details are discussed.

5–10 ■ RECEIVER GAIN AND SENSITIVITY

The superheterodyne receiver block diagram of Figure 5–19 is very generalized and does not suggest the number of RF and IF amplifiers needed. Actually, this is one of the first tasks of a receiver circuit designer. The number of amplifier stages is determined from the required *receiver sensitivity* and the amount of signal power needed to adequately operate the detector. Let's start with the required detector power.

Detector Power Requirement

The power required to operate a detector, or demodulator, obviously depends on the type of detector being used. For a simple passive germanium diode AM detector, we need a couple of milliwatts of carrier power. This is determined as follows: Suppose a 90% AM signal is received. The receiver must amplify this until it is large enough to cause the diode to conduct. Indeed, to prevent negative-peak distortion at the detector, the minimum positive peak V_{min} must cause conduction. A conservative figure for V_{min} when using a germanium detector diode is about 0.2 V, including junction potential and I^2R losses in the diode and detector circuitry. Referring to Figure 5–4, V_c is the average value between A and B; that is

$$V_c = (A + B)/4 \qquad\qquad\qquad (5\text{--}19)$$

Notice also that $B/2 = V_{min}$, so we solve for A in terms of B in

$$m = (A - B)/(A + B)$$

which gives

$$A = [(1 + m)/(1 - m)]B \qquad\qquad (5\text{--}20)$$

Now substitute this into

$$V_c = \frac{1}{4}(A + B)$$

$$= \frac{1}{2}\left\{\frac{1 + m}{1 - m} + 1\right\} V_{min} \qquad\qquad (5\text{--}21)$$

For $m = 0.90$, $V_c = \frac{1}{2}(1.9/0.1 + 1)V_{min} = 10V_{min}$. Hence, we find that $V_c = 10V_{min} = 2$ V pk. Now, $P = V_c^2/2R$ (where V_c is peak volts) and a typical equivalent detector impedance is 1 kΩ. The result is that 2 mW of carrier power is typically required at the detector.

The other piece of information needed to determine the required receiver gain is the desired *receiver sensitivity*. When we talk about the sensitivity of an electronic device, we are talking about the weakest input signal required for a specified output. For example, the sensitivity of a voltmeter might be given as "the minimum input voltage required for full-scale deflection," or "the voltage required for minimum readable deflection." Clearly, for an electronic voltmeter, the more gain built into the meter, the more sensitive it will be. This is true of communication receivers, too. So we might have a receiver sensitivity specification given as the minimum input signal when modulated at, say, 90% AM, to produce a specified audio output voltage or loudness level at the speaker.

Another way high-performance receivers are specified begins with recognizing that the limitation of a communications receiver is noise and that the most important parameter in determining the quality of the received message is the signal-to-noise ratio. This would lead to a definition of receiver sensitivity as "the minimum input signal, when modulated at 90% AM, required to produce a specified signal-to-noise ratio at the audio output." Also, receivers of digital data are specified in terms of the bit error rate (BER) of the output data. The BER, which is easily measured, is directly related to the signal-to-noise ratio at the receiver output, as we will see in the chapter on digital communication systems.

For our purposes here, we will specify the minimum input signal, modulated to 90% AM, required for adequate operation of the detector. In a later chapter on satellite system design, we will determine just how much power reaches the receiver, given transmitted power and distance from transmitter to receiver.

5–11 ■ POWER LEVEL IN dBm AND dBW

A widely used and very useful way to express power levels is to put them in decibels (dB) relative to some reference power level. Two of the most frequently used reference levels in communications are dBm with a reference level of 1 mW and dBW with a reference level of 1 W.

The power level P is converted to dBm using

$$P \text{ (dBm)} = 10 \log(P/1 \text{ mW}) \tag{5–22}$$

and to dBW by

$$P \text{ (dBW)} = 10 \log(P/1 \text{ W}) \tag{5–23}$$

The point of having the power expressed in dBm is that circuit gains and losses are usually expressed in dB and we can operate on the power levels using very simple arithmetic once the conversion has been calculated.

■ ■ ■ ■

EXAMPLE 5–4

A receiver antenna has an output voltage of 10 µV (carrier only) when connected to a 50-Ω receiver.

1. Determine the power level in dBW and dBm.
2. The receiver has one RF amplifier with 10 dB of gain, a mixer with 6 dB of conversion loss, followed by a multipole filter with 1 dB of insertion loss. If available IF amplifiers have 20 dB of gain each, determine the number of IF amplifiers necessary to provide at least 0 dBm (1 mW) to the detector.
3. Sketch a block diagram of a superheterodyne AM receiver showing the power level, in dBm, at each block.

Solution:

1. $P = (10 \times 10^{-6} \text{ V})^2/50 \text{ Ω} = 2 \times 10^{-12} \text{ W (2 pW)}$.
   $P(\text{dBW}) = 10 \log(2 \times 10^{-12} \text{ W/1 W}) = -117 \text{ dBW}$
   $P(\text{dBm}) = 10 \log(2 \times 10^{-12} \text{ W/}10^{-3} \text{ W}) = -87 \text{ dBm}$

2. With a 10-dB gain, the RF amplifier output will be -87 dBm $+ 10$ dB $= -77$ dBm. Following 6 dB of loss due to the mixer and 1 dB of filter insertion loss in the passband, the IF input power will be P (dBm) $= -77$ dBm $+ (-7$ dB) $= -84$ dBm.

   The IF system must provide an overall gain of P_o/P_i or P_o (dBm) $- P_i$ (dBm) $= 0$ dBm $- (-84$ dBm) $= 84$ dB. Notice that gain, being a power *ratio*, is given in dB. At 20 dB/stage, we need five IF amplifiers, one of which requires only 4 dB of gain.

3. The completed block diagram is shown in Figure 5–21.

When you get used to working with dBm power levels, you really do catch on to the convenience of this technique and will quickly learn to recognize typical power levels, such as those of Table 5–1.

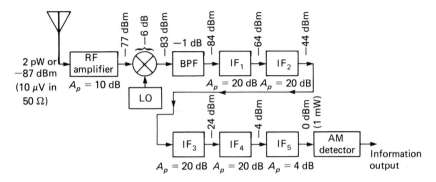

FIGURE 5–21   Signal-level distribution in superhet receiver with 10-μV (−87-dBm) input. The gain of IF₅ has been adjusted to provide 1 mW at the detector.

TABLE 5–1

	Power	
dBm	**mW**	**Comment**
0 dBm	1	
+3 dBm	2	3 dB *above* a milliwatt
+10 dBm	10	
+30 dBm	1 W	also = 0 dBW
−10 dBm	0.1	10 dB below a milliwatt, or
		40 dB below a watt, dBW

5–12 ■ AGC AND DYNAMIC RANGE

So far we have established a sufficiently detailed block diagram to ensure enough gain to demodulate a very weak −87-dBm signal. The next system problem (or challenge) is to consider what would happen if the received signal gets stronger. This does not mean a change in the percent of modulation, because that will naturally vary as the *information* intensity (loudness) varies and is a function of the programming.

What must be considered are the consequences of a substantial increase in the −87-dBm carrier signal level. In an amplitude-modulated system the consequences can range from annoying distortion to a complete wiping off and loss of the information. The annoying distortion occurs if the received signal strength increases to the point at which the IF amplifiers overload and clip the peaks of the AM envelope. An AM signal that has been peak-clipped in a nonlinear IF system is illustrated in Figure 5–22, along with the distorted demodulated output.

It should be obvious that the clipped audio output from the demodulator is going to sound distorted when amplified in the audio system and output through a speaker.

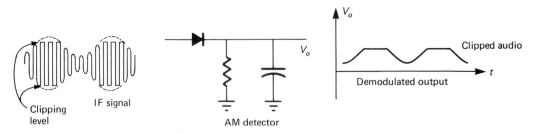

FIGURE 5–22   Effect of clipping due to excessive signal in IF.

Finally, if the received signal gets large enough, all the amplitude variations can be clipped off, resulting in a complete loss of information. This is illustrated in Figure 5–23.

How can such large variations in signal strength occur? One way is to imagine our receiver in an automobile moving through a city. Large buildings and underpasses will greatly attenuate the signal. Such signal-strength variations require the receiver designer to build in automatic signal-level control unless you want to be continually adjusting the receiver gain—which in the preceding situation could be physically dangerous, not to mention a real pain in the neck. The ratio of maximum to minimum signal strength that a receiver is designed to operate over is called the *system dynamic range.*

The point of the preceding discussion is that amplitude-modulated systems that operate in a high dynamic range environment must be designed to maintain linearity in the IF amplifiers by providing automatic gain control (AGC), which keeps the IF signal in the output stages relatively constant.

The usual scheme for AGC is as shown in Figure 5–24, where the gain of the input stage is controlled by controlling the bias of the amplifying device.

Notice that the output of the detector V_o consists of the audio signal (information desired) and a dc voltage that will be proportional to the IF output carrier signal strength. The audio variations are smoothed out with the low-pass filter RC, providing a voltage to control the bias and consequently the gain of Q_1. Notice that Q_1 is a PNP transistor. If the base voltage exceeds -0.2 to -0.3 V, the base-

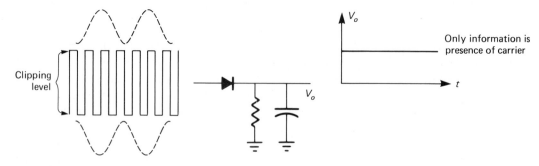

FIGURE 5–23   Information is lost due to severe clipping of the AM signal.

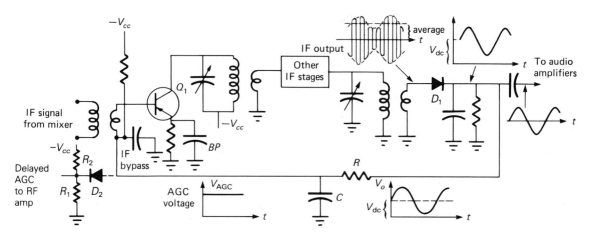

FIGURE 5-24   Typical AGC system.

emitter junction tends to back-bias and the transistor bias current is reduced. This provides the gain control required. If the signal strength from the mixer increases, the detector average (dc) voltage increases and this change is fed back to reduce the gain of the first IF amplifier stage, thereby maintaining a fairly constant IF output.

Since the voltage gain for a common-emitter amplifier is $A_v = -R_c/r_e = -R_cI_c/0.026$, the gain is proportional to collector current and collector current is related exponentially to the base-emitter bias voltage. This is illustrated in Figure 5–25, where I_c (and thus amplifier gain) is essentially zero below V_{ci}, the cut-in voltage. For silicon bipolar transistors, $V_{ci} \approx 0.55$ V. Above this point, gain increases rapidly as the base-emitter voltage increases to 0.6 V and the transistor typically saturates for $V_{BE} > 0.8$ V.

The amount of gain control possible with AGC of a single IF amplifier is limited to a little more than the actual gain of the amplifier itself. This is rarely enough for the typical dynamic range encountered in practice, in excess of 60 dB; therefore, AGC is applied to more than one IF stage. In addition, despite the use of AGC with numerous IF stages, sufficiently strong input signals to the receiver can overload the mixer (and even the RF amplifier) and distort the AM signal. The cure for this is to provide an AGC voltage to the RF amplifier.

FIGURE 5-25   Bipolar transistor bias characteristic.

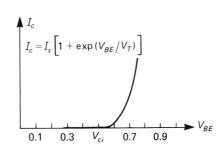

Delayed AGC

AGC applied to an RF amplifier can eliminate distortion produced by an excessively strong received signal. It turns out that there is a price to be paid over and above the added circuit complexity if AGC is applied to the RF amplifier without considering other system parameters.

We haven't studied, as yet, the system design criteria for the reception of extremely weak signals—weak enough to be no larger than the receiver input *noise*. Noise and receiver noise figure (NF) were discussed in Chapter 4. As discussed, the noise figure is a measure of the amount of noise produced by receiver circuits and added to the noise already at the antenna. It was determined that the receiver noise figure (and added noise) can be kept low if the RF amplifier gain is kept *high*.

This means that if we AGC the RF amplifier on weak signals and its gain is reduced, then the signal-to-noise ratio (*S/N*) of the information will be spoiled. The way to a solution is to look at the problem this way: What we want is to have maximum RF amplifier gain for weak signals but, for very strong signals *where S/N is no problem anyway*, reduce the gain, thereby avoiding overload distortion in the mixer and RF amp. The solution, then, is to put a *voltage-level delay circuit* after the regular AGC line to keep the RF amplifier gain high until a sufficiently strong input signal causes the AGC to exceed a set threshold voltage. The voltage-level delay circuit is illustrated in Figure 5–24 by D_2 and R_1 with R_2 connected to $-V_{cc}$. R_1 and R_2 place a negative bias of a few tenths of a volt on the cathode of the silicon diode, D_2. D_2 is reverse biased until the AGC voltage rises to about -0.1 V (Q_1 cuts off at about -0.2 V). If the AGC voltage rises above -0.1 V, D_2 becomes forward biased and conducts to reduce the gain of the RF amplifier. Thus, the delayed-AGC threshold can be set by R_2.

For input signals strong enough to cause the AGC to exceed the threshold value, delayed AGC controls the RF amplifier gain, thereby eliminating high-signal-level distortion while maintaining good *S/N* for weak signals. The circuits described above will be discussed in Chapter 7, so now let's continue the system analysis by studying noise and its effects.

5–13 ■ RECEIVER NOISE

The noise present in electronic components was studied in Chapter 4. Of the various types of noise discussed, most of the noise present in the input of a typical receiver consists of *thermal* and *shot noise*. *Shot noise* is proportional to the bias current in devices such as diodes, transistors, and vacuum tubes.

Receiver *thermal noise* power is proportional to temperature and bandwidth only. This is seen in Equation 4–1, $N_{th} = kTB$. Since receiver power calculations are done in dBm, it is convenient to write an expression for thermal noise in dBm. This is derived from Equation 4–1 as follows:

$$10 \log (kTB) = 10 \log kT + 10 \log B$$

Using $k = 1.38 \times 10^{-23}$ W·s/K and T referenced to 290 K, we obtain Equation 5–24:

$$N_{\text{th}} \text{ (dBm)} = -174 \text{ dBm} + 10 \log B \qquad (5\text{–}24)$$

Noise calculations for receiver circuits get quite complex, so the usual procedure is to calculate the thermal noise from Equation 5–24 and then use the results of semiconductor manufacturer measurements for the other noise contributions such as shot, flicker, and partition.

The circuit noise contributed by a particular transistor or diode is available in the manufacturer's data sheet and is called the *noise figure* of the device. Noise figure is given in units of dB and indicates the amount of noise generated in the device above the kTB thermal noise (referenced to a temperature of 290K). Hence, the total receiver noise power is the sum, in dBm, of the thermal noise and the total receiver noise figure; that is,

$$N = (kTB)(\text{NR})$$

or, in dBm,

$$N \text{ (dBm)} = (-174 \text{ dBm} + 10 \log B) + \text{NF (dB)} \qquad (5\text{–}25)$$

where NF is the total system noise figure.

5–14 ■ SIGNAL-TO-NOISE RATIO (*S/N*)

The quality of the information transmitted through a communications system depends on the amount of distortion of the information signal.

An information system that is common to the experience of each of us is television. When the signal received from a distant transmitter is very weak, noise can degrade the quality of the picture (derived from the video signal) and the sound (from the audio signal). Measurements show that a picture of reasonably good quality requires a video signal-to-noise ratio (S/N) in excess of 40 dB. This means that the signal power S must be more than 40 dB greater than the noise power N.

A typical TV receiver has a noise figure of 12 dB and an IF bandwidth of 6 MHz. Hence, the equivalent noise power referred to the input is, from Equation 5–25, N (dBm) $= (-174 \text{ dBm} + 10 \log 6 \times 10^6) + 12 \text{ dB} = -94.2$ dBm. This is 380 pW of signal power, or 337 μV in a 300-Ω input impedance.

As another and more complete S/N calculation, if only 100 μV is received on the 300-Ω TV antenna, then $S = V^2/R = (100 \ \mu\text{V})^2/300 = 33.3$ pW. This is S (dBm) $= 10 \log (33.3 \times 10^{-12} \text{ W} \div 1 \text{ mW}) = -74.8$ dBm. Consequently, $S/N = S$ (dBm) $- N$ (dBm) $= -74.8 \text{ dBm} - (-94.2 \text{ dBm}) = 19.4$ dB. This would result in a very noisy video output—"snow" on the TV screen.

The video information quality can be improved with an improvement in the receiver noise figure, but the noise bandwidth cannot be reduced much or picture distortion will result due to inadequate signal bandwidth.

5-15 ■ DERIVATION OF THE NOISE FIGURE EQUATION

The derivation of noise figure for a system of cascaded amplifiers is most simply approached in two steps: First derive, in terms of noise figure, an expression for the amount of noise power internal to a single amplifier that is added to the thermal noise normally present at the input of the amplifier, then derive the noise figure for cascaded amplifiers.

Referring to Figure 5–26, let the power gain of the amplifier be G_a and the noise power ratio be NR_a, and let N_a be the internal amplifier noise power that would be measured at the output if absolutely no noise could be present at the input.

Of course, there is always thermal noise

$$N_{th} = kTB \tag{5-26}$$

present at the input to the amplifier by virtue of the fact that conductors at temperature T have N_{th} watts of power as measured in a bandwidth B. Hence, the total noise power at the amplifier output will be

$$N_o = N_{th}G_a + N_a \tag{5-27}$$

where the input thermal noise power is amplified by power gain G_a.

The definition of noise figure NF is

$$NF = 10 \log NR \tag{5-28}$$

where

$$NR = \frac{(S/N)_i}{(S/N)_o} \tag{5-29}$$

Hence,

$$NR_a = \frac{N_o/G_a}{N_{th}} \tag{5-30}$$

That is, the ratio of the total output noise power referred to the amplifier input—that is, divided by amplifier power gain—to input thermal noise power yields the amplifier noise ratio (noise figure). At the output, Equation 5–30 may be written as

$$NR_a = \frac{N_{th}G_a + N_a}{N_{th}G_a} \tag{5-31}$$

That is, total noise power out divided by the output thermal noise power level is the ratio by which the amplifier degrades input signal-to-noise ratio. Equation 5–31 is a more functional definition of noise ratio (noise figure) than 5–30 is.

Now, notice that the right-hand side of Equation 5–31 may be written as the sum of equations

$$NR_a = \frac{N_{th}G_a + N_a}{N_{th}G_a} = 1 + \frac{N_a}{N_{th}G_a} \tag{5-32}$$

FIGURE 5-26 Amplifier with noise.

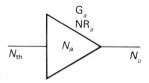

That is,
$$\text{NR}_a - 1 = \frac{N_a}{N_{\text{th}}G_a} \tag{5-33}$$

from which we see that the noise power added to any input noise can be written in terms of amplifier noise ratio as

$$N_a = N_{\text{th}}G_a(\text{NR}_a - 1) \tag{5-34}$$

If multiplied through, Equation 5-34 reexpresses the original notion that

$$N_a = N_{\text{th}}G_a\text{NR}_a - N_{\text{th}}G_a \tag{5-35}$$

That is, input noise times gain subtracted from total output noise equals noise added by the amplifier. Now let us take this notion and Equation 5-34 to derive the noise figure of two amplifiers in cascade. The noise figure of the two-amplifier cascade of Figure 5-27 is derived as follows: The total noise power at the output of amplifier 1, N_{o1}, is, from Equations 5-27 and 5-35,

$$N_{o1} = N_{\text{th}}G_1 + [N_{\text{th}}G_1(\text{NR}_1 - 1)] \tag{5-36}$$

Noise power N_{o1} is now amplified by G_2 and added to the amount of noise, $N_{a2} = N_{\text{th}}G_2(\text{NR}_2 - 1)$, added by amplifier 2; that is,

$$N_{o2} = N_{\text{th}}G_2(\text{NR}_2 - 1) + G_2\{N_{\text{th}}G_1 + [N_{\text{th}}G_1(\text{NR}_1 - 1)]\} \tag{5-37a}$$

$$= N_{\text{th}}G_1G_2 + N_{\text{th}}G_1G_2(\text{NR}_1 - 1) + N_{\text{th}}G_2(\text{NR}_2 - 1) \tag{5-37b}$$

By extending Equation 5-30 to a system of two amplifiers in cascade, it follows that

$$\text{NR}_{\text{sys}} = \frac{N_{o2}/(G_1G_2)}{N_{\text{th}}} = \frac{N_{o2}}{N_{\text{th}}G_1G_2} \tag{5-38a}$$

$$= 1 + (\text{NR}_1 - 1) + (\text{NR}_2 - 1)/G_1 \tag{5-38b}$$

where Equation 5-37b has been substituted in Equation 5-38a to get Equation 5-38b. Hence, the final result for two amplifiers in cascade is

$$\text{NR}_{\text{sys}} = \text{NR}_1 + \frac{\text{NR}_2 - 1}{G_1} \tag{5-39}$$

By extension of this procedure to n amplifiers, the familiar system noise figure,

$$\text{NR} = \text{NR}_1 + \frac{\text{NR}_2 - 1}{G_1} + \frac{\text{NR}_3 - 1}{G_1G_2} + \cdots + \frac{\text{NR}_n - 1}{G_1G_2 \ldots G_{n-1}} \tag{5-40}$$

is obtained. A simple system problem is demonstrated in Example 4-3.

FIGURE 5-27 Two noisy amplifiers cascaded.

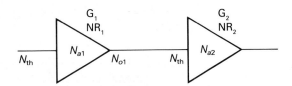

The equivalent noise temperature of a system is often specified instead of the noise figure. The equivalent between noise temperature and noise figure is given by Equation 4–16 as

$$T_{eq} = T_0(\text{NR} - 1) \qquad\qquad (5\text{--}41)$$

where $T_0 = 290$ K (see Section 4–5).

5–16 ■ BANDWIDTH IMPROVEMENT

Improving the signal-to-noise ratio, S/N or SNR, of a received signal is a matter of designing the receiver for low noise figure, eliminating unnecessary noise using the narrowest possible filters while passing the entire information bandwidth and, with modern digital signal processing techniques, sampling and time averaging the received data. In the receiver, before demodulation, there are practical and system-requirement limitations to the improvement achievable by narrowing the bandwidth. The practical problem is in the difficulty of building stable, narrow-band filters. For example, if we are receiving a signal in the standard AM band at 1.5 MHz, building a front-end filter of 10-kHz bandwidth requires an equivalent circuit quality factor of $Q = 1.5$ MHz/10 kHz $= 150$. This is barely achievable. However, it is easily accomplished at the IF amplifier, so that, for a standard $f_{IF} = 455$ kHz, $Q = 455$ kHz/10 kHz $= 45.5$.

The system-requirement constraint is that the circuit bandwidth must exceed the information bandwidth or the information power and spectrum will be reduced, as illustrated in Figure 5–28. Also, if the filter is not symmetrical, additional signal distortion will result.

The best solution is to filter as much as possible at the IF without distorting the information (the sidebands). For critical applications such as space hardware, where temperature changes cause tuned circuits to shift in frequency and symmetry, a temperature-compensated multipole bandpass filter is placed immediately following the mixer, and the IF amplifiers are made broadband with stable gain.

Very often S/N is calculated at the receiver RF input using the RF bandwidth to determine noise power. However, most IF systems have a narrower bandwidth

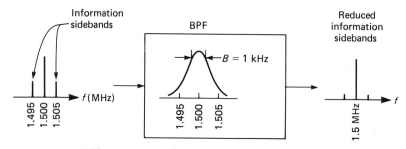

FIGURE 5–28 A filter bandwidth narrower than the information bandwidth, with resulting reduced information power.

than the RF, and consequently the noise power will be relatively lower in the IFs. The simplest method for getting the correct predetection S/N is to use the IF bandwidth as the noise bandwidth in the calculation of $N_{th} = kTB$. The other method is to determine the bandwidth exchange, or *noise-bandwidth improvement* factor, BI.

The bandwidth improvement factor is the ratio by which noise power is reduced by a reduction in bandwidth. As an example, suppose a receiver has an RF bandwidth of 5 MHz and an IF bandwidth of 200 kHz. The noise bandwidth improvement is

$$\text{BI (dB)} = 10 \log (B_{RF}/B_{IF}) \qquad (5\text{--}42)$$

As a consequence, the S/N will be 14 dB better in the IF amplifiers than in the RF input (if the noise figure contribution is ignored).

The last few pages on noise, noise figure, and bandwidth improvement are summarized below in a system design problem to determine the S/N at the input to the demodulator of a superheterodyne receiver block-diagrammed in Figure 5–29.

■ ■ ■ ■

EXAMPLE 5–5

Given that $NF_1 = 2$ dB, NR = 1.6; $NF_{2,3} = 6$ dB, NR = 4.0; $NF_{4,5} = 18$ dB, NR = 63.1; $A_{p1} = 8$ dB, $G_1 = 6.3$; $A_{p2} = 12$ dB, $G_2 = 15.8$; $A_{p3} = -6$ dB, $G_3 = 0.25$; $A_{p4,5} = 20$ dB, $G_{4,5} = 100$:

1. Calculate the system NF (dB).

2. If the RF bandwidth is 5 MHz and the IF bandwidth is 200 kHz, determine the predetection S/N (dB) for a receiver input signal of -80 dBm.

Solution:

1. From Equation 5–40,

$$NR = 1.6 + \frac{4.0 - 1}{6.3} + \frac{4.0 - 1}{(6.3)(15.8)}$$

$$+ \frac{63.1 - 1}{(6.3)(15.8)(0.25)} + \frac{63.1 - 1}{(6.3)(15.8)(0.25)(100)}$$

$$= 1.6 + 0.48 + 0.03 + 2.5^* + 0.025$$

$$= 4.6$$

$$NF \text{ (dB)} = 10 \log 4.6 = \textbf{6.7 dB}$$

2. The most expedient approach here would be to use the IF bandwidth to calculate the thermal noise power and skip noise improvement calculations. However, to illustrate all the concepts, start by determining the receiver input S/N.

---

* Comment: Notice from the fourth term how much the mixer conversion loss and noisy IF amplifier increases the system noise figure. The system NF can be decreased by a couple of dB by either adding another 12-dB gain, RF amp (do the calculation) or by using a lower noise amplifier for the first stage of the IF system.

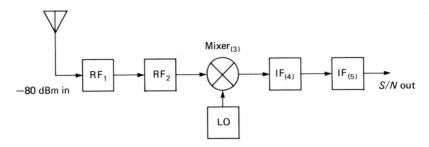

FIGURE 5–29

The received signal is −80 dBm. The thermal noise power at the input is, from Equation 5–24, N_{th} = −174 dBm + 10 log 5 × 10^6 = −107 dBm. Consequently, $(S/N)_i$ dB = −80 dBm − (−107 dBm) = **+27 dB**.

The input S/N is reduced because of system noise (NF = 6.7 dB) but is increased by the effect of reduced noise in a narrow IF bandwidth (BI). The bandwidth-narrowed noise improvement will be, from Equation 5–42, BI (dB) = 10 log(5000/200) = 14 dB. The final result is

$$(S/N)_o \text{ dB} = (S/N)_i \text{ dB} - \text{NF (dB)} + \text{BI (dB)}$$

$$= +27 \text{ dB} - 6.7 \text{ dB} + 14 \text{ dB}$$

$$= \mathbf{+34.3 \text{ dB}}$$

The signal-to-noise ratio at the input to a demodulator (+34.3 dB in Example 5–5) can be a critical system calculation. This is especially true for frequency-modulated (FM) systems because typical FM demodulators exhibit a thresholding effect for input S/N of less than about +13 dB. Frequency modulation is studied in Chapter 9. Chapters 6 and 7 provide circuit details for AM systems.

5–17 ■ BANDWIDTH NARROWING

When designing IF amplifiers and when calculating the bandwidth of a multistage system, remember that the frequency response of each stage affects the overall bandwidth. For the simple case where all single-tuned amplifiers have equal bandwidths, the overall 3-dB bandwidth is computed from

$$\text{BW}_T = \text{BW}_1 \sqrt{2^{1/n} - 1} \qquad \textbf{(5–43)}$$

where BW_1 = bandwidth of each stage

n = number of stages with bandwidth of BW_1

So, for three amplifiers, each of which has BW_1 = 10 kHz, the overall 3-dB bandwidth is

$$\text{BW} = 10 \text{ kHz} \sqrt{2^{1/3} - 1} = 5.09 \text{ kHz}$$

This formula is for single-tuned, *synchronously tuned* amplifier systems only.

Synchronous tuning refers to tuning each IF amplifier stage to the same frequency. While seldom used these days, *stagger tuning* refers to the technique of tuning stages to slightly different frequencies above and below the design value in order to produce a wider bandwidth than for the synchronously tuned system.

■ PROBLEMS

1. Determine from Figure 5–30:
   a. Modulation index.
   b. If the sketch is accurate on the time scale and f_c = 45 kHz, find the modulation frequency from the oscilloscope display shown. (Count cycles.)
   c. Carrier power in 75 Ω.
   d. Power in one sideband.
   e. Total power dissipated by 75 Ω.
   f. The carrier power is what percentage of the total?
   g. How much bandwidth (information bandwidth) is required to transmit this AM signal?
   h. Sketch the time-domain waveform, *include the voltage and frequency,* if *only* the upper sideband is transmitted (assume no carrier).

2. The AM signal of Figure 5–31, with carrier frequency of 1 MHz and modulation frequency of 1 kHz, is transmitted from a 50-Ω antenna.
   a. Calculate the modulation index, in percent.
   b. Determine the power of the unmodulated carrier. (*Hint:* If f_m is removed, the peak carrier amplitude will be the average of the peaks shown in the figure.)
   c. Using P_c from part **b**, calculate the power of each sideband.
   d. The power in one sideband is what percentage of the total transmitted power?
   e. Write the mathematical expression, using cosines, for this AM signal. Include actual numerical values for frequencies and amplitudes.
   f. Suppose two more tones, producing a musical triad chord with the same amplitudes as f_m, are modulated on the carrier. What would be the total modulation index?

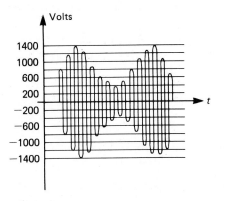

FIGURE 5-30

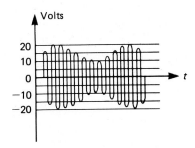

FIGURE 5-31   (Not accurate on time scale.)

3. An AM signal has a 100-V pk, 40-kHz cosine carrier. When modulated by a 10-kHz cosine tone to an index of 50%:
   a. Sketch the signal accurately over a 0.2-ms period.
   b. Sketch the frequency spectrum of this AM signal. Include the frequency and peak voltage of each component.
   c. Write the mathematical expression of this signal in a form that shows the three components—carrier, LSB, and USB. (Again, include the actual numerical frequencies and peak voltages.)
   d. How much bandwidth, in kHz, is required to transmit this AM signal?
   e. Calculate the power delivered to a 50-Ω load for the following:
      (1) P_c
      (2) P_{USB}
      (3) P_{total}
   f. How much power could be saved by transmitting only one sideband (and no carrier)?

4. a. Determine the Fourier transform for the AM signal $e(t)$ of Equation 5–6.
   b. Plot the spectrum determined in part a.

5. For the AM signal of Figure 5–32:
   a. Determine the modulation index.
   b. If the carrier frequency is 45 kHz, what must be the modulation (audio) frequency? (Count cycles.)
   c. If the input impedance of the AM detector circuit shown is 500 Ω, calculate the *carrier* power and *total* power delivered to the detector.
   d. If the diode drop is 0.2 V pk from each RF pulse, sketch the output waveform, complete with peak voltage values.
   e. Determine the average value of V_o, the detector output voltage. (This will be the dc voltage used by the receiver AGC system.)
   f. Calculate the appropriate value of C to demodulate the signal shown. $R = 1$ kΩ.
   g. If $R = 1$ kΩ and $C = 1$ μF are used, does diagonal clipping occur? Why or why not?

6. If the signal of Problem 3 is input to the detector of Figure 5–32,
   a. Sketch the ideal output, $v_o(t)$, and show the voltages and time accurately;
   b. Determine the appropriate value of C for the capacitor if R is 2 kΩ.

7. Use a software package, or write a program and use it, to solve the question of Problem 3.

8. An AM signal at the input to the diode shown in Figure 5–17 has a total power of 2.64 mW and modulation index of 80% with $f_{mod} = 5$ kHz. Determine
   a. Carrier power in mW and in dBm.
   b. Peak voltage of the carrier at the diode input ($Z_i = 1$ kΩ).
   c. Peak voltage of the input AM signal (a sketch might help in thinking about this).
   d. Average (dc) voltage of V_o.
   e. Make a smooth sketch of V_o showing correct voltages and timing.
   f. Sketch the voltage across Z_A (10 MΩ) if R is not a potentiometer. (C_c connects at the top of R.)
   g. Appropriate C for this detector and signal if $R = 2$ kΩ.

9. Calculate the power, in dBm, for 100 μV across 50 Ω. How much is this in dBW?

FIGURE 5-32

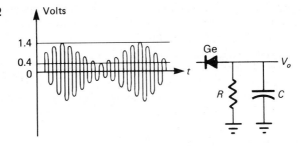

10. The input power to a 50-Ω receiver is 200 pW. Determine the receiver gain required to produce +3 dBm at the detector.

11. The signal received at a 50-Ω antenna is 1 μV pk and is at 108 MHz. For the receiver, assume the following:
   - Use 1 RF amplifier, A_p = +10 dB.
   - Mixer *loss* is A_p = -6 dB.
   - Use LO frequency higher than received carrier (high-side injection).
   - f_{IF} = 10.7 MHz.
   - IF filter following mixer has loss of A_p = -1 dB.
   - We require +3 dBm of power at the detector (2 mW).
   a. Determine the received power in dBm.
   b. Determine LO frequency.
   c. How many IF amplifier stages are needed if A_p = +20 dB/stage?
   d. Draw a block diagram of the receiver (antenna through the detector). Show power level in dBm at each point.
   e. Determine the image frequency. How can we eliminate receiver image response?

12. A 5-μV *peak* carrier is received on a 75-Ω antenna (carrier frequency, 96 MHz). Assume the following for the receiver:
   - One RF amplifier with 12 dB of gain.
   - Mixer conversion efficiency (loss) is 6 dB.
   - LO frequency higher than the received carrier (high-side injection).
   - LO power is 20 mW.
   - f_{IF} = 10.7 MHz.
   - IF filter following mixer has an insertion loss of 1 dB.
   - We require +3 dBm of carrier power at the detector.
   a. Determine the received power, in dBm.
   b. Determine LO frequency.
   c. How many IF amplifier stages are needed if A_p = +25 dB/stage?
   d. Draw a block diagram of the receiver (antenna through detector). Show power level (in dBm) at each input.
   e. Determine the image frequency. How can we eliminate receiver image response?

13. What is the purpose of AGC?

14. Determine the noise power delivered to a receiver input at 300 K and noise bandwidth of 20 kHz.

15. Assume a receiver with the following specifications: 2 RF amps with power gain of 10 dB each and NF (dB) = 4 dB each. Mixer: A_p = −6 dB, NF (dB) = 8 dB. IFs: A_p = 20 dB, NF = 15 dB.

a. Determine the noise figure, in dB.

b. If the overall bandwidth of the receiver is 200 kHz, determine the *S/N* power ratio in the IF for a received signal of −100 dBm.

c. Assuming five IF stages before the detector and the front end (defined above), what will be the signal power at the detector input for a received signal of −100 dBm?

MATHCAD: All from this line through the figure was produced
         on MathCAD. MathCAD executes the equations typed
         below and produces the plot.

   Plot of amplitude modulation Equation 5-2.

$$t := 0, \frac{1}{144} \ ..1 \quad A := 2 \quad Ec := 2 \quad fm := 2 \quad fc := 20$$

(Modulation index = A/Ec)

$Em(t) := A \cdot \sin(2 \cdot \pi \cdot fm \cdot t)$ The modulating signal.

$e(t) := (Ec + Em(t)) \cdot \cos(2 \cdot \pi \cdot fc \cdot t)$ Equation 5-2.

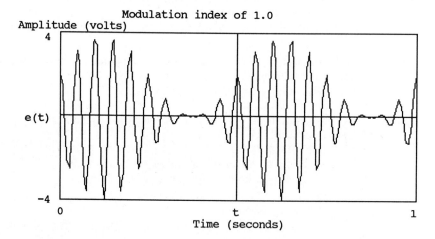

FIGURE 5-33

5–18 ■ COMPUTER ANALYSIS

The analysis for Chapter 5 is as follows:

MATHCAD EXAMPLES: Amplitude-modulation signal equations are analyzed using MathCAD.* The equation for amplitude modulation is entered and then plotted. The effects of modulation index changes can be analyzed by iteration of the process. The modulation index for Figures 5–33 and 5–35 is 1; for Figure 5–34 it is 0.5.

* MathCAD is a trademark of Mathsoft, Inc., Cambridge, Mass.

MATHCAD: All from this line through the figure was produced
         on MathCAD. MathCAD executes the equations typed
         below and produces the plot.

   Plot of amplitude modulation Equation 5-2.

$$t := 0, \frac{1}{144} ..1 \quad A := 1 \quad Ec := 2 \quad fm := 2 \quad fc := 20$$

                    (Modulation index = A/Ec)

Em(t) := A· sin(2·π· fm· t)   The modulating signal.

e(t) := (Ec + Em(t))· cos(2·π· fc· t)   Equation 5-2.

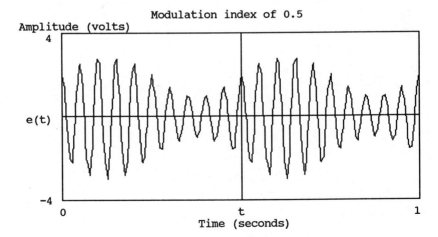

FIGURE 5–34

MATHCAD: All from this line through the figure was produced
         on MathCAD. MathCAD executes the equations typed
         below and produces the plot.

Plot of amplitude modulation Equation 5-3 with a 90
degree phase-shifted sinewave carrier so that the result
looks like Figure 5-33 (Equation 5-2).

$$t := 0, \frac{1}{144} \, ..1 \quad A := 2 \quad Ec := 2 \quad fm := 2 \quad fc := 20$$

(Modulation index = A/Ec)

$$m(t) := A \cdot \sin(2 \cdot \pi \cdot fm \cdot t) \qquad \text{The modulating signal.}$$

$$e(t) := (Ec + m(t)) \cdot \sin\left[2 \cdot \pi \cdot fc \cdot t + \frac{\pi}{2}\right] \qquad \text{Equation 5-3.}$$

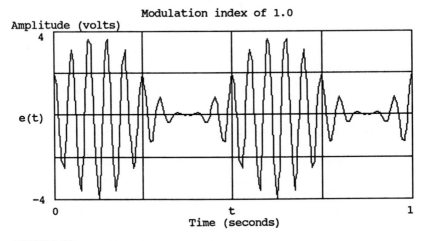

Modulation index of 1.0

FIGURE 5-35

6

Transmitter Circuits

■ INTRODUCTION

The basic block diagram of a radio frequency receiver and several aspects of modulation were discussed in Chapter 5. In this chapter, the elements of transmitters and some circuit concepts specific to RF transmitters will be studied.

In simplest terms, an RF transmitter accepts an electrical signal from some information source and transmits a high-power analog RF carrier signal modulated with the information. Another input to the transmitter is also required—a power supply. The function of transmitter circuits is to convert power from the power supply into the modulated output signal as efficiently as possible.

Usually, system design begins with determining the best type of modulation to use (if not dictated in advance), and a standard block diagram is chosen such as one of those in Figure 6–1. The basic block diagram is then altered as necessary to meet input-output specifications. This is the point in the design process at which experience and good judgment pay off. Making extensive changes in the laboratory is not cost effective.

The final detailed block diagram for a transmitter is usually determined by starting with the output power, efficiency, bandwidth, and linearity requirements;

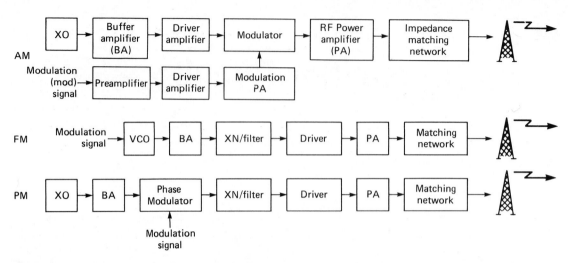

FIGURE 6-1   General block diagrams for transmitters of amplitude, frequency, and phase modulation.

choosing the output amplifier class (A, B, or C) and device; and then repeating this process all the way back to the carrier oscillator. For frequency- and especially phase-modulated transmitters, frequency multipliers (the XN block) may have to be included in order to achieve the required modulation index. This is covered in Chapter 9.   ■

6-1 ■ POWER AMPLIFIERS

Class A, B, and C amplifiers are used in transmitters. They are usually tuned with a bandwidth wide enough to pass all the information sidebands without distortion but as narrow as feasible in order to reduce harmonics and other spurious signals.

Class A amplifiers are used in low-power stages where device dissipation and efficiency are not critical. They have maximum efficiencies approaching 50% when their output is transformer-coupled (25% when RC-coupled) and conduct continuously—that is, over 360° of the input signal cycle.

Class B and C amplifiers are used where high power and efficiency are required. Class B operation is achieved when the active device (transistor or vacuum tube) is biased right at cutoff, so that output current will flow for only one-half of the input signal cycle. Efficiency approaches 78.5% and, for linear system operation, must be used in a push-pull circuit configuration.

Class C cannot be used for audio power amplifiers because the output current flows for less than one-half of the input signal cycle; even in push-pull circuits, the output signal will be too distorted for audio use. Class C is used with tuned circuits, and the short output current pulses make the output "ring" at the circuit resonant frequency. Because the active device is off most of the time, efficiency approaching 100% is theoretically possible but is limited to less than 85% in practice in order to produce high output power.

Examples of class A and B power amplifier analyses are given in the section on audio systems. Here, we will concentrate on the high-power class C final stage of an RF transmitter. While parallel systems of transistorized power amplifiers can achieve hundreds of watts at VHF and UHF, the analysis here will be for the tens of kilowatts required for broadcast. Since this level of power still calls for vacuum-tube amplifiers, a short section is now included to describe the simplest and very often used amplifying tube for class C, the triode.

Triode Vacuum Tube

Triode is the name for a three-electrode (plus heater) vacuum tube. The tube elements are usually formed in a concentric cylindrical arrangement with a lead or straps connected to each element, brought down to a base, and attached to supporting pins (or other socket arrangement) for external connection. The high-voltage plate electrode is usually attached to a connector at the top of the tube to isolate it from the other electrodes. The entire structure is placed in a glass tube or other dielectric material that is then evacuated of air and sealed with only the base socket and plate cap connections exposed.

The *cathode* is typically a metal cylinder wrapped around and electrically isolated from a heater element. The cathode has a specially alloyed metal surface from which electrons can escape with a minimum of thermal energy (heat) supplied by the heater circuit. The heater gets hot enough to boil electrons off the cathode surface, forming an electron cloud much the same as water molecules boiled off the surface of water in a kettle.

The *plate* (anode) is a separate metal cylinder placed around the other tube elements. When a high positive voltage is applied between the plate and cathode, the electrons, liberated by heat from the cathode, will be attracted to the plate and will accelerate across the tube.

The *control grid* is a wire wound in the space between the cathode and plate. Only one end of the grid is electrically connected.

Figure 6–2 illustrates schematically the triode with an electron cloud around the cathode. The two arrows illustrate electrons passing between turns of the

FIGURE 6–2 Triode vacuum tube with bias.

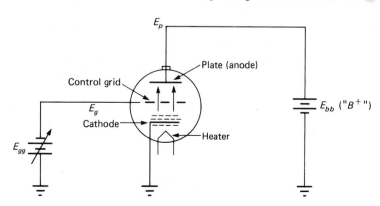

control grid wire on their way to the plate. When a negative voltage is applied between the grid wire and cathode, the negative electric field produces a repulsive force on the negatively charged electrons trying to pass through to the plate.

Indeed, a grid voltage of

$$E_g = -E_p/\mu = E_{CO} \tag{6-1}$$

will completely cut off the electron flow to the plate so that $I_p = 0$. The constant μ is called the *amplification factor* of the tube and is a measure of the effect that the grid voltage has on current flow relative to the plate voltage. As the ac grid voltage varies, the tube current varies accordingly, and an impedance connected in series with the plate lead and power supply (E_{bb}) will allow a plate-voltage variation e_p to develop. The voltage gain of a grounded-cathode, class A vacuum-tube amplifier with plate-load impedance Z_L and internal dynamic plate resistance r_p is

$$A_v = \frac{e_p}{e_g} = \frac{-\mu Z_L}{Z_L + r_p} \tag{6-2}$$

High-power tubes used in class C are usually neutralized* triodes or tetrodes. For high-frequency applications, these triodes are made as small as possible with closely spaced electrodes to minimize cathode-plate transit time. Also, relatively high operating voltages are used in order to minimize transit time and cathode current densities. At UHF and beyond, klystrons are often used for high power.

Figure 6–3 shows a TH382 tetrode used for tens of kilowatts of output power from a UHF television transmitter. Using pyrolitic graphite grids and a water cooling system, this device can provide a video carrier with over 20-kW peak output or a sound carrier output of 13.5 kW at UHF.†

---

* Plate signal is fed back to the grid out of phase with the current in the plate-grid capacitance. The tube is neutralized when the two signals cancel. (Neutralization is discussed later in this chapter.)

† Gerlach and Kalfon, *Proceedings of IEEE Special Issue on UHF TV*, vol. 70, no. 11, November 1982, p. 1335.

FIGURE 6-3 High-power tetrode for UHF television transmitter.

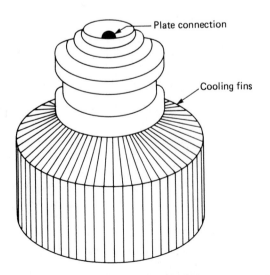

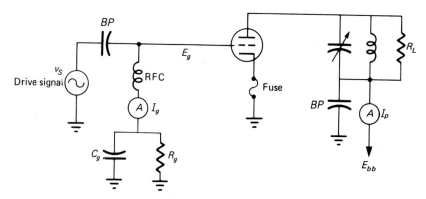

FIGURE 6-4 Class C triode amplifier.

Class C Power Amplifier

The bias of a class C amplifier is set so that the active device (transistor or vacuum-tube) is cut off. The class C vacuum-tube amplifier (Figure 6-4) is typically biased at 2.5 times the cutoff voltage, determined from Equation 6-1 with E_p replaced by the power supply voltage E_{bb}. You can sometimes spot transistorized class C stages by their lack of a base bias resistor, which normally supplies turn-on current from the power supply.*

The active device does not conduct unless the input signal peak exceeds the device cutoff voltage so that output current flows in pulses for *less than* a half-cycle ($<180°$).

The fraction of the input cycle for which output current flows is called the *conduction angle* θ and is given in degrees of phase. In practice a compromise is made between high efficiency (smaller θ) and high power (higher θ). A conduction angle between 120° and 150° is chosen. With this compromise, efficiency is between 80% and 60% with high output power.

Vacuum-tube class C waveforms are shown in Figure 6-5. Notice first that voltages are given in terms of electromotive force (emf), where e is the instantaneous ac voltage and E is a voltage level with respect to ground. The emf nomenclature is traditional for describing vacuum-tube circuits.

As seen in Figure 6-5, the grid bias voltage E_{gg} is about 2.5 times the tube cutoff voltage E_{CO}. The input drive (source) ac peak voltage v_s exceeds the tube cutoff voltage at ①, which is where plate current starts to flow. For optimum operation, the input signal is also large enough to drive the grid positive (at ②), which is where grid current starts to flow (shaded area). The current waveforms are also shown. From the geometry of the grid voltage of Figures 6-5 and 6-6, it is clear that $\phi = (\frac{1}{2})(180° - \theta_p)$. Also, e_g must rise from E_{gg} to $E_{CO} = \phi$ degrees. If, in general, the tube is biased at $E_{gg} = NE_{CO}$ where $N = 2.5$ is typical, then

$$e_g(\text{pk})\sin \phi = E_{CO}(1 - N) \qquad (6-3)$$

---

* As an example, look at the base circuits for the sync separator and especially the horizontal output stage in a TV schematic.

FIGURE 6-5 Triode voltage and current for class C operation.

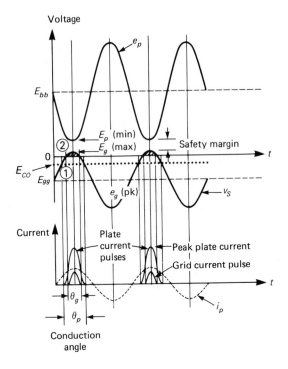

As an example, for $N = 2.5$ and $\theta_p = 120°$, Equation 6-3 yields

$$e_g(pk) = -3E_{CO} \qquad\qquad\qquad (6-4)$$

Class C power stages include an average-reading (dc) milliammeter, which is switched between the grid circuit, to monitor the input drive level, and the cathode or plate, to monitor the output power level. Because the grid is driven positive and current flows in the grid circuit, the input driver stage must supply much more power than for class A operation. The grid current pulses have an average value I_g, which is read on the dc meter in the grid circuit. The required grid drive power is difficult to analyze because of the pulsed nature of the current, but a convenient

FIGURE 6-6 Expanded view of grid voltage showing the relationship between ac grid signal and plate conduction angle.

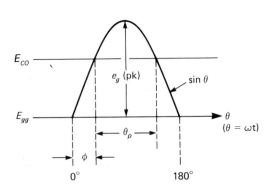

TABLE 6–1 Table of Current Ratios in Class C
Operation

Conduction Angle (degrees)	$\dfrac{i_p(\text{rms})}{i_p(\text{dc})}$	
180	1.11	
170	1.14	
160	1.17	
150	1.20	⎫
140	1.22	Typical
130	1.25	range of operation
120	1.27	⎭
110	1.29	
100	1.31	
90	1.33	

From *Communications Electronics Circuits*, 2nd ed., by J. J. DeFrance. Copyright © 1972 by Saunders College Publishing, a division of Holt, Rinehart and Winston, Inc. Reprinted by permission of Holt, Rinehart and Winston, Inc.

expression developed by H. P. Thomas in the early 1930s is given* as

$$P_g = 0.9 e_g(\text{pk}) I_g \tag{6–6}$$

where $e_g(\text{pk})$ is the peak variation of the sinusoidal grid signal.

For a given value of conduction angle, the rms value of the fundamental component of plate current can be determined by Table 6–1 and the reading of the average-reading (dc) milliammeter in the plate circuit. This allows the calculation of signal power delivered to a high-Q plate-load tuned to the fundamental frequency from

$$P_L = [e_p(\text{pk})/\sqrt{2}] i_p(\text{rms}) \tag{6–7}$$

Whereas currents of higher harmonics will flow from battery to cathode, the high-Q tank circuit provides a high impedance only at the fundamental frequency,

---

* See J. J. DeFrance, *Communications Electronics Circuits*, 2nd ed., pp. 104–105. The analysis of signals other than sinusoids and dc in linear circuits requires a return to the most general definitions. For power, the product of the voltage and current waveforms is averaged. Since most signals in electronics are periodic, the $v(t) \cdot i(t)$ product can be averaged over one cycle. Hence, the average power in a periodic signal can be determined from

$$P = \frac{1}{2\pi} \int_0^{2\pi} v(t) \cdot i(t)\, dt \tag{6–5}$$

Nonlinear signal analysis often requires numerical techniques that are beyond the scope of this text.

so that the plate voltage is sinusoidal at the fundamental frequency, as seen in Figure 6–5. The appropriate impedance for the plate to see, therefore, is

$$R_p = e_p(\text{pk})/[\sqrt{2}i_p(\text{rms})] \tag{6–8}$$

Power and Efficiency

A well-regulated power supply will hold a constant E_{bb} while supplying current pulses whose peaks can get quite large, depending on the plate conduction angle θ_p. For $\theta_p = 120°$, the plate current peaks reach 4.5 times the average supply current. For $\theta_p = 40°$, $i_p(\text{max}) = 13.4I_p(\text{dc})$. With I_p read from the plate circuit dc milliammeter, the plate supply power is found to be

$$P_{\text{dc}} = E_{bb}I_p \tag{6–9}$$

Plate dissipation P_D and efficiency η_p are determined from

$$P_D = P_{\text{dc}} - P_L \tag{6–10}$$

and

$$\eta_p = P_L/P_{\text{dc}} \tag{6–11}$$

■ ■ ■ ■

EXAMPLE 6–1

The class C vacuum-tube (VT) amplifier of Figure 6–4 has the following operating conditions: $E_{bb} = 1$ kV, $I_p = 1$ A dc, $\theta_p = 120°$, $I_g = 20$ mA, $E_{gg} = 2.5E_{CO}$, $\mu = 4.55$. Determine

1. R_g for grid bias.
2. Peak grid voltage and drive power P_g.
3. Peak plate voltage swing for no grid plate safety margin (maximum efficiency).
4. Output power.
5. Power gain in dB.
6. Power dissipated by the VT plate, and efficiency.
7. Load impedance Z_o, L, and C to tune the output with a bandwidth of 1 MHz at 10 MHz.

Solution:

1. $E_{gg} = 2.5E_{CO} = 2.5(-E_{bb}/\mu) = 2.5(-1 \text{ kV}/4.55) = 2.5(-220 \text{ V}) = -550$ V. $I_g = 20$ mA, therefore, $R_g = 550$ V/20 mA = **27.5 kΩ.**
2. From Equation 6–4, $e_g(\text{pk}) = -3E_{CO} = 3(220) = $ **660 V pk.** $P_g = 0.9e_g(\text{pk})I_g = 0.9(660)20$ mA = **11.9 W.**
3. $E_p(\text{max}) = E_{bb} - E_g(\text{max}) - E(\text{safety}) = 1000 - (660 \text{ V} - 550 \text{ V}) - 0 = $ **890 V pk.**
4. From Table 6–1 with $\theta_p = 120°$ and $I_p = 1$ A, $i_p(\text{rms}) = 1.27$ mA. $P_o = (890 \text{ V}/\sqrt{2})(1.27 \text{ A}) = $ **799 W.**
5. $A_p = P_o/P_g = 799$ W/11.9 W = 67.2. $A_p(\text{dB}) = 10 \log 67.2 = $ **18.3 dB.**
6. $P_{\text{dc}} = 1$ kV × 1 A = 1 kW. $P_D = 1000$ W − 799 W = **201 W.** $\eta_p = P_o/P_{\text{dc}} = 799/1$ kW = **79.9%.**

7. $Z_o = e_p/i_p = 890$ V pk$/(1.27\sqrt{2}$ A pk$) = $ **496 Ω.** $Q = f_0/\text{BW} = 10$, and $X_L = X_C = Z_o/Q = $ **49.6 Ω.** At 10 MHz, $L = $ **790 nH** and $C = $ **320 pF.**

6–2 ■ NEUTRALIZATION

The gain and stability of RF amplifiers can be improved by reducing uncontrolled feedback. The feedback is most often internal to the active device. The plate-to-grid interelectrode capacitance of grounded-cathode amplifiers provides a path for feedback; a similar path exists in the drain-to-gate capacitance of field-effect transistors (FETs) in a common-source amplifier configuration. Since vacuum tubes and field-effect transistors have high input and output impedances, the interelectrode feedback path must have a very high impedance to prevent loss of gain when the feedback phase is negative. When the feedback phase is positive, the amplifier will be regenerative and have a very narrow bandwidth.

By adding a parallel path of feedback from output to input, the internal device feedback can be cancelled. This technique is called *neutralization*.

Numerous circuits have been used for providing an appropriately phased current for neutralization. Two of these are shown in Figure 6–7, where the n subscript designates the device that provides a current of just the right phase and magnitude to cancel the current in C_{bc}. In the case of the split inductor circuit, L_a and L_b to ac ground forms a bridge circuit with C_n and C_{bc}. Balancing the bridge requires that

$$C_n = C_{bc}\frac{L_b}{L_a} \qquad (6\text{--}12)$$

FIGURE 6–7   Neutralization networks.

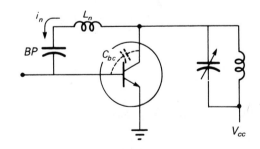

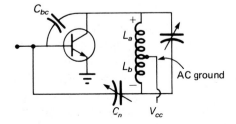

Neutralization tends to be an art as much as a science because the device's internal package reactances and the external circuit phases are not known exactly. However, a first approximation for L_n of Figure 6–7 is a value that will produce resonance with C_{bc}—that is,

$$L_n = 1/[(2\pi f)^2 C_{bc}] \qquad\qquad \text{(6–13)}$$

The result will be replacement of a low-impedance current feedback path through C_{bc} with a high-impedance parallel resonant circuit. This technique is necessarily a narrowband solution.

6–3 ■ INTEGRATED CIRCUIT, RF POWER AMPLIFIERS

Since the early 1970s, RF power modules have been in production. Their small size and reliability have made them especially useful in portable and mobile radio applications. As an example, Motorola's MHW 820 with a 12.5-V power supply will produce 20 W into 50 Ω from 806 to 870 MHz, while achieving 19 dB of gain. The data sheet is shown in Figure 6–8. This 1.4-in. (35.8-mm) × 0.7(18) × 0.35(8.9) hybrid IC is designed to be stable even for highly mismatched loads (VSWR up to 6:1 at all phase angles). Also, harmonic distortion outputs are attenuated by up to 56 dB below the carrier with internal low-pass matching networks of microstrip circuits on alumina circuit board for small size.

The schematic of Figure 6–9 is from a Motorola Applications Note (EB-92A) for a 25-W, 144–148-MHz amplifier for hand-held UHF radios. About all the user has to do with these RF power modules is to provide low-frequency decoupling networks and design an output-matching network/low-pass filter, as shown in Figure 6–9. Small internal components decouple the power supply lines for frequencies above 5 MHz, but the high gain at low frequencies makes the external decoupling networks necessary.

6–4 ■ IMPEDANCE-MATCHING NETWORKS

One of the possible surprises in power amplifier design is the realization that output impedance matching is *not* based on maximum power transfer criteria. One reason for this is the fact that matching the load to the device output impedance results in power transfer at 50% efficiency. More important perhaps are the constraints imposed by the available power supply voltage and the specified output power. To illustrate this point, suppose that a single-ended, solid-state power amplifier must deliver 26 W, and the power supply is specified to be 12.5 V— typical for mobile transmitter systems. The transistor collector voltage can *at best* swing 12.5 V pk (8.84 V rms), so that the transistor must be working into an

MOTOROLA
SEMICONDUCTORS
P.O. BOX 20912 • PHOENIX, ARIZONA 85036

MHW820-1
MHW820-2
MHW820-3

The RF Line

18/20 W — 806–950 MHz

**RF POWER
AMPLIFIER MODULE**

UHF POWER AMPLIFIER MODULE

... designed for 12.5 volt UHF power amplifier applications in industrial and commercial FM equipment operating from 806 to 950 MHz.

- MHW820-1   806–870 MHz
  MHW820-2   806–890 MHz
  MHW820-3   870–950 MHz
- Specified 12.5 Volt, UHF Characteristics
  Output Power = 20 Watts (MHW820-1,2)
                = 18 Watts (MHW820-3)
  Minimum Gain = 19 dB (MHW820-1,2)
                = 17.1 dB (MHW820-3)
  Harmonics   = −58 dBc Max
- 50 Ω Input/Output Impedances
- Guaranteed Stability and Ruggedness
- Features Three Common-Emitter Gain Stages
- Thin-Film Hybrid Construction Gives Consistent Performance and Reliability
- Gold-Metallized and Silicon Nitride-Passivated Transistor Chips
- Controllable, Stable Performance Over More Than 30 dB Range in Output Power

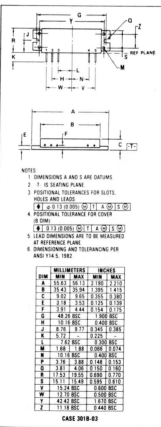

NOTES:
1. DIMENSIONS A AND S ARE DATUMS
2. -T- IS SEATING PLANE
3. POSITIONAL TOLERANCES FOR SLOTS, HOLES AND LEADS:
   | ⊕ | φ 0.13 (0.005) Ⓜ | T | A Ⓜ | S Ⓜ |
4. POSITIONAL TOLERANCE FOR COVER (B DIM):
   | ⊕ | 0.13 (0.005) Ⓜ | T | A Ⓜ | S Ⓜ |
5. LEAD DIMENSIONS ARE TO BE MEASURED AT REFERENCE PLANE.
6. DIMENSIONING AND TOLERANCING PER ANSI Y14.5, 1982.

DIM	MILLIMETERS		INCHES	
	MIN	MAX	MIN	MAX
A	55.63	56.13	2.190	2.210
B	35.43	35.94	1.395	1.415
C	9.02	9.65	0.355	0.380
E	3.18	3.53	0.125	0.139
F	3.91	4.44	0.154	0.175
G	48.26 BSC		1.900 BSC	
H	10.16 BSC		0.400 BSC	
J	8.76	9.77	0.345	0.385
K	5.72	–	0.225	
L	7.62 BSC		0.300 BSC	
M	1.68	1.88	0.066	0.074
N	10.16 BSC		0.400 BSC	
P	3.76	3.88	0.148	0.153
Q	3.81	4.06	0.150	0.160
R	17.53	19.55	0.690	0.770
S	15.11	15.49	0.595	0.610
V	15.24 BSC		0.600 BSC	
W	12.70 BSC		0.500 BSC	
Y	42.42 BSC		1.670 BSC	
Z	11.18 BSC		0.440 BSC	

CASE 301B-03

MAXIMUM RATINGS (Flange Temperature = 25°C)

Rating	Symbol	Value	Unit
DC Supply Voltages	V_{s1}, V_{s2}, V_{s3}	16	Vdc
RF Input Power ($P_{out} \leqslant 25$ W)	P_{in}	400	mW
RF Output Power ($P_{in} \leqslant 400$ mW)	P_{out}	25	W
Storage Temperature Range	T_{stg}	−30 to +100	°C
Operating Case Temperature Range	T_C	−30 to +100	°C

© MOTOROLA INC., 1984   DS5803R2

FIGURE 6–8   (pages 185–89 courtesy of Motorola, Inc.; page 1 of 5)

ELECTRICAL CHARACTERISTICS (Flange Temperature = 25°C, 50 Ω system, and $V_{s1} = V_{s2} = 12.5$ V unless otherwise noted)

Characteristic		Symbol	Min	Typ	Max	Unit
Frequency Range	MHW820-1	BW	806	—	870	MHz
	MHW820-2		806	—	890	
	MHW820-3		870	—	950	
Input Power (P_{out} = 20 W)	MHW820-1, 2	P_{in}	—	200	250	mW
(P_{out} = 18 W)	MHW820-3		—	300	350	
Power Gain (P_{out} = 20 W)	MHW820-1, 2	G_p	19	20	—	dB
(P_{out} = 18 W)	MHW820-3		17.1	17.8	—	
Efficiency (P_{out} = 20 W)	MHW820-1, 2	η	28	32	—	%
(P_{out} = 18 W)	MHW820-3		26	30	—	
Harmonic Output (P_{out} Reference = Rated P_{out})		—	—	—	−58	dBc
Input VSWR (P_{out} = Rated P_{out}, 50 Ω Reference)		—	—	—	2:1	—
Power Degradation (−30 to +80°C) (Reference P_{out} = Rated P_{out} @ T_C = 25°C)		—	—	1.2	1.7	dB
Load Mismatch Stress ($V_{s1} = V_{s2} = V_{s3}$ = 16 Vdc, P_{out} = 25 W, VSWR = 30:1, all phase angles)		—	No degradation in Power Output			
Stability (P_{in} = 0 to 250 mW, [MHW820-1, 2] or 350 mW [MHW820-3] consistent with max, P_{out} = 25 W, V_{s1} = $V_{s2} = V_{s3}$ = 10 to 16 Vdc, Load VSWR = 4:1)		All non-harmonic related spurious outputs ≥ 70 dB below the desired output signal level			
Quiescent Current (I_{s1} with no RF drive applied)		$I_{s1(q)}$	—	—	125	mA

FIGURE 1 — 806-950 MHz TEST SYSTEM DIAGRAM

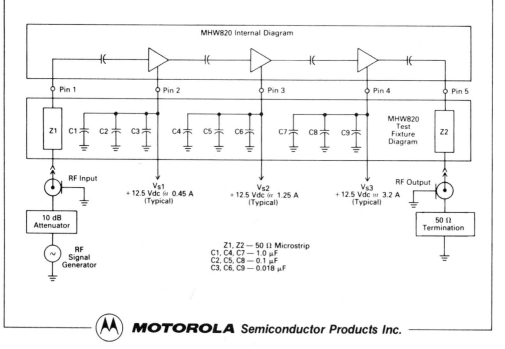

MHW820 Internal Diagram

Pin 1   Pin 2   Pin 3   Pin 4   Pin 5

Z1 C1 C2 C3   C4 C5 C6   C7 C8 C9   MHW820 Test Fixture Diagram   Z2

RF Input

V_{s1}
+ 12.5 Vdc @ 0.45 A
(Typical)

V_{s2}
+ 12.5 Vdc @ 1.25 A
(Typical)

V_{s3}
+ 12.5 Vdc @ 3.2 A
(Typical)

RF Output

10 dB Attenuator

RF Signal Generator

50 Ω Termination

Z1, Z2 — 50 Ω Microstrip
C1, C4, C7 — 1.0 μF
C2, C5, C8 — 0.1 μF
C3, C6, C9 — 0.018 μF

Ⓜ **MOTOROLA** *Semiconductor Products Inc.*

FIGURE 6-8   (page 2 of 5)

MHW820-1 ● MHW820-2 ● MHW820-3

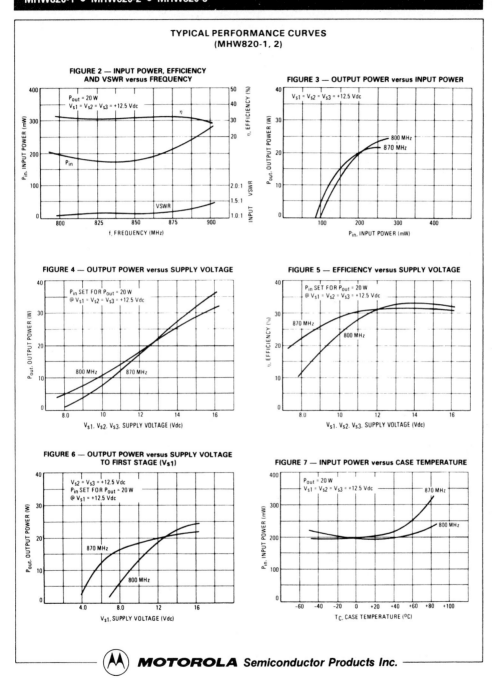

TYPICAL PERFORMANCE CURVES
(MHW820-1, 2)

FIGURE 2 — INPUT POWER, EFFICIENCY
AND VSWR versus FREQUENCY

FIGURE 3 — OUTPUT POWER versus INPUT POWER

FIGURE 4 — OUTPUT POWER versus SUPPLY VOLTAGE

FIGURE 5 — EFFICIENCY versus SUPPLY VOLTAGE

FIGURE 6 — OUTPUT POWER versus SUPPLY VOLTAGE
TO FIRST STAGE (V_{s1})

FIGURE 7 — INPUT POWER versus CASE TEMPERATURE

MOTOROLA *Semiconductor Products Inc.*

FIGURE 6–8   (page 3 of 5)

MHW820-1 ● MHW820-2 ● MHW820-3

TYPICAL PERFORMANCE CURVES
(MHW820-3)

FIGURE 8 — INPUT POWER, EFFICIENCY AND VSWR versus FREQUENCY

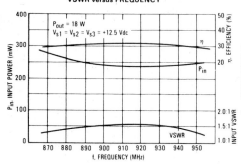

FIGURE 9 — OUTPUT POWER versus INPUT POWER

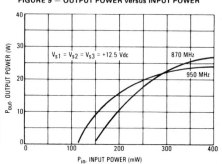

FIGURE 10 — OUTPUT POWER versus SUPPLY VOLTAGE

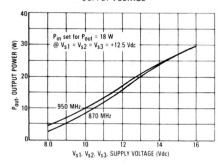

FIGURE 11 — EFFICIENCY versus SUPPLY VOLTAGE

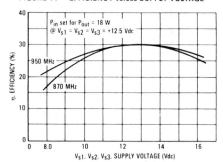

 MOTOROLA *Semiconductor Products Inc.*

FIGURE 6–8   (page 4 of 5)

APPLICATIONS INFORMATION

Nominal Operation

All electrical specifications are based on the following nominal conditions: (P_{out} = Rated, $V_{s1} = V_{s2} = V_{s3}$ = 12.5 Vdc). This module is designed to have excess gain margin with ruggedness, but operation outside the limits of the published specifications is not recommended unless prior communications regarding the intended use has been made with a factory representative.

Gain Control

This module is designed for wide range P_{out} level control. The recommended method of power output control, as shown in Figure 3 and 9, is to fix V_{s1}, V_{s2}, and V_{s3} at 12.5 Vdc and vary the input RF drive level at Pin 1.

A second method of output control is to adjust the supply voltage (V_{s1} independently or V_{s1}, V_{s2}, and V_{s3} simultaneously). However, if any of these voltages fall out of the range from 10 to 16 volts module stability cannot be guaranteed. Typical ranges of power output control using this method are shown in Figures 4, 6, and 10.

In all applications, the module output power should be limited to 25 watts.

outputs at specific operating frequencies and phase angles of input and output VSWR.

FIGURE 13 — TEST FIXTURE CONSTRUCTION
TOP VIEW

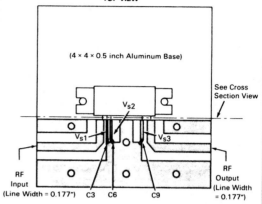

Bring capacitor leads through fiberglass board and solder to V_{s1}, V_{s2}, and V_{s3} lines as close to module as possible.
To insure optimum heat transfer from flange to heatsink, use standard 6–32 mounting screws and an adequate quantity of silicon thermal compound (e.g., Dow Corning 340). With both mounting screws finger tight, alternately torque down the screws to 4–6 inch pounds.

Load Pull

During final test, each module is "load pull" tested in a fixture having the identical decoupling network described in Figure 1. Electrical conditions are V_{s1}, V_{s2} and V_{s3} equal to 16 volts output, VSWR 30:1 and output power equal to 25 watts.

Mounting Considerations

To insure optimum heat transfer from the flange to heatsink, use standard 6–32 mounting screws and an adequate quantity of silicon thermal compound (e.g., Dow Corning 340). With both mounting screws finger tight, alternately torque down the screws to 4–6 inch pounds. The heatsink mounting surface directly beneath the module flange should be flat to within 0.002 inch to prevent fracturing of ceramic substrate material. For more information on module mounting, see EB-107.

FIGURE 12 — TEST FIXTURE ASSEMBLY

Decoupling

Due to the high gain of each of the two stages and the module size limitation, external decoupling networks require careful consideration. Pins 2, 3 and 4 are internally bypassed with 0.018 μF chip capacitors which are effective for frequencies from 5 MHz through 950 MHz. For bypassing frequencies below 5 MHz, networks equivalent to that shown in the test fixture schematic are recommended. Inadequate decoupling will result in spurious

 MOTOROLA *Semiconductor Products Inc.*

FIGURE 6–8 (page 5 of 5)

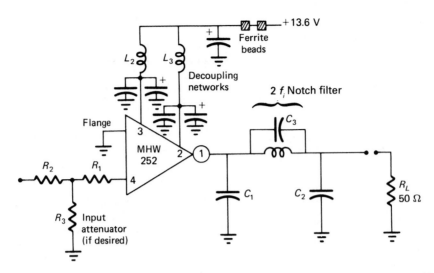

FIGURE 6–9   25-W power amplifier using an integrated circuit RF power amplifier
module.

impedance of 3 Ω ($R = 8.84^2/26 = 3$ Ω). The impedance seen by the active device
must be

$$R_c = [V_{cc} - V_{CE}(\text{sat})]^2/2P \qquad\qquad (6\text{–}14)$$

where $V_{CE}(\text{sat})$ is the transistor saturation voltage, which limits the peak signal
swing. The device break-down voltage (BV) must also be considered. For a collector swing of $V_{pk} = V_{cc}$, the collector voltage will reach $2V_{cc}$. Therefore,

$$V_{cc}(\text{max}) = \text{BV}/2 \qquad\qquad (6\text{–}15)$$

An impedance-matching system may be merely a special wideband *transformer*
which is used for *broadband matching*, or it may be one of a variety of networks
realized with inductor-capacitor combinations or transmission line structures. The
purpose of the impedance-matching network is to transform a load impedance
(which may be complex) to an impedance appropriate for optimum circuit operation. The appropriate circuit impedance is almost always complex, and the network is designed to provide a complex conjugate match over a specified bandwidth. The networks are two-pole (LC) bandpass or low-pass resonant circuits to
minimize noise and spurious signal harmonics.

To illustrate the circuit analysis concepts involved, the equations for the simple
matching network seen in Figure 6–10 are derived next. Extending these techniques to the matching of loads with complex impedances (such as interstage
matching) is a matter of increasing or decreasing the values of the series reactance.

The Q for most matching networks is typically less than 5, thus the *exact*
equations for parallel-to-series conversion are used. These conversion equations
are Equations 1–22a/1–22b and 1–37/1–38.

FIGURE 6–10 Output-matching
network.

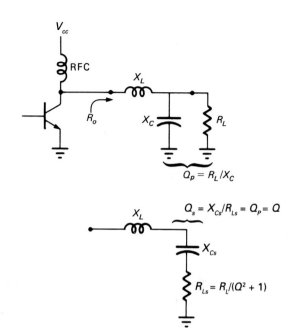

First, the parallel combination of R_L and X_C is converted to its series equiva-
lent. This is done for the series equivalent of R_L with Equation 1–22a and $R_{Ls} = R_L/(Q^2 + 1)$. The matching network will be resonant with low Q for wide bandwidth. Consequently $R_{Ls} = R_o$ to match the load to the desired value of R_o
(Equation 6–14). Therefore, $R_o = R_L/(Q^2 + 1)$, from which

$$Q = \sqrt{(R_L/R_o) - 1} \qquad (6\text{–}16)$$

X_L is the inductive reactance that brings the network to resonance; thus

$$X_L = X_{Cs} \qquad (6\text{–}17)$$

Also, $Q = X_{Cs}/R_{Ls} = X_L/R_o$, which is substituted into $Q = \sqrt{(R_L/R_o) - 1}$ to yield

$$X_L = \sqrt{R_o R_L - R_o^2} \qquad (6\text{–}18)$$

where R_o must be smaller than R_L. Finally, $X_C = R_L/Q = R_L/(X_L/R_o)$ so that

$$X_C = R_o R_L/X_L \qquad (6\text{–}19)$$

The use of this matching network is predicated on the fact that $R_o < R_L$ according
to Equation 6–18. Also, the equations given allow for the matching of pure R_L into
R_o in a way similar to the so-called quarter-wave transformer of Chapter 14 for
transmission lines. However, another component (or more of the preceding
matching sections) will be required to allow for the independent setting of the
overall circuit Q and bandwidth. One such circuit is the Pi- (π-)matching network
of Figure 6–11.

The example to follow includes a check of Equations 6–18 and 6–19 to further
reinforce the circuit analysis concepts that form the basis for deriving all matching
network equations.

FIGURE 6–11   π-matching network showing image resistance
R_I.

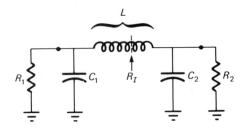

A 50-MHz, integrated circuit, power amplifier is to deliver 26 W to a 50-Ω antenna, $V_{cc} = 12.5$ V, V_{CE}(sat) $= 0$, and the output capacitance of the IC is
internally matched and can be ignored.

1. Determine the values of L and C for the L-matching network of Figure 6–10,
   which will transform the 50-Ω load to the required value of R_o in order to
   achieve the desired output power.

2. Determine the circuit bandwidth.

Solution:

1. **a.** From Equation 6–14, $R_o = (12.5 \text{ V})^2/(2 \times 26 \text{ W}) = 3$ Ω. This is less than
   R_L; therefore, Equation 6–18 is valid. $X_L = \sqrt{(3 \times 50) - 3^2} = 11.9$ Ω. At 50
   MHz, $L = 11.9/2\pi 50 \times 10^6 = $ **0.038 μH.**
   b. Equation 6–19 gives $X_C = 3 \times 50/11.9 = 12.6$ Ω, which at 50 MHz requires
   $C = 1/(2\pi f X_C) = $ **253 pF.**

2. The resonant Q is, from Equation 6–16, $\sqrt{(50/3) - 1} = 4$, and the bandwidth is
   50 MHz/4 = **12.5 MHz.**

   Check the results as follows: Starting at the output parallel combination of
   R_L and X_C, $Q = R/X = 50/12.6 = $ **4.** $R \parallel X_C = [(50)(-j12.6)]/(50 - j12.6) =$
   $630 \underline{/-90°}/(51.5 \underline{/-14°}) = 12.2 \underline{/-76°} = $ **3 − j11.9** Ω. Hence an inductance of
   $+j11.9$ Ω will bring the network to resonance and present the IC output with
   the pure 3 Ω needed to produce 26 W with a 12.5-V power supply.

Pi-Matching Network

One of the most frequently used matching networks for transmitter output stages
is the π-matching network implemented in a 3-pole low-pass filter configuration
with a resonant peak and steep skirts (high-frequency rolloff). The three reactances are determined for an impedance match between terminating impedances
R_1 and R_2 at the resonant frequency ω_0. R_1 and R_2 can be any value.

The network as illustrated in Figure 6–11 is really two L-networks in which the
separate sections transform the terminating impedances to the same *image* resistance R_I. The image resistance is lower than either R_1 or R_2. The value chosen for
R_I will determine the network Q, which is taken as the higher of Q_1 and Q_2 for
convenience. Q_1 is the Q of the parallel combination of R_1X_1 and is $R_1/X_1 = Q_1 =$

$R_1\omega_0 C_1$, where ω_0 is the center frequency of the resonant peak. Also, from Equation 6–16 for the L-section,

$$Q_1 = \sqrt{(R_1/R_I) - 1} \tag{6-20}$$

where $R_1 > R_I$ (also $R_2 > R_I$). Combining, we have

$$C_1 = (1/\omega_0 R_1)\sqrt{(R_1/R_I) - 1} \tag{6-21}$$

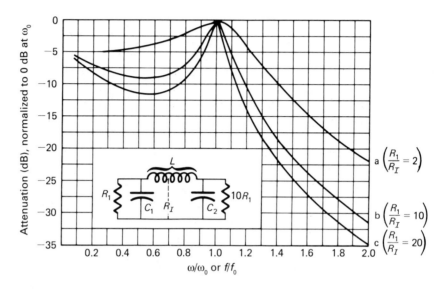

FIGURE 6–12 Unequal terminations ($R_2 = 10R_1$).

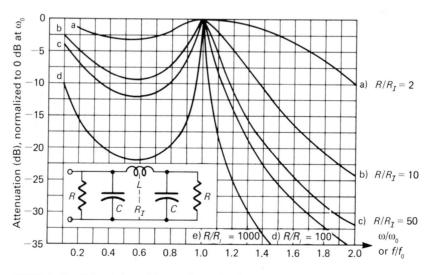

FIGURE 6–13 Selectivity with equal terminations.

Also, $$C_2 = (1/\omega_0 R_2)\sqrt{(R_2/R_I) - 1} \qquad (6\text{--}22)$$

Each part of the inductor will resonate the two L-sections, so the total inductance has a value of $(R_I/\omega_0)(Q_1 + Q_2)$, or

$$L = (R_I/\omega_0)(\sqrt{(R_1/R_I) - 1} + \sqrt{(R_2/R_I) - 1}) \qquad (6\text{--}23)$$

FIGURE 6–14 Loss: unequal terminations ($R_2 = 10R_1$).

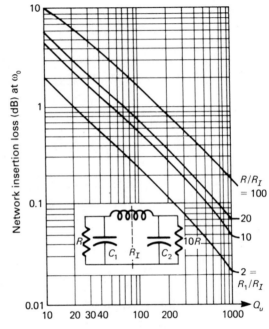

$Q_u = \omega_0\, L/r$ (r = internal resistance of inductor)

FIGURE 6–15 Loss: equal terminations ($R_1 = R_2$).

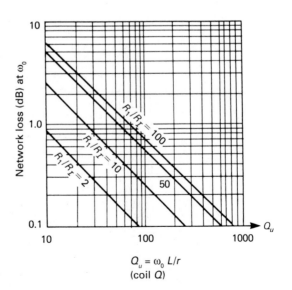

$Q_u = \omega_0\, L/r$
(coil Q)

The choice of R_1 depends on the desired output filtering requirements (selectivity). One of the sets of curves from Figure 6–12 or 6–13 is used, depending on the ratio of terminating impedances. Then R_I is chosen to achieve the desired circuit selectivity. Finally, the matching network insertion loss (attenuation at the resonant frequency) can be determined from Figure 6–14 or 6–15 as a function of the *unloaded Q* of the actual inductor used for L.

■ ■ ■ ■

EXAMPLE 6–3

Design a low-pass π-matching network to match $R_1 = 1$ kΩ and $R_2 = 10$ kΩ at $\omega_0 = 2 \times 10^6$ rad/s. Use the curves for the π-matching network and determine C_1, C_2, and L so that the second harmonic is attenuated by at least 34 dB. Also determine the Q_L and insertion loss at ω_0 if an inductor with $Q_u = 80$ is used.

Solution:

1. From the unequal terminations ($R_2 = 10R_1$) curve of Figure 6–12, the second harmonic will be down by about 34 dB with $R_1/R_I = 20$ (curve c). $R_I = R_1/20 = 50$ Ω.

2. $C_1 = \sqrt{(R_1/R_I) - 1}/(\omega_0 R_1) = \sqrt{20 - 1}/(2 \times 10^6 \times 10^3) = 4.36/(2 \times 10^9) = $ **0.0022 μF.**

3. $R_2/R_I = 10$ k$\Omega/50 = 200$. $C_2 = \sqrt{200 - 1}/(2 \times 10^6 \times 10^4) = (14.1)/(2 \times 10^{10}) = $ **705 pF.**

4. $L = R_I(Q_1 + Q_2)/\omega_0 = 50(4.36 + 14.1)/(2 \times 10^6) = $ **461.5 μH.**

5. The overall loaded Q of the network is about **14.** The low-Q section does not change this much.

6. Q_u of the inductor is given as 80. The insertion loss curves (Figure 6–14) for unequal terminations ($R_2 = 10R_1$) on the $R/R_1 = 20$ curve intersects $Q_u = 80$ at **0.9 dB.** Unfortunately, 0.9 dB is a lot of transmitter power to lose; a 10-kW transmitter would be wasting 1.9 kW. We need a much better inductor—a Q of 200 would cost us only 800 W.

Transistor and Vacuum-Tube Matching Network Equations

All the networks in this section will provide impedance matching with an independently selected Q and bandwidth. The first step in the design procedure is to write Z_i and/or Z_o for the device to be matched. This series equivalent impedance is

$$Z_T = R_T - jX_T \qquad (6\text{–}24)$$

The network equations have been solved to provide a conjugate match between $Z_T = R_T - jX_T$ (capacitive) and system impedance R_L. After the desired bandwidth and Q are determined, compute

$$A = R_T(Q^2 + 1)$$

and

$$B = \sqrt{(A/R_L) - 1} \qquad (6\text{–}25)$$

The network component values for the matching circuit of Figure 6–16 are

$$X_L = QR_T + X_T \qquad\qquad \textbf{(6–26a)}$$

$$X_{C_1} = A/(Q - B) \qquad\qquad \textbf{(6–26b)}$$

$$X_{C_2} = BR_L \qquad\qquad \textbf{(6–26c)}$$

Another excellent matching network is the T network of Figure 6–17. This can be used when R_T is greater or less than R_L. The equations are

$$X_{L_1} = QR_T + X_T \qquad\qquad \textbf{(6–27a)}$$

$$X_{L_2} = BR_L \qquad\qquad \textbf{(6–27b)}$$

$$X_C = A/(Q + B) \qquad\qquad \textbf{(6–27c)}$$

For vacuum tubes where device impedances are greater than 50 Ω, the π network of Figure 6–18 is often used; note that R_T and C_T are in a parallel arrangement. After C_1 is computed, C_T is subtracted in order to determine the appropriate external shunting capacitance.

$$X_{C_1} = R_T/Q, \qquad R_T > R_L \qquad\qquad \textbf{(6–28a)}$$

$$X_{C_2} = R_L \sqrt{\frac{R_T/R_L}{(Q^2 + 1) - (R_T/R_L)}} \qquad\qquad \textbf{(6–28b)}$$

$$X_L = \frac{QR_T - (R_T R_L / X_{C_2})}{Q^2 + 1} \qquad\qquad \textbf{(6–28c)}$$

FIGURE 6–16   Matching network.

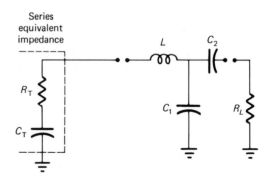

FIGURE 6–17   T matching network.

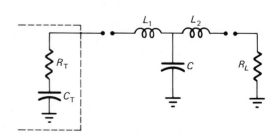

FIGURE 6–18 π-matching network and complex impedance to be matched.

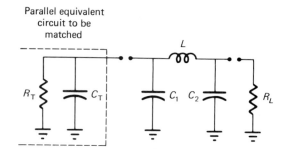

6–5 ■ AM MODULATOR CIRCUITS

Amplitude modulation is produced when the gain of the RF carrier amplifier is varied in accordance with the input information signal. When high output power is required, an efficient class C amplifier is used as the modulator. All-solid-state transmitters exceeding 10 kW are available. When tens of kilowatts of output power are required, the output devices are (liquid-cooled) vacuum tubes.

Solid-State Modulator

Solid-state AM modulators are usually biased class C, and the output RF signal amplitude is modulated by causing the collector supply voltage to vary with the modulation signal. As illustrated in Figure 6–19, the collector voltage is varied by transformer-coupling (T_A) the audio modulation signal in series with V_{cc}. The RF choke (RFC) and RF-bypass capacitor (RF-BP)* isolate the carrier from the audio section and power supply line.

Impedance-matching network L_1-C_3-L_2-C_4 maintains a constant impedance at the collector, but the varying supply voltage effectively shifts this (constant) load-line along the V_{ce} axis. For sinusoidal modulation, the collector supply voltage is

$$V_{ce} = V_{cc} + V_{pk} \sin 2\pi f_m t$$
$$= V_{cc}(1 + m \sin 2\pi f_m t) \qquad \textbf{(6–29)}$$

where $m = V_{pk}/V_{cc}$ is the modulation index.

If 100% modulation ($m = 1$) is to be achieved, the peak audio voltage coupled into the T_A secondary must equal the supply voltage:

$$V_{pk}(\text{audio}) = V_{cc} \qquad \textbf{(6–30)}$$

If the RF amplifier of Figure 6–19 had a constant supply voltage (no modulation), the collector voltage would be V_{cc}. With RF input, the collector voltage can

---

* The symbol shown for the RF-BP indicates a high-capacitance feed-through sometimes referred to as a *filtercon* (filter/connector). This device is screwed or soldered into a metal separator or wall (*bulkhead*) for RF isolation.

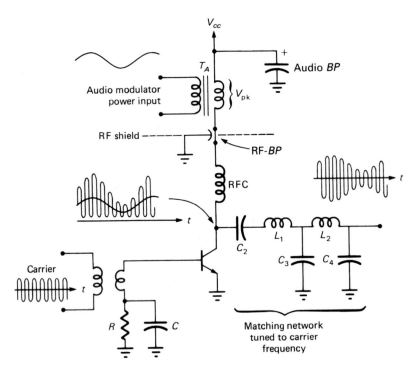

FIGURE 6-19   Transistor AM modulator.

swing almost to ground (less V_{ce} sat) on the negative swing and to $2V_{cc}$ on the positive swing, thereby maintaining an average V_{cc}. For 100% modulation, how-ever, the peak supply voltage reaches $2V_{cc}$, and therefore the positive peak RF swing can reach $4V_{cc}$,* as seen in the collector waveform of Figure 6–19. As a result, the choice of transistor for the AM modulator must include high break-down voltage. In particular,

$$BV \geq 4V_{cc} \qquad\qquad (6–31)$$

Since an amplitude-modulated signal must maintain amplitude linearity, any amplifier following the modulator must be class A or, as is usually the case in transmitters, class B push-pull. Figure 6–20 shows an all-solid-state transmitter output with a tuned push-pull amplifier final. R_1 can be adjusted to bias Q_3 and Q_4 for class A or class B (actually, almost B, called AB) operation. Sometimes R_1 is replaced by a diode that will help track-out temperature variations of V_{BE} for Q_3 and Q_4. Diodes D_1 and D_2 form a network to allow the driver Q_1 to be modulated on positive modulation peaks when V_c exceeds V_{cc}.

---

* The extra drive power required on the positive peaks also can be achieved by collector-modulating the driver amplifier on the positive part of the modulation cycle. (See the network formed by D_1 and D_2 in Figure 6–20.)

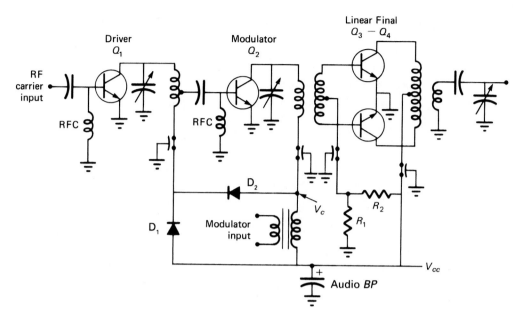

FIGURE 6–20 All-solid-state AM modulator/transmitter.

Vacuum-Tube AM Modulator

High-power TV, AM, and FM broadcast stations transmit 50 kW or more. These HF transmitters require the high plate voltage capability that only vacuum tube technology can provide at this time. The output voltage waveforms and breakdown voltages are ideally the same as for the solid-state AM transmitter. The high-power requirements of broadcast transmitters place very difficult specifications on the audio output stage in the modulator.

If at full modulation the transmitter output is P_t, then from Chapter 5, $P_t = [1 + (m^2/2)]P_c$ and for $m = 1$,

$$P_t = 1.5P_c \qquad (6\text{--}32)$$

where P_c is the unmodulated carrier power. The *peak envelope power* at $m = 1$ is

$$\text{PEP} = 2.66P_t \qquad (6\text{--}33)$$

The total sideband power is $P_{\text{SB}} = (m^2/2)P_c = P_c/2$ for 100% modulation. If the RF amplifier plate efficiency is η_p, then the power supply must deliver $P_{\text{dc}} = P_c/\eta_p$, and the audio system must add $P_c/2\eta_p$ in order to achieve 100% modulation. Hence, the audio power requirement is

$$P_{\text{audio(max)}} = P_{\text{dc}}/2 \qquad (6\text{--}34)$$

The ac impedance that the audio transformer secondary sees is the same that the modulator power supply sees:

$$R_{\text{audio}} = E_{bb}/I_p \qquad (6\text{--}35)$$

■ ■ ■ ■
**EXAMPLE
6–4**

An AM transmitter must transmit 29.93 kW PEP. It maintains an average plate current of 1 A and a 75% plate efficiency. Determine

1. The average output power at 100% modulation.
2. The unmodulated carrier power.
3. The audio power and secondary impedance requirements.
4. The plate supply voltage.

Solution:

1. P_t = 29.93 kW/2.66 = **11.25 kW**
2. P_c = 11.25 kW/1.5 = **7.5 kW**
3. P_{dc} = 7.5 kW/0.75 = 10 kW. P_{audio}(max) = **5 kW.** However, the modulation power transmitted is only 5 kW × 0.75 = 3.75 kW.
4. E_{bb} = P_{dc}/I_{dc} = 10 kW/1 A = **10 kV.** The audio transformer will have a secondary impedance of 10 kV/1 A = **10 kΩ.**

FIGURE 6–21 Grid modulator signals.

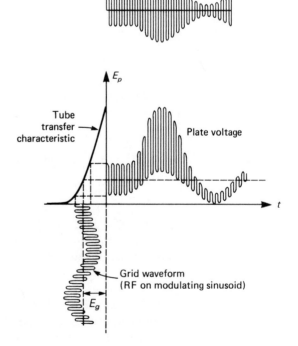

Grid Modulation

The high audio power requirements in plate-modulated AM systems can be greatly reduced by using grid modulation. Grid modulation can be accomplished by transformer-coupling the modulation signal into the grid bias circuit (or base circuit of a bipolar transistor). The advantage of grid modulation is that modulating the RF carrier at the class C amplifier input requires a good deal less modulating power because of the gain of the amplifier. Therefore, grid modulation is called *low-level modulation*, as opposed to high-level plate modulators. This modulator power savings, however, has drawbacks: the peaks of the grid drive signal now have an amplitude variation (see Figure 6–21).

Grid modulation requires very critical adjustments to circuits that still are not as linear as plate modulators. Consequently, they are not often used, with the notable exception of TV broadcasting. The television picture (video) is broadcast as an AM signal with a bandwidth of 4.2 MHz. Because of the difficulty of producing high-power modulation amplifiers with greater than 4 MHz of bandwidth, grid modulation is used.

■ PROBLEMS

1. A power amplifier has 1 kW of RF input drive and 50 kW of output power.
   a. Where does the 50 kW of power come from?
   b. What is the power gain in decibels?

2. A power transistor delivers 50 W at 10 MHz with efficiency η. How much power is wasted as heat for each value of η?
   a. $\eta = 25\%$
   b. $\eta = 50\%$
   c. $\eta = 78.5\%$
   d. $\eta = 85\%$

3. What is the theoretical efficiency for the following types of amplifier operation?
   a. Class A RC-coupled
   b. Class A transformer-coupled
   c. Class B push-pull
   d. Class C (no output power) and class C (practical high-output power operation)

4. Why are class C amplifiers not used for audio power amplification?

5. If the output current of a class C amplifier consists of pulses, how can a continuous sinusoid be transmitted? (This is called *continuous wave* [CW] transmission.)

6. If 50 kW is to be transmitted, why use class C?

7. **a.** How much power must a power-supply supply if a power-supply is supplying power to an 85%, 20-kW power amplifier?
   b. Repeat **a** for class A transformer-coupled operation.

8. Explain the process by which an electron is liberated from the metal cathode and accelerates across the vacuum space of a vacuum tube.

9. A 3CX20000A7 is specified to operate with a heater-element voltage of 6.3 V and 160 A of current at 60 Hz. Determine the heater power for this 25-kW, VHF, FM-transmitter tube.

10. a. The plate voltage of a vacuum tube is 10 kV. If the amplification factor for the tube is 10, what is the cutoff voltage?
   b. If a 5-kΩ load is used and the dynamic plate impedance is $r_p = 50$ kΩ, how much voltage gain can be expected when this tube is used in a class A amplifier?

11. For the class-C RF amplifier shown in Figure 6–22, use four words or fewer to answer each question.
   a. What type of bias is in the grid circuit?
   b. Is plate bias shunt- or series-fed?
   c. What are R_1 and C_1 for?
   d. What is C_2 for?
   e. What is C_3 for?
   f. What is RFC for? What does it do?
   g. Where does the power come from to bias the grid? The plate?
   h. Trace on the schematic the complete RF path of current for the input grid circuit at the instant that the secondary of transformer T_1 is + (as shown). Now trace the plate circuit RF current for this same instant. (Draw complete closed loops of current; ignore current through C_2.)

12. The circuit of Problem 11 has the grid voltage shown in Figure 6–23. If the cutoff voltage is $E_{CO} = -20$ V,
   a. Determine the conduction angle of the plate current.
   b. Sketch the plate and grid instantaneous current waveforms.
   c. If R_1 in the grid circuit is 160 kΩ, what is the average current through it?
   d. If no safety margin is used for the plate-to-grid voltage, calculate the maximum plate voltage swing for optimum operation.
   e. What tank impedance is required to get 200 W output?
   f. The dc plate current is 430 mA. Determine the plate dissipation and efficiency.

13. A tetrode (four electrodes plus heater) is shown in a class C power amplifier in Figure 6–24. With a grid-driving power of 20 W, an oscilloscope connected from grid to ground shows the signal of Figure 6–24b.

FIGURE 6–22

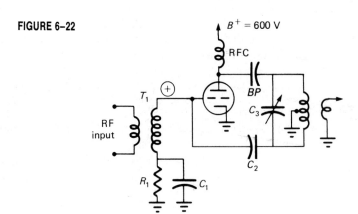

FIGURE 6-23

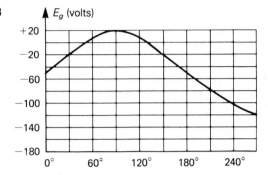

a. Sketch the schematic and the complete dc grid current path. (Show a complete loop.)
b. How much should I_g, the dc grid current, be?
c. What must the value of R_1 be?
d. Calculate the cutoff voltage if $\mu = 10$.
e. If no safety margin is used for the plate-to-grid voltages, calculate the maximum plate voltage swing for optimum operation.
f. What tank impedance is required to get 300 W output?
g. If the required bandwidth is 1 MHz, calculate the required tank inductance ($f_0 = 10$ MHz).
h. If this stage is driving a 50-Ω antenna, what transformer turns ratio is required?
i. If the dc plate current is 430 mA, determine plate dissipation and efficiency.
j. What is the purpose of C_n?

14. A class C vacuum-tube amplifier is operated with a bias of -125 V, $E_{bb} = 600$ V dc, and is driven from a source with 130 V rms. Find
    a. Maximum value of grid swing (above ground).
    b. Optimum plate voltage swing (in rms).

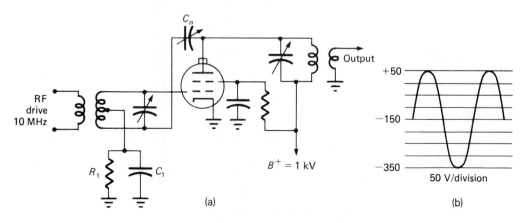

(a)

(b)

FIGURE 6-24

15. If I_p = 200 mA dc and i_p = 350 mA pk for the circuit of Problem 14, determine
   a. Power supplied from the plate power supply.
   b. RF output power.
   c. Plate dissipation.
   d. Efficiency.

16. An RF transistor has a *breakdown* voltage of 56 V, and 100 W must be transmitted. Find
   a. Maximum usable V_{cc};
   b. R'_L (optimum collector load).

17. Show how Figure 6–9 would change if the MHW 820 power amplifier module is used.

18. Determine the L and C values to match a 50-Ω load to a 3.9-Ω amplifier output impedance at 1.6 MHz. Use a series inductor and shunt capacitor to form the low-pass L-matching network.

19. Determine the LC values for a matching network used to match 50 Ω to an impedance of $12 + j12$ Ω at 27 MHz. (*Hint:* Use a capacitor in shunt with the 50 Ω, and adjust the value of the series inductance to achieve the additional 12 Ω reactance.)

20. Figure 6–25 shows an RLC circuit driven from a 1-kΩ-impedance current source.
   a. Determine the equivalent series circuit impedance for the 1-kΩ, 2-μF combination. Sketch the entire (series) network.
   b. Determine the value of L to tune the entire circuit at ω = 1000 rad/s.
   c. Calculate the overall effective Q of the resonant network.
   d. Is this a matching network?
   e. Compute the voltage across the 50-Ω resistor if the input current source is 1A at 1000 rad/s.

21. Use a software package, or write a program and use it, to solve Problem 20e.

22. a. Use circuit analysis to determine the magnitude and angle of the complex impedance Z indicated in Figure 6–26.
   b. What impedance can be matched to 50 Ω using this low-pass L-matching network?

23. a. Design an L-matching network like Figure 6–26 to match a 1-kΩ source to a 50-Ω load at 10 M rad/s.
   b. Show by circuit analysis that the result in part **a** is valid.
   c. Show by computer simulation that the circuit is resonant at 10 M rad/s.

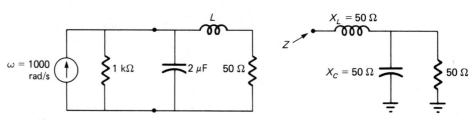

FIGURE 6–25   **FIGURE 6–26**

24. Design a π-matching network to match a 2.4-W, solid-state power amplifier (V_{cc} = 12 V) to a 300-Ω antenna. (The second harmonic output must be more than 20 dB below the 1.59-MHz carrier—that is, −20 dBc.)

25. Use the matching network of Figure 6–16 to match a 50-W solid-state power amplifier (SSPA) to a 50-Ω antenna at 50 MHz with a Q of 10. V_{cc} = 28 V and the SSPA output includes 180 pF of shunt capacitance.

26. Write a high-level-language computer program for solving the following matching network problems of Section 6–4:
   a. The L-matching network (Figure 6–10)
   b. Pi-matching network
   c. Modified-T network (Figure 6–16)
   d. T network (Figure 6–17)
   e. Pi-matching a complex impedance to a real load (Figure 6–18)

27. Figure 6–27 shows a transformer-coupled, class C tuned RF amplifier used as a high-level amplitude modulator (plate).
   a. What peak-to-peak voltage is required of v_A to produce 100% AM?
   b. What bandwidth is required to transmit the modulated signal?
   c. What value must C_2 be in order to tune the RF output?
   d. What is the effective Q of the plate-tuned circuit?
   e. The RFC (a) passes audio but not RF, (b) passes RF but not audio, (c) passes dc only, (d) increases RF gain. (Choose one answer.)

FIGURE 6–27

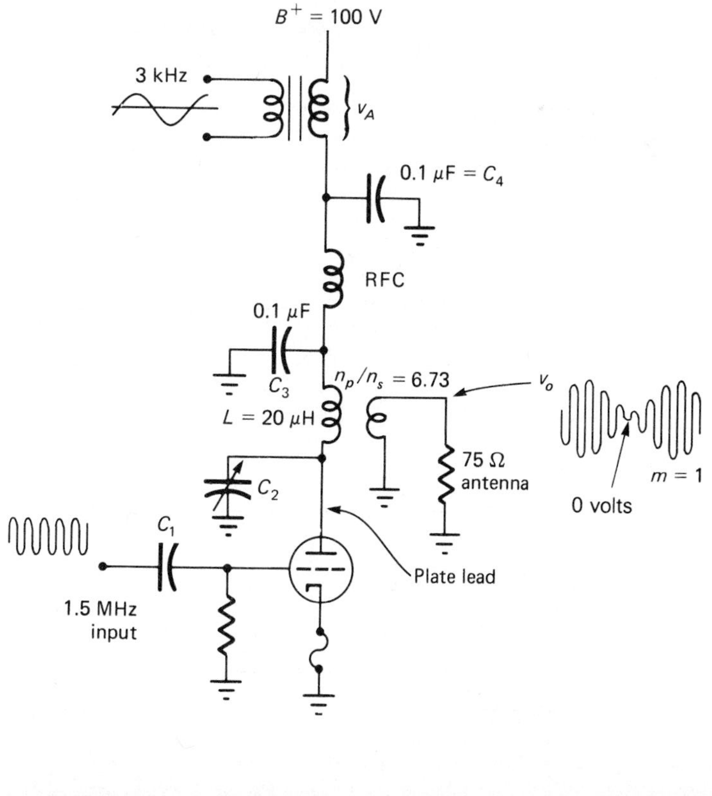

28. An 85% efficient class C amplifier is used as an AM, high-level modulator. The supply voltage is 5 kV and bias current 1.5 A.

   a. Calculate the low-frequency (dc and audio) impedance of the amplifier.

   b. Determine the maximum RF output power.

   c. How much is the total sideband power in a 90% AM output signal?

   d. For a 95% efficient audio transformer, how many watts must the audio system provide to the modulation transformer?

   e. If the audio output stage is class B push-pull and 87% efficient (excluding the transformer), how much power must the dc power supply deliver to get the 90% AM?

6–6 ■ COMPUTER ANALYSIS

Using the computer for communications system and circuit analysis has become a virtual necessity in education and industry. What follows are examples of programs and/or readily available simulation software for solving problems of the type covered in this chapter. The program for Chapter 6 is as follows:

- **Transmitter-matching network:** An impedance-matching network was designed using the techniques of Section 6–4. A 1000-Ω source is matched to a 50-Ω load at 1.59 MHz. The results of a MicroCAP-II simulation of the circuit are shown in Figure 6–28. In particular, the circuit schematic, the Netlist file, and the phase and frequency response are shown.

Micro-CAP II Student Version

XMTRMATC NETLIST

No.	Type	A	B	C	D	Parameter
1	VSIN		0		1	1
2	RESISTOR		1		2	1000
5	RESISTOR		3		0	50
7	INDUCTOR		2		3	21.8UH
8	CAPACITOR		2		0	236PF

Library Elements Used

1

P0= 1000000	P1= .0005	P2= 0	P3= 0
P4= 1	P5= 0	P6= 0	P7= 0

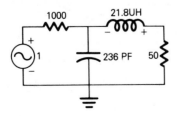

FIGURE 6–28a

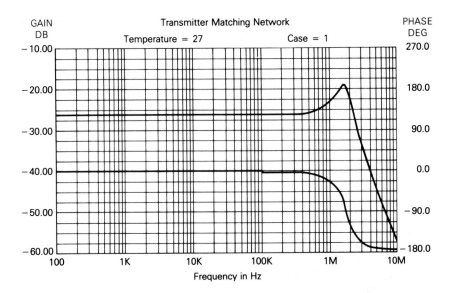

Frequency =   100.00000E+05 Hz   Gain = −57.185 Db
Phase angle = −175.665   Degrees   Group delay =   0.00000E+00
Gain slope = −124.20910E−01 Db/Oct   Peak gain = −19.057 Db/F = 159.75300E+04

FIGURE 6–28b

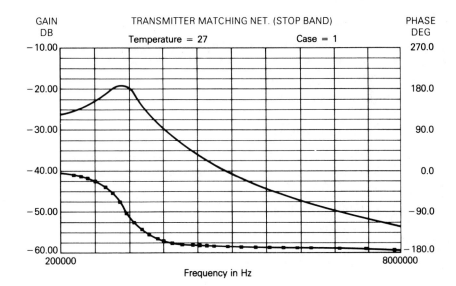

Frequency =   800.00000E+04 Hz   Gain = −53.192 Db
Phase angle = −174.501   Degrees   Group delay =   156.05900E−11
Gain slope = −125.46180E−01 Db/Oct.   Peak gain = −19.068 Db/F = 158.00000E+04

FIGURE 6–28c

7

Receiver Circuits

■ INTRODUCTION

Chapter 5 included introductory discussions on many of the basic concepts of high-frequency receiver systems. Receiver system topics included in Chapter 5 are as follows:

- Information bandwidth
- Multiplexing
- Distortion (see Chapter 3 also)
- Noise, noise figure, and S/N
- Modulation index, power, and bandwidth for AM transmissions

FIGURE 7–1   AM receiver block diagram.

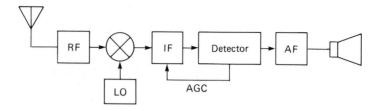

- Superheterodyning (mixing), image response, and RFI, including adjacent channel rejection
- Selectivity, sensitivity, and threshold
- Dynamic range and AGC, including delayed AGC
- AM demodulation

Once the requirements are established for a given system, the circuit performance parameters can be defined and the circuits designed. Some circuit design and analysis techniques needed for receivers have already been developed. These are found in Chapter 1 (tuned circuits, small-signal amplifiers, and transformer coupling), Chapter 2 (oscillators), Chapter 4 (circuit noise analysis), Chapter 5 (AM detectors), and Chapter 6 (impedance-matching circuits).

Following a discussion of low-noise RF amplifiers, circuit analysis and design for the other blocks of the AM receiver of Figure 7–1 will be developed. Receiver circuits specific to modulations other than AM will be covered in appropriate chapters to follow; for example, limiters, FM discriminators, and phase detectors are covered in Chapters 9 and 10. ■

7–1 ■ LOW-NOISE AMPLIFIERS (LNA)

High-performance receivers such as those used for satellite communications require low-noise amplifiers (LNAs) in the input stages to optimize the most critical performance parameter, the noise figure. As seen in the system noise figure equation, Equation 4–15, the first stage of the receiver system is the most important. The first two terms in this equation,

$$NR = NR_1 + (NR_2 - 1)/A_{P1}$$

show that the first stage should have low noise and high gain. Incidentally, the power losses ahead of the first amplifier, such as those due to lossy filters or transmission lines from the antenna, add directly to the NR_1 term. LNAs include two or more stages of amplification, along with matching networks to optimize their performance.

The unfortunate trade-off in low-noise amplifier design is between low noise and high gain because the minimum system noise figure requires that the noise added by the first amplifier circuit—over and above what comes in from the antenna—be as low as possible, whereas the first stage gain should be as high as

FIGURE 7–2 Interdigitated (interleaved finger) geometry of microwave transistor.

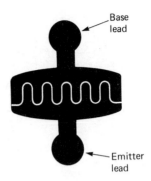

possible. The best compromise is usually achieved by optimizing the first stage for minimum noise at moderate gain and the second stage for higher gain and moderate noise.

The noise sources in RF bipolar transistor amplifiers are shot noise in the emitter-base and base-collector junctions, thermal noise generated by the base-spreading resistance, and the circuit bias resistor noise. Bias resistors can be isolated by transformer coupling at the input. The circuit is identical to the bias-coupling scheme used for IF amplifiers.

Low-noise transistors are manufactured with special attention to reducing the resistance in the base region. One method is to make the base region as narrow as possible; this brings the added benefit of increasing the dc beta of the transistor. Another technique used in RF and microwave transistors is to diffuse multiple-emitter sites, bringing them off the chip with separate leads bonded in parallel to a low-inductance emitter ribbon. The most often used technique for producing a multiple-emitter transistor chip, however, is the geometry illustrated in Figure 7–2, called *interdigitated* or *interleaved fingers*. If N is the number of emitter sites (7 in the 2-GHz transistor of Figure 7–2), then the total base resistance is $R_B = r_b/N$.*

Another thing to be aware of when looking for a low-noise transistor with the right price tag: There is an optimum generator impedance with which to drive the amplifier, and the optimum impedance is a function of frequency. The variation of noise figure with frequency is illustrated in Figure 7–3. This curve shows a "plateau" region for $f \ll f_\alpha$ beyond which the device noise increases, reaching a 6-dB/octave slope. For UHF and below, most transistors operate in the plateau region up to f_c, whereas microwave transistors operate into the region a couple of octaves above f_c. The device specification sheet should include an optimum impedance versus frequency plot in the form of a Smith chart display. The Smith chart is discussed in Chapter 14.

---

* An excellent discussion of high-frequency transistor design, as well as other information for microwave amplifiers and oscillators, can be found in a Microwave Associates Bulletin No. 5210 entitled *Transistor Designers' Guide*, 1978.

FIGURE 7–3 Noise figure variation with frequency.

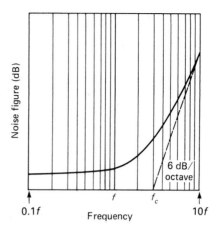

Amplifier Dynamic Range

When optimized for LNA service, the amplifier dynamic range may be limited.
Bipolar transistors make good low-noise devices, but they are more limited than
field-effect transistors in dynamic range.

The dynamic range of an amplifier is loosely defined as the input power range
over which the amplifier is useful. For linear small-signal amplifiers, the useful
range has a low-power limit determined by noise, the *noise floor*. An RF amplifier
can still be useful if the input signal power is lower than the noise power,* but the
noise must be minimized for maximum usefulness.

The high-power limit depends on whether the amplifier is to operate on a single
frequency signal or amplify more than one signal simultaneously. For one signal,
the *1-dB compression point* is usually specified as an upper limit of usefulness for
linear operation. This is shown in Figure 7–4, where the linear gain drops off at the
onset of amplifier saturation.

The linear gain and 1-dB compression point for an amplifier are determined by
plotting the output power versus input and extending the low-level linear region
with a straightedge. The linear gain is found from any point on the straight line.
The 1-dB compression point is given as the output power when the actual amplifier response is 1 dB less than the ideal (straight-line extension) response. It is
often measured directly as the point at which a 10-dB input increase results in a
9-dB output increase.

Third-Order Intercept Point

When two signals (or more) at frequencies f_1 and f_2 are amplified, nonlinearities in
the amplifier will cause second- and third-order mixer-type products to be generated. Second-order products ($2f_1$, $2f_2$, $f_1 + f_2$, and $f_1 - f_2$) are of great concern
only in broadband systems (such as cable TV [CATV]) because these products

* This is not an unusual operating condition for wideband front ends. Noise power is reduced in the
narrower IF bandwidth, but signal power remains high. See ''Bandwidth Improvement'' in Chapter 5.

FIGURE 7-4 Amplifier saturation and third-order distortion characteristics. The 1-dB compression and third-order intercept points are shown.

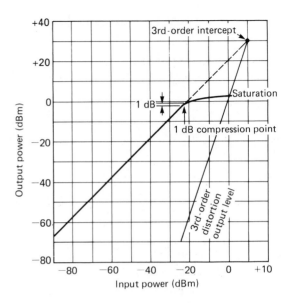

usually fall outside the narrow bandwidth of most systems. The third-order products $(2f_1 + f_2, 2f_1 - f_2, 2f_2 - f_1,$ and $2f_2 + f_1)$* usually fall inside the system bandwidth and cause a distortion called *intermodulation distortion* (IMD). Intermodulation distortion is an especially acute problem in radar systems because the return signal usually has a slightly different frequency than the transmitted signal with which it is compared for range (timing-difference) measurements. The frequency difference is due to doppler effects. Second- and third-order distortion presents major engineering challenges in CATV systems where the beats between various combinations of the (typically) 40–80 RF carriers produce what is called composite second-order (CSO) and composite third-order "triple" beats (CTB). IMD is also a critical design consideration in CATV systems because of the creation of *crosstalk* between channels.

MOSFETs and MESFETs have better dynamic range than bipolar transistors; MOSFETs have second-order (square law) nonlinearities, whereas the bipolars have higher-order nonlinearities associated with their forward- (and reverse-) biased junctions.

■ ■ ■ ■

EXAMPLE 7-1

Determine the gain, 1-dB compression point, and third-order intercept for the LNA of Figure 7–4.

Solution:

1. For an input of −60 dBm, the output is approximately −37 dBm. Hence, the gain is −37 dBm − (−60 dBm) = **+23 dB.**

---

* See the last paragraph of "General Nonlinearity" for more analysis and a cure for third-order products.

2. The 23 dB of gain remains linear until an input of about -22 dBm; 1 dB of gain compression occurs for an input of about **-20 dBm.**

3. The third-order intercept is **$+30$ dBm** (1 W). Notice that this amplifier never outputs $+30$ dBm. The intercept point is merely a point of comparison or a linearity figure of merit for linear amplifiers.

7–2 ■ MIXERS

As stated in the receiver systems discussion of Chapter 5, virtually every high-frequency receiver uses a mixer to *down-convert* the received RF signal to an intermediate frequency (IF) signal. A mixer in RF systems always refers to a circuit with a nonlinear component that causes sum and difference frequencies of the input signals to be generated. Transmitters often use the sum frequency to *up-convert* the carrier to a higher frequency. Audio systems operators refer to mixing two sound tracks. This is *not* a mixer process: no sum and difference frequencies are generated, and no nonlinear components are involved in the circuit. Rather, it is a linear addition of two signals so that the two sound tracks are heard simultaneously. Figure 7–5 gives a comparison for linear combining (addition or summation) and mixing.

An RF mixer requires a nonlinear circuit component. Two broad categories cover the range of nonlinearities: the general nonlinearity that is expressed mathematically by a power series and the nonlinearity produced in switching or sampling mixers.

General Nonlinearity

The output versus input nonlinearity of any device can be expressed mathematically by a power series such as

$$i_o = I_o + av_i + bv_i^2 + cv_i^3 + \cdots + nv_i^n \tag{7-1}$$

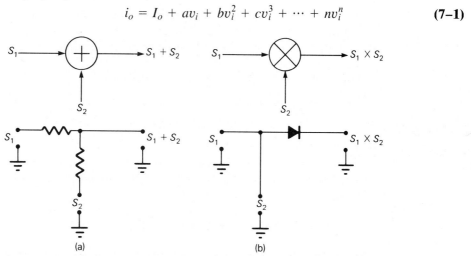

FIGURE 7–5 Summing (a) versus mixing (b).

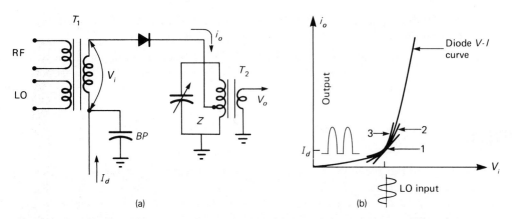

FIGURE 7–6 Mixer circuit (a) and diode characteristic curve (b) showing nonlinear input-output relationship.

illustrated in Figure 7–6. Part (a) shows a mixer circuit with the LO and RF voltages transformer-coupled to a high-frequency diode. The currents produced in the diode produce an output voltage $v_o \propto i_o Z$ in the tuned-output circuit impedance Z.

Figure 7–6b shows the nonlinear input-output V-I curve of the diode. I_d is the bias current developed for the purpose of placing the diode at the best operating point for optimum mixer performance. Curves 1, 2, and 3 show the linear (av_i), second-order (bv_i^2), and third-order (cv_i^3) nonlinearities at the bias point I_d. The constant a is the slope of the i_o curve at the bias point; similarly, constants b and c are scaling factors for the first two nonlinearities.

For a simple second-order nonlinear device such as a FET,

$$i_o = I_d + av_i + bv_i^2 \tag{7–2}$$

The inputs v_{LO} and v_{RF} are linearly added by transformer T_1 to produce $v_i = v_{LO} + v_{RF} = \cos \omega_{LO}t + A \cos \omega_{RF}t$. The magnitude of the LO signal is much greater than the RF and is represented as unity because its primary function is to drive the diode into nonlinear operation.

Substituting for v_i in Equation 7–2 gives

$$\begin{aligned} i_o &= I_d + a \cos \omega_{LO}t + aA \cos \omega_{RF}t + b(\cos \omega_{LO}t + A \cos \omega_{RF}t)^2 \\ &= I_d + a \cos \omega_{LO}t + aA \cos \omega_{RF}t + b \cos^2\omega_{LO}t \\ &\quad + 2bA \cos \omega_{LO}t \cos \omega_{RF}t + bA^2\cos^2\omega_{RF}t \end{aligned} \tag{7–3}$$

It will be helpful to analyze each term of Equation 7–3. The fourth term, which is a result of the second-order (square law) nonlinearity, can be expanded by a trigonometric identity as

$$b \cos^2\omega_{LO}t = \frac{b}{2}(1 + \cos 2\omega_{LO}t) = \frac{b}{2} + \frac{b}{2} \cos 2\omega_{LO}t$$

This shows that some dc and LO second-harmonic currents are produced by distortion. Likewise the last term of Equation 7–3 will yield dc and RF second-harmonic currents. These, together with the scaled input current (terms 2 and 3 of Equation 7–3), yield

$$i_o = I_d + \text{(input signal and second-harmonic current)}$$
$$+ 2bA \cos \omega_{LO}t \cos \omega_{RF}t \qquad (7\text{–}4)$$

where $\qquad\qquad I_d = (I_o + b/2 + bA^2/2)$

The last term of Equation 7–4 is the mixer product of interest,

$$2bA \cos \omega_{LO}t \times \cos \omega_{RF}t$$

Indeed, the mixer derives its circuit symbol from this multiplication. By trigonometric identity,

$$2bA \cos \omega_{LO}t \times \cos \omega_{RF}t = bA \cos(\omega_{LO} + \omega_{RF})t + bA \cos(\omega_{LO} - \omega_{RF})t$$

which shows that the product of two input sinusoidal signals will yield sum and difference frequency currents with amplitudes of one-half of the product term. Replacing the product in Equation 7–4 with this result gives the total circuit current i_o. If the parallel LC circuit of Figure 7–6 is tuned to the difference frequency $\omega_{LO} - \omega_{RF} = 2\pi(f_{LO} - f_{RF})$, then the output voltage developed across the high-impedance tank will be proportional to the RF input voltage but will have an intermediate frequency of $f_{IF} = f_{LO} - f_{RF}$.

It is important to remember that the preceding calculations were limited to a second-order nonlinearity. If the cv^3 of Equation 7–1 had been included, then *third-order intermodulation products*, such as those discussed under ''Third-Order Intercept Point,'' would be generated. The higher-power terms may be treated as a compound of second-power terms. For example,

$$cv^3 = c[(v)^2 \times (v)] = c[(v_{LO} + v_{RF})^2(v_{LO} + v_{RF})]$$

Hence, $\qquad\qquad cv^3 = c[(v_{LO}^2 + 2v_{LO}V_{RF} + v_{RF}^2)(v_{LO} + v_{RF})] \qquad (7\text{–}5)$

The frequencies produced are the triple combinations of $\pm f \pm f \pm f$, where each of the fs can be f_{LO} or f_{RF}. One technique for minimizing the pesky third-order products is to use a *double-balanced mixer*, which has excellent isolation between RF and LO, RF and IF, and LO and IF. An analysis of this circuit, with close attention to the polarities of each cross product

$$f_L \pm f_R, \, 2f_L, \, 2f_R, \, 2f_L \pm f_R, \, 2f_R \pm f_L, \, 3f_L, \, 3f_R$$

as well as f_L and f_R, shows that the only noncanceling outputs are the sum and difference frequencies $f_L \pm f_R$.

Switching or Sampling Mixers

Figure 7–7 shows the mixer circuit symbol and a squarewave LO switching signal. The LO signal does not actually have to be a squarewave, it must merely be strong enough to switch the diode of Figure 7–6 completely on and off. The 50% duty cycle LO waveform shown is the switching function of Equation 8–1 for the balanced mixer/modulator in Chapter 8. Refer to Equations 8–1 through 8–4 for

FIGURE 7-7 Mixer with square-wave local oscillator for switching or sampling.

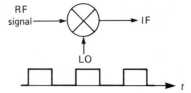

the derivation of the sum and difference frequencies produced in switching mixers and balanced mixers. If the LO produces narrow pulses instead of the squarewave of Figure 7–7, then the mixer output current has stronger mixer products at the higher harmonics of the LO. Since these harmonics will be filtered out anyway, sampling is a perfectly good mixing technique.

Other Frequency Converters

There are numerous aspects of the analyses of mixers to make special note of, but two deserve special mention. First, although a nonlinear circuit component is absolutely necessary to generate the sum and difference frequencies, the overall circuit transfer function v_{IF}/v_{RF} is *linear*; that is, like an amplifier, a wide range exists over which an increase in the RF signal power will result in an equal increase in the power of the IF signal. Second, the output IF voltage is proportional to one-half the input RF voltage; that is, with the second-order nonlinearity for which the mathematical analysis of the previous section was performed, the IF power is at best $20 \log(\frac{1}{2}) = 6$ dB below the RF input.

Figure 7–8 illustrates a few of the almost limitless circuit configurations for mixers (and frequency converters). The simple diode mixer of (1) has been previously discussed, except that here the LO is capacitively coupled. C_1 is always small in value because the LO frequency is almost always higher than the RF; more importantly, the high impedance of the small C_1 allows for isolation of the oscillator circuit from the mixer, whereas T_1 provides for impedance matching of the mixer to the RF circuit.

Mixer (2) is an *active mixer* in which a net conversion gain is achieved. The base-emitter junction is driven into nonlinear or switching operation by the large LO signal. Mixing occurs in the input junction, and the transistor current gain and tuned-collector circuit produce power gain at IF. The overall circuit gain is 6 dB less than the circuit would produce with an IF signal as an input.

Mixer (3) makes use of a dual-gate MOSFET to achieve good isolation between the LO and RF input circuits. Also illustrated in this VHF mixer for a TV tuner (RF front end) is a circuit connection for automatic gain control (AGC).

Integrated circuits, such as the CA (or LM) 3028A which operates to 120 MHz, also give good isolation in this mixer configuration. As seen schematically in mixer (4), the 3028A has a differential amplifier with a (normally) constant current source Q_1 which is modulated by the LO. Thus, sum and difference frequencies are produced. This IC has numerous applications, including AM modulator and cascode amplifier with AGC.

Circuit (5) is the double-balanced mixer that, as mentioned earlier in this section, achieves isolation among each of the three ports. This circuit is seen again

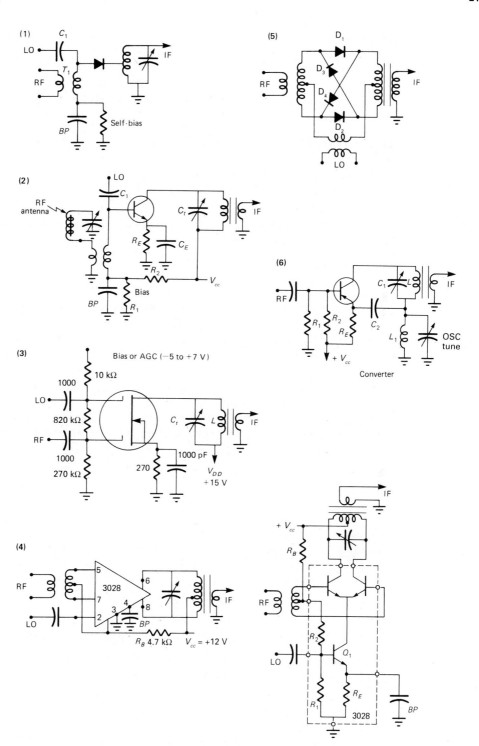

FIGURE 7-8 Mixer circuits.

and again in this book, including its application in high-level digital PSK modems. (See Chapter 8 for an analysis of its operation in single-sideband systems.)

Finally, *converter* is the name given to a circuit that uses a single active device—whether transistor, tube, or IC—to make an oscillator (LO) and an active mixer. Circuit (6) illustrates the frequency converter. The IF tuned circuit has very low impedance at the much higher LO frequency (due to C_1), and C_2 provides a feedback path for the local oscillator. Also, the oscillator tuned circuit is essentially a short circuit to the IF due to L_1; therefore, the emitter circuit has low impedance for good gain.

Mixer Circuit Design

The usual approach taken in mixer design is to follow the rules of thumb for tuned amplifiers in Section 1–6. Then consider the dc bias as if the circuit will operate class A and adjust component values later when the optimum LO power is determined empirically. The ac design has three aspects: (1) the RF signal is transformer-coupled and impedance-matched to optimize the received signal's path. (2) The LO is coupled to the input through a high-impedance circuit component, such as a small-value capacitor, so that this connection does not load down the RF signal path. (We can afford to waste a small amount of local oscillator power for the sake of optimizing the received signal.) (3) The output circuit must be optimum for the intermediate (difference) frequency signal, including the modulation spectrum, while eliminating the f_{RF}, f_{LO}, and $f_{RF} + f_{LO}$. In particular, the strong LO signal must be greatly attenuated in order that the first IF amplifier is not overloaded. Remember, too, that IF current must circulate in the input circuit, so choose input bypass capacitors appropriately; sometimes a low-inductance RF/LO bypass capacitor is placed in parallel with the larger IF bypass capacitor.

A complete circuit design example for a transistor mixer is presented in Example 7–2.

■ ■ ■ ■

EXAMPLE 7–2

Mixer Circuit Design

Design an active mixer using a BJT ($\beta = 50$, $C_{OB} = 1$ pF, $C_{BE} = 30$ pF) to deliver a maximum of 10 mW to a 50-Ω load (collector efficiency = 50%). Assume $f_{RF} = 1.6$ MHz and $f_{IF} = 455$ kHz. $L_1 = 0.25$ mH (infinite Q_u). The IF bandwidth must be 20 kHz with no additional components added to the schematic of Figure 7–9.

1. Determine all values for the R's and C's shown.

2. Determine the turns ratio for T_1 to match the transistor input resistance to 50 Ω, and T_2 turns ratio to achieve the required bandwidth.

3. Determine the LO signal voltage dropped across Q_1 (v_{BE}). Is this enough to drive the transistor nonlinear? Why?

Solution:

1. *DC design:* **Following the amplifier design procedure of Chapter 1, $P_o = 10$ mW, $\eta_c = 50\%$; therefore, $P_{dc} = 10$ mW/0.5 = 20 mW. For $V_E = 10\% V_{cc}$,**

FIGURE 7–9 Active mixer.

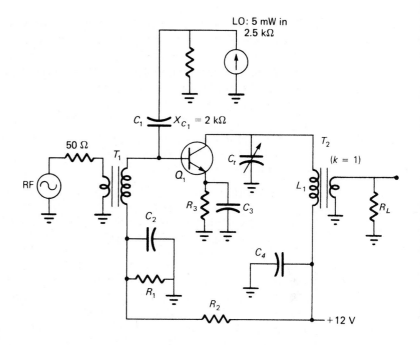

I_c = 20 mW/(12 − 1.2 V) = 1.85 mA and R_3 = 1.2 V/1.85 mA = **648 Ω**. V_B = 1.8 V, and if I_{R1} = 10I_B = 0.37 mA, then R_1 = **4.9 kΩ**. Also R_2 = (12 − 1.8)/0.41 mA = **25 kΩ**.

   AC design: f_{LO} = 1.6 + 0.455 = 2.055 MHz; therefore, C_1 = 1/(2π × 2.055 × 10^6 × 2 kΩ = **39 pF**.

$$X_{C2} = \left(\frac{1}{10}\right)(R_1 \parallel R_2 \parallel R_i) = 61 \ \Omega$$

Therefore, $$C_2 = \frac{1}{(2\pi \times 455 \text{ kHz} \times 61)} = \textbf{5600 pF}$$

   We need to know

$$r_e = \frac{26}{1.85} = 14 \ \Omega$$

and $$R_i(Q_1) = (\beta + 1)r_e = 717 \ \Omega$$

(We will use these values despite the fact that Q_1 will be operating nonlinearly.)

$$X_{C3} = \left(\frac{1}{10}\right)(r_e \parallel R_3) = 1.4 \ \Omega$$

Therefore, $$C_3 = C_4 = \textbf{0.25 } \boldsymbol{\mu}\textbf{F}$$

will provide good ac bypassing. Finally, C_t = 1/[(2π455 kHz)^2(0.25 mH)] = **488 pF** variable or 420 pF fixed, in parallel with a 100-pF variable.

2. T_1: 50 Ω is to match $R_i(Q_1) = 717$ Ω; therefore, $n_p/n_s = \sqrt{50/717} = 0.263$; that is, the secondary has 3.8 times more turns than the primary.

T_2: The only loading is R_L'. Since $X_L = 715$ Ω and $Q_{eff} = 455$ kΩ/20 kΩ = 22.75, then $R_L' = 715 \times 22.75 = 16.3$ kΩ and $n_p/n_s = \sqrt{16.3 \text{ kΩ}/50 \text{ Ω}} = \mathbf{18{:}1.}$

3. For 5 mW in 2.5 kΩ,

$$V_{LO} = \sqrt{2PR} = 5 \text{ V pk}$$

Since T_1 matches the input generator impedance to 717 Ω, we can remove T_1 and place a 717-Ω resistance from base to ground to produce the approximate equivalent ac circuit.

a. If $C_i(Q_1)$ is ignored, the base circuit to ground is $717 \,\|\, R_i(Q_1) = 359$ Ω and the voltage division with C_1 will yield $v_B = (359 \text{ Ω} \times 5 \text{ V pk})/[(359 - j2000)\text{Ω}] = \mathbf{882 \text{ mV pk.}}$ This is more than enough LO to drive Q_1 nonlinear because the minimum required is 50 mV pk.

b. If $C_i(Q_1)$ is not ignored,

$$C_i = C_{BE} + (|A_v| + 1)C_{BC}$$

Actually there is little if any gain at 2.055 MHz because the output is tuned to 455 kHz, and $C_t = 488$ pF provides a low impedance. Even C_{BE} is unpredictable because of the nonlinearity, but let's use $C_i = C_{OB} + C_{BE} = 31$ pF for which $X_C = 2.5$ kΩ. Then $Z_B = 359 \,\|\, j2.5 \text{ kΩ} = 355 \text{ Ω } \underline{/-8°}$ and

$$v_B = \frac{(351.8 - j50.7)5 \text{ V}}{351.8 - j50.7 - j2000} = \mathbf{854 \text{ mV pk}}$$

Since the emitter is well bypassed at the LO frequency, the 0.87 V pk is directly across the base-emitter junction, and 50 mV is about the limit of reasonable linearity. Actually the LO power, or C_1, should be reduced because a maximum of 0.1 V pk is enough to drive Q_1 from high gain to almost cutoff.

Mixing Modulated Signals

When a modulated signal is up- or down-converted in a properly designed mixer, the output will contain the same modulation as at the input. For example, suppose a carrier f_R is amplitude-modulated by a sinusoid of frequency f_m. The time-domain and frequency spectrum for this AM signal are sketched in Figure 7–10 as the input to a mixer. The (ideal) LO spectrum is also sketched with $f_L > f_R$. This is *high-side injection.*

The mixer output spectrum may be sketched by computing the sum and difference frequencies between the LO and each of the RF spectral components. Assigning numerical values to f_R, f_m, and f_L will make this easy, and if you want to portray accurate amplitudes, make the amplitudes of each sum and difference output component one-half that of the RF input spectrum. The results for the IF spectrum are shown at the filter output. Please notice that the time sketch has the same AM index and envelope rate (f_m) and shape. Only the average frequency

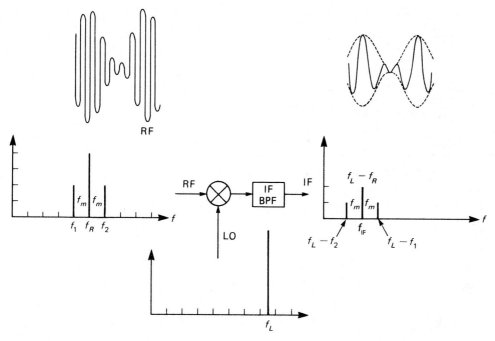

FIGURE 7–10   Down-conversion of AM signal (high-side LO injection to mixer).

(f_{IF}) and amplitudes (½) are different from the RF input. The one-half amplitude (−6 dB) is ideal for a second-order mixer nonlinearity; actual mixer efficiencies will vary.

7–3 ■ TUNING THE STANDARD BROADCAST RECEIVER

When a receiver is to be used for receiving more than one channel (station), either the RF amplifier and LO are tunable, or the LO is fixed and the RF amplifier is broadband enough to handle all the channels while the mixer output feeds separate IFs. Very often the receivers for satellite transponders* and terrestrial microwave repeaters are the broadband RF types with separate IFs for each channel.

Standard broadcast receivers for AM, FM, and television have a relatively narrow but tunable RF input (perhaps with an RF amplifier). The LO frequency and RF amplifier/filter are simultaneously tuned, using a single control, to receive a given station. Figure 7–11 is the schematic of a simple but representative AM superheterodyne receiver available in kit form from Graymark Incorporated. The dashed line associated with C_1 indicates that the tuning capacitors are mechani-

---

* This automatic receiver/transmitter system has transmitter and receiver always on (though at different frequencies) for continuous relay of earth-station information.

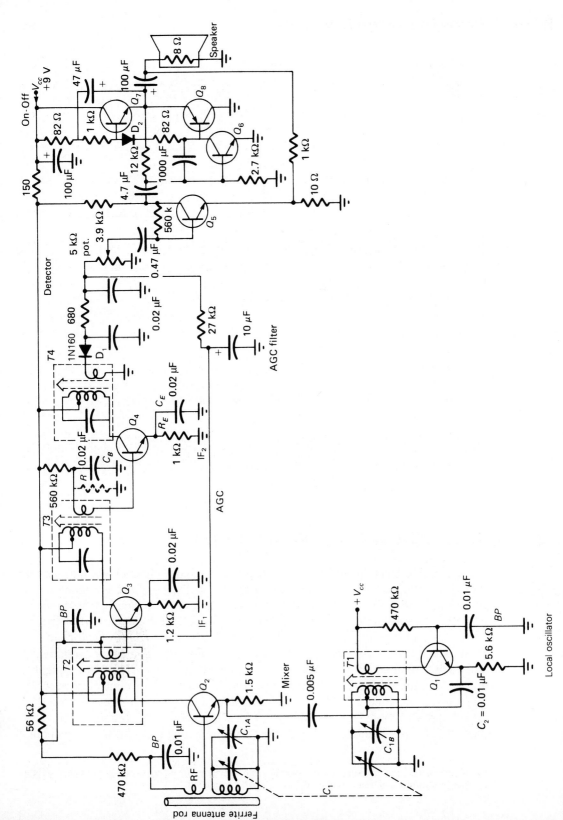

FIGURE 7-11 AM superheterodyne receiver.

cally ganged on one shaft for simultaneous tuning of the RF input and local oscillator. This mechanization is clever enough, but exact tuning of both circuits for all stations does not occur. That is, the separate circuits do not track correctly the RF resonant frequency and the LO frequency to produce exactly 455 kHz. This difficulty is inherent to wide-range superheterodyne systems. The problem is that, while the RF is being tuned from 540 to 1600 kHz (3 : 1 range), the LO must tune from 995 to 2055 kHz (2 : 1 frequency range).

Since the narrowest possible IF bandwidth to pass the AM is desirable for best S/N, a compromise front-end alignment scheme is used in order to avoid sideband (therefore audio) distortion. One procedure used for receiver alignment is as follows (refer to Figure 7–11):

1. Disable the LO with a short to ground from the negative lead of C_2.

2. AC-couple a very low-level 455-kHz signal to the base of Q_2 and tune T_2 through T_4* for a maximum level at the output of T_4. This aligns the narrowband IF amplifiers (use a sweep generator if available).

   Alternatively, AM the input generator and tune for the maximum at the detector output. Keep the output as low as possible by reducing the signal generator power to avoid saturation.

3. Remove the LO short; set the signal generator to 540 kHz and the radio tuning dial (C_1) to the low end of its range. Tune T_1 for 995 kHz using a counter, or maximize the receiver output at T_4 or the detector output (with AM).

4. Set the generator to 1600 kHz and C_1 to the high end of its range. Since C_1 will have a minimum capacitance, the RF and LO trimmers C_{1A} and C_{1B} can be tuned for $f_{LO} = 2055$ kHz and maximum receiver output. Keep the signal generator power low to avoid saturation.

7-4 ■ FILTER REQUIREMENTS FOR INTERFERENCE CONTROL AND SELECTIVITY

Chapter 5 included a discussion of receiver selectivity and RF interference (RFI), including adjacent channel interference. Here, a technique is given for defining the required filter specifications: What is the required attenuation *shape factor*? How many resonators (LC sections) must be used? What is the minimum Q for the resonators? What will be the passband insertion loss and ripple? These questions must be answered in order to specify an appropriate filter. Fortunately, filter selectivity/shape curves are available for those who do not have access to a computer or the time required to solve the polynomials that describe multipole filter response.

---

* The heavy vertical dashed line with an arrowhead in the transformer circuit symbol indicates a variable inductance. Tuning is accomplished by virtue of a threaded ferrite core that can be raised and lowered with respect to the coil windings.

Common AM and FM receivers do not include a separate filter at the mixer output. However, as you probably have already experienced, common AM receivers are notorious for interference from other stations.

Interference is less noticeable in FM systems because of the guardband between stations, the inherent line-of-sight reception, and FM's superior interference suppression using IF limiters and a high modulation index. Receivers for television have had a very complex IF alignment procedure; the better receivers now incorporate *surface acoustic wave* (SAW) filters.

Specialized high-performance receivers such as those designed for satellite and terrestrial microwave links use broadband (low-*Q*), noncritical-tuning IF amplifiers preceded by a multipole bandpass filter. Some of the terminology and jargon used with filters is illustrated in Figure 7–12 for a three-pole Chebyshev filter. Bandpass filter response is more commonly plotted as the inverse of that in Figure 7–12. However, this illustrative plot corresponds to the selectivity curves used in the filter requirement analysis. The *passband* is most commonly defined as the band of frequencies between the 3-dB attenuation points. The band of frequencies between 3-dB attenuation and a prescribed *X*-dB level of attenuation is called the *transition band*. Frequencies beyond the *X*-dB bandwidth (above and below the filter range) are in the *stop band*.

Constant-*k* filter design includes the Chebyshev and Butterworth responses. The *Chebyshev response* is defined as an equal-passband-ripple response; the smooth response of the *Butterworth filter* is defined as maximally flat (0-dB ripple). The Chebyshev filter has faster transition-band rolloff (steep-skirt selectivity) than the Butterworth but poorer group delay characteristics. Group delay causes phase distortion of wideband signals such as pulses.* Another class of filters, including Gaussian and Bessel designs, should be used if group delay is a prime system factor.

---

* Chapter 13 gives more on pulse distortion in PCM systems and modem design. Chapter 3 includes a basic discussion of phase distortion.

FIGURE 7–12 Bandpass filter attenuation characteristic. The region from $BW_{3\,dB}$ to $BW_{X\,dB}$ is often referred to as the transition band.

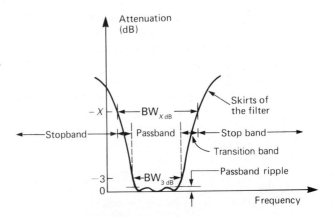

The curves of Figure 7–13 follow the shape of the transition-band attenuation for an equivalent low-pass prototype filter. These curves may be used for a band-pass filter by first computing the value on the horizontal axis, called the attenuation *shape factor*, from

$$SF = \frac{BW_{X\ dB}}{BW_{3\ dB}} \qquad\qquad (7\text{–}6)$$

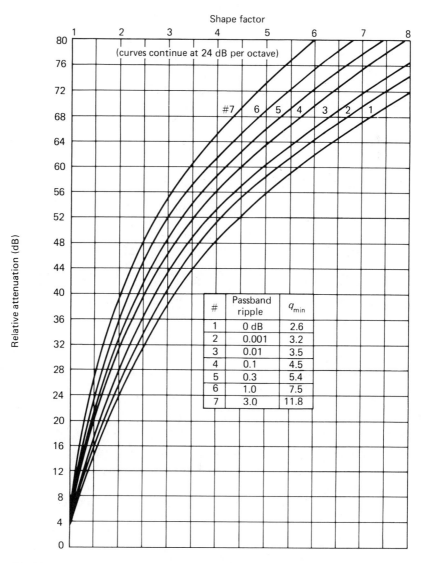

FIGURE 7–13   Relative attenuation for a four-pole low-pass filter network (constant-*k*). (Howard W. Sams & Co. Editorial Staff. ITT Reference Data for Radio Engineers. Indianapolis, 1975. Reproduced with permission of the publisher.)

where $BW_{3\,dB}$ and $BW_{X\,dB}$ are the 3-dB and X-dB bandwidths, respectively. The typical filter specification is for a given amount of attenuation, X dB, at a known, absolute frequency above or below the filter center frequency f_0. The shape factor, given such a specification, is determined from the fact that the center frequency of the filter is the geometric mean of the high and low frequency for a given bandwidth. That is, for the 3-dB bandwidth,

$$BW_{3\,dB} = f_{3H} - f_{3L} \qquad\qquad (7\text{--}7)$$

the center frequency is given by

$$f_0 = \sqrt{f_{3H}f_{3L}} \qquad\qquad (7\text{--}8a)$$

Likewise, for the X-dB bandwidth,

$$f_0^2 = f_{XH}f_{XL} \qquad\qquad (7\text{--}8b)$$

Equation 7–6 may now be written as

$$SF = \frac{BW_{X\,dB}}{BW_{3\,dB}} = \frac{f_{XH} - f_{XL}}{f_{3H} - f_{3L}} \qquad\qquad (7\text{--}9)$$

Substituting Equations 7–7 and 7–8b into Equation 7–9 yields

$$SF = \frac{f_{XH} - (f_0^2/f_{XH})}{BW_{3\,dB}} \qquad\qquad (7\text{--}10a)$$

$$= \frac{f_0}{BW_{3\,dB}}\left[\frac{f_{XH}}{f_0} - \frac{f_0}{f_{XH}}\right] \qquad\qquad (7\text{--}10b)$$

Similarly,
$$SF = \frac{f_0}{BW_{3\,dB}}\left[\frac{f_0}{f_{XL}} - \frac{f_{XL}}{f_0}\right] \qquad\qquad (7\text{--}11)$$

for the lower X-dB frequency point of the bandpass filter response. Comparing Equations 7–10b and 7–11 shows that either the high-side or the low-side stopband frequency may be used if the absolute value is substituted for the rectangular brackets; that is,

$$SF = \frac{f_0}{BW_{3\,dB}}\left|\frac{f_{X\,dB}}{f_0} - \frac{f_0}{f_{X\,dB}}\right| \qquad\qquad (7\text{--}12)$$

Once the filter shape factor has been computed, the curves of Appendix B are searched, starting with the two-pole (low-pass) filter curves, until the X-dB attenuation, shape factor, and passband ripple are all satisfied with the lowest number of circuit poles required. Then the minimum Q required for the circuit components (inductors in particular) is computed from

$$Q_{min} = q_{min}Q_L = q_{min}\frac{f_0}{BW_{3\,dB}} \qquad\qquad (7\text{--}13)$$

where q_{min} is found in the small table inset below the family of curves for each n-pole filter shape. With this and the known or assumed Q_u for available induc-

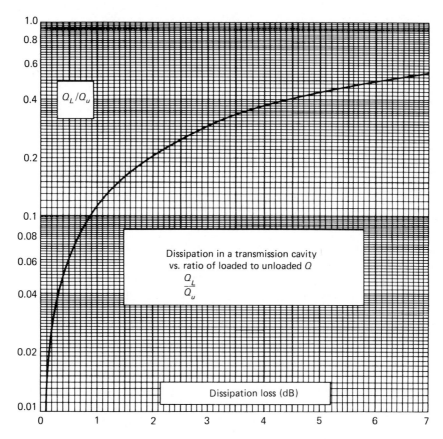

FIGURE 7–14   Insertion loss per pole (low-pass) or resonator (bandpass) for filter with loaded $Q_L = f_0/\text{BW}$, using resonators of unloaded-$Q = Q_u$.

tors, the filter insertion loss due to finite-Q circuit components can be approximated by calculating the Q_L/Q_u ratio, entering the vertical axis of Figure 7–14 and finding the dissipation loss **per pole** for the components. The insertion loss (IL) determined is the minimum passband attenuation if the circuit load and source resistances are equal. For unequal source and load, additional passband loss of $|\,20 \log R_L/R_s\,|$ occurs.

The procedure for determining filter requirements is best illustrated by a working example.

■ ■ ■ ■

**EXAMPLE
7–3**

A receiver with a 10-MHz IF must reject an adjacent channel signal that is 625 kHz away from center frequency. The following bandwidth filter requirements are specified: center frequency 10 MHz, 3-dB bandwidth 500 kHz, selectivity (out-of-band attenuation) 40 dB at 10.625 MHz, passband ripple ≤ 0.5 dB, and 1-dB maximum insertion loss. Determine

1. Number of resonators required.
2. Amount of ripple with chosen filter.
3. Minimum Q per resonator, Q_{min}.
4. Insertion loss (IL).
5. How many resonators would be required in order to produce a Butterworth response with the required selectivity?

Solution:

1. First the attenuation shape factor (SF) is computed using Equation 7–12,

$$\text{SF} = \frac{10 \text{ MHz}}{0.5 \text{ MHz}} \left| \frac{10.625 \text{ MHz}}{10 \text{ MHz}} - \frac{10 \text{ MHz}}{10.625 \text{ MHz}} \right| = 2.43$$

It is helpful to make a sketch like Figure 7–15 to visualize all the frequencies involved in the problem. We want to use a minimum number of resonators for economy and insertion loss. Also, you will appreciate the minimum-pole approach when tuning the filter. Go to the selectivity curves in Appendix B; the horizontal axis is SF, and the vertical axis is attenuation $(V_p/V)_{dB}$. Note that, for all filters, the family of curves intersects SF = 1 at 3 dB. The first family of curves for a 2-pole low-pass network (double-tuned circuit for bandpass) shows that, for SF = 2.43, the best we can do is 21 dB of attenuation at 10.625 MHz (the intersection of 2.43 and 40 dB must be below the curve). So try the three-pole and, finally, the four-pole curves. Now refer to Figure 7–13: Curve 4 won't quite do, but curves 5, 6, and 7 will. Answer: **4 resonators.**

2. Now check the inset. Curve shape number 5 will have a passband ripple (bumps in the response) of 0.3-dB magnitude. Since this is within specifications, we continue with the 4-resonator Chebyshev filter realization; otherwise we would have to use 5 poles.

3. The last column of the inset gives $q_{min} = 5.4$. This will allow us to compute the minimum Q required for the circuit components from Equation 7–13. Thus,

$$Q_{min} = 5.4 \left(\frac{10 \text{ MHz}}{0.5 \text{ MHz}} \right) = \mathbf{108}$$

FIGURE 7–15 Bandpass filter response.

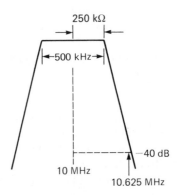

This is easily attained with high-Q toroids or air-core inductors (capacitors are usually much higher Q).

4. Insertion loss is the minimum power loss experienced by the signals we want to pass through the filter. Figure 7–14 gives the circuit component dissipation loss *per resonator*. If the resonators used have just enough unloaded Q (Q_u) to achieve the filter requirement, then $Q_L/Q_u = 20/108 = 0.185$. The transmission loss curve gives 1.8 dB/resonator, so IL = 4 × 1.8 dB = **7.2 dB.** This is *out of spec* (does not meet specifications). IL = 1 dB requires a loss/resonator of 0.25 dB. Figure 7–14 shows that $Q_L/Q_u = 0.032$ is required. Hence we must use resonators with $\boldsymbol{Q_u \geq 625.}$

5. The Butterworth response has 0-dB ripple. This is curve number 1 of each of the figures of Appendix B. A 4-resonator Butterworth filter will attain only 32 dB of selectivity. In order to achieve the required 40-dB attenuation at SF = 2.43, we must use a 6-resonator filter with Q_u/pole of 910 to meet the IL = 1 dB specification.

The matching networks in Chapter 6 are seen to be filters. The π-section network, for example, can be considered the low-pass prototype from which bandpass filters such as those of Figure 7–16 are evolved by the addition of parallel and perhaps series resonating components.

FIGURE 7–16 Bandpass filter configurations. (a) Five-pole, low-pass prototype. (b) Top-coupled or capacitor-coupled 2-resonator. (c) Five-resonator bandpass filter resulting from a 5-pole, capacitor input, $R_{\text{load}} \leq R_{\text{source}}$, low-pass prototype (usually normalized to $\omega_c = 1$ rad/s) to bandpass transformation.

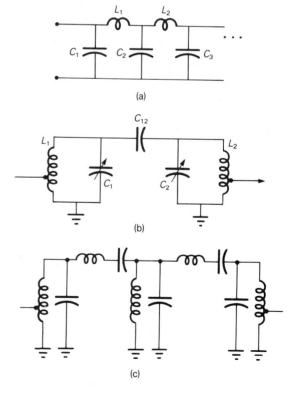

Most of those involved in the technology of radios discover that their greatest challenges are in bandpass filter design and optimization. The subject of bandpass design, including active filters, cannot be done justice in a textbook intended to cover communications. However, design techniques are given for the popular (because of its simplicity) top-coupled bandpass filter of Figure 7–16b. For more complicated structures, including the m-derived end sections for solving the big problem presented to input and output circuits by the radical impedance variations at frequencies near the 3-dB cutoff points, consult a filter designer's or a radio designer's reference book.

7-5 ■ FILTER CIRCUIT DESIGN

The filter networks of Figure 7–16a and b produce the simplest low-pass and bandpass filter circuit implementations. In particular, the capacitor top-coupled, constant-k bandpass filter of Figure 7–16 is especially easy to alter once designed and constructed if the coupling capacitors, C_{ij}, are fixed-value components and the shunt capacitors are variable. In this way the coupling can be well controlled while the variable capacitors are tuned for center frequency and proper filter response shape.

Determination of component values begins by using the filtering requirements/specifications and the procedures in Section 7–4 to select the filter type (Butterworth or Chebyshev), the shape factor, and the number of resonators or poles and q_{min} from the filter curves of Appendix B. Then either the exact equations based on the analysis for locating the polynomial poles, p_k, are used or the simpler method of tables is used.

In the *method of tables*, Table 7–1 or 7–2 is used for obtaining q_1 and k_{xy}. The value q_1 is the required normalized end section resonator Q of the inductor and source (or load) together. The value k_{xy}; $x = 1, 2, 3, \ldots$ and $y = 2, 3, 4, \ldots$ where $y = x + 1$, is a normalized coupling factor between

TABLE 7-1 Circuit design constants for capacitive-coupled Butterworth and Chebyshev filters with equal input and output terminations (page 1 of 2).

Butterworth (Zero Ripple)							
n	q_1, q_n	k_{12}	k_{23}	k_{34}	k_{45}	k_{56}	k_{67}	k_{78}
2	1.414	0.707						
3	1.000	0.707	0.707					
4	0.765	0.841	0.541	0.841				
5	0.618	1.000	0.556	0.556	1.000			
6	0.518	1.169	0.605	0.518	0.605	1.169		
7	0.445	1.342	0.667	0.527	0.527	0.667	1.342	
8	0.390	1.519	0.736	0.554	0.510	0.554	0.736	1.519

TABLE 7–1 (page 2 of 2)

0.01-dB Chebyshev

n	q_1, q_n	k_{12}	k_{23}	k_{34}	k_{45}	k_{56}	k_{67}	k_{78}
2	1.483	0.708						
3	1.181	0.682	0.682					
4	1.046	0.737	0.541	0.737				
5	0.977	0.780	0.540	0.540	0.780			
6	0.937	0.809	0.550	0.518	0.550	0.809		
7	0.913	0.829	0.560	0.517	0.517	0.560	0.829	
8	0.897	0.843	0.567	0.520	0.510	0.520	0.567	0.843

0.1-dB Chebyshev

n	q_1, q_n	k_{12}	k_{23}	k_{34}	k_{45}	k_{56}	k_{67}	k_{78}
2	1.638	0.711						
3	1.433	0.662	0.662					
4	1.345	0.685	0.542	0.685				
5	1.301	0.703	0.536	0.536	0.703			
6	1.277	0.715	0.539	0.518	0.539	0.715		
7	1.262	0.722	0.542	0.516	0.516	0.542	0.722	
8	1.251	0.728	0.545	0.516	0.510	0.516	0.545	0.728

0.5-dB Chebyshev

n	q_1, q_n	k_{12}	k_{23}	k_{34}	k_{45}	k_{56}	k_{67}	k_{78}
2	1.950	0.723						
3	1.864	0.647	0.647					
4	1.826	0.648	0.545	0.648				
5	1.807	0.652	0.534	0.534	0.652			
6	1.796	0.655	0.533	0.519	0.533	0.655		
7	1.790	0.657	0.533	0.516	0.516	0.533	0.657	
8	1.785	0.658	0.533	0.515	0.511	0.515	0.533	0.658

1-dB Chebyshev

n	q_1, q_n	k_{12}	k_{23}	k_{34}	k_{45}	k_{56}	k_{67}
2	2.210	0.739					
3	2.210	0.645	0.645				
4	2.210	0.638	0.546	0.638			
5	2.210	0.633	0.535	0.538	0.633		
6	2.250	0.631	0.531	0.510	0.531	0.631	
7	2.250	0.631	0.530	0.517	0.517	0.530	0.631

From Arthur B. Williams, *Electronic Filter Design Handbook* (New York: McGraw-Hill, 1981).

TABLE 7–2 Circuit design constants for capacitative-coupled Butterworth and Chebyshev filters terminated only at one end.

Ripple (dB)	q_1	k_{12}	k_{23}	k_{34}	k_{45}	k_{56}	k_{67}
2-resonator filter							
0	0.707	1.00					
0.3	0.910	0.904					
1.0	1.11	0.866					
3.0	1.56	0.840					
3-resonator filter							
0	0.500	1.22	0.707				
0.1	0.714	0.961	0.661				
1.0	1.11	0.785	0.645				
3.0	1.68	0.714	0.649				
4-resonator filter							
0	3.83	1.56	0.765	0.644			
0.1	0.667	1.01	0.626	0.618			
1.0	1.10	0.781	0.578	0.614			
3.0	1.72	0.692	0.567	0.609			
5-resonator filter							
0	0.309	1.90	0.900	0.655	0.619		
0.1	0.649	1.044	0.634	0.560	0.595		
1.0	1.105	0.779	0.570	0.544	0.595		
3.0	1.74	0.679	0.554	0.542	0.597		
6-resonator filter							
0	0.259	2.26	1.05	0.732	0.606	0.606	
0.1	0.637	1.06	0.642	0.560	0.539	0.589	
1.0	1.12	0.771	0.566	0.533	0.531	0.589	
3.0	1.75	0.673	0.546	0.529	0.531	0.591	
7-resonator filter							
0	0.223	2.62	1.20	0.824	0.659	0.579	0.598
0.1	0.629	1.08	0.648	0.560	0.531	0.530	0.587
1.0	1.12	0.770	0.564	0.530	0.521	0.527	0.587
3.0	1.76	0.669	0.542	0.523	0.520	0.528	0.588

q_1 is the normalized Q of end sections. The Q of other internal resonators is set by the coupling factors, $k_{(ij)}$. However, all inductors used must have Q_u values exceeding the product $q_{min} Q_L$ or the filter shape cannot be achieved.

consecutive resonators, n. That is, k_{12} is the coupling factor between the first and second resonators—$n = 2$.

Table 7–1 is used for the usual case of equal end (input/output) resistive terminations. Table 7–2 is used when the filter has a termination only at one end. Examples are (1) a 50-Ω source and high-impedance IC input as the filter load, or (2) a high-impedance current source (transistor collector) driving the filter with a 50-Ω filter output load.

Determining the Filter Impedance Level

Because inductors of less than 50 nH are physically very small and difficult to handle, the lower limit of circuit reactance and design impedance R is set as follows:

$$R \geq 2\pi f_0 Q_L q_1 (5 \times 10^{-8}) \qquad (7\text{--}14)$$

where $Q_L = f_0/\text{BW}$ is the overall circuit loaded Q and q_1 is the normalized Q value, taken from Table 7–1, for the first and last resonant nodes in the filter.

R is the *design impedance level* and

$$X_L = \frac{R}{Q_1} \qquad (7\text{--}15)$$

where

$$Q_1 = q_1 Q_L \qquad (7\text{--}16)$$

An important consideration for circuit design is that since inductors in particular have limited Q, the actual circuit source and load resistances must be larger than R. The reason is illustrated by reference to Figure 7–17 for a bandpass filter. The inductor L has finite quality factor, Q_u, the unloaded Q of the coil. Hence, a resistance of R_{coil} given by

$$R_{\text{coil}} = X_L Q_u \qquad (7\text{--}17)$$

already exists in the first resonator of the filter and all other resonators as well. The design value R is then the parallel combination of R_s and R_{coil}; that is,

$$R_s \parallel R_{\text{coil}} = R \qquad (7\text{--}18)$$

The two impedance-level design procedures are as follows:

I. When R is arbitrarily chosen, then R_{coil} is computed from Equation 7–17 using Equations 7–15 and 7–16, and R_s must be determined from

$$R_s = \frac{R R_{\text{coil}}}{R_{\text{coil}} - R} \qquad (7\text{--}19)$$

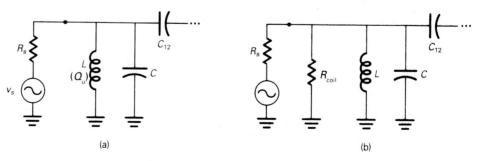

(a)   (b)

FIGURE 7–17 First resonator of a finite-Q, capacitor top-coupled bandpass filter. (a) Ideal circuits. (b) Circuit with finite-Q inductor.

II. Alternatively, if R_s or R_L is specified and inductors of known Q_u are available, then Equations 7–17 and 7–18 are combined for computing X_L as

$$X_L = R_s \left(\frac{1}{Q_1} - \frac{1}{Q_u} \right) \tag{7-20}$$

Now the design value R is determined using Equation 7–15.

Note that if Q_u of the inductors used is ignored, the filter bandwidths will be too wide and the ripple will be less than the design value.

Low-Pass Filter Component Values

The constant-k low-pass filter configuration of Figure 17–16a can be viewed as consisting of end-to-end connections of T-sections in which L_1 is the first component or π-sections in which C_1 is the first component connected to the source. Then the kth component value in the filter comprising a total of n components is computed using Equations 7–21 through 7–23 for the Butterworth case or 7–26 through 7–35 for the Chebyshev case.

For Butterworth filters, the kth normalized component value, g_k, is computed from

$$g_k = 2 \sin[\pi(2k - 1)/2n] \tag{7-21}$$

$$C_k = \frac{g_k}{2\pi f_c R} \tag{7-22}$$

and

$$L_k = \frac{g_k R}{2\pi f_c} \tag{7-23}$$

where $k = 1$ through n, R is the design impedance level, and f_c is the 3-dB cutoff frequency. Equations 7–22 and 7–23 denormalize g_k for the appropriate operating frequency, impedance level, and component type.

Chebyshev low-pass filter component value computations lend themselves to computer implementation, as the following sequence of equations demonstrates. The normalized values, g_k, are

$$g_1 = \frac{2a_1}{S} \tag{7-24}$$

$$g_k = \frac{4(a_{k-1} a_k)}{b_{k-1} g_{k-1}} \qquad k = 2, 3, \ldots \tag{7-25}$$

$$a_k = \sin \left[\frac{\pi(2k - 1)}{2n} \right] \qquad k = 1, 2, \ldots, n \tag{7-26}$$

$$S = \sinh \left(\frac{E}{2n} \right) \tag{7-27}$$

$$E = \ln \left[\coth \left(\frac{r_{dB}}{17.37} \right) \right] \tag{7-28}$$

$$b_k = S^2 + \sin^2\left(\frac{k\pi}{n}\right) \qquad k = 1, 2, \ldots \qquad \text{(7–29)}$$

With the g_k constants computed, the actual circuit component values are denormalized to the design impedance level, R, and the 3-dB cutoff frequency, ω_c, by

$$C_k = \frac{g_k}{\omega_c R} \qquad \text{(7–30)}$$

and

$$L_k = \frac{g_k R}{\omega_c} \qquad \text{(7–31)}$$

where

$$\omega_c = 2\pi f_c \cosh a \qquad \text{(7–32)}$$

and

$$a = \frac{1}{n}\cosh^{-1}\left(\frac{1}{\epsilon}\right) \qquad \text{(7–33)}$$

Bandpass Design

Once the design impedance level, R, has been determined by the procedure involving Equations 7–1 and 7–20, then the inductive reactance is computed from Equation 7–15. In the method of tables design technique, all tuning inductors have the same value:

$$L = \frac{R}{2\pi f_0 q_1 Q_L} \qquad \text{(7–34)}$$

The total capacitance at each filter node must resonate with L at f_0; therefore,

$$C_{\text{node}} = \frac{1}{(2\pi f_0)^2 L} \qquad \text{(7–35)}$$

The coupling capacitors are then computed from

$$C_{xy} = \left(\frac{k_{xy}}{Q_L}\right) C_{\text{node}} \qquad \text{(7–36)}$$

where k_{xy} is taken from Table 7–1 or 7–2.

The actual capacitor, C_i, placed in parallel with a given inductor at node i, must resonate at f_0 with all other *inductors* shorted to ground. This shorting places the coupling capacitors, on each side of the given inductor, in parallel with C_i. Therefore, to compute the actual value of C_i, subtract the value of any coupling capacitor(s) connected directly to C_i.

As an example, consider a 3-resonator filter. There will be three C_i tuning capacitors, C_1 through C_3, and two coupling capacitors, C_{12} and C_{23}. The tuning capacitors are computed as follows:

$$C_1 = C_{\text{node}} - C_{12} \qquad \text{(7–37)}$$

$$C_2 = C_{\text{node}} - C_{12} - C_{23} \qquad \text{(7–38)}$$

and

$$C_3 = C_{\text{node}} - C_{23} = C_1 \qquad \text{(7–39)}$$

Hence, in general, C_{node}, C_{12}, and C_1 must always be computed. Then, for n poles, $C_n = C_1$, $C_{n-1} = C_2$, and so on, only n unique calculations must be performed.

A filter with only one end loaded is designed as before, except that Table 7–2 must be used.

■ ■ ■ ■
EXAMPLE 7–4

Design a 3-resonator, capacitor top-coupled, constant-k bandpass filter centered at 1 MHz, with 1-dB ripple and a 100-kHz 3-dB bandwidth. The source and load resistors are 1.5 kΩ and available inductors have a quality factor of 60.

Solution:

Figure 7–18 illustrates the 3-resonator, capacitor-coupled, constant-k bandpass filter. First, compute the minimum practical design impedance value of R: From Table 7–1, $q_1 = 2.210$ and we compute $Q_L = f_0/BW = 10$. Then, from Equation 7–14, $R \geq 2\pi10^6(10)(2.21)(50 \text{ nH}) \geq 6.9$ Ω, which will be no problem since the source and load are 1500 Ω.

If the inductors had infinite Q, then R would be 1500 Ω. However, we must use procedure II, for which Equation 7–20 yields $X_L = 1500(1/22.10 - 1/60) =$ **42.87 Ω** and $R = X_L Q_1 = 947.5$ Ω, which is the parallel combination of R_s and $R_{coil} = Q_u X_L = 2572$ Ω.

Now, use Equation 7–34 to obtain the value of all three inductors as $L = 947.5/(2\pi10^622.10) =$ **6.823 μH.** Then the C_{node} that resonates with L at f_0 is $C_{node} = 1/[(2\pi10^6)^2\ 6.823\ \mu H] = 3.712$ nF. Table 7–1 gives $k_{12} = k_{23} = 0.645$, and Equation 7–36 yields coupling capacitors, $C_{12} = C_{23} = (0.645/10)3.712$ nF = **240 pF.** Hence, $C_1 = C_{node} - C_{12} = 3712 - 240 =$ **3472 pF** and $C_2 = C_{node} - C_{12} - C_{23} = C_{node} - 2C_{12} = 3472 - 480 =$ **3232 pF.** The design values of all components for the Chebyshev 3-resonator filter have been determined.

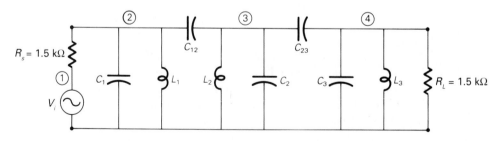

FIGURE 7–18   Three-resonator capacitor top-coupled bandpass filter.

■ ■ ■ ■
EXAMPLE 7–5

Design the 3-resonator Chebyshev filter specified before except that inductors are assumed to have infinite Q. Use PSpice to plot the magnitude and phase of the designed filter circuit to confirm design goals.

Solution:

With infinite-Q inductors, the design value of R is 1500 Ω and the following component values are calculated: L = **10.8 μH,** C_{node} = 2345 pF, C_{12} = C_{23} = **151 pF,** C_1 = C_3 = **2194 pF,** and C_2 = 2044 pF.

This design was simulated on PSpice using the nodes circled in Figure 7–18 for writing the input file. The resulting input and output files are presented in Figure 7–19a–e and careful inspection of the data shows that the ripple is equal, there are three resonant peaks, and valleys are 1.05 dB below the peaks. The 3-dB bandwidth is found to be approximately 103 kHz and, as a point of interest, the 30-dB bandwidth is 246 kHz, so that the 30-dB to 3-dB shape factor is 240/103 = 2.39. Accuracy, even given the need for inspection of data, is within 3% of desired results.

Figure 7-20a shows one of the lab results. Note that, although the 3 resonators are evident, the ripple and flat-top appearance are poor. A major reason for this is that inductor Q was less than 60 (about 40), so there is a smoothing of the ripple toward that expected of a Butterworth design. This is also seen in the 2-resonator results of Figure 7–20b. In addition, inexact component values preclude great accuracy and coupling capacitor values are fairly critical.

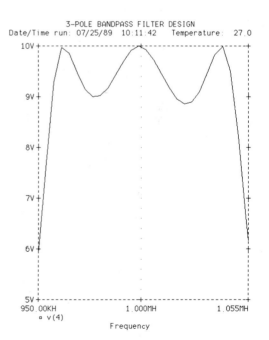

```
* 07/25/89 * Evaluation PSpice
  3-POLE BANDPASS FILTER DESIGN
*****************************
  VIN 0 1 AC 20
  RS 1 2 1.5K
  C1 2 0 2.2NF
  L1 2 0 10.8UH
  C12 2 3 151PF
  C2 3 0 2.05NF
  L2 3 0 10.8UH
  C23 3 4 151PF
  C3 4 0 2.2NF
  L3 4 0 10.8UH
  RL 4 0 1.5K
  .AC DEC 600 0.95MEG 1.055MEG
  .PLOT AC V(4)
  .PROBE
  .END
```

FIGURE 7–19a PSpice circuit file for analyzing the passband ripple for the 3-resonator bandpass filter of Example 7-5.

FIGURE 7–19b Passband ripple for the 3-reso-nator bandpass filter of Example 7–5 using the PROBE feature.

```
*7/25/89*Evaluation PSpice 88
3-POLE BANDPASS FILTER DESIGN
****************************
 VIN 0 1 AC 20
 RS 1 2 1.5K
 C1 2 0 2.2NF
 L1 2 0 10.8UH
 C12 2 3 151PF
 C2 3 0 2.05NF
 L2 3 0 10.8UH
 C23 3 4 151PF
 C3 4 0 2.2NF
 L3 4 0 10.8UH
 RL 4 0 1.5K
 .AC DEC 100 0.7MEG 1.6MEGHZ
 .PLOT AC V(2)
 .PLOT AC V(3)
 .PLOT AC V(4)
 .PROBE
 .END
```

FIGURE 7–19c Circuit file for PSpice
analysis of the 60-dB bandwidth of
the 3-resonator filter of Example 7–5.

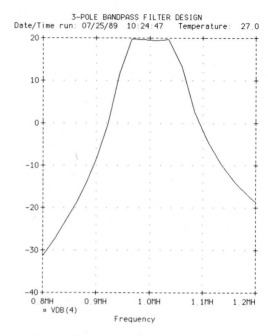

FIGURE 7–19d PSpice plot over the 40-dB
bandwidth of the 3-resonator filter of Example
7–5. Observe the poor high-frequency rolloff.

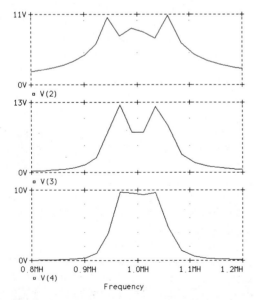

FIGURE 7–19e Filter frequency response at
nodes 2, 3, and output.

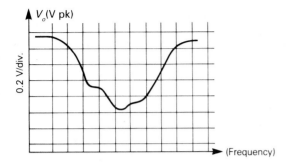

(a) Three-resonator capacitor-coupled
bandpass filter; f_0 = 1.02 MHz
Sweep generator set at 0.7
width, sweep time at 10 ms.
Mode = constant.

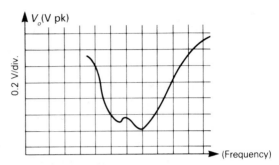

(b) Two-resonator capacitor-coupled
bandpass filter; f_0 = 0.9 MHz
Sweep generator set at 0.7
width, sweep time at 10 ms.
Mode = constant.

FIGURE 7–20   Lab results.

7-6 ■ IF AMPLIFIERS

One of the most popular integrated circuits used in receivers is the CA3028A. (See Figure 7–21 for the data sheet.) This device is shown in Figure 7–8 as a mixer. The RF inputs to the differential amplifier section and the LO switches the diff-amp on and off to produce the nonlinearity for mixing.

In Figure 7–22, the CA3028A is shown as an IF amplifier (or RF amplifier to 120 MHz) with automatic gain control. With the IF input at pin 2 and the output taken from pin 6, the CA3028A is used as a cascode amplifier (pin 5 is at ac ground). The cascode connection is discussed in Chapter 1 (Figure 1–14). When the voltage on pin 1 is equal to the AGC reference voltage on pin 5, Q_1 and Q_2 carry equal currents and the amplifier has high gain. If the AGC voltage on pin 1 is increased, Q_2 current decreases and the IF gain is reduced.

The IF amplifier system for the receiver in Figure 7–11 includes Q_3, Q_4, and the output of Q_2. Transformer coupling is used with the base bias; Q_4 (including R) is typical. The advantages of entering the base bias at the cold end (ac ground) of the transformer secondary are to avoid IF power dissipation in the bias resistors and to avoid the lower-input impedance due to bias resistor shunting. The 0.02-μF bypass capacitor C_B provides the ac ground. For high-frequency IFs, special attention must be given to the physical layout of C_B and C_E because the IF input current passes through these two bypass capacitors.

The dashed line around the transformers indicates that these are specially designed tuned circuits in RFI-tight groundable metal packages for narrow band-width IF applications. They are called *IF cans*. As shown in the 1968 specification sheet of Figure 7–23, this unit includes a 125-pF capacitor, and the arrow between primary and secondary indicates that tuning is attained by tuning-tool (a non-metallic screwdriver) adjustment of the ferrite core. The purpose of the primary

Solid State Division

Linear Integrated Circuits
Monolithic Silicon
**CA3028A, CA3028AF, CA3028AS
CA3028B, CA3028BF, CA3028BS
CA3053, CA3053F, CA3053S**

File Number **382**

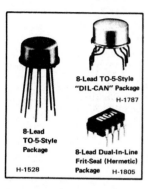

8-Lead TO-5-Style "DIL-CAN" Package
H-1787

8-Lead TO-5-Style Package
H-1528

8-Lead Dual-In-Line Frit-Seal (Hermetic) Package   H-1805

DIFFERENTIAL/CASCODE AMPLIFIERS

For Communications and Industrial Equipment at Frequencies from DC to 120 MHz

FEATURES

- Controlled for Input Offset Voltage, Input Offset Current, and Input Bias Current (CA3028 Series only)
- Balanced Differential Amplifier Configuration with Controlled Constant-Current Source
- Single- and Dual-Ended Operation
- Operation from DC to 120 MHz
- Balanced-AGC Capability
- Wide Operating-Current Range

The CA3028A and CA3028B are differential/cascode amplifiers designed for use in communications and industrial equipment operating at frequencies from dc to 120 MHz.

The CA3028B is like the CA3028A but is capable of premium performance particularly in critical dc and differential amplifier applications requiring tight controls for input offset voltage, input offset current, and input bias current.

The CA3053 is similar to the CA3028A and CA3028B but is recommended for IF amplifier applications.

The CA3028A, CA3028B, and CA3053 are supplied in a hermetic 8-lead TO-5-style package. The "F" versions are supplied in a frit-seal TO-5 package, and the "S" versions in formed-lead (DIL-CAN) packages.

APPLICATIONS

- RF and IF Amplifiers (Differential or Cascode)
- DC, Audio, and Sense Amplifiers
- Converter in the Commercial FM Band
- Oscillator   • Mixer   • Limiter
- Companion Application Note, ICAN 5337 "Application of the RCA CA3028 Integrated Circuit Amplifier in the HF and VHF Ranges." This note covers characteristics of different operating modes, noise performance, mixer, limiter, and amplifier design considerations.

CA3028A, CA3028B, CA3053 Series Differential/Cascode Amplifiers

The CA3028A, CA3028B, and CA3053 are available in the packages shown below. When ordering these devices, it is important to add the appropriate suffix letter to the device.

Package 8-Lead TO-5	Suffix Letter	CA3028A	CA3028B	CA3053
TO-5	T	√	√	√
With Dual-In-Line Formed Leads (DIL-CAN)	S	√	√	√
Frit-Seal Ceramic	F	√	√	√
Beam-Lead	L	√		
Chip	H	√		

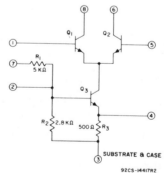

92CS-1441TR2

Fig.1 - Schematic diagram for CA3028A, CA3028B and CA3053.

Information furnished by RCA is believed to be accurate and reliable. However, no responsibility is assumed by RCA for its use; nor for any infringements of patents or other rights of third parties which may result from its use. No license is granted by implication or otherwise under any patent or patent rights of RCA.

Printed in USA/11-73

Supersedes issue dated 3-70

FIGURE 7–21   CA3028A data sheet (page 1 of 2; pages 240–41 courtesy of RCA).

CA3028A, CA3028B, CA3053 _____ File No.382

ELECTRICAL CHARACTERISTICS at T_A = 25°C

CHARACTERISTIC	SYMBOL	TEST CIRCUIT Fig.	SPECIAL TEST CONDITIONS		LIMITS TYPE CA3028A			LIMITS TYPE CA3028B			LIMITS TYPE CA3053			UNITS	TYPICAL CHARACTERISTICS CURVE Fig.
					Min.	Typ.	Max.	Min.	Typ.	Max.	Min.	Typ.	Max.		
DYNAMIC CHARACTERISTICS															
Power Gain	G_P	10a	f = 100 MHz	Cascode	16	20	-	16	20	-	-	-	-	dB	10b
		11a,d	V_{CC} = +9V	Diff.-Ampl.	14	17	-	14	17	-	-	-	-		11b,e
		10a	f = 10.7 MHz	Cascode	35	39	-	35	39	-	35	39	-	dB	10b ✱
		11a	V_{CC} = +9V	Diff.-Ampl.	28	32	-	28	32	-	28	32	-		11b ✱
Noise Figure	NF	10a	f = 100 MHz	Cascode	-	7.2	9	-	7.2	9	-	-	-	dB	10c
		11a,d	V_{CC} = +9V	Diff.-Ampl.	-	6.7	9	-	6.7	9	-	-	-		11c,e
Input Admittance	Y_{11}	-		Cascode				-	0.6 + j 1.6	-				mmho	12
		-		Diff.-Ampl.				-	0.5 + j 0.5	-					13
Reverse Transfer Admittance	Y_{12}	-		Cascode				-	0.0003 - j0	-				mmho	14
		-	f = 10.7 MHz	Diff.-Ampl.				-	0.01 - j0.0002	-					15
Forward Transfer Admittance	Y_{21}	-	V_{CC} = +9V	Cascode				-	99 - j18	-				mmho	16
		-		Diff.-Ampl.				-	-37 + j0.5	-					17
Output Admittance	Y_{22}	-		Cascode				-	0. + j0.08	-				mmho	18
		-		Diff.-Ampl.				-	0.04 + j0.23	-					19
Power Output (Untuned)	P_o	20a	f = 10.7 MHz	Diff.-Ampl. 50 Ω Input-Output	-	5.7	-	-	5.7	-	-	-	-	μW	20b
AGC Range (Max.Power Gain to Full Cutoff)	AGC	21a	V_{CC} = +9V	Diff.-Ampl.	-	62	-	-	62	-	-	-	-	dB	21b
Voltage Gain	A	22a	f = 10.7 MHz	Cascode	-	40	-	-	40	-	-	40	-	dB	22b
at f = 10.7 MHz		22c	V_{CC} = +0V R_L = 1 kΩ	Diff. Ampl.	-	30	-	-	30	-	-	30	-		22d
Differential at f = 1 kHz		23	V_{CC} = +6V, V_{EE} = -6V, R_L = 2 kΩ		-	-	-	35	38	42	-	-	-	dB	
			V_{CC} = +12V, V_{EE} = -12V R_L = 1.6 kΩ		-	-	-	40	42.5	45	-	-	-		
Max. Peak-to-Peak Output Voltage at f = 1 kHz	V_o(P-P)	23	V_{CC} = +6V, V_{EE} = -6V, R_L = 2 kΩ		-	-	-	7	11.5	-	-	-	-	V_{P-P}	
			V_{CC} = +12V, V_{EE} = -12V R_L = 1.6 kΩ		-	-	-	15	23	-	-	-	-		
Bandwidth at -3 dB point	BW	23	V_{CC} = +6V, V_{EE} = -6V, R_L = 2 kΩ		-	-	-	-	7.3	-	-	-	-	MHz	-
			V_{CC} = +12V, V_{EE} = -12V R_L = 1.6 kΩ		-	-	-	-	8	-	-	-	-		
Common-Mode Input-Voltage Range	V_{CMR}	24	V_{CC} = +6V, V_{EE} = -6V		-	-	-	-2.5	(-3.2 - 4.5)	4	-	-	-	V	
			V_{CC} = +12V, V_{EE} = -12V		-	-	-	-5	(-7 - 9)	7	-	-	-		
Common-Mode Rejection Ratio	CMR	24	V_{CC} = +6V, V_{EE} = -6V		-	-	-	60	110	-	-	-	-	dB	
			V_{CC} = +12V, V_{EE} = -12V		-	-	-	60	90	-	-	-	-		
Input Impedance at f = 1 kHz	Z_{IN}		V_{CC} = +6V, V_{EE} = -6V		-	-	-	-	5.5	-	-	-	-	kΩ	-
			V_{CC} = +12V, V_{EE} = -12V		-	-	-	-	3	-	-	-	-		
Peak-to-Peak Output Current	I_{P-P}		V_{CC} = +9V	f = 10.7 MHz	2	4	7	2.5	4	6	2	4	7	mA	-
			V_{CC} = +12V	e_{in} = 400 mV Diff.-Ampl.	3.5	6	10	4.5	6	8	3.5	6	10		

✱ Does not apply to CA3053

FIGURE 7-21   (page 2 of 2)

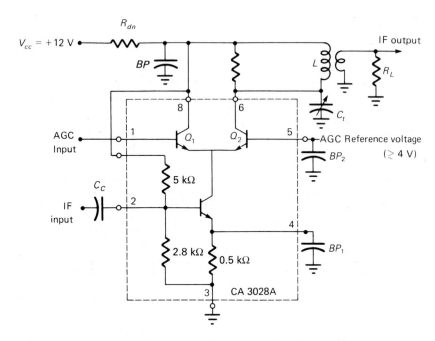

FIGURE 7–22 Cascode IF amplifier with automatic gain control.

winding tap is to increase the effective Q of the collector circuit in the narrowband IF of the standard broadcast receiver.

For instance, suppose the tap is not used. The equivalent circuit is, of course, $Q_{eff} = R_T/X_L$, and the bandwidth is $BW = f_0/Q_{eff}$. If the power supply line (ac ground) is connected to tap point 2 instead of 1, the resulting equivalent circuit is that of Figure 7–24b. Here, $L_1 + L_2 = L$, so the circuit is resonant at the same frequency. However, since $L \propto N^2$, where N is the number of turns for the inductor,

$$X_{L_2} = n^2 X_L \qquad (7\text{–}40)$$

where n is the turns ratio defined by the tap point, $n = n_1/(n_1 + n_2)$. Ignoring finite inductor Q, the effective tapped circuit Q is $Q_T = R_T/X_{L_2} = R_T/(n^2 X_L) = Q_{eff}/n^2$. Since $n < 1$, $Q_T > Q_{eff}$ of the untapped transformer.

■ ■ ■ ■

**EXAMPLE
7–6**

In the circuit of Figure 7–24, $R_T = 2.5$ k and $X_L = 500\ \Omega$. Determine the effective Q of the circuit with point 1 (only) connected to ground, and compare it to the Q with point 2 (only) connected to ground. The tap point is $\frac{1}{3}$ of the inductor turns from the bottom.

Solution:

$Q_{eff} = 2500/500 = 5$. $X_L = n^2 X_{L_2} = (\frac{1}{3})^2 500 = 55.5\ \Omega$. $Q_T = 2500/55.5 = $ **45.** The Q has been increased by $1/n^2 = 9$ times. The bandwidth is $\frac{1}{9}$ of the untapped value.

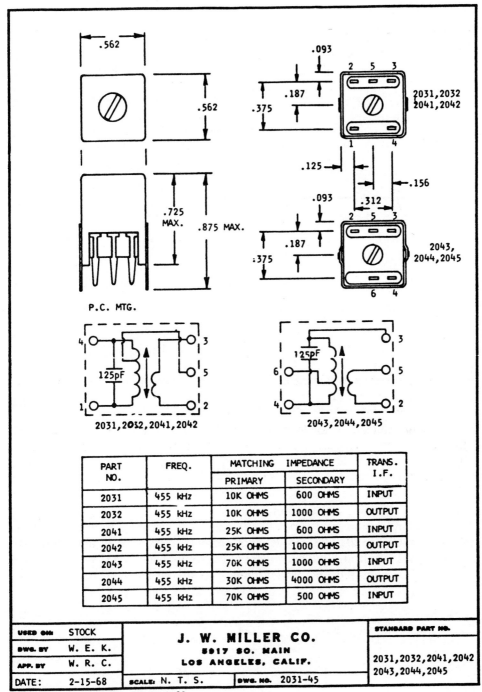

PART NO.	FREQ.	MATCHING IMPEDANCE		TRANS. I.F.
		PRIMARY	SECONDARY	
2031	455 kHz	10K OHMS	600 OHMS	INPUT
2032	455 kHz	10K OHMS	1000 OHMS	OUTPUT
2041	455 kHz	25K OHMS	600 OHMS	INPUT
2042	455 kHz	25K OHMS	1000 OHMS	OUTPUT
2043	455 kHz	70K OHMS	1000 OHMS	INPUT
2044	455 kHz	30K OHMS	4000 OHMS	OUTPUT
2045	455 kHz	70K OHMS	500 OHMS	INPUT

USED ON:	STOCK	J. W. MILLER CO.	STANDARD PART NO.
DWG. BY	W. E. K.	5917 SO. MAIN	2031,2032,2041,2042
APP. BY	W. R. C.	LOS ANGELES, CALIF.	2043,2044,2045
DATE:	2-15-68	SCALE: N. T. S. DWG. NO. 2031-45	

2031,2,41,2,3,4,5,51   5M FH 2-68

FIGURE 7-23   Specification sheet for IF transformers.

FIGURE 7-24 Tapping a coil for higher Q. (a) Tap not used. (b) Ground at tap point.

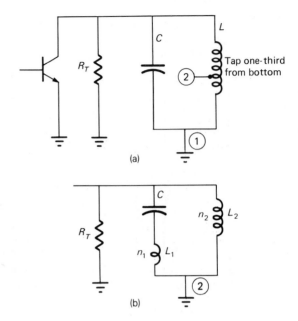

If the secondary of the tapped transformer is loaded with R_L as it is in the receiver application (Figure 7–11), then the total resistance R_T, which is proportional to base-emitter voltage gain, is

$$R_T = r_c \parallel (n_2/n_3)^2 R_L \qquad (7\text{-}41)$$

The dynamic collector impedance is r_c, and n_3 gives the number of secondary turns. Also,

$$Q_{\text{eff}} = \frac{R_T}{X_{LT}}$$

where X_{LT} is the same as Equation 7–40.

The last problem to resolve before studying AGC circuits is how to determine the load impedance that the AM detector presents to the final IF amplifier.

Figure 7–25 illustrates the equivalent circuit of the simple AM peak detector. Diode D_1 is a high-frequency signal diode, usually germanium for low-voltage

FIGURE 7-25 AM peak detector.

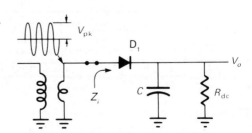

drop and constructed for low capacitance. C is the RF bypass and modulation-smoothing capacitor. R_{dc} is the equivalent dc-path resistance in the detector output to ground. If we assume an ideal diode ($C_D = R_D = V_D$ (junction drop) $= 0$) and an ideal RC time constant, then

$$Z_i = \frac{R_{dc}}{2} \qquad (7\text{–}42)$$

The simplest way to show the validity of Equation 7–42 is to equate the detector input and output power. Assume that the CW carrier is the only input. Then the input power is

$$P_i = \frac{V_{pk}^2}{2Z_i} \qquad (7\text{–}43)$$

where the 2 comes from squaring $V_{rms} = V_{pk}/\sqrt{2}$ for a sinusoid. Because the diode and RC time constant are both perfect, the output voltage is purely dc with magnitude $V_o = V_{pk}$. Since no power is lost in the diode, all the input ac power must be dissipated in R_{dc}. Hence,

$$P_o = \frac{V_{pk}^2}{R_{dc}}$$

and
$$\frac{V_{pk}^2}{2Z_i} = \frac{V_{pk}^2}{R_{dc}}$$

This result shows that $Z_i = R_{dc}/2$ Q.E.D.* Equation 7–42 is used to determine the transformer turns ratio for impedance matching the IF output to the AM detector.

7-7 ■ AGC CIRCUITS

The section on AGC in Chapter 5 introduced the need for automatic gain control in receivers where linearity is required before the demodulator. The voltage waveforms were shown at various points in the AGC system in Figure 5–24. The same system is shown in our prototype AM receiver in Figure 7–11, where the detector diode is reversed. The controlled amplifier is an NPN transistor so that, as the AGC voltage increases (more negative voltages), the Q_3 bias current decreases. When the IF amplifier gain is controlled by a reduction of current, the system is known as *reverse-AGC*. Reverse-AGC is used in satellite and space transponders or battery-operated receivers where minimum power consumption is important. Otherwise, in ground stations and home broadcast systems (AM, FM, and TV), *forward-AGC* is used.

With forward-AGC the amplifier bias point is set at I_F, as shown in Figure 7–26. As the received signal strength increases, the AGC voltage increases. The

---

* *Quod erat demonstrandum,* "which was to be demonstrated."

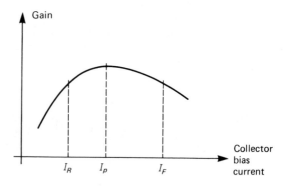

FIGURE 7-26 Transistor gain versus bias current. Bias points for forward- and reverse-AGC.

fed-back AGC voltage increases the already high transistor bias current ($I_p \sim 10$ to 20 mA) so that the base region is flooded with current, thereby spoiling the transistor current gain β. Forward-AGC has the advantage of increasing the amplifier power-handling capability as gain is reduced in order to maintain the best overload characteristic on strong signals when it is most needed.

Figure 7–27 shows an example of a circuit arrangement that would have to operate as forward-AGC. Do not mistake this circuit for a good design—it is merely straightforward to analyze. With no input signal the detector diode will be reverse-biased due to the positive voltage established by the R_1-R_2 voltage divider. These two resistors, along with R_3, set Q_1 for high gain with collector current in the range of I_F in Figure 7–26. As the positive peaks of the received signal at the detector increase, the base bias voltage on Q_1 increases. Hence, the bias must be set near I_F for the gain to be reduced for increasing input signal.

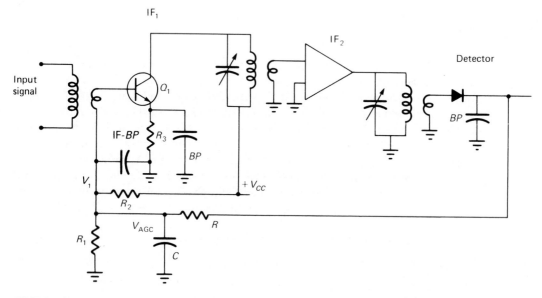

FIGURE 7-27 AGC network.

Assume that the R_1-R_2 voltage divider sets the bias point and AGC line to

$$\frac{R_1 V_{cc}}{R_1 + R_2} = V_1 \qquad (= V_{\text{AGC}})$$

Not only does this establish the Q_1 base bias voltage, but it also sets AGC threshold. This is because the IF input to the detector must reach $V_1 + V_d$ on peaks in order for the diode to start conducting and the AGC voltage to begin rising. V_d is the diode cut-in voltage drop, which we can assume to be approximately 0.2 V for germanium diodes. If V_1 is already established by the choice of R_1, R_2, and V_{cc}, then the detector input power required for the onset of AGC (threshold) must be

$$P_i(\text{threshold}) = \frac{V_{\text{pk}}^2}{2Z_i} \qquad (7\text{–}44)$$

$$= \frac{(V_1 + 0.2)^2}{R_{\text{dc}}}$$

R_{dc} is the dc resistance from the diode cathode to ground, which is, for this circuit,

$$R_{\text{dc}} = R + (R_1 \parallel R_2 \parallel R_{iQ_1}) \qquad (7\text{–}45)$$

where R_{iQ_1} is the dc input resistance of Q_1, approximated by $(\beta + 1)R_3$.

AGC of a transistor is a very nonlinear function. For reverse-AGC the reduction in base voltage reduces emitter current and, therefore, voltage gain due to the increase in r_e. This is approximately linear until V_{BE} goes below about 0.55 V. However, IF amplifiers are designed to amplify the input *power*. At the same time that voltage gain is reduced due to current-starving of the transistor, the current gain is also decreasing—this is in itself a nonlinear function. Forward AGC relies on spoiling the β by flooding the transistor base region with current. Therefore, putting a value on AGC threshold is mostly empirical, which often includes a potentiometer in the circuit. The discussion on receiver sensitivity in Chapter 5 showed that a 2-V peak carrier signal at the detector was a good place for the AGC system to hold the gain when a 90% AM signal is to be demodulated. This was based on a minimum of 0.2 V for the AM envelope.

For the purpose of analyzing an AGC circuit in this section, we will ignore the modulation linearity criteria and determine from circuit values the power level at which AGC action begins; this level is referred to as AGC *threshold*.

At this point, a working example problem is in order. However, since the previous discussion was on forward-AGC, the example problem will concentrate on the reverse-AGC system of our prototype AM receiver.

■ ■ ■ ■

EXAMPLE 7–7

Analyze the circuit of Figure 7–28 for the following:

1. Determine I_B.

2. Determine V_E (assume V_{BE} = 0.6 V), I_E, and β for the transistor.

3. Is the AGC system forward- or reverse-AGC?

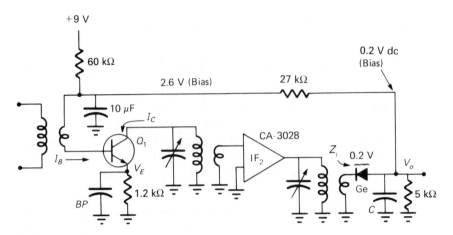

FIGURE 7-28   AGC circuit for analysis.

4. If the total power gain from IF input to detector input is 46 dB, determine the IF input signal power (in dBm) required for AGC threshold. Assume the IF input impedance is equal to Z_i of the detector.

5. If P_i to the detector is 2 mW total for an AM signal with $m = 0.9$, determine each value.

   a. The carrier power P_c
   b. V_c (peak value of carrier)
   c. V_o (dc)
   d. v_o (pk)
   e. Sketch v_o, including the peak voltage at V_{max} and V_{min}.
   f. Determine the value of C if modulation is 5 kHz max.

Solution:

1. $I_{60\ k\Omega} = (9 - 2.6\ \text{V})/60\ k\Omega = 0.107$ mA. $I_{27\ k\Omega} = (2.6 - 0.2\ \text{V})/27\ k\Omega = 0.089$ mA. The difference is $I_B = $ **0.018 mA.**

2. $V_E = 2.6\ \text{V} - 0.6\ \text{V} = $ **2V.** $I_E = 2\ \text{V}/1.2\ k\Omega = $ **1.67 mA.** $\beta + 1 = 1.67/0.018 = 92.6$; therefore, $\beta \approx $ **92.**

3. The detector diode in this circuit is initially biased on with 89 μA of current due to the bias circuit for the IF amplifier. Notice the direction of the detector diode—this is a reverse-AGC system; that is, the signals of increasing signal strength will be rectified and tend to pull V_o negative. This in turn diverts current away from Q_1 so that the gain is reduced, thereby keeping the detector input approximately constant.

4. For the sake of analysis, let's assume that 0.2 V pk is required to initiate AGC action. Then we need to determine only the detector impedance to compute the power. $R_{dc} = 5\ k\ ||\ [27\ k\Omega + (60\ k\Omega\ ||\ R_i(Q_1))]$. $R_i(Q_1) = 92.6(1.2\ k\Omega) = 111\ k\Omega$;

therefore, R_{dc} = 4.7 kΩ. P_d = (0.2 V pk)^2/4.7 kΩ = 8.5 μW(-21 dBm). With 46 dB of gain ahead of the detector, the IF input power need be only P_i = -21 dBm $-$ 46 dB = **-67 dBm.**

5. This is a more appropriate condition for setting the receiver threshold for reasonable demodulator linearity.

    a. From Chapter 5, $P_c = P_t/(1 + m^2/2)$ = **1.42 mW.**

    b. $V_c = \sqrt{2P_c Z_i} = \sqrt{(1.42 \text{ mW})(4.7 \text{ k}\Omega)}$ = **2.58 V pk.**

    c. $V_o = -(V_c - 0.2 \text{ V})$ = **-2.38 V** average.

    d. $V_o(\max) = -[(V_c + V_m) - 0.2 \text{ V}] = -[V_c(1 + m) - 0.2 \text{ V}] = -[2.58 \text{ V}(1 + 0.9) - 0.2 \text{ V}]$ = **-4.70 V pk.**

    e. $V_o(\min) = -(V_o - V_m)$ = **-0.058 V pk.** The results of these calculations are sketched in Figure 7–29.

    f. Also from Chapter 5, an appropriate value of detector smoothing capacitance is $C = \sqrt{(1/m)^2 - 1}/(2\pi \times 5 \text{ kHz} \times 4.7 \text{ k}\Omega)$ = **0.0033 μF.**

FIGURE 7–29 AM detector output.

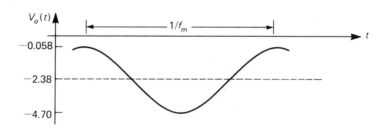

7–8 ■ THE AUDIO SYSTEM

To complete the AM receiver circuit analysis and design, let's analyze the audio system of our prototype receiver (Figure 7–11). The system consists of three amplifier stages with various forms of feedback.

The first stage receives its signal from a 5-kΩ volume (loudness) control potentiometer. The amplifier is a class A common-emitter preamplifier with bias stabilization and signal feedback. There are two completely different forms of negative feedback within this amplifier and feedback comes from the audio system output back to the emitter of Q_5. Since feedback is extremely important in communications systems, these circuits will be analyzed in the next section on distortion and negative feedback.

The second stage is a class A grounded-emitter driver amplifier for the output stage. The output stage is a class AB (almost B) complementary push-pull power amplifier. There are essentially two forms of feedback within these two amplifiers. The 47-μF capacitor at the output provides *bootstrapping*, and bias stability (with a small amount of signal feedback) is provided by the 12-kΩ resistor from output to the base of Q_6. (The bootstrapping technique is described in the next section.) The bias stability technique is virtually identical to that provided by the 560-kΩ

resistor, from collector to base of Q_5, except that both the driver and power amplifier are stabilized.

Before the class B power amplifier is analyzed, you may want to review power and efficiency presented with a transformer-coupled class A power amplifier design example in Appendix 7A. Inclusion of the class A power amplifier will also provide design examples for all three amplifier classifications, since class C was analyzed for transmitters in Chapter 6.

Class B Push-Pull

A simplified version of our prototype receiver audio PA and driver is shown in Figure 7–30.

Resistor R_B is the collector resistor for the common-emitter driver amplifier with Q_1. Diodes D_1 and D_2 are always biased "on" (they do not rectify) and are used to produce a nearly constant voltage drop from the base of Q_2 to Q_3. The diode ac impedance is very low. Diodes, unlike resistors, also help in tracking out V_{BE} changes with temperature. The R_E resistors are often left out of power amplifiers but are included here to make the analysis simpler.

If two power supplies or batteries are used, the capacitor is not required. This has a definite advantage because the value of C is large to minimize frequency and phase distortion at low frequencies ("tilt" or "sag" in squarewave tests). However, for a single power supply the capacitor is required to hold the charge, which supplies current for the negative half of the output signal swing. This is explained as follows: With no input signal the capacitor is charged to approximately $V_{cap} = V_{cc}/2$, whereas Q_2 and Q_3 are biased slightly "on" due to D_1 and D_2 (this is called *class AB*, or "almost B"). When the collector of Q_1 is rising due to ac signal, Q_2 is conducting heavily and the increased voltage drop across R_{E1} has the effect of almost cutting off Q_3. Load current for this positive half of the v_o signal swing comes from V_{cc} via Q_2. For the negative half of the output voltage swing, Q_2 is nearly cut off and Q_3 conducts heavily. The current for this half of the v_o signal swing is supplied by charge stored in C. The capacitance must be very large in order to store enough charge so that V_{cap} in effect replaces a $-V_{cc}$ power supply,

FIGURE 7–30 Complementary push-pull PA/driver.

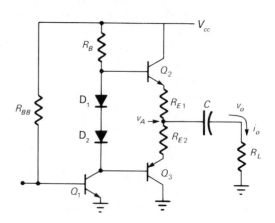

which would be placed from the collector of Q_3 (and emitter of Q_1) to ground in a dual-supply system.

The transformerless class B push-pull power amplifier is analyzed in Example 7–8. From this we will see the reason for the bootstrapping in the prototype receiver audio output.

■ ■ ■ ■

EXAMPLE 7–8

Analyze the class AB (almost B) transformerless push-pull amplifier of Figure 7–30 with $V_{cc} = 9$ V, $R_B = 1$ kΩ, $R_E = 2$ Ω, and $R_L = 8$ Ω. Assume that the silicon diodes drop 0.615 V and the transistors have $V_{BE} = 0.60$ V, $V_{CE}(\text{sat}) = 0$, and $\beta = 200$. Determine the following:

1. DC bias voltages and currents at all points, assuming V_A is set at the optimum bias point. Find R_{BB} to set V_A for the optimum bias
2. Ideal v_o, i_o, and power to the load P_L
3. Efficiency for part 2; also the theoretical efficiency for class B
4. Corrected v_o and P_L; the need for bootstrapping

Solution:

1. V_A (to ground) should be set at about $\frac{1}{2}V_{cc} = 4.5$ V so that the output can have the maximum swing before peak clipping. Because of circuit symmetry the bias voltage between the diodes will equal V_A; consequently, $V_{B2} = \textbf{5.115 V.}$ (V_{B2} is the voltage from the base of Q_2 to ground. The other bias voltages follow this notation.) $V_{C1} = 4.5 - 0.615 = \textbf{3.885 V} = V_{B3}$. $V_{E3} = 3.885 + 0.60 = \textbf{4.485 V,}$ and $V_{E2} = 5.115 - 0.60 = \textbf{4.515 V.}$ $V_{C2} = \textbf{9 V,}$ $V_{E1} = \textbf{0 V,}$ and $V_{B1} = \textbf{0.60 V.}$ $V_o = 0$ on bias (and is the average output voltage). $V_{\text{cap}} = V_A - V_o = \textbf{4.5 V dc.}$

   The voltage across $2R_E$ is $V_{E2} - V_{E3} = 4.515 - 4.485 = 0.03$ V; therefore, $I_{CQ2} = I_{CQ3} = 0.03$ V/4 Ω = $\textbf{7.5 mA.}$ $I_{CQ1} = I_{R_B} = (9$ V $- 5.115$ V)/1 kΩ = $\textbf{3.9 mA.}$ $I_{B1} = 3.9$ mA/$\beta = 19.5$ μA; therefore, $R_{BB} = (9 - 0.60)/19.5$ μA = 431 kΩ.

2. Ideally, when Q_2 conducts at the maximum with $V_{CE}(\text{sat}) = 0$, then $V_{E2} = 9$ V pk. As we will see, this is impossible with such a low-impedance load. Anyway, assuming V_{E2} can reach 9 V on positive peaks, then KVL will yield

$$V_{cc} - V_{CE}(\text{sat}) - V_{\text{cap}} = i_o(\text{ideal}) \times (R_E + R_L).$$

   Hence, $i_o(\text{ideal}) = 4.5$ V pk/10 Ω = $\textbf{450 mA pk,}$ and $v_o = 450$ mA × 8 Ω = $\textbf{3.6 V pk}$ ideally. The power delivered to the load for a full sinewave signal will be $P_L = (0.45\text{A}/\sqrt{2})(3.6$ V$/\sqrt{2}) = \textbf{810 mW}$ ideally.

3. Transistor Q_2 conducts only during the positive half of the output sinusoid. Therefore, the current i_{C2} approximates a half-rectified sinusoid with an average value of i_{pk}/π. Hence, the battery supplies 450 mA/π = 143.2 mA and a voltage of 9 V.

$$P_{dc} = V_{cc} \times \frac{i_o(\text{pk})}{\pi} \qquad\qquad (7\text{--}46)$$

for class B; P_{dc} = 1.29 W, so that the efficiency is η = 810 mW/1.29 W = **62.8%**.

The theoretical efficiency is found for the preceding conditions but no R_E. Then v_o(pk) = $V_{cc}/2$ and i_o(pk) = $V_{cc}/2R_L$. The theoretical efficiency will be

$$\eta = \frac{P_L}{P_{dc}} = \left[\frac{v_o(\text{pk})}{\sqrt{2}} \times \frac{i_o(\text{pk})}{\sqrt{2}} \right] \Big/ \left(V_{cc} \times \frac{i_o(\text{pk})}{\pi} \right)$$

$$= \frac{V_{cc}^2/8R_L}{V_{cc}^2/2\pi R_L} = \mathbf{78.5\%}$$

4. In part 3, i_o was determined to be 450 mA pk. In order for Q_2 to conduct this much current from the battery, the base current would have to be i_{B2} = 450 mA/β = 2.25 mA peak. This has to come through R_B as the current in Q_1 reaches a minimum. However, because R_B is so large, the highest that V_{B2} can actually rise when i_{B2} = 2.25 mA is 9 V − (2.25 mA × 1 kΩ) = 6.75 V pk. As a result Q_2 cannot reach saturation, and indeed the maximum value that i_o can reach is

$$i_o(\text{max}) = \frac{\dfrac{V_{cc}}{2} - V_{BE}}{R_E + R_L + \dfrac{R_B}{\beta}} \qquad\qquad (7\text{--}47)$$

This is determined by assuming that $i_{c1} \rightarrow 0$ on the positive signal peak, and writing KVL through R_B, Q_2, and the output circuit as $V_{cc} - i_B R_B - V_{BE} - i_o(R_E + R_L) - V_{cc}/2 = 0$, with $\beta \gg 1$.

Using Equation 7–47 the maximum output current for this example problem will be i_o(max) = **260 mA pk.** Then v_o(max) = 260 mA × 8 Ω = **2.08 V pk**, and P_L = (2.08 V pk)^2/(2 × 8 Ω) = 270 mW versus 810 mW ideally. Clearly, we need to reduce the ac voltage drop across R_B.

The solution is called *bootstrapping*. In Figure 7–31 (with R_E removed to reduce the voltage loss across it), the signal swing v_o at the base of Q_2 is

FIGURE 7–31 Bootstrap feedback through C_b.

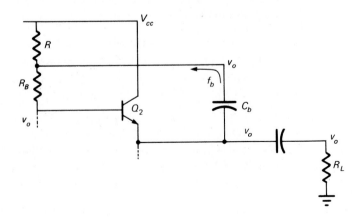

approximately the same in magnitude and phase as at the output because Q_2 (and Q_3) is an emitter-follower configuration. The voltage coupled back to the opposite end of R_B is also v_o, which means that the ac voltage drop across R_B is approximately zero. This means that the (positive) feedback provided by the bootstrap capacitor C_b lifts the base of Q_2 up "by its own bootstraps." The bootstrapping technique allows for a relatively large R_B for the dc voltage drop to give appropriate bias voltages while the ac voltage loss is low.

7–9 ■ DISTORTION AND FEEDBACK

As described in the last section, the prototype AM receiver audio system has three forms of negative feedback plus bootstrapping, which is positive (in-phase) feedback. Each of the three specific examples will be discussed. However, first consider the four general categories for identifying negative-feedback systems.

It is very important to be able to identify and understand the four forms of feedback because each circuit arrangement produces very different results. Figure 7–32 shows in block-diagram form the various circuit arrangements and the effect on input and output impedance.

Notice from Figure 7–32a and b that when the output voltage is sampled, the output impedance is reduced. This is because the effect of the negative feedback is to hold constant that parameter which is sampled. Recall that a constant voltage generator is a low-impedance device. On the other hand, the output current is sampled in c and d, and a constant-current generator is a high-impedance device.

As for the input impedance, shunt feedback will lower the input impedance, whereas a series feedback arrangement will increase the impedance.

In all cases, negative (180° out of phase) feedback *reduces* distortion, phase shifts, and voltage *or* current variations and *increases* circuit bandwidth.

The most common method of reducing circuit-induced distortion is to use negative feedback around the circuit; that is, by taking some of the distorted output and feeding it back 180° out of phase, the original distortion can be reduced. The improvement is *not* due to the reduced gain. The input can be increased until full output power is restored, and the distortion will be reduced by an amount given by the improvement factor T, where

$$T = 1 + \text{(loop gain)} \tag{7–48}$$

The loop gain for Figure 7–33 ia $A_V B$. The feedback factor B is by definition the fraction of the output signal that is fed back to the input. Distortion of amount D, which occurs inside the integrated circuit (IC) of gain A_V, appears at the amplifier output as D_f, the distortion with feedback. A fraction B of the output signal is fed back to the inverting amplifier input. The feedback signal of magnitude BD_f is amplified in the inverting amplifier by $-A_V$ so that $-A_V BD_f$ is fed back to reduce the original distortion D. Thus the distortion seen at the output (with feedback) is

$$D_f = D - A_V BD_f \tag{7–49}$$

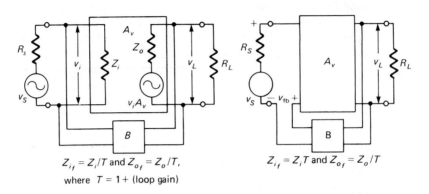

(a) Voltage-sampled/shunt-feedback.   (b) Voltage-sampled/series-feedback.

$Z_{i_f} = Z_i/T$ and $Z_{o_f} = Z_o/T$, $Z_{i_f} = Z_i T$ and $Z_{o_f} = Z_o/T$

where $T = 1+$ (loop gain)

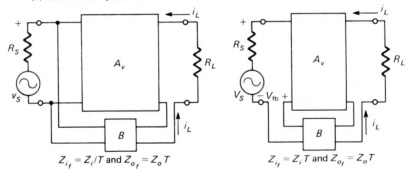

(c) Current-sampled/shunt-feedback.   (d) Current-sampled/series-feedback.

$Z_{i_f} = Z_i/T$ and $Z_{o_f} = Z_o T$ $Z_{i_f} = Z_i T$ and $Z_{o_f} = Z_o T$

FIGURE 7–32 The four feedback arrangements.

Solving Equation 7–49 for D_f yields

$$D_f = \frac{D}{(1 + A_V B)} \tag{7–50}$$

Equation 7–50 shows that the distortion that occurred in the IC is effectively reduced by $1 +$ (loop gain).

FIGURE 7–33 Distortion D in amplifier. The distortion measured at the output is D_f.

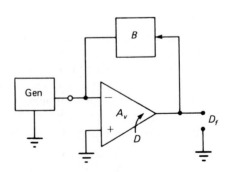

One way of demonstrating the improvement factor is by analyzing a familiar amplifier by the usual techniques and comparing the results to those obtained by using the feedback approach. An example is the first-stage amplifier of the prototype receiver audio system. Example 7–9 will show that, by inserting a resistor from emitter to ground in a common-emitter amplifier producing negative feedback, a reduction in gain will occur that is described by

$$A_{V_f} = \frac{A_V}{1 + A_V B} \qquad (7\text{--}51)$$

and an increase in input impedance described by

$$Z_{i_f} = Z_i(1 + A_V B)$$

■ ■ ■ ■

EXAMPLE 7–9

Current-Sampled/ Series-Feedback

Figure 7–34a is a common-emitter amplifier with no feedback resistor R_E, whereas 7–34b includes feedback due to the 10-Ω emitter resistor.

1. Analyze both circuits for gain and input impedance by the usual circuit technique.

2. Analyze the circuit in (b) by the feedback technique and compare the results to part 1.

3. Determine the output impedance.

Solution:

Since the transistor can affect the results, the exact relationships are written. When $\beta/(\beta + 1) = 0.99$ occurs, the value 1 is used.

1. **a.** No feedback. $A_V = -\beta[R_C/(\beta + 1)r_e] \approx -R_C/r_e = -195 = $ **195** with phase inversion. $Z_i = (\beta + 1)r_e = 100 \times 20 = $ **2 kΩ.**

   b. $A_{V_f} = -\beta[R_C/(\beta + 1)(r_e + R_E)] \approx -3.9 \text{ k}\Omega/30 = $ **−130** (lower with feedback). $Z_{i_f} = (\beta + 1)(r_e + R_E) = 100 \times 30 = $ **3 kΩ** (high because of series feedback).

FIGURE 7–34 (a) Amplifier without feedback. (b) Amplifier with feedback.

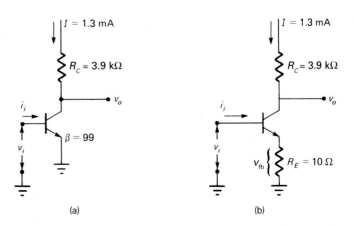

2. The output is taken from the collector. The output current develops a feedback signal v_{fb} that is in series with the input signal current. Hence, the feedback arrangement is current-sampled/series-feedback (Figure 7–32d). The feedback factor for a given circuit can be obscure; however, its definition again is "that fraction of the output signal fed back to the input." So

$$B = \frac{v_{fb}}{v_o} = \frac{i_e R_E}{i_c R_C} = \left[\frac{(\beta + 1)}{\beta} \right] \frac{R_E}{R_C} \approx \frac{R_E}{R_L}$$

Consequently, $B = 10/3.9 \text{ k}\Omega = 0.0026$. Therefore,

$$A_{V_f} = \frac{A_V}{(1 + A_V B)} = \frac{-195}{(1 + 195 \times 0.0026)} = \textbf{-130}$$

and
$$Z_{i_f} = Z_i(1 + A_V B) = (2 \text{ k}\Omega)(1 + 0.5) = \textbf{3 k}\Omega$$

3. For Figure 7–34a (with no feedback) the output impedance is $3.9 \text{ k}\Omega \parallel r_c$. Because feedback with current sampling at the output tends to hold the output current constant, output impedance increases to $Z_{0_f} = Z_0(1 + A_V B) = \textbf{3.9 k}\Omega \parallel$ $\textbf{(1.5 } r_c\textbf{)}$.

Voltage-Sampled/Shunt-Feedback

The first stage of the prototype audio system (Figure 7–11) has bias stabilization and some signal feedback due to the collector-to-base feedback resistor $R_F = 560$ kΩ. The bias stability is achieved because a bias voltage shift at the collector will be fed back by voltage division between the 560 kΩ and dc input impedance of Q_5, which is $R_i(Q_5) = (\beta + 1)(r_e + 10 \text{ }\Omega)$. The ac signal feedback takes the same path with the same output-voltage-sampled/shunt-feedback technique. One difference between the ac and the dc analyses, however, is that the ac feedback current through the 560 kΩ splits between the 5-kΩ potentiometer and a much higher transistor R_i. The ac input impedance is higher than $R_i(Q_5)$ given previously because of the series feedback of signal from the speaker to the emitter of Q_5— around the entire audio system.

Figure 7–35 illustrates, in a more recognizable inverting op-amp form, the amplifier with voltage-sampled/shunt-feedback. R_s is that part of the 5-kΩ potentiometer between the wiper and ground of Figure 7–11. In any case the IC gain is $A_V = 130$ for dc variations (from Example 7–9), and the input impedance is $R_i = 3$ kΩ for dc variations. AC gain is less than 130 because of the input impedance of the next amplifier; however, R_i for the IC is much higher than 3 kΩ because of feedback from the speaker. Since A_V is high, $R_F \gg R_S$ and $R_S \ll R_i$, then

$$B = \frac{V_{fb}}{v_o} \approx \frac{i_F R_1}{i_F R_F} = \frac{R_1}{R_F}$$

where $R_1 = R_i$ for the dc stabilization and $R_1 = R_s \parallel R_i$ for ac. The feedback improvement factor is

FIGURE 7-35 Op-amp configuration for inverting transconductance amplifier.

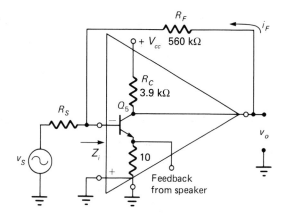

$$T = 1 + A_V\left(\frac{R_1}{R_F}\right) \qquad\qquad (7\text{--}52)*$$

Notice that for the ideal high-gain IC ($A_V B \gg 1$), the ac gain with feedback reduces to approximately

$$A_{V_f} = \frac{-A_V}{A_V B} = \frac{-1}{B} = \frac{-R_F}{R_1}$$

which you recognize for the ideal inverting op-amp.

Voltage-Sampled/Series-Feedback

To analyze the overall audio system with feedback for the reduction of distortion, refer to the simplified sketch of Figure 7–36. This is the audio system of the prototype AM receiver of Figure 7–11 where R_S and v_s vary with the 5-kΩ potentiometer, R_L is an 8-Ω speaker, and the choice of a value for R_C takes into account the effects of loading from the driver input impedance and the feedback from a 470-kΩ resistor (not shown).

First check the feedback phase—is it negative or positive feedback? If the input signal v_s rises, the collector voltage will drop. The driver/PA system is an inverting amplifier, so that v_o will rise. So, if the feedback resistor were connected back to the base, the result would be in-phase or regenerative feedback. By feeding back to the emitter, the rising voltage v_{fb} will tend to reduce the forward bias of the transistor, thereby reducing the voltage variations at the collector (and v_o). This is negative feedback and in fact is voltage-sampled/series-feedback.

---

* An exact analysis (see Appendix C) is based on gain and feedback characterized as conductances for this transconductance amplifier. The exact result is $T = 1 + [A_v R_F R_1/(R_F + R_C)(R_F + R_1)]$, which reduces to Equation 7–52 for the impedance assumptions already stated and $R_F \gg R_C$.

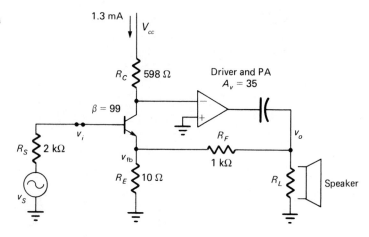

FiGURE 7–36 Audio system with voltage-sampled/series-feedback.

The loop gain is determined as follows:

$$A_V = A_{V_1}A_{V_2} \tag{7-53}$$

where $A_{V_2} = 35$ as shown in Figure 7–36, and

$$A_{V_1} = \frac{-R_C}{r_e + R_E'} \tag{7-54}$$

and

$$R_E' = R_E \parallel R_f \tag{7-55}$$

reflects the loading effect of the feedback network. The feedback factor is the voltage division given by

$$B = \frac{v_{fb}}{v_o} = \frac{R_E}{R_E + R_F} \tag{7-56}$$

So

$$T = 1 + A_V B$$

$$= 1 + \left[\frac{A_{V_1}A_{V_2}R_E}{R_E + R_F}\right] \tag{7-57}$$

■ ■ ■ ■

EXAMPLE 7–10

Voltage-Sampled/ Series-Feedback

Determine the following for the audio system of Figure 7–36 with $R_S = 2\ \text{k}\Omega$:

1. A_{V_f}
2. Z_{o_f}, assuming $Z_o = 5\ \Omega$ (impedance looking into the output and the driver and PA)
3. Z_{i_f}
4. Power to the speaker if $v_s = 20\ \text{mV}$ pk

Solution:

1. $R'_E = 10 \| 1 \text{ k}\Omega = 9.9 \ \Omega$. $r_e = 26/1.3 = 20 \ \Omega$. $A_{V_1} = -598/(20 + 9.9) = \textbf{20.}^*$
   $A_V = 20 \times 35 = 700 \ (56.9 \text{ dB})$. $T = 1 + [(20 \times 35) \times (10/1010)] = 1 +$
   $[(700)(0.0099)] = 7.93 \ (18 \text{ dB})$. $A_{V_f} = (20 \times 35)/7.93 = \textbf{88.3} \ (38.9 \text{ dB})$. This
   solution is also computed from 56.9 dB of forward gain reduced by 18 dB of
   feedback, so $A_{V_f} \ (\text{dB}) = 56.9 \text{ dB} - 18 \text{ dB} = \textbf{38.9 dB.}$

2. $5 \ \Omega/7.93 = \textbf{0.63} \ \Omega.$ This is certainly low enough for driving an 8-Ω speaker.

3. $Z_i = (\beta + 1)(r_e + R'_E) = 100 \times 29.9 = \textbf{2.99 k}\Omega.^*$ For the audio system with
   series feedback, $Z_{i_f} = Z_iT = 2.99 \text{ k}\Omega \times 7.93 = \textbf{23.7 k}\Omega.$

4. Now that the system input impedance is known, we can determine the signal
   loss at the input due to voltage division. $v_i = [Z_{i_f}/(Z_{i_f} + R_S)]v_s$. Therefore, the
   input loss is $v_i/v_s = 23.7/(23.7 + 2 \text{ k}\Omega) = 0.92 \ (-0.7 \text{ dB})$. $v_o = 0.92 \times 88.3v_s =$
   $(81.2)(20 \text{ mV}) = 1.6 \text{ V pk}$, and the average power to the 8-Ω speaker will be
   $P_L = (1.6 \text{ V pk})^2/(2 \times 8 \ \Omega) = \textbf{165 mW.}$

Frequency Effects and Stability with Feedback

In the discussion thus far we have found that the use of negative feedback offers
many improvements over a system without feedback (open-loop). All the assets,
however, do not come without potential hazards, of which we should be aware.

For example, feedback is used in power amplifiers to reduce distortion. Figure
7–37 shows a distorted signal and the inverted feedback signal. 180° of phase
between the original and fed-back signals allows optimum reduction of distortion.
However, suppose the feedback is not 180°, as illustrated in Figure 7–38; the
distortion will not be reduced effectively. If the phase shift is great enough, the
distortion will be worse than without feedback. Even worse, the marginally stable
system with its underdamped behavior can become unstable and oscillate. Either
way the system becomes at least useless, possibly destructive.

How does the extra phase shift occur? The feedback system of Figure 7–39
shows the input and output capacitances of the amplifying device. A multistage IC
or amplifier will possess many of the RC time-constant delays indicated. Each RC
time constant will produce a signal delay (phase shift) in addition to the amplitude
reduction. The sinusoidal input-output relationship (transfer function) for the
R_oC_o voltage-divider network is derived as follows:

$$v_o = (v_s)\left[\frac{-jX_{C_o}}{R_o - jX_{C_o}}\right]$$

$$= v_s\left(\frac{1/j2\pi fC_o}{R_o + 1/j2\pi fC_o}\right)$$

---

* The gain is decreased and the input impedance is increased for this stage due to local feedback—a
current-sampled/series-feedback amplifier.

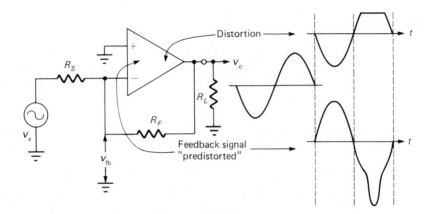

FIGURE 7-37   Feedback 180° for good distortion reduction.

Therefore,
$$\frac{v_o}{v_s} = \frac{1}{(1 + j2\pi fRC)} \tag{7-58a}$$

$$= \frac{1}{\left(1 + \dfrac{jf}{f_c}\right)} \tag{7-58b}$$

where $f_c = 1/(2\pi RC)$. The magnitude reduction with frequency is calculated as

$$\left|\frac{v_o}{v_s}\right| = \frac{1}{\sqrt{1^2 + (2\pi fRC)^2}} \tag{7-59}$$

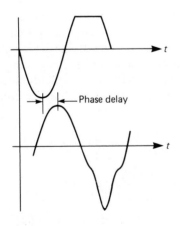

FIGURE 7-38   Additional phase delay will reduce feedback effectiveness.

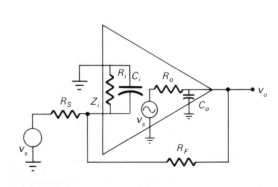

FIGURE 7-39   Internal phase-shifting impedances.

and the phase delay is found from

$$\phi = -\tan^{-1}(2\pi fRC) = -\tan^{-1}\left(\frac{f}{f_c}\right) \qquad \textbf{(7--60)}$$

For multistage, multi-RC-network amplifiers, the phase shifts for each of the corner frequencies are added together. Figure 7–40 illustrates the magnitude and phase-response curves versus frequency for the open-loop amplifier $A_V(f)$. $A_{V_f}(f)$ is the ideal frequency-response Bode plot for the system with feedback, and f_{CO} is the frequency at which the loop gain equals 0 dB. Frequency f_{CO} is called the loop *crossover* frequency.

The phase shift for the system with three RC time constants (Figure 7–40) can be determined at any frequency f by adding up the contributions for each of the three corner frequencies f_1, f_2, and f_3. If the low-frequency phase inversion of the amplifier is not included, then the additional phase shift at the crossover frequency f_{CO} is determined as

$$\phi(f_{CO}) = -\left[\tan^{-1}\left(\frac{f_{CO}}{f_1}\right) + \tan^{-1}\left(\frac{f_{CO}}{f_2}\right) + \tan^{-1}\left(\frac{f_{CO}}{f_3}\right)\right] \qquad \textbf{(7--61)}$$

Phase shifts other than the 180° of the inverting amplifier can result in stability problems. The simple expression for system gain with negative feedback is $A_{V_f} = A_V/(1 + BA_V)$. If we assume that all phase shifts are internal to the IC, B remains constant with frequency. Our concern is with the denominator of this expression,

FIGURE 7–40 System frequency response.

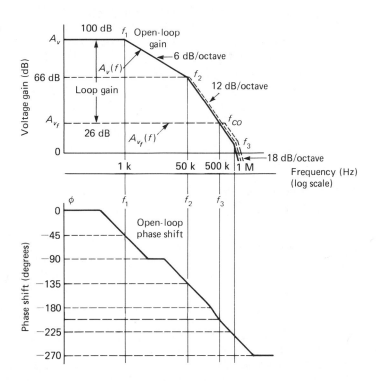

because if $BA_V = -1$, then $A_{V_f} = A_V/(1 - 1)$ goes to infinity; this is the unstable condition that results in oscillations.

The loop gain changes with frequency because of the amplifier rolloff $A_V(f)$. Loop gain $BA_V(f) = -1$ means that the magnitude $|BA_V| = 1$ and the phase shift around the loop has changed a full 180°; that is, loop gain has magnitude and phase $BA_V = |BA_V|\ \underline{/\phi}$. For the system to be stable, $|BA_V|$ must be less than unity (0 dB) before ϕ reaches 180°. The critical frequency to investigate is f_{CO} because this is by definition where $|BA_V(f_{CO})| = 1$. Beyond this frequency loop gain is insufficient to sustain oscillations even if ϕ exceeds 180°; the system will be unstable and have sustained oscillations if $\phi(f_{CO}) \geq 180°$ as calculated by Equation 7–61. The phase response of the Bode plot (Figure 7–40) can also be used to determine the phase shift at f_{CO}.

Another graphical technique is the polar plot of $BA_V(f)$, called a *Nyquist plot*. This is illustrated in Figure 7–41 and is a plot of $|BA_V|\ \underline{/\phi}$ as the system input signal frequency is changed from $f = 0$ to infinity. Notice that the solid curve includes the point $BA_V = -1$, ($|BA_V| = 1\ \underline{/\phi = 180°}$). This system will be unstable and exhibit self-sustained oscillations. The dashed curve, on the other hand, is for the same system except with lower loop gain. Since the point -1 is never inside this curve, this feedback system will be stable; that is, it will not exhibit sustained oscillations.

However, undesirable results will occur if the dashed curve comes too close to point -1. In this case $90° < \phi(f_{CO}) < 180°$ and the system will be *marginally stable*. The marginally stable system will exhibit regenerative peaking in the Bode plot near the natural resonant frequency of the system, and it will exhibit *overshoot* and *ringing* for sudden input signal changes, such as impulses and squarewaves or other voltage steps. Such a system is second order (or higher) like an LC resonant circuit and *underdamped*.

Phase margin (P.M.) is the maximum amount by which the total loop phase shift could be increased before the onset of sustained oscillations. For amplifiers with feedback

$$\text{P.M.} = 180° - |\phi(f_{CO})| \qquad (7\text{–}62)$$

FIGURE 7–41 Nyquist plot.

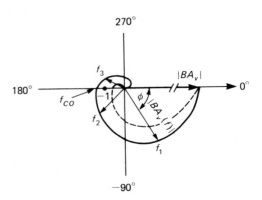

For more on marginally stable second-order systems see Figures 10–27 and 10–30 with the associated discussion. These curves include the amount of ringing and Bode plot peaking for various values of phase margin. For P.M. ≤ 0, the system is unstable and will exhibit oscillations.

Phase margin can be increased and the system stability improved by either reducing the loop gain (turning down the amplifier gain) or moving one or more of the corners higher in frequency (smaller RC). IC manufacturers sometimes include an internal capacitor from collector to base of a high-gain amplifier stage (30 pF in the μA741 or LM101), providing what is called *narrowband* and *dominant-pole* compensation. The RC time constant is so large, due to the Miller-effect capacitance increase for the collector-base capacitor, that the IC acts like it has only one (dominant) corner frequency (pole). For many applications this unnecessarily narrows the amplifier bandwidth, so most ICs have frequency compensation pins to give the circuit designer both gain and damping control. Compensation techniques such as *lag* and *lead-lag* compensation are discussed in Chapter 10.

■ ■ ■ ■

EXAMPLE 7-11

Figure 7–40 shows the open-loop Bode plot for an integrated circuit with three major corner frequencies. The IC gain at low frequencies is A_V (dB) = 100 dB ($A_V = 10^5$), and the corner frequencies are $f_1 = 1$ kHz, $f_2 = 50$ kHz, and $f_3 = 1$ MHz. Determine the system phase margin and stability for the following:

1. 74 dB of negative feedback
2. 34 dB of feedback

Solution:

1. With 74 dB of feedback the system gain with feedback will be $A_{V_{fb}}$ (dB) = 100 dB − 74 dB = 26 dB. This is also called the *closed-loop system gain*. The long dashed line of Figure 7–40 shows the ideal closed-loop frequency response *assuming the loop is stable*. The point of intersection of this line with $A_V(f)$ is the crossover frequency, $f_{CO} = 500$ kHz. The phase margin is investigated at this frequency because, for any frequency less than f_{CO}, there is a positive loop gain. The phase margin is

$$\text{P.M.} = 180° - \left[\tan^{-1}\left(\frac{500 \text{ kHz}}{1 \text{ kHz}}\right) + \tan^{-1}\left(\frac{500 \text{ kHz}}{50 \text{ kHz}}\right) + \tan^{-1}\left(\frac{500 \text{ kHz}}{1000 \text{ kHz}}\right) \right]$$

$$= 180° - (89.9° + 84.3° + 26.6°) = \mathbf{-20.7°}$$

   Since P.M. ≤ 0, the system will oscillate.

2. $A_{V_{fb}}$ (dB) = 100 dB − 34 dB = 66 dB. $f_{CO} = 50$ kHz. Let's see if lowering the loop gain has helped the stability.

$$\text{P.M.} = 180° - \left[\tan^{-1}\left(\frac{50 \text{ kHz}}{1 \text{ kHz}}\right) + \tan^{-1}\left(\frac{50 \text{ kHz}}{50 \text{ kHz}}\right) + \tan^{-1}\left(\frac{50 \text{ kHz}}{1000 \text{ kHz}}\right) \right]$$

$$= 180° - (88.9° + 45° + 2.9°)$$

$$= 180° - 136.8° = \mathbf{43.2°}$$

The system with 34 dB of feedback is stable. (Figure 10–27 shows that the damping factor of a second-order system would be slightly less than 0.5, and there would be slightly more than 15% overshoot—1.15 on the normalized vertical axis. Figure 10–30 shows that the frequency response for a second-order system with $\delta = 0.5$ (P.M. = 45°) will have very little rise. The peak occurs at approximately $0.55\omega_n$, where $\omega_n = 2\pi f_0$ is the natural undamped frequency of oscillation).

The closed-loop frequency response A_{V_f} for our third-order system will look almost identical to the short-dashed line from 66 dB across to $A_V(f)$ at f_{CO} and then following the open-loop response, because there is no more loop gain (and no feedback) for input signals with frequencies greater than f_{CO}.

Please notice also from the closed-loop Bode plot of Figure 7–40 how the bandwidth of the stable system has increased to approximately f_{CO} from the open-loop bandwidth of f_1.

■ PROBLEMS

1. Give two numerical reasons for making the first amplifier in a receiver low noise and high gain.

2. List two important measures for the limits of linearity in receiver circuits.

3. For the amplifier measurements plotted in Figure 7–4, determine
   a. The amount of lost gain at 0 dBm input due to saturation.
   b. The third-order distortion for −23-dBm input (the difference, in dB, between linear and third-order output power).
   c. Amplifier gain for 0-dBm input.

4. An amplifier with 50 Ω output impedance has saturation characteristics as shown in Figure 7–42. Determine
   a. Linear voltage gain in dB.
   b. Output power at the 1-dB compression point.
   c. The third-order intercept point, in volts.

FIGURE 7–42

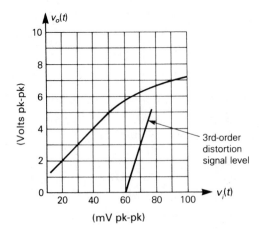

5. Expand cv^3 for $v = \cos \omega_{LO}t + \cos \omega_{RF}t$ to show the third-order intermodulation products.

6. The germanium diode of mixer 1 in Figure 7–8(1) is measured to have $V_{dc} = 0.2$ V. What value of self-bias resistor would allow 200 μA?

7. The dual-gate FET mixer of Figure 7–8 has an AGC voltage of +7 V. Determine
   a. The bias voltage at both gates.
   b. The value of L in order to use a 10-pF (max) variable capacitor at midcapacity if the stray and FET drain capacitances total 3 pF. (IF = 45 MHz for TV application.)
   c. How much bias voltage is measured from source for $I_D = 2$ mA and from the drain?
   d. How much bias power is dissipated by the FET for part c?

8. The resistors in a CA3028 are given in Figure 7–21. For the circuit of Figure 7–8(4)
   a. Determine the bias voltages at all pins shown.
   b. Assuming high-β transistors, what will be the collector bias current in each transistor?

9. The converter of Figure 7–8(6) has $R_1 = 32$ kΩ, $R_2 = 6$ kΩ, and a germanium transistor ($V_{BE} = 0.18$ V). Determine
   a. R_E to set $I_E = 0.8$ mA for a portable AM receiver ($V_{cc} = +9$ V).
   b. L if $C_1 = 125$ pF.
   c. L_1 for tuning the oscillator to 2 MHz with 30 pF. (What frequency will be received?)
   d. What value of reactance will C_1 have at the LO frequency?
   e. Find the reactance of L_1 at 455 kHz.

10. Design a mixer for an FM receiver (IF = 10.7 MHz) to receive 100 MHz. The antenna impedance is 300 Ω, $L_1 = 20$ μH, BW = 500 kHz, and let the LO be the low-side injection. Also, the maximum IF power output will be 6 mW. The schematic is in Figure 7–9, β = 50, $C_{OB} = 1$ pF, $C_{BE} = 30$ pF.

11. The idealized mixer circuit of Figure 7–43 requires design considerations similar to the circuit of Figure 7–9. It is to operate with high-side LO injection and the

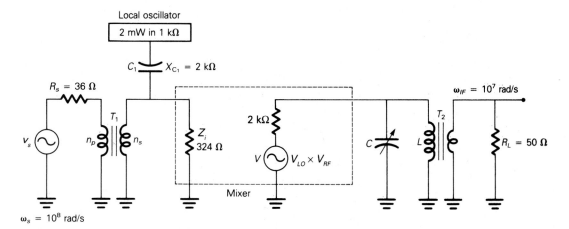

FIGURE 7–43 Idealized mixer circuit.

bandwidth must be 0.75 M rad/s. For the values shown, compute the following:
   a. Turns ratio n_p/n_s for T_1
   b. Turns ratio for T_2 to achieve the required bandwidth if $L = 10\ \mu H$
   c. Peak value of LO voltage across Z_i

12. An imaginary filter response is shown in Figure 7–44. Determine
   a. The attenuation at 2 MHz.
   b. Passband frequencies.
   c. Stopband frequencies (show the 3-dB stopbands on a filter sketch).
   d. 3-dB bandwidth.
   e. Loaded Q of the network.
   f. 60-dB bandwidth.
   g. 60 dB/3 dB shape factor.
   h. The amount of ripple.
   i. The response looks like that of a _____ resonator (low-pass or bandpass or bandstop), (Butterworth or Chebyshev or m-derived) filter.

13. A high-selectivity AM receiver has a 455-kHz IF with a 10-kHz, 3-dB bandwidth requirement. Interference at 427.5 kHz coming through the mixer and RF amplifier must be reduced by 48 dB. Determine the required filter specifications for a Chebyshev and a Butterworth design:
   a. Minimum number of resonators
   b. Expected ripple, if any
   c. Minimum Q per pole for the resonators
   d. Midband insertion loss if Q_{min} is used

14. Design a capacitor top-coupled bandpass filter with a minimum number of resonators, ripple ≤ 0.3 dB, $f_0 = 2$ MHz, $BW_{3\ dB} = 100$ kHz, $R_1 = R_2 = 1\ k\Omega$, and 36 dB of attenuation at 2169.2 kHz.
   a. How many resonators are required?
   b. Will inductors of $Q_u = 60$ be adequate? Why?
   c. Compute the value of L and all capacitors. (Assume $Q_u \to \infty$.)
   d. Compute insertion loss if $Q_u = 100$.
   e. How much attenuation is to be expected at 2.25 MHz?

15. A 2.11-GHz, 28.6-MHz bandwidth satellite receiver must reject an adjacent channel transmitter by 60 dB. Carriers are 36 MHz apart. The filter pass-band ripple

FIGURE 7–44

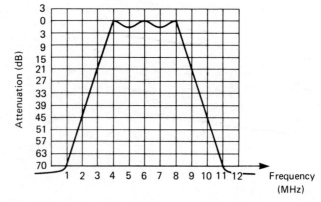

FIGURE 7–45

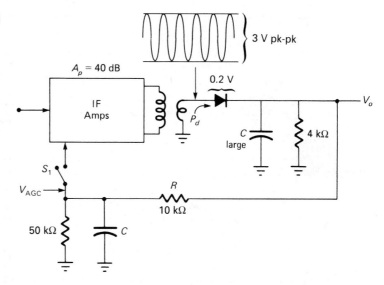

must not exceed 0.5 dB, and the insertion loss shall not exceed 4.4 dB. A cavity filter with $Q_u = 1000$ is to be used.
   a. How many resonators are required?
   b. What will be the actual ripple?
   c. What will be the insertion loss?

16. The effective Q of an IF amplifier must be increased by tapping into the primary with the collector lead. Sketch the circuit and show the turns ratio that will achieve a 5:1 improvement.

17. The IF/AGC system of Figure 7–45 is preceded by 14 dB of gain.
   a. How much power P_d is delivered to the detector (mod. index = 0)?
   b. What is the antenna received signal strength in dBm?
   c. Determine v_o and V_{AGC}, assuming ideal filtering by the large detector capacitor.
   d. Sketch v_o for modulation index of 0.5 and $f_m = 1$ kHz (sine).

18. The AGC detector of Figure 7–46 uses a germanium diode. If AGC threshold is based on the presence of an RF carrier signal with sufficient amplitude to cause the diode to conduct, determine P_i.

FIGURE 7–46   AGC detector.

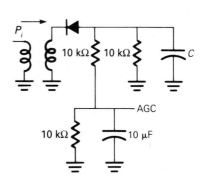

19. For Figure 7–27, what type is the AGC (forward or reverse) if Q_1 is changed to PNP and V_{cc} reversed in polarity? Explain.

20. The class A power amp of Appendix Figure 7A–1 has $V_{cc} = 15$ V, $R_E = 0$, $R_L = 4$ Ω, $r_x = 13$ Ω, and a transistor with $\beta = 100$, $V_{BE} = 0.7$ V, and dissipation (derated) of 1 W.
  a. Draw the power dissipation curve.
  b. Draw accurately the dc and ac load lines.
  c. Calculate the optimum collector load.
  d. Calculate the turns ratio.
  e. Calculate the maximum load power and efficiency (not ideal).
  f. Calculate the load power and efficiency for a 9-V pk collector swing.

21. The push-pull class AB power amp of Figure 7–30 has $V_{cc} = 12$ V, $R_E = 3$ Ω, $R_B = 265$ Ω, $R_L = 10$ Ω, and Q_1 has a 1-Ω emitter-resistor to ground. The diodes have $V_d = 0.70$ V, and the transistors: $\beta = 100$, $V_{BE} = 0.65$, $V_{CE}(\text{sat}) = 0.3$ V. If the bias voltage across C is 6 V,
  a. Determine all bias voltages and currents.
  b. If Q_2 could be driven hard enough to saturate, determine $i_o(\text{sat})_{\text{pk}}$, $v_o(\text{sat})_{\text{pk}}$, $P_o(\text{at sat})$, the power supplied by the battery, and the output circuit efficiency.

22. Refer to Figure 7–47. All ICs are ideal. Both transistors have $\beta = 250$, $r_e \approx 0$.
  a. Show that the system is connected for negative feedback; start with a + at the base of Q_1.
  b. Determine the midfrequency gains of each stage and the total gain (no feedback).
  c. Calculate the closed-loop system gain.
  d. Calculate the system input and output impedances.

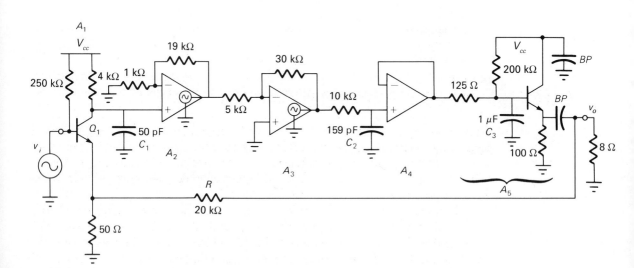

FIGURE 7–47

FIGURE 7–48

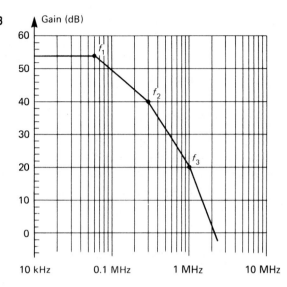

23. In Figure 7–47,
    a. Identify the different types of feedback for each amplifier and for the overall system.
    b. Calculate the time constants associated with C_1, C_2, and C_3.
    c. Sketch the open-loop Bode plot $V_o(f)/V_i(f)$ assuming $A_o = 80$ dB.

24. A system with negative feedback has the open-loop gain response of Figure 7–48. Determine the phase margin, stability, closed-loop system gain (at 10 kHz) and bandwidth, if stable (ignoring regenerative peaking), for the following amounts of feedback $(1 + A_V B)$ dB:
    a. 14 dB
    b. 19 dB
    c. 24 dB
    d. 34 dB

■ APPENDIX 7A: CLASS A POWER AMPLIFIER DESIGN EXAMPLE

Class A power amplifiers (PAs) have the advantage of low distortion over class B and C amplifiers. However, class A is rarely used for high-power output amplifiers because of its relatively poor efficiency. The theoretical efficiencies of 78.5% and 50% for class B versus class A (transformer-coupled) does not tell the whole story, however.

The power rating of the transistors (there must be two of them) for class B operation need to be rated at only 40% of the desired amplifier output power. This is in contrast to transformer-coupled class A, which requires the transistor rating to be twice (200%) the desired amplifier output power. In other words, 1-W transistors can deliver 500 mW to a load in class A, whereas in Class B 2.5 W can be delivered.

Problem:

A power transistor is derated for operation at room temperature with $P_D(\text{max}) =$ 3 W. Its $\beta = 50$ and $V_{BE} = 0.7$ V (high-current). For convenience, assume $V_{CE}(\text{sat}) = 0$. The transformer to be used has power losses equivalent to an $r_x = 3$-Ω primary winding resistance. Referring to the class A audio PA of Figure 7A–1:

1. Sketch the power dissipation curve P_D.
2. Using $V_E = 10\%$ of V_{cc} and optimum class A operation as criteria, determine the quiescent collector bias current I_{CQ}.
3. Determine R_E and R_B.
4. Sketch the dc and ac load lines on the sketch of part 1.
5. Determine the transformer turns ratio and its circuit efficiency.
6. Determine maximum load power P_L and circuit efficiency η.
7. Determine P_L and η for a 10-V pk collector swing.

Solution:

1. $P_D = V_{CE}/I_C = 3$ W. Pick a few values of V_{CE} and calculate the corresponding I_C. The results are shown in Figure 7A–2, and a few values are 333 mA at 9 V, 250 mA at 12 V, 200 mA at 15 V, and 167 mA at 18 V.
2. $V_{CE} = P_D/I_C$. $V_{CE} + I_C R_{dc} = V_{cc} = V_{CE} + (I_C r_x + 0.1 \ V_{cc})$. Substituting for V_{CE} at the Q-point, $(P_D/I_{CQ}) + I_{CQ} r_x + 0.1 \ V_{cc} = V_{cc}$. Multiply both sides by I_{CQ} and collect V_{cc} terms: $r_x I_{CQ}^2 - (0.9 \ V_{cc})/I_{CO} + P_D = 0$. Solving this by the quadratic equation and taking the lower intersection point yields

$$I_{CQ} = \frac{0.9 \ V_{cc}}{2r_x} - \frac{\sqrt{(0.9 \ V_{cc})^2 - 4r_x P_D}}{2r_x} \qquad \textbf{(7A–1)}$$

The result is $I_{CQ} = \textbf{234 mA}$.

FIGURE 7A–1 Class A audio power amplifier.

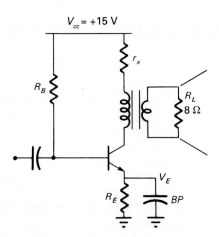

FIGURE 7A-2 Voltage/current relationships in a class A power amplifier.

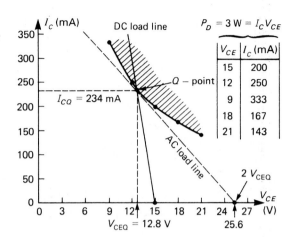

V_{CE}	I_c (mA)
15	200
12	250
9	333
18	167
21	143

3. $R_E = V_E/I_{CQ} = 0.1V_{cc}/I_{CQ} = 1.5\ V/234\ mA = 6.4\ \Omega$; $R_B = (V_{cc} - V_B)/(I_C/\beta) = $ **2.9 kΩ.**

4. **dc:** This is a plot of KVL; $V_{cc} - I_C r_x - V_{CE} - I_C R_E = 0$. Choose a couple of values for V_{CE} within a few volts of V_{cc} and calculate $I_C = 0$ at $V_{CE} = 15\ V$, and $V_{CEQ} = 12.8\ V$ at $I_{CQ} = 234\ mA$; they fall on the P_D curve.

   ac: An optimum load for the transistor will produce maximum output power without overheating the transistor. This must be given by a straight line (linear load) passing through the Q-point (satisfying dc-bias) and tangent to the P_D(max) curve. It must not extend into the shaded region above the maximum dissipation curve (see Figure 7A-2). Graphically and mathematically, the optimum ac load line has a value given by

$$R_C(\text{opt}) = \frac{V_{CEQ}}{I_{CQ}} \qquad (7A-2)$$

   That is, Equation 7A-2 is the slope of the tangent to the P_D curve at the Q-point. From this result, $R_C(\text{opt}) = $ **54.6 Ω.**

5. The transistor should see an ac load of 54.6 Ω. The transformer already has 3 Ω. Therefore, the transformer must transform the 8-Ω speaker impedance to $R'_L = 51.6\ \Omega$.

$$N = \sqrt{\frac{51.6}{8}} = \textbf{2.54:1}$$

   Now the transformer efficiency can be calculated. The signal power division at the transformer is proportional to the resistance ratio $R'_L/(R'_L + r_x) = (R_C - r_x)/R_C$, so the transformer (abbreviated xfmr) efficiency in our application will be

$$\eta_x = 1 - \left(\frac{r_x}{R_C}\right) \qquad (7A-3)$$

   $\eta_x = 1 - (3/54.6) = 94.5\%$, which is about right for a well-designed audio transformer.

6. It is clear from the ac load line of Figure 7A–2 with $V_{CEQ} = 12.8$ V that the maximum ideal collector voltage swing for this circuit will be 12.8 V pk. This ideal condition assumes that $V_{CE}(\text{sat}) = 0$. The actual peak collector swing will be reduced by the collector-emitter saturation voltage $V_{CE}(\text{sat})$.

The maximum signal power developed across the transformer will be $P_C = (12.8 \text{ V pk})^2/(2 \times 54.6\ \Omega) = 1.5$ W, which is 50% of P_D (max). Since $\eta_x = 94.5\%$ and

$$P_L = \eta_x P_C \qquad\qquad\qquad (7A\text{–}4)$$

the load will receive **1.42 W.**

In class A operation the battery supplies a constant average current of $I_{CQ} = 234$ mA to the collector circuit, so that $P_{dc} = V_{cc}I_{CQ} = 3.51$ W. Some of this is dissipated in r_x, R_3, and also as heat from the transistor. The remaining 1.5 W is delivered to R_C. $P_{r_x} = 1.5$ W $-$ 1.42 W $= 0.58$ W and $P_{R_3} = 1.5$ V $\times$ 0.234 A $= 0.351$ W, so $P_D = 3.51$ W $-$ 1.5 W $-$ 0.351 W $= 1.66$ W. You see that with no ac input signal the transistor dissipates at its maximum 3 W rating, but for full drive conditions it runs much cooler because power is delivered to the load. The output efficiency is

$$\eta = \frac{P_L}{P_{dc}}$$

Hence $\eta = 1.42$ W/3.51 W $= \mathbf{40.5\%}$. In fact if the 70 mW of base-bias power is included, the total circuit efficiency is 39.7%. This will be the maximum efficiency for the circuit because when the signal level is reduced, load power is reduced but the circuit dissipations are virtually unchanged (P_{r_x} is reduced slightly).

7. If the input signal drive is reduced so that the collector swing is only 10 V pk, then

$$P_L = \eta_x P_C$$

$$P_L = \frac{(0.945)(10 \text{ V})^2}{2 \times 54.6}$$

$$= \mathbf{866\ mW}$$

and
$$\eta = \frac{P_L}{P_{dc}} = \frac{866 \text{ mW}}{3.51 \text{ W}} = \mathbf{24.7\%}$$

7–10 ■ COMPUTER ANALYSIS

Using the computer for communications system and circuit analysis has become a virtual necessity in education and industry. What follows are examples of programs and/or readily available simulation software for solving problems of the type covered in this chapter.

All the programs are interactive in the sense that an introductory sentence or short paragraph describes the purpose of the program and the user is then prompted for input data.

The programs for Chapter 7 are as follows:

- **Active mixer design program:** A program written in BASIC is presented in Figure 7–49. It is an interactive program for designing an active mixer consisting of input and output impedance-matching transformers, voltage-divider bias resistors for an NPN transistor, and a tuned IF output tank. The local oscillator is RC-coupled to the BJT base. The ac and dc design procedures follow Example 7–2 in the text.

- **PSpice mixer:** The mixer circuit of Figure 7–50b was analyzed using the 1988 version of PSpice called *Interactive PSpice*. The circuit file (Figure 7–50a) was created with the built-in *internal* word processor. The PSpice circuit simulator will work in a nonlinear mode—that is, real-world—for transient analysis simulations. The results of the .TRAN command in the circuit file are shown in Figure 7–50c and d. The two mixer input signals are the RF of Figure 7–50c ($v(4)$ on the Q_1 base) and the slightly lower-frequency LO signal $v(6)$ at the Q_1 emitter; low-side injection is used. The mixer output is seen in Figure 7–50d as $v(9)$: Also seen in this figure is the Q_1 collector signal $v(5)$ where the 15-V bias voltage is obvious and the 455-kHz IF frequency can be estimated using the time axis. The circuit dc voltages and currents as well as the steady-state Fourier distortion analysis results are presented in tabular form in Figure 7–50e. Parts f and g of Figure 7–50 are spectral plots of the mixer output. Note the presence of the LO signal at 975 kHz (down 18.7 dBc) and the RF signal at 1.43 MHz. The two plots in Figure 7–50h show the IF output signal $v(9)$ at 455 kHz with an amplitude of 2.2 V and the RF input signal $v(1)$ at 1.43 MHz with an amplitude of about 245 mV. The mixer conversion gain is therefore determined to be 19 dB.

- **Micro-CAP II mixer analysis:** For comparison, the results of a Micro-CAP II analysis of the preceding PSpice mixer is presented in Figure 7–51a to 7–51f, including the input and output time-domain traces. Also shown are the files for the RF and LO sinusoidal sources, the time-domain analysis (Figure 7–51d), the frequency sweep analysis, and the resulting frequency-domain (Bode) plot.

- **Two-resonator Butterworth and Chebyshev filter design:** The BASIC program of Figure 7–52 computes values for two-resonator capacitor-coupled bandpass filters.

- **Capacitor-coupled bandpass filters:** The PASCAL program of Figure 7–53 computes values for 3- and 4-resonator Chebyshev and Butterworth filters.

```
10 REM    ***           ACTIVE MIXER DESIGN PROGRAM           ***
20 REM    **** EXAMPLE OF STUDENT COMPUTER ASSIGNMENT ****
30 REM    ****          FOR PROFESSOR P. YOUNG. AT ASU        ****
40 CLS
50 REM ** DESIGN OF ACTIVE MIXER (FIGURE 7A-1; ELECTR COMM TECHNIQUES, P. YOUNG)
60 PRINT "********** MIXER CIRCUIT DESIGN PROGRAM ********* "
70 PRINT "THIS IS AN INTERACTIVE PROGRAM FOR DESIGN OF AN ACTIVE MIXER."
80 PRINT "THE CIRCUIT CONSISTS OF INPUT AND OUTPUT Z-MATCHING TRANSFORMERS,"
90 PRINT "VOLTAGE-DIVIDER BIAS RESISTORS FOR THE BJT, AND A  TUNED IF OUTPUT."
100 PRINT "THE  L.O. IS RC-COUPLED TO THE BJT BASE. AC AND DC DESIGN FOLLOW THE"
110 PRINT "TEXT EXAMPLE            ."
120 PRINT "********** DATA INITIALIZATION **************** ":PRINT
130 INPUT "ENTER AMPLIFIER OUTPUT EFFICIENCY (.5 TYP.), THEN PRESS RETURN .";N
140 PRINT:INPUT "ENTER TRANSISTOR'S CURRENT GAIN -- BETA ";BETA
150 PRINT:INPUT "ENTER COB (FARAD)";COB
160 PRINT:INPUT "ENTER CBE (FARAD)";CBE
170 PRINT:INPUT "ENTER VBE (VOLTS)";VBE
180 PRINT:INPUT "ENTER THE MAX OUTPUT POWER 'PO' (WATTS)";PO
190 PRINT:INPUT "ENTER VCC (VOLTS)";VCC
200 REM ********** COMPUTATION OF D.C - DESIGN ***********
210 PRINT " ************ D.C - DESIGN *************"
220 REM FORMULAS
230 PDC=PO/N:VE=.1*VCC:IC=PDC/(VCC-VE):IE=IC:IB=IC/BETA:VB=VBE+VE
240 R1=VB/(10 * IB):R2=(VCC-VB)/(11*IB):R3=VE/IE
250 PRINT TAB(10) "****************"
260 PRINT TAB(10) "* R1 =";R1
270 PRINT TAB(10) "* R2 = ";R2
280 PRINT TAB(10) "* R3 =";R3
290 PRINT TAB(10) "****************"
300 PRINT " ************ A.C. - DESIGN *************"
310 PRINT:INPUT "ENTER RF-FREQUENCY 'FRF' (Hz)";FRF
320 PRINT:INPUT "ENTER IF-FREQUENCY 'FIF' (Hz)";FIF
330 PRINT:INPUT "ENTER IF-OUTPUT LOAD,  RL (OHMS)";RL
340 PRINT: INPUT "ENTER L1 (HENRY)";L1
350 PRINT:INPUT "ENTER LO COUPLING REACTANCE, XC1 (OHMS)";XC1
360 PRINT "ENTER THE TYPE OF L.O. INJECTION 'INJ'"
370 PRINT "1) HIGH "
380 PRINT "2) LOW.  ENTER A 1 OR 2."
390 INPUT INJ
400 IF INJ=1 THEN FLO = FRF+FIF:IF INJ=2 THEN FLO = FRF-FIF
410 PRINT TAB(10)"*******************"
420 PRINT TAB(10)"* FLO =";FLO
430 REM FORMULAS
440 PI=22/7:WLO=2*PI*FLO:WIF=2*PI*FIF:RE=.026/IE:RIN=(BETA+1)*RE
450 PRINT TAB(10) "* RIN =";RIN
460 REM FORMULAS
470 C1=1/(WLO*XC1):XC2=.1*(1/(1/R1+1/R2+1/RIN)):C2=1/(WIF*XC2)
480 XC3=.1*(1/(1/RE+1/R3)):C3=1/(WIF*XC3):C4=C3:CT=1/((WIF*WIF)*L1)
490 PRINT TAB(10)"* C1 =";C1
```

FIGURE 7-49   (page 1 of 2)

```
500 PRINT TAB(10)"* C2 =";C2
510 PRINT TAB(10) "* XC3 =";XC3
520 PRINT TAB(10)"* C3 =";C3
530 PRINT TAB(10) "* C4 =";C4
540 PRINT TAB(10) "* CT =";CT
550 PRINT TAB(10) "********************"
560 PRINT "NOTE: ALL CAPACITORS ARE RATED IN FARADS."
570 PRINT "*** COMPUTATIONS OF TURNS-RATIOS & BASE-TO-GROUND L.O. DRIVE ***"
580 PRINT:INPUT "ENTER BANDWIDTH 'BW' (Hz)";BW
590 PRINT:INPUT "ENTER SOURCE IMPEDANCE 'ZIN' (OHMS)";ZIN
600 REM FORMULAS
610 N1=SQR(ZIN/RIN):XL1=WIF*L1:QEF=FIF /BW:REF=XL1*QEF:N2=SQR(REF/RL)
620 PRINT TAB(10) "*****************"
630 PRINT TAB(10) "* N1 =";N1
640 PRINT TAB(10) "* N2 =";N2
650 PRINT TAB(10) "*****************"
660 PRINT:INPUT "ENTER LOCAL OSCILLATOR'S POWER 'P'(WATT)";P
670 PRINT:INPUT "ENTER LOCAL OSCILLATOR OUTPUT CIRCUIT RESISTANCE (OHMS)";R
680 REM FORMULAS
690 VLO=SQR(2*P*R):RC=RL*N2*N2:AV=RC/RE:CIN=CBE + COB
700 PRINT TAB(10) "*****************"
710 PRINT TAB(10) "* VLO (VOLTS,PEAK) =";VLO
720 PRINT TAB(10) "* AV, at f(IF) =";AV
730 PRINT TAB(10) "*CIN (FARADS) =";CIN;" (VIRTUALLY NO GAIN AT f(LO)!"
740 REM  FORMULAS
750 XCIN=1/(WLO*CIN):ZBG=(XCIN*(RIN/2))/SQR(((RIN/2*(RIN/2))+(XCIN*XCIN))
760 VBG=VLO*(ZBG/SQR((ZGB*ZBG)+(XC1*XC1)))
770 PRINT TAB(10) "* XCIN AT F(LO) =";XCIN;"OHMS"
780 PRINT TAB(10) "* MAGNITUDE OF BJT COMPLEX Z AT F(LO), ZBG =";ZBG;"OHMS"
790 PRINT TAB(10) "* PEAK LO DRIVE ON BJT BASE, VBG =";VBG;"VOLTS"
800 PRINT TAB(10) "*****************"
810 PRINT "IS THE CORRECT DATA FED IN? Y OR N! PLS. CONFIRM"
820 PRINT "IF THE DATA IS WRONG OR YOU WANT TO DO ANOTHER RUN , ANSWER NO"
830 PRINT:INPUT" 1) FOR YES,   2) FOR NO ";Y
840 IF Y = 2 THEN 40:GOTO 830
850 END
```

FIGURE 7-49   (page 2 of 2)

```
**** 06/07/89 *** Evaluation PSpice (Jan. 1988) *** 07:56:40 ***

SPICE MIXER

****      CIRCUIT DESCRIPTION

*****************************************************************

RS1 1 2 300
RS2 7 8 50
RL 9 0 1K
RB1 3 4 120K
RB2 4 0 18K
RE 6 0 1.2K
CS1 2 4 .1UF
CS2 6 7 .1UF
CT 5 0 1748P
L1 5 3 70UH
L2 9 0 10.5UH
K1 L1 L2 .8
VDC 3 0 DC 15
VRF 1 0 SIN (0 .3 1430K)
VLO 8 0 SIN (0 2.25 975K)
Q1 5 4 6 BJT
.MODEL BJT NPN PE=.6 VA=30
.FOUR 455K V(9)
.TRAN .2US 70US .2US
.PLOT TRAN V(9)
.OPTIONS LIMPTS=500 ITL5=10000
.PROBE
.END


**** 06/07/89 *** Evaluation PSpice (Jan. 1988) *** 07:56:40 ***

 SPICE MIXER

****      BJT MODEL PARAMETERS

*****************************************************************

        BJT
        NPN
   IS   100.000000E-18
   BF   100
   NF     1
   VAF   30
   BR     1
   NR     1
   VJE     .6
```

FIGURE 7–50a

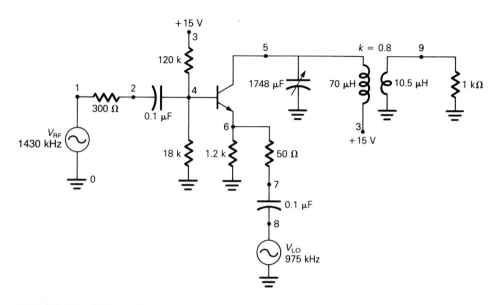

FIGURE 7–50b   Mixer with node numbers for PSpice circuit file.

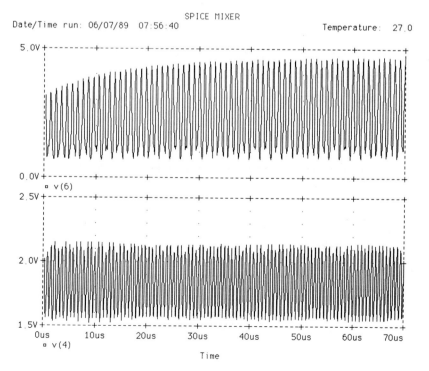

FIGURE 7–50c

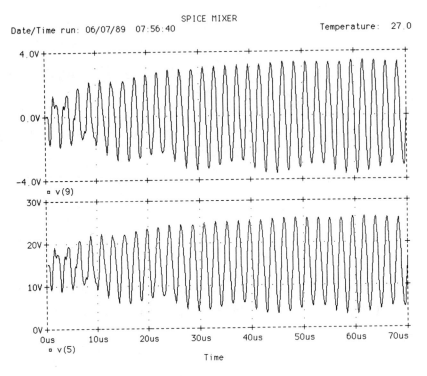

FIGURE 7–50d

```
**** 06/07/89 *** Evaluation PSpice (Jan. 1988) *** 07:56:40 ***

   SPICE MIXER

 ***   INITIAL TRANSIENT SOLUTION      TEMPERATURE = 27.000 DEG C

 ***********************************************************************

 NODE  VOLTAGE    NODE  VOLTAGE    NODE   VOLTAGE    NODE    VOLTAGE

 ( 1)  0.0000     ( 2)  0.0000     ( 3)   15.0000    ( 4)    1.8578

 ( 5)  15.0000    ( 6)  1.0956     ( 7)   0.0000     ( 8)    0.0000

 ( 9)  0.0000

       VOLTAGE SOURCE CURRENTS
       NAME            CURRENT

       VDC             -1.016E-03
       VRF             0.000E+00
       VLO             0.000E+00

       TOTAL POWER DISSIPATION   1.52E-02   WATTS
```

FIGURE 7–50e   (page 1 of 2)

```
**** 06/07/89 *** Evaluation PSpice (Jan. 1988) *** 07:56:40 ***

 SPICE MIXER

****      FOURIER ANALYSIS          TEMPERATURE =    27.000 DEG C

****************************************************************

FOURIER COMPONENTS OF TRANSIENT RESPONSE V(9)


DC COMPONENT =  -4.225532E-03

HARMONIC    FREQ      FOURIER     NORMALIZED     PHASE     NORMALIZED
NO          (HZ)      COMPONENT   COMPONENT      (DEG)     PHASE (DEG)

  1        4.55E+5    3.22E+0     1.00E+0       7.95E+0     0.00E+0
  2        9.10E+5    1.65E-1     5.14E-2      -1.70E+1    -2.49E+1
  3        1.36E+6    4.50E-2     1.40E-2       1.02E+2     9.38E+1
  4        1.82E+6    3.76E-2     1.17E-2       1.24E+2     1.16E+2
  5        2.28E+6    1.87E-2     5.82E03      -6.14E+1    -6.93E+1
  6        2.73E+6    1.16E-2     3.60E-3      -2.26E+1    -3.05E+1
  7        3.18E+6    1.44E-2     4.49E-3       6.36E+1     5.57E+1
  8        3.64E+6    6.73E-3     2.09E-3       6.44E+1     5.65E+1
  9        4.09E+6    3.32E-3     1.03E-3       2.72E+1     1.92E+1

     TOTAL HARMONIC DISTORTION =   5.516834E+00 PERCENT

     JOB CONCLUDED    TOTAL JOB TIME   5566.75 seconds
```

FIGURE 7–50e   (page 2 of 2)

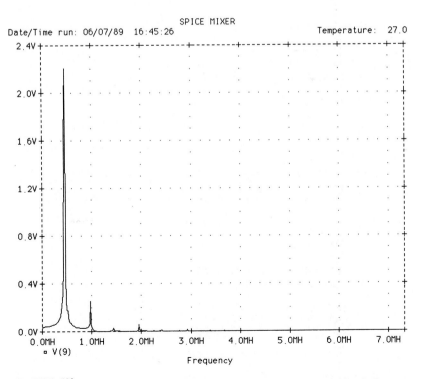

FIGURE 7–50f

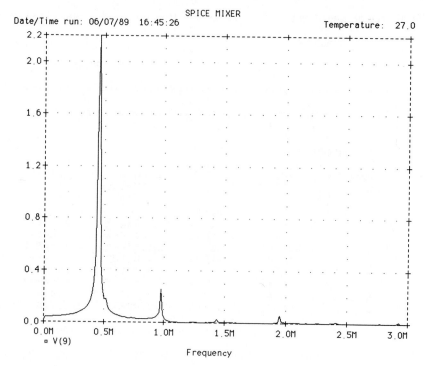

FIGURE 7-50g

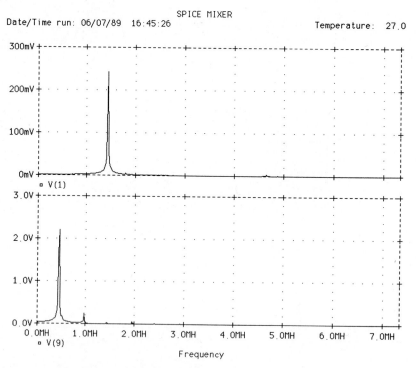

FIGURE 7-50h PSpice frequency spectra.

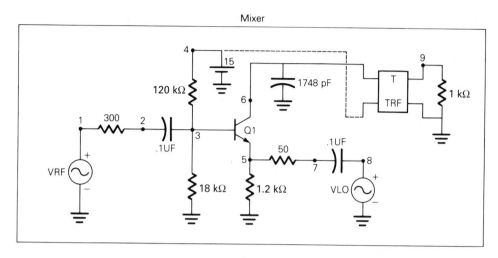

FIGURE 7–51a Mixer with node numbers from Micro-CAP II designer screen.

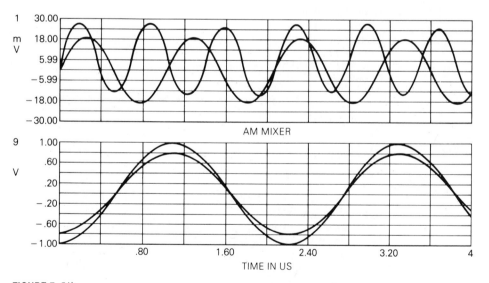

FIGURE 7–51b

Sinusoidal sources Type 0...Alias VRF

	Value	Tolerance(%)
0:Frequency	1430000	0
1:Amplitude/2	.02	0
2:D.C. Voltage level	0	0
3:Phase angle(Radians)	0	0
4:Source resistance	50	0
5:Repetition period	0	0
6:Exponential decay	0	0
7:Frequency shift(Hz/Sec)	0	0

E:Edit   J:Jump   N:Next   L:Last   C:Copy   A:Alter alias   Q:Quit

Sinusoidal sources Type 2...Alias VLO

	Value	Tolerance(%)
0:Frequency	975000	0
1:Amplitude/2	.02	0
2:D.C. Voltage level	0	0
3:Phase angle(Radians)	0	0
4:Source resistance	50	0
5:Repetition period	0	0
6:Exponential decay	0	0
7:Frequency shift(Hz/Sec)	0	0

FIGURE 7–51c

AM MIXER ANALYSIS LIMITS

Simulation time (Tmax/Tmin/Timestep)	4E-6
Display time (Dmax/Dmin/Printstep)	4E-6
Minimum accuracy (%)	5
Upper trace A waveform spec	1
Upper trace B waveform spec	8
Upper trace range (High/Low)	.03/-.03
Lower trace A waveform spec	9
Lower trace B waveform spec	6
Lower trace range (High/Low)	1/-1/30/0
Zero, Read, Leave initial values (Z,R,L)	Z
Edit/review initial conditions (Y,N)	N
Dump node waveform to user file (Y,N)	N
Calculate D.C. operating point (Y,N)	N
Temperature (Low/High/Step)	27
Number of cases	1
Output to Disk, Printer, None (D,P,N)	N
Save, Retrieve, Normal run (S,R,N)	N
Default plotting parameters (Y,N)	N

FIGURE 7–51d

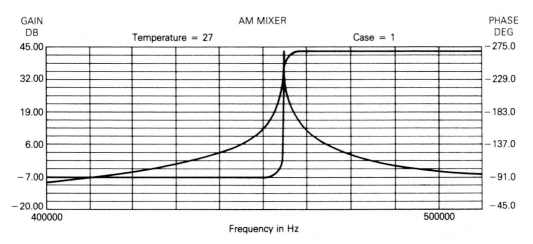

Frequency =   500.00000E+03 Hz   Gain = −6.696 DB
Phase angle = −269.721   Degrees   Group delay = 000.00000E−04 Sec
Gain slope = −645.70260E−01 DB/OCT   Peak gain =   43.626 DB/F = 454.62890E+03

FIGURE 7–51e

AM MIXER ANALYSIS LIMITS

Lowest frequency	400000
Highest frequency	500000
Lowest gain (dB)	-20
Highest gain (dB)	45
Lowest phase shift	-45
Highest phase shift	-275
Lowest group delay	1E-9
Highest group delay	1E-4
Input node number	1
Output node number	9
Minimum accuracy (%)	1
Auto or Fixed frequency step (A,F)	A
Temperature (Low/High/Step)	27
Number of cases	1
Output: Disk, Printer, None (D,P,N)	N
Save, Retrieve, Normal run (S,R,N)	N
Default plotting parameters (Y,N)	N

FIGURE 7–51f   AC frequency response parameters.

```
10 CLS
20 REM * TWO POLE BUTTERWORTH AND CHEBYSHEV FILTER DESIGN *
30 REM *      EXAMPLE OF STUDENT COMPUTER ASSIGNMENT       *
40 REM *****       FOR PROFESSOR P. YOUNG, AT ASU       ****
50 PRINT "THIS PROGRAM COMPUTES THE CIRCUIT COMPONENT VALUES FOR"
60 PRINT "TWO-POLE BUTTERWORTH OR (0.5dB RIPPLE) CHEBYSHEV BANDPASS FILTERS"
70 PRINT :PRINT :PRINT :PRINT : PRINT :
80 INPUT: "       WHAT TYPE OF FILTER? BUTTERWORTH (B), OR CHEBYSHEV (C)"; IS
90 CLS
100 FOR I = 1 TO 8 : PRINT : NEXT I
110 PRINT TAB(19) "PLEASE ANSWER THE FOLLOWING QUESTIONS": PRINT :PRINT
120 INPUT:"              CENTER FREQUENCY Fo (HZ)"; F
130 LET W=2*3.1415927#*F:PRINT :PRINT
140 INPUT:"              FILTER HALF POWER BANDWIDTH BW3dB (HZ)"; B
150 PRINT:PRINT
160 INPUT:"        IMPEDANCE LEVEL OF SOURCE AND LOAD (OHMS)"; R
170 PRINT:PRINT
180 R1=R
190 CLS:PRINT:PRINT
200 IF I$ = "B" THEN GOTO 240 ELSE GOTO 480
210 REM***********************************************************************
220 REM********************BUTTERWORTH**FILTER*****************************
230 REM***********************************************************************
240 Q1 = F/B:Q = 1.4* Q1:K = ((1/Q^2))^.5:B30 = (4/Q)*F:K1 = R/Q:X2 = R1/Q
250 X3 = (1/(K*Q))*(R*R1)^.5:X4 = (X1*X3)/(X1+X3):X5 = X4:CLS
260 PRINT TAB(15) " _____ "
270 PRINT TAB(15) "|                                           |"
280 PRINT TAB(15) "|     ******BUTTERWORTH FILTER******        |"
290 PRINT TAB(15) "|_____|"
300 PRINT:PRINT
310 PRINT TAB(10) "Fo = ";F;" HZ";TAB(40)"WO = ";W;" RAD/S
320 PRINT TAB(10) "BW3dB = ";B;TAB(40)"BW30dB = ";B30
330 PRINT TAB(10) "K = ";K ;TAB(40)"QL = "Q1
340 PRINT TAB(10) "Rs = ";R;" OHMS";TAB(40)"Qmin = ";Q
350 PRINT TAB(10) "R1 = ";R1;" OHMS"
360 C=1/(W*X4):C1=1/(W*X4):C2=1/(W*X3):L1=(X1/W):L2=(X2/W)
370 PRINT TAB(10) "Xa = ";X4;" OHMS";TAB(40)"Ca = "; C;" FARADS"
380 PRINT TAB(10) "Xb = ";X5;" OHMS";TAB(40)"Cb = "; C1;" FARADS"
390 PRINT TAB(10) "Xc = ";X3;" OHMS";TAB(40)"Cc = "; C2;" FARADS"
400 PRINT TAB(10) "X1 = ";X1;" OHMS";TAB(40)"L1 = "; L1;" HENRYS"
410 PRINT TAB(10) "X2 = ";X2;" OHMS";TAB(40)"L2 = "; L2;" HENRYS"
420 PRINT :PRINT "    IF REACTANCES ARE LESS THAN 10 OHMS, IT WILL
430 PRINT "    BE WISE TO USE A HIGHER DESIGN IMPEDANCE."
```

FIGURE 7–52   (page 1 of 2)

```
440 END
450 REM*********************************************************************
460 REM************************CHEBYSHEV**FILTER*****************************
470 REM*********************************************************************
480 Q1 = F/B:Q = 1.95* Q1:K = (2/Q^2)^.5:B30 = (4.85/Q)*F:X1 = R/Q:X2 = R1/Q
490 X3 = (1/(K*Q))*(R*R1)^.5:X4 = (X1*X3)/(X1+X3):X5 = X4:CLS
500 PRINT TAB(15) " _____ "
510 PRINT TAB(15) "|                                        |"
520 PRINT TAB(15) "|      ******CHEBYSHEV FILTER******      |"
530 PRINT TAB(15) "|                                        |"
540 PRINT TAB(15) "|_____|"
550 PRINT:PRINT
560 PRINT TAB(10) "Fo = ";F;" HZ";TAB(40)"WO = ";W;" RAD/S
570 PRINT TAB(10) "BW3dB = ";B;TAB(40)"BW30dB = ";B30
580 PRINT TAB(10) "K = ";K ;TAB(40)"QL = "Q1
590 PRINT TAB(10) "Rs = ";R;" OHMS";TAB(40)"Qmin = ";Q
600 PRINT TAB(10) "R1 = ";R1;" OHMS"
610 C=1/(W*X4):C1=1/(W*X5):C2=1/(W*X3):L1=(X1/W):L2=(X2/W)
620 PRINT TAB(10) "Xa = ";X4;" OHMS";TAB(40)"Ca = "; C;" FARADS"
630 PRINT TAB(10) "Xb = ";X4;" OHMS";TAB(40)"Cb = "; C;" FARADS"
640 PRINT TAB(10) "Xc = ";X3;" OHMS";TAB(40)"Cc = "; C2;" FARADS"
650 PRINT TAB(10) "X1 = ";X1;" OHMS";TAB(40)"L1 = "; L1;" HENRYS"
660 PRINT TAB(10) "X2 = ";X2;" OHMS";TAB(40)"L2 = "; L2;" HENRYS"
670 PRINT :PRINT "    IF REACTANCES ARE LESS THAN 10 OHMS, IT WILL"
680 PRINT "    BE WISE TO USE A HIGHER DESIGN IMPEDANCE."
690 END
```

FIGURE 7-52   (page 2 of 2)

```
Program Capacitor_Coupled_Bandpass_Filter (input,output);

TYPE  LENGTH = STRING[15];

VAR
     FILTER,N : INTEGER ;
     T,AGAIN : LENGTH ;
     R,BW,F,L,C1,C2,C3,C4,Cn,C12,C23,C34 : REAL ;



(*********************************************************************)



Procedure Butterworth (Filter,N : integer ;
                       R,F,BW : real ) ;

   VAR  QL,L,W,K12,K23,K34,Q1 : real ;
        C1,C2,C3,C4,C12,C23,C34,Cn : real ;

begin
     CASE N OF
              3: begin
                   Q1 := 1 ;
                   K12 := 0.707 ;
                   K23 := 0.707 ;
                 end;
              4: begin
                   Q1:= 0.765 ;
                   K12:= 0.841 ;
                   K23:= 0.541 ;
                   K34:= 0.841 ;
                 end;
     end;
     W:= 2*3.14*F;
     QL:= F/BW ;
     L:= R /(W*QL*Q1) ;
     Cn := 1 /(sqr(W)*L);
     C12 := Cn * (K12/QL);
     C23 := Cn * (K23/QL);
     C1 := Cn - C12;
     C2 := Cn - C12 - C23;

     If N = 4 then begin
           C34 := Cn *(K34/QL);
           C3 := Cn - C23 - C34 ;
           C4 := Cn - C34 ;
         end
     Else   C3 := Cn - C23;

 If Filter = 1  then  T := 'Butterworth'
 Else T := 'Chebyshev';

 writeln;
 writeln(' The component values for a ', N ,' pole ', T ,');
 writeln(' capacitor coupled bandpass filter are :');
 writeln;
```

FIGURE 7–53   (page 1 of 4)

```
          writeln('                      L   = ',L:8,'  Henrys');
          writeln('                      C1  = ',C1:8,'  Farads');
          writeln('                      C12 = ',C12:8,'  Farads');
          writeln('                      C2  = ',C2:8,'  Farads');
          writeln('                      C23 = ',C23:8,'  Farads');
          writeln('                      C3  = ',C3:8,'  Farads');

     If N = 4 then Begin
          writeln('                      C34 = ',C34:8,'  Farads');
          writeln('                      C4  = ',C4:8,'  Farads');
        end;



end;




{***********************************************************}




Procedure Chebyshev (Filter,N : integer;
                     R,F,BW : real );

   VAR  QL,L,W,K12,K23,K34,Q1 : real ;
        C1,C2,C3,C4,C12,C23,C34,Cn : real ;
        Ripple : integer ;
begin

     write('  Please enter the Ripple level ( 1 = .1dB, 2 =.5dB,
     write('  3 = 1dB) > ');
     readln(Ripple);


If N = 3 then begin

    CASE Ripple OF
                 1: begin
                        Q1 := 1.433 ;
                        K12 := 0.662 ;
                        K23 := 0.662 ;
                    end;
                 2: begin
                        Q1:=  1.864 ;
                        K12:= 0.647 ;
                        K23:= 0.647 ;
                    end;
                 3: begin
                        Q1 := 2.210 ;
                        K12 := 0.645 ;
                        K23 := 0.645 ;
                    end;
       end;

   end
```

FIGURE 7–53   (page 2 of 4)

```
    Else

        CASE Ripple OF
                      1: begin
                            Q1 := 1.345 ;
                            K12 := 0.685 ;
                            K23 := 0.542 ;
                            K34 := 0.685 ;
                            writeln;

                          end;
                      2: begin
                            Q1:=  1.826 ;
                            K12:= 0.648 ;
                            K23:= 0.545 ;
                            K34:= 0.648 ;

                          end;
                      3: begin
                            Q1 := 2.210;
                            K12 := 0.638;
                            K23 := 0.546 ;
                            K34 := 0.638 ;

                          end;
        end;

        W:= 2*3.14*F;
        QL:= F/BW ;
        L:= R /(W*QL*Q1) ;
        Cn := 1 /(sqr(W)*L);
        C12 := Cn * (K12/QL);
        C23 := Cn * (K23/QL);
        C1 := Cn - C12;
        C2 := Cn - C12 - C23;

        If N = 4 then begin
              C34 := Cn *(K34/QL);
              C3 := Cn - C23 - C34 ;
              C4 := Cn - C34 ;
            end

          Else   C3 := Cn - C23;

        If Filter = 1  then   T := 'Butterworth'
        Else T := 'Chebyshev';

        writeln;
        writeln(' The component values for a ', N ,' pole ', T ,');
        writeln(' capacitor coupled bandpass filter are :');
        writeln;
        writeln('                          L   =  ',L:8,' Henrys');
        writeln('                          C1  =  ',C1:8,' Farads');
        writeln('                          C12 =  ',C12:8,' Farads');
        writeln('                          C2  =  ',C2:8,' Farads');
        writeln('                          C23 =  ',C23:8,' Farads');
        writeln('                          C3  =  ',C3:8,' Farads');
```

FIGURE 7-53   (page 3 of 4)

```
                   If N = 4 then Begin
                         writeln('                     C34 =  ',C34:8,'  Farads');
                         writeln('                     C4  =  ',C4:8,'  Farads');
                     end;

        end;


        BEGIN {+++++++++++++  MAIN PROGRAM  +++++++++++++++++++++++++++}

             writeln;
             writeln;
             writeln ('*Program Capacitor Coupled Bandpass Filter Design*');
             writeln;
             writeln;
             writeln('  This program will calculate the component values');
             writeln('  for either a 3 or 4 pole Butterworth or ');
             writeln('  Chebyshev Capacitor Coupled Bandpass Filter. This');
             writeln('  program will ask the user for a design impedance');
             writeln('  R, center frequency F, and bandwidth BW, of the');
             writeln('  filter to be determined. If a Chebyshev filter is');
             writeln('  desired this program will also ask for the ripple');
             writeln('  level in dB.');

             Repeat
             writeln;
             writeln('  Please enter the type of filter to be determined');
             write('  ( Butterworth = 1, Chebyshev = 2 ) > ');
             readln(filter);

             writeln;
             write('  Please enter the number of poles in the filter');
             write('  (N = 3 or 4) > ');
             readln(N);

             writeln;
             write('  Please enter the Design Impedance (R) in ohms > ');
             readln(R);
             writeln;
             write('  Please enter the center frequency (F) in hertz > ');
             readln(F);
             writeln;
             write('  Please enter the 3dB bandwidth (BW) in kilohertz > ');
             readln(BW);
             writeln;

             If FILTER = 1  then  Butterworth (Filter,N,R,F,BW)
             Else Chebyshev(Filter,N,R,F,BW) ;
             writeln;
             writeln;
             write(' Would you like to determine the components for');
             write(' another Filter? ( Y or N ) >');
             readln(AGAIN);
             Until AGAIN = 'N';
             writeln ;
             writeln('   Goodbye and have a nice day !!');

        END.
```

FIGURE 7–53   (page 4 of 4)

8

Sideband Systems

■ INTRODUCTION

The amplitude-modulated signals discussed in Chapter 5 have a frequency spectrum consisting of a full carrier and two sidebands (DSB-FC). One sideband is the mirror image of the other, and either one may be removed without loss of information; however, removal of either one results in a 3-dB loss of information *power* that would reduce the receiver output S/N by the same amount.

Amplitude-modulated signals with most of one sideband removed are called vestigial sideband (VSB). If all of one sideband and the carrier are eliminated, the result is single-sideband/suppressed-carrier (SSB-SC). Another possibility is to remove only the carrier because the carrier requires most of the transmitted power, and information for its recovery remains with the transmitted spectrum.

When the carrier is suppressed and both sidebands remain, the result is double-sideband/suppressed-carrier (DSB-SC).

In addition to these three AM techniques, this chapter will include a discussion of one of the most important uses of single-sideband on a large scale—the *frequency-division multiplex* of telephone signals. The chapter concludes with a scheme for multiplexing two information channels that has been used in color TV and stereo AM, and now forms the basis for the transmission of digital information. This scheme is called *quadrature multiplexing*. ■

8–1 ■ VESTIGIAL SIDEBAND

The 4.2-MHz video signal of commercial broadcast television is transmitted as a vestigial sideband (VSB) signal, designated A5C by the FCC and International Telecommunication Union. As illustrated in Figure 8–1, the baseband video signal modulates the carrier in a regular double-sideband/full-carrier (DSB-FC) modulator. Before power amplification, this AM signal enters the vestigial sideband filter that eliminates most of the lower sideband. The reason for using VSB is to minimize the transmission spectrum (bandwidth) while maintaining an easily demodulated AM signal; the demodulated low-frequency response of the recovered signal will also be better than for single sideband. As will be discussed in a section to follow, an AM transmission in which the carrier has been suppressed requires special coherent demodulation techniques. By transmitting the carrier along with modulated information, the simple (noncoherent) peak detector of Chapter 5 can be used in the receiver.

Aside from having to use a specially designed vestigial sideband filter in the transmitter, the main disadvantage of VSB is that the receiver must compensate for the part of the sideband that was eliminated at the transmitter.

As seen in Figure 8–2, both sidebands of video signals below 0.75 MHz are transmitted, but only one sideband of the video signals above 0.75 MHz is transmitted. The low-frequency video power will be twice that of the high-frequency signals. If no compensation is provided, the low frequencies will be overempha-

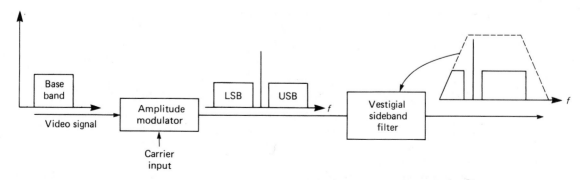

FIGURE 8–1   Generation of vestigial sideband (VSB).

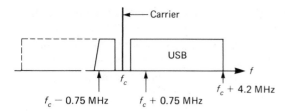

FIGURE 8–2 Television video spectrum.

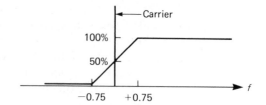

FIGURE 8–3 VSB compensation filter response.

sized in the picture, and the fine details will be of relatively low contrast (washed out). This form of frequency distortion is compensated for in TV receiver IF amplifiers.

Vestigial sideband compensation is accomplished before the demodulation process by providing IF filtering as illustrated in Figure 8–3. The frequency response of the IF amplifier is designed to roll off linearly between ±0.75 MHz of the carrier so that the high video frequencies are emphasized in the IF. The demodulated video will come out with the same relative amplitudes as it had at the studio.

8–2 ■ DOUBLE-SIDEBAND/SUPPRESSED-CARRIER

The amplitude-modulation technique called *double-sideband/suppressed-carrier* (DSB-SC) has an important advantage over regular AM (DSB-FC): The carrier is suppressed during the modulation process. As a result, most of the power in a regular AM transmission, which provides no information, is eliminated.

■ ■ ■ ■

EXAMPLE 8–1

Determine the power savings when the carrier is suppressed in a regular AM signal modulated to an index of 100%.

Solution:

$P_t = (1 + m^2/2)P_c$. $P_{sb} = P_c m^2/2$. The power savings is $(P_t - P_{sb})/P_t = 1/(1 + m^2/2) = 1/1.5 = $ **66.7%** for DSB-SC transmission.

The power savings of Example 8–1 has its price, however. As will be obvious by simple inspection of the waveform of a DSB-SC signal, an AM rectifier cannot be used to demodulate DSB-SC. Demodulation can be achieved only if a locally generated carrier signal is introduced. It must not only have exactly the correct frequency (be frequency-coherent) but also have a phase very close to what the carrier would have if it had been transmitted; that is, DSB-SC demodulation must also be approximately phase-coherent.

If it were not so difficult to demodulate, the excellent power efficiency of DSB-SC would make it very attractive. As it is, double-sideband/suppressed-carrier for information transmission is found in FM stereo and is also the basis for a digital-modulation system called *binary phase-shift key* (B-PSK); otherwise, it is mostly seen as the output of the balanced modulator used in the first step in producing single-sideband.

8–3 ■ BALANCED MODULATOR

The circuit used for producing a double-sideband/suppressed-carrier type of AM signal is shown in Figure 8–4. This circuit is a double-balanced mixer in which the diode pairs D_1-D_2 and D_3-D_4 are alternately switched on and off by the high-frequency carrier signal $v_c(t)$. The carrier signal could be a sinusoid or squarewave at frequency f_c; either way, its amplitude is much larger than that of the information (modulation) signal $m(t)$.

Figure 8–5 shows how the carrier causes alternate reversals* of the polarity of the modulation input signal. In part A the carrier is positive and diodes D_1 and D_2 become low-impedance devices for one-half of the RF cycle, whereas D_3 and D_4 are essentially open-circuited by reverse bias. In part B the modulation signal is coupled to the output with reverse polarity because the carrier signal has switched D_3 and D_4 ''on'' while reverse-biasing D_1 and D_2. The output signal $v_o(t)$ is merely $m(t)$ alternately multiplied by +1 and −1 due to the carrier's switching of the diodes. It should be recognized that due to the balanced output circuit, the carrier signal ideally is not coupled to the secondary of T_2. Consequently, for $m(t) = 0$, then $v_o(t) = 0$. The balance is confirmed by following current from a positive polarity $v_c(t)$ into the center tap of T_1, then splitting and flowing through both D_1

---

* As discussed in Chapter 11 on digital modulation techniques, a double-balanced modulator is used to produce (and demodulate) binary PSK. The carrier input is to T_1 of Figure 8–5, and digital data switches the diodes. The result is most accurately identified as phase-reversal modulation of the input carrier since the data switching period lasts for many carrier cycles.

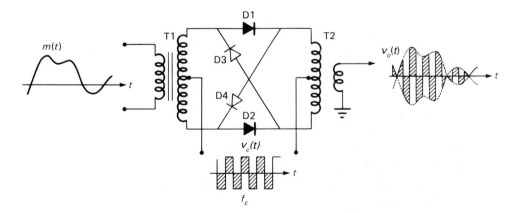

FIGURE 8–4 Balanced ring modulator.

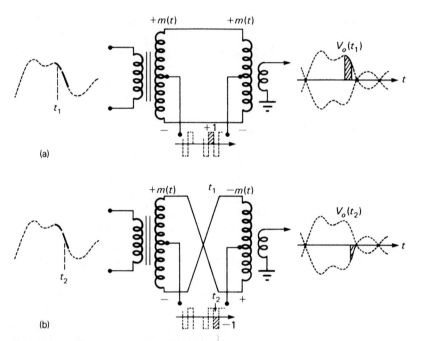

FIGURE 8-5 Balanced modulator phase reversals.

and D_2, converging at the center tap of T_2, and returning to the $v_c(t)$ source. The opposite flowing currents in the primary of T_2 induce voltages of equal magnitude and opposite polarity in the T_2 secondary, which therefore cancel each other.

The squarewave switching function can be written from Table 3–1e with an amplitude of $A = \pi/2$ as

$$v_c(t) = \sin 2\pi f_c t + \left(\frac{1}{3}\right)\sin 2\pi(3f_c)t + \cdots + \left(\frac{1}{n}\right)\sin 2\pi(nf_c)t \qquad \textbf{(8–1)}$$

where n and all previous harmonics are odd only.

As noted, the circuit physically performs a function that is mathematically equivalent to multiplication of time-varying signals $v_c(t)$ and the generalized information signal $m(t)$. Hence, the output is

$$m(t) \times v_c(t) = v_o(t)$$
$$= m(t)\sin 2\pi f_c t + (\frac{1}{3})m(t)\sin 2\pi(3f_c)t$$
$$+ \text{ higher odd harmonics} \qquad \textbf{(8–2)}$$

To illustrate that Equation 8–2 indeed represents a DSB-SC signal, let the modulation signal be a 2-V pk audio tone of frequency $f_m = 5$ kHz so that $m(t) = A \sin 2\pi f_m t = 2 \sin 2\pi(5 \text{ kHz})t$ V. Also, let the carrier frequency be $f_c = 45$ kHz. Substituting into Equation 8–2 yields a modulated output signal of

$$v_o(t) = A \sin 2\pi f_m t \sin 2\pi f_c t + \left(\frac{A}{3}\right) \sin 2\pi f_m t \sin 2\pi (3f_c)t + \cdots$$

$$= 2 \sin 2\pi (5 \text{ kHz})t \sin 2\pi (45 \text{ kHz})t$$

$$+ \frac{2}{3} \sin 2\pi (5 \text{ kHz})t \sin 2\pi (135 \text{ kHz})t + \cdots$$

By the use of the trigonometric identity

$$\sin A \sin B = (\tfrac{1}{2})\cos(A - B) - (\tfrac{1}{2})\cos(A + B)$$

$v_o(t)$ is seen to be

$$v_o(t) = \left(\frac{A}{2}\right)\cos 2\pi (f_c - f_m)t - \left(\frac{A}{2}\right)\cos 2\pi (f_c + f_m)t$$

$$+ \left(\frac{A}{6}\right)\cos 2\pi (3f_c - f_m)t - \left(\frac{A}{6}\right)\cos 2\pi (3f_c + f_m)t + \cdots \qquad \textbf{(8–3)}$$

$$v_o(t) = \cos 2\pi (40 \text{ kHz})t - \cos 2\pi (50 \text{ kHz})t$$

$$+ \frac{1}{3} \cos 2\pi (130 \text{ kHz})t - \frac{1}{3} \cos 2\pi (140 \text{ kHz})t + \cdots \qquad \textbf{(8–4)}$$

Figure 8–6 shows a sketch of Equation 8–4 in both time and frequency domains. If $v_o(t)$ is filtered so that only the first set of sidebands will be transmitted, then the harmonics are missing; the result is shown in Figure 8–7.

For a generalized (nondeterministic) input signal such as music with a maximum audio frequency of $f_A(\max)$, the input and output DSB-SC signals are shown in Figure 8–8.

Balanced modulators for generating DSB-SC with excellent carrier suppression (>50 dB) are also available in integrated circuit technology. The 1496/1596 IC can be used for carrier frequencies up to 100 MHz as a modulator or demodulator.

FIGURE 8–6 Wideband DSB-SC signal.

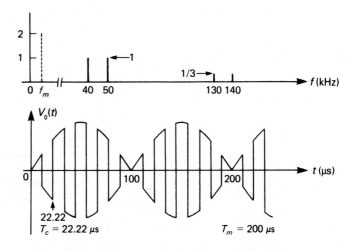

FIGURE 8–7 DSB-SC after filtering higher harmonics.

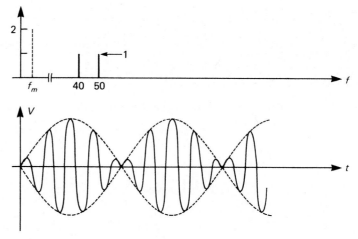

Also, notice the IC internal circuit schematic (Figure 23 of the data sheet) and the typical modulator circuit schematics complete with parts values (Figures 24–29 of the data sheet). Figure 8–9 shows some of the specification sheet, including DSB-SC waveforms and spectra. Also, four-quadrant multipliers such as the 1494L/1594L IC will perform these and numerous other functions.

Equation 8–3 and the spectra shown for DSB-SC justify the introductory comments that double-sideband/suppressed-carrier signals contain no carrier. How-

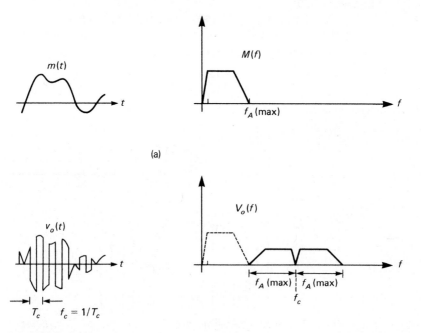

FIGURE 8–8 Time and frequency spectra for (a) the modulation and (b) the DSB-SC.

MOTOROLA
SEMICONDUCTORS
P.O. BOX 20912 • PHOENIX, ARIZONA 85036

MC1496
MC1596

Specifications and Applications Information

BALANCED MODULATOR — DEMODULATOR

SILICON MONOLITHIC INTEGRATED CIRCUIT

BALANCED MODULATOR — DEMODULATOR

. . . designed for use where the output voltage is a product of an input voltage (signal) and a switching function (carrier). Typical applications include suppressed carrier and amplitude modulation, synchronous detection, FM detection, phase detection, and chopper applications. See Motorola Application Note AN-531 for additional design information.

- Excellent Carrier Suppression   65 dB typ @ 0.5 MHz
  50 dB typ @ 10 MHz
- Adjustable Gain and Signal Handling
- Balanced Inputs and Outputs
- High Common-Mode Rejection   85 dB typ

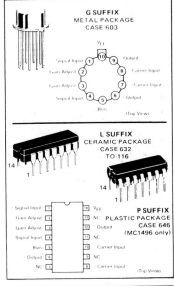

G SUFFIX
METAL PACKAGE
CASE 603

L SUFFIX
CERAMIC PACKAGE
CASE 632
TO-116

P SUFFIX
PLASTIC PACKAGE
CASE 646
(MC1496 only)

(Top View)

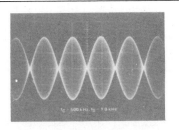

FIGURE 1 –
SUPPRESSED-CARRIER
OUTPUT WAVEFORM

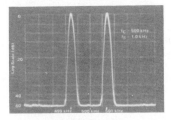

FIGURE 2 –
SUPPRESSED-CARRIER
SPECTRUM

ORDERING INFORMATION		
Device	Temperature Range	Package
MC1496G	0 to +70°C	Metal Can
MC1496L	0 to +70°C	Ceramic DIP
MC1496P	0 to +70°C	Plastic DIP
MC1596G	-55 to +125°C	Metal Can
MC1596L	-55 to +125°C	Ceramic DIP

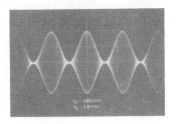

FIGURE 3 –
AMPLITUDE-MODULATION
OUTPUT WAVEFORM

FIGURE 4 – AMPLITUDE-MODULATION SPECTRUM

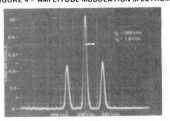

DS 9132 R2

FIGURE 8–9   Integrated circuit balanced modulator (page 1 of 3; pages 297–99 courtesy of Motorola, Inc.).

OPERATIONS INFORMATION

The MC1596/MC1496, a monolithic balanced modulator circuit, is shown in Figure 23.

This circuit consists of an upper quad differential amplifier driven by a standard differential amplifier with dual current sources. The output collectors are cross-coupled so that full-wave balanced multiplication of the two input voltages occurs. That is, the output signal is a constant times the product of the two input signals.

Mathematical analysis of linear ac signal multiplication indicates that the output spectrum will consist of only the sum and difference of the two input frequencies. Thus, the device may be used as a balanced modulator, doubly balanced mixer, product detector, frequency doubler, and other applications requiring these particular output signal characteristics.

The lower differential amplifier has its emitters connected to the package pins so that an external emitter resistance may be used. Also, external load resistors are employed at the device output.

Signal Levels

The upper quad differential amplifier may be operated either in a linear or a saturated mode. The lower differential amplifier is operated in a linear mode for most applications.

For low-level operation at both input ports, the output signal will contain sum and difference frequency components and have an amplitude which is a function of the product of the input signal amplitudes.

For high-level operation at the carrier input port and linear operation at the modulating signal port, the output signal will contain sum and difference frequency components of the modulating signal frequency and the fundamental and odd harmonics of the carrier frequency. The output amplitude will be a constant times the modulating signal amplitude. Any amplitude variations in the carrier signal will not appear in the output.

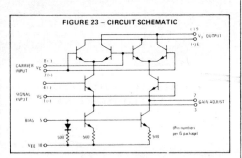

FIGURE 23 – CIRCUIT SCHEMATIC

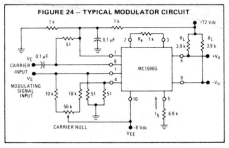

FIGURE 24 – TYPICAL MODULATOR CIRCUIT

TYPICAL APPLICATIONS

FIGURE 26 – BALANCED MODULATOR
(+12 Vdc SINGLE SUPPLY)

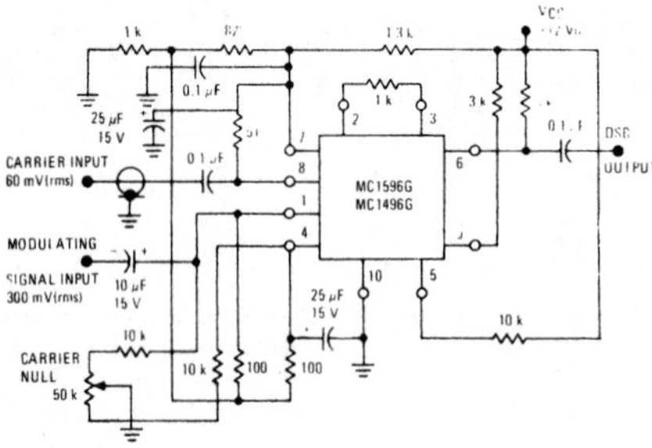

FIGURE 27 – BALANCED MODULATOR-DEMODULATOR

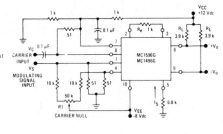

FIGURE 28 – AM MODULATOR CIRCUIT

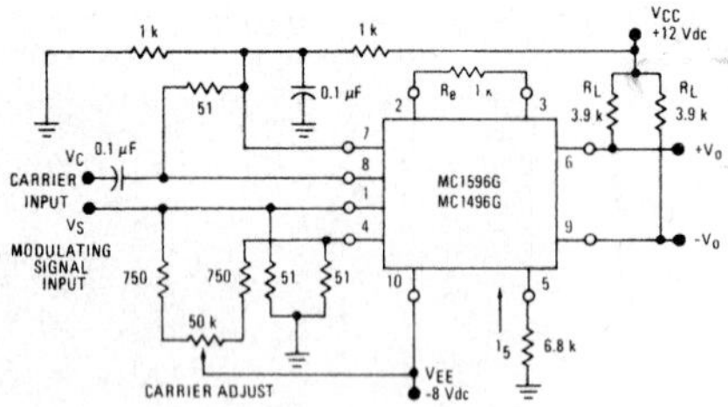

FIGURE 29 – PRODUCT DETECTOR
(+12 Vdc SINGLE SUPPLY)

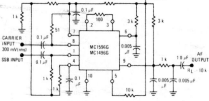

MOTOROLA *Semiconductor Products Inc.*

FIGURE 8-9 (page 2 of 3)

MC1496 ● MC1596

MAXIMUM RATINGS* (T_A = +25°C unless otherwise noted)

Rating	Symbol	Value	Unit
Applied Voltage ($V_6 - V_7, V_8 - V_1, V_9 - V_7, V_9 - V_8, V_7 - V_4, V_7 - V_1,$ $V_8 - V_4, V_6 - V_8, V_2 - V_5, V_3 - V_5$)	ΔV	30	Vdc
Differential Input Signal	$V_7 - V_8$ $V_4 - V_1$	+5.0 $\pm(5+I_5R_e)$	Vdc
Maximum Bias Current	I_5	10	mA
Thermal Resistance, Junction to Air Ceramic Dual In-Line Package Plastic Dual In-Line Package Metal Package	$R_{\theta JA}$	 180 100 200	°C/W
Operating Temperature Range MC1496 MC1596	T_A	 0 to +70 -55 to +125	°C
Storage Temperature Range	T_{stg}	-65 to +150	°C

ELECTRICAL CHARACTERISTICS* (V_{CC} = +12 Vdc, V_{EE} = -8.0 Vdc, I_5 = 1.0 mAdc, R_L = 3.9 kΩ, R_e = 1.0 kΩ, T_A = +25°C unless otherwise noted) (All input and output characteristics are single-ended unless otherwise noted.)

Characteristic	Fig	Note	Symbol	MC1596 Min	MC1596 Typ	MC1596 Max	MC1496 Min	MC1496 Typ	MC1496 Max	Unit
Carrier Feedthrough V_C = 60 mV(rms) sine wave and f_C = 1.0 kHz offset adjusted to zero f_C = 10 MHz	5	1	V_{CFT}	– –	40 140	– –	– –	40 140	– –	µV(rms)
V_C = 300 mVp-p square wave offset adjusted to zero f_C = 1.0 kHz offset not adjusted f_C = 1.0 kHz				– –	0.04 20	0.2 100	– –	0.04 20	0.4 200	mV(rms)
Carrier Suppression f_S = 10 kHz, 300 mV(rms) f_C = 500 kHz, 60 mV(rms) sine wave f_C = 10 MHz, 60 mV(rms) sine wave	5	2	V_{CS}	 50 –	 65 50	 – –	 40 –	 65 50	 – –	dB k

Transadmittance Bandwidth (Magnitude) (R_L = 50 ohms)
Carrier Input Port, V_C = 60 mV(rms) sine wave
  f_S = 1.0 kHz, 300 mV(rms) sine wave
Signal Input Port, V_S = 300 mV(rms) sine wave
  $|V_C|$ = 0.5 Vdc
	8	8	BW_{3dB}	
				–
				
				–
					
					300
					
					80
						
						–
						
						–
							
							–
							
							–
								
								300
								
								80
									
									–
									
									–
										MHz

Signal Gain V_S = 100 mV(rms), f = 1.0 kHz, $|V_C|$ = 0.5 Vdc	10	3	A_{VS}	2.5	3.5	–	2.5	3.5	–	V/V
Single-Ended Input Impedance, Signal Port, f = 5.0 MHz Parallel Input Resistance Parallel Input Capacitance	6		r_{ip} c_{ip}	– –	200 2.0	– –	– –	200 2.0	– –	kΩ pF
Single-Ended Output Impedance, f = 10 MHz Parallel Output Resistance Parallel Output Capacitance	6		r_{op} c_{op}	– –	40 5.0	– –	– –	40 5.0	– –	kΩ pF
Input Bias Current $I_{bS} = \dfrac{I_1 + I_4}{2}$, $I_{bC} = \dfrac{I_7 + I_8}{2}$	7	–	I_{bS} I_{bC}	– –	12 12	25 25	– –	12 12	30 30	µA
Input Offset Current $I_{ioS} = I_1 - I_4$; $I_{ioC} = I_7 - I_8$	7	–	$|I_{ioS}|$ $|I_{ioC}|$	– –	0.7 0.7	5.0 5.0	– –	0.7 0.7	7.0 7.0	µA
Average Temperature Coefficient of Input Offset Current (T_A = -55°C to +125°C)	7	–	$|TC_{Iio}|$	–	2.0	–	–	2.0	–	nA/°C
Output Offset Current ($I_6 - I_9$)	7	–	$|I_{oo}|$	–	14	50	–	14	80	µA
Average Temperature Coefficient of Output Offset Current (T_A = -55°C to +125°C)	7	–	$|TC_{Ioo}|$	–	90	–	–	90	–	nA/°C
Common-Mode Input Swing, Signal Port, f_S = 1.0 kHz	9	4	CMV	–	5.0	–	–	5.0	–	Vp-p
Common-Mode Gain, Signal Port, f_S = 1.0 kHz, $|V_C|$ = 0.5 Vdc	9		ACM	–	-85	–	–	-85	–	dB
Common-Mode Quiescent Output Voltage (Pin 6 or Pin 9)	10	–	V_o	–	8.0	–	–	8.0	–	Vdc
Differential Output Voltage Swing Capability	10	–	V_{out}	–	8.0	–	–	8.0	–	Vp-p
Power Supply Current $I_6 + I_9$ I_{10}	7	6	I_{CC} I_{EE}	– –	2.0 3.0	3.0 4.0	– –	2.0 3.0	4.0 5.0	mAdc
DC Power Dissipation	7	5	P_D	–	33	–	–	33	–	mW

* Pin number references pertain to this device when packaged in a metal can. To ascertain the corresponding pin numbers for plastic or ceramic packaged devices refer to the first page of this specification sheet.

 MOTOROLA *Semiconductor Products Inc.*

FIGURE 8–9 (page 3 of 3)

FIGURE 8–10 Result of non-coherent demodulation of DSB-SC.

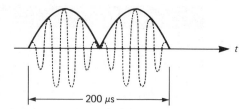

ever, information about the carrier is certainly seen to exist in the transmitted spectrum.

It must be clear from the oscilloscope (time-domain) sketches of Figures 8–6, 8–7, and 8–8 that the peak detector discussed for demodulation of regular AM (DSB-FC) will not yield the correct result for DSB-SC. For instance, when the input is the sinusoidal tone-modulated DSB-SC signal of Figure 8–7, the output of a peak detector will be the "cusp" signal of Figure 8–10. This is certainly not a 5-kHz sinusoid. In fact, a spectral analysis of the waveform of Figure 8–10 will show that no 5-kHz component is present. The fundamental period of the Figure 8–8 waveform is 100 μs, so that the fundamental frequency is 10 kHz. This signal is described in Table 3–1d with $f_0 = 10$ kHz.

8–4 ■ PRODUCT DETECTOR

A DSB-SC signal can be demodulated by reversing the modulation process; that is, the demodulator in Figure 8–4 has its DSB-SC *input at* T_2, a coherent (equal frequency *and* phase) carrier-oscillator signal as shown, and the output taken from T_1. Following Figure 8–5 from right to left, the demodulated output at T_1 is then filtered to eliminate harmonics and noise above $f_A(\text{max})$.

Using a double-balanced mixer and a locally generated carrier signal to demodulate a suppressed carrier signal is commonly referred to as *reinserting the carrier*. The carrier is then referred to as a *product detector* because, as shown for the modulation process, multiplication is the mathematical description for what the circuit does. Another perspective on this coherent demodulation process is to consider the product detector as a mixer and the "local oscillator" as a "beat" frequency oscillator (BFO) at f_c that beats the modulated signal down to an "IF" of 0 Hz. The filtered output is just the original baseband signal. This perspective will be verified by several problems at the end of the chapter.

8–5 ■ PHASE DISTORTION IN THE DEMODULATION OF SUPPRESSED-CARRIER SYSTEMS

Reinserting the carrier for demodulation of a suppressed-carrier signal seems simple enough—tune an oscillator (BFO) to the frequency of the missing carrier. However, even with the correct frequency, severe distortion can result if the reinserted carrier does not have the correct phase relative to the suppressed carrier.

Figure 8–11 shows the transmitted DSB-SC phasors (a), and the correct relationship between the reinserted carrier and sidebands (b). A phase error $\Delta\phi$ will result in the AM phasor signal of (c).

The resultant signal in c is a combination of AM and phase modulation, and the demodulated information which might be that of Figure 8–12a would come out like 8–12b with severe phase distortion. The demodulated signal has the correct fundamental frequency, but the phase distortion has greatly altered the information. The phase distortion problem is worse in DSB-SC than in SSB-SC because of the complication introduced by having the two sidebands. Also, transmission-channel phase shifts, which are not linear between the upper and lower sidebands (envelope-delay distortion), will make the problem even worse.

One technique for solving the problem of producing a coherent carrier for demodulating DSB-SC at a distant receiver is to add a small amount of the carrier oscillator signal to the DSB-SC signal before transmission. This signal, called a *pilot*, can then be used at the receiver to synchronize the local beat-frequency oscillator. The pilot also provides a constant amplitude signal for receiver automatic gain control.

As is done with the DSB-SC part of an FM stereo baseband signal, the transmitted pilot can be a submultiple of the actual carrier, keeping it completely independent of the low-frequency modulation components. However, most systems do not have a bandwidth wide enough to accommodate an octave-frequency offset (see Section 9–8).

Another technique used for carrier recovery at the receiver is to use a circuit called a *Costas loop*, in which information in the demodulated signal is used to control the frequency and phase of the beat-frequency oscillator. The technique is

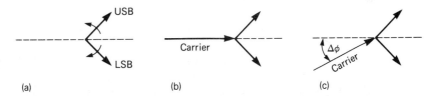

(a)   (b)   (c)

FIGURE 8–11 Phasor representation of DSB-SC. (a) DSB-SC. (b) DSB-SC with carrier "reinserted"—AM. (c) Carrier reinserted with wrong phase.

FIGURE 8–12 Result of phase distortion due to reinserted-carrier phase error. (a) Transmitted. (b) The result of a phase distortion due to phase error of reinserted carrier.

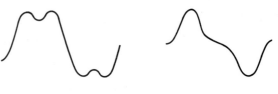

(a)   (b)

described at the end of this chapter. But first let us consider a sideband system that is often used because of its bandwidth and power efficiency, single-sideband/suppressed-carrier.

8–6 ■ SINGLE-SIDEBAND/SUPPRESSED-CARRIER

Single-sideband/suppressed-carrier (SSB-SC) is an amplitude modulation technique used for its outstanding power and bandwidth efficiency. This transmission type is designated as A3J. By eliminating the carrier and one sideband, a power savings of over 83% is realized. Additionally, the bandwidth required for SSB-SC is theoretically one-half that required when both sidebands are transmitted. As is the case for DSB-SC, the advantages are somewhat offset by the need for carrier recovery and reinsertion at the receiver. However, the phase and frequency accuracy requirements are not as critical for single-sideband as they are for DSB-SC. Two techniques for generating SSB-SC are considered: the sideband-filter method and the phasing method.

The Sideband-Filter Method

Figure 8–13 shows a block diagram for an SSB-SC transmitter. The heart of this system is the balanced modulator and sideband filter. The information to be communicated is amplified and fed to the balanced modulator. Also fed to the modulator is an intermediate-frequency (IF) carrier that is frequency- and phase-locked to a stable reference generator in the frequency synthesizer.*

The double-sideband/suppressed-carrier output of the balanced modulator is fed to a sideband filter, where the unwanted sideband is eliminated. The single remaining sideband is at an intermediate frequency and must be up-converted in a mixer to the desired transmission frequency.

After filtering the mixer signal products, the single-sideband/suppressed-carrier AM signal is amplified in linear power amplifiers (LPAs) and coupled to the antenna or perhaps to coaxial transmission lines for multiplexing with other single-sideband signals.

Use of a frequency synthesizer in higher-frequency (HF) radio allows for the generation of a single-sideband signal at a fixed IF frequency and its subsequent up-conversion to one of the many available transmission channels. The synthesized master oscillator (SMO) or up-converter oscillator is variable in discrete steps but has crystal-controlled stability at each of the discrete frequencies selected.

The frequency synthesizer system provides the same highly stable reference frequency to both the IF carrier generator and the variable up-converter oscillator to achieve the frequency stability required of SSB-SC systems. Generally, the

---

* Phase-lock techniques and frequency synthesizers are examined thoroughly in Chapter 10.

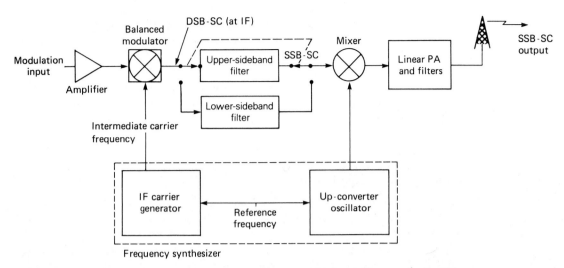

FIGURE 8–13 Single-sideband transmitter block diagram (sideband-filter method). Either upper or lower sideband filtering may be chosen.

total-system frequency drift tolerance for adequate voice quality is considered to be ±50 Hz. However, voice-frequency telegraph (VFTG) and low-speed data systems require less than ±2 Hz.

Typical intermediate frequencies in HF radio range from 50 kHz to 12 MHz. These frequencies allow for the use of mechanical or crystal filters for the attenuation of the unwanted sideband. Mechanical filters* operate to less than 800 kHz and have bandwidths from less than 0.1–10% of the operating frequency (Qs to 20,000). Crystal filters with Qs of 50,000 or more are used at frequencies as high as 20 MHz. Ceramic filters with Qs to 2000 are also available, and at moderate prices. Assessing the number of resonant elements (poles) and minimum Q per resonator required for a given filter application was discussed in Chapter 7. Suffice it to say at this point that a 50-dB minimum suppression of the unwanted sideband when transmitting voice will require at least an eight-resonator filter.

If two filters are available in the sideband filter block of Figure 8–13, the operator may choose either the upper or the lower sideband for transmission. With two filters and another modulator, the unit can also incorporate an independent sideband (ISB) mode of operation, in which two separate voice transmissions occur simultaneously, with one above and the other below the suppressed carrier.

The power amplifier is class AB or B (push-pull) to maintain the required linearity of the amplitude-varying signal. Output power varies from approximately zero during a lull in the conversation to as much as 50 kW peak envelope power in some point-to-point HF radio links.

---

* For an excellent article on mechanical filters, see Robert Johnson, "Mechanical Filters Take on Selective Jobs," *Electronics Magazine*, October 13, 1977.

Automatic level control (ALC) is used to prevent amplifier distortion on peaks. Similar to AGC in receivers, some of the output signal is coupled to a peak detector. The voltage output of the amplitude detector is fed back to control the output level of the driver amplifier. In this system it is important for the ALC voltage to respond quickly because only the peaks are controlled. On the other hand if the amplifier gain is released too quickly, the result is bursts or pops in the output. It is common in ALC design to provide for an attack time on the order of 10 ms and a release time of 100 ms.

The Phase Method of SSB-SC Generation

By properly combining two DSB-SC signals in which either the upper or the lower sidebands are exactly out of phase, a single-sideband signal can be produced. The equal-frequency sidebands that are out of phase will cancel, and the in-phase sidebands reinforce each other to become the transmitted sideband. The block diagram is shown in Figure 8–14. The mathematical proof of this technique is given for a sinusoidal input of frequency ω_m.

The inputs to the top balanced modulator are $\sin \omega_m t$ and $\sin \omega_c t$. The multiplied output is $\sin \omega_m t \sin \omega_c t$, which by trigonometric identity is

$$\sin \omega_m t \sin \omega_c t = 0.5[\cos(\omega_c - \omega_m)t - \cos(\omega_c + \omega_m)t] \qquad \textbf{(8–5)}$$

The inputs to the bottom balanced modulator are $\cos \omega_m t$ and $\cos \omega_c t$. The output of this modulator is

$$\cos \omega_m t \cos \omega_c t = 0.5[\cos(\omega_c - \omega_m)t + \cos(\omega_c + \omega_m)t] \qquad \textbf{(8–6)}$$

Equation 8–5 describes DSB-SC with the upper sideband having opposite polarity to the upper sideband of Equation 8–6. The output of the summing network is the addition of Equations 8–5 and 8–6; that is,

$$\begin{aligned}
\text{SSB-SC output} &= v_o(t) \\
&= 0.5[\cos(\omega_c - \omega_m)t - \cos(\omega_c + \omega_m)t] \\
&\quad + 0.5[\cos(\omega_c - \omega_m)t + \cos(\omega_c + \omega_m)t] \\
&= \cos(\omega_c - \omega_m)t
\end{aligned} \qquad \textbf{(8–7)}$$

The various time waveforms and corresponding frequency spectra for single-tone modulation are shown in Figure 8–15.

The output of the phasing method block diagram of Figure 8–14 is seen in Equation 8–7 to be the lower sideband, with a frequency that is the difference between the carrier and modulation frequencies. Notice that if the output summing network is changed to a difference network, Equation 8–5 will be subtracted from Equation 8–6 and the upper sideband will be transmitted. It can also be shown that reversing the quadrature carrier *or* the audio inputs to the balanced modulators will result in transmission of the opposite sideband.

All this works out very simply on paper. The problems are with the technology. Even with integrated circuitry, keeping the phases of modulator inputs and outputs constant when temperatures and power supply voltages are changing is not

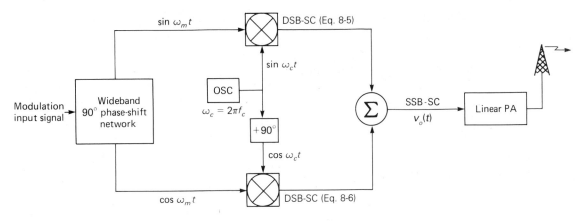

FIGURE 8–14   Single-sideband transmitter block diagram (phase method).

so simple. Furthermore, while the carrier phase-shift network at a single frequency is simple enough, the wideband audio network is required to shift the phase by exactly 90° over the full audio frequency range. The circuit used for this has traditionally been the *all-pass network*, which is implemented with RC branches (Figure 8–16). The desired bandwidth of this network is set between $\omega_2 - \omega_1$, where $\omega_2 = 1/R_2C_2$ is the high-frequency cutoff and ω_1 is set by the other

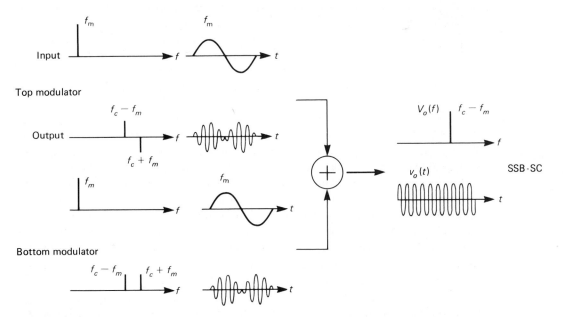

FIGURE 8–15   Time and frequency spectra for phase method of producing SSB-SC.

FIGURE 8–16 Wideband 90°
phase-shift circuit.

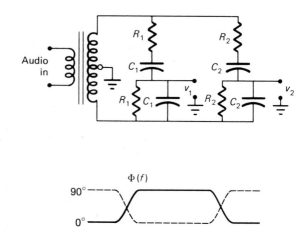

RC time constant. A *sequence discrimination** approach has been shown to be superior for IC implementations.

As integrated circuit technology improves, the phasing method for generation of SSB-SC will allow for the elimination of bulky and expensive filters.

Single-Sideband Receiver

Demodulation of an SSB-SC signal is essentially the same as that described for DSB-SC: The receiver must remain linear and the carrier must be reinserted. The receiver then will be a heterodyned AM receiver with some provision for automatic level control (ALC) to prevent distortion on amplitude peaks. Demodulation is accomplished in a product detector as previously described, and a beat-frequency oscillator (BFO) is used for injecting a carrier.

As mentioned in the section on phase distortion in demodulation of suppressed-carrier systems, SSB-SC is much less phase- and frequency-sensitive than is DSB-SC. In fact, a voice transmission will still be intelligible with a frequency error of ±50 Hz at the demodulator. However, voice frequency telegraph and low-speed data systems require less than ±2 Hz. When you realize that frequency drift is contributed by each of the oscillators in the system—the transmitter IF generator and up-converted oscillator, as well as the receiver local oscillator and BFO—you see that frequency stability is a major concern in suppressed-carrier systems. As an example will show, oscillators with crystal-controlled stability will be required. For multichannel transceivers this means that frequency synthesizers are required.

* E. Daoud and others, "New Active RC-Networks for the Generation and Detection of Single-Sideband Signals," *IEEE Transactions on Circuits and Systems,* vol. CAS-27, no. 12, December 1980, p. 1140.

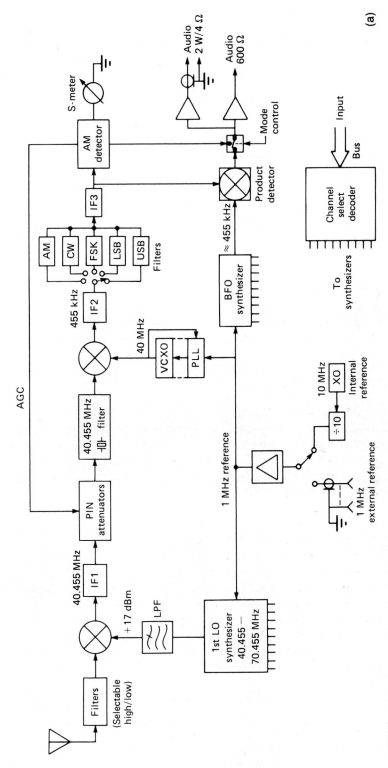

FIGURE 8-17 (page 1 of 2) (a) Block diagram of frequency-synthesized VLF/MF Model 1030 communications receiver. (Cubic Communications)

■ ■ ■ ■

EXAMPLE 8–2

Determine the minimum frequency stability required if a 27.065-MHz oscillator is used to demodulate an SSB-SC voice transmission on Citizen's Band (CB) channel 9. Give the answer in percent and parts per million (ppm).

Solution:

The demodulated voice signal will be barely intelligible if the oscillator drifts by ± 50 Hz. This is ± 50 Hz/27.065 MHz = **± 1.85 ppm** (or 1.85×10^{-6}) $\times 100\%$ = **$\pm 0.000185\%$.** On a short-term basis this is achievable with a crystal oscillator. However, one should be concerned about oscillator and received-signal noise.

Receivers for SSB-SC are usually a double-conversion type; that is, there are two mixers and two IF systems. Also, to achieve the frequency stability required when multichannel operation is employed, the LOs and BFO are synchronized to a highly stable reference oscillator.

(b)

(c)

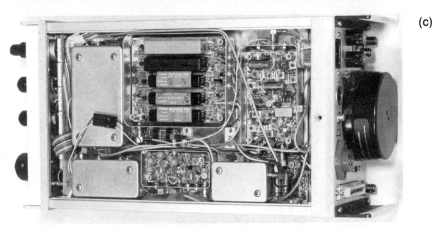

FIGURE 8–17 (page 2 of 2) (b) Front view of hardware. (c) Top, internal view of hardware. (Cubic Communications)

A frequency-synthesized VLF/MF communications receiver block diagram is shown in Figure 8–17. The Cubic Communication Corporation HF1030 model receiver can be used for demodulating AM, CW, SSB-SC, ISB, FM, and FSK. The synthesizer allows reception over a 10-kHz to 30-MHz frequency range in 10-Hz steps and can be programmed via an IEEE-488 BUS or by parallel binary-coded decimal (BCD) control.

The input signal is converted at the first mixer to an IF of 40.455 MHz by the synthesized first LO. The first IF includes an eight-pole crystal filter with a ±4-kHz bandwidth for high selectivity. The first IF includes two PIN-diode attenuators, each having 40 dB of AGC range. The second LO is a 40-MHz crystal oscillator phase-locked to the synthesizer reference source. The BFO can be changed in 10-Hz steps over a range of ±5 kHz and is injected into an active double-balanced-mixer product detector. Figure 8–17b and c shows views of the hardware.

For a discussion of frequency synthesizers, including an example of a phase-lock synthesized CB transceiver, see Chapter 10.

8-7 ■ FREQUENCY-DIVISION MULTIPLEXING IN TELEPHONE SYSTEMS

An excellent example of the application of the single-sideband technique with its bandwidth and power efficiency is in combining analog telephone conversations on long-distance toll trunks.

When introduced in Chapter 5, frequency-division multiplexing (FDM) was suggested as a method by which many radio and TV channels could transmit simultaneously by using different carrier frequencies. In that application, if the carrier frequency separation is at least twice the maximum modulating (voice or music) frequency, then the spectra will not overlap. This allows a receiver to be tuned to one channel without interference from another. By using SSB-SC the transmission bandwidth is cut in half, and twice as many channels of information can be transmitted in the same frequency spectrum. FDM is a technique for simultaneous transmission of many narrow-bandwidth signals over a wideband channel. The individual narrowband signals are separated in the frequency-domain spectrum by modulation on separate carriers. The carriers must have enough frequency separation that the modulated signal spectra do not overlap. By this means the individual information signals can be kept separated by filters.

American Telephone and Telegraph Company (AT&T) has been using frequency-division multiplexing of SSB-SC modulated telephone signals since 1918. In the early years, a maximum of four individual subscribers (telephone users) were combined simultaneously on a single open-wire-pair transmission line. Since that time, coaxial cable and microwave radio links have been used with the equivalent of 13,200 voice-frequency channels transmitted simultaneously on the L5E system by the technique of frequency-division multiplexing. More recently, digital T1 (Chapter 13) and fiber-optic (Chapter 18) transmission have begun to replace SSB techniques.

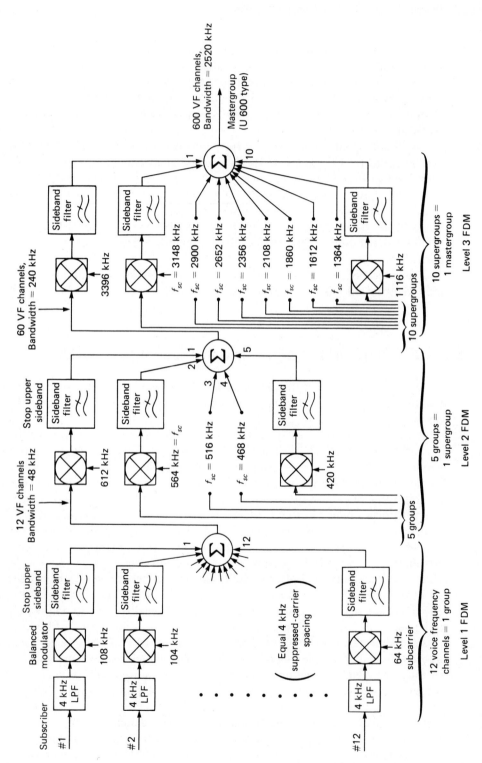

FIGURE 8–18 Analog telephone FDM hierarchy.

The method of multiplexing SSB-SC modulated voice-frequency (VF) telephone conversations in the United States is illustrated in Figure 8–18 based on the AT&T U600 hierarchy. Twelve equally spaced carriers would be used to convert twelve analog incoming calls to lower-sideband-only SSB-SC signals of 4-kHz bandwidth each. The carrier frequencies range from 108 kHz for channel 1 to 64 kHz for channel 12. The result of linearly combining the 12 channels is a 48-kHz-wide frequency-division multiplexed signal called a *group*. Five other groups are produced in the same way, translated up in the frequency spectrum by SSB-SC, and linearly combined to produce a *supergroup*. In the U600 system, ten supergroups are multiplexed to produce a *mastergroup*. The FDM frequency spectrum for this system is illustrated in Figure 8–19.

In each level of the FDM hierarchy, pilot carriers are added for synchronizing and level control for the receiving-end demultiplexing and demodulation processes. The telephone system also uses time-division multiplexing (TDM) to transmit computer data and digitized voice (see Chapters 12 and 13). The mastergroup described above is also multiplexed (FDM) with wideband data* and other

---

* A data modem is used to produce the high-frequency analog signal for multiplexing.

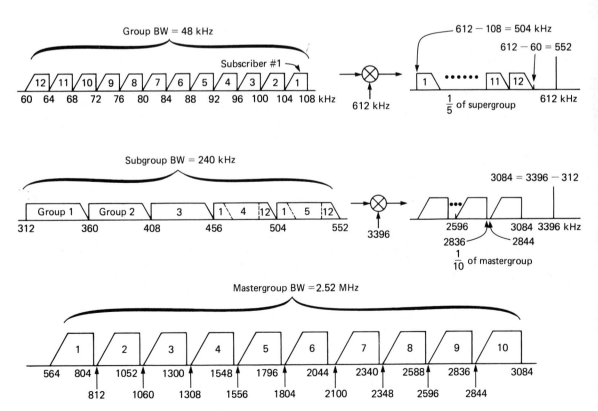

FIGURE 8–19   Analog telephone FDM frequency spectrum.

mastergroups to form higher-level baseband signals. These multiplexed baseband signals can be transmitted directly on coaxial transmission lines or used to modulate a microwave radio transmitter.

The U.S. long-distance L5 system includes the multiplexing of six mastergroups into a *jumbo group*, and there are jumbo groups of 20, 40, and 60 MHz with pilots at 2976, 10992, and 66048 kHz. The main oscillator for this system has a long-term drift rate of less than 1 part in 10^{10} (10^{-4} ppm) per day.* Various hierarchies recommended by CCITT† are in use for internationally compatible telephone transmission using large-diameter coaxial cable with up to a 60-MHz bandwidth and 1-mile repeater spacing. These repeaters are fixed-gain amplifiers, some of which provide transmission system temperature compensation.

■ ■ ■ ■

EXAMPLE 8–3

What is the maximum allowable worst-case drift rate for a 5.12-MHz master oscillator if the frequency must be held to within 1 Hz over 5 years (y) of maintenance-free operation?

Solution:

5 y/Hz × 365 da/y × 5.12 × 10^6 Hz = 0.9344 × 10^{10}, or **9.344 parts in 10^9 per day.**

8–8 ■ QUADRATURE MULTIPLEXING

Frequency-division multiplexing has been discussed. Time-division multiplexing has been mentioned and will be covered with pulse and digital transmissions. Another technique for transmitting two channels of information on a single carrier in a way that will allow the two information signals to be separated at the receiver is *quadrature multiplexing*.

The two information sources (channels) are separately modulated on carriers derived from the same frequency source but with a 90° phase difference; that is, the two carrier signals are in phase quadrature.

Quadrature multiplexing is illustrated in Figure 8–20. The two modulation signals $m_1(t)$ and $m_2(t)$ modulate the quadrature carriers sin $\omega_c t$ and cos $\omega_c t$ in balanced modulators. The modulated signals are filtered (not shown) to eliminate nonlinear mixer products, then linearly added to form the quadrature-multiplexed (QM) signal. As seen at the transmitter output of Figure 8–20 and receiver input of Figure 8–21, the QM signal is merely the sum of two orthogonal (90°) DSB-SC signals with the same suppressed-carrier frequency. To illustrate the principle for a color television application, the color information is transmitted as a vector determined by the amplitude and polarity of the quadrature carriers operating at a

---

* R. L. Freeman, *Telecommunication Transmission Handbook*, 2nd ed. (New York: Wiley-Interscience, 1981), p. 440.

† Consultative Committee for International Telephone and Telegraphy.

FIGURE 8–20 Quadrature multiplexing of two channels. The output is the sum of two orthogonal DSB-SC signals on the same carrier.

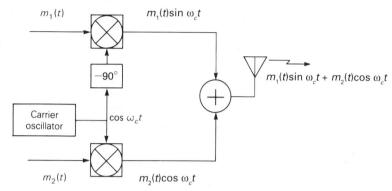

frequency of approximately 3.58 MHz relative to the video carrier. If, for example, the in-phase carrier has an amplitude of $-0.44 V_i$, where V_i is the maximum in-phase carrier voltage of positive polarity, and the quadrature carrier has an amplitude of $-0.9V_q$, then the TV demodulator should interpret the resulting color as green.

The proof that the received orthogonal DSB-SC signals can be separated and demodulated is now demonstrated for the receiver block diagram of Figure 8–21.

The receiver LO or beat frequency oscillator (BFO) is synchronized* to the incoming signal; that is, the oscillator frequency is exactly ω_c. The LO signal $2 \cos \omega_c t$ is split and phase-shifted to give two quadrature LO signals $V_{LO_i} = 2 \cos \omega_c t$ and $V_{LO_q} = 2 \sin \omega_c t$, which are the LO inputs to the in-phase and quadrature mixers, respectively. The mixer outputs are simply the products of their two input signals—the received QM signal and the individual LOs.

---

* The circuit used for deriving the "synchronized" signal is called a Costas (feedback-) loop.

FIGURE 8–21 Quadrature multiplex receiver. The two information signals are $m_1(t)$ and $m_2(t)$.

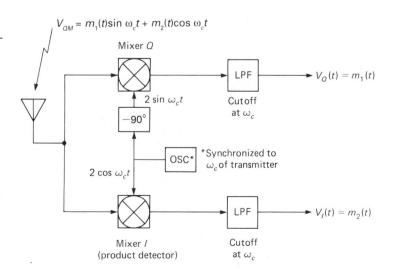

The demodulated outputs after filtering high-order mixer products are derived using the same trig identities as for DSB-SC and in Chapter 7 with mixers. Here, the demodulation of the $m_1(t)$ information signal in the quadrature mixer is shown. The same proof is used to show the demodulation of $m_2(t)$ in the in-phase mixer.

The quadrature mixer output is

$$V_{\text{LO}_q} \times V_{\text{QM}} = (2 \sin \omega_c t)[m_1(t) \sin \omega_c t + m_2(t) \cos \omega_c t] \qquad \textbf{(8–8a)}$$

$$= 2m_1(t)\sin \omega_c t \sin \omega_c t + 2m_2(t)\sin \omega_c t \cos \omega_c t \qquad \textbf{(8–8b)}$$

$$= m_1(t)\cos(\omega_c - \omega_c)t - m_1(t)\cos(\omega_c + \omega_c)t$$
$$+ m_2(t)\sin(\omega_c - \omega_c)t + m_2(t)\sin(\omega_c + \omega_c)t \qquad \textbf{(8–8c)}$$

$$= m_1(t) - m_1(t)\cos 2\omega_c t + m_2(t)\sin 2\omega_c t \qquad \textbf{(8–9)}$$

Please note that the mixer output is the $m_1(t)$ information signal and two second-harmonic mixer products (both DSB-SC), which are easily filtered out with a low-pass filter set just below ω_c. Since the low-pass filters (LPF) of Figure 8–21 have a cutoff frequency of just below ω_c, the output signal from the "quadrature" branch is

$$V_Q(t) = m_1(t) \qquad \textbf{(8–10)}$$

A similar analysis, left for an end-of-chapter exercise, shows that $V_I(t) = m_2(t)$.

A slight extension of the preceding analysis and the block diagram of Figure 8–21 will yield an extremely important result. The result shows that a suppressed-carrier signal does indeed carry information necessary for synchronizing the receiver-system local oscillators.

8–9 ■ COSTAS LOOP FOR SUPPRESSED-CARRIER DEMODULATION

Suppressed-carrier signals such as the telephone SSB-SC FDM signals, the double-sideband/suppressed-carrier transmissions for color in TV, and quadrature-multiplexed stereo, as well as all of the coherent microwave digital-data transmission systems (BPSK, QPSK, QAM, for example), require a coherent local oscillator for demodulation. Coherent in this case means *exact frequency* and *approximate phase*—that is, a small phase error may exist.

When a pilot signal is transmitted (as is the case for telephone FDM, color TV,* and FM stereo) a simple phase-locked loop will lock onto the pilot for "synchronization." Unfortunately, a suppressed-carrier signal with no pilot has no fixed spectral component on which to lock up a phase-locked loop.

As was noted in the section on DSB-SC, the sidebands contain information leading to the whereabouts of the missing carrier. Figure 8–22 shows the additional circuit necessary to extract the required information. It is a dc-coupled product detector (mixer) known as a *phase detector*. The system helps the volt-

---

* The pilot comes in short bursts called the *color burst*.

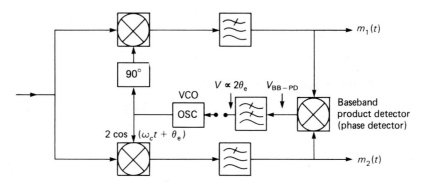

FIGURE 8-22   Costas loop includes BFO-synchronizing for demodulation of all types of suppressed-carrier signals.

age-controlled local oscillator (VCO) to synchronize at a frequency equal to the suppressed-carrier frequency ω_c. Once the frequency of the LO is equal to the suppressed-carrier frequency, the product of $V_I(t)$ and $V_Q(t)$ will produce a voltage proportional to any phase error of the local oscillator. This voltage is used to control the phase and thereby the frequency of the local beat-frequency oscillator. The result described is *carrier recovery*, and the circuit, a Costas loop, is a special type of phase-locked loop.

The analysis of the loop for VCO control is as follows: If the VCO (the receiver local oscillator) is locked to the incoming carrier, then $\omega_{LO} = \omega_c$ and only a small phase error difference (θ_e) between the two signals will exist. Because of the small phase error, Equation 8–8c becomes

$$V_{LO} \times V_{QM} = m_1(t)\cos[(\omega_c - \omega_c)t + \theta_e] - m_1(t)\cos[(\omega_c + \omega_c)t + \theta_e]$$
$$+ m_2(t)\sin[(\omega_c - \omega_c)t + \theta_e] + m_2(t)\sin[(\omega_c + \omega_c)t + \theta_e]$$

$$V_{LO} \times V_{QM} = m_1(t)\cos\theta_e - m_1(t)\cos(2\omega_c + \theta_e)$$
$$+ m_2(t)\sin\theta_e + m_2(t)\sin(2\omega_c + \theta_e) \qquad (8\text{–}11)$$

and Equation 8–10 becomes

$$V_Q(t) = m_1(t)\cos\theta_e \qquad (8\text{–}12\text{a})$$

and

$$V_I(t) = m_2(t)\sin\theta_e \qquad (8\text{–}12\text{b})$$

Now the output of the baseband product detector becomes

$$V_{BB\text{-}PD}(t) = V_Q(t) \times V_I(t) \qquad (8\text{–}13\text{a})$$
$$= m_1(t)m_2(t)\cos\theta_e \sin\theta_e \qquad (8\text{–}13\text{b})$$
$$V_{BB\text{-}PD}(t) = m_1(t)m_2(t)[\sin(\theta_e - \theta_e) + \sin(\theta_e + \theta_e)]/2 \qquad (8\text{–}13\text{c})$$
$$= (\tfrac{1}{2})m_1(t)m_2(t)\sin 2\theta_e \qquad (8\text{–}13\text{d})$$

The low-pass filter preceding the VCO will have a cutoff frequency sufficiently low to integrate the varying information signals $m_1(t)$ and $m_2(t)$ so that the average (dc) voltage applied to keep the VCO tracking any received-carrier frequency drifts will be

$$V_o = (\tfrac{1}{2})\overline{m_1(t)m_2(t)} \sin 2\theta_e \qquad \text{(8–14a)}$$

which is approximately proportional to $2\theta_e$, and Equation 8–14a becomes

$$V_o = K2\theta_e \qquad \text{(8–14b)}$$

■ PROBLEMS

1. What type of modulation is used for transmitting the basic video signal of television? What is the reason?

2. Determine the power savings in percent when the carrier is suppressed in an AM (DSB-FC) signal modulated to
   a. 100%.
   b. 80%.

3. a. What advantage does DSB-SC have over VSB?
   b. What advantage does VSB have over DSB-SC?

4. Write the first five components for the squarewave switching function of Equation 8–1.

5. Write the mathematical expression for the DSB-SC output signal of a balanced modulator with a 15-kHz cosine modulating a 5-V pk sine 38-kHz carrier.

6. Determine the power into 50 Ω for the signal of Problem 5 after all but the first set of sidebands are filtered out.

7. An MC 1496 balanced modulator, when completely unbalanced, produces a 0.5-MHz carrier of 1 V rms. What will be the carrier amplitude when the circuit is brought into balance? (Use the specification sheet of Figure 8–9.)

8. Why is a pilot signal so often transmitted with DSB-SC?

9. A 1-V pk carrier DSB-SC modulated with a 5-kHz sinusoid was demodulated noncoherently—that is, in a regular AM peak detector. Determine the frequencies and amplitudes of the first three harmonics of the demodulated signal (see Figure 8–10).

10. a. Define "coherent" for DSB-SC systems.
    b. If the locally generated carrier frequency differs from the transmitted carrier frequency by 1 Hz, how fast will the phase error be changing in rad/s?

11. a. Name three advantages of SSB-SC over DSB-SC.
    b. Name two advantages of DSB-SC over SSB-SC. (Frequency and phase relationships count as only one.)

12. Determine the power savings in percent when the carrier and one sideband are suppressed in an AM (DSB-FC) signal modulated to
    a. 100%.
    b. 80%. How much better is SSB-SC than DSB-SC in this respect for $m = 1.0$?

13. How is SSB-SC prevented from distorting on peaks? Name the system and describe the circuit.

14. A voice frequency range of 500 Hz to 3.5 kHz is to be transmitted by SSB-SC using a minimum-bandwidth filter. If the modulator carrier frequency is 100 kHz and the lower sideband's upper frequency is to be attenuated by 40 dB, determine the 40- to 3-dB shape factor for the filter skirts. (See Figure 8–23 for reference.)

15. A 100-kHz carrier is modulated by a 3-kHz tone.
    a. What will a spectrum analyzer display for the following types of modulation? Make a sketch and show the frequencies: (**1**) AM (**2**) VSB (**3**) DSB-SC (**4**) SSB-SC.
    b. Sketch the oscilloscope (time-domain) display for the signals of part **a**.

16. Sketch the time-domain signal (oscilloscope) and frequency-domain spectrum (spectrum analyzer) at each point in the block diagram of Figure 8–13 with an upper-sideband pass filter. The inputs are 5-kHz modulation, 2-MHz IF generator, and 32-MHz up-converter oscillator (high-side injection to the mixer). Indicate the frequencies at each point (first-order modulator/mixer products).

17. Continue Problem 14 using the Chebyshev filter curves of Chapter 7 to show that a bandpass filter would have to have 7 resonators with a minimum Q per resonator of at least 543 to achieve the 40 dB of attenuation with less than 1 dB of bandpass ripple.

18. For the SSB-SC filter of Problem 14:
    a. How much attenuation could be achieved with a 7-resonator Butterworth filter?
    b. Design a 7-resonator Butterworth bandpass filter using the procedure of Chapter 7.
    c. Duplicate the results of part **b** with a circuit simulator software package such as PSpice, or Micro-CAP II.

19. Change the summation circuit of Figure 8–14 to a difference circuit, and prove mathematically that the upper sideband will be transmitted.

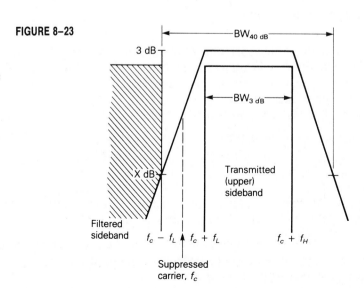

FIGURE 8–23

20. The worst-case oscillator drift in an SSB-SC receiver is to be ±2 Hz. How many parts per million of drift are allowable if the oscillator frequency is 10 MHz?

21. A 5-kHz tone was transmitted from an upper-sideband-only SSB-SC transmitter. Other voice signals are expected to follow. The received upper-sideband signal has a frequency of 455 kHz after passing through a down-converter with high-side injection of LO.
    a. At what frequency must the BFO be set?
    b. What is the maximum BFO drift (in ppm) for intelligible demodulation?
    c. Repeat a and b for telegraph transmission.

22. A 1-kHz squarewave is modulated SSB-SC (upper-sideband transmitted). If the receiver BFO is 10 Hz high, determine the frequencies for the first three demodulated harmonics, and show that they are not harmonically related (consequently, the demodulated output will not be a squarewave).

23. List (a) two advantages and (b) two disadvantages of the phasing method over the sideband-filter method.

24. a. Draw a block diagram for FDM of six telephone signals (4-kHz bandwidth each) whereby the minimum transmitted frequency is 60 kHz. List all the carrier frequencies. Use minimum transmitted bandwidth.
    b. What is the required bandwidth?

25. Showing the endpoint frequencies, sketch the FDM signal spectrum for the CCITT-recommended supergroup in which the suppressed-carrier frequencies are 612, 564, 516, 468, and 420 kHz. Each group is a standard 60–108 kHz SSB-SC signal.

26. Show mathematically that the in-phase mixer (product detector) with filtering will provide $m_2(t)$ information in a quadrature multiplex receiver. Refer to Figure 8–21 and start with $V_{LO_i} \times V_{QM}$, where $V_{LO_i} = 2 \cos \omega_c t$.

8–10 ■ COMPUTER ANALYSIS

Using the computer for communications system and circuit analysis has become a virtual necessity in education and industry. What follows are examples of programs and/or readily available simulation software for solving problems of the type covered in this chapter.

The analysis for Chapter 8 is as follows:

- **MathCAD analysis of DSB-SC:** Figure 8–24 illustrates some of the word processing, mathematical power, and result plotting capability of the software package, MathCAD. The text, the equations, and the plot of the equations evaluated over a time range of 0 to 200 μs in 2-μs steps are all done on the same page. Of course, the plot is computed and drawn automatically by using the command PLOT after the equations are typed in as shown.

The example given is from Equation 8–2 in the text. The plot is seen to correspond to Figure 8–7 for a double-sideband/suppressed-carrier signal produced by a 5-kHz sine modulating a 45-kHz sine carrier.

MathCAD: (Everything from this line through the figure was
        created on MathCAD.)

Plot of wideband double-sideband suppressed-carrier signal;
Equation 8-2 with sinusoidal modulation.

t := 0,.000002 ..0.0002

m(t) := 2· sin(2· π· 5000· t)

Vc(t) := sin(2· π· 45000· t)

Vo(t) := m(t)· Vc(t)

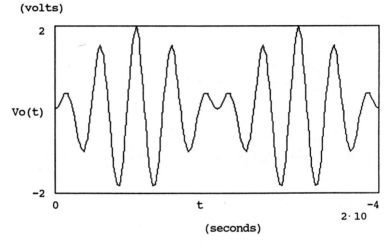

FIGURE 8–24 Example of MathCAD analysis of DSB-SC signal formed by the time-domain product of two sinusoids, one of 45-kHz frequency and the other at 5 kHz. Compare the result with Figure 8–7.

9

Frequency and Phase Modulation

■ INTRODUCTION

The preliminary discussion on modulation in Chapter 5 states that information can be transferred to a high-frequency carrier signal by modulating any of the three parameters that characterize the carrier—amplitude, frequency, and phase. To this point various ways of modulating the carrier amplitude have been presented. In all of the AM schemes the carrier frequency ω_c and phase θ_0 are constant. In this chapter the amplitude will remain constant (ideally) and the information signal will vary either the frequency or the phase angle of the carrier. ■

9-1 ■ FREQUENCY MODULATION

A frequency-modulated signal is any periodic signal whose instantaneous frequency f is deviated from an average value f_c by an information signal $m(t)$. The maximum amount by which the instantaneous frequency is deviated from the carrier frequency f_c is called the *peak deviation*, $\Delta f_c(\text{pk})$.

The block-diagram symbol of a linear frequency modulator, called a *voltage-controlled oscillator* (VCO) or a *voltage-tuned oscillator* (VTO), is illustrated in Figure 9–1. The VCO is an oscillator with a voltage-variable reactance that controls the oscillator output frequency. The circuit is dc-coupled and considered to be very broadband and linear. k_o, with units of Hz/V, is the proportionality constant that gives the amount by which the input voltage causes the output frequency to deviate from the average (carrier) value f_c. Hence the instantaneous output frequency is expressed as

$$f = f_c + k_o v_m(t) \tag{9-1}$$

It is assumed, of course, that when $v_m = 0$, $f = f_c$.* Otherwise,

$$\Delta f_c = k_o v_m(t) \tag{9-2}$$

and $$f = f_c + \Delta f_c$$

Because the circuit is dc-coupled, broadband, and linear, the oscillator output frequency variations about f_c will follow exactly the input information voltage variations $v_m(t)$; that is, as illustrated in Figure 9–2, an input sinusoidal modulation signal will produce an identical output sinusoidal variation of the oscillator frequency, an input squarewave or digital (data) signal will produce an output squarewave or digital (data) frequency variation, and so forth. This says nothing about what the output phase angle is doing—an issue that will be dealt with when FM and phase modulation (PM) are compared. The phase-angle variations also must be determined in order to compute the modulation index.

Modulation Index

The modulation index is used in communications as a measure of the relative amounts of information-to-carrier amplitude in the modulated signal. This param-

---

* Many actual VCO circuits have an input bias offset voltage. Its value is unimportant, and we define v_m with respect to this bias offset reference.

FIGURE 9-1 VCO block diagram.

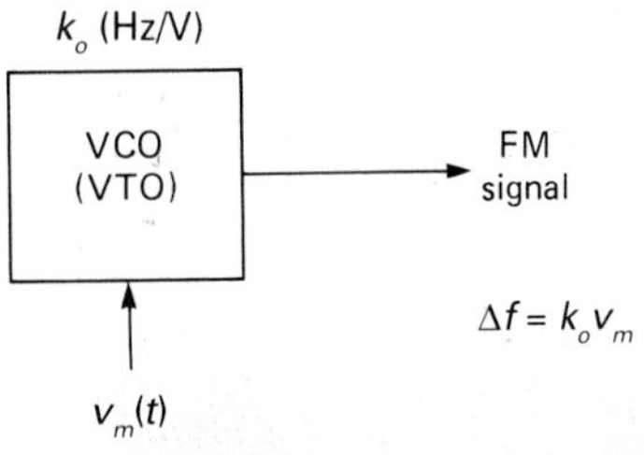

k_o (Hz/V)

VCO
(VTO)

FM
signal

$\Delta f = k_o v_m$

$v_m(t)$

FIGURE 9–2 The frequency variations in FM are directly proportional to the modulating voltage.

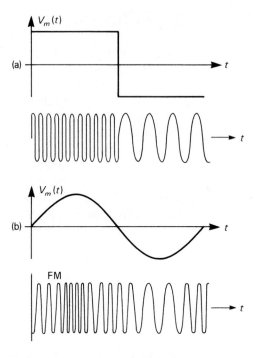

eter is also used to determine the spectral power distribution. For angle modulation by a sinusoid, the modulation index is defined as the peak phase angle deviation of the carrier. For phase modulation, this is easy to determine. However, for FM the modulated signal must be analyzed to determine the phase variations.

In general a signal with angle modulation is expressed simply as

$$s(t) = A \cos \theta(t) \qquad\qquad (9\text{--}3)$$

where $\theta(t)$ is the instantaneous angular displacement of the signal from a zero phase reference. Angle modulation of a high-frequency carrier will result by varying the phase of

$$s(t) = A \cos(\omega t + \phi) \qquad\qquad (9\text{--}4)$$

where either the angle ϕ or the angle ωt is varied. When ϕ is varied by the information signal, $\phi(t) \propto m(t)$, the result is called *phase modulation* (PM). If ϕ is held constant and the angle ωt is modulated by the information signal deviating the carrier frequency, the result is called *FM*.

Angular frequency ω is the rate of change of phase—that is,

$$\omega \equiv \frac{d\theta}{dt} \qquad\qquad (9\text{--}5)$$

This gives the fundamental relationship between frequency and phase. Equation 9–1 gives the instantaneous frequency of an FM signal as

$$f = f_c + k_o v_m(t)$$

where $v_m(t) \propto m(t)$ is the time-varying modulation (information) voltage applied to a linear VCO of k_o Hz/V circuit sensitivity. Since $\omega = 2\pi f$, then

$$\frac{d\theta}{dt} = 2\pi f_c + 2\pi k_o v_m(t) \qquad (9\text{–}6)$$

This differential equation is solved for θ by separating the phase and time variables as

$$d\theta = 2\pi f_c \, dt + 2\pi k_o v_m(t) \, dt$$

and integrating both sides to yield

$$\theta(t) = \int 2\pi f_c \, dt + \int 2\pi k_o v_m(t) \, dt \qquad (9\text{–}7)$$

where the limits of integration are from 0 to t. The first integral is easy to evaluate because $2\pi f_c$ contains all constants and

$$2\pi f_c \int dt = 2\pi f_c t + \theta_0$$

where θ_0 is the arbitrary starting phase at $t = 0$. Therefore, the instantaneous phase of an FM signal is determined from

$$\theta(t) = 2\pi f_c t + \theta_0 + 2\pi k_o \int v_m(t) \, dt \qquad (9\text{–}8)$$

where the modulation signal waveform must be integrated in order to evaluate the peak phase deviations of the carrier.

The modulation index for angle modulation by a sinusoidal input signal such as $v_m(t) = V_{pk}\cos 2\pi f_m t$ is defined as the peak phase deviation of the carrier. For FM the phase deviation of the carrier must be determined from Equation 9–8, where the integral of the third term is evaluated as follows:

$$2\pi k_o \int V_{pk}\cos(2\pi f_m t) \, dt = 2\pi k_o V_{pk}\left(\frac{1}{2\pi f_m}\right)\sin(2\pi f_m t + \phi_0) \qquad (9\text{–}9)$$

The arbitrary constant phase at $t = 0$, ϕ_0, can be combined with θ_0 in Equation 9–8 and set to zero. After canceling the 2π terms of Equation 9–9, Equation 9–8 is written as

$$\theta(t) = 2\pi f_c t + \left(\frac{k_o V_{pk}}{f_m}\right)\sin 2\pi f_m t$$

or
$$\theta(t) = 2\pi f_c t + \left[\frac{\Delta f_c(pk)}{f_m}\right]\sin 2\pi f_m t \qquad (9\text{–}10)$$

The *cosine-modulated* FM signal is written, using Equation 9–3, as

$$s_{FM}(t) = A \cos(2\pi f_c t + m_f \sin 2\pi f_m t) \qquad (9\text{–}11)$$

where $m_f = \Delta f_c(pk)/f_m$ is the peak of the sinusoidally varying carrier phase angle.

The *modulation index*, in units of radians, for sinusoidal FM is calculated from

$$m_f = \frac{\Delta f_c(\text{pk})}{f_m} \tag{9-12}$$

■ ■ ■ ■

EXAMPLE 9-1

A 1-MHz VCO with a measured sensitivity of k_o = 3 kHz/V is modulated with a 2-V peak, 4-kHz sinusoid. Determine the following:

1. The maximum frequency deviation of the carrier.
2. The peak phase deviation of the carrier and therefore the modulation index.
3. m_f if the modulation voltage is doubled.
4. m_f for $v_m(t) = 2 \cos[2\pi(8 \text{ kHz})t]$, volts.
5. Express the FM signal mathematically for a cosine carrier and the cosine-modulating signal of part 4. The carrier amplitude is 10 V pk.

Solution:

1. $\Delta f_c(\text{pk}) = k_o V_m(\text{pk}) = (3 \text{ kHz/V})(2 \text{ V pk}) = \textbf{6 kHz.}$
2. From Equation 9–12 the peak phase deviation by the modulating sinusoid is $\Delta f_c(\text{pk})/f_m$ = 6 kHz/4 kHz = **1.5 rad** = m_f. The radian units are usually dropped. For emphasis, this example problem will include radians in parentheses.
3. For a linear modulator $\Delta f \propto \Delta V_m$ so m_f = 3.0 (rad) is double the answer of part 2. Of course this can be computed from Equation 9–12. Since V_m increases to 4 V, then $m_f = \Delta f_c(\text{pk})/f_m$ = (4 V × 3 kHz)/4 kHz = **3** (rad).
4. The modulation is sinusoidal with 2 V pk so the carrier-frequency deviation is the same as part 1—$\Delta f_c(\text{pk})$ = 6 kHz. Since f_m = 8 kHz, the peak carrier phase deviation and modulation index is m_f = 6 kHz/8 kHz = **0.75.**
5. The modulating signal is a cosine (so is the carrier). Therefore, using the value of m_f = 0.75, we have $v_{\text{FM}} = \textbf{10 cos}(2\pi10^6t + 0.75 \sin 2\pi8 \times 10^3t)$.

Sidebands and Spectrum

The frequency spectrum of an FM signal is very different than that of AM. For low deviations (m_f < 0.25), called *narrowband FM* (NBFM), there is a carrier and a set of sidebands much like AM. However, this is where the similarity ends. Even for narrowband FM the amplitude of the carrier decreases as m_f increases. Also, and of more fundamental importance, the two sidebands are shifted 90° relative to the carrier (as compared to AM). This comparison is illustrated in Figure 9–3.

Figure 9–4 shows a comparison of AM and NBFM phasor variations. Notice that the amplitude for the NBFM seems to vary slightly in the illustration. This is

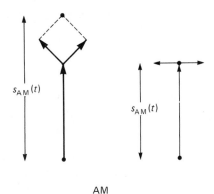

AM

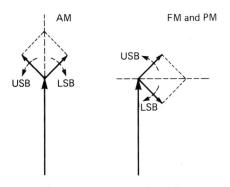

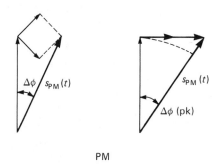

PM

FIGURE 9–3 Comparison of phasor relationships in AM and FM/PM.

FIGURE 9–4 Comparison of instantaneous transmitted phasors—$s_{AM}(t)$ and $s_{PM}(t)$.

because the angle drawn for the illustration is greater than 14.3° (0.25 rad). For $m_f > 0.25$, other sets of sidebands are produced such that the phasor sum of carrier and all the sidebands results in a *resultant signal vector* $s_{FM}(t)$ with constant amplitude. This is discussed later for wideband FM.

The spectrum for angle-modulated signals (FM and PM) may be determined from the FM Equation 9–11, written here as an instantaneous voltage (A is peak carrier voltage) and with angular frequency for convenience:

$$v_{FM} = A \cos(\omega_c t + m_f \sin \omega_m t) \qquad (9\text{–}13)$$

By the trigonometric identity

$$\cos(a + b) = \cos(b + a) = \cos b \cos a - \sin b \sin a$$

Equation 9–13 may be written as

$$v_{FM} = A[\cos(m_f \sin \omega_m t)]\cos \omega_c t - A[\sin(m_f \sin \omega_m t)]\sin \omega_c t \qquad (9\text{–}14)$$

where the magnitudes of the quadrature carrier components are bracketed.

For low-deviation FM ($m_f < 0.25$) the angular variations of the cosine and the sine in the brackets of Equation 9–14 can be approximated as $\cos \Delta\phi \approx 1$ and

$\sin \Delta\phi \approx \Delta\phi$. Therefore, narrowband FM can be approximated as

$$v_{\text{NBFM}} = \underbrace{A \cos \omega_c t}_{\text{discrete carrier}} - \underbrace{A(m_f \sin \omega_m t)\sin \omega_c t}_{\text{sidebands term}} \qquad (9\text{--}15)$$

Also,

$$Am_f \sin \omega_c t \sin \omega_m t = \left(\frac{Am_f}{2}\right)\cos(\omega_c - \omega_m)t - \left(\frac{Am_f}{2}\right)\cos(\omega_c + \omega_m)t$$

This shows clearly the constant amplitude carrier and the quadrature sidebands, as illustrated in Figure 9–4. The sidebands term is, from Chapter 8, double-sideband/suppressed-carrier (DSB-SC), with peak amplitude $Am_f/2$ varying sinusoidally at the modulation rate ω_m.

As seen from Equation 9–15, the bandwidth required to transmit NBFM must include the carrier and a single set of sidebands $\pm\omega_m$ from the carrier frequency.

Wideband FM

Wideband FM (WBFM), or simply FM in general with no approximations, is expressed mathematically by Equation 9–14 for sinusoid modulation. Let us agree that the amplitude variations of the quadrature carrier components become very complicated. In point of fact the amplitudes of the frequency components [sinusoid($m_f \sin \omega_m t$)] cannot be evaluated in closed form but are tabulated in the table of Bessel functions of the first kind of order n (Table 9–1). The continuous Bessel functions $J_n(m_f)$ are shown in Figure 9–5, where

$$V_c(\text{pk}) = A J_0(m_f) \qquad (9\text{--}16)$$

gives the peak voltage amplitude of the carrier for any modulation index m_f up to and including $m_f = 10$. The peak voltage of sidebands on either side of the carrier is given by

$$V_n(\text{pk}) = A J_n(m_f) \qquad (9\text{--}17)$$

where $n = 1, 2, \ldots$ is the order of the sideband—that is, the number of the particular sideband pair. So the carrier, $n = 0$, would be

$$V_o(\text{pk}) = A J_0(m_f)$$

However, $V_c(\text{pk})$ is more easily associated with the carrier (see Equation 9–16).

Notice that for $m_f < 0.25$, the carrier drops by less than 2% and only one set of sidebands, the first-order sidebands $J_1(m_f)$, have a value exceeding 1% (0.01) of the unmodulated carrier. These values justify the approximations used for NBFM. The first-order set of sidebands are separated from the carrier by f_m.

Notice also that for $m_f \geq 0.25$ there are additional sets of sidebands. This is the wideband FM case, and the sideband sets are separated from the carrier by f_m, $2f_m, 3f_m, \ldots, nf_m$ corresponding to the order of the Bessel function $J_n(m_f)$. This can also be seen in the example FM signal spectra of Figure 9–6. Notice, too, that all sets of sidebands start at zero for no modulation, and the carrier J_0 starts at 1.0

TABLE 9-1 Bessel Functions of the First Kind, $J_n(m_f)$

m_f	J_0	J_1	J_2	J_3	J_4	J_5	J_6	J_7	J_8	J_9	J_{10}	J_{11}	J_{12}	J_{13}	J_{14}
0.00	1.00	—	—	—	—	—	—	—	—	—	—	—	—	—	—
0.25	0.98	0.12	—	—	—	—	—	—	—	—	—	—	—	—	—
0.5	0.94	0.24	0.03	—	—	—	—	—	—	—	—	—	—	—	—
1.0	0.77	0.44	0.11	0.02	—	—	—	—	—	—	—	—	—	—	—
1.5	0.51	0.56	0.23	0.06	0.01	—	—	—	—	—	—	—	—	—	—
2.0	0.22	0.58	0.35	0.13	0.03	—	—	—	—	—	—	—	—	—	—
2.4	0	0.52	0.43	0.20	0.06	0.02	—	—	—	—	—	—	—	—	—
2.5	−0.05	0.50	0.45	0.22	0.07	0.02	0.01	—	—	—	—	—	—	—	—
3.0	−0.26	0.34	0.49	0.31	0.13	0.04	0.01	—	—	—	—	—	—	—	—
4.0	−0.40	−0.07	0.36	0.43	0.28	0.13	0.05	0.02	—	—	—	—	—	—	—
5.0	−0.18	−0.33	0.05	0.36	0.39	0.26	0.13	0.05	0.02	—	—	—	—	—	—
6.0	0.15	−0.28	−0.24	0.11	0.36	0.36	0.25	0.13	0.06	0.02	—	—	—	—	—
7.0	0.30	0.00	−0.30	−0.17	0.16	0.35	0.34	0.23	0.13	0.06	0.02	—	—	—	—
8.0	0.17	0.23	−0.11	−0.29	−0.10	0.19	0.34	0.32	0.22	0.13	0.06	0.03	—	—	—
9.0	−0.09	0.25	0.14	−0.18	−0.27	−0.06	0.20	0.33	0.31	0.21	0.12	0.06	0.03	0.01	—
10.0	−0.25	0.05	0.25	0.06	−0.22	−0.23	−0.01	0.22	0.32	0.29	0.21	0.12	0.06	0.03	0.01

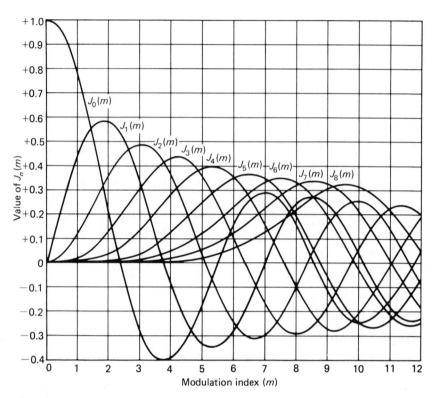

FIGURE 9-5 Amplitudes of carrier and sidebands for FM (and PM) relative to the un-modulated carrier.

(100% of the unmodulated carrier amplitude). Finally, notice that the carrier and each set of sidebands reach a maximum amplitude and then decrease, one after another, going through a null.* Then they become negative (meaning a 180° phase reversal), go through a maximum with that phase, rise again, and continue this cyclical peak-null-peak-null with ever-decreasing peak values.

Bessel function values can be computed (easily with a computer) using the infinite series approximation

$$J_n(m_f) = \left(\frac{m_f}{2}\right)^n \left[\frac{1}{n!} - \frac{(m_f/2)^2}{1!(n+1)!} + \frac{(m_f/2)^4}{2!(n+2)!} - \frac{(m_f/2)^6}{3!(n+3)!} + \cdots\right] \quad (9\text{-}18)$$

where the quantity $(N)!$ is the *factorial*, $1 \times 2 \times 3 \times \cdots \times (N)$.

The Bessel function solutions do not show, but can be used to prove, that FM (unlike AM) is a constant power signal irrespective of modulation index. This also means that the amplitude seen on an oscilloscope will stay constant with increas-

---

* The first carrier null at $m_f = 2.40$ is an important measurement point for a WBFM system. For example, if f_m and $V_m(\text{pk})$ are measured when the modulation index is set at 2.40 using a spectrum analyzer or narrowband filter, the deviation and modulator sensitivity can easily be calculated.

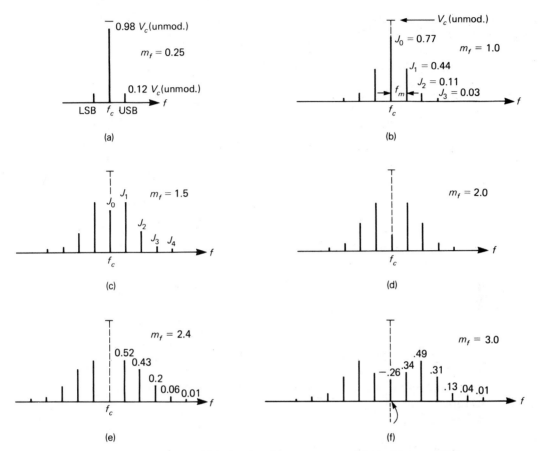

FIGURE 9-6 Frequency spectra for FM signals with constant modulation frequency f_m but different carrier-frequency deviations.

ing m_f as long as the power in all the sidebands is taken into account—which means that the measurement bandwidth must be wide enough to include all sidebands. Try a calculation for $m_f = 1.5$; the relative power in the signal is

$$P/P(\text{unmod.}) = [J_0(m_f)]^2 + 2[J_n(m_f)]^2$$
$$= (0.51)^2 + 2[0.56^2 + 0.23^2 + 0.06^2 + 0.01^2] \qquad \textbf{(9–19)}$$
$$= 1.0005$$

This has a round-off error of 0.05%.

If you start drawing phasor (vector) diagrams like Figure 9–4 for higher m_f (and therefore larger ϕ), you will see very quickly that with a single set of sidebands, the resultant signal vector cannot maintain a constant length. In order to end up with a constant resultant amplitude, the additional sidebands must be included with magnitudes adjusted according to the Bessel table, *and* each sideband must be shifted an additional 90° from the previous one. Figure 9–7 shows an example for $m_f = 1.0$.

FIGURE 9-7 Carrier and sidebands add up to transmitted FM (or PM) phasor.

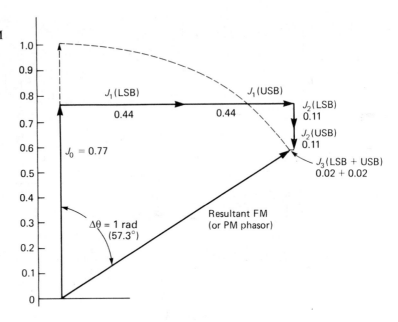

As seen in Figure 9-5 and in the spectra of Figure 9-6 with a constant f_m, increasing the carrier deviation by increasing the modulating signal voltage amplitude will result in a greater m_f and more sidebands, thereby requiring more bandwidth.

Information Bandwidth and Power

The total power in an angle-modulated signal is a constant as long as all of the sidebands are accounted for. In theory this means an infinite bandwidth is required. However, 99% of the signal is accounted for if all of the "significant" sidebands are present. The Bessel function from Table 9-1 includes sidebands with voltage amplitudes of 1% or more of the unmodulated carrier $[J_n(m_f) > 0.01]$. If A is the peak voltage of the unmodulated carrier, then $A^2/2R$ gives the power in the FM signal as measured in real impedance R. The power may also be determined by adding the power in individual components. This is expressed by Equation 9-19, where

$$P(\text{unmod}) = \frac{A^2}{2R} = P_{\text{total}} \qquad (9\text{-}20)$$

The power in just one of the nth-order sidebands is

$$[J_n(m_f)]^2 P(\text{unmod}) \qquad (9\text{-}21)$$

The 99% bandwidth determined by including all of the sideband *pairs* of Table 9-1 is the *information bandwidth* for a given modulation index. In practice a convenient rule of thumb used for determining the *circuit* bandwidth is based on the approximation called *Carson's rule*,

$$BW = 2(f_m + \Delta f_c) \qquad \textbf{(9-22a)}$$
$$= 2f_m(1 + m_f) \qquad \textbf{(9-22b)}$$

The actual 3-dB circuit bandwidth is made larger than the Carson's rule bandwidth if low distortion is desired.

■ ■ ■ ■

EXAMPLE 9-2

An FM signal expressed as $v_{FM} = 1000 \cos(2\pi 10^7 t + 0.5 \cos 2\pi 10^4 t)$ is measured in a 50-Ω antenna. Determine the following:

1. Total power
2. Modulation index
3. Peak frequency deviation
4. Modulation sensitivity if 200 mV pk is required to achieve part 3
5. Spectrum
6. Bandwidth (99%) and approximate circuit bandwidth by Carson's rule
7. Power in the smallest sideband (just one sideband) of the 99% bandwidth
8. Total information power

Solution:

1. $P_T = (V \text{ pk})^2/2R = 1000^2/100 = \textbf{10 kW.}$
2. The $0.5 \cos 2\pi 10^4 t$ term gives the peak phase variation and $m_f = \textbf{0.5.}$
3. $m_f = \Delta f_c(\text{pk})/f_m$; therefore, $\Delta f_c(\text{pk}) = 0.5 \times 10^4 \text{ Hz} = \textbf{5 kHz pk.}$
4. $k_o = \Delta f_c/\Delta V_m = 5 \text{ kHz}/0.2 \text{ V} = \textbf{25 kHz/V.}$
5. The spectrum is determined from Table 9-1 with $m_f = 0.5$, $A = 1000$ V pk, and $f_m = 10$ kHz. The carrier is $AJ_0(0.5) = 940$ V pk at 10 MHz. The first sidebands are $AJ_1(0.5) = 240$ V pk at 9.990 MHz and 10.010 MHz, and the second (and least significant) sidebands are $AJ_2(0.5) = 30$ V pk at 9.98 MHz and 10.020 MHz.
6. From 5, the 99% information bandwidth is $2 \times 20 \text{ kHz} = \textbf{40 kHz,}$ and from Carson's rule, BW $= 2(10 \text{ kHz} + 5 \text{ kHz}) = \textbf{30 kHz}$ for reasonably low circuit distortion.
7. There are two ways to arrive at the solution: From part 5 each sideband is 30 V pk, so $P_{1sb} = (30 \text{ V pk})^2/100 = \textbf{9 W.}$ Using Equation 9-18 with $m_f = 0.5$ and using Equation 9-21 with $P = 10$ kW, $P_{1sb} = (0.03)^2 10 \text{ kW} = \textbf{9 W.}$
8. The power in the information part of the signal is $P_T - P_c$, where the carrier power is $P_c = (940 \text{ V pk})^2/100 = 8.836$ kW. $P_T - P_c = 10 \text{ kW} - 8.836 \text{ kW} = \textbf{1.164 kW.}$ Low-index modulation is not very efficient (11.64% in this case). Incidentally, the total power must add up as follows: $P_T = P_c + P_1 + P_2 = 8.836 \text{ kW} + 2[(240^2/100) + 9 \text{ W}] = 10.006$ kW, a 0.06% round-off error.

9–2 ■ PHASE MODULATION

Phase modulation (PM) is a very important form of angle modulation. It is the modulation of choice for satellite and deep-space missions because, like FM, its noise properties are superior to AM, but unlike FM, it can be produced in a simple circuit driven from a frequency-stable, crystal-controlled carrier oscillator. A VCO is intentionally made very frequency-variable to produce high deviations and a high modulation index. As a consequence the average (carrier) frequency tends to drift.

In fact, a stable carrier frequency is such a valuable performance criterion that phase modulation with a trick is often used for producing FM. This is discussed further in the treatment of "FM from PM" in Section 9–4. Called *phase-shift key* (PSK), phase modulation is used extensively in transmitting digital information (Chapter 17).

Phase modulation of a high-frequency carrier is expressed mathematically by varying the phase ϕ of the general Equation 9–4 with a constant carrier frequency f_c. Thus PM is expressed by

$$v_{PM} = A \cos[\omega_c t + k_p v_m(t)] \tag{9-23}$$

where the phase variation $\phi(t)$ is proportional to the modulating voltage signal $v_m(t)$. For a linear phase modulator,

$$\phi(t) = k_p v_m(t) \tag{9-24}$$

where the proportionality constant k_p, in rad/V, is called the *modulator sensitivity*.

For a sinusoidal modulation signal of $v_m(t) = V_{pk} \cos 2\pi f_m t$, the PM signal becomes

$$v_{PM} = A \cos(2\pi f_c t + k_p V_{pk} \cos 2\pi f_m t) \tag{9-25a}$$

$$= A \cos(2\pi f_c t + \Delta\phi(pk) \cos 2\pi f_m t) \tag{9-25b}$$

Since

$$\Delta\phi(pk) = k_p V_{pk} \tag{9-26}$$

is the peak phase deviation of the sinusoidally varying carrier phase angle, the *modulation index* in units of radians for this PM signal is

$$m_p = \Delta\phi \,(pk) \tag{9-27}$$

Equation 9–25 can now be written in terms of the modulation index as

$$v_{PM}(t) = A \cos(2\pi f_c t + m_p \cos 2\pi f_m t) \tag{9-28}$$

Please compare Equations 9–28 and 9–11 for FM by the same modulating cosine. The comparison will be elaborated on in the section on "FM from PM." When Equation 9–28 is expanded by trigonometric identity and the narrowband

FIGURE 9–8   NBPM using a balanced AM modulator and sideband phase shift.

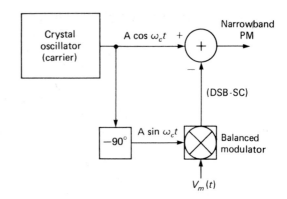

approximation ($\Delta\phi < 0.25$) applied, the result written as a signal voltage is

$$V_{\text{NBPM}}(t) = \underbrace{A \cos \omega_c t}_{\text{discrete carrier}} - \underbrace{A(m_p\cos \omega_m t)\sin \omega_c t}_{\text{sidebands term}} \qquad (9\text{–}29)$$

Compare this with Equation 9–15 for NBFM.

Equation 9–29, with its constant-amplitude carrier and quadrature DSB-SC signal of sinusoidal modulation with peak amplitude Am_p, suggests a simple circuit implementation. This is shown in block diagram form in Figure 9–8.

We will see this scheme with a slight modification (trick) for generating FM. Note also that the scheme of Figure 9–8 with an additional balanced modulator (in series with $A \cos \omega_c t$) for producing a quadrature multiplexed signal (QM, Chapter 8) is used for transmitting the color (chrominance) vector in television.

Diode Modulator for PM

Figure 9–9 shows the most straightforward method of producing a PM signal. A voltage-variable capacitance diode called by various names (Varicap, Varactor, and tuning diode) is used to vary the tuning of a parallel-resonant circuit. The

FIGURE 9–9   Phase modulation system.

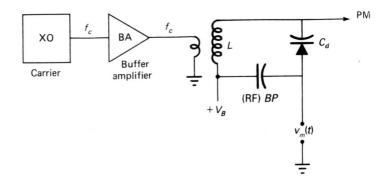

stable crystal-controlled-frequency carrier signal will thus be phase-shifted in accordance with the modulation voltage.

The most graphically straightforward method for determining the amount of phase shift is to use the diode voltage-capacitance curve of Figure 9–10. Notice that the $C(V)$ curve is plotted on log-log paper. $C(V)$ is very nonlinear (approximately a square root function), and consequently voltage variations must be kept small to minimize distortion of the transmitted PM signal. Even the 1-V peak of Example 9–3 is too much, but it is convenient for demonstrating the principles.

The vertical scale will yield the fractional change $\Delta C_d / C_d$ in diode capacitance for small input voltage changes. The capacitance changes can be estimated by determining the slope of the $C(V)$ curve at the bias point V. Thus

$$\frac{dC_d}{dV} = \frac{d}{dV}\,[C_0(1 + 2\mid V\mid)^{-1/2}]$$

$$\frac{dC_d}{dV} = \frac{-C_d}{1 + 2\mid V\mid} \tag{9–30a}$$

or

$$\frac{\Delta C_d}{C_d} = \frac{-\Delta V}{1 + 2\mid V_R\mid} \tag{9–30b}$$

where V_R is the reverse bias across C_d, and ΔV is the peak value of V_m, the modulation signal voltage. Equation 9–30 is then used to determine the *tuned-circuit* frequency shift from Equation 2–24,

$$\frac{\Delta f_0}{f_0} = -0.5\,\frac{\Delta C_d}{C_d} \tag{9–31}$$

With this the resulting peak phase shift $\Delta\phi(\text{pk})$ can be calculated. For sinusoidal angle modulation $m_p \equiv \Delta\phi(\text{pk})$, and knowing the peak input voltage swing, the modulator sensitivity can be calculated from

$$k_p = \frac{\Delta\phi}{\Delta V_m} \tag{9–32}$$

An exact formula for the phase shift is derived from the phase-versus-frequency curve for the parallel tank impedance,

$$Z = \frac{R_p}{1 + jQ\rho}$$

where

$$\rho = \frac{f}{f_0} - \frac{f_0}{f} \tag{9–33a}$$

$$= \left(1 + \frac{\Delta f_0}{f_0}\right) - \left(1 + \frac{\Delta f_0}{f_0}\right)^{-1} \tag{9–33b}$$

Δf_0 is a small deviation, less than 10% from center frequency, f_0. The phase-versus-frequency curve, $\phi(f)$, follows a tangent curve that is the phase of impedance, Z; that is, $\phi(f) = -\tan^{-1}Q\rho$ for which the magnitude provides the FM modulation index given by

$$\Delta\phi = m_p = \tan^{-1}Q\rho \tag{9–34}$$

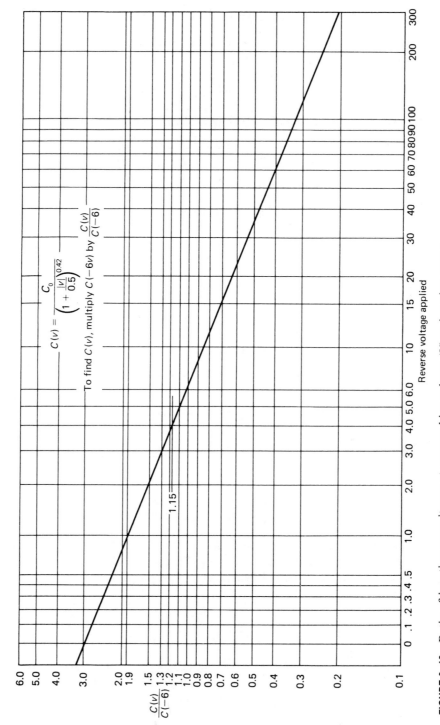

$$C(v) = \frac{C_0}{\left(1 + \dfrac{|v|}{0.5}\right)^{0.42}}$$

To find $C(v)$, multiply $C(-6v)$ by $\dfrac{C(v)}{C(-6)}$

Reverse voltage applied

FIGURE 9-10 Ratio of junction capacitance at reverse-bias voltage (V) to junction capacitance at 6 V reverse-bias versus applied reverse-bias voltage.

Derivation of Modulator Equations

To derive an expression for modulation sensitivity, k_p, using component values in the circuit of Figure 9–9, we begin with the definition from Equation 9–32, $k_p = d\phi/dV_m$. The parallel tuned circuit consists of R, L, and $C_d(V)$—the diode tuning capacitor that varies with bias voltage variations. We combine the change of capacitance with voltage, the corresponding change of frequency with capacitance, and the resulting change of phase with frequency. Hence,

$$k_p = \frac{d\phi}{dV_m} = \frac{dC}{dV} \cdot \frac{df}{dC} \cdot \frac{d\phi}{df} \tag{9–35}$$

For the diode of Figure 9–10,

$$C_d(V) = C_0(1 + 2 \mid V_R \mid)^{-0.42} \tag{9–36}$$

Hence,
$$\frac{dC_d}{d \mid V_R \mid} = -0.42(2)C_0(1 + 2 \mid V_R \mid)^{-0.42-1} \tag{9–37a}$$

$$= \frac{-0.84}{(1 + 2 \mid V_R \mid)} \frac{C_o}{(1 + 2 \mid V_R \mid)^{0.42}}$$

$$= \frac{-0.84C_d}{1 + 2 \mid V_R \mid} \tag{9–37b}$$

In general, for
$$C_d(V) = C_o(1 + 2 \mid V_R \mid)^{-p} \tag{9–38}$$

the rate of diode capacitance change with voltage is computed by differentiating Equation 9–38 as

$$\frac{dC_d}{d \mid V_R \mid} = \frac{-2pC_d}{1 + 2 \mid V_R \mid} \tag{9–39}$$

where, for various diode junction doping profiles, $0.3 < p < 0.6$. For convenience, let the power be $p = 0.5$, resulting in a square-root capacitance-voltage characteristic. Then

$$\frac{dC_d}{d \mid V_R \mid} = \frac{-C_d}{1 + 2 \mid V_R \mid} \tag{9–40}$$

which is the same as Equation 9–30a.

The second term in Equation 9–35, df/dC, is, from Equations 2–24 and 9–31 for a tuned circuit with the only capacitance being C_d,

$$\frac{df}{dC} = \frac{df_0}{dC_d} = -0.5 \frac{f_0}{C_d} \tag{9–41}$$

For small frequency deviations, $f \approx f_0$, then the phase change is also small, so that ρ in Equation 9–33a becomes

$$\rho = \frac{f^2 - f_0^2}{f_0 f} = \frac{(f + f_0)(f - f_0)}{ff_0} \approx \left(\frac{2f}{f}\right)\left(\frac{\Delta f}{f_0}\right) \tag{9–42a}$$

$$\approx \frac{2 \, \Delta f}{f_0} \tag{9–42b}$$

which, when combined with $\phi = -\tan^{-1}Q\rho \approx -Q\rho$ from Equation 9–34, yields the approximation

$$\frac{d\phi}{df} = -\frac{2Q}{f_0} \tag{9–43}$$

where the differential operator d has been substituted for the discrete delta. Substituting Equations 9–40, 9–41, and 9–43 into Equation 9–35 for k_p yields

$$\frac{d\phi}{dV_m} = \frac{-C_d}{1 + 2\mid V_R \mid} \cdot \frac{-f_0}{2C_d} \cdot \frac{-2Q}{f_0} = \frac{-Q}{1 + 2\mid V_R \mid} \tag{9–44}$$

and the magnitude of Equation 9–44 gives the phase modulator sensitivity at the circuit bias point, V_R, as

$$k_p = \frac{Q}{1 + 2\mid V_R \mid} \tag{9–45}$$

Also, the modulation index m_p resulting from a change in voltage, dV_m, applied to the tuning diode is, from Equation 9–24, $\Delta\phi(\text{pk}) = k_p V(\text{pk})$, and from Equation 9–27, $m_p = \Delta\phi(\text{pk})$; therefore,

$$m_p = \Delta\phi(\text{pk}) = \frac{Q\,\Delta V_m(\text{pk})}{1 + 2\mid V_R \mid} \tag{9–46}$$

for the circuit of Figure 9–9. Equations 9–45 and 9–46 are *exact slope* equations for a simple phase modulator.

■ ■ ■ ■

EXAMPLE 9–3

A 1-V pk sinusoid is to phase-modulate a 5-MHz carrier using the circuit of Figure 9–9 with the following circuit values: $V_B = +6$ V, a 3.3-kΩ resistor shunts L, a 44-pF capacitor is in parallel with C_d, which is rated at 56 pF for a reverse bias of 6 V. Determine the following:

1. Value of L to set resonance at $f_0 = f_c$ with ($V_m = 0$ V)
2. Circuit Q and peak frequency shift of f_0
3. Modulation index
4. Modulator sensitivity
5. Use a linearized phase-slope model to demonstrate graphically the circuit phase shifts.

Solution:

1. Solve the resonant circuit formula for inductance. $L = 1/(2\pi f_0)^2 C$. For $f_0 = 5$ MHz and total $C = 100$ pF, $L = \mathbf{10\ \mu H.}$
2. $Q = R/X_L = 3.3$ k$\Omega/2\pi(5$ MHz$)(10\ \mu$H$) = \mathbf{10.5.}$ At the positive peak of the modulating sinusoid, the reverse diode voltage will be $6 - 1 = 5$ V. The vertical axis of Figure 9–10 shows that the relative capacitance change from 6 V to 5 V is approximately 1.10—that is, a 10% increase in C_d. The total circuit capacitance will increase to 100 pF + 10% of 56 pF = 105.6 pF, which gives $\Delta C_t/C_t =$

5.6%. Using Equation 9–31, the resonant frequency will decrease by $\Delta f_0/f_0 =$ (½)5.6% = 2.8%.

3. From part 2, $\Delta f_0/f_0 = 0.028$, so using Equation 9–33b, $\rho = 1.028 - (1/1.028) =$ 0.055, and therefore $\Delta\phi = m_p = \tan^{-1}Q\rho = \tan^{-1}(10.5 \times 0.055) = $ **0.53 rad.**

4. $k_p = \Delta\phi/\Delta V_m = 0.53$ rad/1V = **0.53 rad/V.** (This is also about 30°/V.)

5. A *graphical method* for finding the phase shift of the tuning-diode modulator is illustrated in Figure 9–11. For the sake of simplicity, assume that the phase-shift curve is linear across the 3-dB bandwidth, BW. The phase at the 3-dB points (±BW/2 from f_0) will be 45° ($\pi/4$ rad), and the problem reduces to similar triangles. Then

$$\frac{\Delta\phi}{\pi/4} = \frac{-\Delta f_0}{(BW/2)}$$

from which

$$\Delta\phi \approx -\left(\frac{\pi}{2}\right) Q \frac{\Delta f_0}{f_0} \qquad\qquad \textbf{(9–47)}$$

where $Q = f_0/$BW. Observe that when Δf is negative, then $\Delta\phi$ is positive. Hence, $\Delta\phi/\Delta f$ is a negative slope. For $\Delta f_0/f_0 = 0.028$, $\Delta\phi \approx (\pi/2) \times 10.5 \times 0.028 = 0.46$ rad for a 13% error. The approximation is not very good, but the technique is instructive.

FIGURE 9–11 Phase shift to carrier due to shift in circuit resonant-frequency.

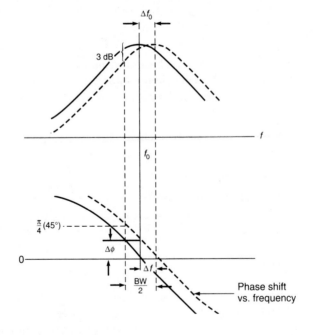

Determining the Limits of Linearity

The procedure developed here is easily generalized for any linear circuit used beyond the limits of linearity. Here an equation is derived for the maximum resonant frequency shift $\Delta f_0(\text{max})$ beyond which the phase modulator of Figure 9-9 will produce distortion exceeding a maximum fractional error ϵ. The distortion is usually given as a percentage, such as "5% accuracy," for which $\epsilon = 0.05$.

Phase as a function of frequency for a single tuned circuit such as that in a phase modulator is a tangent function. $\text{Tan}^{-1}x$ deviates from a $-45°$ straight line V_2 near $f = f_0$ by an amount ϵ, where

$$\epsilon = \frac{V_2 - V_1}{V_2} \tag{9-48}$$

is the fractional error and 100ϵ yields the percent error. To determine the maximum deviation for a given linearity specification, find the mathematical expressions for V_1 and V_2.

The infinite series approximation to the arctangent function is

$$-\tan^{-1}x = -x + \frac{x^3}{3} + \cdots \tag{9-49}$$

and the slope at $x = 0$ is most easily approximated by differentiating the right side of Equation 9-49, with the result being -1. Thus, the straight-line approximation of Equation 9-49 is $V_2 = -x$, where $x = Q\rho \approx 2Q(\Delta f/f_0)$ for the phase modulator.

Approximating the actual function, $V_1 = -\tan^{-1}x$, by the first two terms of Equation 9-49, ϵ becomes

$$\epsilon = \frac{-x - (-x + x^3/3)}{-x} = \frac{x^2}{3} \tag{9-50}$$

so that, in terms of fractional error from linearity,

$$x = \sqrt{3\epsilon} \tag{9-51}$$

For the phase modulator of Figure 9-9,

$$x = Q\left(\frac{f}{f_0} - \frac{f_0}{f}\right) \tag{9-52a}$$

$$\approx 2Q\frac{\Delta f_0(\text{max})}{f_0} \tag{9-52b}$$

in radians. Consequently, the frequency deviation of the tuned circuit from f_0 should be held to less than a value given by

$$\Delta f_0 \leq \frac{\sqrt{3\epsilon}f_0}{2Q} \tag{9-53a}$$

or

$$\Delta f_0 \leq \frac{\sqrt{3\epsilon}BW}{2} \tag{9-53b}$$

where $BW = f_0/Q$.

The maximum allowable phase deviation for the modulator to maintain a linearity specification of ϵ is, from Equation 9-51 where $x = \Delta\theta$,

$$\Delta\theta_{\text{max}} = \sqrt{3\epsilon} \tag{9-54}$$

Equation 9–54 also gives the maximum phase modulation index m_p(max), and the allowable voltage applied to the phase modulator should be held below a maximum given, from Equation 9–32, by

$$\Delta V_m(\text{max}) = \frac{\sqrt{3\epsilon}}{k_p} \qquad\qquad (9\text{–}55)$$

■ ■ ■ ■

**EXAMPLE
9-4**

If the phase modulator for Figure 9–9 is to maintain better than 1% linearity, determine the following:

1. Maximum frequency deviation of the tuned circuit from center frequency, as a percentage of a half-bandwidth ($BW/2$)
2. Maximum modulation index
3. Maximum peak voltage applied if the modulator has a sensitivity of $k_p = 0.577$ rad/V

Solution:

1. 1% linearity means that $\epsilon = 0.01$, so that $\sqrt{3\epsilon}$ in percent is $\sqrt{0.03} \times 100 = 17.3\%$. Therefore, Δf_0(max) should be held to less than **17.3%** of the circuit half-bandwidth, determined by R and tuning reactance.
2. From Equation 9–54, m_p(max) $= \Delta\theta_{\text{max}} = $ **0.173 rad.**
3. From Equation 9–55, ΔV_m(max) $= 0.173$(rad)/0.577 rad/V $= $ **0.3 V pk.**

Wideband PM

An analysis of the linearity of the tangent function and Equation 9–34 for phase-versus-frequency shifts of an LC-tuned circuit indicates that the modulation index must be limited to less than 0.55 rad (31.5°) in order to keep linearity within 10%. On the other hand, the modulation index must be many more times 0.55 in order for the noise-reducing qualities of angle modulation to be effective. While modulator linearity for the tuning-diode modulator can be nearly doubled by adding a second diode in a series back-to-back configuration, the real solution is to use frequency multipliers to increase the phase (and frequency) deviation.

Multiplier circuits are discussed later in this chapter. They are routinely used in PM systems for increasing narrowband PM to wideband PM ($m_p > 0.25$). The PM equation (9–28) is seen to be the same as for FM (Equation 9–9), except for the integration of the modulating signal in FM to show the phase-angle variations. For a given modulation index the sidebands, spectra, and power relationships of sinusoidal PM and FM are identical. The Bessel function tables are used with m_p replacing m_f. The only way to see the difference between PM and FM with a spectrum analyzer is to increase (or decrease) the modulation frequency f_m: The PM index will remain constant, but the FM index will change. This is because $m_p = \Delta\phi$(pk) for PM is independent of modulation *frequency*, but $m_f = \Delta f_c$(pk)/f_m is inversely proportional to the modulating frequency.

Demodulation of PM

The circuit for demodulating a phase-modulated signal is appropriately called a *phase detector*. What is necessary for this process is a nonlinear circuit element and a phase reference. Figure 9–12 shows a typical analog phase detector in which the nonlinear element is a diode gated by the phase-reference signal, and the current pulses are integrated by the capacitor and resistor network. Two such coherent gate/integrator networks are combined in the balanced circuit of Figure 9–12.

FIGURE 9–12   Phase detector.

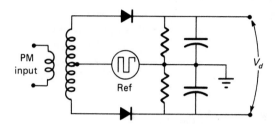

If the reference signal is to provide a fixed phase reference, then it must be obvious that the frequency of this signal must be *exactly* equal to the received carrier frequency. Thus PM requires frequency-coherent demodulation.

The details of how the phase-detector circuit converts phase-difference variations into output voltage variations are given in Chapter 10, along with the derivation of the coherent reference signal from the received carrier signal. In short, a phase-locked loop is used.

9–3 ■ ADVANTAGES OF PM OVER FM

FM and PM certainly have their similarities, but their technological differences have historically dictated the choice between them. For mass applications such as broadcast, phase modulation has been disadvantageous because it requires coherent demodulation, which requires a phase-locked loop. FM can use a noncoherent demodulator, which was simpler and less expensive. With today's ICs, the phase-locked loop circuitry is less expensive because neither a transformer nor LC tuned circuit is required (see the IC phase detector of Chapter 10).

An important difference exists between the modulation index of PM and FM. Specifically,

$$m_p = \Delta\phi(\text{pk})$$

$$m_f = \Delta\omega_c t(\text{pk}) = \frac{\Delta f_c(\text{pk})}{f_m}$$

Because PM has a constant index with modulation frequency changes, it offers better demodulated S/N performance than FM and also does not require preemphasis. These issues are discussed later for FM.

As a final advantage, PM is generated after the carrier oscillator has developed a very stable (crystal-controlled) frequency. Consequently carrier drift, and therefore the necessity for receiver bandwidth, is reduced for PM.

This technology leaves FM with only two advantages over PM. First, the FM VCO can directly produce high-index FM, whereas high-index PM requires multiplier circuits. Second, a long-term change, such as extremely slow telemetry state changes, can be detected using FM, whereas the phase-locked loop has relatively short-term phase memory, so PM is ineffective in this application.

FM broadcast will probably never be replaced by PM because of its traditional use and not because of any technological advantages. Since it is here to stay, we now turn to the technology of FM, beginning with FM transmitter circuits with automatic carrier-frequency control, FM receiver circuits, and FM from PM.

9–4 ■ FM TRANSMITTER CIRCUITS

Some of the specifications for commercial FM broadcast in North America are given in Table 9–2. These requirements will provide some design goals to help put circuits into perspective.

A standard FM transmitter will use a VCO to produce the FM signal. The VCO is followed by a buffer amplifier to provide isolation and therefore better oscillator frequency stability. The FM signal is then amplified up to its final output power by power amplifiers and their driver amplifiers. The power amplifier is almost always operated class C and near saturation for high efficiency and with some peak-limiting to level the transmitted signal envelope.

Two solutions to the poor frequency stability exhibited by VCOs are common. One uses PM with a trick to produce low-index FM followed by frequency multipliers to achieve high indices. This technique is referred to as the *Armstrong method*, and the NBFM from PM is discussed after VCO design.

TABLE 9–2 FCC Requirements for FM Broadcast in North America	**Assigned Carrier Center Frequency, f_c**
	f_c range: 88.1–91.9 MHz (noncommercial) 92.1–107.9 MHz (commercial)
	f_c spacing: 200 kHz between stations
	Transmitted carrier stability: $f_c \pm 2$ kHz
	f_m range: 50 Hz to 15 kHz with 75 µs preemphasis
	Max. deviation: $\Delta f_c(\text{max}) = 75$ kHz $\equiv 100\%$ modulation
	Max. transmitter power: 100 kW horizontal polarization 100 kW vertical polarization

The second technique for achieving carrier frequency stability is to use the wideband VCO and stabilize the carrier center frequency by the use of feedback. This technique is referred to as the *Crosby AFC method* and is analyzed in Section 9–5.

Since power driver amplifiers and class C finals (final power amplifiers) were covered in Chapter 6 on transmitters, we need to consider only VCOs, FM from PM, frequency multipliers, and AFC here.

Voltage-Controlled Oscillators (VCOs)

The simplest technique for controlling the frequency of an oscillator with a voltage is to use a tuning diode. An ac equivalent of a VCO is illustrated in Figure 9–13. The voltage-variable capacitance device was described under "Diode Modulator for PM" in Section 9–2, and the diode capacitance versus tuning voltage is illustrated in Figure 9–10. The equations developed in that section for the frequency shift of an LC_d resonant circuit using only the diode capacitance C_d are applicable here, except that the approximations used are not accurate for VCOs with a large tuning range. Instead of using the approximation of Equation 9–31, it is better to derive an exact expression for frequency deviations, where for convenience we use the normalizing capacitance $C_d = C(-6 \text{ V})$ for the curve of Figure 9–10. If the only circuit capacitance is C_d, then $f_0 = 1/2\pi\sqrt{LC_d}$. At any other voltage V, the diode capacitance is $C(V)$ and the new frequency will be $f = 1/2\pi\sqrt{LC(V)}$.

The deviation from center (carrier) frequency will be $\Delta f_0 = f - f_0$. The fractional (percentage in decimal form) deviation is

$$\frac{\Delta f_0}{f_0} = \frac{f - f_0}{f_0} = \frac{f}{f_0} - 1 \qquad (9\text{–}56)$$

which, using the previous tuned circuit equations, yields

$$\frac{\Delta f_0}{f_0} = \sqrt{\frac{C_d}{C(V)}} - 1 \qquad (9\text{–}57a)$$

FIGURE 9–13 Varicap VCO.

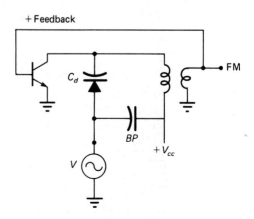

In terms of the vertical axis of the diode C_d versus bias voltage curve of Figure 9–10, Equation 9–57a becomes

$$\frac{\Delta f_0}{f_0} = \sqrt{\frac{C(-6)}{C(V)}} - 1 \qquad \text{(9–57b)}$$

where $C(-6) = C_d$ is the diode capacitance at -6 V reverse bias.

Equation 9–57b can be used with tuning diodes rated at other than 6 V where the value $C(-6)$ is determined from Figure 9–10 or Equation 9–38 by using specified values for p and C_d at given reverse bias voltages. If another capacitor is in parallel or series with the tuning diode, then the circuit equation for total capacitance is modified, which will modify the derivation for Equation 9–31. Also, in Equations 9–57a and 9–58 for k_o, C_d must be replaced by the effective tuning capacitance of the circuit. The VCO sensitivity k_o can be derived for the single tuning diode circuit of Figure 9–13 using some of the results from the derivation for the PM modulator circuit, Figure 9–9. The defining equation for k_o is

$$k_o = \frac{df_0}{dV_i} = \frac{df_0}{dC} \frac{dC_d}{dV_d} \qquad \text{(9–58)}$$

Using Equations 9–41 and 9–40,

$$k_o = \frac{-f_0}{2C_d} \frac{-C_d}{1 + 2V_R} = \frac{f_0}{2 + 4V_R} \qquad \text{(9–59)}$$

where V_R is the actual reverse bias voltage across C_d.

■ ■ ■ ■

EXAMPLE 9-5

The VCO of Figure 9–13 has a capacitance given by $C_d = C_0(1 + |V_d|/\phi)^{-1/2}$, where $\phi = 0.5$ is the forward-bias junction voltage and $C_d = 100$ pF at $V_d = 6$ V. Also $L = 1$ mH and $V_{cc} = +6$ V. Determine the following:

1. f_0 at $V = 0$
2. Δf_0 for $V = +1$ V using (a) graphical method and (b) exact slope method
3. The VCO sensitivity using (a) and (b) of part 2
4. The modulation index if the input signal is $V_m(t) = \sin 6.28 \times 10^4 t$
5. Compare k_o values from the graphical and exact slope methods.

Solution:

1. $f_0 = 1/2\pi\sqrt{10^{-3}(10^{-10})} = $ **503.3 kHz.**
2. **a.** From Figure 9–10 with $V = 6 - 1 = 5$ V, $C(V)/C(-6) = 1.1$, and from Equation 9–57b, $\Delta f_0/f_0 = (\sqrt{1.1})^{-1} - 1 = -0.0465$. Then $\Delta f_0 = -0.0465(503.3 \text{ kHz}) = $ **−23.4 kHz.**

   b.
   $$\Delta f_0 = \frac{df_0}{dV_m} \Delta V_m = \frac{f_0 \Delta V_m}{2 + 4|V_R|}$$
   $$= 503.3 \text{ kHz} \left(\frac{-1V}{2 + 24}\right) = \textbf{−19.4 kHz}$$

3. **a.** $k_o = \Delta f_0/\Delta V = -23.4$ kHz$/-1$ V = **23.4 kHz/V.**
   b. Again, since $\Delta V = -1$ V, $k_o =$ **19.4 kHz/V.**

4. $f_m = \omega/2\pi = 6.28 \times 10^4/2\pi = 10$ kHz. The peak voltage is 1 V; therefore,
   a. $\Delta f_0 = k_o \Delta V = 23.4$ kHz pk $= \Delta f_c$(pk). $m_f = \Delta f_c$(pk)$/f_m =$ **2.34 rad;**
   b. $m_f = 19.4/10$ kHz = **1.94 rad.**

5. The discrepancy between the graphical and exact slope methods is [(2.34 − 1.94)/2.34] (100%) = **17.1%.**

Integrated Circuit VCO

A stable, linear VCO can be formed by combining an MC1648 oscillator IC with an external tuning diode such as the MV1401. The circuit illustrated in Figure 9–14 shows the tuning diode D_1 reverse biased at 4.6 V by setting the V(bias) supply to 6 V. V(bias) can, of course, be set for best linearity or used as a variable dc voltage to determine the oscillator output frequency versus modulation voltage from which the slope at 4.6 V is the oscillator FM sensitivity, k_o; see Figure 7 of the MC1648 data sheet, Figure 9–15.

The modulation signal V_m is coupled by C_c to vary the diode voltage, and C_1 is an RF bypass of value low enough not to unduly limit the modulation high-frequency bandwidth. If modulation with low-frequency binary is desired, replace C_c with a 10-kΩ resistor. R_1 isolates V(bias); C_2, C_3, and C_4 are RF bypasses; and L tunes with D_1 at the desired carrier frequency, f_0. R_2 will reduce the harmonics of f_0 if a squarewave carrier is not desired, and C_5 couples the FM to the output via R_3 and R_4. R_4 provides a 50-Ω output impedance, whereas R_3 minimizes distortion by increasing the output load resistance.

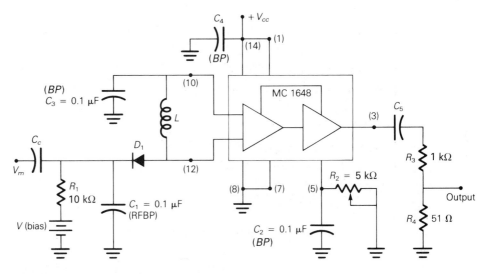

FIGURE 9–14 Integrated circuit VCO.

The MC1648 is an emitter-coupled oscillator, constructed on a single monolithic silicon chip. Output levels are compatible with MECL III logic levels. The oscillator requires an external parallel tank circuit consisting of the inductor (L) and capacitor (C).

A varactor diode may be incorporated into the tank circuit to provide a voltage variable input for the oscillator (VCO). The MC1648 was designed for use in the Motorola Phase-Locked Loop shown in Figure 9. This device may also be used in many other applications requiring a fixed or variable frequency clock source of high spectral purity (See Figure 2).

The MC1648 may be operated from a +5.0 Vdc supply or a -5.2 Vdc supply, depending upon system requirements.

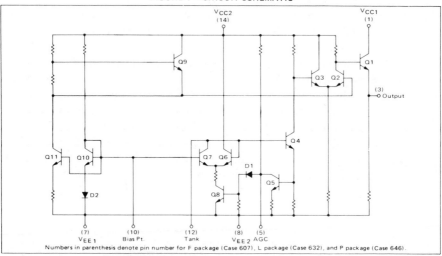

Numbers in parenthesis denote pin number for
F package (Case 607) L package (Case 632), and
P package (Case 646).

Input Capacitance = 6 pF typ
Maximum Series Resistance for L (External Inductance) = 50 Ω typ
Power Dissipation = 150 mW typ/pkg (+5.0 Vdc Supply)
Maximum Output Frequency = 225 MHz typ

SUPPLY VOLTAGE	GND PINS	SUPPLY PINS
+5.0 Vdc	7, 8	1, 14
-5.2 Vdc	1, 14	7, 8

FIGURE 1 – CIRCUIT SCHEMATIC

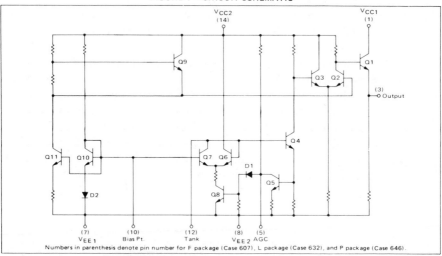

Numbers in parenthesis denote pin number for F package (Case 607), L package (Case 632), and P package (Case 646).

FIGURE 2 – SPECTRAL PURITY OF SIGNAL AT OUTPUT

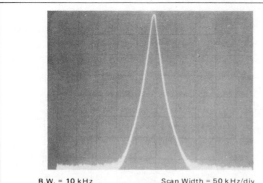

B.W. = 10 kHz　　　　Scan Width = 50 kHz/div
Center Frequency = 100 MHz　　Vertical Scale = 10 dB/div

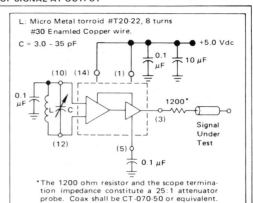

L: Micro Metal torroid #T20-22, 8 turns
#30 Enamled Copper wire.

C = 3.0 - 35 pF

*The 1200 ohm resistor and the scope termination impedance constitute a 25:1 attenuator probe. Coax shall be CT-070-50 or equivalent.

FIGURE 9–15　(page 1 of 5; pages 346–50 courtesy of Motorola, Inc.)

Supply Voltage = +5.0 volts

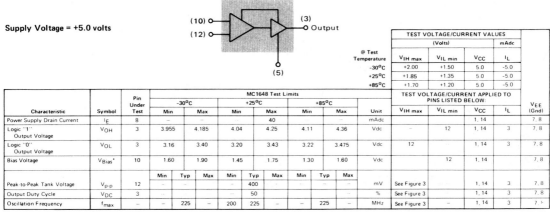

	@ Test Temperature	TEST VOLTAGE/CURRENT VALUES			
		(Volts)			mAdc
		$V_{IH\ max}$	$V_{IL\ min}$	V_{CC}	I_L
	-30°C	+2.00	+1.50	5.0	-5.0
	+25°C	+1.85	+1.35	5.0	-5.0
	+85°C	+1.70	+1.20	5.0	-5.0

Characteristic	Symbol	Pin Under Test	MC1648 Test Limits						Unit	TEST VOLTAGE/CURRENT APPLIED TO PINS LISTED BELOW:				V_{EE} (Gnd)
			-30°C		+25°C		+85°C			$V_{IH\ max}$	$V_{IL\ min}$	V_{CC}	I_L	
			Min	Max	Min	Max	Min	Max						
Power Supply Drain Current	I_E	8	–	–		40			mAdc			1, 14		7, 8
Logic "1" Output Voltage	V_{OH}	3	3.955	4.185	4.04	4.25	4.11	4.36	Vdc	–	12	1, 14	3	7, 8
Logic "0" Output Voltage	V_{OL}	3	3.16	3.40	3.20	3.43	3.22	3.475	Vdc	12		1, 14	3	7, 8
Bias Voltage	V_{Bias}*	10	1.60	1.90	1.45	1.75	1.30	1.60	Vdc		12	1, 14		7, 8

Characteristic	Symbol	Pin Under Test	Min	Typ	Max	Min	Typ	Max	Min	Typ	Max	Unit	$V_{IH\ max}$	$V_{IL\ min}$	V_{CC}	I_L	V_{EE} (Gnd)
Peak-to-Peak Tank Voltage	V_{p-p}	12	–	–	–	–	400	–	–	–	–	mV	See Figure 3		1, 14	3	7, 8
Output Duty Cycle	V_{DC}	3	–	–	–	–	50	–	–	–	–	%	See Figure 3		1, 14	3	7, 8
Oscillation Frequency	f_{max}	–	–	–	225	–	200	225	–	–	225	–	MHz	See Figure 3	–	1, 14	3	7, 8

*This measurement guarantees the dc potential at the bias point for purposes of incorporating a varactor tuning diode at this point.

V_{p-p} output is typically 500 mV @ 225 MHz.

ELECTRICAL CHARACTERISTICS

Supply Voltage = -5.2 volts

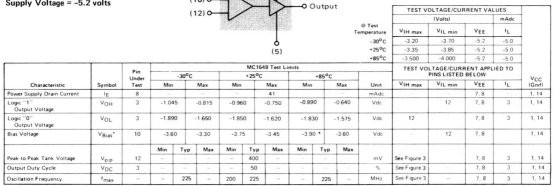

	@ Test Temperature	TEST VOLTAGE/CURRENT VALUES			
		(Volts)			mAdc
		$V_{IH\ max}$	$V_{IL\ min}$	V_{EE}	I_L
	-30°C	-3.20	-3.70	-5.2	-5.0
	+25°C	-3.35	-3.85	-5.2	-5.0
	+85°C	-3.500	-4.000	-5.2	-5.0

Characteristic	Symbol	Pin Under Test	MC1648 Test Limits						Unit	TEST VOLTAGE/CURRENT APPLIED TO PINS LISTED BELOW:				V_{CC} (Gnd)
			-30°C		+25°C		+85°C			$V_{IH\ max}$	$V_{IL\ min}$	V_{EE}	I_L	
			Min	Max	Min	Max	Min	Max						
Power Supply Drain Current	I_E	8	–	–		41			mAdc		–	7, 8		1, 14
Logic "1" Output Voltage	V_{OH}	3	-1.045	-0.815	-0.960	-0.750	-0.890	-0.640	Vdc	–	12	7, 8	3	1, 14
Logic "0" Output Voltage	V_{OL}	3	-1.890	-1.650	-1.850	-1.620	-1.830	-1.575	Vdc	12		7, 8	3	1, 14
Bias Voltage	V_{Bias}*	10	-3.60	-3.30	-3.75	-3.45	-3.90 *	-3.60	Vdc	–	12	7, 8		1, 14

Characteristic	Symbol	Pin Under Test	Min	Typ	Max	Min	Typ	Max	Min	Typ	Max	Unit	$V_{IH\ max}$	$V_{IL\ min}$	V_{EE}	I_L	V_{CC} (Gnd)
Peak-to-Peak Tank Voltage	V_{p-p}	12	–	–	–	–	400	–	–	–	–	mV	See Figure 3		7, 8	3	1, 14
Output Duty Cycle	V_{DC}	3	–	–	–	–	50	–	–	–	–	%	See Figure 3		7, 8	3	1, 14
Oscillation Frequency	f_{max}	–	–	–	225	–	200	225	–	–	225	–	MHz	See Figure 3	–	7, 8	3	1, 14

*This measurement guarantees the dc potential at the bias point for purposes of incorporating a varactor tuning diode at this point.

V_{p-p} output is typically 500 mV @ 225 MHz.

FIGURE 3 – TEST CIRCUIT AND WAVEFORMS

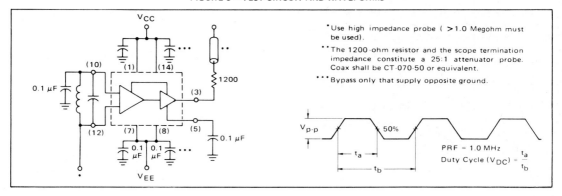

* Use high impedance probe (>1.0 Megohm must be used).

** The 1200-ohm resistor and the scope termination impedance constitute a 25:1 attenuator probe. Coax shall be CT-070-50 or equivalent.

*** Bypass only that supply opposite ground.

$$\text{PRF} = 1.0 \text{ MHz}$$
$$\text{Duty Cycle } (V_{DC}) = \frac{t_a}{t_b}$$

FIGURE 9–15   (page 2 of 5)

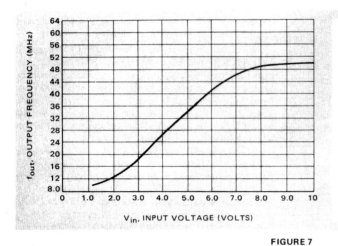

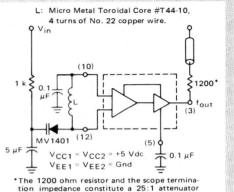

64
60
56
52
48
44
40
36
32
28
24
20
16
12
8.0
0
1.0
2.0
3.0
4.0
5.0
6.0
7.0
8.0
9.0
10
f_{out}, OUTPUT FREQUENCY (MHz)
V_{in}, INPUT VOLTAGE (VOLTS)
L: Micro Metal Toroidal Core #T44-10,
4 turns of No. 22 copper wire.
V_{in}
(10)
1 k
0.1 μF
L
1200*
f_{out}
(3)
MV1401
(12)
(5)
5 μF
$V_{CC1} = V_{CC2} = +5$ Vdc
$V_{EE1} = V_{EE2} =$ Gnd
0.1 μF
*The 1200 ohm resistor and the scope termination impedance constitute a 25:1 attenuator probe. Coax shall be CT-070-50 or equivalent.

FIGURE 7

18
17
16
15
14
13
12
11
10
9.0
8.0
0
1.0
2.0
3.0
4.0
5.0
6.0
7.0
8.0
9.0
10
f_{out}, OUTPUT FREQUENCY (MHz)
V_{in}, INPUT VOLTAGE (VOLTS)
L: Micro Metal Toroidal Core #T44-10,
4 turns of No. 22 copper wire.
V_{in}
C = 500 pF
(10)
1 k
0.1 μF
L
C
1200*
f_{out}
(3)
MV1401
(12)
(5)
5 μF
$V_{CC1} = V_{CC2} = +5$ Vdc
$V_{EE1} = V_{EE2} =$ Gnd
0.1 μF
*The 1200 ohm resistor and the scope termination impedance constitute a 25:1 attenuator probe. Coax shall be CT-070-50 or equivalent.

FIGURE 8

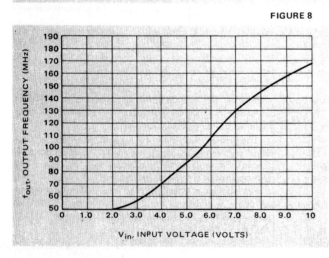

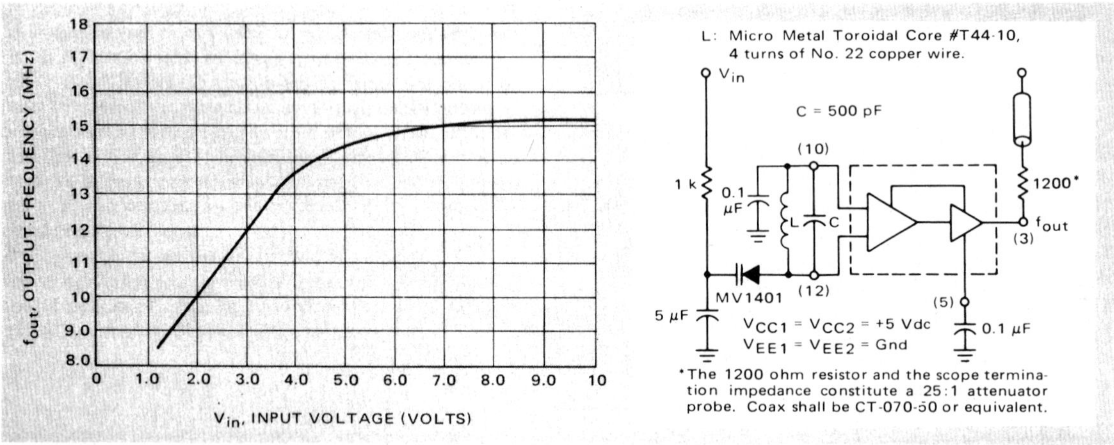

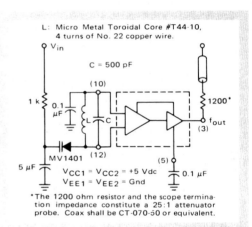

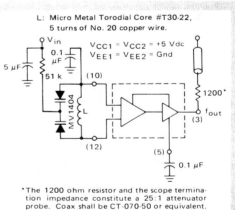

190
180
170
160
150
140
130
120
110
100
90
80
70
60
50
0
1.0
2.0
3.0
4.0
5.0
6.0
7.0
8.0
9.0
10
f_{out}, OUTPUT FREQUENCY (MHz)
V_{in}, INPUT VOLTAGE (VOLTS)
L: Micro Metal Torodial Core #T30-22,
5 turns of No. 20 copper wire.
V_{in}
$V_{CC1} = V_{CC2} = +5$ Vdc
$V_{EE1} = V_{EE2} =$ Gnd
0.1 μF
5 μF
51 k
(10)
MV1404
L
1200*
f_{out}
(3)
(12)
(5)
0.1 μF
*The 1200 ohm resistor and the scope termination impedance constitute a 25:1 attenuator probe. Coax shall be CT-070-50 or equivalent.

FIGURE 9–15 (page 3 of 5)

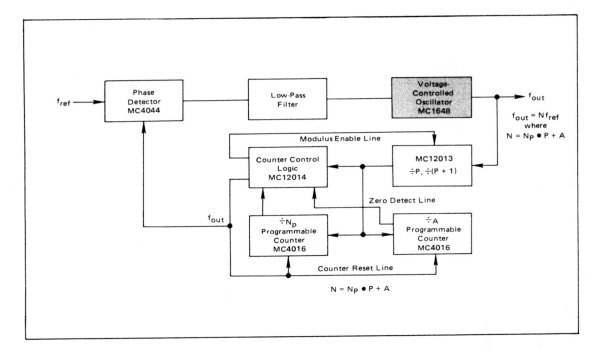

$$f_{out} = N f_{ref}$$
where
$$N = N_p \bullet P + A$$

N = N_p • P + A

Figure 10 shows the MC1648 in the variable frequency mode operating from a +5.0 Vdc supply. To obtain a sine wave at the output, a resistor is added from the AGC circuit (pin 5) to V_{EE}.

Figure 11 shows the MC1648 in the variable frequency mode operating from a +5.0 Vdc supply. To extend the useful frequency range of the device a resistor is added to the AGC circuit at pin 5 (1 kohm minimum).

Figure 12 shows the MC1648 operating from +5.0 Vdc and +9.0 Vdc power supplies. This permits a higher voltage swing and higher output power than is possible from the MECL output (pin 3). Plots of output power versus total collector load resistance at pin 1 are given in Figures 13 and 14 for 100 MHz and 10 MHz operation. The total collector load includes R in parallel with Rp of L1 and C1 at resonance. The optimum value for R at 100 MHz is approximately 850 ohms.

FIGURE 10 – METHOD OF OBTAINING A SINE-WAVE OUTPUT

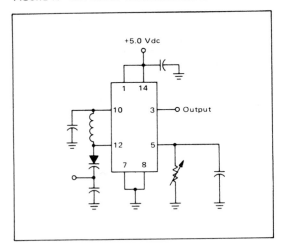

FIGURE 11 – METHOD OF EXTENDING THE USEFUL FREQUENCY RANGE OF THE MC1648

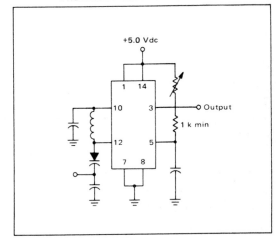

FIGURE 9–15 (page 4 of 5)

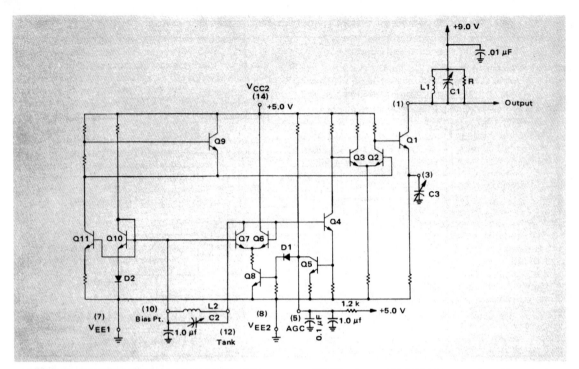

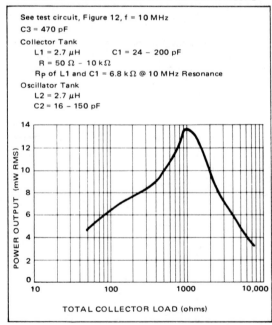

FIGURE 13 – POWER OUTPUT versus COLLECTOR LOAD

See test circuit, Figure 12, f = 100 MHz

C3 = 3.0 – 35 pF

Collector Tank
  L1 = 0.22 μH   C1 = 1.0 – 7.0 pF
  R = 50 Ω – 10 kΩ
  Rp of L1 and C1 = 11 kΩ @ 100 MHz Resonance

Oscillator Tank
  L2 = 4 turns #20 AWG 3/16″ ID
  C2 = 1.0 – 7.0 pF

FIGURE 14 – POWER OUTPUT versus COLLECTOR LOAD

See test circuit, Figure 12, f = 10 MHz

C3 = 470 pF

Collector Tank
  L1 = 2.7 μH   C1 = 24 – 200 pF
  R = 50 Ω – 10 kΩ
  Rp of L1 and C1 = 6.8 kΩ @ 10 MHz Resonance

Oscillator Tank
  L2 = 2.7 μH
  C2 = 16 – 150 pF

FIGURE 9–15 (page 5 of 5)

FIGURE 9-16 High-index FM from a phase modulator preceded by an integrator (Armstrong method).

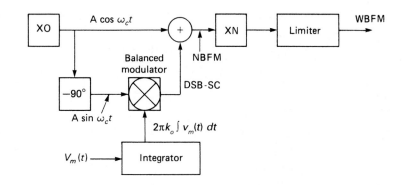

FM from PM

Equation 9–8 for FM, when substituted into the general angle-modulation expression (Equation 9–3), yields

$$v_{FM}(t) = A \cos[2\pi f_c t + \theta_0 + 2\pi k_o \int v_m(t) \, dt] \qquad (9\text{–}60)$$

This equation reveals the trick needed for using a phase modulator to produce FM: integrate the modulating signal $v_m(t)$ before modulating the phase modulator. Any phase modulator will do, and a properly designed RC or RL network, or an op-amp integrator, may be placed in series with the modulation input. One example, popular in FM broadcast, is shown in Figure 9–16. Low-index FM is produced in the narrowband phase modulator (same as Figure 9–8) preceded by an integrator. To achieve the high-modulation indices required by the FCC, frequency multipliers follow the FM modulator.

As an example, NBPM and FM have an index less than 0.25 rad; in order to produce 5 rad ($\Delta f_c(pk)/f_m$ = 75 kHz/15 kHz = 5), a multiplication of at least 20 is required.

Two important system design considerations will be expanded upon in the next section for multiplier circuits. One is that the crystal carrier oscillator frequency must be $f_{XO} = f_c/N$, where f_c is the assigned carrier frequency and N is the multiplication factor. The second is that all frequency deviations are increased by a factor N. This means that crystal-oscillator instabilities (drifts and noise) are increased by N.

■ ■ ■ ■

EXAMPLE 9-6

An FM broadcast station uses the Armstrong method with a times-24 multiplication. If f_c = 100.1 MHz, compute

1. The crystal-oscillator frequency.

2. Maximum allowable oscillator drift.

3. Peak phase deviation at the NBFM output to produce 100% modulation with f_m = 15 kHz.

4. Peak phase deviation of the NBFM if f_m is reduced to 5 kHz; also determine the peak frequency deviation of the transmitted carrier.

Solution:

1. f_{XO} = 100.1 MHz/24 = **4.170833 MHz.**

2. From Table 9–2 the maximum drift at the output is ±2 kHz. At the oscillator the drift must be less than ±2 kHz/24 = **±83⅓ Hz** or **19.98 parts/million** (same ppm as at output).

3. 100% modulation in broadcast FM is Δf_c(pk) = 75 kHz. At the modulator output (NBFM) the frequency deviation must be 75 kHz/24 = 3.125 kHz. The modulation index and peak phase deviation is 3.125 kHz/15 kHz = **0.2083 rad.** Of course, this is 5 rad at the transmitter output.

4. A phase modulator by itself will maintain a constant index as f_m varies. *However*, when f_m is lowered from 15 kHz to 5 kHz, the peak modulating voltage at the integrator output will rise because of the $1/f_m$ (6 dB/octave) slope of the integrator circuit (see Figure 9–17). (The integrator also delays all inputs by 90°.) As a result of the voltage rise, the peak phase deviation of the phase modulator circuit will rise to $\Delta\phi$(pk) = (15 kHz/5 kHz) × 0.2083 rad = **0.625 rad** = m_f. This NBFM index is increased by the multiplier(s) to m_f(out) = 0.625 × 24 = 15. The peak carrier frequency deviation will be Δf_c(pk) = $m_f f_m$ = 15 × 5 kHz = **75 kHz pk.**

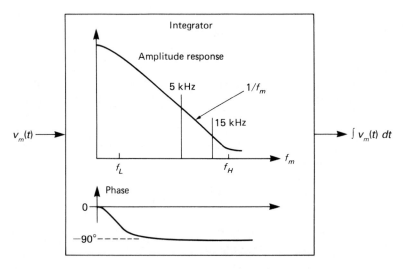

FIGURE 9–17   Gain slope and phase shift of an integrator circuit.

Reactance Modulator for FM

Figure 9–18 shows a circuit for which the output impedance is a highly capacitive reactance, and this reactance varies with the input signal voltage. The equivalent output capacitance C_{eq} is the result of feedback via a large reactance $X_C = 1/2\pi fC$, where $X_C \gg R$. The derivation of C_{eq} is as follows:

$$i_d = g_m v_g \tag{9-61}$$

Also, by voltage divider action,

$$v_g = \frac{v_d R}{R - jX_C} \approx \frac{v_d R}{-jX_C} \tag{9-62}$$

where $X_c \gg R$. For a very high value of X_C such that $i_{fb} \ll i_d$ and $Z_o = v_d/i_d$, Equations 9–61 and 9–62 are combined to yield $Z_o = (-jX_C/R)v_g/g_m v_g$, or

$$Z_o = -j\left[\frac{1}{2\pi f(g_m RC)}\right] \tag{9-63}$$

The output impedance is seen to be equivalent to a capacitance of value approximating

$$C_{eq} = g_m RC \tag{9-64}$$

Reactance Modulator Sensitivity

The circuit output reactance varies with voltage because the FET transconductance g_m varies as

$$g_m = g_{mo}\left(1 - \frac{V_{gs}}{V_p}\right) \tag{9-65}$$

where g_{mo} is the transconductance when the gate-to-source voltage is equal to

FIGURE 9–18 Reactance modulator.

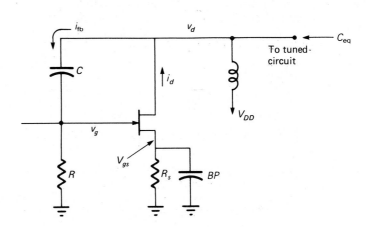

0 V, and V_p is the device pinch-off voltage. $V_{gs} = I_d R_S$ is the self-bias voltage. However, small changes due to the signal at the input Δv_g will cause g_m to vary as

$$\Delta g_m = \left(\frac{-g_{mo}}{V_p}\right) \Delta V_g \qquad (9-66)$$

This is the result of differentiating g_m in Equation 9–65 with respect to V_{gs}.

The variations of C_{eq} with g_m are found by differentiating Equation 9–64 with respect to g_m. That is, $\Delta C_{eq} = RC \, \Delta g_m$. Therefore, with Equation 9–66,

$$\Delta C_{eq} = -\frac{g_{mo} RC}{V_p} \Delta V_g$$

or substituting Equation 9–64 for C and $g_m/g_{mo} = (1 - V_{gs}/V_p)$ yields the fractional change in C_{eq},

$$\frac{\Delta C_{eq}}{C_{eq}} = \frac{\Delta v_g}{V_p - V_{gs}} \qquad (9-67)$$

If the reactance circuit is connected as the total tuning capacitance of a VCO, then Equation 9–57b or 9–58 may be used to determine the VCO sensitivity. For example, Equation 9–58 with Equation 9–67 will give the peak frequency deviation as

$$\Delta f_0 \text{ (pk)} = -\frac{V_g(\text{pk})f_0}{2(V_p - V_{gs})}$$

$$= \frac{V_g(\text{pk})f_0}{2(V_{gs} - V_p)} \qquad (9-68)$$

and the VCO sensitivity is $k_o = \Delta f_0(\text{pk})/V_g(\text{pk})$ or

$$k_o = \frac{f_0}{2(V_{gs} - V_p)} \qquad (9-69)$$

■ ■ ■ ■

EXAMPLE 9–7

Determine the (1) equivalent capacitance C_{eq} and (2) sensitivity for a 4-MHz reactance modulator with $C = 56$ pF, $R = 100$, $R_s = 500$, $I_D = 4$ mA. The FET has a pinch-off voltage of -4 V and $g_{mo} = 4$ mS.

Solution:

1. $|V_{gs}| = V_s = 4$ mA $\times$ 0.5 kΩ = 2 V dc. $g_m = (4 \text{ mS})(1 - 2/4) = 2$ mS, and $C_{eq} = 2$ mS $\times$ 0.1 kΩ $\times$ 56 pF = **11.2 pF.**

2. $k_o = 4 \text{ MHz}/2[-2 - (-4 \text{ V})] = $ **1 MHz/V.**

Frequency Multipliers

Frequency multipliers are used in FM and PM to increase the frequency and phase deviation, thereby increasing the modulation index.

The most straightforward method of frequency multiplication is to drive a high-frequency amplifier or diode nonlinear; the nonlinear current will contain harmonics of the input signal (see Table 3–1). The output circuit is then tuned to present a high impedance at the desired harmonic N to develop maximum power at $f_o = Nf_i$.

If the input signal is $A \sin[(\omega_c + \Delta\omega_c)t + \Delta\phi]$ and the nonlinearity produces a squarewave current, then from Table 3–1e the signal current will be *proportional to*

$$\left(\frac{2A}{\pi}\right)\sin[(\omega_c + \Delta\omega_c)t + \Delta\phi] + \left(\frac{1}{3}\right)\sin[(3\omega_c + 3\Delta\omega_c)t + 3\Delta\phi] + \cdots$$

$$= \left(\frac{2A}{\pi}\right) \sum_{N,odd}^{\infty} \left(\frac{1}{N}\right)\sin[(N\omega_c + N \Delta\omega_c)t + N \Delta\phi] \qquad \textbf{(9–70)}$$

The output circuit is tuned to $N\omega_c$ and should have a proportionately ($\times N$) increased bandwidth to include the N times frequency and N times phase deviations; the percent bandwidth and circuit Q remain unchanged.

Notice from Equation 9–70 that a squarewave nonlinearity is extremely inefficient if an even harmonic (2, 4, . . .) multiplication is desired. To minimize power loss (and maximize efficiency) the nonlinearity should produce current impulses as nearly as possible. This is because all the harmonics (even and odd) of the input signal are produced and because all the harmonics will have nearly the same amplitude. Such a circuit would be called a *comb generator*.

One device made specifically for very efficient, high-multiplication applications is the *step recovery diode* (SRD), or *snap diode*. This device accumulates current for a short part of each input cycle before it suddenly releases it with a "snap." The circuit efficiency or power loss is proportional to approximately $1/N$, as opposed to $1/N^2$ for a good transistor multiplier. Of course, the transistor's current gain will make up for some of the loss by providing power gain over and above the multiplication efficiency.

Transistor multiplier circuits are used for low multiples of N—doublers, triplers, and quadruplers. For high-power triplers and especially for quadruplers, current "idler" networks are used to improve the efficiency. Three series-resonant idlers are illustrated in the quadrupler of Figure 9–19. The idlers help in the output filtering problem, but more importantly they improve the circulation of harmonic currents to enhance the nonlinearity. The idlers in Figure 9–19 are series tuned to f_i, $2f_i$, and $3f_i$, whereas the parallel output tank is tuned to $f_o = 4f_i$. Also, for high-efficiency power transistor multipliers, it is important to realize that most of the nonlinearity is produced in the collector-base junction (varactor behavior), and not the base-emitter, in order to maintain a high β.

Varactor is the name for a (usually high-power) diode operated with high voltage swings in order to take advantage of the nonlinear capacitance of the reverse-biased junction. Varactors do not have the high-multiplication efficiency of SRDs, but they have higher impedance, making them more attractive for high-power applications.

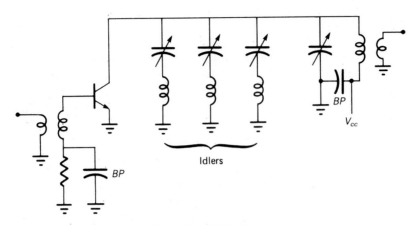

FIGURE 9–19   Quadrupler (times-4 multiplier).

9–5 ■ FREQUENCY STABILITY AND AFC

The FCC specification for the assigned carrier frequency of a standard broadcast FM station is, from Table 9–2, $f_c = f$(assigned) $\pm$ 2 kHz. As calculated in Example 9–5 for a station assigned the carrier frequency of 100.1 MHz, the frequency stability $\Delta f_c/f_c$ must be better than 19.98 ppm (0.001998%).

A frequency stability of 20 ppm is no problem for a well-designed crystal-controlled oscillator. However, high-frequency deviation is not practical with crystal-controlled VCOs even with multiplication. And the stability is reduced by the multiplication factor N.

LC-tuned VCOs have good deviation sensitivity but poor stability with respect to frequency drifts due to aging effects and temperature changes.

One solution to these dilemmas is to use PM with a broadband audio integrator and then multiply. Another solution is to modulate the LC VCO and then lock the average value of the VCO frequency to a crystal reference through a narrow-bandwidth phase-locked loop. This technique is shown in Figure 9–20.

The frequency divider of Figure 9–20 is useful for minimizing interference from the crystal oscillator (XO). The low-pass filter (LPF) prevents feedback of modulation frequencies (FM feedback) and eliminates the possibility of the loop locking to a sideband.

FIGURE 9–20   Phase-locked FM generator.

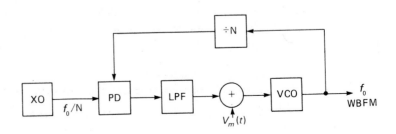

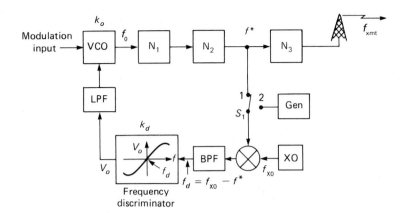

FIGURE 9–21 Transmitter automatic frequency control (AFC) loop.

One more technique for stabilizing an oscillator frequency is the automatic frequency control (AFC) system shown in Figure 9–21. This system is used for stabilizing the carrier center frequency against drift in FM transmitters. AFC without the multiplier and down-converter (mixer and XO) is used in receivers for stabilizing the first local oscillator. In many television receivers it is called AFT—*automatic fine tuning.*

The Crosby AFC loop of Figure 9–21 is an excellent system to analyze for understanding feedback. Suppose the VCO frequency f_0 drifts (*d*) by an amount df_0. Then with switch S_1 open (position 2), there will be no feedback and the transmit frequency will be

$$f_{xmt} = (N_1 \times N_2 \times N_3)(f_0 + df_0)$$
$$= 12f_0 + 12df_0$$

with $N_1 = N_2 = 2$ and $N_3 = 3$. To meet FCC specifications,

$$df_{xmt} = N_1 N_2 N_3 \, df_0 < 2 \text{ kHz} \tag{9–71}$$

With the generator set exactly to

$$f^* = N_1 N_2 f_0$$

the frequency at the frequency discriminator input is $f_d = f_{XO} - f^*$, where high-side injection $f_{XO} > f^*$ is almost always used. The frequency discriminator is a special LC tuned circuit that changes frequency variations to amplitude variations (FM to AM) and is followed by peak AM detectors in a balanced configuration. When properly aligned the circuit is tuned to f_d and the output voltage is 0 V.

---

* Very high index transmission is used in systems with severe multipath distortion problems. FM feedback can be used in a high-index receiver to intentionally reduce m_f and bandwidth while simultaneously stabilizing the frequency/phase characteristics of the receiver—typical feedback improvements.

The $v_o(t)/f_d(t)$ transfer characteristic of the discriminator is linear over a wide frequency range, with a slope (gain) of

$$\frac{\Delta V_o}{\Delta f_d} = k_d \tag{9--72}$$

An input frequency shift of $\pm df_d$ will result in an output voltage change of $\pm\Delta V_o = k_d df_d = \pm V_o$.

The purpose of the low-pass filter (LPF) in Figure 9--21 is to prevent any modulation signals from being fed back (FM feedback), thereby reducing the modulation index. FM feedback has its useful applications, but standard broadcast transmission is not one of them.* The very slow dc changes in V_o due to df_d drifts detected by the frequency discriminator are fed back to the VCO with the correct polarity for forcing the VCO back to nearly f_0. There will, of course, always be a small frequency error in order to produce the correction voltage. The higher the loop gain, the smaller the frequency error necessary to produce the appropriate correction voltage.

In order to complete the system analysis, let us go back to switch S_1. The signal at pin 1 of S_1 has an open-loop frequency error of $df_{ol}^* = N_1 N_2\, df_0$. When the switch is closed to pin 1, the system with feedback will reduce this frequency error to df_{fb}^*. After down-conversion in the mixer, the frequency error at the discriminator input will be $df_d = -df_{fb}^*$. The discriminator output voltage will be $V_{ofb} = -k_d\, df_{fb}^*$, so the VCO frequency will be corrected to $df_{ofb} = k_o V_{ofb} = -k_o k_d\, df_{fb}^*$. This frequency correction, when multiplied by N_1 and N_2, is the correction signal that reduces the open-loop error of f^*; that is,

$$df_{fb}^* = df_{ol}^* - k_o k_d N_1 N_2\, df_{fb}^* \tag{9--73}$$

Solving for df_{fb}^* in Equation 9--73 yields the amount of drift at f^* for the system with feedback

$$df_{fb}^* = \frac{df_{ol}^*}{1 + (k_o k_d N_1 N_2)} \tag{9--74}$$

The error in the transmitted carrier frequency will be

$$df_{xmt}(fb) = N_3 \times df_{fb}^* \tag{9--75}$$

Equation 9--75 shows the expected result for a system with feedback, the open-loop drift is reduced by 1 + (loop gain), where the magnitude of the gain once around the loop is $k_o N_1 N_2 k_d$, or

$$\text{loop gain} = k_o k_d N_1 N_2 \tag{9--76}$$

The reason for going to the trouble and expense of down-converting f^* to the intermediate frequency f_d will be revealed following an example demonstrating the preceding analysis. The reason for separating the times-4 multiplier into two doublers with output filtering ($\times 2$/BPF) is to make a technological point: The higher the multiple N, the more difficult it is to reduce the output harmonic content by 60 dB to meet spectral purity specifications while maintaining wide information bandwidths. In this regard the FCC is much less demanding than the military, for example, because of the great quantities of electromagnetic radiations filling the frequency spectrum and the greater need for interference-free transmissions. The rule of thumb for such applications is to break down the overall multiplication factor to include small submultiples, then start the multiplier string with the smallest submultiple, and filter heavily. A doubler, for example, gives almost an octave to work with for the filtering.

■ ■ ■ ■

EXAMPLE 9–8

A VCO with a long-term frequency stability of $df_0/f_0 = 200$ ppm is used in the AFC system of Figure 9–21. The transmit carrier frequency is to be 108 MHz, and the circuit component gains are as follows: $k_o = 10$ kHz/V, $N_1 = 2$, $N_2 = 2$, $N_3 = 3$, $k_d = 2$ V/kHz; also, $f_{XO} = 37$ MHz. All circuits except the VCO are considered to be perfectly stable. Determine

1. f_0, f^*, and f_d.
2. df_{xmt} open-loop; is it within FCC specs?
3. Transmitted carrier frequency and frequency error closed-loop.

Solution:

1. $f_0 = 108$ MHz/12 = **9 MHz**. $f^* = 4f_0 =$ **36 MHz**. $f_d = f_{XO} - f^* =$ **1 MHz**.
2. Without feedback, the VCO will drift by $(df_0/f_0) \times f_0 = 200$(Hz/MHz) $\times$ 9 MHz = 1800 Hz. At the transmitter output this will be $df_{xmt} = 12 \times 1800$ Hz = **21.6 kHz**. This is also the same as 200 ppm $\times$ 108 MHz = 21.6 kHz. In any case, it is way beyond the specified ±2 kHz and f_{xmt} = 108.0216 MHz.
3. $df_{ol}^* = 4\ df_0 = 4 \times 1800$ Hz = 7200 Hz. The loop gain is (10 kHz/V) $\times$ (2 $\times$ 2) (2 V/kHz) = 80. The drift at f^* is reduced by 81 so that $df_{fb}^* = 7200$ Hz/81 = 88.9 Hz, which at the transmitter output becomes 3×88.9 Hz = **266.7 Hz** = $df_{xmt}(fb)$. The transmitted carrier, instead of being 108 MHz, is **108,000,266.7 Hz**—well within ±2 kHz.

The reason for down-converting f^* before sampling by the frequency discriminator is that the discriminator uses an LC tuned circuit to set the zero reference. The discriminator circuit is also subject to drift.

For example, suppose the discriminator of the Example 9–8 AFC system is stable to 100 ppm. At 1 MHz input, this is only 100 Hz of drift, but without the down-conversion provided by the mixer and XO, the input will be $f^* = 36$ MHz and the discriminator drift will be 3600 Hz. Any drift in the discriminator itself— call it df_{discr}—will result in an output voltage change of

$$\Delta V_o = k_d df_{discr} \tag{9–77}$$

Assuming the worst situation, in which this drift is in the same direction as the VCO drift, the voltage change will push the VCO frequency by the additional amount of $\Delta f_0 = k_o k_d df_{discr}$. The additional drift due to the discriminator as measured at f^* is

$$df_{discr}^* = N_1 N_2 k_o k_d df_{discr} \tag{9–78}$$

Notice that feedback will reduce this by very little. Therefore, the absolute frequency of the discriminator must be as low as practicable. Crystal oscillators have relatively excellent frequency stability. However, it is easy to show that the XO drift contribution at f^* is

$$df^*_{XO} = k_o k_d N_1 N_2 df_{XO} \qquad (9\text{–}79)$$

Using the same notation for the VCO drift contribution at f^*, let us write it as

$$df^*_0 = N_1 N_2 df_0 \qquad (9\text{–}80)$$

Then the total open-loop drift at f^* is

$$df^*_{ol} = df^*_0 + df^*_{discr} + df^*_{XO} \qquad (9\text{–}81)$$

■ ■ ■ ■

EXAMPLE 9–9

The transmitter of Example 9–8 has a discriminator with 200 ppm stability and an XO with 5 ppm. Determine the following for the worst case of all drifts in an additive direction:

1. ΔV_o with S_1 in position 2
2. ΔV_o with no mixer for frequency down-conversion
3. df^* with S_1 in position 2
4. Will the closed-loop transmitter meet FCC specifications?

Solution:

1. ΔV_o due to discriminator drift is (200 ppm $\times$ 1 MHz) $\times k_d$ = (200 Hz) $\times$ (2 V/kHz) = **0.4 V.** ΔV_o due to XO drift is (5 ppm $\times$ 37 MHz) $\times k_d$ = (185 Hz) $\times$ (2 V/kHz) = 0.37 V. The total is ΔV_o = **0.77 V.**

2. With no mixer, f_d = 36 MHz and df_d = 7200 Hz. The ΔV_o = 7.2 kHz $\times$ 2 kHz/V = **14.4 V.** This would push the VCO 144 kHz, and the closed-loop system will not meet FCC specs even with feedback.

3. df^*_0 = 1.8 kHz $\times$ 4 = 7.2 kHz, the same as in Example 9–8. df^*_{discr} = 0.4 V $\times$ 10 kHz/V $\times$ 4 = 16 kHz. df^*_{XO} = 0.367 V $\times$ 10 kHz/V $\times$ 4 = 14.8 kHz. The total from Equation 9–81 is 7.2 + 16 + 14.8 = **38 kHz.**

4. Loop gain is 80 from Example 9–8, so df^*_{fb} = 38 kHz/81 = 469 Hz and df_{xmt} = 3 $\times$ 469 Hz = **1.41 kHz,** which meets FCC specs.

9–6 ■ FM RECEIVERS

Receivers for FM and PM are superheterodyne receivers like their AM counterparts. However, with the exception of delayed-AGC to prevent mixer saturation on strong signals, the AGC of AM receivers is not used for FM and PM because there is no information in the amplitude of the transmitted signal. Constant amplitude into the FM detector is still desirable, however, so additional IF gain is used and the final IF amplifiers are allowed to saturate. The harmonics of the IF signal produced by saturation are often reduced substantially by tuned circuits used to preserve only the required information bandwidth. The IF amplifiers used in the output stages require special design consideration for good saturation characteristics and are called *limiters*, or *passband limiters* when the output is filtered to preserve the information bandwidth.

Since the gain of FM receivers is very high for good limiter action, a special circuit is used to squelch the audio output to eliminate loud noise. When a signal with sufficient S/N is received, the squelch is deactivated for normal reception.

FM Demodulators

An FM demodulator produces an output voltage that is proportional to the instantaneous frequency of the input. The circuit symbol used in Figure 9–21 is repeated in Figure 9–22 to indicate that the frequency-to-voltage transfer function can be nonlinear, but over the linear range of operation

$$v_o(t) = k_d f(t) \qquad\qquad (9\text{--}82)$$

where k_d, in V/kHz, is the gain slope of the overall circuit and $f(t)$ denotes frequency versus time. k_d is measured as indicated by Equation 9–72.

There are three general categories of FM demodulator circuits:

- Phase-locked loop (PLL) demodulator
- Slope detection/FM discriminator
- Quadrature detector

They all produce an output voltage proportional to the instantaneous input frequency.

PLL Demodulator

The phase-locked loop used as an FM demodulator is conceptually the simplest. As seen in Figure 9–23, the PLL consists of a phase detector (PD), low-pass filter

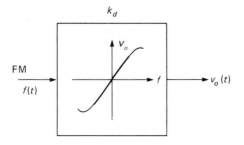

FIGURE 9–22 Frequency discriminator circuit symbol.

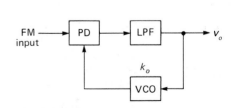

FIGURE 9–23 Phase-locked loop FM detector.

(LPF), and voltage-controlled oscillator (VCO). The PLL VCO, with voltage-to-frequency conversion sensitivity of k_o, is tuned to the FM carrier frequency with $V_o = 0$.* The behavior of the loop (described in detail in Chapter 10) produces a voltage V_o that forces the VCO to track the input frequency. Hence, v_o will be directly proportional to the input frequency variations of the FM carrier that, from Equation 9–2, will be

$$f_c(t) = k_o(\text{xmtr}) \times v_m(t)$$

With limiters ahead of the demodulator and a properly compensated loop, the circuit gain will be constant, so that

$$v_o(t) = k_d f_c(t) k_d [k_o(\text{xmtr}) \times v_m(t)]$$
$$= [k_d k_o(\text{xmtr})] v_m(t)$$

Therefore, $\qquad\qquad\qquad v_o(t) \propto v_m(t) \qquad\qquad\qquad$ **(9–83)**

where $v_m(t)$ is the transmitted information signal.

Integrated circuits such as the LM 1310 and the 1800 PLL FM stereo demodulator use phase-locked techniques.

Slope Detection/FM Discriminator

FM demodulators that use the frequency-versus-amplitude characteristic of tuned circuits for frequency discrimination form this class. The tuned portion of the circuit performs an FM-to-AM conversion, followed by AM peak detection.

The demodulators to be discussed in this class are the traditional FM detectors: slope detector, balanced slope detector, Foster-Seeley phase-shift discriminator, and ratio detector.

The simplest of these circuits, the *slope detector*, is rarely used because of poor linearity. It is illustrated in Figure 9–24a. The center frequency f_0 of the tuned circuit is set such that the input carrier signal falls on the slope of the resonance curve. Either the high- or the low-frequency slope may be used. With f_c on the low side of f_0 and the diode in the direction shown in Figure 9–24a, a positive S curve transfer characteristic is realized. As frequency varies from f_c, the RF amplitude varies. These RF amplitude variations are peak-detected with the diode and RF low-pass filter.

Figure 9–24b illustrates the *balanced slope detector*—literally a balanced version of Figure 9–24a. It provides better linearity than the single slope detector, but it still has a somewhat limited frequency range, and the two variable capacitors increase the complexity.

The mathematical development begins at the modulator where, in Equations 9–8 and 9–9, it was demonstrated that a sinusoidal information signal

$$v_m(t) = V_{\text{pk}} \cos \omega_m t \qquad\qquad\qquad \textbf{(9–84)}$$

---

* For single-power-supply implementations the VCO is tuned to f_0 with no input signal, and then $v_o(t)$ can be taken from a series-coupling capacitor. As with all RC-coupled systems, there will be a low-frequency corner limitation.

FIGURE 9–24 (a) Slope detector.
(b) Balanced slope detector.

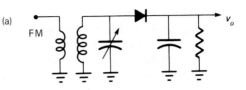

(a)
FM

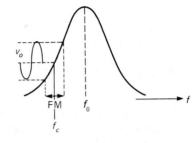

(b)
FM

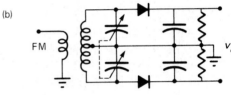

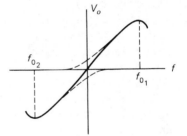

frequency-modulated on a carrier of frequency, ω_c, results in Equation 9–11,

$$s_{\text{FM}}(t) = A \cos(\omega_c t + m_f \sin \omega_m t) \tag{9–85}$$

where

$$m_f = \frac{\Delta f_c(\text{pk})}{f_m} \tag{9–86a}$$

$$= \frac{2\pi k_o V_{\text{pk}}}{\omega_m} \tag{9–86b}$$

To recover $v_m(t)$ at the demodulator output, the off-resonant slope of the tuned circuit differentiates $s_{\text{FM}}(t)$ as follows:

$$v_d(t) \propto \frac{d}{dt} [s_{\text{FM}}(t)] \tag{9–87a}$$

$$= \frac{d}{dt} [A \cos(\omega_c t + m_f \sin \omega_m t)] \tag{9–87b}$$

$$= -A(\omega_c + m_f \omega_m \cos \omega_m t) \sin(\omega_c t + m_f \sin \omega_m t) \tag{9–87c}$$

Equation 9–87c has high-frequency variations around the carrier and amplitude (magnitude) variations of $-A\omega_c - Am_f\omega_m \cos \omega_m t$.

The output of the envelope detector is the magnitude of $v_d(t)$, with polarity depending on the diode(s) direction. Hence,

$$v_o(t) \propto A\omega_c + Am_f\omega_m \cos \omega_m t \qquad \text{(9–88a)}$$

$$= V_{dc} + A2\pi k_o V_{pk} \cos \omega_m t \qquad \text{(9–88b)}$$

using Equations 9–88a and 9–88b and noting that the dc voltage is proportional to carrier amplitude and frequency. After coupling through a series capacitor, the information signal is

$$v_o(t) = k_d V_{pk} \cos \omega_m t \qquad \text{(9–89)}$$

where $k_d = 2\pi k_o A$ plus any circuit losses is the demodulator sensitivity.

Foster-Seeley Phase-Shift Discriminator

The FM demodulator pioneered by Foster and Seeley uses a modified double-tuned circuit to change the FM to AM, that is, to differentiate the input FM signal. This is then followed by a balanced AM detector network to recover the information signal. The Foster-Seeley discriminator is rarely used, but the circuit analysis is very instructive.

Figure 9–25 shows the circuit and the vectors (phasors) formed by the special coupling arrangement used. The input signal voltage across the transformer primary winding v_p is coupled by capacitor C_c directly across the RF choke. A resistor can be used in place of the RFC, but losses occur in the peak detectors. (RFC provides the low-frequency current return path.)

The secondary induced-voltage v_s is divided equally between v_1 and v_2 by a secondary-winding center-tap. The secondary (as well as the primary) circuit is tuned to resonance at the carrier frequency. $L_2 C_2$ forms a high-Q resonant circuit which at resonance circulates a current i_s with the same phase as v_s. Our model has v_s as a point-source voltage in series with the secondary winding inductance. As will be derived, $\mathbf{v}_s = (\mathbf{v}_1 + \mathbf{v}_2)$ is in phase quadrature (90°) to $\mathbf{v}_p$ at resonance,

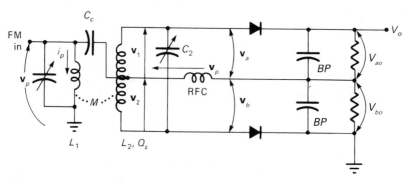

FIGURE 9–25 Foster-Seeley phase-shift discriminator.

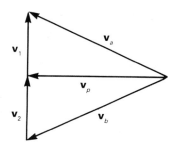

FIGURE 9-26 Vector diagram for phasors in Foster-Seeley discriminator for f_i at circuit resonance.

where the boldface indicates a vector quantity. The resultant vector diagram is shown in Figure 9-26, where $\mathbf{v}_a$ and $\mathbf{v}_b$ are the phasor signals applied to the upper and lower peak detectors, respectively. The detector output voltage V_o will be equal to the difference of the peak value V_{ao} and V_{bo} for the balanced peak detectors; that is,

$$V_o = V_{ao} - V_{bo} \qquad (9-90)$$

where $\qquad V_{ao} = \sqrt{2}\,|\,v_a(\text{rms})\,| \qquad$ and $\qquad V_{bo} = \sqrt{2}\,|\,v_b(\text{rms})\,|$

The important input signal transfer function relating the vectors of Figure 9-26 is derived for the air-coupled transformer as follows: Equation 1-46 is $M = k\sqrt{L_1 L_2}$, where M is the transformer mutual inductance and k is the coefficient of coupling; Equation 1-49b is $v_2 = -j\omega M i_p$; and for a series resonant circuit, impedance is $Z = r_s(1 + jQ\rho)$, where $\rho = f/f_0 - f_0/f$.

Current in the transformer primary is $i_p = v_p/j\omega L_1$, $v_2 = -j\omega M i_p = -(M/L_1)v_p$. Current circulating in $L_2 C_2$ will be $i_s = v_s/Z_{ss}$ and, at resonance, $Z_{ss} = r_s$ where $r_s = \omega L_2/Q_s$. At any frequency ω, $Z_{ss} = r_s + j(\omega L_2 - 1/\omega C_2) = r_s(1 + jQ_s\rho)$. Since $i_s = v_s/Z_{ss}$, then

$$i_s = \frac{-(M/L_1)v_p}{r_s(1 + jQ_s\rho)} \qquad (9-91)$$

For a high-Q secondary, $Q_s > 5$, the voltage across one-half of L_2 will be

$$v_1 = i_s\left(\frac{j\omega L_2}{2}\right) = \frac{-(j\omega L_2 M/2 r_s L_1)}{1 + jQ_s\rho}\,v_p$$

$$v_1 = v_2 = \frac{K}{1 + jQ_s\rho}\,v_p\underline{/-90^\circ} \qquad (9-92)$$

where $K = \omega L_2 M/2 r_s L_1$. Note that at resonance, $\rho = 0$ because $f/f_0 = f_0/f = 1$, so that $\mathbf{v}_1 (= \mathbf{v}_2) = K\mathbf{v}_p$, except for the 90° phase difference between the primary and secondary circuits. Also, since $\omega_0 L_2/r_s = Q_s$,

$$K = \frac{Q_s M}{2 L_1}$$

$$= \frac{k Q_s}{2}\sqrt{\frac{L_2}{L_1}} \qquad (9-93)$$

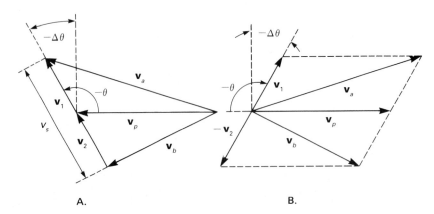

FIGURE 9–27 Identical vector diagrams for discriminator with f_i above circuit resonance. (a) corresponds to Figure 9–26.

where k is the transformer coupling coefficient. The transformer and secondary Q will produce an optimum FM detector in terms of sensitivity/low distortion trade-offs if $kQ_s = 1.5$ and $L_2/L_1 = 1.77$. Thus, good results are obtained for $K = 1$.

When the FM signal frequency reaches its peak deviation $f = f_c + \Delta f_c(\text{pk})$, then one calculates $\rho(\text{max})$ and the phase shift of v_1 from the argument of $1/(1 + jQ_s\rho)$; that is,

$$\Delta\theta = -\tan^{-1}Q_s\rho \qquad (9\text{--}94)$$

Then the vector diagram of Figure 9–26 is modified as illustrated in Figure 9–27 (for $f > f_0$). The peak output $V_o(\text{pk})$ and the detector sensitivity

$$k_d = \frac{\Delta V_o}{\Delta f}$$

for this *phase-shift discriminator* can then be determined graphically or by calculation using the law of cosines. An example will illustrate both techniques.

■ ■ ■ ■

EXAMPLE 9–10

A Foster-Seeley phase-shift discriminator has circuit values such that K of Equation 9–93 is $K = 0.5$. Also, $|v_1| = 4$ V rms and $Q_s = 5.77$. If the FM peak frequency deviation is 5% of the resonant frequency, 10.7 MHz, determine the (1) peak output voltage and (2) discriminator sensitivity.

Solution:

1. $\mathbf{v}_1 = \mathbf{v}_2 = 4$ V$\underline{/-90°}/[1 + j5.77(1.05 - 1/1.05)] = 3.45$ V$\underline{/-120°}$ or, for convenience, $v_1 = v_2 = 4$ V, $\Delta\theta = 30°$. This produces the 30–60–90° right triangle of Figure 9–28, where $|v_p| = |v_1|/K = 8$ V rms. The graphical solution can be obtained from an accurate sketch of the vector diagram. However, for a 30–60° right triangle with a hypotenuse of 4, the short side is 2 and the other side

FIGURE 9–28 Solution for Example 9–10.

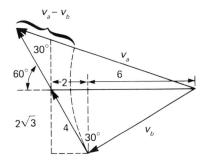

must be $2\sqrt{3}$. The large triangles are also 30–60–90°, therefore, $|\mathbf{v}_a| = \sqrt{10^2 + (2\sqrt{3})^2} = 10.58$ V rms, and $|\mathbf{v}_b| = \sqrt{6^2 + (2\sqrt{3})^2} = 6.93$ V rms. $V_{ao} = \sqrt{2}\,|\mathbf{v}_a| = 14.96$ V pk, and $V_{bo} = 9.80$ V pk. $V_o = 14.96 - 9.80 = $ **5.16 V dc** (or peak for an ac modulation signal).

By the law of cosines:

$$v_b = \sqrt{4^2 + 8^2 - 2 \times 4 \times 8 \cos 60°} = 6.93 \text{ V rms}$$

$$v_a = \sqrt{4^2 + 8^2 - 2 \times 4 \times 8 \cos 120°} = 10.58 \text{ V rms}$$

$$V_o = \sqrt{2}(10.58 - 6.93) = \textbf{5.16 V dc}$$

2. 5% of 10.7 MHz = 535 kHz pk. (Commercial FM never uses this much deviation.) The discriminator sensitivity is $k_d = 5.16$ V/535 kHz = **9.6 mV/kHz.** This is very low for a discriminator. However, the values given are convenient for drawing the phasor diagram.

Ratio Detector

Equation 9–92 shows that the magnitude of the vectors involved in the FM to AM conversion (by phase-shift differentiation) will vary if the circuit input signal v_p changes. The vector diagrams clearly indicate that amplitude variations of v_p and/or v_s will cause a proportional variation of detector sensitivity and therefore demodulated information amplitude. The use of limiters will reduce AM noise of all sorts, and the FM demodulation will be very accurate. Coherent demodulation preceded by limiting will be even more accurate (better S/N).

The ratio detector is a circuit very similar to the Foster-Seeley discriminator, except for the balanced peak-detector configuration that makes it much less sensitive to AM. As seen in Figure 9–29, diode D_b is reversed, a large-C ($\sim$10-μF) capacitor has been connected from diode D_a to D_b, and the demodulated output information is taken from between the resistors as shown. Reversing the direction of D_b allows for a current path, which keeps the voltage V_c across the large-capacity smoothing capacitor constant; that is, capacitor C absorbs AM.

V_c is approximately equal to the value of v_s(pk-pk) across the transformer secondary. This average voltage is maintained, and $2RC$ is the time constant. If the FM signal frequency increases, v_s remains constant, but the vectors shift as in

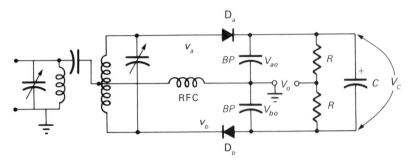

FIGURE 9-29   Ratio detector.

Figure 9–27. Then D_a will conduct more heavily and V_{ao} will increase, but D_b conducts less and V_{bo} decreases—the same as for the Foster-Seeley circuit. Thus, the *sum* of V_{ao} and V_{bo} stays constant, but the *ratio* varies with input frequency.

The voltage V_R across either resistor remains constant, $V_R = v_s(\text{pk-pk})/2$, and as a result of this voltage division for the ratio detector,

$$V_o = 0.5(|\ V_{ao}\ | - |\ V_{bo}\ |)$$
$$V_o(\text{ratio detector}) = \tfrac{1}{2}V_o(\text{Foster-Seeley}) \qquad \qquad \textbf{(9–95)}$$

That is, for otherwise equal circuits, the ratio detector problem is worked the same way as the Foster-Seeley discriminator, but the circuit sensitivity and output voltage are one-half those of the Foster-Seeley.

Receivers using the ratio detector should have a bandpass limiter stage ahead of the detector. This will improve predetection filtering because ratio detectors have low-input Q owing to the constant voltage across C, and the Q varies if v_p tries to change. Also, the less sensitive the FM receiver is to AM, the better the noise improvement and capture ratio (to be discussed later).

If the limiter is not used, very slow received-signal changes cause the voltage across C to change. This voltage can be used for receiver gain control by taking the AGC voltage across either of the output resistors, depending on the polarity desired. Some receivers used in a high dynamic-range environment use this AGC for delayed-AGC of the RF amplifier to prevent saturation of the mixer or even a second RF stage.

Quadrature FM Detectors

Quadrature FM detectors use a reactance, capacitor C_1 in Figure 9–30, for example, to produce two signals with a 90° phase difference. The phase-shifted signal is then applied to an LC-tuned circuit resonant at the carrier frequency. Frequency changes will then produce an additional leading or lagging phase shift that is detected by comparing zero crossings (coincidence detector) or by analog phase detection.

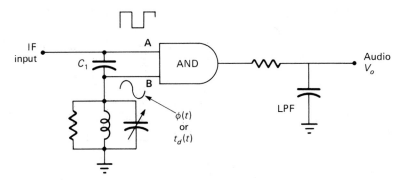

FIGURE 9-30   FM quadrature (coincidence) detector.

The quadrature FM detectors of Figures 9–30 and 9–32 are the *coincidence* and *analog phase-shift* detectors, respectively. These circuits are popular in integrated circuit implementations because there is no complicated and bulky transformer involved.

The *coincidence detector* provides two equal-frequency, phase-quadrature signals to a digital AND gate. As seen in Figure 9–31 the pulse width of the AND gate output varies as a function of the resonant-circuit phase shifts due to frequency changes at pin B. The RC output circuit provides the pulse integration for deriving the average output voltage. The AND gate can also be thought of as a digital multiplier producing the product *AB*.

FIGURE 9-31   Coincidence detector operation. Input *B* is shown with two different phase shifts relative to input *A*. The resulting AND gate outputs are also shown.

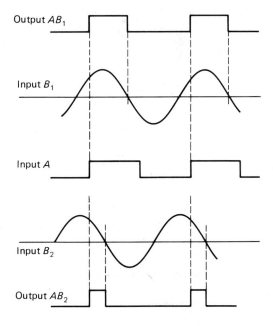

FIGURE 9-32   Quadrature FM
detector.

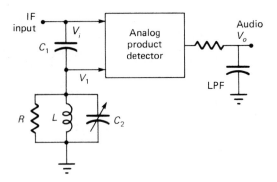

Figure 9–32 shows the same circuit as Figure 9–30 except for the analog product detector. As previously derived for coherent detection of AM signals and phase detection, the multiplication of two periodic signals with the same frequency produces a dc voltage that is directly proportional to the signal phase difference. This phase-shift detector is the analog equivalent of the digital AND gate coincidence detector of Figure 9–30. For small phase shifts (narrowband FM), the output will be reasonably linear and expressed approximately by

$$V_o = \left(\frac{V_1 V_i}{2}\right) \sin Q\rho \qquad (9\text{--}96a)$$

$$\propto Q\rho \qquad (9\text{--}96b)$$

$$= K_{vc}Q\rho \qquad (9\text{--}96c)$$

where Q is the resonant-circuit effective Q and $\rho = f/f_0 - f_0/f$ is the fractional deviation of frequency from resonance, and the phase-to-voltage constant for the circuit is K_{vc}. For the design at frequency $\omega_0 = 2\pi f_0$,

$$Q = \frac{R}{\omega_0 L} \qquad (9\text{--}97)$$

$$\omega_0^2 = \frac{1}{L(C_1 + C_2)} \qquad (9\text{--}98)$$

Also, choose $C_1 \approx C_2$ and L as large as practical while maintaining a high Q. An excellent analysis by Edward Richley is provided in the May 1991 issue of *RF Design Magazine*.

Both the quadrature detectors require prelimiter circuits to maintain constant input amplitudes. A commercially available IC for FM using the quadrature detector is the LM 3089 (Figure 9–33), which includes IF amplifiers through audio preamplification. A noise squelch (muting) circuit is also included.

IF Amplifiers and Limiters

The IF system for broadcast FM commonly operates at 10.7 MHz and requires a bandwidth of at least 180 kHz (Carson's rule). The input stages can be similar to

LM3089

Audio/Radio Circuits

LM3089 FM Receiver IF System

General Description

The LM3089 has been designed to provide all the major functions required for modern FM IF designs of automotive, high-fidelity and communications receivers.

Features

- Three stage IF amplifier/limiter provides 12μV (typ) −3 dB limiting sensitivity
- Balanced product detector and audio amplifier provide 400 mV (typ) of recovered audio with distortion as low as 0.1% with proper external coil designs

- Four internal carrier level detectors provide delayed AGC signal to tuner, IF level meter drive current and interchannel mute control
- AFC amplifier provides AFC current for tuner and/or center tuning meters
- Improved operating and temperature performance, especially when using high Q quadrature coils in narrow band FM communications receivers
- No mute circuit latchup problems
- A direct replacement for CA3089E

Block and Connection Diagram

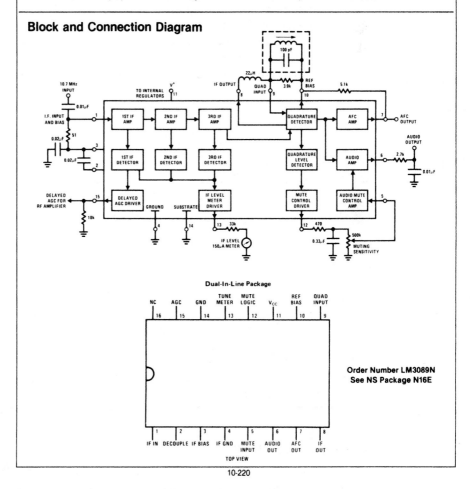

10-220

FIGURE 9-33   (page 1 of 3; pages 371–73 courtesy of National Semiconductor, Inc.)

Typical Performance Characteristics (Continued)

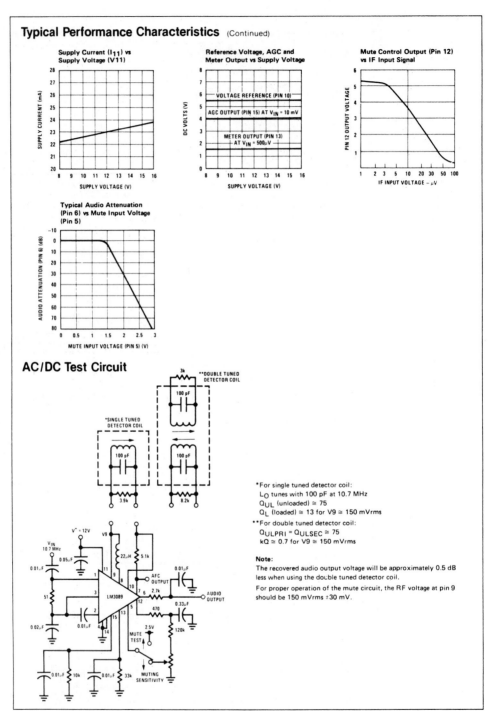

AC/DC Test Circuit

*For single tuned detector coil:
 L_O tunes with 100 pF at 10.7 MHz
 Q_{UL} (unloaded) ≅ 75
 Q_L (loaded) ≅ 13 for V9 ≅ 150 mVrms

**For double tuned detector coil:
 $Q_{ULPRI} = Q_{ULSEC}$ ≅ 75
 kQ ≅ 0.7 for V9 ≅ 150 mVrms

Note:
The recovered audio output voltage will be approximately 0.5 dB less when using the double tuned detector coil.

For proper operation of the mute circuit, the RF voltage at pin 9 should be 150 mVrms ±30 mV.

FIGURE 9-33   (page 2 of 3)

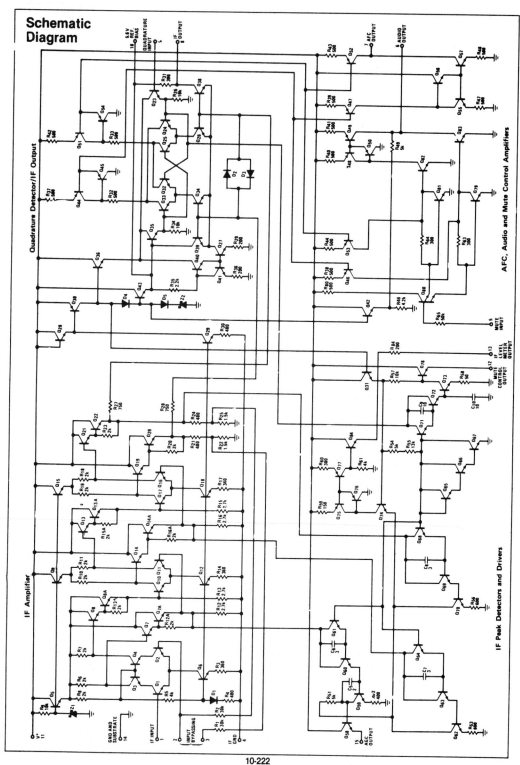

Schematic Diagram

FIGURE 9–33 (page 3 of 3)

373

FIGURE 9–34 Bandpass limiter/
amplifier.

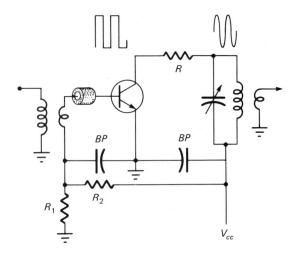

the AM IF amplifiers of Chapter 7, but the output stages are bandpass (tuned) limiter/amplifiers.

Bandpass limiters (BPL) require special design considerations. The reason is that high-frequency tuned amplifiers may exhibit undesirable behavior when driven into saturation. Radical changes in transistor characteristics and impedances along with inappropriately long circuit component leads (especially in the emitter where feedback exists) can cause very high frequency oscillations called *parasitic oscillations*. Parasitics are typically at too high a frequency to be observed as a burst of oscillation, but they manifest themselves as sudden changes (snaps) of output power or as very peculiar distortions (glitches) on oscilloscope displays. This phenomenon is certainly not unique to amplifiers in saturation. Parasitics are as common in medium-frequency RF work as the common cold. The most commonly applied preventative is a low-Q ferrite bead (Z-bead), usually placed on the base lead.

Another practical design consideration is a resistor placed in series with the collector circuit. Values between 100 and 500 Ω are common. They sometimes help with the parasitic problem, but their main function is to maintain at least 100 Ω of impedance across the tank when the collector—normally considered a high-impedance, constant-current source—is in saturation. Too large a value of resistance will unnecessarily reduce the output power. Too low a value will allow the circuit Q to drop excessively.

Figure 9–34 illustrates a typical bipolar BPL stage. Notice the resistor in series with the collector, as previously mentioned. Also, note that R_2 may be eliminated entirely. This would result in class C operation, which is appropriate in this application because the input signal is rectified and the remaining input circuit will provide a self-adjusting bias for constant amplifier gain. Care must be taken with the RC time constant and the transformer/transistor selection to ensure that the reverse bias on the base does not exceed the base-emitter breakdown voltage of around 3 V for small-signal devices.

9–7 ■ NOISE PERFORMANCE

Under special circumstances the noise performance of angle-modulated (FM and PM) transmissions can be much improved (more than 20 dB) over AM and baseband transmissions. There are various special circumstances, but the major ones are (1) predetection (IF) $S/N > 10$ dB, (2) use of IF amplitude *limiters*, and (3) use of *high-modulation index*. Also, for broadband FM, an additional 12 dB of improvement may be realized by the use of *preemphasis*.

The use of limiter circuits reduces AM noise at the demodulation input and keeps the discriminator sensitivity constant. The use of sufficient receiver gain and hard limiting helps to ensure that the noise peaks at the design threshold are almost never larger than the signal. As seen in Figure 9–35, phase deviations due to noise are less than 45° (0.79 rad) and signal phase deviations greater than this will have improved S/N at the demodulator output.

Even without modulation, a *noise-quieting* or *capture effect* is exhibited when using limiters. How can this occur when, due to limiting, the carrier amplitude into the demodulator is not increasing, even though the received signal strength increases? Figure 9–36 illustrates the effect. Figure 9–36a shows the IF limiter output when the noise power is greater than the carrier. We see only noise impulses at the demodulator input because noise is saturating the limiter and the signal is too weak to do anything about it—the receiver is captured by the noise. Figure 9–36b, however, shows the receiver captured by the carrier to the extent that the carrier peaks have the IF system so far into saturation that the weaker noise cannot cause a deflection except possibly at zero-crossover points. This means that the greater the amount of IF limiting, the steeper the crossover slopes and the lesser the effect of noise on the output. The result is that the signal has captured the receiver and the noise is suppressed.

Even without limiting but with modulation, increasing the modulation index will result in a greater demodulator output voltage swing ($v_o \propto m_f$) so that, as long as the input carrier-to-noise ratio C/N is sufficient, the demodulator output S/N

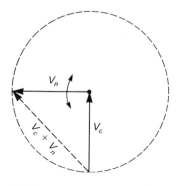

FIGURE 9–35 Carrier and noise phasors (no modulation) at limiter output.

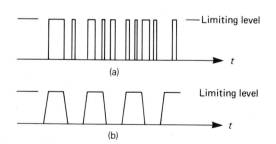

FIGURE 9–36 Limiter circuit output with receiver (a) captured by noise, and (b) captured by signal (carrier only).

FIGURE 9–37 FM thresholding
for C/N less than 13 dB (low-
modulation index).

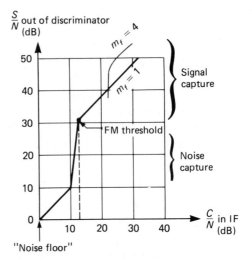

will improve as m_f^2; that is, for strong-carrier conditions, doubling the modulation
index will improve the demodulated S/N by 6 dB. The *FM-thresholding effect* and
modulation index improvement are shown in Figure 9–37. Notice that for low-
modulation index, FM threshold occurs at approximately $C/N = 13$ dB. This is
the "knee" of the linear operating range for the discriminator.

Preemphasis for FM

The main difference between FM and PM is in the relationship between frequency
and phase,

$$f = \frac{1}{2\pi} \frac{d\theta}{dt}$$

FIGURE 9–38 Detector noise
output spectra for (a) PM and
(b) FM.

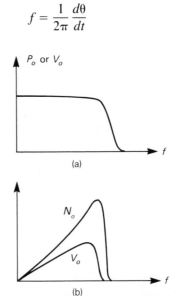

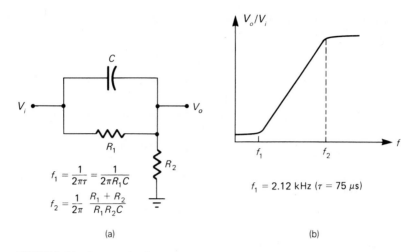

FIGURE 9-39 Preemphasis network (a) and frequency response (b).

A PM detector has a flat noise power (and voltage) output versus frequency (power spectral density). This is illustrated in Figure 9-38a. However, an FM detector has a parabolic noise power spectrum, as shown in Figure 9-38b. The output noise voltage increases linearly with frequency. If no compensation is used for FM, the higher audio signals would suffer a greater S/N degradation than the lower frequencies. For this reason compensation, called *emphasis*, is used for broadcast FM.

A preemphasis network at the modulator input provides a constant increase of modulation index m_f for high-frequency audio signals. Such a network and its frequency response are illustrated in Figure 9-39.

With the RC network chosen to give $\tau = R_1C = 75$ μs in North America (150 μs in Europe), a constant input audio voltage level with frequency will result in a nearly constant rise in the VCO input voltage for frequencies above 2.12 kHz. The larger-than-normal carrier deviations and m_f will preemphasize high-audio frequencies.

At the receiver demodulator output, a low-pass RC network with $\tau = RC = 75$ μs will not only decrease noise at higher audio frequencies but also deemphasize the high-frequency information signals and return them to normal amplitudes relative to the low frequencies. The overall result will be nearly constant S/N across the 15-kHz audio baseband and a noise performance improvement of about 12 dB over no preemphasis. Phase modulation systems do not require emphasis.

9-8 ■ STEREO FM

The broadcast of high-fidelity stereo began in 1961 on FM radio. The transmission of two channels of sound information (stereophonic) on a single carrier required compatibility with existing high-fidelity monophonic FM receivers. This communication system design challenge was solved by using frequency-division multi-plexing (FDM) within the already established 200-kHz station bandwidth.

Figure 9–40 illustrates the approach used at the transmitter, including the composite baseband spectrum which modulates the carrier. Audio signals from both left and right microphones are combined in a linear matrixing network to produce a left-plus-right (L + R) signal and left-minus-right (L − R) signal. The sum L + R provides the total monaural audio signal for compatibility, and L − R is produced by inverting the R signal and adding it to the L signal.

Both L + R and L − R are signals in the audio band and must be separated before modulating the carrier for transmission. This is accomplished by translating the L − R audio signal up in the spectrum. As seen in Figure 9–40, the frequency translation is achieved by amplitude-modulating a 38-kHz subsidiary carrier in a balanced modulator to produce DSB-SC. This form of modulation is used in order to eliminate deviation of the FM carrier by the subcarrier. (A DSB-FC AM carrier would be at least twice the amplitude of the sidebands.) However, the stereo receiver will need a frequency-coherent 38-kHz reference signal to demodulate the DSB-SC. To simplify the receiver, a frequency- and phase-coherent signal is derived from the subcarrier oscillator by frequency division (÷2) to produce a *pilot*. The 19-kHz pilot fits nicely between the L + R and DSB-SC L − R signals in the baseband frequency spectrum. All three signals are added together to form the composite stereo baseband signal, which frequency-modulates the main station carrier. As indicated by its relative amplitude in the baseband composite signal, the pilot is made small enough so that its FM deviation of the carrier is only about 10% of the total 75-kHz maximum deviation. After the FM stereo signal is received and demodulated to baseband, the 19-kHz pilot is used to phase-lock an oscillator, which provides the 38-kHz subcarrier for demodulation of the L − R signal.

The composite stereo baseband signal can look very interesting on an oscilloscope. A simple example using equal frequency but unequal amplitude audio

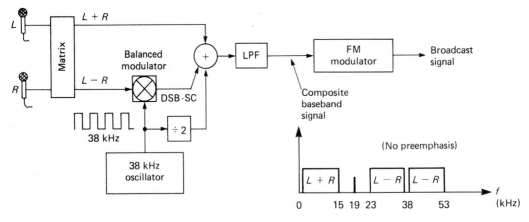

FIGURE 9–40   FM stereo generation block diagram.

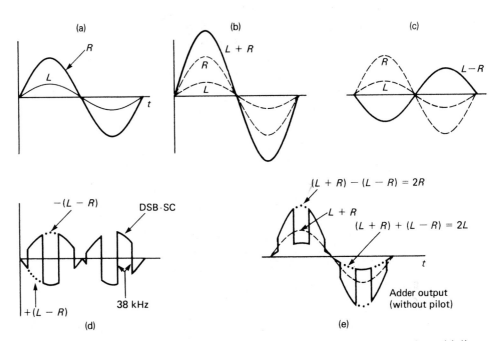

FIGURE 9-41 Development of composite stereo signal. The 38 kHz alternately multiplies the L − R signal by +1 and −1 to produce the DSB-SC in the balanced AM modulator (part d). The adder output (shown in part e without pilot) will be filtered to reduce higher harmonics before FM modulation.

tones in the L and R microphones is used to illustrate the formation of the composite stereo (without pilot) in Figure 9-41.

Figure 9-42 shows the block diagram of an RCA CA-3090 IC stereo decoder for recovering the left and right audio signals at the receiver. A similar inductorless IC made by Motorola, the MC-1310, also uses a phase-locked loop to achieve subcarrier frequency coherency for demodulating the L − R signal.

A 76-kHz VCO is used and a divide-by-2 (÷2) flipflop provides a symmetrical 38-kHz subcarrier to the synchronous L − R product detector for demodulation of the DSB-SC. Another ÷2 flipflop divides the 38 kHz down to 19 kHz for input to the phase detector. The PLL loop gain is high enough to achieve a very low phase error difference between the pilot reference and local VCO. It can be shown that a phase error of 26° between the local signal and the suppressed 38-kHz subcarrier will result in a degradation of stereo separation down to 26 dB. Thus the high-gain PLL has largely replaced the older technique of filtering and doubling the 19-kHz pilot to reconstruct the subcarrier. Both the CA-3090 and MC-1310 achieve 40 dB of channel separation.* The output D of the L − R detector is combined in the linear matrix network with the L + R of the composite baseband signal to produce

---

* That is, a stereo generator with only a right-channel tone will be at least 40 dB down in the left output.

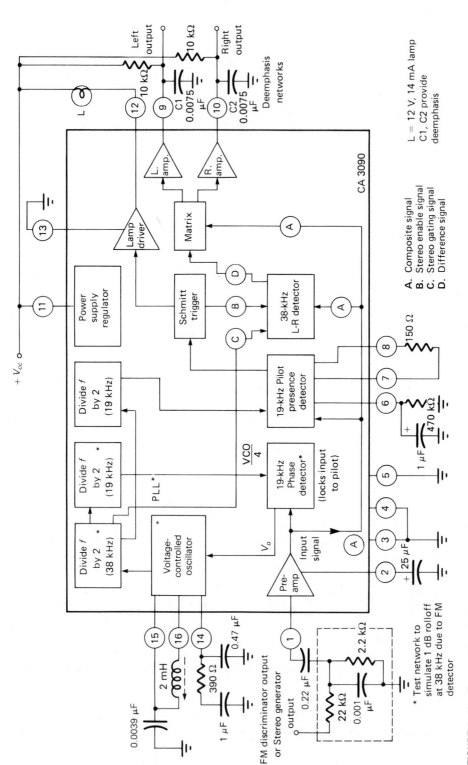

FIGURE 9–42 Stereo FM decoder using the CA3090 IC with PLL pilot synchronizing. (From Leonard Feldman, "New FM Stereo Decoder," *Radio Electronics Magazine,* August 1973.)

(L + R) − (L − R) = 2R and (L + R) + (L − R) = 2L. The 2 in 2R and 2L indicates twice the amplitude and is not a frequency multiplier. The deemphasis networks are external to the decoder chip and have time constants of 75 μs.

■ PROBLEMS

1. **a.** Describe an FM signal.
   b. What is meant by carrier deviation Δf_c?

2. A linear VCO produces a 2-V pk unmodulated 100-MHz carrier and has a modulation sensitivity of $k_o = 10$ kHz/V. If the VCO is modulated by a 707.1-mV rms, 5-kHz cosine signal, determine
   a. Peak deviation of the carrier frequency.
   b. Modulation index.

3. Repeat Problem 2 for the following separate conditions:
   a. Only the modulation voltage (VCO input) is doubled.
   b. Only the modulation frequency is doubled.
   c. Only the carrier amplitude is doubled.

4. **a.** Express, mathematically, the VCO output of Problem 2.
   b. Repeat for Problem 3. (Assume a sine carrier.)

5. The modulation voltage of Problem 2 is reduced to 100 mV pk.
   a. Express this NBFM transmitted signal mathematically using actual voltages and frequencies.
   b. Make a phasor sketch of the NBFM signal with the actual peak voltage amplitudes indicated on the sketch.

6. **a.** Determine the relative amplitudes of all significant spectral components in the FM signal of Problem 2 ($m_f = 2$).
   b. Determine the actual peak voltages for part **a**.
   c. Sketch the frequency spectrum accurately. Include actual frequencies for each component.

7. **a.** Determine the power of the carrier transmitted on a 75-Ω antenna if $m_f = 2$ and V_c (unmodulated) is 2 kV pk.
   b. Repeat for V_c(unmod) = 2 V pk.
   c. Find the power of one J_4 sideband in part **a**.

8. Prove that the relative powers of all sidebands and carrier for $m_f = 2.0$ add up to unity (within 1%).

9. A spectrum analyzer shows that a transmitted FM signal ($f_c = 100$ MHz) has frequency components separated by 10 kHz, and the carrier amplitude is equal to the amplitudes of the first set of sidebands.
   a. Determine the modulation index.
   b. If $k_o = 30$ kHz/V, find the modulation input voltage and frequency.
   c. If the unmodulated carrier power was 10 kW, determine the power transmitted at 100.01 MHz when modulated.
   d. What is the minimum amount of input signal voltage to the VCO required to make the carrier power zero?

10. Compare the information bandwidth (99%) with Carson's rule bandwidth for each problem.
   a. 2
   b. 3a and b
   c. 5
   d. 9a and d

11. **a.** What type of modulator is shown in Figure 9–43?
   b. If $C_D = 100$ pF at 6 V and decreases by 10% at the modulation peak, where $V_m(t) = 1.3 \sin 628t$ V, calculate the output modulation index. (Use the exact or the linear, phase-slope approximation.)
   c. Determine the modulator sensitivity. (Assume linearity holds over the voltage range.)

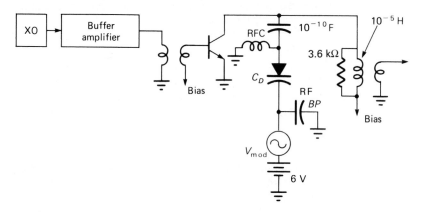

FIGURE 9–43

12. A phase modulator with sensitivity of 0.5 rad/V has a modulation input of $4 \cos 2\pi 10^4 t$ V. Determine the peak phase deviation and modulation index m_p.

13. Use the capacitance-voltage curve to determine the resonant frequency of Figure 9–44 for the following voltages:
   a. $V = -1$ V
   b. $V = -4$ V
   c. $V = 4$ V
   d. Plot f_o versus V.

FIGURE 9–44

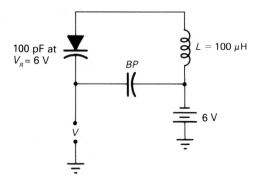

14. a. Determine the modulation sensitivity for a VCO using the circuit of Figure 9–44 with $V = \pm 1$ V.

   b. Find the FM modulation index for $V = \sin 75.4 \times 10^3 t$ V.

15. What is the major difference in demodulators for PM and FM?

16. a. Why is PM preferred over FM for moderate- to narrow-bandwidth space communications?

   b. Why does high-index PM require frequency multipliers?

17. a. The temperature coefficient (TC) for a 98-MHz VCO is 100 ppm/°C. If the VCO is used in a satellite orbiting the moon and the temperature increases by 40°C during an orbit, by how many hertz will the carrier frequency drift?

   b. Repeat for a crystal-controlled oscillator with TC = 1 ppm/°C.

18. A J-FET reactance modulator is shown in Figure 9–45. $I_s = 3$ mA (dc), $g_m = 6$ mS, and $V_p = -5$ V.

   a. What value of L is required to resonate the oscillator tuned circuit to 4 MHz if $V_i = 0$?

   b. What will be the peak percentage change of capacitance for the variable reactance section if $V_i = 0.5$ V pk?

   c. Determine the modulator sensitivity.

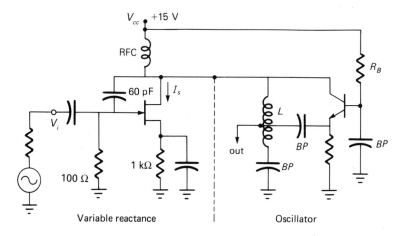

FIGURE 9–45

19. a. Discuss the circuit requirements for multiplying a 12.5-kHz peak deviation broadcast FM signal to 100% modulation index.

   b. What is the best system arrangement for meeting FCC specifications on out-of-band emissions?

   c. Name three multiplier device/circuits and comment on their relative efficiencies.

20. An FM discriminator has a slope (S-curve) of 2 V/kHz and is tuned to 10.7 MHz. Determine the output voltage and frequency for the following input FM signal: $v(t) = 10 \sin[67.23 \times 10^6 t - \cos 31.42 \times 10^3 t]$ V. (*Hint:* Compare $v(t)$ to Equation 9–11.)

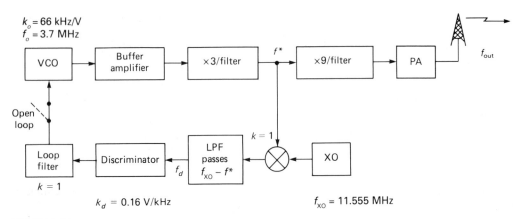

$k_o = 66$ kHz/V
$f_o = 3.7$ MHz

FIGURE 9–46

21. The Crosby AFC system of Figure 9–21 has $N_1 = 3$, $N_2 = 1$, $N_3 = 1$, $k_o = 5$ kHz/V, $k_d = 1$ V/kHz, and an XO with $f = 34$ MHz.
  a. Determine VCO, f^*, and discriminator frequencies for transmitting a 99-MHz carrier.
  b. Find the transmitter output frequency drift, open-loop, if the VCO drifts a total of 300 ppm. Is it within FCC specs?
  c. Repeat part **b** for the closed-loop system. Is this within FCC specs?
  d. Add a 10-dB dc amplifier between the LPF and VCO. Repeat part **c.**

22. An FM transmitter (see Figure 9–46) uses AFC (Crosby method) to stabilize the carrier frequency from the following long-term frequency drifts: VCO 200 ppm, discriminator 200 ppm, XO 5 ppm. Determine
  a. Transmitter output frequency f_{out}, without drift.
  b. Output frequency drift open-loop. Is this within FCC specs?
  c. Total long-term drift at f^* with the loop closed but before correcting for feedback improvement.
  d. Transmitter closed-loop output drift. Is this within FCC specs?

23. A Foster-Seeley discriminator is operating at $\omega_0 = 1.703 \times 10^6$ rad/s. The circuit components are $C_1 = 70$ pF, $L_1 = 3$ μH, $C_2 = 150$ pF, $L_2 = 1.5$ μH (total secondary), Q of secondary $= 20$, $M = 1$ μH. Use a primary rms voltage of 4 V.
  a. Determine v_1 (voltage across ½ of secondary), v_A (rectifier input), V_{AO} (rectifier output), and V_o (net discriminator output) at resonance (ω_0).
  b. Draw the phasor diagram for part **a.**
  c. Repeat **a** and **b** for a frequency decrease of 1%.

24. For the Foster-Seeley discriminator of Figure 9–26, $C_2 = 101.3$ pF, $R = 4.7$ kΩ, and the transformer has $L_1 = L_2 = 10$ μH and a coupling coefficient of 0.10. If $V_p = 8$ V rms at the circuit resonant frequency, determine the following:
  a. V_o at 4.6729 MHz. (*Hint*: Use $Z_i = 0.5 R_{dc}$ for a peak detector to compute Q_s.)
  b. The detector sensitivity (V/Hz).

25. What will be V_o for Problem **23c** if a ratio detector is used with the same component values and input as Problem 23?

26. Determine the output voltage from the quadrature FM detector receiving a 10.715-MHz CW signal. The voltage proportionality constant is 10 V, the resonant circuit center frequency is 10.700 MHz, and the bandwidth is 200 kHz.

27. **a.** The signal-to-noise ratio at the input of an FM ratio detector is +20 dB. What will the demodulated S/N be, in dB?
    b. If the rms noise level at the detector output is 20 mV, what will be the output signal level?

28. Repeat Problem 27 for $(S/N)_i = 10$ dB.

29. Explain the "capture effect" in an FM receiver with hard limiting.

30. Why is preemphasis used in broadcast FM systems?

31. **a.** By how many dB is an 8-kHz modulation signal preemphasized assuming a 6-dB-per-octave gain slope starting at 2 kHz?
    b. Repeat for a 15-kHz signal. (*Hint*: The exact number of octaves n may be determined from 2^n.)

32. The stereo FM system of Figure 9–40 has a 10-kHz tone in the left channel, whereas the right channel is silent. Draw the frequency spectrum and oscilloscope display (time-domain) at each point shown in the system. Indicate each frequency component in the spectrum.

33. What is the purpose of the "pilot" in FM stereo?

34. Explain all the steps from the FM discriminator output to the L and R speakers in recovering a stereo broadcast. What time constant is used for FM deemphasis?

9–9 ■ COMPUTER ANALYSIS

Using the computer for communications system and circuit analysis has become a virtual necessity in education and industry. What follows are examples of programs and/or readily available simulation software for solving problems of the type covered in this chapter.

All the programs are interactive in the sense that an introductory sentence or short paragraph describes the purpose of the program and the user is then prompted for input data.

The programs for Chapter 9 are as follows:

- Three plots of time-domain FM signals with varied modulation parameters demonstrate the use of MathCAD (Figure 9–47).

- One time-domain plot of a PM signal using MathCAD allows for a comparison to the FM signal with the same parameters (Figure 9–48).

- An FM and PM modulation circuit solver: Figure 9–49 is a program in BASIC to solve for modulation index and sensitivity for the circuits of Figures 9–9 and 9–13. In addition, the maximum allowable modulation voltage is given if 10% distortion is not to be exceeded.

MATHCAD: All of this was created on MathCAD.

Plot of a frequency modulated signal; Equations 9-3 and 9-8.

	Modulator sensitivity	Peak amplitude	Frequency of the mod. sig. & carrier

$t := 0, \dfrac{1}{144} \, .. \, 1 \qquad ko := 4 \qquad Vpk := 3 \qquad fm := 2 \qquad fc := 20$

$A := 2 \qquad\qquad Vm(t) := Vpk \cdot \cos(2 \cdot \pi \cdot fm \cdot t) \qquad$ The mod. index is:

$$Mf := ko \cdot \frac{Vpk}{fm}$$

$$s(t) := A \cdot \cos\left[2 \cdot \pi \cdot fc \cdot t + 2 \cdot \pi \cdot ko \cdot \int_{0}^{t} Vm(t) \ dt \right]$$

Amplitude (volts)

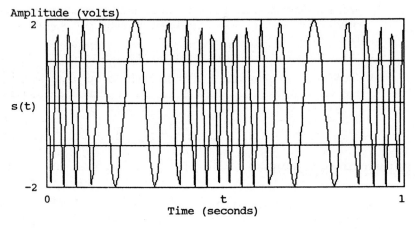

FIGURE 9-47a

MathCAD:

Plot of a frequency modulated signal; Equations 9-3 and 9-8.

	Modulator sensitivity	Peak amplitude	Frequency of the mod. sig. & carrier	

$t := 0, \dfrac{1}{144} ..1$ $ko := 4$ $Vpk := 3$ $fm := 4$ $fc := 20$

$A := 2$ $Vm(t) := Vpk \cdot \cos(2 \cdot \pi \cdot fm \cdot t)$ The mod. index is:

$$Mf := ko \cdot \frac{Vpk}{fm}$$

$$s(t) := A \cdot \cos\left[2 \cdot \pi \cdot fc \cdot t + 2 \cdot \pi \cdot ko \cdot \int_0^t Vm(t)\ dt\right]$$

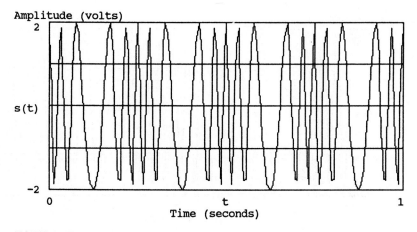

Amplitude (volts)

Time (seconds)

FIGURE 9-47b

MathCAD:

Plot of a frequency modulated signal; Equations 9-3 and 9-8.

	Modulator sensitivity	Peak amplitude	Frequency of the mod. sig. & carrier

$t := 0, \dfrac{1}{144} ..1 \quad$ ko := 4 $\quad$ Vpk := 1.5 $\quad$ fm := 2 $\quad$ fc := 20

A := 2 $\qquad$ Vm(t) := Vpk$\cdot$ cos(2$\cdot \pi \cdot$ fm$\cdot$ t) $\qquad$ The mod. index is:

$$Mf := ko\cdot \dfrac{Vpk}{fm}$$

$$s(t) := A\cdot \cos\left[2\cdot \pi \cdot fc\cdot t + 2\cdot \pi \cdot ko\cdot \int_0^t Vm(t)\ dt\right]$$

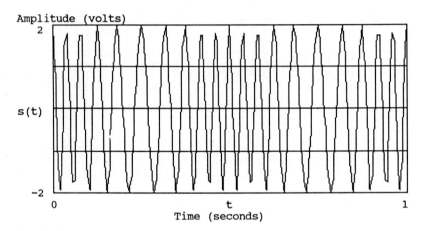

Amplitude (volts)

FIGURE 9–47c

MathCAD:

Plot of a phase modulated signal; Equation 9-28.

	Delta θ	Peak amplitude	Frequency of the mod. sig. & carrier

$t := 0, \dfrac{1}{144} \; ..1 \quad Mp := 6 \qquad A := 2 \qquad fm := 2 \quad fc := 20$

$$Spm(t) := A \cdot \cos(2 \cdot \pi \cdot fc \cdot t + Mp \cdot \cos(2 \cdot \pi \cdot fm \cdot t)) \hspace{2cm} (9\text{-}28)$$

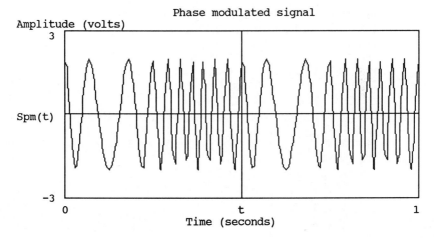

Phase modulated signal

FIGURE 9-48

```
10 REM  ***   FM & PM  VCO Sensitivity and Modulation index   ***
20 REM * Example of student computer assignment*
30 REM *******  for Prof. P. YOUNG , at ASU  **********
40 CLS
50 PRINT "                    FM & PM MODULATION CIRCUIT SOLVER "
60 PRINT:PRINT:PRINT
70 PRINT "       Please choose type of circuit to solve :"
80 PRINT:PRINT
90 PRINT "                    1. Frequency Modulation ( FM )."
100 PRINT
110 PRINT "                    2. Phase Modulation ( PM )."
120 PRINT
130 PRINT "                    3. Exit the program."
140 PRINT:PRINT
150 INPUT "             Enter transaction :";SELECT
160 IF SELECT < 1 THEN GOTO 40
170 IF SELECT > 3 THEN GOTO 40
180 IF SELECT = 1 THEN GOTO 210
190 IF SELECT = 2 THEN GOTO 540
200 IF SELECT = 3 THEN GOTO 920
210 CLS:PRINT:PRINT
220 PRINT "          The following calculations are based on figure 9-13"
230 PRINT
240 PRINT "          of Young's 'Electronic Communication Techniques'"
250 PRINT
270 PRINT:PRINT:PRINT:PRINT:PRINT
280 INPUT "          Press the Return to continue";X
290 CLS:PRINT:PRINT:PRINT
300 PRINT "                    Frequency Modulation ( FM )"
310 PRINT
320 PRINT
330 PRINT "          To solve for the modulator sensitivity enter the"
340 PRINT
350 PRINT "          following circuit values :":PRINT:PRINT:PRINT
360 INPUT "          Varicap diode capacitance at 6volts, Cd ( in farads ) : ";CD1
370 PRINT
380 INPUT "             L  ( in henries ) : ";L1
390 FO = 1/(2*3.14*SQR(L1*CD1)):KO = FO/(2*(1+2*6)):PRINT:PRINT
400 PRINT "          The sensitivity (gain), Ko = "KO " Hz/volt."
410 PRINT:PRINT:INPUT "          Press the Return Key to continue ";Y
420 CLS:PRINT:PRINT
430 PRINT "                    Frequency Modulation ( FM )"
440 PRINT:PRINT
450 PRINT "          To solve for the modulation index (Mf ) enter "
460 PRINT
470 PRINT "          the following :":PRINT:PRINT
480 PRINT "          The Peak input modulation voltage in volts :"
490 PRINT:INPUT "               Vm = ";VM
500 PRINT:PRINT "          The modulation frequency in hertz :"
```

FIGURE 9-49   (page 1 of 2)

```
510 PRINT:INPUT "                    fm = ";FM:DFO = KO * VM:MF = DFO / FM
520 PRINT:PRINT:PRINT "          The Modulation Index Mf = "MF "radians"
530 PRINT:INPUT "       Press the Return key to continue " ;Y:GOTO 40
540 CLS:PRINT:PRINT
550 PRINT "               The following calculations are based on figure 9-9"
560 PRINT
570 PRINT "           of Young's 'Electronic Communication Techniques'"
580 PRINT
600 PRINT:PRINT:PRINT:PRINT:PRINT
610 INPUT "            Press the Return key to continue";Y
620 CLS:PRINT:PRINT:PRINT
630 PRINT "                    Phase Modulation ( PM )"
640 PRINT:PRINT
650 PRINT "           To solve for the modulator sensitivity enter the"
660 PRINT
670 PRINT "              following circuit values :":PRINT:PRINT
680 INPUT "              R ( in ohms ) : ";R:PRINT
690 INPUT "              L ( in henries ) :    ";L2:PRINT
700 INPUT :        Varicap diode capacitance at 6volts, Cd ( in farads ) : ";CD2
710 FO2 = 1/(2*3.14*SQR(L2*CD2)):Q = R / (2*3.14*FO2*L2):KP = Q / (1+2*6)
720 PRINT:PRINT:PRINT "          The sensitivity  (gain), Kp = ";KP;
730 PRINT " rad/volt.":PRINT:PRINT
740 INPUT "   Press the Return key to continue";Y
750 CLS:PRINT:PRINT:PRINT
760 PRINT "                    Phase Modulation ( PM )"
770 PRINT:PRINT
780 PRINT "           To solve for the modulation index ( Mp ) enter"
790 PRINT
800 PRINT "           the following :":PRINT:PRINT
810 PRINT "           The peak modulation input voltage in volts :"
820 PRINT:INPUT "                  Vm = ";VM2
830 DFO2 = ( VM2 * FO2 ) / (2+4*6):MPL = (3.14/2) * Q * ( DFO2/FO2 )
840 E = ( 1 + ( DFO2/FO2 ) ) - ( 1 / ( 1 + ( DFO2/FO2 ))):MPE = ATN(Q*E)
850 PRINT:PRINT:PRINT "     'Linearized' Model : Mp = "MPL "radians"
860 PRINT:PRINT "        ' Exact' Model     : Mp = "MPE " radians."
870 VMX = .55/KP   :REM   MAX  VM FOR 10 PERCENT LINEARITY
880 IF MPE > .548 THEN GOSUB 900:PRINT:PRINT
890 INPUT "      Press return to continue";Y:GOTO 40
900 PRINT "(For less than 10% distortion, Vm must be held below";VMX;" volts for this modulator.)"
910 RETURN
920 END
```

FIGURE 9–49   (page 2 of 2)

10

Phase-Locked Loops

■ INTRODUCTION

The phase-locked loop was described for the synchronous reception of radio signals as far back as 1932 by H. deBellescizi in France and has found use in numerous applications. Its widespread use today in all areas of electronics is due to the convenience and small size of the integrated circuit. Some of the communications applications are shown in the block diagram configurations of Figure 10–1 and will be described in appropriate sections and chapters which follow. Although the phase-locked loop (PLL) is nonlinear over some of its operating range, the linearized noiseless model used in this chapter provides the basis for most PLL applications. A treatment of noise and other important topics can be found in various books on PLLs and frequency synthesizers (using the phase-lock principle).

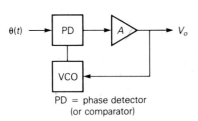

PD = phase detector
(or comparator)

(a) Coherent *phase detector*
or *FM demodulator.*

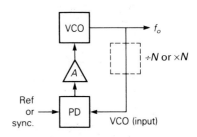

(b) *Stabilized VCO* (voltage-controlled oscillator).
Applications: frequency synthesizer;
TV horizontal AFC.

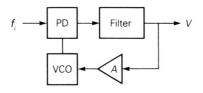

(c) *Tracking filter.*

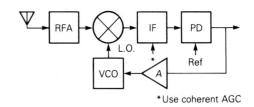

*Use coherent AGC

(d) *Phase-locked receiver* to receive deep
space signals that are buried in noise
(negative *S/N*); also as *Doppler detector*
or to solve Doppler shift problems in earth
satellites or fast-moving systems.

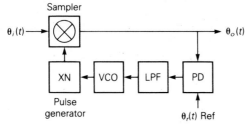

(e) *Sampled PLL.*

Other applications: "cleaning up" a noisy
oscillator with a wideband PLL;
pulse width modulator or demodulator;
bit sync of PCM; touch-tone decoder.

FIGURE 10–1   Phase-locked loop applications.

As illustrated in Figure 10–1c for the frequency-tracking filter application, the phase-locked loop consists of a phase detector (PD), loop filter, a voltage-controlled oscillator (VCO), and, very often, a low-frequency op-amp. The loop is said to be *phase-locked* when the VCO frequency f_o and the loop input signal frequency f_i are identical ($f_o = f_i$), and only a phase difference between the VCO and input signal exists.

Without an input signal the VCO is free-running at frequency f_{FR}. When a signal arrives, the system enters an *acquisition* (prelock) *mode*. When an input signal is present and the loop is locked, then $f_o = f_i$, by definition, and the VCO will track frequency changes of the input signal. This is called the *tracking* (locked) *mode.* The range of frequencies over which the VCO will track the input signal is called the *hold-in range* and is determined by the PLL loop gain and linearity. Like any feedback system, the loop gain is the product of the gains for each component in the loop. ■

10-1 ■ LOOP COMPONENTS

The Phase Detector

A phase detector is a dc-coupled mixer. A simple analog phase detector is shown in Figure 10–2a. Consider the phase detector as a simple switch, as illustrated in Figure 10–2b. The signal with frequency f_o simply opens and closes the (diode) switch. If $f_i \neq f_o$, then the circuit behavior is that of a mixer producing the sum and difference frequencies. The capacitors shown are chosen to bypass f_i, f_o, and $f_i + f_o$, and therefore only the beat $(f_i - f_o)$ signal is seen at v_d. This will be the case prior to phase lock during the acquisition mode of operation.

After the loop is locked, f_o will be exactly equal to $f_i (f_o = f_i)$ by definition of "locked," and only a phase difference can exist between the two phase-detector input signals. A phase difference between the two input signals results in a dc voltage V_d, which is proportional to the phase difference, $\theta_e = \theta_i - \theta_o$. This is first shown mathematically, then graphically.

A mixer performs the mathematical function of multiplication. Thus for sinusoidal inputs,

$$v_d = A \, \sin(\omega_i t + \theta_i) \times 2 \, \cos(\omega_o t + \theta_o)$$
$$= A \, \sin[(\omega_i t + \theta_i) - (\omega_o t + \theta_o)] + A \, \sin[(\omega_i t + \theta_i) + (\omega_o t + \theta_o)]$$

FIGURE 10-2 (a) Analog phase detector. (b) Simplified model.

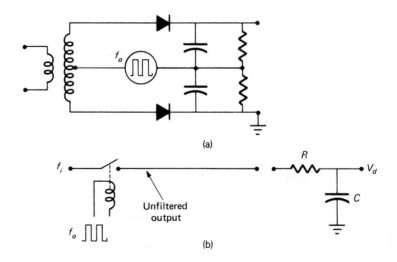

(a)

Unfiltered output

(b)

FIGURE 10-3 Analog phase detector characteristic—output voltage versus input phase difference.

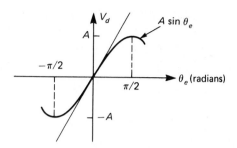

When phase-locked, $\omega_o = \omega_i$, the second harmonic term, $A \sin[2\omega_0 t + \theta_o + \theta_i]$, is filtered out, leaving

$$V_d = A \sin(\theta_i - \theta_o) \tag{10-1}$$

This voltage is directly proportional to the input signal amplitude and, more importantly, the phase error θ_e if the signal amplitude is held constant. Indeed, for *small* θ_e, this transfer function is linear as seen in Figure 10-3. The comments of the last two sentences assume that the VCO voltage is much larger in amplitude than the input signal. If the VCO amplitude is not strong enough to be considered a switching function, or the signal level is so strong as to be comparable in level to the VCO voltage, then A of Equation 10-1 will be proportional to one-half the product of the VCO and input signal levels.

Figure 10-4 helps to show the results graphically as oscilloscope measurements. The f_o signal is shown as a squarewave because its function is to open and

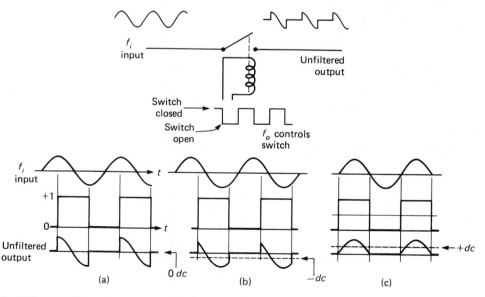

FIGURE 10-4 Phase detector waveforms.

close the switch at a rate exactly equal to f_i. For the waveforms shown, f_o and f_i are equal and the loop is locked. When the signals are out of phase by 90° as in part (a), a zero dc output results; if the phase is slightly advanced as in part (b), a small negative dc output is produced; and when the signals are exactly in phase as in part (c), the result is a dc output proportional to the f_i signal level—exactly the kind of signal that is needed for a lock indication in telephone touch-tone decoders or AGC in coherent receivers.

You can see here that the average (dc) value of the unfiltered phase detector output is proportional to the phase difference between the f_i and f_o signals. With the low-pass filter (integrator) at the output, only the dc voltage is measured as V_d. To determine the actual value of V_d, you must integrate the unfiltered waveform over one cycle and divide by the period. If you think you are not into integration, look again. Do you see that there is a negative average in part (b), a zero average in part (a), and a positive average in part (c)? Yes? Congratulations, your eye and brain integrate just fine.

Phase-Detector Gain

The phase detector just discussed is a mixer, which mathematically is a multiplier, with a direct-coupled output. When the loop PLL is locked and has no stress in it (zero static phase error), the voltage applied to the VCO must be zero, so the phase detector output must also be zero. Thus the phase detector (locked loop) operates with a 90° offset, but this is really no different than a circuit-bias offset. The locked-loop phase detector input-output (transfer) characteristic is shown in Figure 10–5. This is seen to be similar to the FM discriminator S-curve. However, the PD characteristic is a continuous sinusoid repeating every 2π radians. Also, during the tracking mode, operation is limited to the portion of the curve between $\pm \pi/2$ where $|\theta_e| < \pi/2$. This is because, for θ_e between $\pi/2$ and $3\pi/2$, the slope of the PD characteristic is negative, resulting in an unstable (regenerative feedback) loop. For sinusoidal inputs it is clear from Figure 10–5 that the slope of the phase detector characteristic curve,

$$V_d = A \sin \theta_e \qquad \qquad (10\text{--}2)$$

is not constant. In fact, it rises with a maximum slope at $\theta_e = 0$, and levels off to a slope of zero (no gain) at $\theta_e = \pi/2$ rad. Since we are usually interested in loop

FIGURE 10–5 Analog phase detector characteristic.

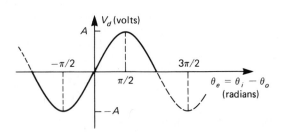

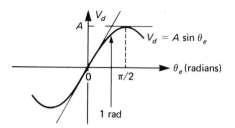

FIGURE 10–6 Sinusoidal characteristic of analog phase detector.

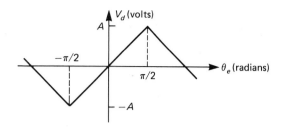

FIGURE 10–7 Phase comparator characteristic for squarewave inputs.

behavior for small θ_e, let's find the slope (gain) at $\theta_e = 0$. A proof of this is given,* but it turns out that the peak voltage A is the volts-per-radian gain of this phase detector because the tangents to the peak and the PD curve at $\theta_e = 0$ intersect at 1 rad, as seen in Figure 10–6. Therefore, the gain of the analog phase detector is

$$k_\phi = A \quad \text{(volts/radian)} \quad \quad \quad (10\text{–}3)$$

If the input signals are both squarewaves, the phase detector characteristic will be linear, as illustrated in Figure 10–7. It is clear from Figure 10–7 that the gain of this circuit is constant over the range of inputs $\theta_e = \pm\pi/2$ and is given by $k_\phi = V_d/\theta_e = A/(\pi/2)$; that is,

$$k_\phi = \frac{2A}{\pi} \quad \text{(volts/radian)} \quad \quad \quad (10\text{–}4)$$

The phase detectors of Figures 10–8 and 10–9 also produce the characteristic of Figure 10–7, although—as shown for the exclusive-OR circuit of Figure 10–9— there will be a dc-bias offset to V_d if a single supply voltage is used. The circuit of Figure 10–8 is typical of balanced integrated circuit implementations. This circuit is also used as a balanced AM modulator for producing double-sideband/suppressed-carrier signals and consists of differential amplifiers. The oscillator input polarity determines which differential pair conducts, and the signal input determines whether R_{C1} or R_{C2} receives the current. The output voltage will be the difference between $i_1 R_{C1}$ and $i_2 R_{C2}$.

Digital Phase (Timing) Comparators

Digital phase detectors can be realized using an exclusive-OR (Figure 10–9) or an edge-triggered set-reset flipflop (RS-FF) circuit. The exclusive-OR output Y is low when both inputs are high or low; otherwise Y is high, indicating "or." The output

---

* The slope of the PD curve, $V_d(\theta) = A \sin \theta_e$, at the origin is

$$\frac{d}{d\theta} [A \sin \theta_e] \,|_{\theta_e=0} = A \cos 0 = A$$

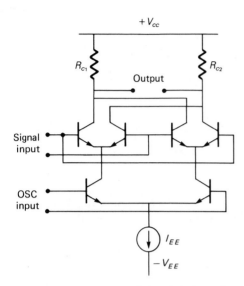

FIGURE 10-8 Integrated circuit balanced detector.

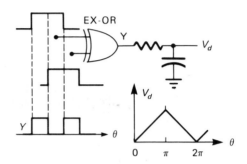

FIGURE 10-9 Digital implementation of phase detector using an exclusive-OR gate.

is smoothed (integrated) to produce V_d. The exclusive-OR requires symmetrical squarewave inputs, which may become a system problem, whereas the edge-triggered RS-FF works well with pulses. Furthermore, as illustrated for the circuit of Figure 10–10, the RS-FF phase detector can produce a linear PLL over a full θ_e range of 2π rad, which is twice that for the other phase detectors. The problem with using digital phase detectors in sensitive communication receiver applications is in the difficulty of filtering the sharp impulses and their harmonics to prevent radio-frequency interference (RFI).

Amplifiers

The second loop component is an amplifier commonly referred to as the *dc amp*. Its function is to increase the loop gain by amplifying the phase detector output voltage. Figure 10–11 shows three voltage amplifiers and their gain parameter $k_A = A_V$ (volts out/volts in). The bandwidth of the dc amp must be very high compared to the loop bandwidth or loop instability will result—even to the point

FIGURE 10-10 Digital implementation of phase detector using a set-reset flipflop.

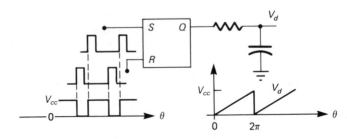

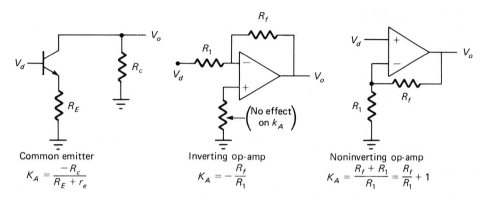

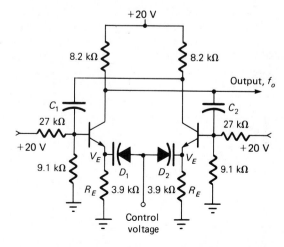

FIGURE 10–11   Operational amplifiers increase PLL loop gain.

of oscillation due to excessive phase shift around the loop, which would produce positive (regenerative) feedback.

Voltage-Controlled Oscillator (VCO)

The voltage-controlled oscillator was studied in its role as a frequency modulator. Various circuits were shown for controlling the frequency of the oscillator. The simplest approach is to use a voltage-variable capacitor (varicap diode) in a multi-vibrator so that when the varicap bias voltage changes, the capacitance changes, thereby changing the oscillator frequency. The frequency of the free-running multi-vibrator circuit of Figure 10–12 is controlled by the variable reactance of D_1 and D_2. In IC implementations, D_1 and D_2 are realized by reverse-biased collector junctions. It should be noted that the control voltage must not exceed $V_E + 0.5$ V $+ v_e$, where v_e is the positive peak of the oscillator signal across R_E and 0.5 V causes forward bias of the silicon diodes. The input-output characteristic for

FIGURE 10–12   Tuning diode control of free-running multivibrator—voltage-controlled oscillator.

FIGURE 10–13　VCO character-istic.

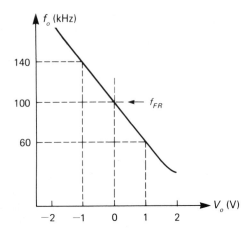

the VCO is shown in Figure 10–13. The VCO should be operated within the linear range between 60 and 140 kHz. Over this range the VCO loop-gain parameter is constant and determined from

$$k_o = \frac{\Delta f_o}{\Delta V_o} \tag{10-5}$$

$$= \frac{(60 - 140)\ \text{kHz}}{[1 - (-1)]\ \text{V}} = -40\ \text{kHz/V}$$

10–2 ■ BASIC LOOP BEHAVIOR

Locking the Loop

The concept of a locked loop is best understood from a description of what you would experience in the electronics lab with a generator and an oscilloscope.

Start with switch S_1 open (Figure 10–14) and a signal generator with frequency f_i connected to the input. With f_i not equal to the free-running frequency (f_{FR}), the

FIGURE 10–14　PLL block dia-gram.

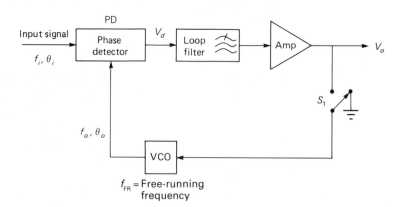

FIGURE 10–15 Beat-frequency output at V_o with loop open. The input generator frequency is being varied from $f_i < f_{FR}$ to $f_i > f_{FR}$.

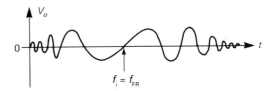

phase detector will produce the sum and difference frequencies. The loop (low-pass) filter filters out the sum frequency ($f_i + f_{FR}$), f_i, and f_{FR}, but the difference ($f_i - f_{FR}$)—the beat between the signal generator and VCO—is allowed to pass through. The beat is amplified and seen as V_o on an oscilloscope. As the generator frequency is varied to bring f_i closer to f_{FR}, the beat frequency gets lower and lower, and if you have a steady enough hand on the generator control, you will see V_o on the scope as a voltage that you can vary from $+V$, through 0, to $-V$ and back as the frequencies go through a zero beat ($f_i = f_o$). You have, in fact, traced out the phase detector characteristic amplified by the dc amp and the loop-filter characteristic. This is illustrated in Figure 10–15.

The loop-filter characteristic can be observed by continuing to increase the generator frequency and noticing that the *amplitude* of the beat decreases at higher (beat) frequencies; this "rolloff" will be observed when f_i is above or below f_{FR}.

Acquisition

In Figure 10–14, with the VCO input grounded and $V_o = 0$, measurements will show that $f_i = f_{FR}$. However, if $f_i \neq f_{FR}$, then the beat is observed at V_o. When the switch is closed, the beat-frequency signal at V_o will cause the VCO frequency f_o to change. If the voltage is large enough (high loop gain) and the filter bandwidth wide enough, then the VCO will be deviated from f_{FR} and lock at the instant that $f_o = f_i$. The amount by which the VCO frequency must be changed is $\Delta f = f_i - f_{FR}$. The time required for the loop to lock depends on the type of loop and loop dynamics. For the simplest PLL with no loop filter, this *acquisition time* is on the order of $1/k_v$ seconds—where k_v is defined in Equation 10–8. Also, the range of f_i over which the loop will lock, the *lock range*, is equal to the hold-in range for the simple PLL.

Locked Loop: The Tracking Mode

When the loop is locked we know that $f_o = f_i$. This is the unique feature of the phase-locked PLL: there is *no* frequency error. Only a phase difference between the signal and the VCO can exist. This phase difference $\theta_i - \theta_o = \theta_e$ is called the *static* (dc) *phase error*. θ_e is the input to the phase detector when the loop is locked and is required in order for the phase detector to produce a dc output voltage V_d that, when amplified by the dc amplifier, will produce exactly enough V_o to keep the VCO frequency deviated by Δf. If f_i increases, then Δf increases and θ_e must increase in order to provide for more V_o to keep the VCO tracking f_i.

The definition of *locked* is that $f_i = f_o$ and the loop will track any change in f_i. Any subsequent shift of θ_i or θ_o will be tracked out so that only θ_e, as defined by Equation 10–7, remains.

Loop Gain and Static Phase Error

The locked PLL is seen in Figure 10–16, where the loop filter is shown with dashed lines because the static condition holds and the loop filter has a gain of unity. Each block in the loop has its own gain parameter. The phase comparator develops an output voltage V_d in response to a phase difference between the reference input and the VCO. The transfer gain k_ϕ has units of volts/radian of phase difference. The amplifier shown is wideband with a voltage gain of k_A volts/volt (dimensionless). Thus, $V_o = k_A V_d$.

The VCO free-running frequency is f_{FR}. The VCO frequency f_o will change in response to an input voltage change. The transfer gain k_o has units of hertz/volt. At this point in the development, the only function of the loop filter is to eliminate the high frequencies, namely, f_i, f_o, and the sum of these. The *loop gain* for this system, like that for any other feedback system, is simply the gain of each block multiplied around the loop; thus

$$k_L = k_\phi \cdot k_A \cdot k_o \tag{10–6}$$

The units of k_L are (V/rad · V/V · Hz/V) = Hz/rad.

Assume that a signal with frequency f_i is an input to the phase detector, and the loop is locked. If the frequency difference before lock was $\Delta f = f_i - f_{FR}$, then a voltage $V_o = \Delta f/k_o$ is required to keep the VCO frequency equal to f_i. So the phase comparator must produce $V_d = V_o/k_A = \Delta f/(k_o k_A)$, and the static phase error $\theta_e = \theta_i - \theta_o$ must be $\theta_e = V_d/k_\phi$. Combining gives $\theta_e = \Delta f/(k_o k_A k_\phi) = \Delta f/k_L$. This is a fundamental equation for the PLL in phase lock;

$$\theta_e = \frac{\Delta f}{k_L} \tag{10–7}$$

In many computations the loop gain must be in units of radians/second rather than

FIGURE 10–16 PLL in tracking mode (locked).

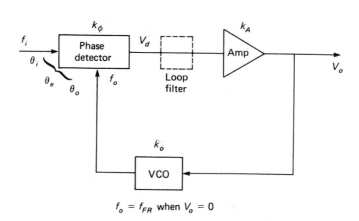

$f_o = f_{FR}$ when $V_o = 0$

in hertz/radian. The conversion is made using 2π rad/cycle. Hence, loop gain is also given by

$$k_v = 2\pi k_\phi k_A k_o \qquad (10\text{–}8)$$

in units of second^{-1}, or radians/second.

Hold-In Range

The range of frequencies for f_i over which the loop can maintain lock is called *hold-in* range. Assuming that the amplifier does not saturate and the VCO has a wide frequency range, the phase detector characteristic limits the hold-in range. It should be clear from the phase-detector characteristics (Figures 10–6 and 10–7) that, as the static phase error increases due to increasing f_i, a limit for V_d is reached beyond which the phase detector cannot supply more voltage for VCO correction. The phase detector simply cannot produce more than A volts. The total range of V_d is $\pm A = 2A$, so that the total range of θ_e is π rad. From Equation 10–7, the minimum to maximum input frequency range, $f_i(\text{max}) - f_i(\text{min}) = \Delta f_H$, will be

$$\Delta f_H = \pi k_L$$

or

$$\Delta f_H = \frac{k_v}{2} \qquad (10\text{–}9)$$

The edge-triggered RS flipflop phase comparator of Figure 10–9 can provide twice this, $\Delta f_H = k_V$. Let's stop here and work through an example that will demonstrate quantitatively what has been covered so far.

■ ■ ■ ■

EXAMPLE 10–1

Figure 10–17 provides enough information to analyze the static behavior of a phase-locked loop.

1. Determine k_A for the op-amp.
2. Calculate the loop gain in units of seconds^{-1} and in decibels (at $\omega = 1$ rad/s).
3. With S_1 open as shown, what is observed at V_o with an oscilloscope?
4. When the loop is closed and phase-locked, determine
    a. the VCO output frequency.
    b. The static phase error at the phase comparator output.
    c. V_o. (Is this rms, pk-pk, or what?)
5. Determine the hold-in range Δf_H.
6. Determine A, the maximum value of V_d.

Solution:

1. $k_A = (R_f/R_1) + 1 = 4$ kΩ/1 kΩ $+ 1 = $ **5.**
2. $k_L = k_\phi k_A k_o = 0.1$ V/rad $\times 5 \times -30$ kHz/V $= -15 \times 10^3$(Hz/rad). Then, $k_v = 15 \times 10^3$ cycles/s-rad $\times (2\pi$ rad/cycle$) = $ **94,300 s^{-1},** and k_v(dB) $= 20$ log $k_v = 20$ log$(94.3 \times 10^3) = $ **99.5 dB** at 1 rad/s.

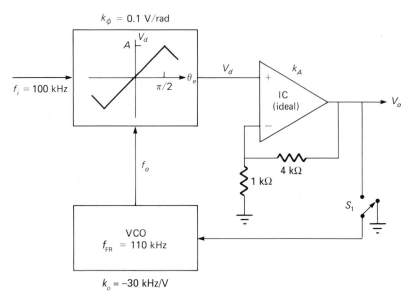

FIGURE 10–17   Example PLL.

3. V_0 will be a sinusoidally varying voltage with a frequency of $|f_i - f_{FR}| = 10$ **kHz.** This assumes that a very small capacitor internal to the phase comparator filters out f_o, f_i, and $f_o + f_i$.

4. **a.** When the loop is locked, $f_o = f_i = 100$ kHz by the definition of locked, and only a phase difference can exist between the input signal and VCO. This phase difference θ_e is the loop-error signal (static phase error), which results in V_d at the detector output and, when amplified by k_A, provides enough voltage V_o to make the VCO frequency be exactly equal to f_i.

   b. The free-running frequency of the VCO is 110 kHz. In order for the VCO to equal 100 kHz, the VCO input voltage must be $V_o = (100\ \text{kHz} - 110\ \text{kHz})/k_o = -10\ \text{kHz}/-30\ \text{kHz/V} = $ **0.33 V dc.** Then, because $k_A = 5$, V_d must be $V_d = 0.33\ \text{V}/5 = 0.0667$ V. Finally $\theta_e = V_d/k_\phi = 0.0667\ \text{V}/0.1$ V/rad = **0.667 rad.**

   Once again, we have derived the basic relationship, $\theta_e = \Delta f/k_L = (f_i - f_{FR})/k_L = -10\ \text{kHz}/(-15 \times 10^3\ \text{Hz/rad}) = $ **0.667 rad.**

   c. V_o was calculated in (b), but let's do it another way. The input to the phase detector (loop-locked) was determined from $\theta_e = \Delta f/k_L = 0.667$ rad. Since $V_d = k_\phi \theta_e$, we have $V_d = 0.1$ V/rad × 0.667 rad = 0.0667 V dc. Now, we are assuming Z_i of the op-amp is much larger than R of the loop filter, so there is no voltage drop across R. The input to the op-amp is 0.0667 V dc, so that $V_o = k_A V_d = 5 \times 0.0667$ V dc = **0.33 V dc.** This is enough to keep the VCO at 100 kHz when, in fact, its rest frequency is 110 kHz.

5. The question is, when the loop is locked, how much can f_i change in frequency before the loop just cannot provide enough V_o to keep the VCO at $f_o = f_i$? Assuming that the VCO and dc amplifier don't saturate, we look at the phase

detector characteristic. Clearly V_d can increase with θ_e until $V_d \rightarrow V_{max} = A$, at which point $\theta_e = \pi/2$. Beyond this, V_d decreases for increasing static phase error, and the phase detector simply cannot provide more output voltage to continue increasing f_o, and the loop breaks lock. The total hold-in range is $\pm\pi/2$, or π rad. The frequency difference between these break-lock points will be $\Delta f_H = \theta_e(max) \times k_L = \pi \times 15$ kHz/rad = **47.1 kHz.**

6. At the frequency where $\theta_e = \pi/2$, we have $V_{d(max)} = A$. Therefore, $V_d = k_\phi\theta_e = 0.1$ V/rad $\times \pi/2$ rad = **0.157 V dc.**

10-3 ■ LOOP FREQUENCY RESPONSE AND BANDWIDTH

In previous sections steady-state behaviors of the PLL and its individual components were described. In this section, the frequency response and bandwidth for the locked-loop condition is discussed in order to complete the background necessary to describe PLL applications such as analog FM demodulators and others. This discussion is also a necessary prerequisite for the development of concepts such as loop dynamic behavior and system stability.

Frequency Response of Individual Loop Components

All individual loop components discussed so far are dc-coupled except for the phase detector input and VCO output. The bandwidth of a well-designed phase detector with high-frequency diodes or transistors for switching can be extremely wide. Commercially available detectors have a bandwidth (overall) in excess of 500 MHz, so the real frequency limitation will be the R_oC low-pass circuit at the phase detector output; this is used to filter out the two input signals and the mixer sum (f_i, f_o, and $f_o + f_i$). The R_oC corner frequency is designed to be well above the loop bandwidth (which is defined later).

The bandwidth of the loop amplifier must be much wider than the loop bandwidth. This requirement often precludes the use of such popular ICs as the 741 operational amplifier with its 2-MHz gain-bandwidth product. The reason for use of wideband detectors and amplifiers is loop stability. Each high-frequency corner (RC time constant—also called a *pole*) can increase the negative-feedback phase delay by up to 90°. Furthermore, each pole within or near the loop bandwidth can increase the "order" of the loop, as will be demonstrated in Section 10-6, "Improving Loop Response."

The VCO frequency response is the primary component limiting the loop bandwidth. You should remember from the study of the VCO in its use as an FM modulator that the modulation index is inversely proportional to the frequency at which the input voltage varies (f_m); that is, $m_f \propto 1/f_m$. Hence for sinusoidal variations of VCO inputs $V_o(t)$, the loop gain is effectively reduced as the frequency of $V_o(t)$ increases. This gain rolloff with frequency is characteristic of integrators and is sketched in Figure 10-18.

FIGURE 10–18   VCO gain frequency response, $k_o(f)$.

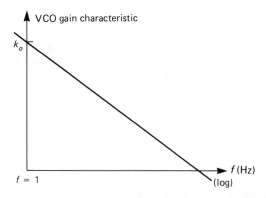

   The derivation of the VCO as an integrator in the PLL is now given. A sinusoidal VCO output, like any sinusoid, can be written as

$$e(t) = E \sin[\theta_o(t)]$$

where $\theta_o(t)$ is the instantaneous phase at any time t. The VCO frequency at any instant is the rate of change of phase; that is,

$$f_o = \frac{1}{2\pi} \frac{d\theta_o(t)}{dt} \qquad\qquad (10\text{--}10)$$

Since we can measure the VCO frequency and how it varies with the input control voltage V_o, we can determine how the VCO phase varies with V_o. From the preceding, observe that $d\theta_o(t) = 2\pi f_o\, dt$; integrating both sides gives

$$\theta_o(t) = 2\pi \int f_o\, dt$$

If the frequency doesn't change, then the integral is simply

$$\theta_o(t) = 2\pi f_o t + \theta_o$$

where θ_o, the integration constant, represents any initial phase offset at $t = 0$. Hence the VCO output becomes the familiar

$$e(t) = E \sin(2\pi f_o t + \theta_o)$$

   If the VCO frequency changes due to a changing input control voltage $V_o(t)$, then for a linear VCO characteristic, the change is

$$\Delta f_o = k_o V_o(t)$$

and the instantaneous frequency is

$$f(t) = f_o + k_o V_o(t)$$

The VCO phase at any instant is, therefore,

$$\theta_o(t) = 2\pi f_o t + 2\pi k_o \int V_o(t)\, dt + \theta_o \qquad\qquad (10\text{--}11)$$

In terms of phase-locked loop behavior, it is clear from this derivation that the VCO integrates the input voltage waveform $V_o(t)$. Indeed, if the VCO input is sinusoidal,

$$V_o(t) = V_o\cos 2\pi ft$$

then the variation of $\theta_o(t)$ is

$$2\pi k_o \int V_o\cos 2\pi ft \; dt = \left(\frac{k_o}{f}\right) V_o\sin 2\pi ft$$

$$= \left(\frac{k_o}{f}\right) V_o(t)\underline{/-90°} \qquad \textbf{(10–12)}$$

This shows that the VCO's ability to control the loop phase response is inversely proportional to the frequency of V_o, and the 90° lag at all frequencies indicates a pure integrator (pole at $f = 0$). The loop frequency response of the VCO is sketched in Figure 10–18, where its gain is theoretically infinite at $f = 0$ but is k_o at $f = 1$ Hz or $2\pi k_o$ at $\omega = 1$ rad/s.

PLL System Response

In the last section the frequency response of each component in the loop was discussed except for the loop compensation filter (see Figure 10–19). The phase detector is broadband relative to the frequencies (rates of change) within the loop. The cutoff frequency of the phase detector is determined by its output resistance and the value of the bypass capacitor. This cutoff frequency is high for reasonably high VCO frequencies. Also, a very broadband amplifier is assumed. Given these conditions, then the loop frequency response will be limited by the VCO and its integrating effect. For a phase-locked loop with no compensation filter (uncompensated loop), the overall open-loop frequency response is determined by the product of the individual component responses. The Bode plot for this loop versus

FIGURE 10–19

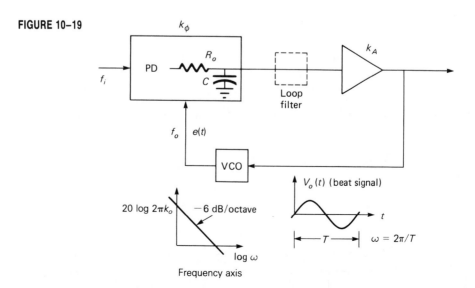

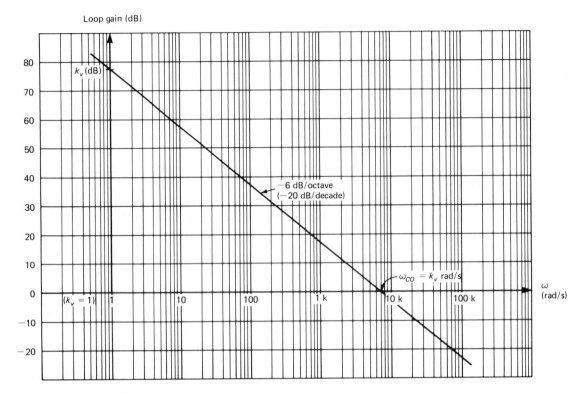

FIGURE 10-20 Open-loop frequency response (Bode) plot, $k_v(\omega)$.

the log of frequency ω is sketched in Figure 10–20. The loop gain in decibels is derived from the product $2\pi k_\phi k_A k_o$ at $\omega = 1$ rad/s, and the loop gain decreases as $1/\omega$—that is, at -6 dB/octave (-20 dB/decade). Because of the one-to-one decrease in loop gain with frequency, the loop gain will be 0 dB ($k_v = 1$) at a frequency of k_v rad/s.

The frequency where the open-loop gain crosses through 0 dB is referred to as the loop *crossover* frequency ω_{CO}. For the uncompensated loop,* the crossover frequency is equal to k_v, that is,

$$\omega_{CO} = k_v \qquad (10\text{–}13)$$

Closed-Loop Frequency Response and Bandwidth

The closed-loop Bode plot for the uncompensated PLL is shown in Figure 10–21. Notice that when the system is operating closed-loop, the system gain is unity (0

---

* The uncompensated PLL is sometimes referred to as the "natural" loop response because no circuit components other than the VCO, PD, and perhaps an amplifier are included in the "natural" loop. This term will not be used, however, in order to avoid confusion with the "natural" loop frequency of a second-order PLL.

FIGURE 10-21 Open- and closed-loop frequency response for uncompensated PLL.

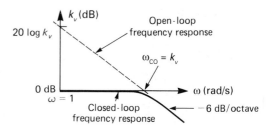

dB) up to frequencies approaching ω_{CO}, but it then decreases at the open-loop rate of 6 dB/octave; that is, unlike most negative feedback amplifier systems* that you have studied, a closed-loop PLL always has 100% feedback.

As illustrated in Figure 10-21 for the PLL of Figure 10-19, the closed-loop Bode plot is flat for low frequencies in the loop, is down 3 dB at ω_{CO}, and then decreases at 6 dB/octave for high loop signal frequencies. Hence, for the uncompensated PLL, the bandwidth (one-sided) is

$$\text{BW} = \omega_{CO} = k_v \qquad (10\text{--}14)$$

This is the one-sided bandwidth because the PLL can track input signal frequencies that are above and below f_{FR} of the VCO. The closed-loop Bode plot of the uncompensated PLL to input frequency deviations is thus seen to be identical to that of a low-pass filter (LPF). In point of fact, the PLL is a "tracking filter," and the VCO will track variations in the input frequency f_i; these input frequency variations produce a varying V_o with internal-loop frequency ω.

So the uncompensated PLL, in which all corners from loop components are well outside of the loop bandwidth k_v, responds to input frequency changes like that of a first-order low-pass RC filter and we can write the closed-loop gain versus frequency as

$$k_{CL} = \frac{1}{1 + j\omega/\omega_{CO}} \qquad (10\text{--}15)$$

The input frequency-to-output voltage transfer function is derived in Appendix D. Equation D-3, repeated here, is

$$\frac{V_o}{f_i} = \frac{2\pi k_\phi k_A}{j\omega + k_\phi k_A k_o} = \frac{2\pi k_\phi k_A / k_\phi k_A k_o}{\dfrac{j\omega}{k_v} + 1}$$

$$= \frac{2\pi/k_o}{1 + j\dfrac{\omega}{\omega_{CO}}} \qquad (10\text{--}16)$$

where $\omega_{CO} = k_v$ and f_i is the input signal (carrier) frequency, whereas ω is the rad/s frequency of phase detector outputs (demodulation signals).

---

* Except for the voltage-follower op-amp.

Derivations of the equations for the frequency response of first- and second-order PLLs are given in Appendix D and will be helpful before the following graphical description of the conclusions from that analysis. Incidentally, the conclusions are nicely confirmed using a 565 integrated circuit PLL in the circuit of Figure 10–35.

■ ■ ■ ■
EXAMPLE 10–2

Sketch the open-loop and closed-loop Bode plots for the loop of Example 10–1. Make the sketch on semilog graph paper with frequency in radians/second on the log axis.

Solution:

1. Since $k_\phi = 0.1$ V/rad, $k_A = 5$, and $k_0 = -30$ kHz/V, the loop gain is $k_v = 2\pi(15$ kHz/V$) = 94,300$ s^{-1}. Then k_v(dB) $= 20\log 94.3 \times 10^3 = 99.5$ dB at $\omega = 1$ rad/s. The open-loop gain, sketched in Figure 10–22, is seen to intersect $\omega = 1$ rad/s at 99.5 dB, decreasing at 20 dB/decade (6 dB/octave) and passing through 0 dB at $\omega_{CO} = k_v = 94,300$ rad/s.

2. The closed-loop gain is also plotted in Figure 10–22. Notice that the system gain is 0 dB for low ω, -3 dB at $\omega = k_v$, and follows the open-loop gain curve at 6 dB/octave for high frequencies.

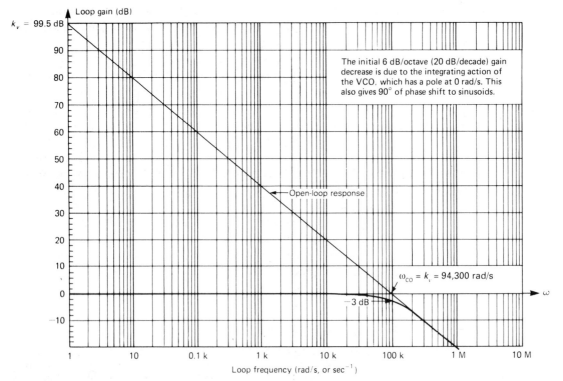

FIGURE 10–22   Frequency response Bode plots.

10-4 ■ FM AND FSK APPLICATIONS OF PLLs

When a PLL has locked to an input signal, the VCO will follow slow changes in the input signal frequency f_i. Suppose f_i increases by an amount Δf_i. In order for the loop to remain locked ($f_o = f_i$), the VCO voltage must increase by $\Delta V_o = \Delta f_i/k_o$. This voltage change is produced by the amplified change in V_d, which is produced by an increased phase difference, $\Delta\theta_e = 2\pi\Delta f_i/k_v$.

As a specific example, suppose that an FM signal with carrier frequency f_i is modulated to an index of $m_f = 4$ by a 1-kHz sinusoid. From Chapter 9 we know that the carrier frequency will be deviated above and below f_i by an amount $f_i = m_f f_m = 4 \times 1\text{ kHz} = 4\text{ kHz}$ pk. If this FM signal is the input to a PLL with a VCO gain of $k_o = 10\text{ kHz/V}$ and loop bandwidth much greater than 1 kHz, then the VCO input voltage V_o will be a 1-kHz sinusoid with a peak amplitude of $\Delta V_o = \Delta f_i/k_o = (4\text{ kHz pk})/(10\text{ kHz/V}) = 400\text{ mV}$ pk. Incidentally, this 400-mV (peak) sinusoid will be centered at a voltage determined by the difference between the VCO free-running frequency f_{FR} and the input carrier frequency f_i as follows: $V_o = (f_i - f_{FR})/k_o$. Care must be taken to ensure that the peak frequency deviations do not exceed the loop hold-in range or the loop will break lock on peak excursions. The result will *not* be simple peak clipping but will be a partial sinusoid with extreme distortion where the loop breaks lock and slips* cycles until the input frequency returns to within the lock range of the loop. For the uncompensated PLL, the lock range is the same as the hold-in range.

If the modulation rate f_m is increased, the demodulated output from the PLL will be attenuated and phase-shifted just as if it had passed through a low-pass filter with cutoff frequency $f_{CO} = k_v/2\pi$.

Frequency-shift key (FSK) is the name for FM of an analog carrier by digital binary (two-state) pulses. The carrier frequency is instantaneously keyed from f_S, called *SPACE*, to f_M, called *MARK*. As discussed in Chapter 13, f_S is usually a higher frequency than f_M and is transmitted to convey a digital 0. The rate at which the carrier frequency is switched between f_S and f_M is the same as the digital data rate (in bits/second) and is referred to as the *baud* rate, after J. M. E. Baudot.

The PLL can be used for digitally modulating FSK (using the VCO) and for demodulating FSK. Hence, the PLL can be used as an *FSK modem*—modulator/ demodulator.

As previously discussed, an uncompensated PLL behaves as a low-pass filter to input frequency shifts. Consequently, only low baud rate FSK signals will be accurately demodulated by an uncompensated PLL because the loop will be slow to respond to the input step-change of carrier frequency. Indeed, the time constant for the voltage (V_o) is given by the inverse of the loop bandwidth,

$$\tau = \frac{1}{\omega_{CO}} = \frac{1}{k_v} \tag{10-17}$$

in seconds. An example of this is given as Example 10-3.

---

* The phase-detector characteristic (Figure 10–5, for example) repeats every cycle (2π rad). Also, the loop cannot lock between $\pi/2 < \theta_e < 3\pi/2$ because the loop will have the wrong phase (regenerative). Consequently, when the loop breaks lock due to a transient or other excessive input change, the system must slip one or more full cycles before relocking in a stable mode. This is called *cycle slipping*.

■ ■ ■ ■

**EXAMPLE
10–3**

A PLL with $k_o = -0.75$ kHz/V, $f_{FR} = 3.5$ kHz, $k_\phi = 0.3184$ V/rad, and $k_A = 5$ is used as an FSK demodulator. The input signal has $f_S = 4$ kHz, $f_M = 2$ kHz, and the modulation is shown in Figure 10–23a. As seen, the baud rate is 1333 bits/s and the data is $\cdots 1\ 0\ 0\ \cdots$.

Sketch accurately the PLL output $V_o(t)$.

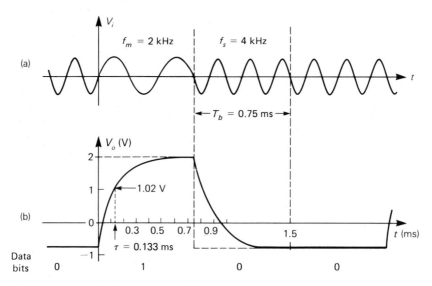

FIGURE 10-23   FSK input (a) and demodulated output (b) of PLL.

Solution:

For $f_i = f_M = 2$ kHz, $V_o = \Delta f_o / k_o = (f_i - f_{FR})/k_o = (2$ kHz $- 3.5$ kHz$)/(-0.75$ kHz/V$) = +2$ **V**. For $f_i = f_S = 4$ kHz, $V_o = (4$ kHz $- 3.5$ kHz$)/(-0.75$ kHz/V$) = -0.67$ **V**. The loop time constant is $\tau = 1/k_v = 1/[(0.75$ kHz/V$) (0.3184$ V/rad$)(5)(2\pi$ rad/cycle$)] = 1/7502 = \tau = 0.133$ ms. It takes 0.133 ms for V_o to rise from -0.67 V to 63% of the total voltage range $2 - (-0.67) = 2.67$ V. 63% of 2.67 V is 1.69 V, so at time τ, $V_o = 1.69$ V $- 0.67$ V $= 1.02$ V (see the plot of V_o in Figure 10–23b).

10-5 ■ NOISE

The problem of noise in phase-locked loops will not be analyzed in any detail here, but a few comments will perhaps spawn interest in further study.

The *noise margin* or maximum allowable peak noise deviation is determined by the PLL's ability to remain locked. If a noisy input signal has frequency or phase deviations that exceed the hold-in range of the loop, the loop will break lock. The uncompensated (first-order) loop will reacquire quickly after the noise transient diminishes, but a great deal of distortion will be introduced if the PLL is being used as a demodulator.

To get a quantitative idea of the loop noise margin, consider the results of Example 10–3 as seen in Figure 10–23b. The output voltage V_o is 2 V for a transmitted MARK. How high can V_o rise on a noise transient caused by a deviation of the MARK frequency or circuit variations of V_o before the loop breaks lock? The static phase error when $f_i = f_M = 2$ kHz is $\theta_e = \Delta f/k_L = (2 \text{ kHz} - 3.5 \text{ kHz})/(1.19 \text{ kHz/rad}) = 1.26$ rad. However, for typical phase detectors, the loop will break lock if θ_e exceeds $\pi/2 = 1.57$ rad. Consequently loop transients that would cause θ_e to increase by 0.31 rad will result in a loss of lock—temporarily, one hopes. In terms of voltages, $V_d(\text{max}) = k_\phi\theta_e(\text{max}) = (0.3184 \text{ V/rad}) \times (1.57 \text{ rad}) = 0.5$ V and $V_o(\text{max}) = 5 \times 0.5 = 2.5$ V max. Since, as seen in Figure 10–23b, $V_o(\text{MARK}) = 2$ V and we have calculated $V_o(\text{max}) = 2.5$ V, we see that the noise margin for V_o will be $V_o(\text{NM}) = 0.5$ V (peak). This can result from noise in the PLL itself, from the noise input signal amplitude if no limiter precedes the PLL, or from an input signal frequency deviation (due to noise) of $\Delta f_i(\text{NM}) = k_o \times V_o(\text{NM}) = (0.75 \text{ kHz/V})(0.5 \text{ V pk}) = 375$ Hz, peak noise.

10–6 ■ IMPROVING LOOP RESPONSE

The results of the FSK demodulation example (see Figure 10–23) point out the disadvantage of an uncompensated PLL. The response is that of a first-order RC low-pass filter. For many applications, including the FSK demodulator, the uncompensated loop is too slow, and the response to a sudden change in input frequency (step response) produces a very rounded V_o.

The loop response can be improved so that V_o will have a shorter rise time. This is accomplished by making the loop slightly regenerative and is called *loop compensation*.

The simplest technique for compensating a PLL is to insert a phase-delay network into the loop. The PLL of Figure 10–24 shows an RC lag network inserted in the loop between the phase detector and amplifier. The RC time constant is chosen for optimum loop response in a given system application.

The result of adding an RC integrator to a negative feedback loop that already has an integrator (the VCO) is to produce a second-order system. A second-order system with which you are already familiar is shown in Figure 10–25. This circuit consists of a first-order RC network with series inductor added to produce a second-order network. If R is not too large, the circuit has a series resonance at a frequency $f_o = 1/(2\pi\sqrt{LC})$—called the *undamped natural frequency*—and as illustrated in Figure 10–25, a sudden step-input-voltage change will cause an initially sluggish V_o to rise rapidly, overshoot, and ring. The ringing will eventually dampen out after a time determined by the circuit Q, where $Q = 2\pi f_o L/R$. Second-order systems are also characterized by a quantity called the *damping factor* (or damping ratio) δ, related to Q by

$$\delta = \frac{1}{2Q} \tag{10–18}$$

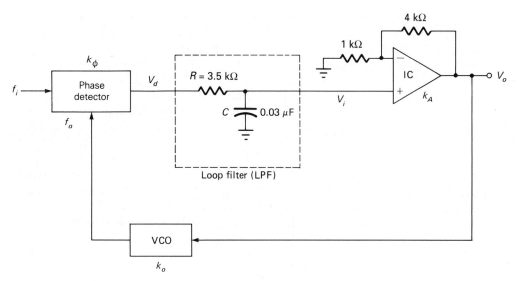

FIGURE 10-24 PLL with RC (lagging phase) compensation network.

FIGURE 10-25 Step response in underdamped second-order circuit (RLC low-pass).

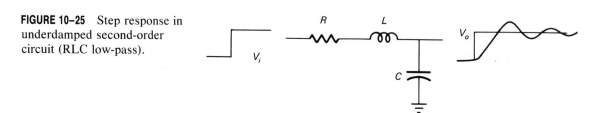

If R is large enough so that $\delta > 1$, there will be no ringing, but the rise time will be poorer than the first-order circuit. A damping factor of unity is the critical damping value; for values greater than this, the system is *overdamped*.

Second-order feedback systems such as the PLL shown in Figure 10–24 must be carefully designed to produce an appropriate dynamic response to a sudden input change. For example, an underdamped FSK demodulator can overshoot the desired output voltage value and ring violently during each digital binary transition. Such a response would make the PLL system useless for this application. On the other hand, if the values of R and C in the loop filter are chosen properly, the compensated PLL will be designed to rise rapidly, have no overshoot, and settle at the final steady-state voltage without ringing, thereby producing a much more satisfactory demodulator than the uncompensated PLL.

Determining Loop Stability and Dynamic Behavior Using the Bode Plot

The sinusoidal frequency response plot (Bode plot) yields a great deal of information for determining the dynamic behavior of feedback systems. For instance, the open-loop Bode plot of the uncompensated first-order PLL of Example 10–3 (Figure 10–26) shows a single 6-dB/octave slope, and the crossover frequency is

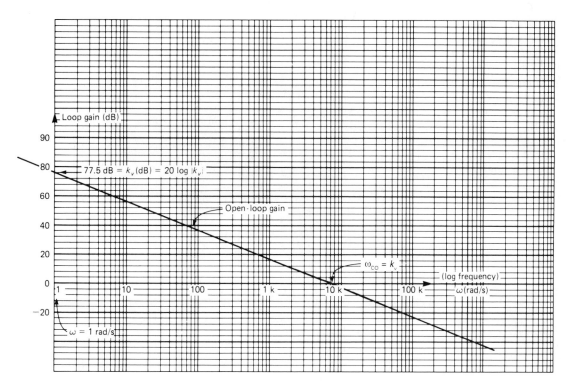

FIGURE 10-26

the same as $\omega_{CO} = k_v = 7.5$ k rad/s. The corner responsible for the 6-dB/octave rolloff is at $f = 0$ and is due to the integrating effect of the VCO. Such a pure integrator also causes a 90° phase lag.

A system with negative feedback has 180° of phase delay built into the loop. If any of the circuits in the loop, such as a loop compensation network, increase phase shift close to 180° at frequencies for which positive loop gain exists, the loop will have marginal stability and be regenerative. If the additional phase shift is greater than or equal to 180° within the loop bandwidth, the system is unstable and will oscillate.

The uncompensated PLL, with its 90° additional phase delay due to the integrating effect of the VCO, will have a phase margin of 180° − 90° = +90° and is not at all regenerative, as seen by the step response of Figure 10–23. However, a loop-compensating lag network will produce additional phase delay, and the phase margin can be considerably less than 90°. Indeed, reducing the phase margin is precisely what the compensating network is designed to do. However, the phase margin must not be allowed to be less than about 40° for most applications or the system will be too underdamped and ring excessively after sudden input changes.

Figure 10–27 shows the overshoot of the final (steady-state) value $V_o(SS)$, and ringing for various values of loop-phase margin.

FIGURE 10–27 Second-order loop response to step-change of input frequency. $V_o(SS) = \Delta f/k_o$ is the final (steady-state) value of VCO input voltage. The loop-damping factor is δ, P.M. is loop phase margin based on the Bode asymptotic approximation, and ω_n is the natural undamped frequency of the loop in radians/second.

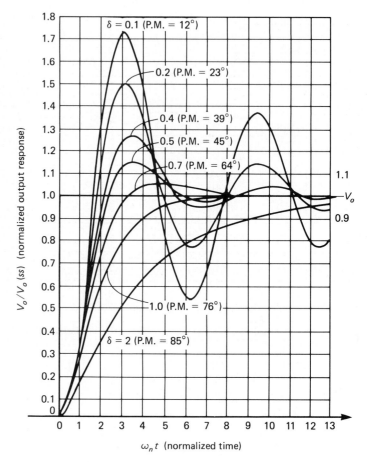

Phase Margin

Phase margin (P.M.) is determined at the loop crossover frequency ($k_v = 1$, that is, 0 dB) and is the maximum amount by which the total loop phase shift could be increased before the onset of oscillations. Hence P.M. is a measure of loop stability; Figure 10–27 shows the extent of overshoot and ringing for a loop with known phase margin.

The phase margin is computed for a second-order loop with lag compensation as follows:

$$\text{P.M.} = 180° - \left[90° + \tan^{-1}\left(\frac{\omega_{CO}}{\omega_c}\right)\right] \qquad \textbf{(10–19)}$$

where $\omega_c = 1/RC$ is the cutoff frequency of the lag network, and ω_{CO} is the loop crossover frequency determined from the Bode plot (see Figure 10–28).

The quantity inside the brackets of Equation 10–19 is the phase shift due to loop components. The 90° is due to the VCO, and the arctangent of ω_{CO}/ω_c gives

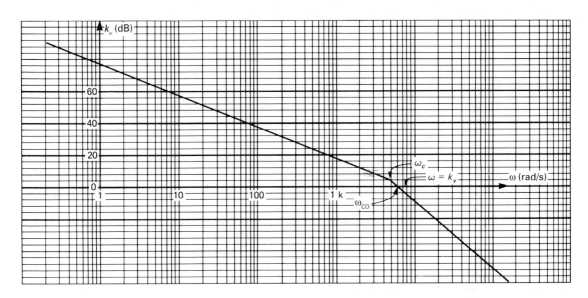

FIGURE 10–28   Open-loop Bode plot, PLL with RC filter lag network.

the phase lag at the loop crossover frequency due to the RC loop compensating filter: Examine related circuit-equation derivatives in Equations 7–58 through 7–61.

■ ■ ■ ■

EXAMPLE 10–4

Determine the phase margin for the loop of Example 10–3 if an RC low-pass filter is inserted between the phase detector and op-amp. $R = 6.8$ kΩ and $C = 0.03$ μF.

Solution:

First make the Bode plot. From Example 10–3, $k_v = 0.75$ kHz/V $\times$ 0.3184 V/rad $\times 5 \times 2\pi$ (rad/cycle) = **7500 s^{-1}**. k_v(dB) = 20 log k_v = 20 log 7.5 $\times 10^3$ = 77.5 dB at 1 rad/s. Also, $\omega_{CO} = k_v =$ **7500 rad/s.** This is plotted in Figure 10–28 on semilog paper. The points k_v (dB) = 77.5 dB and $k_v = 7.5$ k rad/s are joined with a straight line. This gives the Bode plot of the uncompensated loop and is seen to drop at 20 dB/decade (6 dB/octave). Next, $\omega_c = 1/RC = 1/(6.8$ kΩ $\times 0.03 \times 10^{-6}) = 4900$ rad/s is marked on the uncompensated Bode plot, and the loop rolls off at 40 dB/decade (12 dB/octave) from this point. This line passes through the crossover point (k_v(dB) = 0 dB) at approximately $\omega_{CO} =$ **5800 k rad/s.** From Equation 10–19, P.M. = 90° − tan^{-1}(5.8 k/4.9 k) = **40°**. For the second-order loop with P.M. = 40°, Figure 10–27 indicates that $\delta \approx 0.4$. The maximum amount of overshoot (25% or 2 dB) and the ringing are evident from the curve. However, an accurate sketch of the demodulated FSK signal requires working with the time scale, and ω_n has not yet been defined.

Damping and Undamped Natural Frequency

For the second-order PLL with a simple lag network such as that shown in Figure 10–24, the loop damping factor can be determined by sketching the open-loop Bode plot, calculating the phase margin, and getting δ from the curves of Figure 10–27. However, a more direct approach is to use the results of the PLL system analysis of Appendix D. The equations developed for lag compensation yield (Equation D–8)

$$V_o(s) = \frac{\Delta f k_v \omega_c / k_o}{s(s^2 + \omega_c s + k_v \omega_c)}$$

Comparing this to the general second-order system frequency response (Equation D–9),

$$F(s) = \frac{\omega_n^2}{s(s^2 + 2\delta\omega_n s + \omega_n^2)}$$

shows that
$$\omega_n^2 = k_v \omega_c$$

and
$$\delta = \frac{\omega_c}{2\omega_n}$$

from which
$$\delta = 0.5 \sqrt{\frac{\omega_c}{k_v}} \qquad \qquad \textbf{(10–20)}$$

where $\omega_c = 1/RC$ is the cutoff frequency of the loop filter.

It is clear from Equation 10–20 that the loop damping is determined by the location of ω_c with respect to $\omega = k_v$. Indeed, if ω_c is set by R and C to be exactly equal to k_v, then $\delta = 0.5$. Also, to achieve $\delta = 0.707$, which gives optimum damping for most applications, ω_c must be set exactly one octave above k_v—that is, $\omega_c/k_v = 2$.

If the loop phase margin is reduced to zero, the damping factor becomes zero and the loop will oscillate continuously. The frequency of oscillation is called the *undamped natural frequency* (ω_n) and is found to be the geometric mean of ω_c and k_v. Thus, from the preceding development,

$$\omega_n = \sqrt{\omega_c k_v} \qquad \qquad \textbf{(10–21)}$$

This undamped natural frequency has the same meaning as $\omega_0 = 1/\sqrt{LC}$ for a tuned circuit.

Care must be taken in the system design to choose a loop damping factor that prevents the overshoot of V_o from exceeding a value that would cause the loop to break lock. For instance, Figure 10–29 illustrates the results for the demodulator of Example 10–3 if the PLL has a damping factor of $\delta = 0.40$. Note that the initial overshoot has caused V_o to exceed 2.5 V, which was previously determined to be the maximum value of V_o for the hold-in range. Consequently, the loop will break lock and behave in a very complex manner as a combination of nonlinear FM with

FIGURE 10-29 Result of overshoot causing the loop to exceed the hold-in range.

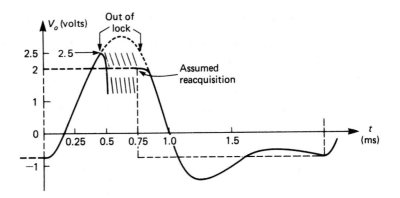

beating at $f = 3.5$ kHz $- 2$ kHz $= 1.5$ kHz and attempts at loop acquisition in an
underdamped second-order system.

The loop filter time constant must be decreased in order to increase the damping factor so that the peak overshoot will not exceed 2.5 V at V_o. Recall from
Example 10–3 that the full step in V_o is from -0.67 V to $+2$ V. With V_o rising to
$+2.5$ V, the overshoot is $(2.67$ V $+ 0.5)/2.67$ V $= 1.19$, or a 19 percent overshoot.
The ringing curves of Figure 10–27 indicate that for $V_o/V_o(\text{SS}) = 1.19$, the damping must be greater than approximately 0.46. Indeed, a damping factor of $\delta = 0.5$
would show V_o rising only to 2.4 V (a 15% overshoot), and therefore the system
will be able to follow the swings of V_o.

Rise in Loop Frequency Response

In addition to the overshoot and ringing for a step change in input frequency, as
seen in Figure 10–27, an underdamped second-order loop also shows its regenerative nature for sinusoidal variations of input frequency.

Figure 10–30 shows a rise peaking near ω_n in the otherwise flat frequency
response for underdamped second-order feedback loops. The peak is found to
occur at a frequency given by

$$\omega = \omega_n\sqrt{1 - 2\delta^2} \qquad (10\text{–}22)$$

and the relative rise of the closed-loop gain reaches a peak value of

$$\frac{1}{2\delta\sqrt{1 - \delta^2}} \qquad (10\text{–}23)$$

Thus, these curves give the output frequency response of a PLL used as an FM
demodulator.

There will be undesirable high-frequency peaking if the loop phase margin is
less than 64°, which corresponds to a damping factor of $\delta = 0.707$.

An example is now provided to demonstrate the use of Figures 10–27 and
10–30 for determining the closed-loop sinusoidal frequency response and dynamic
behavior of a second-order PLL.

FIGURE 10–30 Closed-loop frequency response of second-order loop. Frequency axis is normalized to ω_n, the undamped natural frequency of the loop.

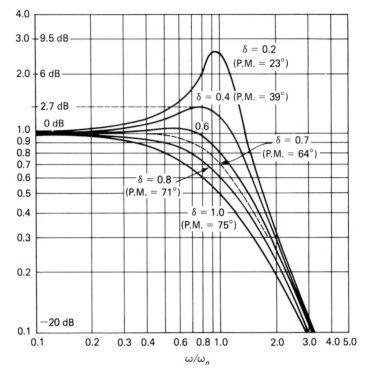

■ ■ ■ ■

**EXAMPLE
10–5**

The loop of Examples 10–3 and 10–4 has $k_o = -0.75$ kHz/V, $k_v = 7500$ s^{-1}, and a loop filter of $R = 6.8$ kΩ, $C = 0.03$ μF ($\omega_c = 4900$ rad/s).

1. Determine the damping factor and the undamped natural frequency (in radians per second and hertz).

2. Plot the open-loop and closed-loop frequency response of this loop used as an FM demodulator.

3. Make an accurate sketch of V_o for input step changes of frequency 375 Hz below and above f_{FR} to show the dynamic behavior for an FSK demodulator.

4. Determine the value of R that will compensate the loop for $\delta = 0.707$, and sketch, using dots, the step response on the same sketch as part 3.

Solution:

1. $\delta = \frac{1}{2}\sqrt{\omega_c/k_v} = 0.5\sqrt{4.9\ \text{k}/7.5\ \text{k}} = \mathbf{0.4.}$ $\omega_n = \sqrt{\omega_c k_v} = \sqrt{(4.9\ \text{k})(7.5\ \text{k})}$ = **6.1 k rad/s.**

2. k_v (dB) = 77.5 dB. Since ω_n is known, the horizontal axis can be denormalized for our specific problem. The peak of the closed-loop response is $20 \log[1/(2\delta\sqrt{1 - \delta^2})] = 2.6$ dB at $\omega = 0.82\omega_n = 5.0$ k rad/s. The closed-loop response showing the peaking is seen in Figure 10–31.

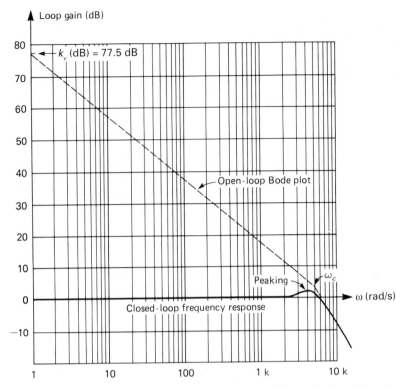

FIGURE 10-31 Frequency response of FSK demodulator showing the peaking due to slightly regenerative loop ($\delta = 0.4$).

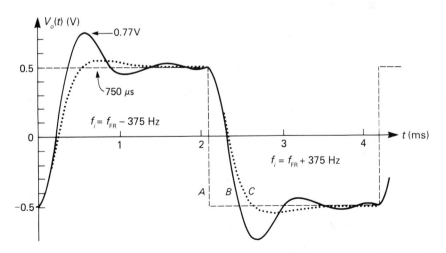

FIGURE 10-32 Response of second-order PLL to input frequency step changes. (a) is ideal, (b) $\delta = 0.4$, and (c) $\delta = 0.707$.

3. For $\Delta f = \pm 0.375$ kHz, $V_o(\text{SS}) = \Delta f / k_o = -0.375/-0.75 = +0.5$ V. The axes of Figure 10–27 are denormalized as follows: *Vertical*—1 is now 0.5 V; therefore, each number on the vertical axis is divided by 2 and read in volts. *Horizontal*— $\omega_n = 6.1$ k rad/s; therefore, $\omega_n t = 1$ gives $t = 1/6100 = 0.164$ ms, and so forth, with other values of $\omega_n t$. The sketch for $\delta = 0.4$ is shown in Figure 10–32 for a 4.5-ms-long, unity duty-cycle (squarewave), FSK signal. The peak of the 0.4 damping factor curve of Figure 10–27 is seen to overshoot the full step value (1 V total) by 27% at $\omega_n t = 3.5$. Hence, V_o will reach 0.77 V (1.27 V − 0.5 V) at $t = 0.57$ ms.

4. The R must be changed to produce a cutoff frequency determined from Equation 10–20 with $\delta = 1/\sqrt{2}$. Hence, $\omega_c = 4\delta^2 k_v = 4(0.707)^2 \times 7500$ rad/s = 15 k rad/s. Since $\omega_c = 1/RC$, then $R = 1/(15 \times 10^3) \times (0.03 \times 10^{-6}) = 2.2$ kΩ. The resultant sketch of $V_o(t)$ for $\delta = 0.707$ is shown dotted in Figure 10–32.

Switching and Settling Time

The results of Example 10–5 (Figure 10–32) show the overshoot and ringing that occur for the PLL compensated for $\delta = 0.4$ and 0.7. The greater rise time associated with a small damping factor makes underdamping attractive for some applications. However, most applications in which the PLL is switched from one condition to the next will also have a required maximum settling time.

The *settling time* is the time it takes for the system to settle to within a specified percentage of the final steady-state value. For example, suppose that the application for which the PLL of Example 10–5 is used requires that the system must be 0.5 V ± 10% within 750 µs of the input frequency transition. The plot of $V_o(t)$ in Figure 10–32 shows that, although the $\delta = 0.4$ response rises to 450 mV in about 320 µs (compared to 475 µs for the $\delta = 0.7$ response), the underdamped PLL will then exceed the upper specification limit of 550 mV, overshoot all the way to 770 mV, and not be settled to within 500 mV + 50 mV until 1.1 ms after the input switching instant. Meanwhile the $\delta = 0.7$ response does not exceed +50 mV at any time after 475 µs.

Clearly, the PLL with $\delta = 0.7$ will easily meet the specifications, but $\delta = 0.4$ is too lightly damped and the R or C of the loop compensation network must be decreased in value to improve the damping.

Use of the universal second-order response ringing curves of Figure 10–27 will allow the determination of the range of damping factors which will be suitable to meet overshoot and settling-time requirements. Use Figure 10–27 for Example 10–5 to show that the damping factor must be greater than 0.6. Also notice that an overdamped loop may not meet specifications either. Figure 10–27 will show that $\delta = 1.0$ will just meet the settling-time requirement, but loops more overdamped will be too slow. If the 0.03-µF capacitor of Example 10–5 ($k_v = 7500$ rad/s) is used for compensation and Equation 10–20 is combined with $\omega_c = 1/RC$, then R is determined to be between 1100 and 3100 in order to meet the settling-time requirements.

10-7 ■ PHASE-LOCKED LOOP COMPENSATION

Improving the response of PLLs has been discussed for the technique called *lag compensation*. The effect of lag compensation was to produce a second-order system whose phase margin could be controlled with the *RC* time constant (pole compensation). There is a fundamental limitation to lag compensation because, for some applications, not enough system characteristics can be defined independently; that is, if the loop gain and damping factor (overshoot and ringing characteristics) are defined, then the loop bandwidth is out of our control—it cannot be set independently of k_v and δ.

There is a loop compensation technique called *lead-lag compensation*, which permits the independent control of all three characteristics, k_v, δ, and loop bandwidth.

Lead-Lag Compensation

Lead-lag compensation allows us to control all the loop parameters—k_v, δ, and bandwidth B—independently. This is necessary in applications such as demodulators, tracking filters, and coherent receivers,* where the noise bandwidth as well as the loop response are important. Since most applications call for $\delta = 0.707$, a simple graphical solution is given for determining the values of R_1, R_2, and C in the lead-lag network shown in Figure 10–33. The complete design procedure is as follows:

1. Determine the loop gain parameters k_ϕ, k_A, and k_o from IC manufacturer's data sheets; or better yet, measure them.

2. After k_v is calculated, draw the uncompensated loop Bode plot on semilog paper with frequency ω on the log scale, starting with k_v (in dB) on the linear vertical scale at $\omega = 1$ rad/s. Mark k_v (rad/s) on the frequency axis, and draw a straight line connecting this point and k_v (dB). Continue this 20-dB/decade line to the bottom of the page. (This procedure is more accurate than moving down 20 dB and over 1 decade from k_v (dB).)

3. Mark the *desired* loop bandwidth B (in radians/second) on the 0-dB axis. Remember that this is the one-sided bandwidth. Since the loop operates for f_i on either side of the loop center frequency f_{FR}, the system 3-dB noise bandwidth is $2B$ rad/s.

4. From the point of part 3 marking the loop bandwidth B, construct a −6-dB/octave line and mark ω_2 on this line at $(B/2, + 6$ dB$)$; that is, ω_2 is at +6 dB loop gain and one octave below B. This frequency, read on the horizontal axis, determines the values of R_2 and C while establishing $\delta = 0.707$. The procedure is to choose an arbitrary, but convenient, value for C (0.01–1.0

* See Figure 10–1.

FIGURE 10–33 Lead-lag compensation network.

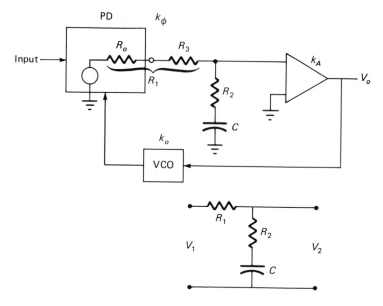

μF) and compute R_2 from

$$\omega_2 = \frac{1}{R_2 C}$$

or
$$R_2 = \frac{1}{\omega_2 C} = \frac{2}{BC} \tag{10–24}$$

5. Now, from the point on the Bode plot of part 4 (ω_2), construct a straight line up at -40 dB/decade (12 dB/octave) until it intersects the line drawn for the uncompensated loop response. The point of intersection is ω_1, which is read from the horizontal axis.

   The lead-lag compensation design for $\delta = 0.707$ is completed by calculating the value of R_1 from

$$\omega_1 = \frac{1}{(R_1 + R_2)C} \tag{10–25}$$

or $R_1 = 1/\omega_1 C - R_2$. Finally, it must be realized that R_1 includes the output impedance R_0 of the phase detector, so the value of resistor to add in series is $R_3 = R_1 - R_0$.

   This completes the fully compensated phase-locked loop. An example is shown in Figure 10–34 for $k_v = 126{,}000$ rad/s (102 dB at 1 rad/s) and $B = 1000$ rad/s.

   A circuit example for demonstrating lead-lag compensation in the laboratory is shown in Figure 10–35 using the LM565 PLL IC of Figure 10–36. Compute the phase margin for the lead-lag compensated loop by adapting Equation 10–19 to produce

$$\text{P.M.} = 90° - \tan^{-1}\left(\frac{B}{\omega_1}\right) + \tan^{-1}\left(\frac{B}{\omega_2}\right) \tag{10–26}$$

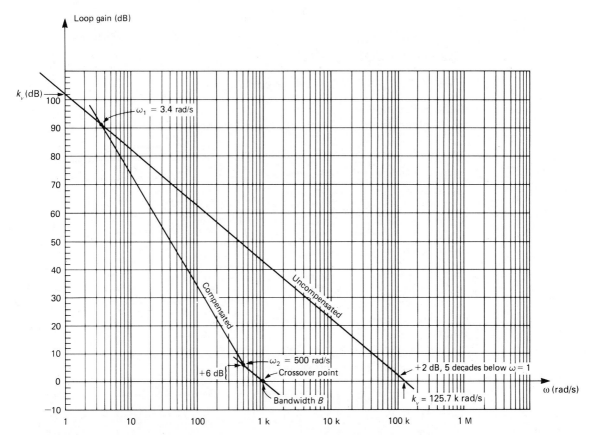

FIGURE 10-34 Bode plots of uncompensated and lead-lag compensated phase-locked loop.

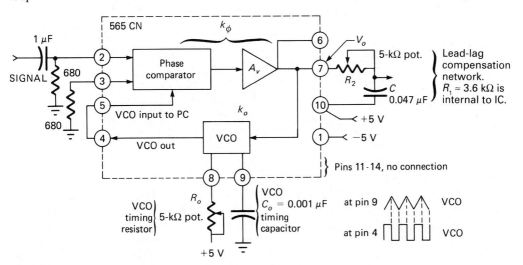

FIGURE 10-35 PLL measurement circuit using LM565CN integrated circuit.

Industrial Blocks

LM565/LM565C Phase Locked Loop

General Description

The LM565 and LM565C are general purpose phase locked loops containing a stable, highly linear voltage controlled oscillator for low distortion FM demodulation, and a double balanced phase detector with good carrier suppression. The VCO frequency is set with an external resistor and capacitor, and a tuning range of 10:1 can be obtained with the same capacitor. The characteristics of the closed loop system—bandwidth, response speed, capture and pull in range—may be adjusted over a wide range with an external resistor and capacitor. The loop may be broken between the VCO and the phase detector for insertion of a digital frequency divider to obtain frequency multiplication.

The LM565H is specified for operation over the –55°C to +125°C military temperature range. The LM565CH and LM565CN are specified for operation over the 0°C to +70°C temperature range.

Features

■ 200 ppm/°C frequency stability of the VCO

■ Power supply range of ±5 to ±12 volts with 100 ppm/% typical

■ 0.2% linearity of demodulated output

■ Linear triangle wave with in phase zero crossings available

■ TTL and DTL compatible phase detector input and square wave output

■ Adjustable hold in range from ±1% to > ±60%.

Applications

■ Data and tape synchronization
■ Modems
■ FSK demodulation
■ FM demodulation
■ Frequency synthesizer
■ Tone decoding
■ Frequency multiplication and division
■ SCA demodulators
■ Telemetry receivers
■ Signal regeneration
■ Coherent demodulators.

Schematic and Connection Diagrams

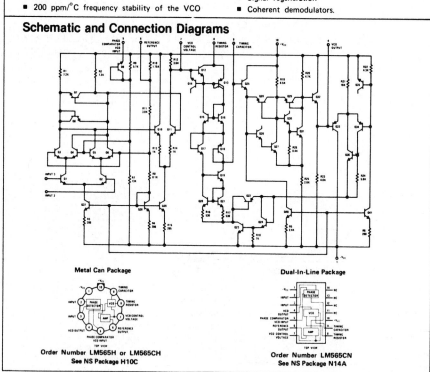

Metal Can Package

Order Number LM565H or LM565CH
See NS Package H10C

Dual-In-Line Package

Order Number LM565CN
See NS Package N14A

9-42

FIGURE 10–36   (page 1 of 4; pages 426–29 courtesy of National Semiconductor, Inc.)

Applications Information

In designing with phase locked loops such as the LM565, the important parameters of interest are:

FREE RUNNING FREQUENCY

$$f_o \cong \frac{1}{3.7 \, R_0 C_0}$$

LOOP GAIN: relates the amount of phase change between the input signal and the VCO signal for a shift in input signal frequency (assuming the loop remains in lock). In servo theory, this is called the "velocity error coefficient".

$$\text{Loop gain} = K_0 K_D \left(\frac{1}{\text{sec}}\right)$$

$$K_o = \text{oscillator sensitivity} \left(\frac{\text{radians/sec}}{\text{volt}}\right)$$

$$K_D = \text{phase detector sensitivity} \left(\frac{\text{volts}}{\text{radian}}\right)$$

The loop gain of the LM565 is dependent on supply voltage, and may be found from:

$$K_o K_D = \frac{33.6 \, f_o}{V_c}$$

f_o = VCO frequency in Hz

V_c = total supply voltage to circuit.

Loop gain may be reduced by connecting a resistor between pins 6 and 7; this reduces the load impedance on the output amplifier and hence the loop gain.

HOLD IN RANGE: the range of frequencies that the loop will remain in lock after initially being locked.

$$f_H = \pm \frac{8 \, f_o}{V_c}$$

f_o = free running frequency of VCO

V_c = total supply voltage to the circuit.

THE LOOP FILTER

In almost all applications, it will be desirable to filter the signal at the output of the phase detector (pin 7) this filter may take one of two forms:

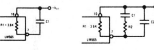

Simple Lag Filter   Lag-Lead Filter

A simple lag filter may be used for wide closed loop bandwidth applications such as modulation following where the frequency deviation of the carrier is fairly high (greater than 10%), or where wideband modulating signals must be followed.

The natural bandwidth of the closed loop response may be found from:

$$f_n = \frac{1}{2\pi} \sqrt{\frac{K_o K_D}{R_1 C_1}}$$

Associated with this is a damping factor:

$$\delta = \frac{1}{2} \sqrt{\frac{1}{R_1 C_1 K_o K_D}}$$

For narrow band applications where a narrow noise bandwidth is desired, such as applications involving tracking a slowly varying carrier, a lead lag filter should be used. In general, if $1/R_1 C_1 < K_o K_d$, the damping factor for the loop becomes quite small resulting in large overshoot and possible instability in the transient response of the loop. In this case, the natural frequency of the loop may be found from

$$f_n = \frac{1}{2\pi} \sqrt{\frac{K_o K_D}{\tau_1 + \tau_2}}$$

$$\tau_1 + \tau_2 = (R_1 + R_2) \, C_1$$

R_2 is selected to produce a desired damping factor δ, usually between 0.5 and 1.0. The damping factor is found from the approximation:

$$\delta \simeq \pi \tau_2 f_n$$

These two equations are plotted for convenience.

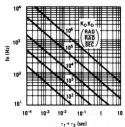

Filter Time Constant vs Natural Frequency

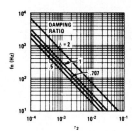

Damping Time Constant vs Natural Frequency

Capacitor C_2 should be much smaller than C_1 since its function is to provide filtering of carrier. In general $C_2 \leq 0.1 \, C_1$.

FIGURE 10-36 (page 2 of 4)

Absolute Maximum Ratings

Supply Voltage	±12V
Power Dissipation (Note 1)	300 mW
Differential Input Voltage	±1V
Operating Temperature Range LM565H	$-55°C$ to $+125°C$
LM565CH, LM565CN	$0°C$ to $70°C$
Storage Temperature Range	$-65°C$ to $+150°C$
Lead Temperature (Soldering, 10 sec)	$300°C$

Electrical Characteristics (AC Test Circuit, $T_A = 25°C$, $V_C = ±6V$)

PARAMETER	CONDITIONS	LM565			LM565C			UNITS
		MIN	TYP	MAX	MIN	TYP	MAX	
Power Supply Current			8.0	12.5		8.0	12.5	mA
Input Impedance (Pins 2, 3)	$-4V < V_2, V_3 < 0V$	7	10			5		kΩ
VCO Maximum Operating Frequency	$C_o = 2.7$ pF	300	500		250	500		kHz
Operating Frequency Temperature Coefficient			-100	300		-200	500	ppm/$°C$
Frequency Drift with Supply Voltage			0.01	0.1		0.05	0.2	%/V
Triangle Wave Output Voltage		2	2.4	3	2	2.4	3	V_{p-p}
Triangle Wave Output Linearity			0.2	0.75		0.5	1	%
Square Wave Output Level		4.7	5.4		4.7	5.4		V_{p-p}
Output Impedance (Pin 4)			5			5		kΩ
Square Wave Duty Cycle		45	50	55	40	50	60	%
Square Wave Rise Time			20	100		20		ns
Square Wave Fall Time			50	200		50		ns
Output Current Sink (Pin 4)		0.6	1		0.6	1		mA
VCO Sensitivity	$f_o = 10$ kHz	6400	6600	6800	6000	6600	7200	Hz/V
Demodulated Output Voltage (Pin 7)	±10% Frequency Deviation	250	300	350	200	300	400	mV_{pp}
Total Harmonic Distortion	±10% Frequency Deviation		0.2	0.75		0.2	1.5	%
Output Impedance (Pin 7)			3.5			3.5		kΩ
DC Level (Pin 7)		4.25	4.5	4.75	4.0	4.5	5.0	V
Output Offset Voltage $|V_7 - V_6|$			30	100		50	200	mV
Temperature Drift of $|V_7 - V_6|$			500			500		$\mu V/°C$
AM Rejection		30	40			40		dB
Phase Detector Sensitivity K_D		0.6	.68	0.9	0.55	.68	0.95	V/radian

Note 1: The maximum junction temperature of the LM565 is 150°C, while that of the LM565C and LM565CN is 100°C. For operation at elevated temperatures, devices in the TO-5 package must be derated based on a thermal resistance of 150°C/W junction to ambient or 45°C/W junction to case. Thermal resistance of the dual-in-line package is 100°C/W.

FIGURE 10–36 (page 3 of 4)

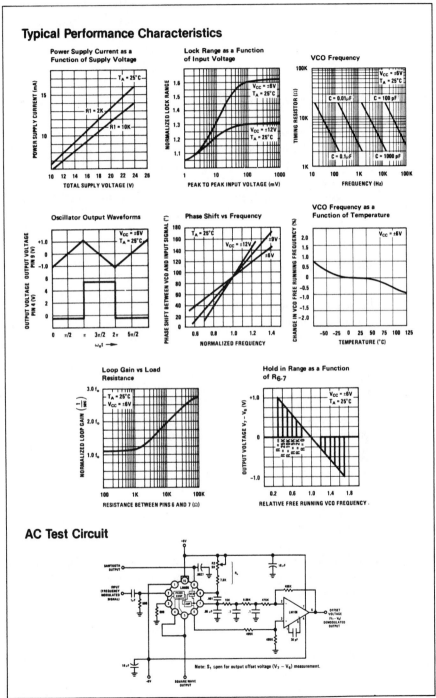

FIGURE 10-36   (page 4 of 4)

Compare the results to the values in Figure 10–30; $\delta = 0.7$, P.M. $\approx 64°$. It is clear that the lead phase of the *zero* (ω_2) overcomes the increasing phase lag of the *pole*, ω_1.

10–8 ■ FREQUENCY SYNTHESIZERS

A *frequency synthesizer* is a variable-frequency generator with crystal-controlled stability. Synthesizers are used for instrumentation, automatic test equipment, and communications systems, as well as the more common uses—push-button TV channel selection, synthesized amateur transceivers, CB radios and scanners.

Frequency synthesizers are used in multichannel communication links for generating stable transmitter carrier frequencies and stable local oscillator frequencies for receivers. Typically, channel selection is guided by microprocessor control of digital logic gating circuits or synchronous counters.

There are two basic approaches to frequency synthesis—direct and indirect. For *direct* frequency synthesis, the output signal is derived by combining multiple crystal-controlled oscillator outputs or a single crystal oscillator with multiple-divider/comb-generator sections. Various combinations are digitally switched in, and the unwanted subharmonics and other spurious frequencies are filtered out. This approach can be very fast, limited only by the digital switching logic and pulse response of the output filters.

The *indirect synthesis* approach relies on a spectrally pure VCO and programmable phase-locked loop circuitry. While slower than the direct approach and susceptible to FM noise on the VCO, indirect frequency synthesis using the phase-lock principle is less expensive, requires much less filtering, and offers greater output power with lower spurious subharmonics.

Direct Frequency Synthesizers

Two examples are given to illustrate the direct synthesis approach for synthesizing a variable-frequency signal source from oscillators with quartz-crystal frequency stability—that is, with less than ±10 ppm over a wide environmental temperature range.

Figure 10–37 shows a block diagram of an approach that uses a single-crystal reference oscillator and dividers to provide a fine resolution of the discrete output frequencies. *Resolution* is the smallest difference possible between two frequencies at the synthesizer output.

Figure 10–37 shows a high-stability, 64-kHz master reference oscillator followed (horizontally) by a *comb generator*, which is a circuit used to produce a pulse rich in harmonics of the 64-kHz input signal. A harmonic-selector filter controlled by tuning logic is tuned to the desired harmonic and rejects all other spurious outputs. If the synthesizer consisted only of this group of blocks, the resolution would be 64 kHz because the output can be switched only to the various harmonics of the 64-kHz master reference.

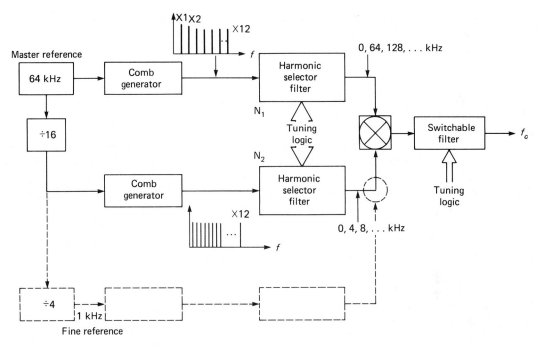

FIGURE 10-37 Direct synthesizer.

In order to improve the resolution and thereby achieve a finer separation between the possible output frequencies, a divide-by-16 circuit is used with a comb generator and selector filter to produce 4-kHz frequency steps. The selected frequencies from the upper and lower harmonic-select filters are mixed to produce sum and difference frequencies, and the output (switchable) filter passes the desired output frequency. The resolution, or smallest possible discrete frequency step, is now seen to be 4 kHz. For instance, with $N_1 = 2$, $N_2 = 2$, and the output filter passing the mixer sum, then $f_0 = 128 \text{ kHz} + 8 \text{ kHz} = 136 \text{ kHz}$. The next higher output frequency would be 140 kHz.

The addition of more divider/comb-generator/harmonic-filter/mixer combinations will provide for finer resolution of the output frequency. This is illustrated in the block diagram with dashed lines, producing a synthesizer with 1-kHz resolution and an output frequency range from 1 kHz to perhaps 828 kHz, assuming that all of the comb generators can operate up to times-12. The resolution is seen to be equal to the fine reference frequency.

A multicrystal, direct-synthesis scheme for producing the transmit carrier and two receiver local oscillators for a 23-channel citizens band transceiver is shown in Figure 10-38. This synthesizer is a 6-4-4 crystals/oscillator scheme (a 6-4-2 scheme is also used) and, when tuned to emergency CB channel 9 (27.065 MHz carrier frequency), crystals 3, 7, and 11 are used. With the receiver oscillator off, crystals 3 and 11 produce the transmit carrier: $37.700 - 10.635 = 27.065 \text{ MHz}$. With the receiver oscillator on and the transmit oscillator off, the first and second

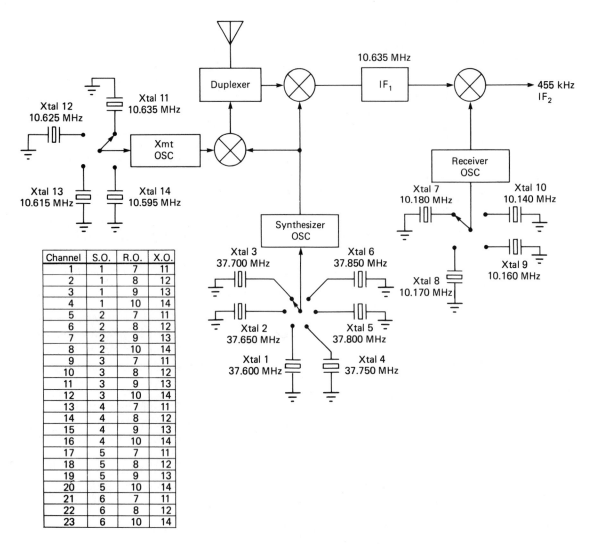

Channel	S.O.	R.O.	X.O.
1	1	7	11
2	1	8	12
3	1	9	13
4	1	10	14
5	2	7	11
6	2	8	12
7	2	9	13
8	2	10	14
9	3	7	11
10	3	8	12
11	3	9	13
12	3	10	14
13	4	7	11
14	4	8	12
15	4	9	13
16	4	10	14
17	5	7	11
18	5	8	12
19	5	9	13
20	5	10	14
21	6	7	11
22	6	8	12
23	6	10	14

FIGURE 10–38   Frequency synthesis (6-4-4) for 23-channel CB transceiver.

local oscillators for this double-conversion receiver are produced by the synthesis and receiver oscillators as follows:

37.700 MHz from the synthesis oscillator with Xtal-3 is the first-LO frequency. Thus, the first-IF frequency is $37.700 - 27.065$ MHz $= 10.635$ MHz, so that FM receiver IF transformers can be used. The receiver oscillator with Xtal-7 (10.180 MHz) is the second-LO, and the second-IF frequency is 10.635 MHz $- 10.180$ MHz $= 455$ kHz, so that AM receiver IF transformers can be used.

This method of synthesis has not been popular for a number of years. It has been replaced by PLL synthesizers but has been offered to show the evolutionary nature of the synthesizer technology.

Phase-Locked Synthesizers

The most frequently used technique for frequency synthesis is the indirect method utilizing a voltage-controlled oscillator in a programmable PLL. The simplest system is the one-loop synthesizer of Figure 10–39, consisting of a digitally programmable divide-by-N circuit used to divide the VCO output frequency for comparison with a stable reference source.

The VCO provides the synthesizer output signal at frequency f_o and is phase-locked to the stable reference signal. The reference frequency f_{ref} is usually derived by dividing down the frequency of a stable crystal- (or even cesium-) controlled oscillator. This ensures that the reference is extremely stable because frequency division reduces absolute frequency deviations just as multiplication will increase them.

The digitally programmable frequency-divider output, f_o/N, is determined by the value of N selected by the user and is compared to the reference signal in the phase detector (PD). When the loop is locked for a specific value of N, then $f_o/N = f_{ref}$ by definition of phase-locked; therefore the synthesizer output is

$$f_o = Nf_{ref} \qquad\qquad (10\text{–}27)$$

The divider can be a simple integer divider such as the 74192 programmable up-down counter, or noninteger divider systems* such as the fractional-N method (producing $f_o = (N + 1/M)f_{ref}$) and the two-modulus prescaler circuit of Figure 10–44, the MC12012, for example, using a technique called *pulse swallowing*. For our purposes, only integer dividers are considered.

The loop gain for the simple PLL synthesizer of Figure 10–39 is

$$k_L = \frac{k_\phi k_o}{N} \qquad\qquad (10\text{–}28)$$

It is important to realize that the frequency-divider circuit reduces the loop gain so

* For an excellent presentation of divider techniques see Chapter 6 of W. F. Egan, *Frequency Synthesis by Phase Lock* (New York: Wiley-Interscience, 1981).

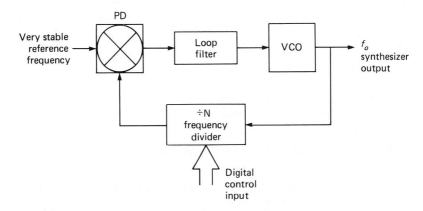

FIGURE 10–39 Phase-locked frequency synthesizer.

FIGURE 10–40 Microprocessor-
controlled LO synthesizer for
TV.

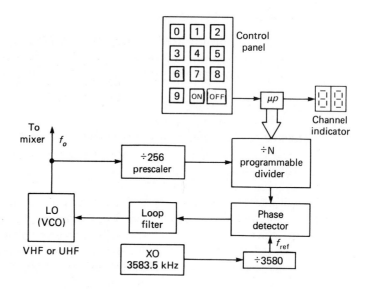

that the other loop components need to have relatively higher gain than the
conventional PLL. A more troublesome design problem, however, is that, as N
changes, so does the loop gain. There are linearizer circuits to ameliorate this
problem.

Figure 10–40 shows the use of a very high frequency prescaler in a one-loop
synthesizer used for push-button TV channel selection. The VHF local oscillator
(LO) frequency is greater than 100 MHz, so high frequency emitter-coupled logic
(ECL) dividers are used to prescale the VHF signal below 1 MHz, where low-cost
TTL or CMOS technology can be used. The prescaler will reduce the resolution
by an amount equal to the prescale division ratio P. Hence,

$$\text{resolution} = Pf_{\text{ref}} \qquad\qquad \textbf{(10–29)}$$

with a prescaler.

■ ■ ■ ■

**EXAMPLE
10–6**

The microprocessor-controlled VHF LO synthesizer of Figure 10–40 has a phase
comparator with $k_\phi = 1$ V/rad and an output impedance of 3.5 kΩ. Determine the
following:

1. f_{ref}

2. N for the TV to receive channel 5 ($f_{\text{LO}} = 123.000$ MHz)

3. The synthesizer frequency resolution

4. Loop gain and value of capacitor to compensate the loop to $\delta = 0.5$ and have
   the VCO frequency within $\pm10\%$ of its specified value in less than 10 ms after
   selection of channel 5 (Assume that the maximum frequency step at the phase
   detector is within the loop bandwidth.)

5. What value must the VCO sensitivity be?

Solution:

1. $f_{ref} = f_{XO}/3580 = 3583.5$ kHz/3580 = **1.00098 kHz.**

2. $f_o = 256Nf_{ref}$; therefore, N = 123.000 MHz/[(256)(1.00098 kHz)] = **480.**

3. With the prescaler, the resolution will be $Pf_{ref} = 256 \, f_{ref} = 256.25$ kHz. To prove this, change the programmable divider to N + 1 = 481, and compare the new f_o to the old.

$$f_o(N + 1) = 1.00098 \text{ kHz} \times 256 \times 481 = 123,256.67 \text{ kHz}$$

$$f_o(N) = 1.00098 \text{ kHz} \times 256 \times 480 = 123,000.42 \text{ kHz}$$

$$\text{resolution} = f_o(N + 1) - Nf_o = \textbf{256.25 kHz}$$

4. With the assumption stated, the loop will lock up unaided (without frequency-sweep circuitry); hence, we use the universal overshoot and ringing curves of Figure 10–27. V_o must stay within relative values 0.90 and 1.1 (+10%) on the $\delta = 0.5$ curve. This is satisfied by $\omega_n t = 4.6$. With $t = t_s = 10$ ms, we need the loop to have $\omega_n = 4.6/10$ ms = **460 rad/s**. Since $\delta = 0.5\sqrt{\omega_c/k_v}$, then $\omega_c = k_v$ for $\delta = 0.5$. Also $\omega_n = \sqrt{\omega_c k_v}$, so that $\omega_n = \omega_c = k_v = $ **460 rad/s.**

   A capacitor is placed across the phase detector output ($R_o = 3.5$ kΩ) to form the lag-compensation network. $C = 1/\omega_c R = 1/(460 \times 3500) = $ **0.62 μF.**

5. $k_v = 2\pi k_\phi k_o/N = 460$ rad/s. Therefore, k_o is required to be $k_o = Nk_v/2\pi k_\phi = (256 \times 480)(460)/2\pi(1 \text{ V/rad}) = $ **9 MHz/V.** This figure is not unrealistic for a 123 MHz VCO.

Translation Loops and Multiple-Loop Synthesizers

One technique used to reduce a high-frequency VCO output to reasonable frequencies without a prescaler, and to provide a frequency offset, is shown within the dashed area of Figure 10–41. Figure 10–41 is the receiver block diagram for a PLL synthesized 40-channel citizen band transceiver with delta tuning for fine-frequency adjustments. The mixer and 35.42-MHz crystal oscillator translate the VCO frequency range from 37.66–38.10 MHz down to 2.24–2.68 MHz for input to the programmable divider. Notice that the reference oscillator also provides the second LO for the double-conversion receiver.

Multiple-loop synthesizers combine all the techniques discussed thus far. The addition of more loops increases the resolution and frequency coverage; the individual loops also act as tracking filters to reduce unwanted mixer products and spurious output components. Ideally, double-balanced mixers are used throughout. Figure 10–42 illustrates the basic multiple-loop synthesizer for n loops. If the mixer outputs are filtered to pass the difference frequency and the output frequency of each VCO is lower for higher n, the synthesizer output frequency can be shown to be

$$f_o = f_{ref}\left[N_1 + \frac{N_2}{M_2} + \cdots + \frac{N_n}{(M_2 M_3 \cdots M_n)}\right] \qquad \textbf{(10–30)}$$

and $\qquad \text{resolution} = \dfrac{f_{ref}}{(M_2 M_3 \cdots M_n)} \qquad$ **(10–31)**

FIGURE 10–41 PLL frequency
synthesizer for a 40-channel CB
transceiver.

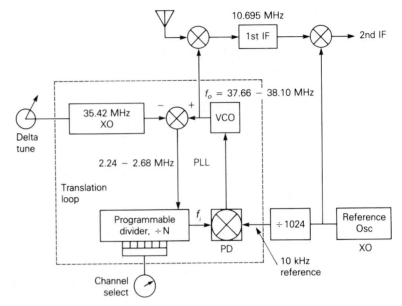

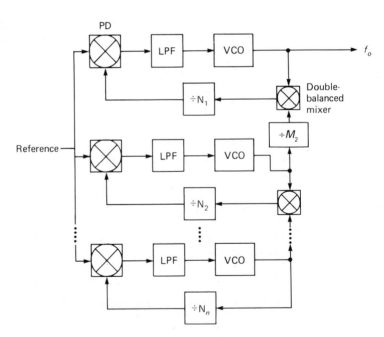

FIGURE 10–42 An *n*-loop, multiloop synthesizer.

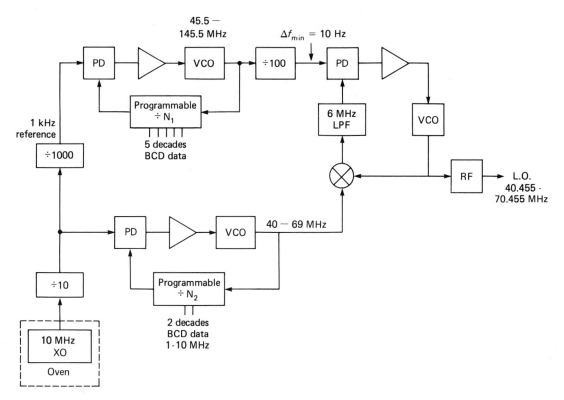

FIGURE 10–43 Multiloop main synthesizer for HF-1030 communications receiver. (Courtesy of Cubic Communications.)

A variation of the multiple-loop synthesizer shown in Figure 10–43 is used in a Cubic Communications HF-1030 AM and single-sideband receiver. The figure is incomplete but it is clear that the resolution is 10 Hz, and 3 million discrete local oscillator frequencies can be synthesized. The oven temperature control of the crystal-reference oscillator for the portable HF-1030 allows the unit to maintain frequency stability specifications of 1 ppm/month and less than 1 Hz/°C for environmental temperature changes.

As observed in the above discussion, many frequency dividers are used in signal synthesizers. High frequency dividers require special circuit considerations. Figure 10–44 is an example of an emitter-coupled-logic (ECL) divider/prescaler integrated circuit.

SEMICONDUCTORS

P.O. BOX 20912 • PHOENIX, ARIZONA 85036

MC12012

MECL PLL COMPONENTS

TWO-MODULUS PRESCALER

TWO-MODULUS PRESCALER

The MC12012 is a two-modulus prescaler which consists of three functional blocks: 1) a controllable divide-by-5/divide-by-6 prescaler; 2) a divide-by-2 prescaler, and 3) a MECL to MTTL translator. When used with the MC12014 Counter Control Logic function and the MC4016 programmable counter, a divide-by-N programmable counter can be constructed for operation to 200 MHz. This arrangement is especially useful in frequency synthesizer applications.

- $\div 2, \div 5/\div 6, \div 10/\div 11, \div 10/\div 12$
- MECL to MTTL Translator on Chip
- +5.0 or –5.2 V Operation*
- 200 MHz (Typ) Toggle Frequency

*When using +5.0 V supply, apply +5.0 V to pin 16 (V_{CC}) and ground pin 8 (V_{EE}). When using –5.2 supply, ground pin 16 (V_{CC}) and apply –5.2 V to pin 8 (V_{EE}).

L SUFFIX
CERAMIC PACKAGE
CASE 620

FIGURE 1 – TYPICAL FREQUENCY SYNTHESIZER APPLICATION

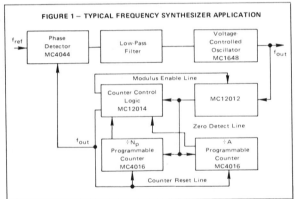

PIN ASSIGNMENT

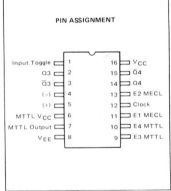

Input Toggle	1	16	V_{CC}
Q3	2	15	$\overline{Q}4$
$\overline{Q}3$	3	14	Q4
(–)	4	13	E2 MECL
(+)	5	12	Clock
MTTL V_{CC}	6	11	E1 MECL
MTTL Output	7	10	E4 MTTL
V_{EE}	8	9	E3 MTTL

FIGURE 2 – LOGIC DIAGRAM

To obtain an MTTL output connect 5 and 4 to 2 and 3 or 14 and 15 respectively. The MECL outputs (2, 3, 14, 15) require terminating resistors. When used, the translator (4 and 5) will provide the proper termination for connection to it.

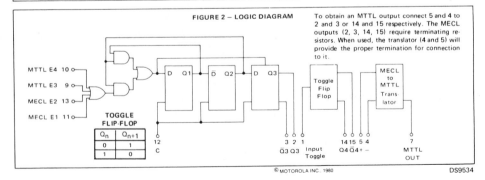

Q_n	Q_{n+1}
0	1
1	0

DS9534

FIGURE 10–44 (page 1 of 6; pages 438–43 courtesy of Motorola, Inc.)

MC12012

FIGURE 3 – AC TEST CIRCUIT

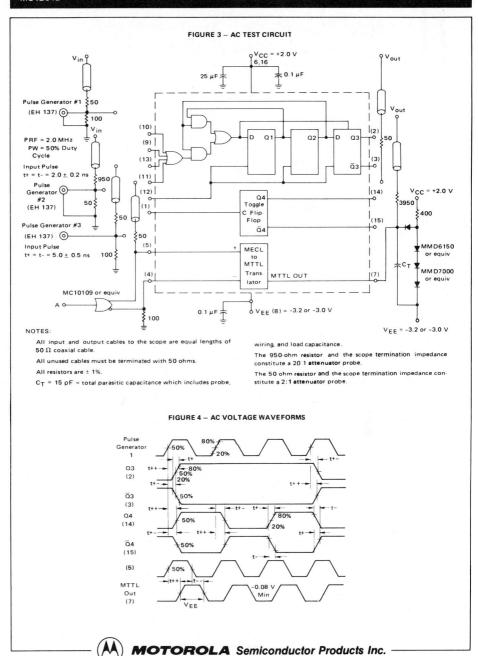

FIGURE 4 – AC VOLTAGE WAVEFORMS

NOTES:

All input and output cables to the scope are equal lengths of 50 Ω coaxial cable.

All unused cables must be terminated with 50 ohms.

All resistors are ± 1%.

C_T = 15 pF = total parasitic capacitance which includes probe,

wiring, and load capacitance.

The 950-ohm resistor and the scope termination impedance constitute a 20:1 attenuator probe.

The 50 ohm resistor and the scope termination impedance constitute a 2:1 attenuator probe.

Ⓜ **MOTOROLA** *Semiconductor Products Inc.*

FIGURE 10–44   (page 2 of 6)

MC12012

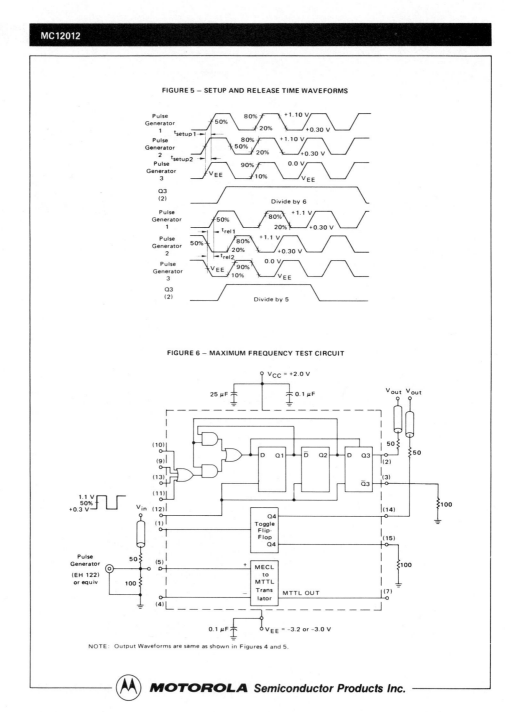

FIGURE 5 — SETUP AND RELEASE TIME WAVEFORMS

FIGURE 6 — MAXIMUM FREQUENCY TEST CIRCUIT

NOTE: Output Waveforms are same as shown in Figures 4 and 5.

MOTOROLA *Semiconductor Products Inc.*

FIGURE 10–44   (page 3 of 6)

MC12012

FIGURE 9 – ÷5/6

MTTL E4 10
MTTL E3 9
MECL E2 13
MECL E1 11

D Q1   D̄ Q2   Q3

Toggle Flip Flop

MECL to MTTL Trans-lator

12
C

3 2 1
Q̄3 Q3 C4

14 15 5 4
Q4 Q̄4 + –

7
MTTL OUT

Divide by 2 use Toggle Flip-Flop

E1 + E2 + E3 + E4 = 1

COUNT	Q1	Q2	Q3
6	0	1	1
7	1	1	1
5	1	0	1
1	1	0	0
0	0	0	0
2	0	1	0

E1 + E2 + E3 + E4 = 0

To obtain an MTTL output connect 5 and 4 to 2 and 3 or 14 and 15, respectively.

FIGURE 10 – ÷10/11

MTTL E4 10
MTTL E3 9
MECL E2 13
MECL E1 11

D Q1   D̄ Q2   D Q3

Togle Flip Flop

ECL to TTL Trans-lator

C 12

3 2 1
Q̄3 Q3 C4

14 15 5 4
Q4 Q̄4 + –

7
MTTL Out

E1 + E3 + E4 = 1

COUNT	Q1	Q2	Q3	Q4
14	0	1	1	1
15	1	1	1	1
13	1	0	1	1
9	1	0	0	1
8	0	0	0	1
10	0	1	0	1
7	1	1	1	0
5	1	0	1	0
1	1	0	0	0
0	0	0	0	0
2	0	1	0	0

E1 + E3 + E4 = 0

To obtain an MTTL output connect 5 and 4 to 14 and 15, respectively.

Ⓜ **MOTOROLA** *Semiconductor Products Inc.*

FIGURE 10–44 (page 4 of 6)

MC12012

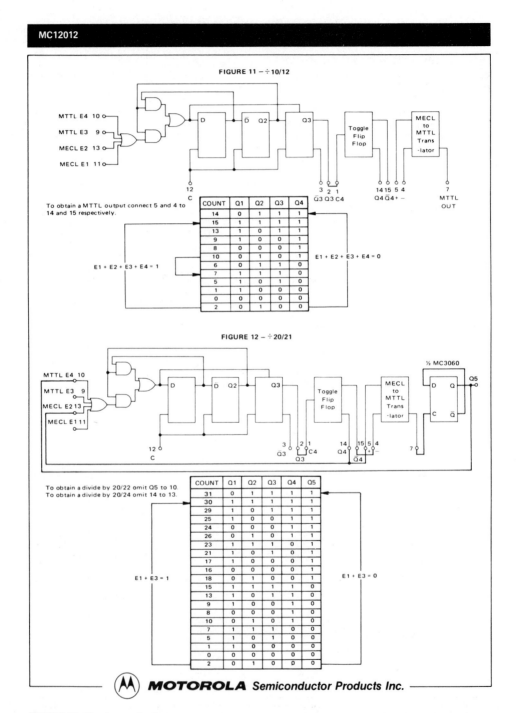

FIGURE 11 – ÷ 10/12

To obtain a MTTL output connect 5 and 4 to 14 and 15 respectively.

E1 + E2 + E3 + E4 = 1

COUNT	Q1	Q2	Q3	Q4
14	0	1	1	1
15	1	1	1	1
13	1	0	1	1
9	1	0	0	1
8	0	0	0	1
10	0	1	0	1
6	0	1	1	0
7	1	1	1	0
5	1	0	1	0
1	1	0	0	0
0	0	0	0	0
2	0	1	0	0

E1 + E2 + E3 + E4 = 0

FIGURE 12 – ÷ 20/21

½ MC3060

To obtain a divide by 20/22 omit Q5 to 10.
To obtain a divide by 20/24 omit 14 to 13.

E1 + E3 = 1

COUNT	Q1	Q2	Q3	Q4	Q5
31	0	1	1	1	1
30	1	1	1	1	1
29	1	0	1	1	1
25	1	0	0	1	1
24	0	0	0	1	1
26	0	1	0	1	1
23	1	1	1	0	1
21	1	0	1	0	1
17	1	0	0	0	1
16	0	0	0	0	1
18	0	1	0	0	1
15	1	1	1	1	0
13	1	0	1	1	0
9	1	0	0	1	0
8	0	0	0	1	0
10	0	1	0	1	0
7	1	1	1	0	0
5	1	0	1	0	0
1	1	0	0	0	0
0	0	0	0	0	0
2	0	1	0	0	0

E1 + E3 = 0

Ⓜ **MOTOROLA** *Semiconductor Products Inc.*

FIGURE 10–44   (page 5 of 6)

FUNCTION DESCRIPTION

THE MC12012 TWO MODULUS PRESCALER

Three functional blocks are contained in the MC12012 variable modulus prescaler: 1) a controllable ÷5/÷6 prescaler; 2) a ÷ 2 prescaler; and 3) an ECL to TTL translator (for single power supply operation).

Selection of division by 5 or by 6 is made by inputs to E1 through E4. If all E inputs are low before the transition of the clock pulse driving Q3 high, Q3 will stay high for 3 clock pulses, then will go low for 3 clock pulses. This provides a divide by 6 function.

On the other hand, if any one or all of the E inputs are high prior to the positive transition of the clock pulse driving Q3 high, Q3 will stay high for only 2 clock pulses, then will go low for 3 clock pulses. The result is division by 5.

For the ÷5 operation, at least one of the E inputs must go high sometime before the clock pulse. This time is referred to as the "setup time." Specifications for setup time are given in the electrical characteristics table: t_{setup1} and t_{setup2} for E1 and E2 (MECL inputs), and E3 and E4 (MTTL inputs).

For the divide by 6 operation all E inputs must be low for some time prior to the clock pulse. This time is referred to as the "release time." Data for release time is given in the electrical characteristics table; t_{re1} and t_{re2} for E1, E2, E3, E4.

The data given in the tables for setup and release times are referenced to the positive transition of the clock pulse causing Q3 to go high. If it is necessary to reference the setup and release times to the positive transition of Q3, add t++ (specified for Q3) to the setup/release times given. It should be noted that the logic states for the enable inputs are important only for only one clock pulse which causes Q3 to go high (within the limits specified by setup and release times).

The ÷ 5/÷ 6 prescaler may be connected externally to the ÷ 2 prescaler to form a ÷ 10/÷ 11 prescaler (Figure 10) or a ÷ 10/÷ 12 prescaler (Figure 11).

By way of an example showing how a ÷ 10/÷ 11 prescaler operates, note that if E1, E3, and E4 (Figure 10) are held in a low state, the counter divides by 11. To do this, a feedback connection is established from Q4 to E2 (or to E1). With this feedback, the ÷ 5/÷ 6 prescaler divides by 5 when Q4 is high, and by 6 when Q4 is low.

Since Q4 changes state with each positive transition of Q3, the prescaler alternates between ÷ 5 and ÷ 6 resulting in a ÷ 11 at Q4.

If any one or all of the E inputs are high (Figure 10), the 5/6 prescaler always divides by 5 and a divide by 10 results at Q4.

With the addition of external flip-flops and counters (MECL or MTTL) various other modulus prescalers may be produced (20/21, 20/22, 20/24, 40/41, 50/51, 100/101, etc.).

OUTLINE DIMENSIONS

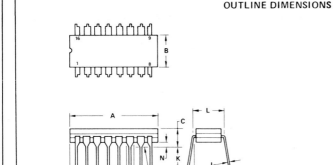

DIM	MILLIMETERS		INCHES	
	MIN	MAX	MIN	MAX
A	19.05	19.81	0.750	0.780
B	6.22	6.99	0.245	0.275
C	4.06	5.08	0.160	0.200
D	0.38	0.51	0.015	0.020
F	1.40	1.65	0.055	0.065
G	2.54 BSC		0.100 BSC	
H	0.51	1.14	0.020	0.045
J	0.20	0.31	0.008	0.012
K	3.18	4.06	0.125	0.160
L	7.37	7.87	0.290	0.310
M	–	15⁰	–	15⁰
N	0.51	1.02	0.020	0.040

CASE 620

NOTES:
1. LEADS WITHIN 0.13 mm (0.005) RADIUS OF TRUE POSITION AT SEATING PLANE
2. AT MAXIMUM MATERIAL CONDITION'
   PKG. INDEX: NOTCH IN LEAD
   NOTCH IN CERAMIC OR INK DOT'
3. DIM "L" TO CENTER OF LEADS WHEN FORMED PARALLEL'

FIGURE 10–44 (page 6 of 6)

■ PROBLEMS

1. Define *phase-locked*.

2. An analog phase detector with two input generators (no VCO) has a beat-frequency output of 4 V pk-pk at 100 Hz. Determine the phase detector gain (sensitivity) in volts/radian and volts/degree of phase.

3. A phase comparator with triangular transfer characteristic has a maximum output
   voltage of 4 V. Determine the gain in volts/radian and volts/degree of phase.

4. A VCO is linear between 260 kHz and 300 kHz. The corresponding input voltages
   are 200 mV and −200 mV, respectively.
   a. Determine the VCO gain (sensitivity).
   b. The free-running frequency.

5. What will be the loop gain of a PLL with f_{FR} = 200 kHz and component gains of
   k_ϕ = 0.5 V/rad, k_A = −4, and k_o = −30 kHz/V? The phase-detector characteristic
   is triangular.
   a. In kHz/radian.
   b. In radians/second.
   c. In dB at 1 rad/s

6. a. Determine the VCO input voltage V_o for the PLL of Problem 5 locked to a 180-
      kHz input signal.
   b. What will be the static phase error?

7. Determine the maximum voltage possible from the phase detector of Problem 5.

8. Determine the hold-in frequency range for the PLL of Problem 5.

9. In Figure 10–45 switch S_1 is at position 2 and f_i = 110 kHz. Capacitor C is part of a
   low-pass filter with a cutoff frequency of 70 kHz. If the phase detector is a
   (double-balanced) mixer,

FIGURE 10–45

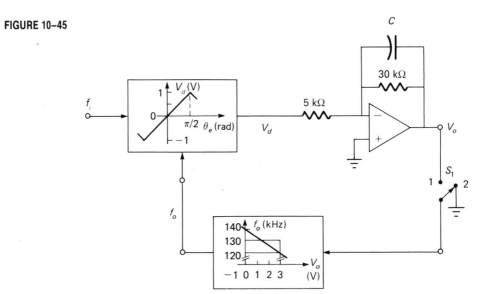

   a. What frequencies are present at V_d and why?
   b. What frequency is present on pin 1 of S_1?
   c. Describe the system behavior when S_1 is switched to position 1.

10. After the PLL of Figure 10–45 locks up with f_i = 110 kHz, determine the following:
    a. f_o
    b. V_o
    c. V_d
    d. θ_e

11. Determine the hold-in frequency range of the PLL of Figure 10–45.

12. **a.** If f_i is changed to 145 kHz, repeat Problem 10.
    b. If f_i varies sinusoidally between 133⅓ kHz and 145 kHz at a rate of 2 kHz, sketch V_o accurately in time and amplitude.

13. The PLL of Figure 10–46 has the following parameters: k_o = 10 kHz/V, k_ϕ = 2 V/rad (triangular characteristic), and f_{FR} = 50 kHz.
    a. What is the value of the reference input f_{ref} if V is measured at 500 mV dc?
    b. What is the phase difference between the VCO output and the reference signal?
    c. What improvement would have to be done to this system to reduce the static (dc) phase error to 2°?
    d. Maximum V for *any* input frequency.
    e. Make an open-loop Bode plot for this PLL.
    f. Calculate the rise-time (time constant) and steady-state V_o if the input signal frequency suddenly changes to 60 kHz.

FIGURE 10–46

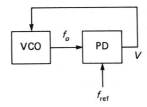

14. A PLL demodulating FSK has the following parameters: k_ϕ = 1 V/rad [triangular], k_A = 5, k_o = 2 kHz/V, f_{FR} = 1.7 kHz. The FSK input signal is shown in Figure 10–47 with $f(space)$ = 2.2 kHz and $f(mark)$ = 1.2 kHz.

FIGURE 10–47

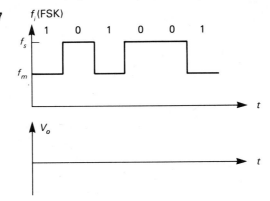

a. If the data rate is 1000 bits/s and the PLL is perfectly compensated, sketch V_o (the VCO input voltage) showing amplitude and time.
b. Compute the rise time (time constant) for the uncompensated loop.

15. a. Make an open-loop Bode plot for the PLL of Problem 14.
b. Use a dashed line to show the closed-loop frequency response of the loop on the plot of part a.
c. Determine the closed-loop bandwidth of this system (in radians/second and hertz).
d. Sketch V_o, relative to the steady-state V_o, accurately for the FSK input shown in Figure 10–48—a 5-kHz peak rectangular pulse above the VCO free-running frequency f_{FR}.

FIGURE 10–48

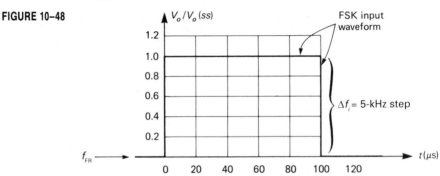

16. A PLL has a loop gain of 10^5 s^{-1} and an RC compensation filter with $\omega_c = 0.5 \times 10^5$.
a. Determine the loop damping factor.
b. What percent of overshoot will occur for an input FSK signal?

17. A second-order loop has $\delta = 0.4$ and $\omega_n = 10,000$ rad/s. Use Figures 10–27 and 10–30 to determine
a. The time required for the loop to settle to within 10% of steady-state.
b. Time of occurrence and maximum amount of overshoot (in dB).
c. Maximum amount in dB of the closed-loop frequency response rise and the frequency at which the peak occurs.

FIGURE 10–49

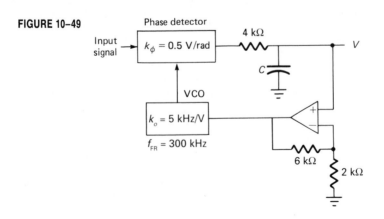

18. The circuit of Figure 10–49 is used as a demodulator for FSK. An input carrier of 310 kHz is received.
    a. Determine V after lock-up.
    b. If $C = 0.04$ μF in the filter shown, will the loop be overdamped or under-damped? Why?
    c. What value of C will produce $\delta = 0.7$?
    d. Sketch V for $\Delta f_i = \pm 12$ kHz from carrier center frequency at a 500-Hz rate for $\delta = 0.7$ loop compensation. (The time scale doesn't matter, but the *volts do*.)

19. A PLL with loop gain of 10^5 s^{-1} rings when *pulsed* because of a poorly designed lag compensation network.
    a. If the complex reactance of the capacitor in Figure 10–24 is $1/sC$, use the voltage-divider rule to show that $V_i/V_d = 1/(RCs + 1) = (1/RC)/(s + \omega_c)$.
    b. Sketch the Bode plot of a lead-lag compensated PLL and determine R_1 and C in order to achieve a closed-loop bandwidth (one-sided) of 318.3 Hz and damping factor of $\delta = 0.707$ with $R_2 = 500$ Ω.

20. Lead-lag compensation: Change the compensation network of Problem 18 to give the loop an open-loop crossover frequency of 3000 rad/s and a damping factor of 0.707.
    a. Determine the values of any components used in addition to the 4-kΩ resistor shown.
    b. Calculate the phase margin for this loop.

21. A loop has $k_\phi = 0.8$ V/rad, $k_A = 12.5$, and $k_o = 2$ kHz/V.
    a. If a 0.1-μF capacitor is used, determine R_1 and R_2 for a lead-lag network to compensate the loop for a 0.707 damping factor and a bandwidth of 636.62 Hz.
    b. If the complex reactance of the capacitor in Figure 10–33 is $1/sC$, use the voltage-divider rule to show that $V_2/V_1 = [R_2/(R_1 + R_2)][(s + \omega_2)/(s + \omega_1)]$.

22. A squarewave FSK signal is applied to a 565 PLL. The VCO control voltage V_o is seen on an oscilloscope as shown in Figure 10–50.
    a. Determine the percentage of overshoot.
    b. Use the ringing curves of Figure 10–27 to estimate the loop damping factor.
    c. What is the frequency of the FSK-modulating squarewave?
    d. What is the bit rate if the demodulated signal is binary?

FIGURE 10–50

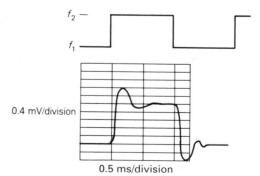

0.4 mV/division

0.5 ms/division

23. For the PLL frequency synthesizer of Figure 10–51, N is equal to the decimal equivalent of the binary input.

a. Determine f_o.
b. Increase the binary input by 1 (LSB = 1). Determine f_o.
c. How fine is the *resolution* (minimum frequency change)?
d. Determine f_o for a binary input of 1100.

FIGURE 10–51

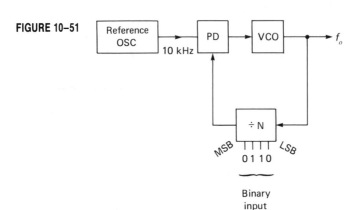

24. For the synthesizer of Figure 10–37 with a master reference generator of 1 MHz and a fixed divider ratio of 200,
    a. What is the minimum frequency step (resolution)?
    b. If $N_1 = 5$ and $N_2 = 6$, determine the synthesizer output frequency. (Assume that the switchable output filter selects the upper sideband out of the balanced mixer.)
    c. Repeat **b** for $N_1 = 6$ and $N_2 = 6$.
    d. Repeat **b** for $N_1 = 5$ and $N_2 = 7$.
    e. Give two possible methods for increasing the resolution.

25. a. Which crystals are used to receive CB channel 7 with the synthesizer of Figure 10–38?
    b. Determine the following frequencies when tuned to channel 7: 1st LO, 2nd LO, transmit, and receive frequencies.

26. For the phase-locked CB transceiver of Figure 10–41:
    a. Determine f_i when the loop is locked.
    b. If the loop is locked and the VCO frequency is 37.67 MHz, what division ratio ($\div N$) is programmed into the programmable divider?
    c. Determine the frequency of the channel to be received. (Assume high-side injection to the first mixer.)
    d. Determine the second-IF frequency.

27. The VCO of Figure 10–52 is phase-locked to a reference oscillator as shown. The system $\omega_n = 1000$ s^{-1}.
    a. Determine N for the programmable divider shown if f_o is to be 10 MHz.
    b. If N must be an integer, what is the resolution for this synthesizer? Prove it.
    c. Determine the minimum damping factor which will allow f_o to be within $\pm 8\%$ of its final value after 11 ms when switching from one frequency to the next.

FIGURE 10–52 $k_o = 100 \text{ kHz/V}$

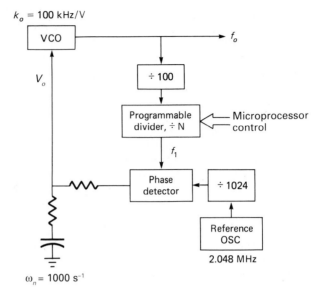

d. Determine the maximum damping factor to do part **c** if the loop is to be overdamped. (Estimate by reading between the lines on the second-order response curves.)

10–9 ■ COMPUTER ANALYSIS

Using the computer for communications system and circuit analysis has become a virtual necessity in education and industry. What follows are examples of programs and/or readily available simulation software for solving problems of the type covered in this chapter.

All the programs are interactive in the sense that an introductory sentence or short paragraph describes the purpose of the program and the user is then prompted for input data.

The programs for Chapter 10 are as follows:

- **Phase-locked loop basics:** The BASIC program of Figure 10–53 calculates the voltages at the VCO and amplifier inputs, as well as the static phase error and the PLL hold-in frequency range, when the loop component gains, the VCO free-running frequency, and the loop input signal frequency are specified.

- **Phase-locked loop analysis with lag compensation:** The BASIC program of Figure 10–54 does the preceding plus, if user desires, the program will compute the uncompensated loop time constant (from which the bandwidth is easily computed) and the resistor or capacitor value for compensating the loop with any desired damping factor.

```
10 REM *****            PHASE LOCK LOOP BASICS            *****
20 REM *    EXAMPLE OF STUDENT COMPUTER ASSIGNMENT    *
30 REM *****        FOR PROFESSOR P. YOUNG, AT ASU        *****
40 CLS
50 PRINT "   PHASE-LOCKED LOOP BASICS;   FROM CHAPTER 10 :":PRINT
60 PRINT "CALCULATES THE CIRCUIT LEVELS AND GAIN OF A PHASE-LOCKED LOOP;":PRINT
70 PRINT "VCO INPUT VOLTAGE IS Vo, OUTPUT OF PHASE DETECTOR IS Vd,":PRINT
80 PRINT "PHASE ERROR IS Oe":PRINT
90 PRINT "----------------------------------------------":PRINT
100 PRINT:INPUT "PLEASE INPUT FREQUENCY Fin , IN HZ. Fin = ";F1:PRINT
110 PRINT:INPUT "FREE RUNNING FREQUENCY, IN HZ,    F(FR) = ";F2:PRINT
120 PRINT:INPUT "GAIN OF VCO (HZ/V), Ko = ";KO:PRINT
130 PRINT:INPUT "AMPLIFIER VOLTAGE GAIN, KA = ";K1:PRINT
140 PRINT:INPUT "PHASE DETECTOR GAIN, RADIANS/V ,KO = ";K2
150 PRINT:PRINT:PRINT
160 REM FORMULAS
170 F3=F1-F2:VO=F3/KO:V1=VO/K1:E1=V1/K2:K3=K1*K2*KO:FH =ABS(3.1415*K3)
180 PRINT "SOLUTIONS:"
190 K5=2*3.1415*K3:K4=ABS(K5):E2=F3/K3
200 PRINT "Vo = ";VO;"VOLTS":PRINT
210 PRINT "Vd = ";V1;"VOLTS":PRINT
220  PRINT "STATIC PHASE ERROR IS, Oe = ";E1;" RADIANS":PRINT
230 PRINT "LOOP GAIN, KL = ";ABS(K3);" HZ/RAD":PRINT
240 PRINT "LOOP GAIN, Kv = ";K4;" RAD/S":PRINT
250 PRINT "HOLD-IN FREQUENCY RANGE IS, DELTA-FH = ";FH;"HZ.":PRINT
260 PRINT:INPUT "COMPUTE ANOTHER ? (Y/N)";Q$
270 IF Q$ ="Y" THEN 100:END
```

FIGURE 10–53

```
10 REM ****  PHASE LOCK LOOP ANALYSIS WITH LAG COMPENSATION ****
20 REM ****        EXAMPLE OF STUDENT COMPUTER ASIGNMENT        ****
30 REM ****            FOR PROFESSOR P. YOUNG, AT ASU            ****
40 CLS
50 PRINT "          -PHASE LOCKED LOOP ANALYSIS AND LAG COMPENSATION-"
60 PRINT:PRINT
70 INPUT "The Phase Detector Gain (volts/radian) is: ";KO:PRINT
80 INPUT "The Voltage Controlled Oscillator Gain Parameter (Hz/volt) is: ";KO
90 PRINT
100 INPUT "The DC Amplifier Gain is: ";KA:PRINT
110 INPUT "The VCO Free Running Frequency is: ";FR
120 REM FORMULAS
130 PI=3.14159:KL=KO*KO*KA 'Loop Gain in Hz/radian
140 KV=ABS(2*PI*KL) 'Loop Gain in radians/second
150 FH=ABS(PI*KL) 'Hold In Frequency Range in Hz
160 F1=FR+FH/2 'Upper Frequency Limit
170 FO=FR-FH/2 'lower frequency limit
```

FIGURE 10–54   (page 1 of 3)

```
180 DB=20*(LOG(KV)/LOG(10)) 'open loop gain at 1 radian/second
190 VO=(FO-FR)*KO/KL 'phase detector output voltage at FO
200 V1=(F1-FR)*KO/KL 'phase detector output voltage at F1
210 V2=KA*VO 'amplifier output voltage at FO
220 V3=KA*V1 'amplifier output voltage at F1
230 T=1/KV 'uncompensated loop time constant
240 CLS:GOSUB 800:PRINT:PRINT
250 INPUT "The PLL Input Frequency (Hz) is: ";FI
260 IF FO>F1 THEN 300 'find out which is greater FO or F1
270 IF FI>F1 THEN 750 'check input frequency FI for out of range
280 IF FI<FO THEN 750 'if out of range give error message
290 GOTO 320 'do calculations
300 IF FI>FO THEN 750 'check input frequency for out of hold in range
310 IF FI<F1 THEN 750 'if out of range give error message
320 VD=(FI-FR)*KO/KL 'phase detector output voltage
330 VO=VD*KA 'amplifier output voltage
340 PE=VD/KO 'static phase error
350 CLS
360 GOSUB 870:PRINT:PRINT
370 PRINT "Fi =";FI;"Hz":PRINT
380 PRINT "Static Phase Error:";PE;"radians":PRINT
390 PRINT "Phase Detector Output Voltage:";VD;"volts":PRINT
400 PRINT "Amplifier Output Voltage:";VO;"volts" :PRINT:PRINT
410 INPUT "Do You Wish To Calculate for another Input Frequency? 'Y' or 'N'";Y$
420 IF Y$="N" THEN 430:IF Y$="n" THEN 430:GOTO 240
430 CLS
440 INPUT "Do LAG COMPENSATION? 'Y' or 'N'";Y$
450 IF Y$= "N" THEN END:IF Y$="n" THEN END
460 CLS
470 GOSUB 800:PRINT:PRINT
480 PRINT "LAG COMPENSATION":PRINT
490 INPUT "What is the Desired Damping Factor";DF
500 WC=(DF/.5)^2*KV 'cutoff frequency
510 RC=1/WC 'value of resistance times capacitance for lag network
520 WN=(WC*KV)^.5 'undamped natural frequency
530 CLS
540 PRINT "LAG COMPENSATION":PRINT
550 PRINT "DF =";DF:PRINT
560 PRINT "Cutoff Frequency (wc):";WC;"radians/second":PRINT
570 PRINT "Undamped natural Frequency (wn):";WN;"radians/second":PRINT
580 PRINT "RC =";RC:PRINT
590 INPUT "Calculate for 'C' or 'R' ";Y$:PRINT
600 IF Y$="C" THEN 630:IF Y$="c" THEN 630
610 IF Y$="R" THEN 680:IF Y$="r" THEN 680
620 GOTO 590 'input not r or c
630 INPUT "What is Your Chosen Value for 'R' (ohms)";R
640 C=RC/R
650 PRINT
660 PRINT "The Capacitor Value is:";C;"farads"
```

FIGURE 10–54   (page 2 of 3)

```
670 GOTO 710
680 INPUT "What is Your Chosen Value for 'C' (farads)";C
690 R=RC/C:PRINT
700 PRINT "The Resistor Value is:";R;"ohms"
710 PRINT:PRINT
720 INPUT "Do LAG COMPENSATION for another Damping Factor 'Y' or 'N';Y$
730 IF Y$="N" THEN END:IF Y$="n" THEN END
740 GOTO 460 'do lag compensation again
750 CLS
760 PRINT "The Frequency You Input is Out of the Hold In Range   TRY AGAIN!"
770 PRINT
780 PRINT "Frequency Hold In Range:";FH;"Hz",FO;"Hz  to  ";F1;"Hz"
790 GOTO 300
800 PRINT "            -PLL PARAMETERS-":PRINT
810 PRINT "KO =";KO;"volts/radian","Ko =";KO;"Hz/volt","KA =";KA,"FR =";FR;"Hz"
820 PRINT:PRINT
830 PRINT "LOOP GAIN:","KL =";KL;"Hz/radian","Kv =";KV;"radians/second"
840 PRINT
850 PRINT "Open Loop Gain at 1 radian/second:";DB;"dB":PRINT
860 PRINT "Uncompensated Loop Time Constant (1/Kv):";T;"seconds":PRINT
870 PRINT "Frequency Hold In Range:";FH;"Hz,",FO;"Hz  to   ";F1;"Hz"
880 PRINT
890 PRINT "Phase Detector Output Voltage Range:";VO;"volts   to   ";V1;"volts"
900 PRINT
910 PRINT "Amplifier Output Voltage Range:";V2;"volts   to   ";V3;"volts"
920 RETURN
```

FIGURE 10-54   (page 3 of 3)

11

Pulse and Digital Modulation

■ INTRODUCTION

Until a few decades ago, most information signals were in a continuous analog form. Then came digital computers and integrated circuits. The proliferation of computers, storing and processing information in discrete forms of energy and measured in *bits*, has produced a strong demand for digital data transmission equipment.

Whereas amplitude, frequency, and phase modulation of piecewise continuous carriers is used for transmission of digital data, a fundamentally different modulation technique called *pulse code modulation* (PCM) is used for converting analog signals into digital bits. Digital electronic circuits, processing equipment, and PCM transmission systems have some real advantages over their analog counterparts:

- Digital circuits are relatively inexpensive and simple to interface.
- Combining computer data with video and voice signals allows maximum flexibility when all are in a digital format.

- Non-real-time digital techniques allow signal processing that would be impossible with only analog circuits.

- Noise does not accumulate from repeater to repeater along a digital transmission link because the pulses can be completely regenerated in the repeaters.

In addition to constant-amplitude, constant-width digital signals, the continuous-wave modulation techniques already studied can be adapted to pulsed carriers. This type of analog/pulsed hybrid signal is considered next.

Analog/Pulse Modulation

The modulation of pulses is very similar to analog modulation; that is, the amplitude of pulses can be varied to produce pulse amplitude modulation (PAM). The duration or width of pulses can be varied to produce pulse duration or pulse width modulation (PDM or PWM). And, finally, the position of pulses can be varied to produce pulse position modulation (PPM).

These three modulation schemes mentioned so far are not digital and in fact require analog channels for transmission. PCM, on the other hand, is a truly digital signal and can be processed by digital computer circuitry. Pulse/analog modulated signals—PCM is not one—are more often used in instrumentation and control systems than for telecommunications. PAM, PWM, and PPM are actually the AM, FM, and PM of pulse carriers, with all the advantages and disadvantages of their purely analog cousins. PAM does, however, have the capability, like PCM, of being time-division multiplexed and is used in the voice/PCM interface. It also is used to produce higher-level modulation schemes for data modems and digital radio. Consequently, PAM is an important pulse modulation technique in communications systems. ■

11–1 ■ SAMPLED DATA AND THE SAMPLING THEOREM

A long-used technique for remote monitoring of such parameters as temperature, power consumption, motion, flow rates in such applications as satellites, ocean buoys, mountaintop repeaters, and radio transmitters is to transmit periodic samples of these parameters on a telemetry link. A *transducer* is used to produce an electrical voltage proportional to the parameter magnitude, and the voltage amplitude is periodically sampled. Voltage sampling of the analog voice signal from a telephone produces a PAM signal as the first step in the development of PCM for long-distance telecommunications.

Figure 11–1 shows an example of a sampler. The switch S_1 is closed instantaneously every T_s seconds to allow the instantaneous amplitude of v_A to appear at the output. The output is, in fact, a pulse amplitude modulated signal.

The sampling process can be described mathematically as multiplication of the two input time functions $v_A(t)$ and $s(t)$; that is, a sampler is a mixer with a train of very narrow pulses as the local oscillator input. The remarkable thing about this

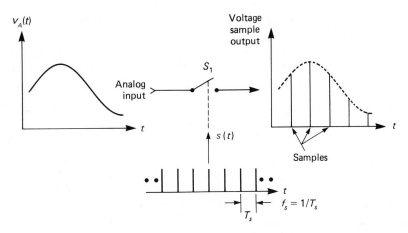

FIGURE 11-1   Impulse sampling of an analog voltage.

process, as stated in the sampling theorem, is that if the analog input is sampled instantaneously at regular intervals at a rate that is at least twice the highest analog frequency

$$f_s \geq 2f_A(\text{max}) \tag{11-1}$$

then the samples contain *all* of the information of the original signal. In practice $f_s = 2kf_A(\text{max})$, where $k > 1$ is the oversampling factor. Hence, a continuous analog voltage could be sampled at $f_s > 2f_A(\text{max})$, a list of the sample values could be made and sent by mail along with the value of f_s, and at the receiving end the entire analog signal could theoretically be reconstructed exactly.

A look at the frequency spectrum of the signals involved in sampling will illustrate an important concept that can also be demonstrated mathematically. The frequency spectrum of the sampled analog signal, called ideal PAM and illustrated in Figure 11-2, will show why it is necessary to sample at or above the *Nyquist rate* $f_s \geq 2f_A(\text{max})$, why all the information is present in the sampled output signal, and what is necessary to recover the original analog signal.

The analog signal $v(t)$ has a signal spectrum represented by the Fourier transform $V(f)$, and the sampling signal

$$s(t) = \sum_{n=-\infty}^{\infty} \delta(t - nT_s) \tag{11-2}$$

consists of instantaneous impulses every nT_s seconds, where $n = 0, \pm 1, \pm 2, \ldots$. The Fourier transform of $s(t)$ is

$$S(f) = \frac{1}{T_s} \sum_{n=-\infty}^{\infty} \delta(f - nf_s) \tag{11-3}$$

which represents a spectrum consisting of all the harmonics of the sampling frequency f_s.

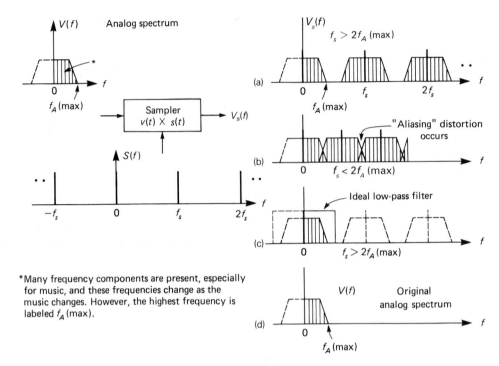

FIGURE 11–2 Sample spectra and their outputs. (a) $f_s > 2f_A(\text{max})$. Nyquist criteria met. (b) $f_s < 2f_A(\text{max})$. Frequency foldover or "aliasing" distortion occurs. (c) $f_s > 2f_A(\text{max})$ and recovery of original information with low-pass filter. (d) The original analog signal spectrum following recovery as in (c).

The time-domain product performed by the sampler produces a sampled output spectrum given by

$$V_s(f) = \frac{1}{T_s} \sum_{n=-\infty}^{\infty} V(f - nf_s) \qquad\qquad (11\text{–}4)$$

where this spectrum consists of replicas of the analog signal spectrum $V(f)$, translated in frequency by each of the sampling frequency harmonics. This is the same process, then, as modulating n harmonically related carriers with frequencies nf_s with a single analog information signal spectrum represented generally as $V(f)$.

Figure 11–2 shows the frequency spectrum for an arbitrary but typical analog signal to be sampled. Ideal sampling impulses are also shown. Notice that the frequency spectrum of the periodic impulses is a series of equal-amplitude harmonics of the sampling rate f_s. The sampler is a wideband (harmonic) mixer producing upper and lower sidebands at each harmonic of the sampling frequency. Figure 11–2a illustrates the correct way to sample: if sampling is done at $f_s > 2f_A(\text{max})$, the upper and lower sidebands do not overlap each other, and the original information can be recovered at the receiver by merely passing the signal through a low-pass filter (see parts (c) and (d)). However, if the sampling rate is less than the Nyquist rate, $f_s < 2f_A(\text{max})$, the sidebands overlap, as shown in Figure 11–2b. The result is *frequency-folding* or *aliasing* distortion, which makes it impossible to recover the original signal without distortion. The Nyquist criterion for sampling is identical to the criterion for the minimum possible carrier frequency for analog AM.

A visual example of aliasing distortion is seen when watching an old-time western with a spoked-wheel wagon or stagecoach. The camera is sampling the scene (the information). If the camera sampling (shutter) rate falls behind the speed of the turning wheel, the spokes and wheel seem to stop and even go backward; the rate of apparent backward motion is the aliasing frequency. In any case, it is impossible to recover the true information (wheel speed and direction) once the image is cast in celluloid (without some sleight of hand). A final comment on this scene is noteworthy: Most of the picture is unaffected; only the high-frequency parts are unrecoverable.

11–2 ■ PULSE AMPLITUDE MODULATION— NATURAL AND FLAT-TOP SAMPLING

The circuit of Figure 11–3 is used to illustrate pulse amplitude modulation (PAM). The FET is the switch used as a sampling gate. When the FET is on, the analog voltage is shorted to ground; when off, the FET is essentially open, so that the analog signal sample appears at the output. Op-amp 1 is a noninverting amplifier that isolates the analog input channel from the switching function. Op-amp 2 is a high-input-impedance voltage follower capable of driving low-impedance loads (high "fanout"). The resistor R is used to limit the output current of op-amp 1 when the FET is "on" and provides a voltage division with r_d of the FET. (r_d, the drain-to-source resistance, is low but not zero.)

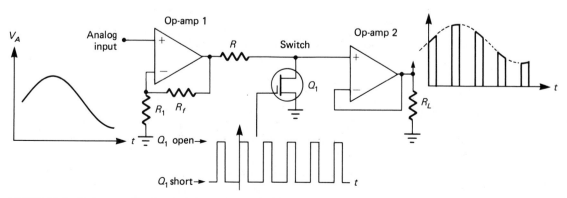

FIGURE 11–3   Pulse amplitude modulator, natural sampling.

Notice that the tops of the PAM pulses of Figure 11–3 retain the shape of the original analog waveform; this is called *natural sampling*. The frequency spectrum of the sampled output is different than that for ideal (impulse) sampling because the spectrum for the finite-width sampling pulses is different. As shown in Chapter 3, the amplitudes of frequency components for narrow finite-width pulses decreases for higher harmonics in a (sin x)/x fashion. This alters the information spectrum, which requires the use of frequency compensation (called *equalization*) before recovery by a low-pass filter.

The most common technique for sampling voice in PCM systems is to use a sample-and-hold circuit. As seen in Figure 11–4, the instantaneous amplitude of the analog (voice) signal is held as a constant charge on a capacitor for the duration of the sampling period T_s. This technique is useful for holding the sample constant while other processing is taking place, but it alters the frequency spectrum and introduces an error, called *aperture error*, resulting in an inability to recover exactly the original analog signal. The amount of error depends on how much the analog signal changes during the holding time, called *aperture time*. To estimate the maximum voltage error possible, determine the maximum slope of the analog signal and multiply it by the aperture time ΔT in Figure 11–4. The analysis is done for a delta modulator later in this chapter and is useful for determining the sampling rate required to keep below a specified amount of error.

A sample-and-hold circuit produces a rectangular pulse of A volts for a duration of T_s seconds starting at $t = 0$. The (Laplace) transfer function for this circuit is derived as follows: The rectangular pulse produced can be described by

$$h(t) = Au(t) - Au(t - T_s) \tag{11-5a}$$

$$= \begin{cases} 0 & t < 0 \\ A & 0 \le t \le T_s \\ 0 & t > T_s \end{cases} \tag{11-5b}$$

where $u(t)$ is the unit step function; see Example 1–10 and Table 1–2, transform 2.

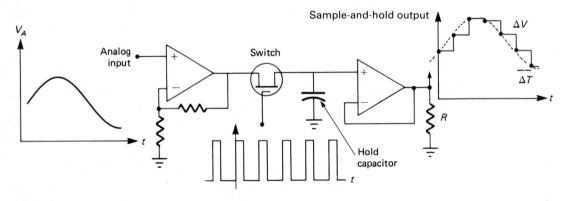

FIGURE 11-4 Sample-and-hold circuit and flat-top sampling.

The transfer function is found by taking the Laplace transform of $h(t)$ as

$$H(s) = \int_0^{T_s} A e^{-st}\, dt \tag{11-6a}$$

$$= A \left(\frac{e^{-st}}{-s} \right) \Big|_0^{T_s} = A \frac{e^{-sT_s} - e^0}{-s}$$

$$= A \frac{1 - e^{-sT_s}}{s} \tag{11-6b}$$

Time-Division Multiplexing

Observe the natural and flat-top PAM pulse trains of Figures 11–3 and 11–5. Notice that narrow pulses occurring at a rate of $f_s \gg 2f_A(\text{max})$ leaves quite a bit of space between pulses. PAM transmissions allow for the use of a single transmission line by more than one information source operating at the same time. As illustrated in Figure 11–6, by sampling many signals at the same rate and interleaving the pulses into separate time slots, many information sources can share the same transmission line "simultaneously." This technique of dividing the time and combining many different signals into one signal path is called *time-division multiplexing* (TDM). At the receiving end of the TDM system, the individual analog signals can be recovered by circuits similar to those at the transmitter. The receiver commutator demultiplexes the PAM input, and the low-pass filters demodulate the three individual PAM signals to compensate and recover the analog voltages.

In the three-channel multiplexed PAM system of Figure 11–6, each channel is filtered and sampled once per revolution (cycle) of the commutator. Notice that the commutator is performing both the sampling and the multiplexing. The commutator must operate at a rate that satisfies the sampling theorem for each channel. Consequently, the channel of highest cutoff frequency determines the commutation rate for the system of Figure 11–6.

As an example, suppose the maximum signal frequencies for the three input channels are $f_{A1}(\text{max}) = 4\,\text{kHz}, f_{A2}(\text{max}) = 20\,\text{kHz}$, and $f_{A3}(\text{max}) = 4\,\text{kHz}$. For the TDM system of Figure 11–6, the multiplexing must proceed at $f \geq 2f_A(\text{max}) = 40$ kHz to satisfy the worst-case condition.

FIGURE 11–5 Flat-top PAM signals.

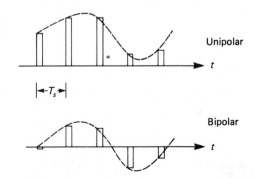

Unipolar

$\leftarrow T_s \rightarrow$

Bipolar

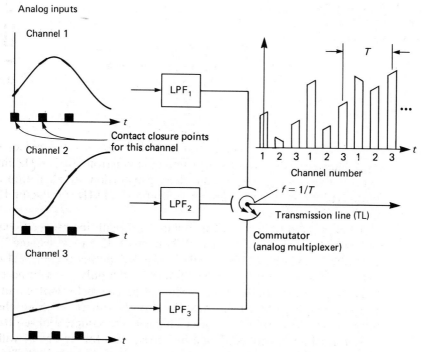

FIGURE 11-6 Time-division multiplex of three information sources.

With this, we can calculate the transmission line pulse rate as follows: The commutator completes one cycle, called a *frame*, every 1/40 kHz = 25 μs. Each time around, the commutator picks up a pulse from each of the three channels. Hence, there are 3 pulses/frame × 40 k frames/s = 120 k pulses/s.

This example illustrates that multiplexing of many channels will require relatively high-frequency transmission systems. A little creativity, however, can help minimize the transmission system bandwidth. For example, notice that the 4-kHz channel is being sampled at five times the rate required by the sampling theorem. But if we slow down the commutator, the 20-kHz channel will be inadequately sampled. One thought might be to multiplex at 8 k frames/s and sample the 20-kHz channel 5 times per frame. If you sketch this, as is done in Figure 11-7, you

FIGURE 11-7 Possible TDM solution.

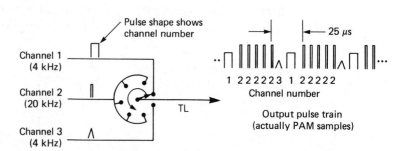

FIGURE 11-8 TDM solution for minimum transmission line pulse rate (and bandwidth).

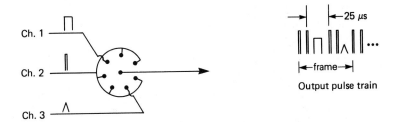

Output pulse train

discover that there are 7 pulses/frame × 8 k frames/s = 56 k pulses/s, which looks good. However, the two missing samples stolen from the 20-kHz channel result in inadequate sampling and periodic aliasing distortion. A calculation will show that, for no errors, the commutation rate must be 17.14 kHz, producing 120 k samples/s on the transmission line.

A better scheme is shown in Figure 11–8 with insertion of channels 1 and 3 between two samples of channel 2. Now, with 12.5 μs/pulse and 7 pulses/frame, the multiplexing rate can be (2 pulses/25 μs)/(7 pulses/frame) = 11.428 k frames/s and 11.428 k frames/s × 7 pulses/frame = 80 k pulses/s with no errors.

There are other ways of interspersing the channel samples, but the minimum transmission-line pulse rate—and bandwidth—for these three channels is 80 k pulses/s. In fact, while the slower channels are sampled more often than necessary, the minimum transmission-line pulse rate for no errors will be twice the required sampling rate for the highest-frequency input channel. This is because at least one pulse of the fast channel must be dropped in order to insert a sample from the slower channels. Realistically, the only way to solve a problem like this is to sketch the multiplexed samples and the maximum allowable time between samples.

Intersymbol Interference (ISI)

Another consideration concerning the transmission of pulses over circuits with limited bandwidth and nonlinear phase response is intersymbol interference. As noted in Chapter 3, rectangular pulses will not remain rectangular in less than infinite bandwidth systems. As the transmission bandwidth narrows, the pulses round off, and what is worse, phase distortion will cause tilt and even affect the next pulse. Figure 11–9 is a sketch from Chapter 3 of a pulse distorted in a typical transmission channel (like a telephone line). Notice how the next pulse cannot help but be affected because the line is still active with channel 1 energy when the channel 2 signal comes on. This effect is *intersymbol interference* (ISI). As more

FIGURE 11-9 Intersymbol interference; transmission-channel distortions.

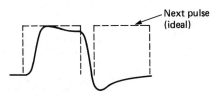

Next pulse (ideal)

FIGURE 11–10 Low-pass data channel with brick wall ($r = 0$) and raised cosine ($0 < r \leq 1$) rolloff characteristics.

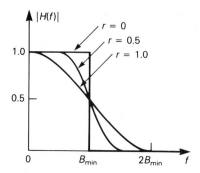

pulsed channels are multiplexed, high frequency response and linear phase response become more critical. Also, when the distortion on one pulse affects the next pulse, cross talk between channels results.

Transmission channels can be compensated (called *equalization*) to make the channel response closer to the ideal channel. Perhaps the most common technique is to use a channel filter that has a raised cosine rolloff characteristic with rolloff factor r, where $0 < r \leq 1$. The amplitude frequency response is illustrated in Figure 11–10, where B_{min} is the absolute minimum (Nyquist) bandwidth. As will be demonstrated under "Bandwidth Requirements" in Section 11–4, the maximum data rate of rectangular-shaped pulses that can be supported without ISI in a channel having the transfer function $H(f)$ is given by

$$f_B(\text{max}) = \frac{2B_{min}}{1 + r} \tag{11-7}$$

where B_{min} is in units of hertz and $f_B(\text{max})$ is in pulses per second.

11–3 ■ PULSE WIDTH AND PULSE POSITION MODULATION (PWM AND PPM)

PWM and PPM are pulsed carrier modulation techniques in which varying voltages, such as telemetry signals, are used to vary the duration or the relative position of carrier pulses. In pulse width modulation, the width of each pulse is made directly proportional to the amplitude of the information signal. In pulse position modulation, constant-width pulses are used, and the position or time of occurrence of each pulse from some reference time is made directly proportional to the amplitude of the information signal. PWM and PPM are compared and contrasted to PAM in Figure 11–11.

Figure 11–12 shows a PWM modulator. This circuit is simply a high-gain comparator that is switched on and off by the sawtooth waveform derived from a very stable frequency oscillator. Notice that the output will go to $+V_{cc}$ the instant the analog signal exceeds the sawtooth voltage. The output will go to $-V_{EE}$ the instant the analog signal is less than the sawtooth voltage. With this circuit the average value of both inputs should be nearly the same. This is easily achieved with equal-

FIGURE 11–11   Analog pulse modulation signals.

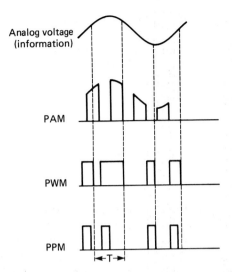

value resistors to ground. Also, the $+V$ and $-V$ values should not exceed the V_{cc} or $-V_{EE}$ rails. A 710-type IC comparator can be used for positive-only output pulses that are also TTL compatible. PWM can also be produced by modulation of various voltage-controllable multivibrators. One example is the popular 555 timer IC. Other (pulse output) VCOs, like the 566 and that of the 565 phase-locked loop IC, will produce PWM. This points out the similarity of PWM to continuous analog FM. Indeed, PWM has the advantages of FM—constant amplitude and good noise immunity—and also its disadvantage—large bandwidth.

Close comparison of the PWM and PPM signals of Figure 11–11 shows that the trailing edge of each PWM pulse occurs at a regular (clock) interval. This is achieved by using the sawtooth waveform as a switching signal in the modulator of Figure 11–12. A triangular switching signal would not produce the same result because both edges are a function of the analog input. When the trailing (or

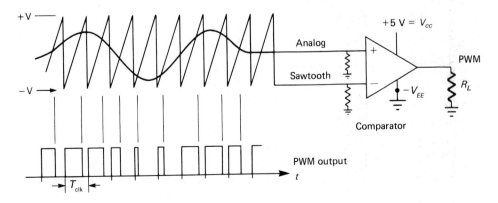

FIGURE 11–12   Pulse width modulator.

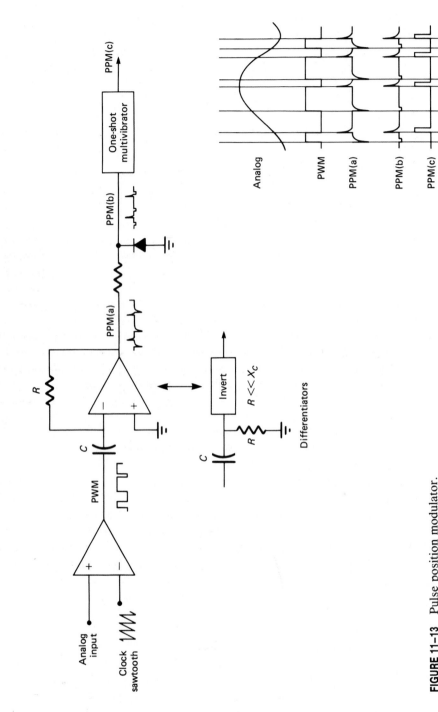

FIGURE 11-13 Pulse position modulator.

leading) edge does provide a fixed-reference time of occurrence, the PWM signal can be used to generate pulse position modulation.

As shown in Figure 11–13, a PPM signal is produced by differentiating a PWM signal. Functionally, the differentiation is realized by an op-amp differentiator or a well-designed RC high-pass filter. A well-designed RC differentiator has a very short time constant so that the decaying edge dissipates before the arrival of the next pulse.

As shown in Figure 11–13, a diode can be used to eliminate the negative-polarity clock pulse if desired, and a one-shot multivibrator will provide a constant-width, flat-top PPM signal.

Demodulation

Since the width of each pulse in the PWM signal shown in Figure 11–13 is directly proportional to the amplitude of the modulating voltage, and the trailing edge is at regular clock intervals, demodulation can be achieved by time-averaging (integrating) the received pulses. In another similar approach, the signal is differentiated as shown in Figure 11–13 (to PPM in part (a)), then the positive pulses are used to start a ramp, and the negative clock pulses stop and reset the ramp. The effect of this procedure is to produce frequency-to-amplitude conversion (or equivalently, pulse width–to-amplitude conversion). The variable-amplitude ramp pulses are then time-averaged (integrated) to recover the analog signal.

The last approach suggests a technique often used in phase meters to demodulate PPM. As illustrated in Figure 11–14, a narrow clock pulse sets an RS flipflop output high, and the next PPM pulse resets the output to zero. The resulting signal, PWM, has an average voltage proportional to the time difference between the PPM pulses and the reference clock pulses. Time-averaging (integration) of the output produces the analog variations. This circuit is the RS-FF phase detector introduced in Chapter 10.

It is seen that PPM has the same disadvantage as continuous analog phase modulation: a coherent clock reference signal is necessary for demodulation. The reference pulses can be transmitted along with the PPM signal. This is achieved by full-wave-rectifying the PPM pulses of Figure 11–13a, which has the effect of reversing the polarity of the negative (clock-rate) pulses. Then an edge-triggered flipflop (J-K or D-type) can be used to accomplish the same function as the RS flipflop of Figure 11–14, using the clock input. But transmitting the reference

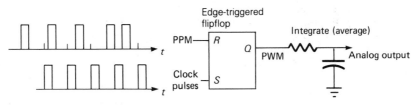

FIGURE 11-14   PPM demodulator.

signal along with the PPM does not come without penalty: More pulses per second will require greater bandwidth, and the width of the pulses limit the pulse deviations for a given pulse period.

11-4 ■ DIGITAL MODULATION

Pulse Code Modulation (PCM)

As we have seen, pulse modulation methods such as PAM, PWM, and PPM are basically pulsed versions of AM, FM, and PM. These are not digital modulations because digital systems use pulses of constant amplitude, width, and duration. The information in digitally modulated signals is contained in the presence or absence of a pulse and consequently is less susceptible to noise than analog/pulse modulated signals.

There is an extremely important, truly digital modulation technique called *pulse code modulation* (PCM), in which the amplitude of an analog signal is sampled and a sequence of digital 1s and 0s are produced, which represent the amplitude of the analog signal at the sampling instant. Thus, PCM is produced by an analog-to-digital conversion process. As in the case of other pulse modulation techniques, the rate at which samples are taken and encoded must conform to the Nyquist sampling rate: The sampling rate must be greater than or equal to twice the highest frequency in the analog signal; $f_s \geq 2f_A(\max)$.

A simple example to illustrate the pulse code modulation of an analog signal is shown in Figure 11–15. Here, an analog input sample becomes three binary digits (bits) in a sequence which represents the amplitude of the analog sample.

At time $t = 1$, the analog signal is 3 V. This voltage is applied to the encoder for a time long enough that the 3-bit digital sequence (''word''), 011, is produced. The second sample at $t = 2$ has an amplitude of 6 V, which is encoded as 110. This particular example system is conveniently set up so that the analog voltage value (decimal) is encoded with its binary equivalent. The decimal/binary equivalence is given in Table 11–1.

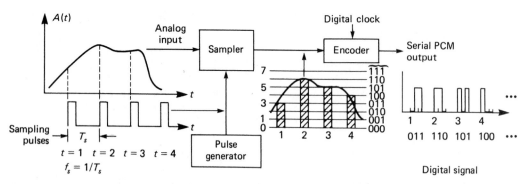

FIGURE 11–15 A 3-bit PCM system showing analog-to-3-bit-digital conversion.

TABLE 11–1 Decimal-to-Binary Equivalents

Decimal Value	Binary Code
0	000
1	001
2	010
3	011
4	100
5	101
6	110
7	111

Notice that, with a 3-bit code, $2^3 = 8$ different levels, called *quantization levels*, can be distinguished by the encoder. One more bit in the digital code allows twice as many quantization levels ($2^4 = 16$) to be distinguished. The added complexity in the encoding circuitry to produce a 4-bit code would allow finer resolution and greater accuracy in reproducing the analog signal at the demodulator.

The demodulator must take the digital code words and generate an analog voltage. This is digital-to-analog conversion. Analog-to-digital (A/D) and digital-to-analog (D/A) converter circuits are discussed in more detail in the next chapter.

Multiplexing PCM Channels

The serial PCM output of Figure 11–15 shows that there is time between the last bit of one encoded sample and the first bit of the next sample. This time can be used for multiplexing another digital signal of another analog channel. Figure 11–16 shows an example of time-division multiplexing in which two PCM channels are combined (multiplexed) on a single transmission line.

In Figure 11–16, the analog signals of channels 1 and 2 are converted to 3-bit/sample digital words. These 3-bit samples are then picked off by the multiplexer circuit and serially (one at a time, in series) transmitted down the transmission

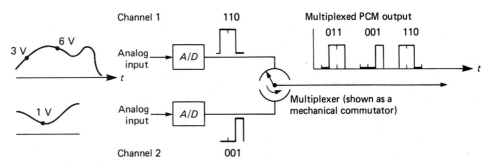

FIGURE 11–16 Two-channel multiplexed 3-bit PCM system.

line. This time-sharing of a single transmission line (TDM) allows many channels, such as telephone conversations, to be transmitted over a single transmission line channel.

The channel 1 and 2 inputs could very well be carrying the signals from two telephone conversations. Compared to frequency-division multiplexed analog techniques, PCM is increasingly used for transmitting telephone conversations over long distances because of its superior noise rejection and pulse-regeneration capabilities. The digital pulses of a multiplexed PCM telephone system (digitized voice traffic) can be transmitted between switching centers from city to city via digital microwave radio or optical fiber systems. The received noisy pulses can be regenerated into clean (high S/N) pulses for noise-free telephone connections or retransmitted to other switching centers for long-haul links.

Bandwidth Requirements

Suppose that the two-channel, multiplexed PCM system of Figure 11–16 is being sampled and multiplexed at 6000 samples per second with essentially no deadtime between the three-bit samples. Figure 11–17 shows several "frames" (complete turns of the commutator). Since each frame is made up of 2 (channels/frame) × 3 (bits/channel), there are 6 bits/frame. The commutator completes 6000 frames per second, so the transmission line is carrying 6 bits/frame × 6 k frames/s = 36 k bits/s. If digital *TTL signals* are used, then the worst-case digital signal, in terms of transmission line changes, occurs when alternate 1s and 0s are transmitted; this situation is seen as the first four bits of Figure 11–17 or the squarewave signal of Figure 11–18.

Since there are 36 k bits/s for our two-channel system, each bit requires

$$T_b = \frac{1}{f_b} = \frac{1}{36 \text{ k bits/s}} = 27.8 \ \mu\text{s/bit}$$

Also, we see from Figure 11–18 that the basic line rate is

$$f = \frac{1}{T} = \frac{1}{2T_b} = \frac{1}{2}f_b \qquad\qquad (11\text{–}8)$$

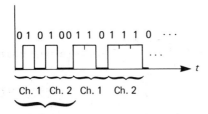

FIGURE 11-17 Multiplexed serial output of two-channel, 3-bit PCM systems.

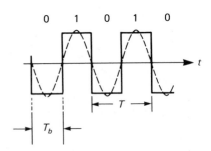

FIGURE 11-18 Squarewave and fundamental frequency (dashed line).

The basic line rate is the fundamental frequency of a binary squarewave and is the absolute minimum bandwidth required to transmit the basic information of the digital TTL signal of Figure 11–17. That is,

$$BW_{min} = \frac{1}{2} f_b \qquad (11-9)$$

where f_b is the transmission-line bit rate (baud rate). That is, an ideal noiseless channel can carry up to 2 bits per cycle of bandwidth for transmitted binary signals. Thus, for our 3-bit PCM system multiplexing two channels at 6 k frames/s, the absolute minimum bandwidth required is

$$BW_{min} = \frac{1}{2} f_b = \frac{1}{2} \ 36 \text{ k bits/s} = 18 \text{ kHz}$$

Refer to Equation 11–5 for intersymbol interference (ISI) considerations.

Delta Modulation (DM)

In pulse code modulation, the analog signal is sampled at regular intervals, and the amplitudes of these samples are coded in a multiple-bit digital format. A variation on this A/D technique, called *delta modulation*, compares successive signal samples and transmits their difference with one bit, thus reducing the number of bits required for coding. This savings can be very important for fast-changing signals like full audio (music) and, in particular, television video, which has a 4-MHz bandwidth.

Like PCM, a delta modulation system consists of an encoder and a decoder; unlike PCM, however, a delta modulator generates single-bit words that represent the difference (delta) between the actual input signal and a quantized approximation of the preceding input signal sample. This is represented in Figure 11–19 with a sample-and-hold, comparator, up-down counter staircase generator, and a D-type flipflop (D-FF) to derive the digital pulse stream.

The operation of the simplified delta modulator of Figure 11–19 is as follows: The continuous analog signal is band-limited in the low-pass filter (LPF) to pre-

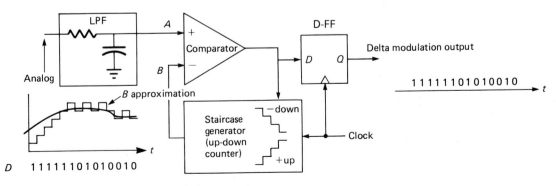

FIGURE 11–19   Possible delta modulation encoder.

FIGURE 11-20   DM demodulator.

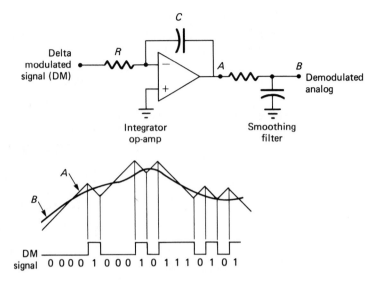

vent aliasing distortion, as in any sampling system. The analog signal V_A is then compared to its discrete approximation V_B. If the amplitude of V_A is greater than V_B, the comparator goes high, calling for positive-going steps from the staircase generator; if, however, V_B exceeds V_A, the comparator goes low, calling for negative-going steps from the staircase generator. The comparator also sets the D-FF and the output will be properly clocked because the edge-triggered D-FF can change state only at rising edges of the input clock: 1s will correspond to positive steps and 0s to negative steps of the staircase generator.

Decoding of the delta modulation (DM) signal can be accomplished with an up-down staircase generator and a smoothing filter or simply by integrating the DM pulses, as shown in Figure 11–20. The resulting demodulated signal is illustrated as curve B.

A practical implementation of a delta modulator is shown in Figure 11–21, where the up-down counter and D/A converter (sometimes called a DAC) com-

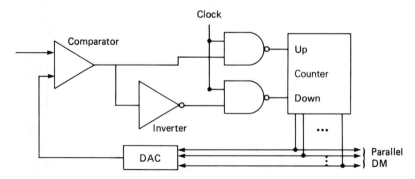

FIGURE 11-21   Up-down staircase generator for delta modulator.

FIGURE 11–22 Critical design parameters in constant step-size linear delta modulation.

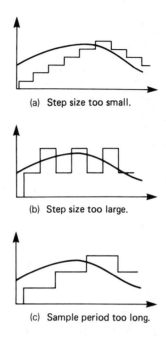

(a) Step size too small.

(b) Step size too large.

(c) Sample period too long.

prise the staircase generator of Figure 11–19. The delta modulator of Figure 11–21 is usually referred to as a *tracking* or *servo* A/D converter.

A close look at the discrete version of the analog signal in Figure 11–19 (curve *B*) should help you to realize that low-distortion delta modulation depends on the ability of the system to closely follow the changes of the analog signal. Clearly, as seen in Figure 11–22, the critical parameters determining the quality of a system using a constant step size are the designer's choice of step size and sampling period length.

With too small a step size, the analog signal changes cannot be followed closely enough; this is called *slope overload* (Figure 11–22a). With too large a step size, two problems arise: poor signal approximation (resolution) and large quantization noise (Figure 11–22b). This condition is called *granular noise*.

Too long a period has the same problem as too small a step size and poor resolution (Figure 11–22c). When the period is too short, too much transmission bandwidth is required.

■ ■ ■ ■

EXAMPLE 11–1

A 5-V pk, 4-kHz sinusoid is to be converted to a digital signal by delta modulation. The step size must be 10 mV. Determine the minimum clock rate that will allow the DM system to follow exactly the fastest input analog signal change, that is, to avoid slope overload.

Solution:

The fastest rate of change of a sinewave, $v(t) = V \sin \omega t$, is the slope at the zero

FIGURE 11-23

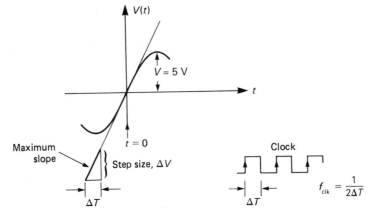

crossover points ($t = 0$ in Figure 11–23). The slope at $t = 0$ is found as follows:

$$\text{slope} = \frac{dv(t)}{dt} = \frac{d}{dt}\,(V\,\sin\,\omega t) = \omega V\,\cos\,\omega t$$

At $t = 0$ the slope is $\omega V\,\cos\,0 = \omega V$ or, since $\omega = 2\pi f$, the maximum slope of the sinewave is

$$\frac{\Delta v}{\Delta t} = 2\pi f V \qquad\qquad\qquad \textbf{(11–10)}$$

where f is the frequency of the analog sinusoid.
    Solving for Δt,

$$\Delta t = \frac{\Delta v}{2\pi f V} = \frac{10\text{ mV}}{2\pi(4000\text{ Hz})\,5\text{ V}} = 0.079\ \mu\text{s/cycle}$$

Δt is half the clock period because steps occur only at positive transitions of the clock in a practical system.
    Thus, $T_{\text{clk}} = 2\Delta t = 1/f_{\text{clk}}$, so that

$$f_{\text{clk}} = 1/2\Delta t = 1/(2 \times 0.079 \times 10^{-6}\text{ s/cycle}) = \textbf{6.3 MHz}$$

Practical DM.   A circuit configuration in present use for telecommunication applications involving filters, speech scramblers, instrumentation, and remote motor control is shown in Figure 11–24.
    To demodulate the digital signal simply integrate the pulses. In fact, if the integrator has the same RC time constant as the preceding modulator, the integrated demodulator output is the same as the B curve of Figure 11–24. The integrator output is then put through a sharp cutoff low-pass filter to smooth out the final analog signal. The circuit of Figure 11–24 could be modified to include a variable-gain amplifier (VGA) and decision logic as indicated in Figure 11–25 for adaptive delta modulation.
    Example 11–1 points out that linear DM is not viable for transmitting voice accurately because too high a pulse rate is required. Two changes can reduce the

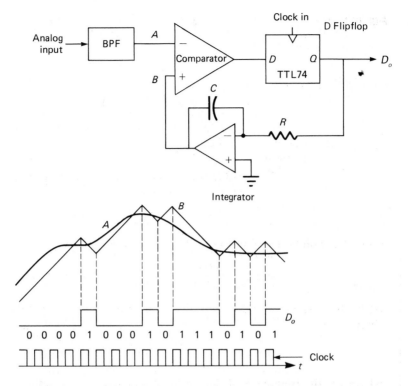

FIGURE 11-24   Integrating linear delta modulator block diagram and signals.

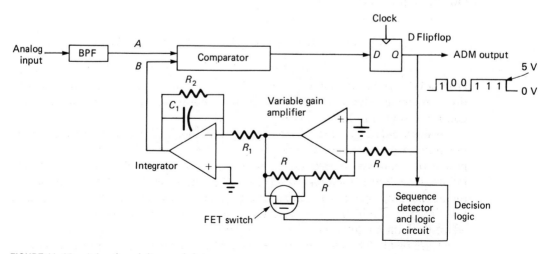

FIGURE 11-25   Adaptive delta modulator.

transmission rate considerably: One is to use companding as PCM does (companding is discussed in Chapter 12); the other, more important, solution is to use adaptive delta modulation.

Adaptive DM (ADM). A solution to the trade-offs and compromises of the simple delta modulation system just given is to have a variable-step-size system. This could be accomplished as in Figure 11–25 with a variable-gain amplifier at the output of the D-type flipflop and a decision circuit that counts the number of + and − steps taken over a given period of time and decides whether the step size should be increased and by how much.

As an example of how this circuit might operate, suppose that the adaptive algorithm (decision criteria) will be as follows: The preceding four bits of ADM output are counted. If an equal number of 1s and 0s occur in this interval (the last 4 bits), then the VGA gain will be $A_v = -1$ (the FET switch will be a short). If more 1s than 0s or more 0s than 1s are counted, the step size is doubled, $A_v = -2$ (the FET switch will be open-circuited). With the VGA gain doubled, the input to the integrator will be doubled. The results are constructed for an analog input and compared to the results for the linear DM in Figures 11–26 and 11–27.

One interesting application of the delta modulator is in a telephone PCM *codec* (coder/decoder). Typical PCM requires expensive, precision analog components. One approach to producing an inexpensive all-digital PCM codec that could be installed in every telephone is to pass the analog signal through a low-pass analog filter to limit the baseband, then convert this analog signal to the 1-bit-per-sample DM. The DM is converted to linear PCM, then to the A-law (Europe) or μ255-law (U.S.) PCM before serial transmission on telephone networks.

A block diagram of the dual modulation codec just described is shown in Figure 11–28.

An integrated circuit for producing adaptive delta modulation with a variable-slope integrator is shown in Figure 11–29.

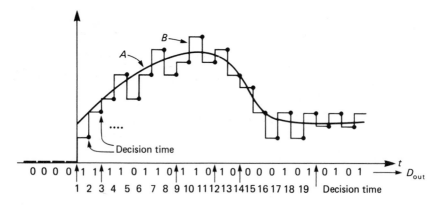

FIGURE 11–26 ADM with step or double step sizes. Arrows are shown along the horizontal axis to indicate where the step size changes. These changes are based on the number of 1s and 0s of the previous 4 bits.

FIGURE 11-27 Different DM results depending on step size used.

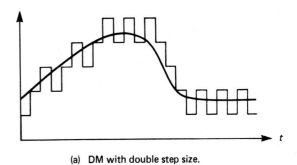

(a) DM with double step size.

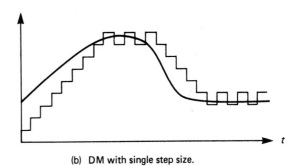

(b) DM with single step size.

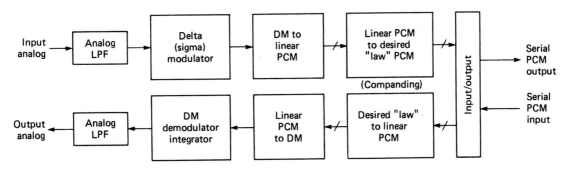

FIGURE 11-28 Dual modulation codec. Deriving PCM from delta modulation with an all-digital IC.

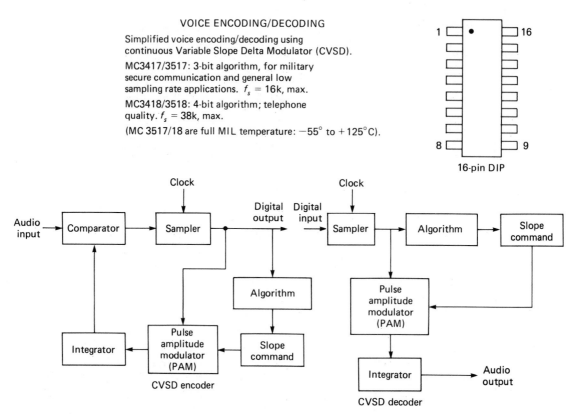

VOICE ENCODING/DECODING

Simplified voice encoding/decoding using continuous Variable Slope Delta Modulator (CVSD).

MC3417/3517: 3-bit algorithm, for military secure communication and general low sampling rate applications. $f_s = 16k$, max.

MC3418/3518: 4-bit algorithm; telephone quality. $f_s = 38k$, max.

(MC 3517/18 are full MIL temperature: $-55°$ to $+125°$C).

16-pin DIP

FIGURE 11-29 Adaptive delta modulation IC from Motorola.

11-5 ■ QUANTIZATION NOISE

It is appropriate to mention here an unavoidable limitation of digital modulation. The sharp signal variations above the average, inherent in delta modulation, and the discrete variations in demodulated PCM constitute a form of noise unique to digital modulation systems. This kind of noise, called *quantization noise*, is due to the discrete nature of digital signals and sets a lower limit on the minimum recoverable analog signal variation. Quantization noise and system resolution design are covered in Chapter 12.

11-6 ■ QUALITATIVE COMPARISONS OF PULSE AND DIGITAL MODULATION SYSTEMS

Like PAM, PCM can be time-division multiplexed because the modulated samples maintain a fixed position (slot) and duration in time. However, PCM is less noise-sensitive than PAM, and PCM can use digital constant-amplitude circuitry, unlike

FIGURE 11-30 Pulse and digital modulation waveforms.

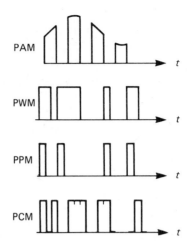

PAM, which requires linear circuits. A disadvantage of PCM is its greater bandwidth requirement. For example, in a simple 3-bit PCM system, three pulses must be transmitted, whereas only one is transmitted for the PAM sample (see Figure 11–30).

PWM and PPM are rarely used in multiplexed communication systems because of the large bandwidths required; frequency-division multiplexing, which requires a more complex system than TDM, must also be used. While PWM and PPM have better noise performance than PAM (like FM and PM over AM), PWM and PPM are not easily regenerated, and therefore noise accumulates over long-haul networks.

An additional disadvantage for PPM over all the other techniques is that PPM, like continuous-phase modulation, requires coherent demodulation. This usually means that a phase-locked loop and its acquisition circuitry are required.

In addition to these advantages for PCM over other pulse modulation techniques, the use of digital terminal equipment makes PCM more desirable in today's communications marketplace.

■ PROBLEMS

1. A motor tachometer registers variations as high as 10,000 rev/s. What is the minimum rate required for sampling this signal?

2. A voice channel has frequency components of up to 3.5 kHz.
   a. What minimum sampling rate is required if this signal is to be pulse amplitude modulated?
   b. What oversampling factor produces 8400 samples per second?

3. A voice channel has a frequency range of 1 to 3 kHz.
   a. How slowly can it be sampled for PAM?
   b. What kind of distortion occurs if it is sampled at 4.2 kHz?

4. Suppose a sampler produces the ideal PAM spectrum of Figure 11–2a. This signal is then ac-coupled to filter out everything below f_A(max). How could the original analog signal be recovered? (This is what happens in most applications where PAM is transmitted.)

5. What is required to demodulate ideal (impulse-sampled) PAM? Natural-sampled PAM?

6. Sketch a simple four-channel TDM system; include transmit and receive ends.
   a. If all input channels are limited to 4 kHz, what is the minimum multiplexed output pulse rate?
   b. How wide (in milliseconds) is each pulse if 50% duty cycle pulses, as shown in Figure 11–6, are used?

7. Show a multiplexed PAM system that can properly multiplex the following analog channels to achieve 120 k pps transmission rate:
   a. Five channels of 4 kHz and one channel of 10 kHz.
   b. Change the scheme to achieve 64 k pps. At what rate does the commutator operate?

8. To observe the time-domain effects of rectangular pulse transmission through bandlimited systems:
   a. Make an accurate sketch of the first three odd harmonics of $6/n$ sin $n\theta$, where θ varies from 0° to 180° in 10° intervals ($n = 1$ for fundamental or first harmonic). Also, add the amplitudes (of the three curves) at each interval and plot the sum of the three amplitudes. The sum is the narrowband approximation of a squarewave whose Fourier components add up as

$$\sum_{n=1(n\ \text{odd})}^{5} (6/n)\ \sin\ n\theta$$

   b. By replotting points or using tracing paper, shift the fundamental ($n = 1$) curve to the left by 30°. Continue the shifted (phase-delayed) curve out to $\theta = 180°$. Plot the new sum (resultant). Notice what will happen to any squarewave pulse occurring before or after this "squarewave" pulse. What happens is called *intersymbol interference* (ISI) distortion. The resultant (sum) curve also shows an approximation to the "tilt" or "sag" caused by phase distortion.

9. What is intersymbol interference and how can it cause interchannel cross talk?

10. **a.** Name the pulse modulations of Figure 11–31.
    b. Sketch a circuit to demodulate each signal of Figure 11–31.

FIGURE 11–31

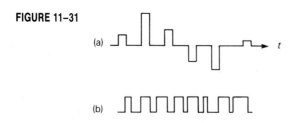

11. Sketch the pulse modulation that results from differentiating signal (b) of Figure 11–31. Name the resultant signal.

12. **a.** Sketch the output of the PWM modulator of Figure 11–32. Show correct voltage magnitude.
   b. Sketch a circuit, without repeating Figure 11–13, to produce PPM from the analog signal of Figure 11–32. Sketch the resulting PPM signal.
   c. What is the sampling rate for the PWM signal?

FIGURE 11–32

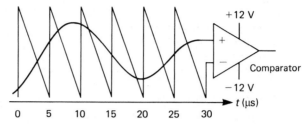

13. **a.** Sketch a PAM output for the analog signal of Figure 11–32. Use 2-μs-wide pulses occurring at 0, 5 μs, 10 μs, 15 μs, and so on.
   b. Sketch a demodulator circuit and its output.

14. What pulse modulated signal is most suited to demodulation by a phase-locked loop?

15. The VCO of Figure 11–33 is modulated by the data pulse shown. Use the information in the figure to determine the following.
   a. What type of pulse modulation is produced?
   b. Find the VCO frequency for $V = 0$ (f_{FR}).
   c. What value of V is required at the input to produce the output shown?

FIGURE 11–33

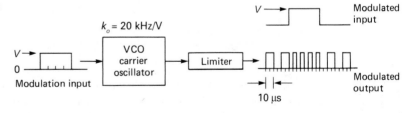

16. **a.** Show that the integrator of Figure 11–34 produces a constant ramp with a peak output voltage of $V_o = -(VW/RC)$. V is a constant input voltage applied for W seconds.

FIGURE 11–34 Integrate-and-dump circuit.

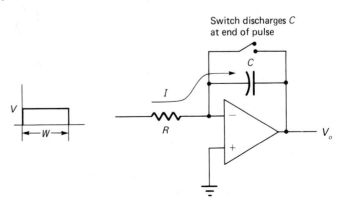

b. Show that the average of V_o over the interval W is $-VW/2RC$. (For PWM, the pulse width is proportional to the analog peak value—by definition—and the *integrate-and-dump* circuit of Figure 11–34, when followed by an averaging circuit, demodulates PWM. The "averaging circuit" averages over a clock period, not just over W.)

17. Sketch the Q output of the RS-FF demodulator of Figure 11–14.

18. Show that the average output voltage of Q for Problem 17 is proportional to the time between a clock pulse and a PPM pulse. (The output voltage is averaged over a full clock period.)

19. What advantage does PPM have over PWM?

20. a. Give one advantage of PCM over each of the pulse/analog modulations.
   b. Give a disadvantage of PCM.

21. How many bits are needed to quantize an analog signal into at least 250 levels?

22. If 6 bits are used in a PCM system, how many separate input levels are distinguishable?

23. A digital signal (binary) is to be transmitted at 9 k bits/s. What absolute minimum bandwidth is required to pass the fastest information change undistorted?

24. Four channels of 5-bit PCM are multiplexed at 8 k frames/s (with no additional pulses added). Determine the output pulse rate and pulse width.

25. What advantage does delta modulation have over PCM?

26. A ramp is to be modulated with delta modulation. The ramp starts at time zero with 1.25 V and continues with a slope of 5 V/ms [$v(t) = 5t + 1.25$ V (t in ms)]. The delta modulator takes 10 samples/ms (10 k samples/s) of 1 V/step. The first 1-V step starts at time zero.
   a. Sketch the ramp and DM estimate as it follows the ramp.
   b. Determine the transmitted binary sequence.

27. Repeat Problem 26 for an adaptive delta modulator (ADM) with an adaptation decision circuit algorithm as follows: (1) Look at the last four samples (assume all 0s before $t = 0$), (2) If the number of 1s is 0 or 4, the step is 1.5 V; if the number of 1s is 1 or 3, the step is 1.0 V; if the number of 1s is 2, the step is only 0.5 V.

28. A 12-V pk, 4-kHz sinusoid is to be modulated in linear DM with step size 10 mV.
   a. Determine the minimum clock rate.
   b. Change the step size to 100 mV and determine the minimum clock rate.
   c. Change the sinusoid to 15 kHz and determine the minimum clock rate.
   d. Comment on linear DM as a viable technique for transmitting voice; compare it to PCM, for instance.

11–7 ■ COMPUTER ANALYSIS

Using the computer for communications system and circuit analysis has become a virtual necessity in education and industry. What follows are examples of programs and/or readily available simulation software for solving problems of the type covered in this chapter.

```
10 REM ****              DELTA MODULATION               ****
20 REM *   EXAMPLE OF STUDENT COMPUTER ASSIGNMENT   *
30 REM ****         FOR PROFESSOR P. YOUNG, AT ASU       ****
40 CLS
50 REM VARIABLES: VS = SINUSOIDAL VOLTAGE
60 REM                 FS = FREQUENCY OF SINUSOID
70 REM                 SS = STEP SIZE
80 REM                 FC = MINIMUM CLOCK RATE
90 VS=O:FS=O:SS=O:FC=O
100 PRINT "THIS DELTA-MODULATOR (DM) PROGRAM ASKS FOR INFORMATION ABOUT"
110 PRINT "THE PEAK VOLTAGE AND FREQUENCY OF AN ANALOG INPUT SINEWAVE TO"
120 PRINT "BE CONVERTED TO A DIGITAL SIGNAL BY DELTA MODULATION.  IT ALSO"
130 PRINT "ASKS FOR THE DM STEP SIZE.  GIVEN THIS INFORMATION THE PROGRAM"
140 PRINT "COMPUTES THE MINIMUM CLOCK RATE THAT WILL ALLOW THE DM SYSTEM"
150 PRINT "TO EXACTLY FOLLOW THE FASTEST INPUT ANALOG SIGNAL CHANGE."
160 PRINT
170 PRINT:INPUT "ENTER THE SINUSOIDAL PEAK VOLTAGE: ";VS
180 PRINT:INPUT "ENTER THE FREQUENCY OF THE SINUSOID, HZ: ";FS
190 PRINT:INPUT "ENTER THE DELTA MODULATOR, VOLTAGE STEP SIZE: ";SS
200 LET FC=(2*3.14159*FS*VS)/(2*SS):PRINT
210 PRINT "THE MINIMUM CLOCK RATE OF THE DM SYSTEM MUST BE";FC;"HERTZ"
220 END
```

FIGURE 11–35

All the programs are interactive in the sense that an introductory sentence or short paragraph describes the purpose of the program and the user is then prompted for input data.

The program for Chapter 11 is as follows:

- **Delta modulator:** Given the amplitude, frequencies, and step size, the program of Figure 11–35 computes the minimum clock rate that will allow the specified DM system to exactly follow the highest-frequency analog input signal.

12

Digital Communication Concepts

■ COMMUNICATION CONCEPTS

In designing an efficient communication system, the notion of information content and the efficiency of its transfer must be considered. *Information theory* as used by electronic system designers is a scientifically established body of knowledge about the quantity and quality of information transmitted through electronic channels.

The quantity aspect of information involves determining the amount, speed, and efficiency of information transfer through a finite bandwidth channel. The quality aspect involves the determination of how precisely the received information represents what was transmitted. It should be noted that neither quantity nor quality has anything to do with the "meaning" of a message.

When writing in the English language, the physical symbol called a *letter* is used to construct *words*. The information contained in each word is determined by the choice and position of letters from the set of 26 called the *alphabet*. And the information in the set of words (sentence, paragraph, chapter, book, library) is determined by the choice and position of words (and other characters) out of the total possible words and their positions. As a specific example, the information contained in the English words *stop* and *tops* is determined by the choice and position of letters (symbols) from the available set of 26. But the "meaning" that a given word has for you is quite beyond the scope of this book.

So, "information" comes down to a specific choice or set of choices from all possible choices. Good efficiency in the transfer of information is based on transmitting the smallest number of symbols necessary to complete the message. ■

12–1 ■ DIGITAL INFORMATION

Information requires a change; no information is transmitted in a continuous symbol. Keeping in mind that zero is a very definite quantity—none, as opposed to any other quantity—the simplest information system would be binary. In a binary information system, the information is contained in the choice of one state out of two possible states—one "something" out of two possible "somethings." If the states are assigned the symbols 1 and 0, then 1 and 0 are our binary digits. (A decimal system has 10 digits.) The two words *binary digit* have been contracted into the new English word *bit*, and 1 bit is the *smallest division of binary information*.

A *bit* is a division of information, and the choice of a symbol, 1 or 0, constitutes the information.

Coding

Now, suppose that 1 or 0 are the only symbols we want to use in our information system—they constitute the entire "alphabet"—and we are faced with distinguishing between 16 possible objects. A code will accomplish this. We assign to each object a name (or word) using a specific arrangement of symbols from our alphabet.

Let us specify that each code word (a complete set of bits) is to be made up of binary digits placed side by side and further that every word will contain the same number of bits. How many bits are required to distinguish 16 different objects? The bits will be side by side, ___ & ___ & ___ & ___ . . . , and each bit requires a choice from two symbols (1 or 0). So one bit can define 2 objects; 2 bits can define 2 & $2 = 2 \cdot 2 = 2^2 = 4$ objects; 3 bits can define 2 & 2 & $2 = 2 \cdot 2 \cdot 2 = 2^3 = 8$ objects; 4 bits can define $2^4 = 16$ objects. Thus, for a binary (base 2) system, the number of bits n required to distinguish M different things (objects, voltage levels, keys on a terminal keyboard, and so on) is found from

$$2^n = M \tag{12–1}$$

The number of bits can be found by taking the logarithm of both sides:

$$\log_B 2^n = \log_B M$$

$$n \log_B 2 = \log_B M$$

Therefore, the number of bits required is

$$n = \frac{\log_B M}{\log_B 2} \qquad (12\text{--}2)$$

If $B = 2$ is used, then since $\log_2 2 = 1$, the exact number of bits required is given by

$$n = \log_2 M \qquad (12\text{--}3a)$$

for binary. If your calculator doesn't have $\log_2$, then use $B = 10$ so that the minimum number of bits required is

$$n = \frac{\log_{10} M}{\log_{10} 2} \approx 3.32 \log_{10} M \qquad (12\text{--}3b)$$

■ ■ ■ ■

EXAMPLE 12–1

How many bits would be required to completely distinguish the 88 keys of a piano?

Solution:

1. We could get the answer by trying powers of 2. Thus, $2^6 = 64$ is not enough. But $2^7 = 128$ is more than enough. Therefore, a 7-bit binary code can be used. Of course more than 7 bits could be used, but this would reduce system efficiency.

2. Another way to solve the problem is to use the exact formula. Thus the exact number of bits required $= \log_2 88 \approx 3.32 \log_{10} 88 = 6.46$ bits. Since a fraction of a bit cannot be implemented in practice, we must use the next highest whole number, namely, **7 bits.**

Coding Efficiency

Example 12–1 leads directly to the concept of coding efficiency; that is, if exactly 6.46 bits of information are required but 7 bits must be used,

$$\text{coding efficiency} = \frac{\text{exact number of digits required}}{\text{actual number of digits used}} \qquad (12\text{--}4)$$

Thus for the preceding sample,

$$\text{coding efficiency} = \frac{6.46 \text{ bits}}{7 \text{ bits}} = 0.923 = 92.3\%$$

Coding systems other than binary could be used to specify the 88 piano keys of the preceding example. If a decimal ($n = 10$) system is used, the system will require at least $\log_{10} 88 = 1.944$ "dits" (decimal digits). Thus, two dits are necessary—this, of course, will allow up to 99 keys. And the efficiency of this system will be

$$\frac{1.944 \text{ dits}}{2 \text{ dits}} = 0.972 = 97.2\%$$

For the 88-key piano example, a decimal number system would be more efficient than binary. However, binary computers do not operate internally with 10 symbols.

12–2 ■ INFORMATION TRANSFER RATE

The bit has been defined as the basic unit of information in a digital binary system. The speed at which information is transferred from one computer or terminal to another in a digital system, the *information transfer rate*, is measured in bits per second. Thus, if the serial digital word 101001 is transferred from one computer to another in 6 ms, there are 6 bits per 6 ms, or an information transfer rate of

$$f_i = \frac{6 \text{ bits}}{6 \text{ ms}} = 1000 \text{ bits/s}$$

In the literature, you will see 1000 bits per second expressed as 1 kbps or 1 kb/s.

12–3 ■ SIGNALING (BAUD) RATE

Later we will study higher-order (M-ary, where $M > 2$) transmission systems wherein one signal change on the transmission line will represent a group of two *or more* binary bits. Thus, a distinction must be made between the end-to-end information transfer rate f_i and the rate at which transmission-line changes are occurring—the signaling rate, or baud rate, f_B, after J. M. E. Baudot.

The signaling rate specifies how fast signal states are changing in the communications channel and is measured in *symbols*/second, or baud. If the communications equipment is transmitting the binary signal as shown in Figure 12–1,* there is one level per millisecond, or a transmission line signaling rate of 1000 symbols/s = 1 k baud. In a purely binary system, the bit rate f_b and baud rate f_B are equal.

However, suppose that instead of each interval T_B allowing a choice of only two levels (1 V or 0 V), we have a choice of four different voltages (say, 0 V, 1 V, 2 V, and 3 V). This four-level *quaternary* system provides twice as many choices as the two-level *binary* system and can transmit twice as much information.

---

* Notice that 101 are the first 3 bits transmitted since $t = 0$ is the origin of the time scale shown. This will be a convention used throughout this book.

FIGURE 12–1 Binary transmission.

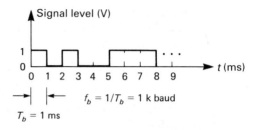

$T_b = 1$ ms

TABLE 12–1 Quaternary System Transmission Levels

Binary Pair	Transmission-Line Voltage (V)
00	0
01	1
10	2
11	3

Binary message   1 0   1 0   0 1   1 1 $\begin{cases} f_i = 1 \text{ kb/s (k bits/s)} \\ f_B = 1 \text{ k baud (symbols/s)} \end{cases}$

Quaternary transmission   2 V   2 V   1 V   3 V

(a) Equivalent quaternary transmission signal

f_i (transfer rate) = 8 bits/4 ms = 2 kb/s   (computer clocks must run twice as fast as for Figure 12–1)

f_B (baud rate) = 4 symbols/4 ms = 1 k baud

(b) Quaternary transmission signal (slower rate)

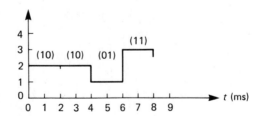

f_i = 8 bits/8 ms = 1 kb/s

f_B = 4 symbols/8 ms = 0.5 k baud

FIGURE 12–2 Four-level transmission of a binary message (computer data), at both (a) the same transmission-line rate and (b) one-half the binary transmission-line rate.

For example, let the binary (computer data) message be 10100111, as shown in Figure 12–1. We could build digital circuits to separate the bits into groups of two as they come serially (one after the other) out of the computer and assign one of four voltage levels to each pair. One possible arrangement might be as shown in Table 12–1, where a pair of 0s (00) would be assigned to the voltage 0 V, and so forth for the other three possible combinations—01 = 1 V, 10 = 2 V, and 11 = 3 V. Thus, to send the binary message 10100111, our quarternary system transmits 2 V, 2 V, 1 V, and 3 V, as shown in Figure 12–2. Part a shows that the total message could be sent in 4 ms with the same baud rate as the binary system and twice the information transfer rate. (The computer clock rate would have to be doubled.) On the other hand we could slow the signaling rate, as in part b, and get the same end-to-end information transfer rate (the same number of bits/second from computer to computer) while using one-half the transmission-line band-width. The price paid for this improvement is system complexity, because now we have to store 2 bits in a register and generate a multilevel signal on the transmis-sion line; then, at the receiver, the multilevel signal must be decoded back into the original binary voltages.

12–4 ■ SYSTEM CAPACITY

By system capacity we mean the maximum *rate* at which *information* can be transferred through a system of given design. In this section the expression is developed for system capacity (the Shannon-Hartley theorem)

$$C = \frac{\text{information}}{T_m} = \frac{1}{T_B} \log_2 M$$

$$= 2f_c(\text{min})\log_2 M \qquad \textbf{(12–5)}$$

where T_m is the message time, $1/T_B$ is the signaling rate, and $\log_2 M$ is the number of bits from Equation 12–3a, with M the number of levels per information interval. The parameters are illustrated in Figure 12–3. The absolute minimum bandwidth (cutoff frequency, low-pass) required will be

$$f_c(\text{min}) = \frac{1}{2T_B} \qquad \textbf{(12–6)}$$

FIGURE 12–3 Multilevel digital signaling. M symbols may be represented by $\log_2 M$ binary symbols.

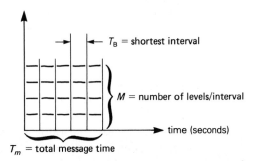

T_B = shortest interval

M = number of levels/interval

time (seconds)

T_m = total message time

Let us say that during each interval of T_B seconds duration, the information will consist of any one of four possible (voltage) levels. Then during interval 1 there are 4 ($M = 4$) possible voltages, each of which is likely to occur during interval 1. *And*, in interval 2, there are also $M = 4$ equally likely possibilities. Therefore, in two intervals, there are 4 *and* $4 = 4 \cdot 4 = 4^2$ possibilities. (The exponent indicates the number of intervals.)

If all the information we want to communicate takes a total time T_m, where T_m is made up of many intervals of length T_B seconds each, then the number of intervals during T_m is T_m/T_B.

Combining what we have done so far brings us to the conclusion that the number of possibilities (possible signal levels) during the message is proportional to M^{T_m/T_B}, where T_m provides more possibilities through *more time* and T_B provides more possibilities through shorter intervals within T_m. To solve for T_m, take the logarithm

$$\log(M^{T_m/T_B}) = \left(\frac{T_m}{T_B}\right) \log M$$

This expression is proportional to the total amount of information transmitted during the message period T_m. In order to have the above quantity in bits—the basic unit of information—the logarithm base is 2. The result, then, is information $= (T_m/T_B)\log_2 M$ (bits). This is the maximum information content of the multilevel discrete signal described graphically in Figure 12–3. The first part of Equation 12–5 is obvious from this as we solve for C, *system capacity*:

$$\frac{\text{information}}{T_m} = C = \frac{1}{T_B} \log_2 M$$

which gives us the maximum rate (in bits/second) at which information can be transmitted through a system designed to communicate a T_m-seconds-long message divided up into T_B-seconds-long intervals, wherein during each interval, the signal can assume any one of M distinct, equally probable, levels. "Levels" can be voltages, frequencies, or phases. If the probability p of transmitting a given level (symbol) is not equal to that of any other level ($p_i \neq p_j$), then the information content of the ith level is given by

$$I_i = \log_2 \left(\frac{1}{p_i}\right) \tag{12–7}$$

and the average information, called entropy, transmitted is

$$H = \sum_{i=1}^{m} p_i I_i \quad \text{(bits)} \tag{12–8a}$$

$$= -(p \log_2 p + q \log_2 q) \tag{12–8b}$$

for binary with state probabilities of p and q, where $q = 1 - p$.

To complete Equation 12–5 and expand on the theory, expect that M, the number of different levels, is related to system S/N (signal-to-noise ratio) and that T_B is related to bandwidth. As illustrated in Figure 12–4, the fastest changes would be for alternate high-low level changes. Hence, the channel requires an

FIGURE 12–4   For an absolute minimum bandwidth channel, only the fundamental frequency (solid curve) will be received even though rectangular data (dashed curve) is transmitted. The basic binary information is still received, however.

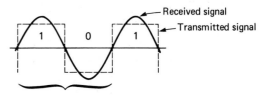

1 cycle of f_c (min), where f_c (min) is the minimum channel cutoff frequency.

absolute minimum bandwidth of 1 cycle in $2T_B$ seconds so that the minimum channel cutoff frequency is $f_c(\text{min}) = 1/(2T_B)$ for an ideal noiseless channel. Since

$$\frac{1}{T_B} = 2f_c(\text{min})$$

then

$$C = 2f_c(\text{min})\log_2 M \qquad\qquad \textbf{(12–9)}$$

As an example, for binary $M = 2$, and $C = 2f_c(\text{min})\log_2 2 = 2f_c(\text{min})$. Therefore, for binary,

$$C = 2f_c(\text{min}) \qquad\qquad \textbf{(12–10)}$$

For a channel or system with noise—and they are all noisy—the *Shannon limit* gives an expression for maximum channel capacity in terms of bandwidth (BW) and signal plus noise power–to–noise power ratio as

$$C = \text{BW} \log_2\left(\frac{S + N}{N}\right) \qquad\qquad \textbf{(12–11a)}$$

or

$$C = \text{BW} \log_2\left(\frac{S}{N} + 1\right) \qquad\qquad \textbf{(12–11b)}$$

in bits/second, where BW is the *actual* upper cutoff frequency *of the channel* $f_c(\text{min})$, S = signal power, and N = noise power. BW in Equation 12–11 is the effective *used* bandwidth, not just the bandwidth available.

This mathematical expression is usually misinterpreted. For example, consider an ideal 3-kHz bandwidth telephone circuit with $S/N = 30$ dB. We calculate $S/N = \text{antilog}_{10}(30/10) = 10^3$ and

$$C = 3 \times 10^3 \log_2(1 + 10^3) = 29.92 \text{ kb/s}$$

However, $C = 2f_c(\text{min}) = 6$ kb/s for an ideal *noiseless* channel! How can an ideal, noiseless, channel have less capacity than a noisy one? The answer is that $C = 6$ kb/s is for a 3-kHz channel carrying *binary*. The preceding Shannon limit tells us that we can transmit information at a rate of 29.9 kb/s, but we obviously cannot do it by transmitting binary over a 3-kHz channel. We would have to transmit over this channel using a 32-level code system,

$$C = 2f_c(\text{min})\log_2 M$$

$$\log_2 M = \frac{C}{2f_c} = \frac{29.9 \text{ kb/s}}{2 \times 3\text{kHz}}$$

so $M = 31.6$. Such a system might use PAM in which each bit of information is assigned 1 of 32 possible signaling levels. On the other hand, we could use binary PCM with five bits ($2^5 = 32$, which is greater than 31.6), but it would take five times longer to get the information transferred through the 3-kHz telephone line. Thus the Hartley expression,

$$H = CT_m \qquad\qquad \text{(12–12)}$$

gives the maximum total information H that can be transferred in T_m seconds over a channel whose information capacity is C.

This discussion provides an idea of why it took 30 min to construct a single TV frame transmitted from the Mariner IV at Mars—which has nothing to do with the time it took for the RF signal to travel that 124 million miles: $d = vt$, $t = d/v = 2 \times 10^8$ km/(3×10^8 m/s) = 11 min. The point is that when the signal reached earth antennas, it was so weak that it was buried in noise. Reducing the noise required narrowing the bandwidth, which in turn required slowing the signaling rate and/or using higher-level coding.

Multilevel transmission systems are covered in Chapter 17. However, the primary reason that multilevel signaling is being designed into digital transmission systems is to conserve a dwindling natural resource—bandwidth.

12–5 ■ BANDWIDTH CONSIDERATIONS

A major concern for digital communication system designers is the bandwidth requirement of the transmission channel; or, from another perspective, if we are to transmit digital data through a channel of known bandwidth such as a telephone line, the maximum signaling rate (for a given amount of distortion) must be limited.

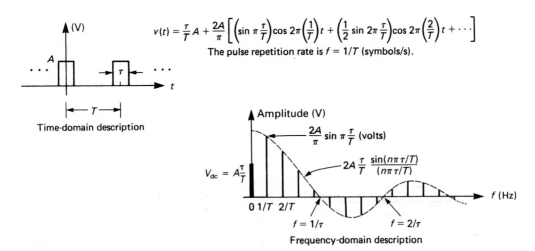

$$v(t) = \frac{\tau}{T} A + \frac{2A}{\pi} \left[\left(\sin \pi \frac{\tau}{T} \right) \cos 2\pi \left(\frac{1}{T} \right) t + \left(\frac{1}{2} \sin 2\pi \frac{\tau}{T} \right) \cos 2\pi \left(\frac{2}{T} \right) t + \cdots \right]$$

The pulse repetition rate is $f = 1/T$ (symbols/s).

Time-domain description

Amplitude (V)

$\frac{2A}{\pi} \sin \pi \frac{\tau}{T}$ (volts)

$V_{dc} = A\frac{\tau}{T}$

$2A \frac{\tau}{T} \frac{\sin(n\pi\tau/T)}{(n\pi\tau/T)}$

$0 \ 1/T \ 2/T$

$f = 1/\tau$ $f = 2/\tau$

f (Hz)

Frequency-domain description

FIGURE 12–5   Time and frequency description of a rectangular pulse train.

FIGURE 12–6 Return-to-zero
(RZ) data stream.

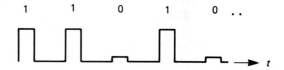

In order to determine the amount of frequency spectrum used up by discrete (digital) signals, we review here pulse spectra from Chapter 3.

As seen in Figure 12–5, sharp digital pulses require a tremendous amount of frequency spectrum (bandwidth). Since transmission channels have a limited bandwidth, the communication system designer needs to know (1) the minimum possible bandwidth required for a given pulse rate and (2) how pulses can be shaped to minimize the bandwidth and distortion of the data pulses.

Minimum Possible Bandwidth

Digital data transmitted in the format shown in Figure 12–5 is called *return-to-zero* (RZ) because the transmission line returns to 0 V between data pulses; a data stream would look like Figure 12–6. The first component in the frequency spectrum of Figure 12–5 is the average value of the signal, the dc component. Its amplitude is determined from the pulse peak value V and the duty cycle τ/T. Thus, for pulses with a 20% duty cycle and +5-V amplitude, $\tau/T = 0.2$, $V = 5$ V, and the dc component is 5 V × 0.2 = 1 V.

The second component in the frequency spectrum has a frequency equal to the pulse repetition rate, $f = 1/T$, and is shown in the time domain in Figure 12–7a. Its

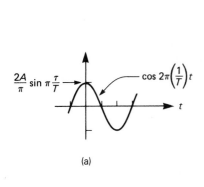

(a)

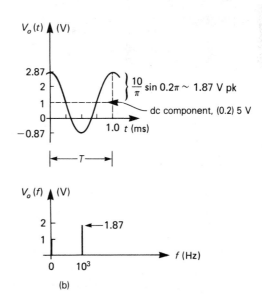

(b)

FIGURE 12–7 Output of ideal minimum-bandwidth channel with a rectangular-shaped input digital signal. (a) Fundamental frequency signal, time-domain description. (b) Time (upper) and frequency (lower) description of output.

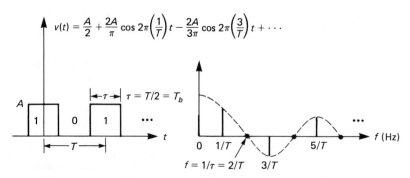

FIGURE 12–8 Squarewave signal in the time and frequency domains.

peak amplitude is determined from the duty cycle τ/T and the pulse amplitude A. For 5-V pulses with a duty cycle of 20% and pulse repetition frequency (PRF) of $f = 10^3$ pulses per second, the "fundamental" signal is

$$\left(\frac{10}{\pi}\sin 0.2\pi\right)\cos(2\pi \times 10^3 t) = 1.87\cos(2\pi \times 10^3 t)\ \text{V}$$

If the (noiseless) transmission channel can be represented by an ideal low-pass filter with cutoff frequency just above the PRF (1000 pps), passing the pulses of Figure 12–6 through this transmission channel will result in the signal of Figure 12–7b. Notice that although we no longer have squared-off pulses, the basic information is present; that is, the presence or absence of voltage at $t = T$ is clearly shown in the time domain. Digital circuits (such as Schmitt triggers) are available to reshape the cosine pulses into rectangular pulses.

An important conclusion of this discussion is that the absolute minimum bandwidth required to transmit rectangular pulses, not including squarewaves, in an ideal (Nyquist) noiseless channel is equal to the pulse repetition rate.

If squarewaves (rectangular pulses with a 50% duty cycle) are transmitted, the channel cutoff frequency is not as critical because the second harmonic is at a null, as are all even harmonics of the fundamental (see Figure 12–8).

However, notice that when binary data is transmitted in this format (called *nonreturn-to-zero*, NRZ), the fastest signal change in the channel will be for alternate 1s and 0s; that is, the sequence 101010 . . . represents the maximum rate of signal change on the transmission line. This signal format has two bits (a 1 and a 0) occurring during one squarewave period T; that is, $T = 2T_B$. Consequently, the minimum channel cutoff frequency must be more than $f_c \geq 1/(2T_B) = \frac{1}{2}f_B$. For example, if 1000 bits/s are transmitted NRZ, the minimum cutoff frequency is 500 Hz.

How close to $\frac{1}{2}f_B$ the bandwidth can be made depends on how much intersymbol interference (ISI) can be tolerated, which in turn depends on how closely the overall channel frequency response resembles the ideal Nyquist channel. Channel equalization (compensation) approximating a raised-cosine characteristic near the band edges allows the smallest possible bandwidth in practice; see Equation 11–7.

Pulse Shaping for Minimizing Bandwidth and Pulse Distortion

Advanced digital transmission systems are designed to use pulses whose shapes maximize the percentage of total signal power within the first "lobe" of the spectrum. Figure 12–9 shows pulses and the first "lobe," and indicates how power is more concentrated for some pulse shapes as compared to others. As an example, the raised-cosine pulse is seen to have a faster spectral rolloff than rectangular pulses. Thus, for a channel with a bandwidth wide enough to include the first lobe, transmitting raised-cosine pulses will result in the reception of more power with less pulse distortion than for rectangular pulses.

FIGURE 12–9 Frequency spectra for three different digital transmission signal formats.

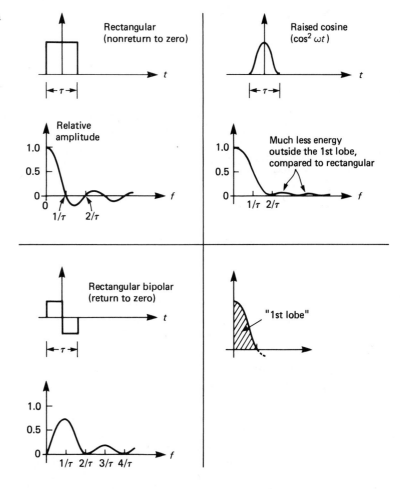

Digital Transmission Formats

Most digital data is generated by computers and terminals that usually operate internally on a parallel TTL-level signal format. TTL (transistor-transistor logic) levels are 0 to 0.8 V for a logic 0 and 2.0 to 5.0 V for a logic 1 input level. Also, output current levels are typically less than 16 mA.

Transmitting binary computer data over distances greater than a few meters requires circuitry other than TTL because capacitance and line resistance will limit the frequency response and distort the pulses. In addition, it is often undesirable or impossible to transmit the dc component of typical TTL signals; this is the case with transformer-coupled telephone lines. Where direct connection to a transmission line is desirable, various types of linedrivers and signal formats are used.

Some of the more common digital formats are illustrated in Figure 12–10. The NRZ signal is the same as the common TTL format, and NRZ-B is a bipolar version.

The advantage of the RZ format will be seen in the study of synchronous data systems, where the increase in signal transitions will help in system synchronization.

Bipolar transmission formats have the advantage of a zero dc component, assuming an equal number of 1s and 0s occur* during a message. The dc component is an important consideration in noisy systems because dc changes due to

* This condition is often ensured by scrambling and bit-stuffing techniques.

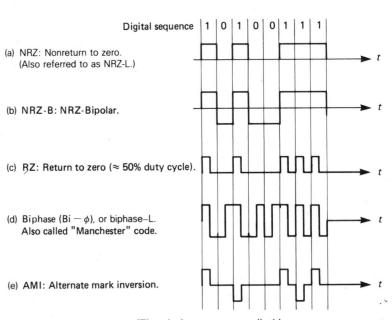

FIGURE 12–10 A few digital transmission formats.

Digital sequence 1 0 1 0 0 1 1 1

(a) NRZ: Nonreturn to zero.
    (Also referred to as NRZ-L.)

(b) NRZ-B: NRZ-Bipolar.

(c) RZ: Return to zero (≈ 50% duty cycle).

(d) Biphase (Bi — φ), or biphase–L.
    Also called "Manchester" code.

(e) AMI: Alternate mark inversion.

(The telephone company calls this
3-level code "Bipolar.")

short bursts of continuous 1s or 0s will change the decision threshold and can result in more errors.

An additional advantage of bipolar over polar (on/off) formats is that, for the same S/N, polar requires twice the average power (four times the peak power) compared to bipolar. This is shown in the example in the next section.

The biphase (Bi–ϕ) format uses a $+/-$ squarewave cycle for a MARK and a $-/+$ for a SPACE. Each bit period contains one full cycle, thereby eliminating dc wander problems inherent in all the above signal formats. A disadvantage of biphase, and RZ, is the requirement for twice the bandwidth of NRZ and NRZ-B. The biphase format is also referred to as a "Manchester" code.

The AMI format (Figure 12–10e), called bipolar with a 50% duty cycle by the telephone industry, is similar to RZ except that alternate 1s are inverted. The dc component is less than for RZ, the minimum bandwidth is less than for RZ and biphase. An additional advantage of AMI is that, by detecting violations of the alternate-one rule, transmission errors can be detected.

12–6 ■ POWER IN DIGITAL SIGNALS

If a digital transmission has an equal number of 1s and 0s during a message, then the power can be averaged over the message period and the signal modeled as a continuous pulse stream. The generalized pulse stream is shown in Figure 12–11. The normalized ($R = 1$) average power is derived for a signal $f(t)$ from

$$\overline{P} = \lim_{T \to \infty} \frac{1}{T} \int_{-T/2}^{T/2} [f(t)]^2 \, dt \qquad \textbf{(12–13a)}$$

where T is the period of integration. If $v(t)$ is a periodic signal with period T_0, then

$$\overline{P} = \frac{1}{T_0} \int_{-T_0/2}^{T_0/2} [v(t)]^2 \, dt \qquad \textbf{(12–13b)}$$

Of course, the period can start at any point in time.

FIGURE 12–11 Pulse streams. (a) Rectangular pulses. (b) Rectangular pulses with $\tau/T = 0.5$ (squarewave).

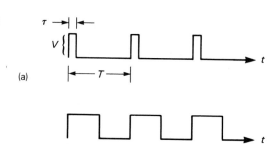

If the rectangular pulses of amplitude V in Figure 12–11a start at $t = 0$, then

$$v(t) = \begin{cases} V & 0 \le t \le \tau \\ 0 & \tau < t \le T \end{cases} \qquad (12\text{--}14)$$

and, from Equation (12–13b),

$$\overline{P} = \frac{1}{T_0} \int_0^\tau [V]^2 \, dt = \frac{1}{T_0} V^2 t \Big|_0^\tau$$

$$= \frac{\tau}{T_0} V^2 \qquad (12\text{--}15)$$

from which

$$P = \left(\frac{\tau}{T}\right) \frac{V^2}{R} \qquad (12\text{--}16)$$

Incidentally, since the rms (effective value) of a periodic wave is found from $P = (V_{rms})^2/R$, it follows that the rms voltage for rectangular pulses is

$$V_{rms} = \sqrt{\frac{\tau}{T}} \, V \qquad (12\text{--}17)$$

because $P = (V_{rms})^2/R = (\sqrt{\tau/T} \, V)^2/R = (\tau V^2)/(TR)$.

In the squarewave case of Figure 12–11b, $\tau/T = 0.5$ so that $P = V^2/2R$ (squarewaves; that is, NRZ signals). So the rms voltage for the unipolar squarewave is $V_{rms} = V/\sqrt{2}$—just as it is for sinusoids.

■ ■ ■ ■

EXAMPLE 12–2

Compare the power of an NRZ squarewave to NRZ-bipolar (NRZ-B) where the peak-to-peak amplitudes are equal (in order for the two signals to have the same S/N when received over a noisy channel). Refer to Figure 12–12.

Solution:

The power in an NRZ signal is $P_{NRZ} = V^2/2R$.

For the NRZ-B signal, $V \to V/2$ and, since there are pulses in each half-period (instead of a large pulse followed by a zero amplitude pulse),

$$P_{NRZ\text{-}B} = 2\,\frac{(V/2)^2}{2R} = \frac{V^2}{4R}$$

FIGURE 12–12 Comparison of NRZ and NRZ-bipolar.

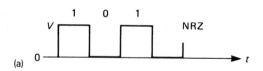

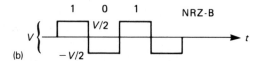

Comparing results, it is seen that the on/off NRZ signal has twice the power of the NRZ-B signal.

Also, the instantaneous (peak) power for NRZ is V^2/R, whereas the peak power for NRZ-B is $(V/2)^2/R = V^2/4R$, for a 4:1 difference in peak power.

A final comment on these two important digital formats is that the dc power for rectangular RZ and NRZ signals is found from $(\tau V/T)^2/R$, whereas for the bipolar signal it is zero.

12–7 ■ PCM SYSTEM ANALYSIS

The digital communication concepts developed thus far can be used to determine the requirements of PCM systems.

Pulse code modulation, introduced in Chapter 11, is a technique for converting analog (including pulsed) information into a specific binary code for transmission through digital communication channels. In PCM, the information signal to be transmitted is sampled at regular intervals, and each sample is given an n-bit digital code corresponding to the amplitude of the sample. For telephone systems, sampling and encoding are performed in the transmit section of a *codec* (coder-decoder). After transmission, the n-bit encoded samples are demodulated in the receiver section of another codec circuit.

The PCM system consists of many information sources, both analog and digital, *concentrated* into a *few* digital channels or multiplexed into a single digital data stream for transmission between data communications equipment (DCE). The transmissions between DCEs can be current or voltage pulses on transmission lines, or for longer distances, the PCM pulses can modulate an analog carrier for transmission by ASK (amplitude-shift key), FSK (frequency-shift key), or PSK (phase-shift key). Such digital carrier systems allow the transmission of digital data on transmission lines, including fiber-optic lines, and by satellite and terrestrial (land) microwave radio links.

Previously, the pulse code conversions were by high-speed A/D and D/A converters multiplexed over many analog channels. But now, large-scale integration (LSI) has made it economical to use a codec in each channel and then concentrate or multiplex the resulting digital signals.

Figure 12–13 shows the transmit/multiplex part of a PCM system in which each channel of information is digitally encoded, multiplexed with other similarly encoded channels, and then transmitted after a framing or frame-synchronizing bit is added.

To analyze this PCM system, consider the AT&T 24-channel, D3 (also GTE Lenkurt's 9002B) "long-distance" (toll) telephone bank.

The analog voice signal is transformer-coupled and frequency band-limited to less than 4 kHz by the low-pass filter. Typical uncompensated telephone lines have a 3-dB frequency response between 300 Hz and 3.3 kHz, but the LPF is used to ensure that no aliasing distortion will occur in the sampling process due to the presence of signals with frequencies greater than 4 kHz.

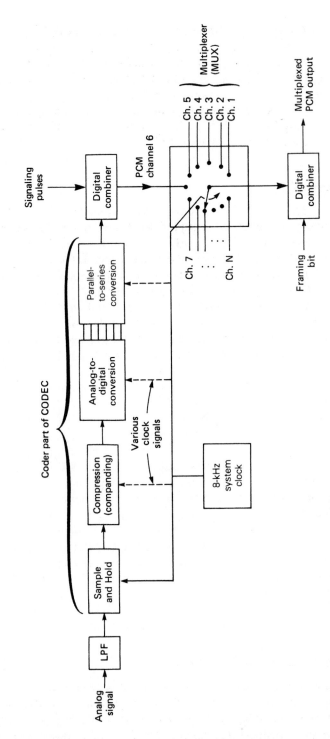

FIGURE 12–13 Multiplexed PCM transmitter system.

The sampling theorem states that if a signal is sampled at a rate greater than twice the highest input signal frequency, $f_s \geq 2f_A(\text{max})$, so that no aliasing (frequency foldover) occurs, then the original signal can be recovered without distortion. From this, it is clear that we could sample at 2×4 kHz = 8 kHz or greater. The telephone system samples at 8 kHz. The samples must be held long enough to be encoded but short enough to multiplex the other 24 channels and one framing bit—all within $\frac{1}{8000} = 125$ μs. The encoding time will be calculated after other system requirements are determined.

Digitizing/Encoding

The critical part of a PCM system is the analog-to-digital conversion process. This is where the analog voltage sample is compared to the discrete levels (*quanta*, or quantum levels) defined in the A/D converter, and a digital code word corresponding to the closest quantum level is assigned. The digital code word then represents the PCM approximation of the particular analog signal sample.

The A/D converter (ADC) designer will be concerned with accuracy, linearity, speed, and a list of specifications too numerous to explain here. A short listing follows. For our purposes it will be enough to consider resolution/dynamic range, number of bits required, noise, and conversion rate. Conversion rate is considered in the discussion of converter circuit schemes.

Some ADC Specifications

- Accuracy
- Linearity
- Precision
- Monotonicity
- Resolution/dynamic range
- Aperture time, delay, and jitter
- Quantization error, or number of bits required
- Settling time
- Conversion rate
- Temperature coefficient
- Noise
- Bandwidth

Resolution and Dynamic Range

Dynamic range (see Figure 12–14) is the ratio of largest-to-smallest analog signal that can be transmitted. The resolution, or quantization (step) size q, is the smallest analog input voltage change that can be distinguished by the A/D converter.

Clearly, from the basic relationship $2^n = M$ (Equation 12–1), the more bits used, the more discrete quantization levels M can be distinguished (resolved).

FIGURE 12-14 Dynamic range of analog signal.

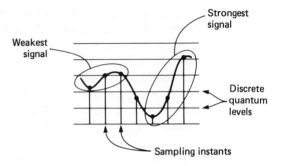

From the linear ADC transfer characteristic of Figure 12–15, it can be seen that $q = V_{max}/M = V_{FS}/M$, which, when combined with Equation 12–1, yields the basic linear ADC relationship

$$q = \frac{V_{FS}}{2^n} \qquad\qquad \textbf{(12–18)}$$

where q = resolution (smallest analog voltage change that can be distinguished)

   n = number of bits in the digital code word

   V_{FS} = full-scale voltage range for the analog signal

FIGURE 12–15 Linear ADC characteristic and quantization noise.

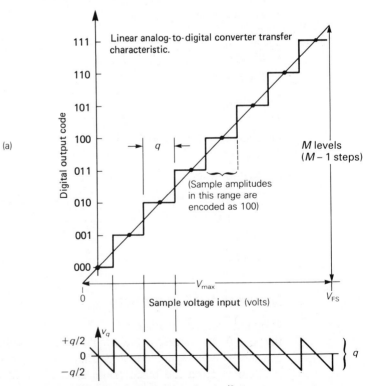

(a)

(b) Quantum uncertainty or quantization noise, $\pm q/2$.

The bits of the code word are usually written from left to right, as illustrated in Figure 12–15, where the leftmost bit has the most weight (most significant bit, MSB) and the rightmost bit has the least weight (least significant bit, LSB, not to be confused with lower sideband). The LSB, smallest digital change, determines the PCM resolution and has the same analog value as q.

The analog dynamic range capability of a PCM system V_{max}/V_{min} is the same as the ADC parameters V_{FS}/q. And, since $M = 2^n = V_{FS}/q$, the dynamic range (DR) is expressed mathematically as

$$DR \equiv \frac{V_{max}}{V_{min}} = \frac{V_{FS}}{q} = 2^n \qquad (12\text{–}19)$$

or, in decibels,

$$DR\ (dB) = 20 \log \frac{V_{max}}{V_{min}} \qquad (12\text{–}20)$$

Note that $20 \log 2^n = 20n \log 2 = 6.02n$, so we can write

$$DR\ (dB) \approx 6n$$

which means that there are approximately 6 dB/bit of dynamic range capability for a linearly encoded PCM system. Table 12–2 summarizes resolution, dynamic range, and accuracy for up to 16 bits, where accuracy is the system resolution in percent of full-scale, or the maximum possible accuracy of the demodulated analog signal.

TABLE 12–2 Linear ADC

Resolution, in Bits n (bits)	Number of Levels $M = 2^n$	Dynamic Range (dB)	Accuracy $\frac{1}{M} \times 100(\%)$
0	1	0	—
1	2	6.02	50
2	4	12	25
3	8	18.1	12.5
4	16	24.1	6.2
5	32	30.1	3.1
6	64	36.1	1.6
7	128	42.1	0.8
8	256	48.2	0.4
9	512	54.2	0.2
10	1024	60.2	0.1
11	2048	66.2	0.05
12	4096	72.2	0.02
13	8192	78.3	0.01
14	16384	84.3	0.006
15	32768	90.3	0.003
16	65536	96.3	0.0015*

* 15 ppm (parts per million)

■ ■ ■ ■

**EXAMPLE
12–3**

Determine the dynamic range capability of an 8-bit linear (noncompanded) PCM system.

Solution:

$DR(dB) \approx 6n = 6$ dB/bit $\times$ 8 bits $\approx$ **48 dB.** Said another way, an analog system for which the dynamic range is 48 dB ($V_{max}/V_{min} \approx 251$) requires an 8-bit PCM system.

Quantization Noise

Quantization noise is unique to digital (and some pulsed) transmission systems and is produced in the digitization process.

As seen in Figure 12–15b, a digitally encoded analog sample will have an exact-amplitude uncertainty of $\pm q/2$. Only analog values corresponding to the code center points will have no error (called *quantization error*). Demodulated PCM can be thought of as the analog input with quantization noise added. The quantization noise voltage V_{qn} is sawtoothed with a peak value of $q/2$ and can be calculated from the average normalized power of Equation 12–13b. On a time basis over a period T_0 of the quantization noise voltage waveform (Figure 12–15b), $v_q(t) = -(q/T_0)t$, then

$$\overline{N}_q = \frac{1}{T_0} \int_{-T_0/2}^{T_0/2} \left[\frac{-q}{T_0} t \right]^2 dt \qquad \textbf{(12–21a)}$$

$$= \frac{q^2}{T_0^3} \int_{-T_0/2}^{T_0/2} t^2 \, dt = \frac{q^2}{12} \qquad \textbf{(12–21b)}$$

Hence, the effective voltage is from $V_{qn} = \sqrt{N_q}$,

$$V_{qn} = \frac{q}{2\sqrt{3}} \qquad \textbf{(12–22)}$$

in volts rms. The noise power will be

$$N_q = \frac{q^2}{12R} \qquad \textbf{(12–23)}$$

for a linear ADC characteristic.

Signal-to–Quantization Noise Ratio

Let us calculate the S/N_q for a sinusoid of maximum level. As seen in Figures 12–14 and 12–15, the maximum peak-to-peak sinusoid voltage will be equivalent to qM; that is, V_s(pk-pk) $= qM$ and $V_s = qM/2$ volts peak. The rms value is

$$\frac{V_s(\text{pk})}{\sqrt{2}} = \frac{qM}{2\sqrt{2}} \qquad \textbf{(12–24)}$$

volts rms. The average power of a sinusoidal signal is

$$S = \frac{V_{rms}^2}{R}$$

$$= \frac{q^2 M^2}{8R} \tag{12-25}$$

Combining Equations 12–23 and 12–25 yields the maximum signal-to–quantization noise for a sinusoid quantized into M levels,

$$S/N_q = \frac{q^2 M^2/8}{q^2/12} = \frac{3}{2} M^2 \tag{12-26}$$

Keep in mind that this is the maximum S/N for a linear analog-to-digital converter. For analog input signals of voltage V_s that are less than the maximum, substitute $S = V_s^2$(pk-pk) for $q^2 M^2$ in Equation 12–26.

Companding

Linear quantizing in PCM systems has two major drawbacks. First, the uniform step size means that weak analog signals will have a much poorer S/N_q than the strong signals. Second, systems of wide dynamic range require many encoding bits and consequently wide system bandwidth.

A technique to improve S/N_q for weak signals is to decrease the step size q for weak signals and increase it for strong signals, as illustrated in Figure 12–16. This nonlinear encoding/decoding, called *companding*, tends to equalize the S/N_q (or signal-to-digitizing distortion) over the expected range of analog amplitudes and requires rather complex digital hardware.

Another technique to accomplish the same result is to use a linear encoder preceded by an analog voltage compressor, which also helps to prevent high-level signals from saturating the system. After decoding, a complementary expander restores the original dynamic range. A companding curve is sketched in Figure 12–17.

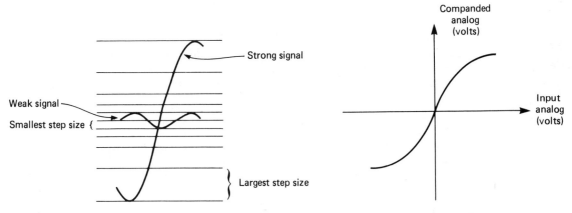

FIGURE 12-16 Nonlinear step-size quantizing.　　　　**FIGURE 12-17** Companding curve.

FIGURE 12–18   Sketch of μ-255 transfer characteristic.

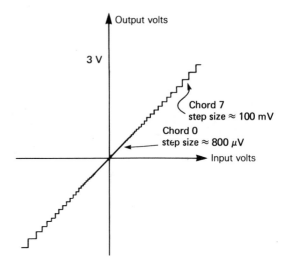

The actual shape of the curve depends on the system designer or, in the case of standard telephone systems, it depends on where you live. North America employs AT&T's so-called μ-255 companding shape known as the "mu" law, whereas in Europe, the CCITT* regulatory board specifies the "A" law. The two companding curve shapes are usually obtained by a series of short linear segments, called *chords*, each decreasing in slope as distance from the origin increases.

The North American "μ" law has 16 linear segments, 8 for positive and 8 for negative voltages. There are 16 steps of equal size in each chord. A simplified sketch is shown in Figure 12–18. As a practical matter, companding laws are quite similar, and despite the complex circuitry, LSI codecs are companding with digital techniques and make provisions for the use of either law.

To summarize the PCM fundamentals developed here, the AT&T T1 (GTE 9002B) system is used as an example.

■ ■ ■ ■

EXAMPLE 12–4

Summary of PCM System Analysis

The 24-channel (D1) AT&T T1 PCM carrier telephone system band-limits input voice frequencies in each channel to 4 kHz. The ratio of maximum-to-minimum voice level gives a dynamic range of 72 dB. Following the multiplexing of the 24 digitized voice signals, a framing bit is inserted to synchronize the system and to identify each channel's data. Reference to Figure 12–19 should aid in visualizing this analysis.

1. The *minimum sampling rate* is

$$f_s \geq 2f_A(\text{max}) = 2 \times 4 \text{ kHz} = \textbf{8000 samples/second}$$

2. The voice dynamic range for long (toll-grade) lines is 72 dB, so the ratio of *maximum to minimum analog signal levels* to be resolved is found from Equa-

---

* International Consultative Committee for Telephony and Telegraphy.

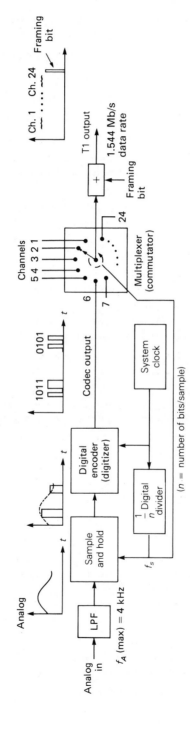

FIGURE 12-19 24-channel PCM T1 system with individual codec per-channel multiplexing.

tion 12–20: 72 dB $= 20 \log_{10} V_{max}/V_{min}$, and by the definition of logarithms, $V_{max}/V_{min} = 10^{72/20} = 10^{3.6} = \mathbf{3981} = M$.

3. The *number of bits* required to quantize 3981 equal levels (linear ADC) is, from $M = 2^n$, using Equation 12–3b,

$$n \approx 3.32 \log 3981 = 3.32 \, (3.6) = 11.95, \qquad \text{or } \mathbf{12 \ bits}$$

This is also determined as follows: 6 dB/bit requires 72 dB/(6 dB/bit) = **12 bits.**

4. By compressing the voice signal, the μ-255 companding codec provides the same dynamic range with 256 discrete levels using 8 bits. Eight bits are nearly worldwide practice for digital voice transmission.

5. If only one channel (telephone conversation) is transmitted, the bit (and baud) rate would be 8000 samples/second × 8 bits/sample = 64 kbps. Thus, a 4-kHz voice channel digitized to 8 bits would require an absolute minimum *bandwidth* of BW $= \frac{1}{2}f_b = $ (64 kb/s)/2 bits/cycle = **32 kHz.**

6. 24 8-bit samples plus 1 framing bit are multiplexed at 8000 frames per second. The T1 bit (and baud) rate is (24 channels × 8 bits/channel) + 1 framing bit = 192 + 1 = 193 bits/frame. 193 bits/frame × 8000 frames/s = 1.544 Mb/s. So the T1 line carries 1.544 million bits/s, which requires a baseband bandwidth of at least BW $= \frac{1}{2}f_b = \frac{1}{2}$(1.544 Mb/s) = 772 kHz. However, AT&T typically uses closer to 1.5 MHz for the T1 line bandwidth.

7. Figure 12–20 illustrates in a unipolar-RZ format the 8-bit samples for one 24-channel frame and the frame (synchronizing) bit. Notice that each frame takes $\frac{1}{8000}$ s = 125 μs and that each *bit* gets a slot of *time* equal to 125 μs/193 bits = 648 ns. The transmission format used for a T1 line is AMI with a 50% duty cycle (not illustrated in Figure 12–20): consequently, each pulse is only 324 ns wide.

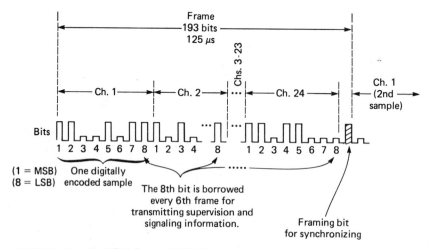

FIGURE 12-20 T1 PCM frame, 50% RZ polar format.

8. Also indicated in Figure 12–20 is the fact that during every sixth frame, the 8th bit (LSB) of each channel is borrowed by the microprocessor-controlled CPU, allowing numerous loop supervision signals, including off-hook conditions, rotary-dial pulses, call charging information, and so forth.

The *signaling rate* can be determined as follows:

$$6 \text{ frames} \times 125 \ \mu\text{s/frame} = 750 \ \mu\text{s}$$

for a signaling rate of $1/750 \ \mu\text{s} = $ **1⅓ kb/s.** During the sixth frame, the LSB is borrowed from each channel, resulting in more quantization distortion; but this occurs only 17% (⅙) of the time.

12–8 ■ PCM TELEPHONE CIRCUITRY

The circuits that perform the coding and decoding (codecs) in PCM systems are A/D and D/A converters. Since most A/D converters incorporate a D/A circuit, digital-to-analog converters are discussed first.

Digital-to-Analog Converters (DACs)

The receiving end of a PCM system accepts a digitally encoded serial data stream. Framing bits help to separate the encoded samples into their respective channels and then the receive side of each codec clocks the encoded samples into a register for short-term storage (buffering) and serial-to-parallel conversion. A D/A converter circuit will convert the parallel digital bits (d_1–d_4 in Figure 12–21) to an analog voltage equal to the original sample with some quantization error.

The Binary-Weighted Resistor Converter. A binary-weighted resistor-type DAC is illustrated in Figure 12–22a. Here, Q_1–Q_4 are MOS transistors used as switches activated by the parallel data word. Each closed switch sets up an amount of current determined by the reference voltage and series resistance $2^{(n-i)}R$. A high-gain IC is used in a current-summing op-amp configuration, and the currents from each of the high data bits are summed in feedback resistor R_f to produce an output voltage V_A. As an example, Figure 12–22a shows that if d_4 is high and MOSFET Q_4

FIGURE 12–21   Serial-to-parallel and digital-to-analog conversion.

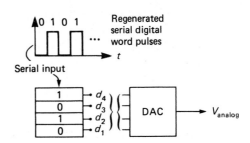

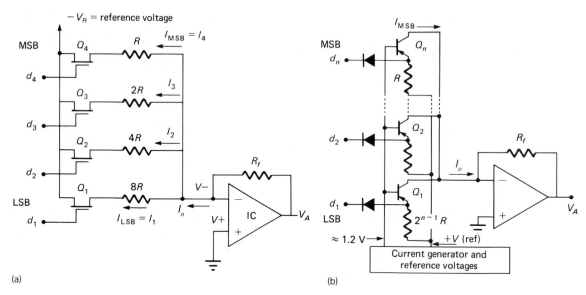

FIGURE 12-22 Binary-weighted resistor digital-to-analog (D/A) converter. (a) Weighted resistor D/A converter. (b) Weighted current source implementation of binary-weighted resistor D/A converter.

conducts (assume $r_{FET} \ll R$), $I_{MSB} = V/R$ flows through R_f.* If, say, d_1 is also high and Q_1 conducts, $I_{LSB} = I_1 = V/(2^3 R) = V/8R$. Thus the output current is

$$I_o = \sum_{i=1}^{n} d_i I_i \qquad (12\text{–}27\text{a})$$

where $d_i = 1$ or 0, corresponding to the input data, and

$$I_i = \frac{V}{2^{n-i}} R \qquad (12\text{–}27\text{b})$$

The analog output voltage is $V_A = R_f I_o$.

Figure 12–22b shows an LSI implementation of the weighted-resistor DAC. The bipolar transistors are current sources for each weighted bit and are switched on or off by means of the control diodes connected to each emitter. The base of each transistor remains biased to $+1.2$ V, so that when an input bit is high, the transistor current source is on.

Difficulties arise in implementing high-order, weighted-resistor D/A converters due to the large values of resistors. For instance, if $R = 10$ kΩ, then the LSB resistor for an 8-bit D/A converter is $2^{8-1} \times 10$ kΩ $= 1.28$ MΩ. Not only does such a large resistance consume a large amount of space in an integrated circuit, but

---

* Because of the high-gain IC and feedback resistor R_f, $V^- \approx V^+ = 0$, and the voltage across R is only V. Assuming an ideal IC with $Z_i \to \infty$, all the current, $I = V/R$, must flow through R_f, thus maintaining $V^- \approx V^+$.

problems also arise with temperature stability and the differences in switching speed for each section. The ladder-type, or R-$2R$, D/A converter described in the following section avoids these problems.

R-2R Ladder-Type D/A Converter. The D/A converter shown in Figure 12–23 maintains constant impedance level R at any node 1 through n and constant I_R (except during switching transitions). Note that each switch, actuated by a parallel-data input bit, is connected to ground or to "virtual ground" V^-. Consequently $I_R = V_R/R$ is a constant. However, I_R divides into $1/2^n$ binary weighted currents at each node. These binary weighted currents are either shunted to ground or summed into I_o to determine the output analog voltage V_A, depending on the digital word status of d_1–d_n.

I_{MSB} is easily calculated to be $V_R/2R$, and I_{n-1} is determined by calculating the voltage at node $n - 1$. Summing currents at node n requires that $I = I_R - I_{MSB}$. Thus $V_{n-1} = V_R - IR = V_R - (I_R - I_{MSB})R = V_R - (V_R/R - V_R/2R)R = V_R/2$, so that $I_{n-1} = V_{n-1}/2R = V_R/4R$. Proceeding down the ladder, $I_1 = V_R/2^nR$. The sum of currents is, then,

$$I_o = \frac{V_R}{R}\left(\frac{d_n}{2} + \frac{d_{n-1}}{4} + \frac{d_{n-2}}{8} + \cdots + \frac{d_1}{2^n}\right) \qquad (12\text{–}28)$$

where d is either 1 or 0, depending on the digital word to be decoded. As usual, for the inverting op-amp,

$$V_A = R_f I_o \qquad (12\text{–}29)$$

The advantages of the R-$2R$ ladder D/A converter are that only two resistor values (laser-trimmed) are used, impedance levels are constant at all nodes for constant switching speed, and except for I_{MSB}, the weighted currents are determined by resistor ratios (rather than absolute resistor values) for improved temperature tracking.

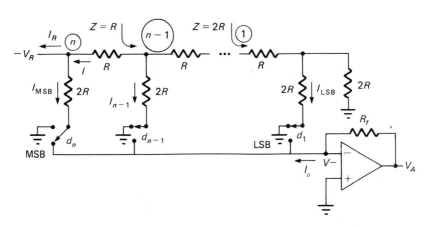

FIGURE 12–23 R-$2R$ ladder-type digital-to-analog converter circuit.

Analog-to-Digital (A/D) Converters

Numerous circuit configurations exist for converting an analog signal into digitally encoded signals. The different circuits can be grouped into a few general converter types, and the choice of a particular converter usually comes down to limitations of resolution and speed, versus complexity and cost. The most used A/D converters are the counter/ramp, tracking or servo, integrating (single- and dual-slope), successive approximation, the parallel (flash) converter, charge-balancing (voltage-to-frequency), and the more recent switched-capacitor converter.

To illustrate the basic concepts, only the counter/ramp and successive approximation converters are discussed.

Counter/Ramp A/D Converter. The analog signal sample is applied to a comparator while a staircase voltage or ramp builds up to the analog voltage value, at which point the ramp (and conversion process) stops and resets for the next sample.

A circuit configuration for a continuous, 4-bit, counter/ramp A/D converter is shown in Figure 12–24. In this circuit, the D/A converter converts the binary counter output to a staircase "ramp." The ramp voltage V_D starts at V_{min} and is compared in the comparator to the analog input voltage V_A. As long as V_A is greater than V_D, the comparator output is positive, thus enabling the AND gate to clock the binary counter. The 4-bit binary counter counts up from 0000, corresponding to $V_D = V_{min}$, until V_D exceeds V_A, whereupon the comparator output goes low, thus disabling the AND gate which, in turn, stops the counter (and ramp). The expanded region on the left of Figure 12–24 indicates an example in which the counter output contains the digital code (0101), which is proportional to V_A at the sampling instant t_s. If a parallel input data register (or latch) is connected as shown, the encoded sample is quickly entered at t_s, and after a short delay, the counter is reset to zero, which returns V_D to V_{min}. The comparator output then goes high, enabling the AND gate to allow the sampling process to start up again.

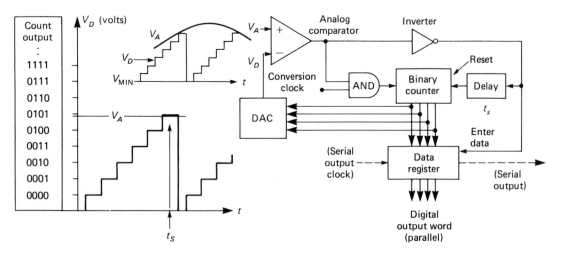

FIGURE 12–24   Counter/ramp A/D converter.

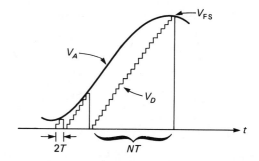

FIGURE 12–25 Three samples taken on a full-scale, V_{FS}, analog signal. Note the unequal sampling periods.

Incidentally, the register, which is not necessary for the A/D conversion process, could merely latch the data for parallel output or could be a serial-shift register for parallel-to-serial data conversion.

A major drawback of this type of encoder is its unequal conversion times, as illustrated in Figure 12–25. If $V_A \approx V_{min}$, the conversion is completed in one clock period T, whereas, for $V_A = V_{FS}$, the conversion takes

$$MT = \frac{M}{f_{clk}} \qquad\qquad (12\text{–}30)$$

where M is the number of clock periods or steps. $M = 2^n$, where n is the number of bits in the counter and DAC.

■ ■ ■ ■

EXAMPLE 12–5

A 4-bit counter and D/A converter are used with a 10-kHz clock.

1. Determine the maximum conversion time if the circuit is adjusted so that $V_{max} = qM$, where $q = 100$ mV is the minimum step voltage of the D/A converter.

2. How long will it take to encode a 560-mV input signal, and what will the digital word be for this sample?

Solution:

1. $M = 2^4 = 16$ steps are possible with a 4-bit system, and $V_{max} = qM = 100$ mV $\times$ 16 = 1.6 V. (Obviously an amplifier at the D/A converter output can extend this range. Or if better resolution is desired, the bit capacities of the digital counter and D/A converter must be increased.) $T = 1/10$ kHz = 0.1 ms, so the conversion time is $MT = 16 \times 0.1$ ms = **1.6 ms.**

2. The first step would have been to calculate the minimum step size from $q = V_{FS}/M = 1600$ mV/16 = 100 mV/step, but this information was given. The sampled analog value of 560 mV requires 560 mV/100 mV/step = 5.6 steps. The conversion will be completed at the sixth step, as illustrated in Figure 12–26. The time required is 0.1 ms/step $\times$ 6 steps = **0.6 ms**. Since the counter stops at step 6, the digital word will be **0110** ($6_{10} = 0110_2$).*

* $6_{10} = 0110_2$ says, "Six, base 10 (decimal), is equal to **0110** base 2 (binary)."

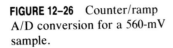

FIGURE 12–26 Counter/ramp
A/D conversion for a 560-mV
sample.

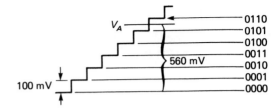

■ ■ ■ ■

**EXAMPLE
12–6**

At what rate must a 4-bit counter/ramp A/D converter run in order to sample at
4 kHz?

Solution:

$MT = 2^4T = 16/f_{clk}$, so each sample must be completed in $1/4000$ s. Therefore, the
A/D converter clock rate is computed as $16/f_{clk} = 1/4000$, $f_{clk} = 16 \times 4$ kHz =
64 kHz.

Successive-Approximation A/D Converter. This very popular analog-to-digital converter
has a block diagram similar to the counter A/D converter, but the binary counter
has been replaced in the feedback loop by a digital programmer circuit called a
successive-approximation register (SAR). (See Figure 12–27.) With the programmer (SAR) controlling the DAC, a 1 is applied to the most significant bit (MSB)
and $V_D \rightarrow 1/2$ full-scale (FS). If the analog signal is greater than $1/2$FS, the comparator output is still high, so the SAR applies a 1 to the next significant bit. However, if the analog input is less than $1/2$FS, as shown in Figure 12–28, the comparator output will go low and, in the next clock period, the SAR will remove the MSB
1 (MSB $\rightarrow$ 0) and apply a 1 to the next most significant bit position. The SAR,
getting feedback from the comparator, continues its routine of applying successive 1s to lower (finer) bit positions of the DAC. After all bits have been tried, the
digital output of the SAR indicates the closest approximation to the analog signal.
A good analogy for the SAR routine is putting successive half-weights on a balance scale. If the first applied weight ($1/2$FS) is too much, you remove it and put a
smaller weight ($1/2 \times 1/2$FS = $1/4$FS), and so forth, until the best balance is
achieved. The good conversion efficiency of this technique means that high-reso-

FIGURE 12–27 Successive-
approximation A/D converter.

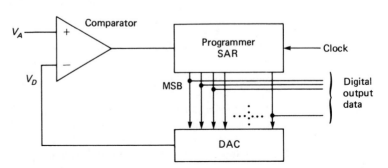

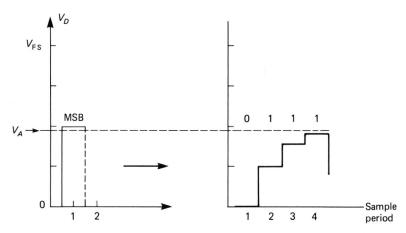

FIGURE 12–28   One sample taken with a 4-bit SAR.

lution conversions can be made in very short times. An additional advantage for the SAR over the counter/ramp converter is that each analog sample is completed in the same amount of time; that is, conversion time remains constant.

As an example, a 4-bit SAR clocked at 10 kHz will convert the smallest and the largest analog input signals into 4 bits at a rate of 0.1 ms/bit, thereby requiring a total conversion time of 0.4 ms. Recall that, for the equivalent counter/ramp ADC, a full-scale analog signal took 1.6 ms.

A final A/D conversion technique, the servo or tracking converter, is quite different from the other techniques because single-bit digital words are generated that represent the difference (delta) between the actual input signal and a quantized approximation of the preceding input signal sample. In other words, this is a delta modulator. Delta modulation and adaptive delta modulation were discussed in Chapter 11.

Codec

An example of a per-channel PCM codec (COder/DECoder) for voice is Motorola's MC14407/4 24-pin CMOS device shown in Figure 12–29. As seen in the block diagram, the analog voice input goes to a sample-and-hold circuit, followed by analog-to-digital conversion by the successive approximation technique. You can get μ- or A-law companding by specifying the 14407 or the 14404 device. As is typical for codecs, the DAC is part of the A/D conversion loop and handles the D/A conversion for decoding input PCM as well. Typical dc power consumption for this LSI chip is 80 mW, whereas 1 mW is used in the power-down mode.

Telephone Dialing Tones—Synthesizing and Decoding

The telephone system is still in the process of converting dialing signals from 10-Hz dc interrupt pulses to dual-tone pairs. Since only seven tones (and an eighth, 1633 Hz, for special functions) have to be generated, a simple synthesis technique is used. Typically, an off-chip 3.5795-MHz (TV chroma oscillator) crys-

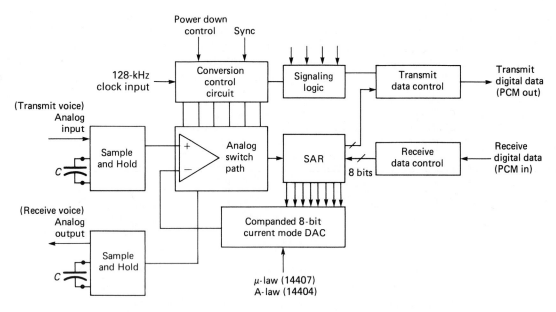

FIGURE 12–29   Codec block diagram. Derived from Motorola's MC 14407/4 full-duplex, 8-bit companded PCM codec.

tal is used for generating the reference. The reference is divided in two separate counters controlled by the telephone touch-tone keyboard logic. The output sinewaves are digitally synthesized in ROMs (or a programmable logic array) and converted to the low-distortion sinewaves in separate D/A converters. The two tones, one from the high-frequency (1209–1477 Hz) group and the other from the low-frequency (697–941 Hz) group, are linearly summed in an op-amp and low-impedance line driver.

Each number touched for dialing causes two tones to be transmitted over the telephone line. As an example (see Figure 12–30), when ☐1 is touched, 697 Hz and 1209 Hz are simultaneously transmitted. Various decoding techniques are avail-

FIGURE 12–30   Two-tone frequency grid for multifrequency "touch-tone" dialing.

Low group ↓	1209 Hz	1336 Hz	1477 Hz	1633 Hz ← High group
697 Hz	1	ABC 2	DEF 3	
770 Hz	GHI 4	JKL 5	MNO 6	(Special function keys)
852 Hz	PRS 7	TUV 8	WXY 9	
941 Hz	*	OPER 0	#	

able for receivers. One technique for touch-tone decoding consists of seven phase-locked loops, each of which can lock to one of the dial tones; if the appropriate sequence of numbers is decoded, your phone rings. An example of an IC with a single touch-tone decoder PLL is the 567 circuit of Figure 12–31. The VCO is set by R_1 and C_1 to the tone frequency to be decoded.

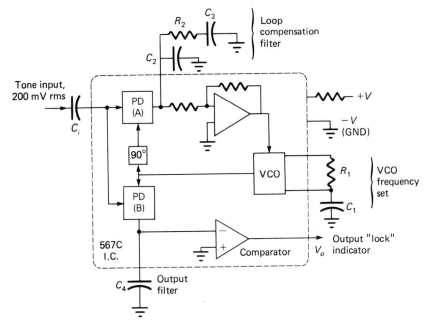

FIGURE 12–31 567 IC touch-tone decoder. This circuit decodes (detects) only one telephone "ringing" tone.

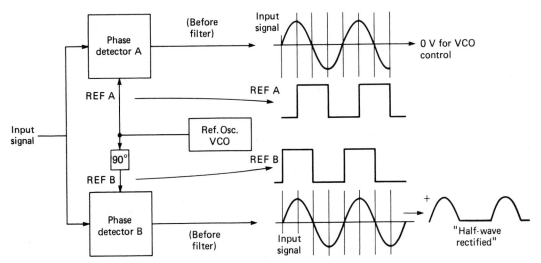

FIGURE 12–32 Phase relationships for quadrature phase detectors of the 567 IC.

The PLL consisting of PD(A), loop compensation filter $R_2 C_2 C_3$, op-amp, and VCO will lock to the input tone and provide a coherent reference to the quadrature detector PD(B). Because PD(B) operates in quadrature (90°) to the PLL detector, it becomes a coherent half-wave rectifier at the instant of tone acquisition (lock) and produces a positive dc voltage, which drives the inverting comparator low to indicate "lock-up." Figure 12–32 shows the phase relationships for these detectors. (These phase relationships and the resultant outputs were discussed in Chapter 10.)

■ PROBLEMS

1. How many bits would be required to define all the characters for a 54-key computer terminal? (Assume a 1 : 1 correspondence—that is, no character/figure shifts.)

2. We want to quantize a signal into 35 different levels.
   a. How many bits of binary are required?
   b. Calculate the coding efficiency for this system.
   c. Repeat a and b for a decimal system (base 10 versus base 2).

3. A 5-bit binary signal is transmitted in 100 μs.
   a. Calculate the information rate.
   b. What is the absolute minimum bandwidth required for regular NRZ signals in an ideal noiseless transmission system?

4. a. If the bandwidth of a telephone line is 4 kHz, what is the highest information rate that can be transmitted as binary, assuming an ideal noiseless channel?
   b. What is the highest information rate for a 4-level encoded PAM transmission?

5. Determine the channel capacity of a 4-kHz channel with S/N = 12 dB.

6. How many levels must be used to achieve full transmission capacity for Problem 5?

7. a. Determine the power in a 100-mV peak, NRZ squarewave on a 600-Ω channel (in μW and dBm).
   b. Repeat for the NRZ signal narrowed to a 20% duty cycle pulse.

8. a. Calculate the amplitude of the second harmonic for the signals of Problem 7a and b.
   b. How many decibels below the fundamental (first harmonic) is the second harmonic for Problem 7b?

9. A baseband (transmission line) transmission system with an overall 30.1-dB signal-to-noise ratio is to be used for delivering an information rate of up to 45 Mb/s.
   a. What is the absolute minimum bandwidth?
   b. How many levels of encoding must be used?
   c. The resulting system has a resolution equivalent to an n-bit PCM system. What is the value of n?

10. A 48-channel PCM system samples voice at 8 kHz and encodes the samples with

8-bit words (characters). If 2 bits are added (for framing) after the 48 channels are multiplexed, determine the following:
   a. Number of bits/frame
   b. Number of frames/second
   c. Number of bits/second
   d. What kind of multiplexing (FDM, TDM, CDM, or FBM) is being used?
   e. Calculate the absolute minimum (Nyquist) bandwidth required.

11. Do Problem 10 for 40 channels and one framing bit.

12. An analog voice signal is to be linearly quantized into 64 discrete levels; 5 V are available.
   a. How many bits are required?
   b. The resolution q is how many volts?
   c. The quantization noise (noise margin) is how many volts?
   d. Calculate the quantization noise power in 600 Ω.
   e. Calculate the dynamic range of this system.

13. The full-scale analog voltage for a 6-bit PCM encoder is 10 V pk-pk. Determine the signal-to-quantization noise ratio, in decibels, for an input signal of $v(t) = 2 \sin 18850t$.

14. The dynamic range capability of a binary transmission system is 50 dB.
   a. What is the minimum number of bits to use for PCM (noncompanding; that is, linear) in this system?
   b. What will be the coding efficiency?

15. A 5-V pk sinusoid is quantized into 64 levels.
   a. Calculate the maximum signal-to–quantization noise ratio for this system in decibels.
   b. If S/N_q must exceed 40 dB, determine the number of levels and bits required for PCM.

16. A 30-channel multiplexed PCM telephone system has an expected analog input of 40 V max with a 42-dB dynamic range and a max bandwidth of 5 kHz. If one bit is used for framing and the output binary signal is bipolar NRZ, determine the following:
   a. Minimum expected analog input voltage
   b. The minimum number of bits to use for encoding
   c. The system resolution, in percent
   d. The multiplexed PCM output baud rate, if multiplexing and sampling are done at the same rate
   e. The absolute minimum output bandwidth required

17. Which companding law is a French engineer most likely to specify for a codec? What does companding improve?

18. What is a codec and what does it do?

19. The weighted resistor D/A converter of Figure 12–22 has a reference voltage of 5 V and $R_f = 1$ kΩ. If the LSB is to produce 25 mV at V_A, determine R for the 4-bit D/A converter.

20. The LSB resistor for a 7-bit, weighted-resistor D/A converter would have what value, if $R = 10$ kΩ?

FIGURE 12–33

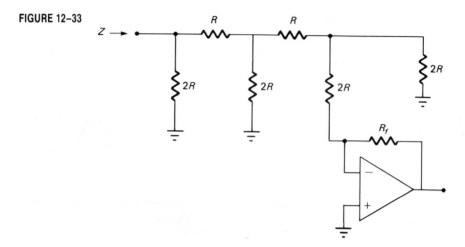

21. **a.** Show that $Z = R$ for the R-2R ladder network of Figure 12–33.
    b. If $V_R = 2$ V, $R = 1$ kΩ and $R_f = 2$ kΩ in Figure 12–23 configured for 3 bits and with the appropriate d_n connections made, find: (a) $V_A(101_3)$, (b) I_{LSB} and I_{MSB}.
    c. If $V_R = -5$ V and $R_f = 2$ kΩ, find R to make $V_A(LSB) = 10$ mV.

22. A 0- to 5-V max analog signal is to be applied to an 8-bit SAR A/D converter running at 8 k samples/s. The input analog is 4 V dc.
    a. Sketch the voltage applied to the analog approximation side of the comparator (V_D, Figure 12–27) for the full sampling period. (Use at least a 2-in., full-scale grid.)
    b. How long does it take for the analog approximation of part **a** to be within 5% of full scale (5 V)?
    c. What is the digital output word transmitted? (Write the final 1s and 0s.)

23. The D/A converter and op-amp of a 4-bit counter/ramp A/D converter are capable of producing a maximum voltage range of 0 to 5.1 V. Determine the output binary word (1s and 0s pattern, with LSB on the right) for an analog input of 3.5 V dc.

24. What two tones are transmitted when the ⑧ key is touched (DTMF)?

25. **a.** What will be the polarity and magnitude of the dc voltage from the unfiltered output of detector PD(B) of Figure 12–32? Assume a 200-mV rms sinewave.
    b. Repeat for a large capacitor on the detector output.

26. A touch-tone decoder has the following parameters: $f_{FR} = 1633$ Hz $\pm$ 1%, $k_\phi = 0.01$ V/rad, $k_A = 20$, and $k_o = 1$ kHz/V.
    a. If the VCO drifts to its spec limit, how many cycles is it from center frequency?
    b. Assume the VCO is off center by 8 Hz when an exactly correct touch-tone arrives. Calculate the static phase error after lockup.
    c. Calculate the VCO input voltage after lockup.

12–9 ■ COMPUTER ANALYSIS

Using the computer for communications system and circuit analysis has become a virtual necessity in education and industry. What follows are examples of programs and/or readily available simulation software for solving problems of the type covered in this chapter.

All the programs are interactive in the sense that an introductory sentence or short paragraph describes the purpose of the program and the user is then prompted for input data.

The programs for Chapter 12 are as follows:

- **Digital communication concepts/PCM system analysis:** The BASIC program of Figure 12–34 solves six different basic datacom problems, including analog-to-digital converter resolution and analog-input dynamic range, system capacity, quantization noise, and multiplexed data rate.

- **Binary weighted-resistor D/A converter:** Given specified values for the number of bits and reference voltage, and assuming 10-kΩ series and feedback op-amp resistors, the analog output voltage is computed. Changes in this BASIC program (Figure 12–35) will allow other choices of resistors.

- **Counter/ramp A/D converter:** This BASIC program solves for typical design trade-offs such as step size, maximum conversion time, conversion time for a given analog signal amplitude, and the minimum clock rate (Figure 12–36).

- **Successive-approximation A/D:** For specified reference voltage and number of bits, the digital (binary) output word is given along with its accuracy for a given analog input voltage (Figure 12–37).

- **Simulation of a counter/ramp A/D:** The BASIC program of Figure 12–38 simulates the counter/ramp circuit and provides a "dual-trace scope" graphic display (real time!) of the signals input to the comparator. The two inputs are an analog sinewave and the D/A converter ramp output.

```
10 REM *      DIGITAL COMMUNICATION CONCEPTS        *
20 REM *   EXAMPLE OF STUDENT COMPUTER ASSIGNMENT   *
30 REM ****   FOR PROFESSOR P. YOUNG, AT ASU    ****
40 CLS:PRINT
50 PRINT "*****************************************************************"
60 PRINT "*                                                               *"
70 PRINT "*           DIGITAL  COMMUNICATION  CONCEPTS                     *"
80 PRINT "*                                                               *"
90 PRINT "*****************************************************************"
100 PRINT:PRINT:PRINT:PRINT
110 PRINT "[1]  Solving for bits required to distinguish M different things"
120 PRINT
130 PRINT "[2]  Solving dynamic range/resolution problems of an n-bit PCM system
140 PRINT
150 PRINT "[3]  Solving system capacity of an M-level discrete PAM system"
160 PRINT
170 PRINT "[4]  Solving maximum channel capacity (channel or system with noise)
180 PRINT
190 PRINT "[5]  Solving Quantization Noise problems"
200 PRINT
210 PRINT "[6]  PCM system analysis"
220 PRINT:INPUT "Please choose one of the above numbers ",A:PRINT
230 ON A GOTO 270,380,550,630,730,850
240 PRINT
250 INPUT "Do you want to do another job? Please type [Y]es or [N]o.",Y$
260 IF (Y$="n") OR (Y$="N") THEN 1070 ELSE CLS:GOTO 100
270 CLS :PRINT "Number of bits required to distinguish M different things:"
280 INPUT "Input M, M = ",M
290 X=(LOG(M)/LOG(2))
300 PRINT:IF (X-INT(X))>.00001 THEN Y=FIX(X)+1 ELSE GOTO 350
310 PRINT "The number of bits required is, n = ";X;"."
320 PRINT "We must use the next highest whole number,namely ";Y;"bits."
330 PRINT "Therefore, the coding efficiency will be ";100*X/Y;"%."
340 GOTO 240
350 X=INT(X)
360 PRINT "Number of required bits is ";X;" ,and coding efficiency is 100%."
370 GOTO 240
380 CLS
390 PRINT "Solving dynamic range and resolution problems of an n-bit linear"
400 PRINT "(noncompanded) PCM system. Either n is known, or DR(dB) is known."
410 PRINT:INPUT "How many bits does the digital encoder have? n = ",N:PRINT
420 PRINT "The dynamic range capability is DR(dB)=";20*LOG(2^N-1)/LOG(10);"dB"
430 PRINT
```

FIGURE 12–34   (page 1 of 3)

```
440 INPUT "To find n when DR(dB) is known, key in DR (in dB), ",DB:PRINT
450 Z=DB/6.0206  :IF (Z-INT(Z))>.00001 THEN Z=INT(Z)+1 ELSE Z=INT(Z)
460 IF (N-INT(N))>.00001 THEN N=INT(N)+1 ELSE N=INT(N)
470 PRINT "If the system dynamic range range is ";DB;"dB, then the number"
480 PRINT "of bits required in a linear ADC is ";Z;" bits."
490 PRINT
500 INPUT "PCM system resolution for an n-bit system, where n=";N
510 INPUT "and the analog input full voltage range is Vfs (Volts,pk-pk)=",V
520 PRINT
530 PRINT "The resolution (smallest distinguishable analog input signal) of a(n)
    ";N;"bit linear PCM encoder with full-scale analog input level of ";V;"volts
    (pk-pk) is ";V/(2^N);"volts,pk-pk."
540 PRINT:GOTO 240
550 CLS  :PRINT "System capacity---Shannon-Hartley theory:":PRINT
560 INPUT "Key in the system transmission rate (1/Tb, or 2*bandwidth,Hz) ",A
570 PRINT
580 INPUT "Key in the number of discrete PAM levels used in the system, M =",M
590 PRINT:X=(LOG(M)/LOG(2))
600 IF (X-INT(X))>.00001 THEN X=INT(X)+1 ELSE X=INT(X)
610 PRINT "The system data rate capacity =";A*X;"bits/sec"
620 GOTO 240
630 CLS
640 PRINT "Solving for the system capacity for a channel or system with noise;"
650 PRINT
660 INPUT "Key in the system bandwidth (Hz) ",BW:PRINT
670 INPUT "Key in the channel signal to noise ratio, S/N (in dB) = ",SN:PRINT
680 C=BW*LOG(10^(SN/10)+1)/LOG(2)
690 PRINT "The maximum channel capacity, C = ";C;"bps":PRINT
700 A=C/2/BW:IF A-INT(A)>.00001 THEN A=INT(A)+1 ELSE A=INT(A)
710 PRINT "The number of required levels in a M-ary transmission system to achieve
    this system capacity is ";2^A
720 GOTO 240
730 CLS  :PRINT "Solving quantization noise problems for a linear ADC:"
740 PRINT
750 INPUT "Key in the the number of bits used for encoding ,n = ",N
760 PRINT :INPUT "The encoder input full-scale voltage (pk-pk) is, Vfs = ",VFS
770 Q=VFS/(2^N):PRINT
780 INPUT "Key in the system resistance level (ohms) ",R:PRINT
790 PRINT "The quantization noise voltage is Vqn = ";Q/2/3^.5;" Vrms."
800 PRINT "and the noise power is Nq = ";Q^2/12/R;" watts."
810 PRINT :PRINT "The maximum possible S/Nq = ";10*LOG(1.5*2^(N*2));"dB."
820 INPUT "Input a specified pk-pk analog signal amplitude less than Vfs:",VS
825 IF VS > VFS THEN GOTO 820
```

FIGURE 12–34   (page 2 of 3)

```
830 PRINT :PRINT "S/N =";10*LOG(1.5*VS^2/Q^2);" dB, assuming sinusoids."
840 GOTO 240
850 CLS  :PRINT "PCM System Analysis:":PRINT
860 INPUT "Enter the number of channels ",N:PRINT
870 INPUT "Enter the maximum frequency of the input voice channels, (in Hz)",F
880 PRINT
890 INPUT "Enter the dynamic range desired, in dB, (linear encoding!) ",DR
900 PRINT
910 INPUT "Enter the number of framing bits used for sync and supervision ",FB
920 PRINT
930 PRINT "The minimum sampling rate is ";2*F;"samples/sec":PRINT
940 M=10^(DR/20)
950 PRINT "The ratio of maximum to minimum analog signal levels Vmax/Vmin =M =";M
960 PRINT
970 PRINT "The number of bits required to quantize M equal levels is ";LOG(M)/LOG(2);"bits"
980 PRINT
990 INPUT "Key in the number of bits that are used for digital voice transmission ",B
1000 PRINT:NBF=N*B+FB
1010 PRINT "The number of bits that a frame contains is ";NBF;"bits/frame "
1020 PRINT
1030 PRINT "The max transmission line data rate is, C = ";NBF*2*F;"bits/sec"
1040 PRINT
1050 PRINT "The minimum baseband bandwidth for NRZ data is ";NBF*F;"Hz"
1060 PRINT:GOTO 240
1070 END
```

FIGURE 12–34   (page 3 of 3)

```
10 REM *        BINARY WEIGHTED RESISTOR DAC      *
20 REM * EXAMPLE OF STUDENT COMPUTER ASSIGNMENT *
30 REM **** FOR PROFESSOR P. YOUNG, AT ASU   ****
40 CLS
50 REM *************    PRINT    *******************
60 PRINT" THIS PROGRAM SOLVES FOR THE OUTPUT VOLTAGE "
70 PRINT" OF A BINARY-WEIGHTED RESISTOR TYPE DAC (SEE "
80 PRINT" FIGURE 12-22 ECT BY YOUNG).  THE PROGRAM "
90 PRINT" ASKS THE USER TO ENTER THE 4 BIT BINARY "
100 PRINT" WORD AND THEN CONVERTS IT TO AN ANALOG VOLTAGE. "
110 PRINT" CIRCUIT VALUES ARE Vref = -5V.  R = Rf = 10K OHMS "
120 PRINT"*****************************************************"
130 REM ********* INITIALIZATIONS ***************
140 LET R=10000:LET RF=10000:LET V1=0:LET V2=0:LET V3=0:LET V4=0:LET VA=-5
145 PRINT
150 PRINT" THE BINARY BITS ARE WRITTEN AS 1 OR 0 :":PRINT
160 INPUT" THE LSB IS "; X1
170 IF X1=0 GOTO 190
180 V1= -(RF/(8*R))*VA
190 INPUT" THE 2nd LSB IS "; X2
200 IF X2=0 GOTO 220
210 V2= -(RF/(4*R))*VA
220 INPUT" THE 2nd MSB IS "; X3
230 IF X3=0 GOTO 250
240 V3= -(RF/(2*RF))*VA
250 INPUT" THE MSB IS "; X4
260 IF X4=0 GOTO 280
270 V4= -(RF/R)*VA
280 VOUT= V1+V2+V3+V4
290 PRINT:PRINT:PRINT"THE OUTPUT VOLTAGE = "; VOUT "V"
300 END
```

FIGURE 12–35

```
10 REM ****      SUCCESSIVE APPROXIMATION ADC     ****
20 REM *  EXAMPLE OF STUDENT COMPUTER ASSIGNMENT  *
30 REM ****     FOR PROFESSOR P. YOUNG, AT ASU    ****
40 CLS
50 PRINT"THIS PROGRAM COMPUTES THE BINARY OUTPUT AND ACCURACY (Fig 12-28) OF"
60 PRINT "A SUCCESSIVE-APPROXIMATION ADC FOR A GIVEN ANALOG INPUT VOLTAGE."
70 PRINT:INPUT; "INPUT REFERENCE VOLTAGE Vr IN VOLTS.  Vr=";R:PRINT "V"
```

FIGURE 12–36   (page 1 of 2)

```
80 INPUT;"INPUT ANALOG SIGNAL VOLTAGE Va IN VOLTS.  Va=";S:PRINT"V"
90 INPUT;"INPUT NUMBER OF BITS OF RESOLUTION FOR THE SAR";B:PRINT
100 PRINT:PRINT "BIT","V OUT","BINARY","PERCENT OF INPUT SIGNAL"
110 PRINT "                              OUTPUT"
120 FOR N=1 TO B
130 LET G =H+(R/(2^N))
140 IF G<S THEN LET H = G
150 IF G<S THEN LET D = 1 ELSE LET D = 0
160 P=((G-S)/S)*100
170 PRINT N,G,D,P
180 NEXT N
190 END
```

FIGURE 12–36   (page 2 of 2)


```
10 REM *        COUNTER/RAMP A-TO-D CONVERTER        *
20 REM *  EXAMPLE OF STUDENT COMPUTER ASSIGNMENT     *
30 REM ****     FOR PROFESSOR P. YOUNG, AT ASU    ****
40 CLS
50  PRINT "THIS PROGRAM ANALYZES A `COUNTER/RAMP´ A-TO-D CONVERTER"
60  PRINT "BY DETERMINING THE STEP SIZE FOR AN N-BIT COUNTER AND DAC AT "
70  PRINT "CLOCK RATE, F. IT WILL ALSO  1) CALCULATE THE MAX CONVERSION"
80  PRINT "TIME AS WELL AS THE CONVERSION TIME OF A GIVEN ANALOG"
90  PRINT "INPUT SIGNAL, 2) THE RATE AT WHICH THE ADC MUST RUN IN ORDER"
100  PRINT "TO SAMPLE AT FS SAMPLES/S.":PRINT:PRINT
110  INPUT "ENTER V, IN VOLTS, THE FULL-SCALE ANALOG VOLTAGE RANGE"; V
120 INPUT "ENTER N, THE COUNTER BIT-SIZE"; N
130 INPUT "ENTER F, IN HZ, THE CLOCK FREQUENCY"; F
140 INPUT "ENTER V1, THE INSTANTANEOUS ANALOG INPUT SIGNAL VOLTAGE"; V1
150 INPUT "ENTER FS, IN HZ, THE SAMPLING FREQUENCY ( >= 2*FaMAX )"; F1
160 PRINT:PRINT
170 T=1/F:M=2^N:Q=V/M:T1=M*T:S=V1/Q:K=S
180 IF K=INT(S) GOTO 200
190 K=INT(S)+1
200 T2=T*K:D=K:T4=1/F1:P=M/T4:PRINT:PRINT
210 PRINT "THE MAX CONVERSION TIME IN SECONDS (FULL SCALE) =";T1:PRINT
220 PRINT "THE CONVERSION TIME (SECONDS) REQUIRED TO ENCODE THE "
230 PRINT "ANALOG INPUT SIGNAL IS ";T2:PRINT
240 PRINT "THE DIGITAL WORD (IN DECIMAL) IS ";D:PRINT
250 PRINT "THE RATE AT WHICH THE CLOCK MUST RUN IN ORDER TO"
260 PRINT "SAMPLE AT FS IS, IN HZ, ";P;"."
270 PRINT "REPEAT THE PROBLEM USING THIS VALUE, OR HIGHER, FOR F."
280 END
```

FIGURE 12–37

```
 10 '================================================================
 20 '
 30 '        COUNTING ANALOG TO DIGITAL SIMULATION PROGRAM
 40 '
 50 '     CREATED BY (THOSE WHIZ KIDS) KASANG AND WEINSTEIN
 60 '
 70 '     FOR CET 473 UNDER PROFESSOR PAUL YOUNG, FALL 1987
 80 '
 90 '================================================================
120 '
140 'This program simulates an A/D converter using the counter
150 'ramp technique. The user is asked to enter analog sinewave
160 'parameters as well as the parameters for the counter circuit.
170 'The program then calculates the appropriate scaling factors
180 'for amplitude and time scales, followed by a superimposed
190 'plot of both the sinewave input and the digital ramp output
200 'from the D/A. See Figure 12-24 of "Elec. Comm. Tec." by Young.
210 '
240 '*************************************************************
250 '
260 '                 DEFINITION OF VARIABLE NAMES
270 '
280 '    VARIABLE NAME                        DEFINITION
290 '
300 '    USER DEFINED INPUTS:
310 '
320 '    SWF            SINEWAVE FREQUENCY (ANALOG INPUT)
330 '    SWA            SINEWAVE PEAK TO PEAK AMPLITUDE
340 '    DCO            DC OFFSET OF SINEWAVE
350 '    CR             CLOCK RATE FOR DIGITAL COUNTER
360 '    VREF           MAXIMUM D/A VOLTAGE
370 '    CYC            NUMBER OF SINEWAVE CYCLES TO BE DISPLAYED
380 '    NOB            NUMBER OF BITS IN THE BINARY COUNTER
390 '
400 '    PROGRAM GENERATED:
410 '    LPV            NUMBER OF PIXELS PER VOLT (VERTICAL)
420 '    VMAX           PEAK POSITIVE VOLTAGE
430 '    LAZ            NUMBER OF PIXELS FROM ZERO TO VMAX
440 '    ZP             GRAPHICAL POSITION OF ZERO VOLTS
450 '    NODEG          NUMBER OF ELECTRICAL DEGREES TO BE DISPLAYED
460 '    LPD            NUMBER OF PIXELS PER DEGREE (HORIZ)
470 '    SPL            NUMBER OF SECONDS PER PIXEL
480 '    LPST           NUMBER OF PIXELS PER COUNTER STEP (HORIZ)
490 '    NOST           NUMBER OF VERTICAL STEPS
500 '    VPST           VOLTS PER PIXEL (VERTICAL)
510 '    SSIZE          STEP SIZE IN PIXELS (VERTICAL)
520 '    DELTA          INSTANTANEOUS HORIZONTAL STEP WIDTH (PIXELS)
530 '    THETA          INSTANTANEOUS ELECTRICAL ANGLE IN DEGREES
540 '    THETAR              "              "         "    " RADIANS
550 '    SINW           AMPLITUDE OF SINEWAVE AT THETAR
560 '    XS             CURRENT GRAPHICAL POSITION (HORIZ) IN PIXELS
570 '    XT             PREVIOUS     "            "        "    "    "
580 '    YS             CURRENT SINEWAVE AMPLITUDE IN PIXELS
590 '    YT             PREVIOUS     "           "      "    "
600 '    CNTR           INTERVAL COUNTER FOR REPETITIVE CALCULATIONS
610 '    YRAMP          CURRENT RAMP AMPLITUDE IN PIXELS
620 '    YTRAMP         PREVIOUS "        "       "   "
623 '    MAXRAMP        MAXIMUM RAMP POSITION IN PIXELS (VERTICAL)
625 '*************************************************************
```

FIGURE 12–38   (page 1 of 3)

```
627 '
628 '                            Program starts here
630 KEY OFF
640 SCREEN 2
650 CLS
660 GOSUB 910                    'GET INITIAL DATA
670 CLS
680 GOSUB 990                    'TO SET UP SCREEN PARAMETERS
690 XT=60:YT=ZP-DCO*LPV          'SET ORIGIN OF SINE WAVE
700 IF LPST >= 3 THEN INC=1 :GOTO 1220  'Check step size bounds
710 '===============================================================
720 '                  Plot routine for small step size
730 DELTA=-1:INC=.1
740 FOR THETA = 0 TO NODEG STEP INC
750 CNTR=CNTR+INC
760 DELTA=DELTA +INC
770 IF DELTA >= LPST THEN YRAMP=YRAMP-SSIZE:DELTA=-INC
780 THETAR=THETA*3.14159/180
790 SINW=.5*SWA*SIN(THETAR)
800 YS=ZP-(SINW+DCO)*LPV
810 XS=LPD*THETA+60
820 IF CNTR>.99 THEN PSET(XS,YS)
830 YT=YS
840 IF CNTR>.99 THEN PSET(XS,YRAMP):CNTR=0
850 YTRAMP=YRAMP
860 XT=XS
870 IF (YRAMP<=YS) OR (YRAMP<=MAXRAMP) THEN LINE (XS,YRAMP)-(XS,ZP)
875 YRAMP=ZP:YTRAMP=YRAMP:XT=XS
880 NEXT THETA
890 LOCATE 12,60
900 END
905 '===============================================================
907 '                          Data entry routine
910 INPUT"Enter sine wave frequency in hertz ";SWF
920 INPUT"Enter peak to peak amplitude in volts ";SWA
930 INPUT"Enter the DC offset for the sine wave in volts ";DCO
940 INPUT"Enter the clock rate for the counter in hertz ";CR
950 INPUT"Enter the peak value of reference voltage in volts ";VREF
960 INPUT "Enter the number of cycles for the sinewave ";CYC
970 INPUT "Enter the number of bits in the counter ";NOB
980 RETURN
985 '===============================================================
987 '                    Screen setup routine
990 LINE (60,0)-(60,199)               'PRINT Y-AXIS
1000 'FIND ZERO ON SCREEN
1010     LPV=(199-19)/SWA              '# OF LINES PER VOLT
1020     VMAX=.5*SWA+DCO               'MAX VOLTAGE ABOVE ZERO
1030     LAZ=VMAX*LPV                  '# OF LINES ABOVE ZERO
1040     ZP=19+LAZ                     'ZERO POSITION
1050     LOCATE INT(ZP/8),4
1060     PRINT "0"
1065 '
1067 '              Calculate various screen parameters
1070 LINE (55,19)-(65,19)
1080 LINE (0,ZP)-(639,ZP)              'PRINT X-AXIS
1090 NODEG=CYC*360                     '# OF DEGREES TO BE DISPLAYED
1100 LPD=580/NODEG                     '# OF LINES PER DEGREE HORIZ.
1110 SPL=1/(SWF*360*LPD)               'SECONDS PER LINE HORIZ.
1120 LPST=1/(CR*SPL)                   'LINES PER COUNTER STEP HORIZ.
```

FIGURE 12-38   (page 2 of 3)

```
1130 NOST=2^NOB-1                          '# OF VERTICAL STEPS
1140 VPST=VREF/NOST                        'VOLTS PER STEP VERTICAL
1150 SSIZE=LPV*VPST                        'STEP SIZE VERTICAL LINES
1160 LOCATE 4,4
1170 PRINT VMAX;
1175 MAXRAMP = ZP - LPV * VREF
1180 RETURN
1190 END
1200 '=============================================================
1210 '                      Plot routine for large step size
1220 DELTA=-1
1230 FOR THETA = 0 TO NODEG STEP INC
1240 DELTA=DELTA +INC
1250 IF DELTA >= LPST THEN YRAMP=YRAMP-SSIZE:DELTA=-INC
1255 SOUND 8000-YRAMP*40,1.5
1260 THETAR=THETA*3.14159/180
1270 SINW=.5*SWA*SIN(THETAR)
1280 YS=ZP-(SINW+DCO)*LPV
1290 XS=LPD*THETA+60
1300 LINE(XT,YT)-(XS,YS)
1310 YT=YS
1320 LINE (XT,YTRAMP)-(XS,YRAMP)
1330 YTRAMP=YRAMP
1340 XT=XS
1350 IF (YRAMP<=YS) OR (YRAMP<=MAXRAMP) THEN YRAMP=ZP
1355 LINE(XS,YTRAMP)-(XS,YRAMP):YTRAMP=YRAMP:XT=XS
1360 NEXT THETA
1370 END
```

FIGURE 12–38   (page 3 of 3)

13

Data Communication Techniques

■ INTRODUCTION

Much of the quantitative groundwork for digital communication systems was laid in Chapter 12. In particular, analog-to-digital conversion was discussed, using the voice PCM system as a model. Telephone PCM systems are also used for transmitting computer information.

This chapter begins to explore data communications. While voice communications are important to all of us, the world of digital communication is primarily computer/terminal data transmissions.

A short listing of data services will give an idea of the expanding services involving data communications:

- Teletype mail exchange (Telex and TWX); Western Union, facsimile
- Direct Distance Dialing and WATS (AT&T), SPRINT (GTE), Dataphone

(Switched) Digital Service (AT&TIS), and the developing international integrated-services digital network (ISDN)

- TELENET and TYMNET (U.S.), DATAPAC (Canada), TRANSPAC (France), and AT&T's Advanced Information Systems (AIS)/Net 1000
- U.S. satellite networks (Satellite Business System [SBS], MCI, Western Union, RCA, SP Communications)

Data services provide multiplexed terminal transactions over distances of a few feet via cable to thousands of miles via satellite repeaters. One of the reasons that digital equipment continues to replace analog is because experience has shown that time-division multiplexing equipment is less expensive than frequency-division multiplexing equipment. Another reason is security: Digital signals can be processed in encoders, scramblers, and so forth in such a way as to encrypt messages; the receiver, programmed with the code, can decipher the transmission to recover the message. Finally, digital signaling such as PCM can operate virtually error-free if the system signal-to-noise ratio can be maintained in excess of about 20 dB.

Standardization of the signals used for communications between terminal equipments preceded the tremendous growth in digital information transfer. We now turn to a discussion of the standardized digital bit patterns (codes) used in communication links between terminals. ■

13-1 ■ STANDARD DIGITAL CODES

Digital communication codes go back in history well before the harnessing of electricity. The first well-known electrical code was Samuel Morse's code of long and short bursts, which could be (and still is) adapted to electrical and light bursts. The Morse code is a binary code using two unequal-length symbols—dash and dot, or long/short.

The codes used in computers and data communication machines today are more structured, having equal-length characters (digital words) made up of equal-duration symbols (bits). The earliest of these codes was a 5-bit code of J. M. E. Baudot, a variation of which is the CCITT-2 code of Table 13–1; it is still used in some teletype (TTY) systems. One example is Western Union's Telex,* operating at 66 English words/min (50 bits/s). Notice how each 5-bit code word is used to represent one of the 26 letters of the alphabet, as well as another keyboard character (figure) by using the "figure shift." The 58 characters shown would not have been possible with a five-bit code ($2^5 = 32$).

The most frequently used digital communications code today is the ASCII code.† This is a 7-bit code, and an eighth bit may be added for parity. With 7 bits,

---

* Telex is an international message service using telegraph circuits. Another Western Union service, called TWX ("Twix"), is a U.S.-based teletype message service that uses the dial-up public telephone network.

† American Standard Code for Information Interchange.

TABLE 13-1 Two Early Standard Digital Codes

"ARQ" 1234567	CCITT-2 12345	Character	
0001110	11111	*(letter shift)*	
0100110	11011		*(figure shift)*
0011010	11000	A	–
0011001	10011	B	?
1001100	01110	C	:
0011100	10010	D	(WRU)
0111000	10000	E	3
0010011	10110	F	%
1100001	01011	G	@
1010010	00101	H	£
1110000	01100	I	8
0100011	11010	J	(BELL)
0001011	11110	K	(
1100010	01001	L	)
1010001	00111	M	.
1010100	00110	N	,
1000110	00011	O	9
1001010	01101	P	0
0001101	11101	Q	1
1100100	01010	R	4
0101010	10100	S	'
1000101	00001	T	5
0110010	11100	U	7
1001001	01111	V	=
0100101	11001	W	2
0010110	10111	X	/
0010101	10101	Y	6
0110001	10001	Z	+
0000111	00000		(blank)
1101000	00100		(space)
1011000	01000		(line feed)
1000011	00010		(carriage return)

Characters within parentheses are telegraph function signals and are not printed.

$2^7 = 128$ different symbols can be defined. As seen in Table 13–2, ASCII includes numbers, uppercase and lowercase letters of the alphabet, and numerous typewriter symbols (comma, colon, and so on), as well as data-link control codes such as STX (Start of Text) and ACK (Acknowledge). The parity bit is used for detecting transmission errors and is discussed in Section 13–2. When used in asynchronous transmissions, stop and start bits are used to separate one digital word from the next. With ASCII, the first bit transmitted is a 0 for the start bit, followed by

TABLE 13–2 The U.S. ASCII Code: American Standard Code for Information Interchange

Bit	7	0	0	0	0	1	1	1	1	
Position	6	0	0	1	1	0	0	1	1	
1234	5	0	1	0	1	0	1	0	1	HEX 0
0000		NUL	DLE	SP	0	'	P	@	p	0
1000		SOH	DC1	!	1	A	Q	a	q	1
0100		STX	DC2	"	2	B	R	b	r	2
1100		ETX	DC3	#	3	C	S	c	s	3
0010		EOT	DC4	$	4	D	T	d	t	4
1010		ENQ	NAK	%	5	E	U	e	u	5
0110		ACK	SYN	&	6	F	V	f	v	6
1110		BEL	ETB/ EOB	'	7	G	W	g	w	7
0001		BS	CAN	(	8	H	X	h	x	8
1001		HT	EM	)	9	I	Y	i	y	9
0101		LF	SS	*	:	J	Z	j	z	A
1101		VT	ESC	+	;	K	[	k	{	B
0011		FF	FS	,	<	L	\	l	⌐	C
1011		CR	GS	–	=	M	]	m	}	D
0111		SO	RS	.	>	N	¬	n	|	E
1111		SI	US	/	?	O	—	o	DEL	F
	HEX1	0	1	2	3	4	5	6	7	

Meaning of nondisplayable data link control characters:

NUL = Blank
SOH = Start of header
STX = Start of text
ETX = End of text
EOT = End of transmission
ENQ = Inquiry
ACK = Acknowledgment
BEL = Bell
BS = Back space
HT = Horizontal tab
LF = Line feed
VT = Vertical tab
FF = Form feed
CR = Carriage return
SO = Shift out
SI = Shift in
DLE = Data-link escape

DC1 = Device control #1
DC2 = Device control #2
DC3 = Device control #3
DC4 = Device control #4
NAK = Negative acknowledgment
SYN = Synchronous idle
ETB = End transmitted block
CAN = Cancel
EM = End of medium
SUB = Start special sequence
ESC = Escape or break
FS = File separator
GS = Group separator
RS = Record separator
US = Unit separator
SP = Space
DEL = Delete

FIGURE 13-1   ASCII format for
asynchronous systems.

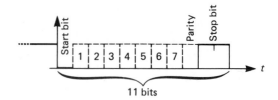

11 bits

the 7-bit character plus parity, and closed by a 2-bit-long 1 for stop. The total
asynchronous (nonsynchronized clocks) ASCII character format, then, requires
11 bits, as seen in Figure 13–1.

Another code used in computers and instrumentation control, especially where
numerical calculations or decimal number displays are involved, is the binary-
coded decimal (BCD) system. An extension of BCD for communications use is an
8-bit code called EBCDIC, for Extended BCD Interchange Code (see Table 13–3).
This code, like ASCII, includes the alphabet, decimal numbers, control codes,
and other symbols. However, parity is not included.

Codes and Error Control Techniques

Numerous techniques have been applied to the task of achieving the channel data
capacity with low error rates promised by the information theory begun by Claude
Shannon.* A major division exists between types of error control techniques.
First, there is the *automatic request for retransmission* (ARQ), in which a short
message is returned to the sender when an error has been detected, and then there
is the *forward error correction* (FEC), in which redundant bits built into the code
allow detected errors to be corrected at the decoder. Then a fundamental distinction must be made between the *standardized alphanumeric codes* and the *redundancy codes* for error detection and correction.

The standardized alphanumeric digital codes such as ASCII, EBCDIC,
CCITT-2, and others have been developed primarily for the purpose of providing
a universally agreed-on set of fixed-length binary patterns to represent standard
alphanumeric characters. The most common application of these codes is for
interfacing between computers that use internal machine codes and multivendor
peripheral equipment such as monitors, printers, and also serial communication
links. Listings showing the one-to-one correspondence between the binary and
alphanumeric characters for the most commonly used codes are given in Tables
13–1 through 13–3.

In contrast to standardized codes (tables), there are literally hundreds of coding
techniques that insert redundancy (extra bits) into digital data for the purpose of
providing error-detection capability or error-detection-with-correction capabilities. Some of the most common encoding techniques in electronic communica-

* C. E. Shannon, "A Mathematical Theory of Communication," *Bell System Technical Journal*, vol.
38, 1948, pp. 623–656.

TABLE 13-3 The EBCDIC Code

Code Bit 8		0	0	0	0	0	0	0	0	1	1	1	1	1	1	1	1
Bit 7		0	0	0	0	1	1	1	1	0	0	0	0	1	1	1	1
Bit 6		0	0	1	1	0	0	1	1	0	0	1	1	0	0	1	1
Bit 5		0	1	0	1	0	1	0	1	0	1	0	1	0	1	0	1
Position 1234	HEX	0	1	2	3	4	5	6	7	8	9	A	B	C	D	E	F
0000	0	NUL	DLE	DS		SP	&	-									0
1000	1	SOH	DC1	SOS				/		a	j			A	J		1
0100	2	STX	DC2	FS	SYN					b	k	s		B	K	S	2
1100	3	ETX	DC3							c	l	t		C	L	T	3
0010	4	PF	RES	BYP	PN					d	m	u		D	M	U	4
1010	5	HT	NL	LF	RS					e	n	v		E	N	V	5
0110	6	LC	BS	EOB ETB	UC					f	o	w		F	O	W	6
1110	7	DEL	IL	PRE ESC	EOT					g	p	x		G	P	X	7
0001	8		CAN							h	q	y		H	Q	Y	8
1001	9	RLF	EM							i	r	z		I	R	Z	9
0101	A	SMM	CC	SM		¢	!		:								
1101	B	VT				.	$	,	#								
0011	C	FF	IFS		DC4		*	%	@								
1011	D	CR	IGS	ENQ	NAK	(	)	―	'								
0111	E	SO	IRS	ACK		+	;		=								
1111	F	SI	IUS	BEL	SUB			?	"								
	HEX	0	1	2	3	4	5	6	7	8	9	A	B	C	D	E	F

TABLE 13–4 Gray Code

Bit Position 1234	Character or Value
0000	0
1000	1
1100	2
0100	3
0110	4
1110	5
1010	6
0010	7
0011	8
1011	9

tions applications are convolutional and trellis encoding, and Reed-Solomon, Hamming, Golay, and Bose-Chaudhuri-Hocquenhem (BCH) codes.

Digital codes have been devised that are specifically designed to reduce the effects of noise. One of these is an example of exact-count or constant-ratio codes in which the number of 1s is the same for each code word.

The first of the exact-count codes is the "ARQ." It is listed in Table 13–1 next to the CCITT-2 because the 7-bit ARQ code defines the same set of figures and letters as the CCITT-2 code. However, each 7-bit digital word of the ARQ code has exactly three 1s, so that a simple count can detect transmission errors.

Another code in which errors are easily detected is the Gray code of Table 13–4. The Gray code is especially useful for encoding slowly changing telemetry data. Notice that from one data word to the next, there is only a 1-bit change. Thus, if a normally slow-changing parameter, such as temperature in a satellite package, is being monitored or the fine control of a positioning motor is being relayed by a telemetry link, short-term errors due to noise can easily be detected. An additional benefit of a 1-bit-change code for this application is the reduced signaling rate and consequently narrower bandwidth required.

As an example, suppose a thermistor used as a transmitter temperature monitor is transmitting $1000_2 = 8_{10}$ to the satellite control station. The earth station chart recorder is expected to be varying between 7 and 9 or, perhaps, indicating a steady rise or fall in the temperature reading. The CCITT-2 code reading 01100 will equivalently vary between 11100 (a 1-bit change) and 00011 (a 4-bit change) from the nominal. A fundamentally important principle in electronic communications is that slowly changing signals require less channel bandwidth, and reduced channel bandwidth permits less noise to enter the system. In addition, short bursts of 01110 = "colon" or 01101 = "zero" can be ignored as transients.

13–2 ■ ERROR DETECTION

One way of overcoming the effects of noise and interference in data transmission systems, other than by increasing the transmitted power, is by detecting the

FIGURE 13-2 Parity generator/ checker. Used as a parity-bit generator, ground P_i for odd parity or tie high (1) for even parity. As an even parity checker, P_o will go high (true) for a valid check.

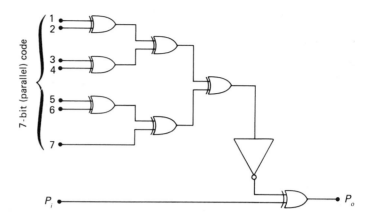

occurrence of errors and by getting a retransmission or by using codes that allow the data to be corrected at the receiver.

Besides the purely random thermal and shot noise encountered in communication systems, we find that transient noise can be devastating in data transmission. For instance, a 10-ms transient in a voice system (telephone) will sound like a "click." Enough of these will be annoying but the message can be transmitted. The same 10-ms "glitch" in a 1200-bit/s data transmission will obliterate 12 data bits; this could wipe out a few people's bank accounts!

We must design into our data systems methods for detecting and correcting errors. The simplest error detection scheme is parity. *Parity* is a simple technique for detecting the occurrence of errors in a digital signal transmission. A parity circuit like that of Figure 13–2 can examine the 1 bits of each transmitted character (digital word) and produce an additional 1 or 0 to make the total come out even (even parity) or odd (odd parity). The same circuit can be used to check parity at the receiver. If the parity check is false, an error has occurred.

As an example of determining the parity for a single character, take a character from the ASCII code and format, as in Figure 13–3. Suppose the 7-bit digital word is (LSB) 0011010 (MSB). If we decide that the system should be based on even parity, then the parity generator looks at the 7 bits, finds that there are an odd number of 1s, so that the parity bit must be a 1 in order to make the total (8 bits) come out even. For even parity then, the parity generator inserts the 1 into the data stream, and the transmitted 8 bits are 00110101.

At the receiving-end computer or terminal, the 8 bits are examined by a hardware- or software-implemented parity checker that counts the number of 1s (actually, modulo-2 addition) and outputs a digital pulse to indicate whether or not the received 8-bit character, in fact, has the correct parity.

FIGURE 13-3 ASCII character format. Note that b_1, the LSB, is the first bit of the digital word to be transmitted.

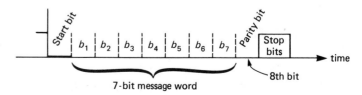

FIGURE 13–4 Exclusive-OR
circuit symbols and truth table.

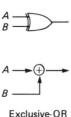

Inputs		Outputs
A	B	EX-OR
0	0	0
1	1	0
0	1	1
1	0	1

Exclusive-OR

Truth table

(even for first two rows, odd for last two rows)

If the parity check is negative (wrong parity received), then, depending on the system protocol, the transmitted character is either simply rejected or the transmission is delayed while the receiver signals the transmitter to repeat that character. If a repeat request is to be made, there must be a return transmission channel available to carry the request for retransmission. It must also be realized that, for such a system, the actual rate of data transmission will vary with the number of repeats so that the actual system data rate will be error-rate dependent. Automatic repeat request (ARQ) techniques are discussed with "Protocols."

Let us look at the operation of the parity generator using the preceding character sequence: The exclusive-OR adds two bits and produces an output indicating if the sum is even (0) or odd (1). A truth table is included with Figure 13–4. The exclusive-OR circuit performs modulo-2 arithmetic. All the ordinary rules of addition or, equivalently, subtraction are obeyed except that in modulo-2 addition $1 + 1 = 0$, and there are never any "carries."

With this instant review, follow the 7-bit digital word, 0011010, through the parity generator of Figure 13–5 to see if it produces a 1. Since we want even parity, P_i is tied high ($P_i = 1$).

Now try an even 7-bit digital word like 1011010 and confirm that P_o will be a 0. You can also show that the inverter is not required. If it is deleted, make P_i the opposite polarity.

FIGURE 13–5 Parity generator
(even). Even parity is being
generated for a 7-bit digital
word. For an ASCII format
transmission, this parity bit is
added as the eighth bit.

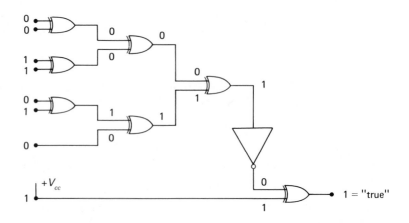

As a parity checker, the circuit of Figure 13–5 has the 8-bit word (7-bit character + parity) as the input. If the output is 1, an even parity word was received. If the output is 0, an odd parity word was received.

Perhaps the greatest drawback to the parity check as a form of coding for overcoming noisy transmission environments is that an even number of errors will go undetected. For instance, if an even (or odd) character has two bits inverted during transmission, the received character will still be even (or odd).

Another limitation to the simple parity being discussed is that an error occurring between transmitter and receiver can be detected, but there is no provision for determining which bit was in error, only which character. A scheme has been devised to determine which bit of a character is in error so that the bit can be corrected at the receiving terminal.

Error Detecting and Correcting

One scheme for correcting detected errors is based on the generation of a parity word, or character, for a group (block) of data characters. Figure 13–6 shows five consecutive characters, including their parity bits (even), to form a block of data. The block check character (BCC) is easily determined as the parity of the individual columns. Thus, for an even-parity system, we see that the BCC is 00111100. This block check character is transmitted after the block of five data characters, and circuitry (or software) checks the parity of the rows and columns. The intersection of the row and column with a parity error is the location of the incorrect bit. Correction is achieved by merely inverting the bad bit. This scheme is no more foolproof than other parity schemes because parity in the columns (the LR check) is defeated if a double error occurs in a column. There are numerous possibilities for an even number of errors in rows and/or columns for a system in a noisy environment, but this scheme is useful in a low-noise environment in which system security depends on correct data and when duplexing for retransmission of data is not feasible.

Cyclic Redundancy Checks

One of the most effective techniques for detecting multiple or singular errors in data transmissions with a minimum of hardware is the cyclic redundancy check (CRC).

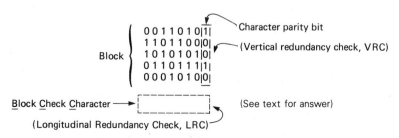

FIGURE 13-6 Vertical and longitudinal checks.

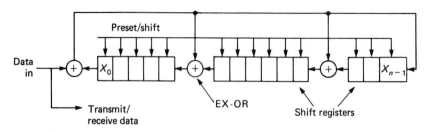

FIGURE 13-7   Cyclic redundancy block check register.

A typical block diagram for a CRC system is shown in Figure 13–7. There are shift-register/exclusive-OR arrangements like this at the transmitter and the receiver. The shift registers are initialized to zero and the message is transmitted; simultaneously, it is passed into the transmitter's CRC block check register. At the end of the message, the transmitter sends the contents of its block check register (BCR). The receiver initializes its identical BCR to zero and receives the message followed by the cyclic redundancy check. After the message and CRC have passed through the receiver BCR, the register contents must equal zero. This is true regardless of the particular placement of the exclusive-ORs, in the two BCRs.

Sometimes the initializations are not zero (zero is not always the best choice for a final word), but the thing to realize is that, because of the feedback arrangement, the exact state of the register is dependent upon a lot of past history.

The most common cyclic redundancy check registers are the CRC-12, CRC-16, and internationally, the CCITT's recommended V41. The V41 system, shown in Figure 13–8, is a 16-stage shift register intended for low-speed data transmission with data split into blocks of either 220, 460, 940, or 3820 bits, depending on the transmission link. This CRC system has a 99.95% chance of detecting bursts of errors up to 12 bits in length.

The calculations performed by CRC systems such as Figure 13–8 involve a division of the message signal, called the *message polynomial M(x)*, by the specific system generator polynomial $G(x)$. The quotient is discarded, and a remainder is transmitted as the block check character.

Polynomials of Binary Coefficients.   The polynomial representation of the 8-bit binary word or message 10111_2 is

$$M(x) = 1 \cdot x^4 + 0 \cdot x^3 + 1 \cdot x^2 + 1 \cdot x^1 + 1 \cdot x^0 \qquad \text{(13–1a)}$$
$$= x^4 + x^2 + x + 1 \qquad \text{(13–1b)}$$

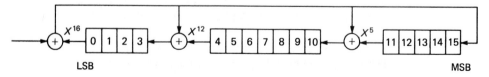

FIGURE 13-8   CCITT-V41 CRC system. The generator polynomial, $G(x)$, is $X^{16} + X^{12} + X^5 + 1$.

where the coefficients are chosen from one of the binary symbols 1 or 0. The order (highest power) of the polynomial in Equation 13–1 is 4; that is, an n-bit binary word or message can be represented by a polynomial of order $n - 1$. Modulo-2 (MOD-2) addition is performed in exclusive-OR gates with results as follows:

$$0 + 0 = 0 \qquad \text{(13–2a)}$$

$$0 + 1 = 1 \qquad \text{(13–2b)}$$

$$1 + 0 = 1 \qquad \text{(13–2c)}$$

$$1 + 1 = 0 \qquad \text{(13–2d)}$$

MOD-2 subtraction produces the same results as MOD-2 addition.

In cyclic redundancy check mathematics, the order of the generator polynomial is made one *more* than the order of the block check character, BCC. Stated another way, if the number of *bits* in the BCC is c, then the order of the generator polynomial $G(x)$ is c. To construct a transmitted data sequence $D(x)$ that is evenly divided by $G(x)$ with no remainder, first shift the message to the left by the number of bits in the BCC, $C(x)$, to make room for adding $C(x)$ to the right of the message. Expressed mathematically, this is the same as multiplying the message polynomial $M(x)$ by x^c. That is, the message, shifted left by c bits, is expressed mathematically as

$$M'(x) = x^c M(x) \qquad \text{(13–3)}$$

The transmitted data sequence $D(x)$ is the BCC added to the shifted message. This is expressed mathematically as

$$D(x) = M'(x) + C(x) \qquad \text{(13–4)}$$

The BCC generators of CRC systems are usually implemented in hardware from exclusive-OR gates and serial shift registers; see Figure 13–7. The message is input and the result is that the left-shifted message $M'(x)$ is divided by the generator polynomial $G(x)$. The binary states remaining in the register cells after the last message bit followed by three zeros is the remainder of the division process, expressed as

$$\frac{M'(x)}{G(x)} = Q(x) + C(x) \qquad \text{(13–5)}$$

where $Q(x)$ is the quotient of the division process that is not physically captured during the process and $C(x)/G(x)$, the remainder, provides the BCC, $C(x)$. Remember that the subtractions involved in the long-division process are done modulo-2. The block check character $C(x)$ is serially shifted out (to the left in Figure 13–7) of the CRC system and appended to the end of the message on the transmit data line by a logic arrangement that connects the x_0-cell output to the transmit line.

At the receiver, if $D(x)$ is divided by $G(x)$ and the remainder is zero, then no errors have occurred. This is demonstrated as follows:

$$\frac{D(x)}{G(x)} = \frac{M'(x)}{G(x)} + \frac{C(x)}{G(x)} \qquad \text{(13–6)}$$

which, from Equation 13–5, is

$$\frac{D(x)}{G(x)} = \left[Q(x) + \frac{C(x)}{G(x)} \right] + \frac{C(x)}{G(x)} \tag{13–7}$$

However,

$$\frac{D(x)}{G(x)} = Q(x) + (1 + 1) \left(\frac{C(x)}{G(x)} \right) \tag{13–8}$$

and, in modulo-2 arithmetic, $1 + 1 = 0$. Therefore,

$$\frac{D(x)}{G(x)} = Q(x) \tag{13–9}$$

and the division at the receiver has no remainder. That is, the receiver CRC registers contain all zeros, indicating no transmission errors. An example will illustrate the CRC calculations and the polynomial division process.

■ ■ ■ ■

EXAMPLE 13–1

The simple binary message 10111_2 is to be transmitted with cyclic redundancy check (CRC) error detection. The CRC generator polynomial is $G(x) = x^3 + x^2 + x$.

1. Determine the BCC, $C(x)$.
2. Determine $D(x)$ and the transmitted data sequence in bits.
3. Show that $D(x)/G(x)$ has a remainder of zero.

Solution:

From the given, $M(x) = x^4 + x^2 + x + 1$ and $G(x)$ is of order 3, the highest power of $G(x)$. From Equation 13–3,

$$M'(x) = x^c M(x) = x^3(x^4 + x^2 + x + 1) = x^7 + x^5 + x^4 + x^3.$$

$C(x)$ will be the remainder from $M'(x)/G(x)$ in Equation 13–5.

$$\frac{M'(x)}{G(x)}: \quad x^3 + x^2 + x \,\overline{\smash{\big)}\, x^7 + x^5 + x^4 + x^3}$$

$$
\begin{array}{r}
x^4 + x^3 + x^2 + x\ \ + 1 \\
x^3 + x^2 + x + 0\ \overline{\smash{\big)}\ x^7 + 0\ + x^5 + x^4 + x^3 + 0\ + 0 + 0} \\
x^7 + x^6 + x^5 \\
\hline
x^6\ \ \ \ \ \ \ + x^4 + x^3 \\
x^6 + x^5 + x^4 \\
\hline
x^5\ \ \ \ \ \ \ + x^3 \\
x^5 + x^4 + x^3 \\
\hline
x^4 \\
x^4 + x^3 + x^2 \\
\hline
x^3 + x^2 \\
x^3 + x^2 + x \\
\hline
x\ \ \ \ \ = C(x)
\end{array}
$$

1. $C(x) = x$.
2. The BCC is 010_2. Hence, from Equation 13–4, the answers to **(2)** are $D(x) = x^7 + x^5 + x^4 + x^3 + x$ and, in binary, $D_o = \mathbf{10111010}$.
3. At the receiver, if no errors have occurred in the transmission link, then the remainder from $D(x)/G(x) = \mathbf{0}$.

$$
\frac{D(x)}{G(x)}: \quad x^3 + x^2 + x \ \overline{\big)\
\begin{array}{l}
x^4 + x^3 + x^2 + x\ + 1 \\[2pt]
\hline
x^7 \qquad\ + x^5 + x^4 + x^3 \qquad\quad + x \\
\underline{x^7 + x^6 + x^5} \\
\qquad x^6 \qquad\ + x^4 + x^3 \\
\qquad \underline{x^6 + x^5 + x^4} \\
\qquad\qquad x^5 \qquad\ + x^3 \\
\qquad\qquad \underline{x^5 + x^4 + x^3} \\
\qquad\qquad\qquad x^4 \qquad\qquad + x \\
\qquad\qquad\qquad \underline{x^4 + x^3 + x^2} \\
\qquad\qquad\qquad\qquad x^3 + x^2 + x \\
\qquad\qquad\qquad\qquad \underline{x^3 + x^2 + x} \\
\qquad\qquad\qquad\qquad\ \ 0\ +\ 0\ +\ 0 \qquad\quad = R
\end{array}}
$$

As demonstrated, the remainder is zero when there are no errors.

More details on these calculations as well as examples can be found in McNamara,* as well as other telecommunications sources.

Some of the first work in error correction coding was done by R. W. Hamming at Bell Labs. Codes developed by adding check bits within simpler nonredundant codes, then performing multiple parity checks at the receiver are called *Hamming codes*. The *Hamming distance* is the number of bits by which any two code words differ. To provide for error detection, the hamming distance must exceed 1 bit; to provide for error correction, the hamming distance must exceed 2 bits. The addition of redundant (noninformation) bits is a common technique for improving the reliability of data communications in a noisy environment. Indeed, system S/N improvement, called *coding gain*, has been computed for several digital modulation techniques and coding schemes including convolutional, Golay, BCH, and Hamming codes.†

Since the name of the game in communication systems is to overcome noise, we next turn our attention to the characterization of noise in digital data systems.

---

* John E. McNamara, *Technical Aspects of Data Communication* (Bedford, Mass.: Digital Equipment Corporation, 1977).

† R. L. Freeman, *Telecommunication Transmission Handbook*, 2nd ed. (New York: Wiley-Interscience, 1981), pp. 623, 624.

13-3 ■ NOISE AND DATA ERRORS

As stated in the section on data codes for error detection and correction, noise transients can be devastating in data communication systems. The example of 12 bits being obliterated by a 10-ms noise transient in a 1200-baud transmission illustrated the point. Also, in Chapter 12, we found that quantization noise is an unavoidable part of the analog-to-digital conversion process in PCM systems.

In this section, the effects of random thermal and shot noise on the quality of digital information transfer is discussed.

As illustrated in Figure 13–9, if a noise peak exceeds one-half of the peak-to-peak signal voltage at a sampling instant, the sampled data bit will be in error. If the noise distribution is known, the probability of error occurrence can be analyzed and characterized in terms of the average number of errors versus the average signal-to-noise power ratio.

The thermal and shot noise in receivers tends to be statistically random *white noise*, and the noise peaks, around the average data voltage V, tend to have a Gaussian or normal distribution, as seen in the familiar bell-shaped curve of Figure 13–10.

The curve of Figure 13–10 shows that noise is clustered around the value V, which would be measured in the absence of noise. Hence, this curve gives a relative measure of the occurrence of different values of voltage during the time that a digital 1 is being received. Notice that values near $+V$ are more likely to be measured at the receiver than other values. Also notice that, when the randomly occurring voltage values drive the received voltage below the decision level of 0 V, the receiver might record a data error. The dashed curve is the case when noise is present—that is, when $\bar{v}_n$, the rms noise voltage, is reduced.

The probability of a data bit error occurring is determined by calculating the area under the noise distribution curve (called a *probability density function*, pdf).

Probable values of received voltage when $+V$ has been transmitted.

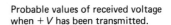

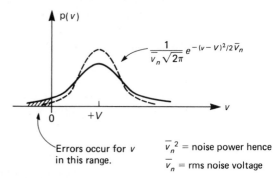

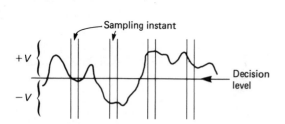

FIGURE 13–9 Noisy data at the receiver. 1011 was transmitted. An error is registered in the first bit.

FIGURE 13–10 Gaussian "normal" noise distribution. V is the value of signal without noise and $\bar{v}_n$ is the rms noise voltage.

The area to be calculated, shaded in Figure 13–10, is for all possible noise pulses with amplitude greater than V with a negative polarity. Assuming a decision level of 0 V and equal probabilities of transmitting a $+V$ or $-V$, then the probability of bit errors, P_e, or the theoretically expected bit error rate, is found from Equation 13–10:

$$P_e = \int_{-\infty}^{0} p(v) \; dv \qquad (13\text{–}10)$$

The integration necessary to determine the (shaded) area has no closed-form solution for a Gaussian pdf $[p(v)]$, but because of its importance in many areas of engineering design, it has been solved by numerical methods and put into tables and graphical form.

Figure 13–11 shows the final results of the analysis for binary transmissions in a Gaussian noise environment. The probability of error P_e, or theoretical bit error rate (BER), is shown as a function of signal-to–noise power ratio in decibels. The signal power is proportional to V^2 in each bit interval, and $\bar{v}_n$, the rms noise voltage, when squared, is proportional to noise power N and is a statistical quantity called the *variance*. Hence, the S/N in decibels can be calculated from

$$\frac{S}{N} \text{ (dB)} = 20 \log \frac{V}{\bar{v}_n} \qquad (13\text{–}11)$$

where V = peak received voltage (without noise) and $\bar{v}_n$ = rms voltage of the noise only. The value V is *one-half* the peak-to-peak received signal voltage without noise.

Bit error rate (BER) is a measured quantity. When BER is measured and adequately documented for a data link, the user can calculate the number of errors

FIGURE 13–11 Theoretical bit error rate versus S/N for binary transmissions in white Gaussian noise.

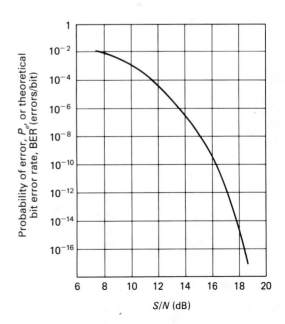

Probability of error, P_e, or theoretical bit error rate, BER (errors/bit)

S/N (dB)

to expect in a received binary message. Examples of how the error function*
curve is used follow.

■ ■ ■ ■
**EXAMPLE
13–2**

Output measurements on a binary data receiver indicate a signal-to-noise ratio of
about 10 dB. Determine the number of errors expected in a message 100,000 bits
long.

Solution:

The probability-of-error curve of Figure 13–11 gives $P_e \approx 1 \times 10^{-3}$ (average
number of errors/bit): an average of 1 error for every 1000 bits transmitted. Thus,
we can expect about 100 errors in a 10^5-bit message; that is,

$$\text{no. errors} = \left(\frac{\text{no. bits}}{\text{message}}\right) \times P_e = \frac{100,000 \text{ bits}}{\text{message}} \times \frac{1 \times 10^{-3} \text{ errors}}{\text{bit}} = \textbf{100 errors}$$

■ ■ ■ ■
**EXAMPLE
13–3**

1. Determine the minimum S/N required to achieve a 10^{-5} BER for a binary
   transmission.
2. If the received signal has a peak signal voltage of 100 mV (no noise), what is the
   maximum value of rms noise for the 10^{-5} BER?

Solution:

1. The error rate curve for 10^{-5} shows $S/N \approx \textbf{12.5 dB.}$
2. 12.5 dB = 20 log 100 mV/$\bar{v}_n$, so

$$\bar{v}_n = \frac{100 \text{ mV}}{10^{12.5/20}} = \textbf{23.7 mV rms}$$

For S/N greater than 9.5 dB (BER $< 1.5 \times 10^{-3}$), the theoretical BER curve of
Figure 13–11 may be approximated by

$$\text{BER} = \frac{1}{z\sqrt{2\pi}} \, e^{-z^2/2} \qquad\qquad (13\text{–}12)$$

where $z = V/\bar{v}_n$ is the signal voltage–to–noise voltage ratio.
    The curve of Figure 13–11 can be used for both NRZ and NRZ-bipolar signal
formats. BER is used to compare various transmission systems and signal formats. What you will notice about these comparison curves is that they have a
similar shape—either the complementary error function of Figure 13–11 or an
exponential function.

---

* Figure 13–11 is actually a plot of a complementary error function (erfc), which gives the probability
that an instantaneous voltage measurement will have a value greater than V when the Gaussian noise
has an rms value of $\bar{v}_n$.

13–4 ■ INTERCOMPUTER COMMUNICATIONS

One of the first considerations in determining how to get one computer or terminal communicating with another is the distance separating the units. Internally, computers and terminal devices use parallel transistor-transistor logic (TTL) circuits that operate at 0 to 5 V or MOS levels. But TTL-level connections are severely distance-limited so that, for interconnecting devices separated by more than one meter, the parallel data are converted to a serial (series) data stream for transmission of all data on a single wire-pair circuit. The circuit that performs the parallel-to-serial and serial-to-parallel conversions in a computer terminal is called a *programmable communications interface* (PCI). More generalized circuits are called *universal data receiver/transmitters* (UART and USART).

Once the data is in a serial format, the serial data, along with data-link control signals, is brought to an interface connector, usually on the rear of the terminal chassis. However, TTL-level signals are not usually brought out of the terminal. The TTL-level signals are usually converted to a higher-voltage bipolar format for interfacing equipment separated by up to 50 ft or so. The mechanics of interfacing teletype systems are considerably different from the above and will be discussed in the section on 20-mA loop systems.

For very short distances between terminals or other computer peripherals, the communications interface could be hardwired with a multiwire ribbon or cable to carry the serial data and the control signals necessary to start and maintain a rationally organized and reasonably error-free flow of traffic using a prescribed protocol. Figure 13–12 illustrates a simple interface between a terminal and computer.

The interface shown in Figure 13–12 is limited to approximately 60 ft (20 m) due to external interference (RFI), cross talk between wires, and pulse-degrading capacitance. Such limitations exist even when using low-impedance line driver/receiver circuits such as the 1488 and 1489 ICs.

Medium- and long-distance data communications require an additional unit with the general name of *data communications* or *data channel equipment* (DCE); this is usually a modem. Data terminal equipment (DTE) will interface with the communications equipment, which transmits and receives digital data over vari-

FIGURE 13–12 Data terminal equipment (DTE) interface. Three ways of indicating multiple-lead interconnections are shown.

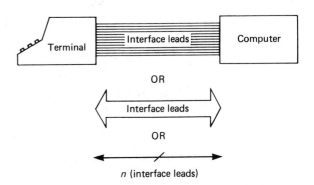

ous transmission media, from copper or optical fiber cables to microwave radio and satellite links.

13–5 ■ SERIAL TRANSMISSION INTERFACING

A generalized single-line interface (computer/terminal to transmission line) is shown in Figure 13–13. The next sections will cover more specific systems and the system-level hardware design decisions necessary to determine appropriate interfacing.

Data may be transmitted in either a synchronous or asynchronous mode. Synchronous transmissions require the receiving DCE to synchronize its data clock to the received signal before actual data can be transferred. In asynchronous transmission, the start and stop bits frame each data character.

In general, data signals will be flowing in both directions on the transmission line. This is called *duplex* operation. It is *full-duplex* if communication signals can move in both directions *simultaneously*, and *half-duplex* if signals can flow in only *one direction at a time* on the transmission line. *Simplex* is the name given to a system restricted to signal flow in one direction only. The transmission line shown in Figure 13–13 can be any one of many different communication media such as a hard wire pair or coaxial cable, telephone lines, microwave radio, or satellite links. The signals can be binary dc-level changes (current or voltage) or digitally modulated carrier schemes operating synchronously or asynchronously by the protocols (rules) commonly used in the industry or developed specifically for a particular application. A breakdown of the major system-level decisions which must be made are listed in Table 13–5.

For a complete discussion of data communication interfacing, we need to cover each of the topics in Table 13–5. But, as you will see, there are so many combinations and permutations among the listings preceding that reading various literature and equipment data sheets can leave your head spinning. So study Figure 13–13 and Table 13–5 and discussion will begin with the process of converting the computer data from its usual parallel form to a serial data stream.

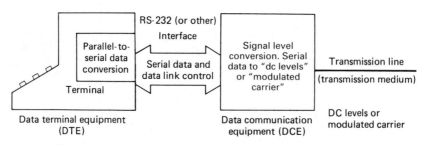

FIGURE 13–13 Generalized interfacing for medium- to long-distance data communication.

TABLE 13–5 System Design Choices for Serial Interfaces

Transmission Network (Line)	Transmission Line Signals	Protocols
1. Hardwire	1. Levels	1. Asynchronous
2. Private dedicated telephone lines	a. 20 mA loop, TTY-type	a. RS-232-C, CCITT, etc.
3. Switched (telephone) network	b. voltages, RS-232-C, standards, etc.	2. Synchronous
a. acoustic coupled	2. Carrier	a. character oriented: Bisync
b. direct connection	a. FSK	b. bit-oriented (BOP):
(i) auto answer	b. Multilevel PSK and QAM (APSK)	(i) X.25
(ii) auto dial		(ii) HDLC
4. System multiplexing (trunks)	**Modes**	(iii) SDLC
5. Microwave radio link	1. Asynchronous	
6. Satellite link	2. Synchronous	

UART for Asynchronous Communication

The circuit used to implement the parallel-to-serial conversion and handle some of the terminal input/output controls of Figure 13–13 is given different names by various manufacturers. There are also functional variations between the different devices, but in general, the circuit is an LSI (large-scale integration) IC and is functionally a programmable communications interface (PCI). When the circuit is capable of universal application to many different computer terminals, the names UART (universal asynchronous receiver/transmitter) or USART (universal synchronous/asynchronous receiver/transmitter) are used. Other names for circuits that perform similar or complementary functions are *data-link controller, communications controller,* and *asynchronous communications interface adapter.*

The most popular binary code for asynchronous communication is the ASCII code (see Section 13–1 on codes). In addition to having a serial data stream (typically seven bits), start, stop, and parity bits are included. These are usually added in the UART. Some UART chips include the clock circuitry as well, and the clock rate is computer-controllable. Some of the control leads for various UARTs are shown in the block diagrams of Figures 13–14 and 13–15.

Figure 13–14 shows the transmitter section and receiver section of a UART capable of handling up to 8-bit codes (such as EBCDIC). Control inputs select whether or not parity is to be computed and transmitted, whether or not one or two stop bits are to be added to the transmitted character, odd or even parity (if used), and the number of data bits in the character code (the choice of 5- to 8-bit coded word). Various strobe inputs and flag outputs give the terminal control over timing along with the clock input. f_b is the serial data output rate, and the clock

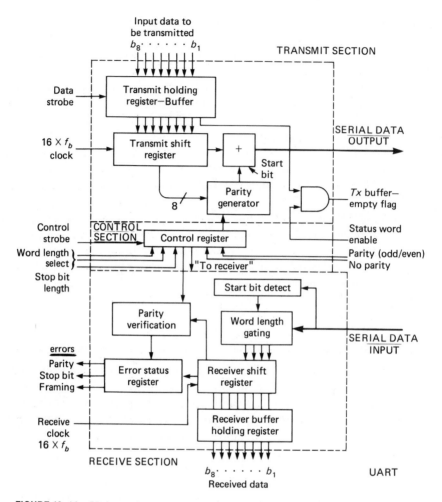

FIGURE 13–14   Universal asynchronous receiver/transmitter (UART).

driving the timing generator typically operates at $16 \times f_b$ hertz. The "to receiver" signals set the receiver section format for data received from another compatible UART. The holding and shift registers in the transmitter and receiver provide double buffering of this and the AMI UART of Figure 13–15. A *buffer* is a temporary data storage device (register).

    A 16-times-the-bit-rate clock allows the receiver section of the UART to sample the transmission line very rapidly, waiting for the stop-to-start bit transition to occur. When this transition is detected, eight clock pulses are counted before the start bit is sampled, at the center of the start bit. If the sampled signal is low, a valid start bit is assumed to exist (and not noise), whereupon a divide-by-16 circuit is enabled so that subsequent bits are sampled approximately in the center of the bit to reduce the effects of transmission line distortions of the data.

Features

- [ ] **Full or Half Duplex Operation Transmits and Receives Serial Data Simultaneously or Alternately**
- [ ] **Automatic Internal Synchronization of Data and Clock**
- [ ] **Automatic Start Bit Generation**
- [ ] **Buffered Receiver and Transmitter Registers**
- [ ] **Completely Status Circuitry**
- [ ] **TTL I/O Compatible**
- [ ] **Single + 5 Volt Supply**
- [ ] **Fully Programmable**
  - [ ] Work Length
  - [ ] Baud Rate
  - [ ] Even/Odd Parity
  - [ ] Parity Inhibit
  - [ ] One, One and One Half or Two Stop Bit Generation
- [ ] **Automatic Data Received/Transmitted Status Generation**
  - [ ] Transmission Complete
  - [ ] Parity Error
  - [ ] Framing Error
  - [ ] Overrun Error
- [ ] **Ceramic, CerDIP, or Plastic Packages Available**

Applications

- [ ] **Peripherals**
- [ ] **Terminals**
- [ ] **Mini Computers**
- [ ] **Facsimile Transmission**
- [ ] **Concentrators/Modems**
- [ ] **Asynchronous Data Multiplexers**
- [ ] **Modems/Data Sets**
- [ ] **Printers**
- [ ] **Remote Data Acquisition Systems**
- [ ] **Card and Tape Readers**

General Description

The AMI S1602 is a programmable Universal Asynchronous Receiver/Transmitter (UART) fabricated with N-Channel silicon gate MOS technology. All control pins, input pins and output pins are TTL compatible, and a single + 5 volt power supply is used. The UART interfaces asynchronous serial data from terminals or other peripherals, to parallel data for a microprocessor, computer, or other terminal. Parallel data is converted by the transmitter section of the UART into a serial word consisting of the data as well as start, parity, and stop bit(s). Serial data is converted by the receiver section of the UART into parallel data. The receiver section verifies correct code transmission by parity checking and receipt of a valid stop bit. The UART can be programmed to accept word lengths of 5, 6, 7, or 8 bits. Even or odd parity can be set. Parity generation checking can be inhibited. The number of stop bits can be programmed for one, two, or one and one half when transmitting a 5-bit code.

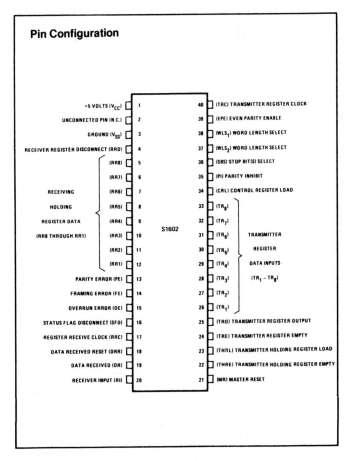

Pin Configuration

Pin	Signal
1	+5 VOLTS (V$_{CC}$)
2	UNCONNECTED PIN (N.C.)
3	GROUND (V$_{SS}$)
4	RECEIVER REGISTER DISCONNECT (RRD)
5	(RR8)
6	(RR7)
7	(RR6) RECEIVING
8	(RR5) HOLDING
9	(RR4) REGISTER DATA
10	(RR3) (RR8 THROUGH RR1)
11	(RR2)
12	(RR1)
13	PARITY ERROR (PE)
14	FRAMING ERROR (FE)
15	OVERRUN ERROR (OE)
16	STATUS FLAG DISCONNECT (SFD)
17	REGISTER RECEIVE CLOCK (RRC)
18	DATA RECEIVED RESET (DRR)
19	DATA RECEIVED (DR)
20	RECEIVER INPUT (RI)
40	(TRC) TRANSMITTER REGISTER CLOCK
39	(EPE) EVEN PARITY ENABLE
38	(WLS$_1$) WORD LENGTH SELECT
37	(WLS$_2$) WORD LENGTH SELECT
36	(SBS) STOP BIT(S) SELECT
35	(PI) PARITY INHIBIT
34	(CRL) CONTROL REGISTER LOAD
33	(TR$_8$)
32	(TR$_7$)
31	(TR$_6$) TRANSMITTER
30	(TR$_5$) REGISTER
29	(TR$_4$) DATA INPUTS
28	(TR$_3$) (TR$_1$ – TR$_8$)
27	(TR$_2$)
26	(TR$_1$)
25	(TRO) TRANSMITTER REGISTER OUTPUT
24	(TRE) TRANSMITTER REGISTER EMPTY
23	(THRL) TRANSMITTER HOLDING REGISTER LOAD
22	(THRE) TRANSMITTER HOLDING REGISTER EMPTY
21	(MR) MASTER RESET

S1602

FIGURE 13–15 UART data sheet (pages 549–50 courtesy of Gould AMI Semiconductors). (page 1 of 2)

Block Diagram

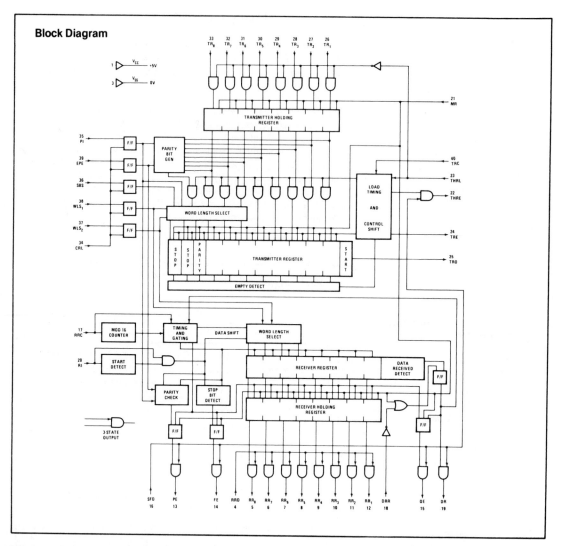

FIGURE 13–15   (page 2 of 2)

The 20-mA Loop—Teleprinters

Historically and presently, teletypewriters (TTY) "make" or "break" a 20-mA current loop connecting other teleprinters. The keyboard and printer are both electromechanical marvels of cams, latches, shafts, electromagnetic solenoids, and a common motor. A simplistic sketch is shown in Figure 13–16 to give an idea of how current pulses might be encoded according to the opening and closing of contacts by cams set by mechanical latches when a key on the keyboard is pressed. At the receiver, regenerated current pulses actuate the appropriate solenoids to set latches, allowing the transmitted character to be printed.

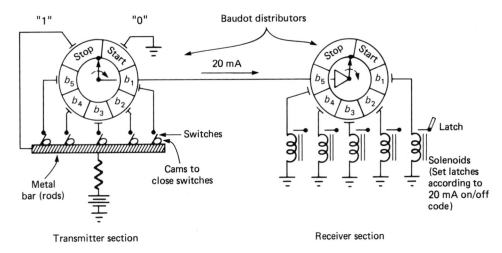

FIGURE 13-16 Simplified version of the electromechanical teleprinter.

Today's ASR 33 and 35 teleprinters allow automatic send and receive to and from a paper-tape punch and reader. Transmission speeds are in the range of 45–220 baud with the ASR 33 operating at 110 baud. The 110 baud rate can be converted to *words per minute* (wpm) by assuming an average of 5 letters (characters) and a space per English word. Then the rate in words/minute, R_{wpm}, is computed from

$$R_{\text{wpm}} = R_{\text{b/s}} \times \left(\frac{60 \text{ s}}{\text{min}}\right) \times \left(\frac{\text{bits}}{\text{character}}\right)^{-1} \times \left(\frac{1}{6}\right) \qquad \textbf{(13–13)}$$

If the 11-bit, full ASCII format is used, the conversion is

$$110 \frac{\text{bits}}{\text{s}} \times 60 \frac{\text{s}}{\text{min}} \times \frac{\text{character}}{11 \text{ bits}} \times \frac{\text{English word}}{6 \text{ characters}} = 100 \text{ words/min}$$

Modern 20-mA loop interface circuitry includes active and passive transmitters and receivers, and optical couplers for isolation. The active circuit is one that supplies a high-impedance, 20-mA current, whereas the passive circuit only sinks current. An example of an active transmitter circuit is shown in Figure 13–17, along with an optical coupler-isolator.

The voltages given in Figure 13–17, +1.3 V (= 5 V − 20 mA × 150 Ω − 0.7 V) and −11.3 V (= −15 V + 20 mA × 150 Ω + 0.7 V), exist when the digital input is low and Q_1 is conducting. The voltage drop across the 1-kΩ emitter resistor of Q_1 causes 3.7 mA through Q_1. This is calculated as follows:

$$I_{Q_1} = \frac{5 \text{ V} - 1.3}{1 \text{ k}\Omega} = 3.7 \text{ mA}$$

The transistors have fairly high β so that the current through the 1-kΩ base resistor of Q_2 is about 3.7 mA, causing the base voltage of Q_2 to be 3.7 V above

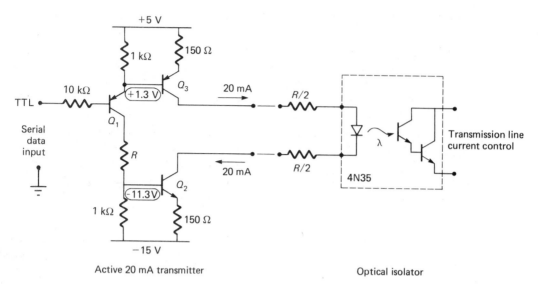

FIGURE 13-17 An active 20-mA transmitter and optical isolator for teletype.

−15 V—i.e., −11.3 V. The collector currents I of Q_2 and Q_3 will be the same and are calculated from Q_3 as follows: If $V_{BE} = 0.7$ V, then

$$I_{EQ_3} \approx I = \frac{5 \text{ V} - (1.3 \text{ V} + .7 \text{ V})}{0.15 \text{ k}\Omega} = \frac{3 \text{ V}}{0.15 \text{ k}\Omega} = 20 \text{ mA}$$

When the digital input goes high, $I \approx 0$ as follows: The base voltage of Q_1 goes high, thereby cutting off Q_1. With no current through Q_1, $I_{Q_1} = 0$, the emitter of Q_1 and base of Q_3 will be at +5 V, cutting off Q_3. Also, the base of Q_2 will be at −15 V so that Q_2 is cut off. R must be kept below 3.5 kΩ to prevent Q_1 from saturating and spoiling the switching speed of the circuit.

When great distances are required, teletype and telegraph services like TWX and Telex are interfaced with the telephone system through modems using FSK carriers. Just dial the TWX or Telex number of the company to which you want to send electronic mail and let the ASR unit run the punched tape to transmit the data in block mode. In fact, you can even start the punched tape with the TWX (Telex) number and get auto-dial. Since transmitting computer data is done primarily by modems, computer interfacing is now discussed.

EIA and CCITT Voltage Interfacing of Computer Devices

As mentioned in the introduction of intercomputer communications, the serialized data of computer terminals, along with data link control signals, are brought out to an interface connector on the rear of the terminal cabinet. As computers, terminals, printers, and other types of data terminal equipment came on the market, all with incompatible interface connections, the EIA (Electronics Industries Association) in the United States and CCITT in Europe drew up interface standards. The early standard developed by the EIA, called RS-232 which is now in the ''C'' revision (RS-232-C), generally outlines the electrical characteristics, connector, and pinout, as well as control lead functions.

Most terminal devices come equipped with RS-232-C-compatible connectors (DB25P/DP25S) having pin numbering and voltage levels reasonably close to those specified (see Table 13–6). In addition to levels and pin designations, the standards establish a simple protocol called *handshaking*.

The RS-232-C (CCITT V.28) voltage levels are $+3$ to $+15$ V dc for a 0 logic level and -3 to -15 V dc for a logic 1 with 25 V maximum, open-circuited. Note the negative logic used at the communications interface. Other electrical characteristics are given in Figure 13–18. The RS-232-C standard is intended for use in interfaces of up to 50 ft and baud rates less than 20 kb/s.

TABLE 13–6 RS-232-C Signal Function Connector Pin Designations

Pin	Meaning	Modern Mnemonic
1	Protective ground	PG
2	Transmitted data	TD
3	Received data	RD
4	Request to send	RTS
5	Clear to send	CTS
6	Data set ready	DSR
7	Signal ground	SG
8	Data carrier detect	DCD
9	(unassigned)	
10	(unassigned)	
11	(unassigned)	
12	Secondary data carrier detect	S-DCD
13	Secondary clear to send	S-CTS
14	Secondary transmitted data	S-TD
15	Secondary transmitter clock	S-TC
16	Secondary received data	S-RD
17	Receiver clock	RC
18	Divided clock receiver	DCR
19	Secondary request to send	S-RTS
20	Data terminal ready	DTR
21	Signal quality detect	SQ
22	Ring indicator	RI
23	Data rate selector	
24	External transmitter clock	
25	Busy	

FIGURE 13–18 Electrical characteristics of RS-232-C and
CCITT V.28 interfacing.

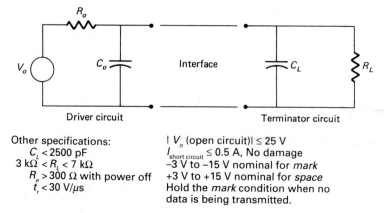

Driver circuit   Terminator circuit

Other specifications: $| V_o$ (open circuit)$| \le 25$ V
  $C_L < 2500$ pF $I_{short\ circuit} \le 0.5$ A, No damage
  3 k$\Omega < R_L < 7$ kΩ -3 V to -15 V nominal for *mark*
  $R_e > 300\ \Omega$ with power off $+3$ V to $+15$ V nominal for *space*
  $t_r < 30$ V/μs   Hold the *mark* condition when no
       data is being transmitted.

The control circuits specified in RS-232-C are basically compatible with
CCITT's V.24 standard and are shown along with the newer EIA RS-449 standard
in Table 13–7. Figure 13–19 is a pictorial of the leads for a typical terminal-to-
modem interface.

TABLE 13–7 Interface Lead Designations for EIA-449, RS-232-C, and CCITT's V.24
(page 1 of 2)

EIA RS-449		EIA RS-232-C		CCITT's V.24	
SG	Signal ground	AB	Signal ground	102	Signal ground
SC	Send common			102A	DTE common
RC	Receive common			102B	DCE common
IS	Terminal in service				
IC	Incoming call	CE	Ring indicator	125	Calling indicator
TR	Terminal ready	CD	Data-terminal ready	108/2	Data-terminal ready
DM	Data mode	CC	Data-set ready	107	Data-set ready
SD	Send data	BA	Transmitted data	103	Transmitted data
RD	Receive data	BB	Received data	104	Received data
TT	Terminal timing	DA	Transmitter signal-element timing (DTE source)	113	Transmitter signal-element timing (DTE source)
ST	Send timing	DB	Transmitter signal-element timing (DCE source)	114	Transmitter signal-element timing (DCE source)
RT	Receive timing	DD	Receiver signal-element timing	115	Receiver signal-element timing (DCE source)
RS	Request to send	CA	Request to send	105	Request to send
CS	Clear to send	CB	Clear to send	106	Ready for sending

TABLE 13–7 (page 2 of 2)

EIA RS-449		EIA RS-232-C		CCITT's V.24	
RR	Receiver ready	CF	Received-line signal	109	Data-channel received-line-signal detector
SQ	Signal quality	CG	Signal-quality detector	110	Data-signal-quality detector
NS	New signal				
SF	Select frequency			126	Select transmit frequency
SR	Signaling-rate selector	CH	Data-signal-rate selector (DTE source)	111	Data-signaling-rate selector (DTE source)
SI	Signaling-rate indicator	CI	Data-signal-rate selector (DCE source)	112	Data-signaling-rate selector (DCE source)
SSD	Secondary send data	SBA	Secondary transmitted data	115	Transmitted backward-channel data
SRD	Secondary receive data	SBB	Secondary received data	119	Received backward-channel data
SRS	Secondary request to send	SCA	Secondary request to send	120	Transmit backward-channel line signal
SCS	Secondary clear to send	SCB	Secondary clear to send	121	Backward-channel ready
SRR	Secondary receiver ready	SCF	Secondary received-line-signal detector	122	Backward-channel received-line-signal detector
LL	Local loopback			141	Local loopback
RL	Remote loopback			140	Remote loopback
TM	Test mode			142	Test indicator
SS	Select standby			116	Select standby
SB	Standby indicator			117	Standby indicator

In order to accommodate higher interfacing speeds and reduce ground noise effects, the EIA developed the RS-422 electrical specification. RS-422 provides for a two-wire balanced line with differential output at the transmitter and differential input at the receiver. This is shown in Figure 13–20.

As is typical of differential systems, twisted-wire pairs help reduce common-mode noise interference and system ground currents, particularly the ever-present 60-Hz noise. A possible circuit could be constructed using a twisted-wire pair (about six turns/inch) between an AM 78/8830 dual differential line driver and the

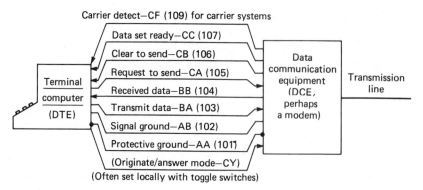

FIGURE 13-19 Control lines ("circuits") for RS-232-C and the similar CCITT V.24 standards. Circuit designations are XX for the EIA RS-232 standard and (XXX) for the CCITT international standard.

FIGURE 13-20 RS-422 balanced interfacing. Outputs and inputs are differential circuits and the transmission line is balanced (typically a twisted-wire pair).

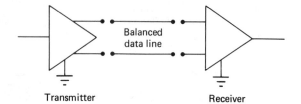

AM 78/8820 DDL receiver. Differential interfaces are capable of accommodating up to 100 kb/s over short lines or 20 kb/s over 200-ft lines.

13-6 ■ CARRIER SYSTEMS AND MODEMS

In order to transmit data over long distances it seems natural to consider the already existing telephone network. Unfortunately, transmitting dc levels beyond the local exchange office is not possible without special dedicated lines, and some form of modulation is required to use these common-carrier analog facilities.

The equipment used to modulate and demodulate digital data for transmission over analog facilities is called a *modem*, from MOdulator/DEModulator. Most modems are RS-232-C-compatible for interfacing with the terminal/computer equipment and can be coupled directly or by acoustical coupling to the telephone line.

Modems

The type of modulation used in modems depends on the information bit rate because standard voice-grade telephone lines are bandwidth-limited, extending from about 300 Hz to about 3300 Hz. Modems operating up to 1200 b/s use binary

FSK (frequency-shift key), but for higher rates, up to 9600 b/s, multilevel PSK (phase-shift key) and QAM (quadrature amplitude modulation) are used. Clearly, a 3300-Hz channel cannot respond to binary line changes of 9600 b/s, so a 9600-b/s modem must use at least an eight-level modulation scheme.

Frequency Shift Key (FSK) Modems

Low- to medium-speed modems (up to 1200 bits/s) use FSK. FSK is binary (digital-data) frequency modulation of an analog carrier. Thus, a voltage-controlled oscillator (VCO) can be used with the two voltage levels of the data, shifting the carrier to two discrete frequencies. This is illustrated in Figure 13–21.

In the AT&TIS type-202* modem, a 1700-Hz (±1%) VCO is deviated ±500 Hz (±1%) by data in a bipolar format. (Recall that an RS-232-C interface is bipolar.) A transmitted frequency of 1200 Hz is called a *mark* and corresponds to a digital 1; 2200 Hz is called a *space* and corresponds to a digital 0. *Mark* and *space* are old telegraph terms. Data can be sent by the 202-type modem on a standard two-wire voice-grade telephone line at up to 1200 bits/s. Signal strength into the 600-Ω telephone line should be between −6 and −9 dBm, but definitely within 0 and −12 dBm.

The demodulation of FSK signals can be accomplished with a standard FM discriminator or a phase-locked loop. When a terminal device is to transmit data, a request-to-send (RTS) is relayed to the modem which, when the telephone connection is completed, transmits the carrier clamped to the *mark* frequency. The receiving modem, using a phase-locked discriminator, will lock to (acquire) the transmitted carrier, and a quadrature detector will activate the carrier detect (CD) flag. When the transmitting computer gets clear-to-send (CTS) from its modem, data is transmitted. For direct-connect modems there is up to 200 ms of delay between the reception of RTS and the sending of CTS back to the computer. The data is transmitted asynchronous mode and usually ASCII with even parity.

The *mark* and *space* frequencies within the 3-kHz telephone bandwidth for the 202 modem are shown in Figure 13–22. Also in the figure is a narrow channel at 387 Hz that is used for automatic requests for retransmission of erroneous data.

---

* Most modem manufacturers reference the "Bell" standard models, now owned by AT&T Information Systems. The 202 modem is similar to CCITT's V.23 recommendation.

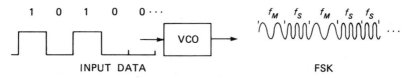

FIGURE 13–21 FSK generation by voltage-controlled oscillator.

FIGURE 13–22 202-type modem frequencies.

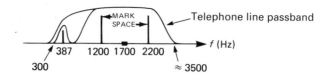

Automatic Retransmission Request (ARQ)

The receiving station checks parity of each transmitted character, and in the case of batch data transmissions, a block check character is computed and, in effect, compared to the one transmitted at the end of each block of data.

If the receiving station detects an error, it must transmit this information to the sending station. Rather than using all the time required for the receiving station to seize control of the channel and request a retransmission of the erroneous block or character, the 202-type modem provides a very low-speed reverse or secondary channel. This channel, used for supervision, is frequency-division multiplexed below the main channel on a 387-Hz carrier that can be on/off keyed (amplitude-shift key, ASK) at up to 5 b/s (baud). The CCITT's V.23 provides for a 75-baud reverse channel.

In *stop-and-wait ARQ,* the block of data is transmitted and simultaneously stored in a buffer at the transmitter. The transmitter then waits until the receiving station responds positively with the code character ACK before transmitting a new block. However, if an error is detected, a negative acknowledgment, NAK, is received by the transmitter via the reverse channel, and the transmitter must retransmit the last block which it stored in the buffer.

There are continuous ARQ techniques that avoid the stop-and-wait routine, but these require a full-duplex channel. The 202-type modem should not be operated full-duplex on a two-wire line because distortion and insufficient isolation between the forward and reverse channels can result in transmission errors in both directions.

A Full-Duplex, Two-Wire Line, FSK Modem

Figure 13–23 shows the frequencies used in the 103A-type asynchronous modem. A single two-wire transmission line can be used in a simultaneous two-way communication system with this modem. The trick, of course, is to reduce the frequency deviation so that two full channels will fit within the 300–3300-Hz telephone bandwidth.

FIGURE 13–23 Full-duplex modem frequencies for type-103A modems.

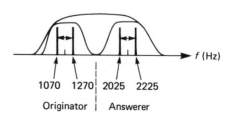

To get an idea of the relationship between *mark-space* frequency separation and baud rate, use Equation 13–14. A ratio of 0.7 between these two parameters is about optimum. As an example, let

$$h = \frac{f_S - f_M}{f_B} \tag{13–14}$$

where f_S = *space* frequency, f_M = *mark* frequency, and f_B = baud rate. The relationship of Equation 13–14 is also written in the literature as $h = 2\Delta f_d T_B$, where Δf_d is the *mark* (or *space*) frequency deviation from center frequency and $2T_B$ is the period of the fundamental frequency of a squarewave modulating signal at baud rate f_B. The relationship yielding h is that of an FM modulation index.

The 103-type modem uses a *mark-space* frequency separation of 200 Hz, so a baud rate of about 286 b/s is appropriate with $h = 0.7$. In fact, the signaling rate for the 103-type modem is specified up to 300 baud. The *mark-space* frequencies used are as follows: The modem originating the transmission uses 1270 Hz/1070 Hz, and the answering modem uses 2225 Hz/2025 Hz.

A typical modem and simplified interfacing is shown in Figure 13–24 using the Motorola MC 6850 ACIA (UART) and MC 6860 modem chip, which is like the 103-type. Not included on the 6860 modem chip, but very important to the communications system design, are the transmit and receive filters.

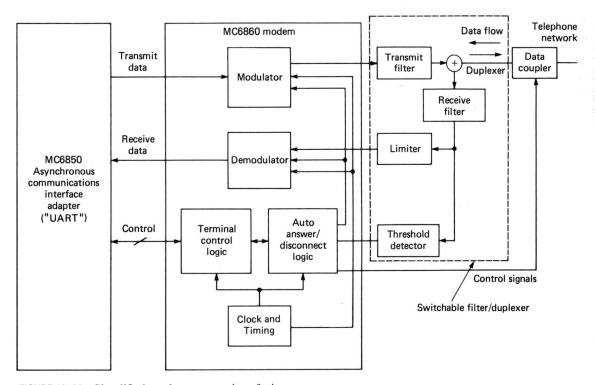

FIGURE 13–24 Simplified modem system interfacing.

FIGURE 13-25 Two-pole bandpass active filter.

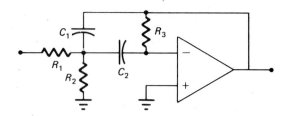

The receive filter should provide greater than 35 dB of attenuation to the adjacent channel. In addition, phase linearity or group delay in the passband is a very important design consideration. Group delay for the filter in the 300-Hz bandwidth between *mark* and *space* frequencies should not exceed 0.8 ms. Group delay is defined by

$$t_d = \frac{\Delta\phi}{\Delta f} \tag{13-15}$$

where $\Delta\phi$ = change in phase and Δf = change in frequency. Notice that it is not necessary that there be no phase shift, only that the phase slope in the passband be fairly constant over the used bandwidth. Motorola Application Note AN-747 gives a detailed design procedure for a 0.5-dB ripple, Chebyshev, active filter using two-pole sections, as shown in Figure 13-25.

The resultant six-pole filter is included in the schematic of the switchable filter/duplexer of Figure 13-26, along with the frequency responses for both the answer (receive) and originate (transmit) modes.

Continuous-Phase FSK and MSK

FSK modems should use some form of continuous phase FSK (CP-FSK) to minimize phase discontinuities and the resultant frequency transients.* This allows for the simplification of demodulator circuitry and improved BER due to the minimization of line transients.

Continuous-phase FSK is achieved by selecting the difference between the two signaling frequencies (*mark* and *space*) to be a multiple of one-half of a bit cycle (180° phase difference). As seen in Figure 13-27, at each data bit transition the carrier phase is continuous due to the exact relationship between *mark/space* frequencies and the bit period.

In the case of Figure 13-27, there is exactly one-half of a bit cycle between the *mark* and *space* frequencies. Consequently, there is a maximum phase difference (180°) between a transmitted 1 and 0, while at the same time, there is a minimum possible frequency difference (one-half of a bit cycle) between the *mark* and *space* frequencies. This particular choice of CP-FSK signaling is called *minimum shift key* (MSK).

---

* Recall that $\omega = d\phi/dt = (2\pi)f$; that is, frequency is determined by the rate of change of phase. Consequently, sudden changes in phase result in large frequency deviations and signal transients in narrowband circuits.

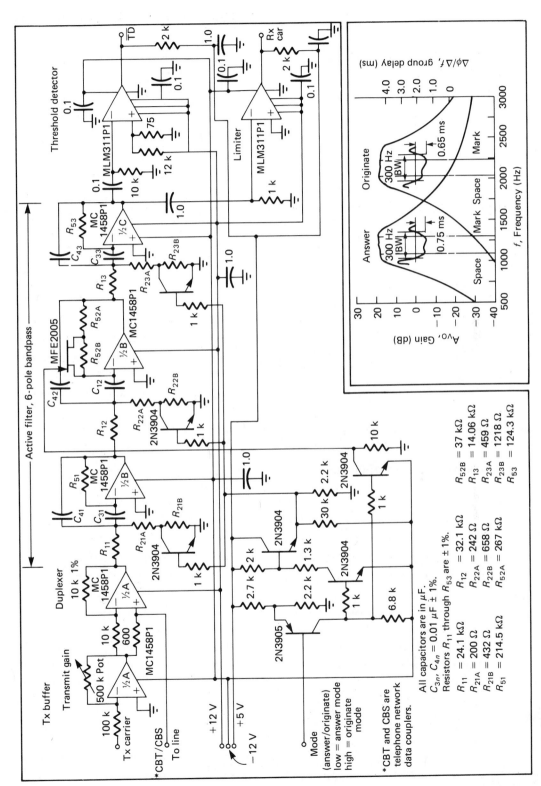

FIGURE 13-26 Application Note AN-747. Low-speed modem system design using the MC 6860. (Courtesy of Motorola Semiconductor Products, Inc., Box 20912, Phoenix, AZ 85036.)

FIGURE 13–27 FSK signals.
(a) Noncontinuous phase.
(b) Continuous-phase FSK
(CP-FSK), also called MSK.

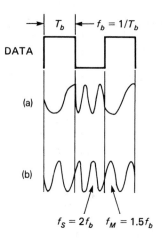

The relationship of frequencies in MSK is $f_S - f_M = 0.5 f_B$. Hence from Equation 13–14, $(f_S - f_M)/f_B = h$, we see that MSK has a modulation index of $h = 0.5$. With its continuous-phase feature, MSK produces a frequency spectrum that falls off faster than binary PSK and has a first-lobe null at $0.75 f_B$, as opposed to BPSK's null at f_B. However, MSK's spectrum is not as good in terms of bandwidth utilization as QPSK. PSK techniques for synchronous data modems are discussed in Chapter 17.

13–7 ■ SYNCHRONOUS COMMUNICATION TECHNIQUES

Modems used for transmission of digital data on standard telephone lines at data rates exceeding 1200 b/s use phase-shift key and high-level modulation techniques. These systems also use synchronous data transmission formats and protocols in order to increase the coding efficiency and overall system throughput. For instance, eliminating the start and stop bits of an 11-bit asynchronous ASCII code format improves the coding efficiency by 3/11 = 27.3%. This, of course, means that a given set of ASCII-coded message characters can be completed 27.3% sooner than by asynchronous transmission.

Since the start and stop bits are not used in synchronous data transmission, it is necessary to distinguish one character (and, perhaps, a parity bit) from another by synchronizing the receiver to the transmitter data clock before the message is transmitted. Synchronous protocols differ in how this is accomplished but the general procedure is for the transmitter to initially send an 8-bit sync character (or frame start flag) that the receiver recognizes. Circuitry in the receiver uses the bit pattern to synchronize the clock, a procedure called *bit timing recovery*. Bit timing recovery is covered along with PSK and high-level digital modulation techniques in Chapter 17 for digital radio. However, keep in mind that modems for high-speed synchronous data transmission use these techniques for transmitting over telephone networks. For reference purposes, Table 13–8 lists the characteristics of standard telephone dial-up and leased-line modems.

TABLE 13-8 Standard Modem Characteristics

CCITT (V.XX), AT&T (Bell) Specification	Bit Rate (bits/s)	Modulation Scheme	Wires/ Duplex	Sync/ Async	Remarks Dial-up (D) Leased (L)
Bell 103	0–300	FSK	2/full	async	D,L
Bell 202	1200	FSK	4/half	async	D,L
V.22, Bell 212	1200/600	QPSK/FSK	2/full	either	D,L
V.22bis	2400	16-QAM	2/full	either	(fallback: 212/103);D,L
V.26, Bell 201	2400	QPSK	2/half, 4/full	sync	D
V.26bis	2400	QPSK	4/full	sync	L
V.26ter	2400	QDPSK	2/full	sync	L
Bell 208	4800	8-PSK	2/half, 4/full	sync	L
V.27	4800/2400	8-DPSK/QDPSK	4/full	sync	L
V.27ter	4800/1200	8-DPSK/QDPSK	2/half	sync	L
Bell 209	9600/2400	16-QAM	4/half	sync	L
V.29	9600	16-QAM (modified)	4/half	sync	L
V.32	9600	32-QAM*	2/full	sync	L
V.33	14400	32-QAM*	4/full	sync	L
V.35	56000		4/full	sync	L

* With trellis-coded modulation (TCM) and echo cancellation.

Data Networks and Synchronous Transmission Protocols

A data network consists of nodes or communications stations and the arrangement of transmission lines and facilities interconnecting them. The two basic network configurations, shown in Figures 13–28 and 13–29, are the centralized and the distributed networks, respectively. The centralized network has a single

FIGURE 13-28 Centralized network.

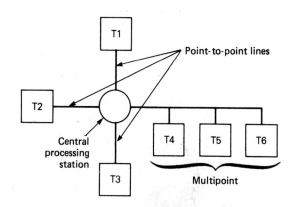

FIGURE 13–29   Distributed net-work.

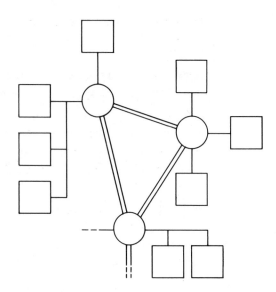

processing station that controls all traffic between point-to-point connections and multipoint connections. The distributed network is an interconnection of more than one centralized network.

Point-to-Point

Point-to-point connections are those between a fixed pair of terminals such as T1 and T2 in Figure 13–28. Point-to-point lines may be leased (dedicated) connections so that they are permanently connected whether or not the line is in use. Point-to-point connections may also be made in a switched telephone network, in which case a dial-up connection is made through the central processing facility, but only for the duration of a single call. The switched network connection allows subsequent transmissions to or from any other station in the network by standard dialing procedures, manual or automatic.*

Multipoint Lines

A multipoint line is shared, one at a time, between more than two stations (see Figure 13–28) under control of the central processing station. The central processing unit (CPU) polls and selects the other (tributary) stations. *Polling* is an "invitation to send" transmitted in a predetermined sequence to each tributary station (terminal). *Selection* is a "request to receive" notification from the central processor to one of the tributary stations.

Multiplexers and concentrators may also be part of a multipoint (also called *multidrop*) network. In this case, a few stations can be multiplexed onto one

* Automatic calling units like the AT&T type 801 offer dial pulses (801A) or touch tones (801C).

transmission line under control of the CPU, or many stations can be concentrated onto a few lines (all under central control).

Switched Networks—Synchronous Transmissions

Large networks like the public telephone network are typically distributed networks in which voice and/or digital data can be routed through electronic or electromechanical switches under the control of line signals and computers located in central or remote (toll) offices. Since the start and stop bits define each character in asynchronous data transmissions, timing information is taken care of at the receiver. However, synchronous transmission of digital information—which more and more includes digitized (PCM) voice—can be switched through the network by three different methods: circuit switching, message switching, and packet switching.

Circuit switching is used in ordinary analog telephone connections. The circuit path is established for the duration of the call, whether or not anyone is communicating.

In message switching, illustrated in Figure 13–30, the message is sent to a switching center, where it is stored until it can be forwarded to a closer center. This store-and-forward technique is typical of telegram and electronic mail transmissions and is more efficient than circuit switching because a line is busy only when actual data is being transferred. Hence, message switching is more efficient and more cost effective than circuit switching.

Packet Switching

Packet switching is another technique finding applications in large systems with limited transmission facilities such as satellite and land microwave radio links.

FIGURE 13–30   Message (M) switched network.

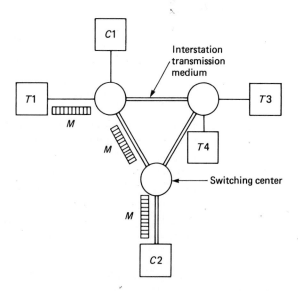

FIGURE 13–31   Packet-switched
network.

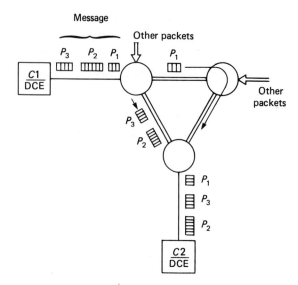

The message is divided into variable-length units called *packets* that are switched
through the network with other packets in a statistically concentrated* fashion in
order to maximize system utilization time. A typical packet length is 1000 bits.

Each packet is given a destination address and other protocol information, and
switched through the system on the first available path. As shown in Figure 13–31,
a prior commitment of a data link may require that packets take different paths to
their destination (computer C_2 in this case). Clearly, the individual packets must
be numbered sequentially so that the receiving station can reassemble the message. The numbering and control for packet-switched networks are part of the
system protocol.

A packet received at a node is *stored* in a buffer for *forwarding* on queue. When
the packet reaches the head of the queue it is presented to the processor, which
determines the next node to which it will be sent. Then the packet is switched to
an output queue to await its turn for transmission.

The delays, or service times, for the various operations are analyzed by Mar-
kov statistics, meaning that Poisson or exponential arrival and queuing times are
assumed. Much of the pioneering work in queuing theory was performed by the
Danish mathematician Erlang in the early 1900s. The most commonly employed
queuing theory model is the single-server queue. The problem then is to determine
the message arrival statistical process and the service completion time process for
a single server, such as one cashier at the grocery store.

---

* Statistical concentrating works as follows: The processor continuously samples input buffers and
output lines. Any pause in data transmission detected on a transmission line will cause the processor to
redirect input data packets (in buffer storage) to that line *on a priority basis*. The efficient use of
transmission lines in this way allows the use of only enough line capacity to handle the *average* data
rate instead of the *peak* data rate.

The message arrival process is assumed to obey a Poisson statistical distribution and the probability of a packet arrival is the same for any interval of time. The probability Pr that exactly k packets arrive at the node in t seconds is given by

$$\text{Pr}(k,t) = \frac{(rt)^k}{k!} e^{-rt} \qquad (13\text{–}16)$$

where the average rate of packet arrivals is r per second. The probability of k or fewer arrivals in t seconds is found by adding up the probability of zero arrivals, one arrival, and so on, up to and including k; that is,

$$\text{Pr}(K \le k) = \sum_{i=0}^{k} \frac{(rt)^i}{i!} e^{-rt} \qquad (13\text{–}17)$$

where K is the random variable for which k is a particular value.

For the single-server queue with a Poisson service time distribution, the probability of exactly c service completions (c packets transmitted) in t seconds is given by

$$\text{Pr}(c,t) = \frac{(\mu t)^c}{c!} e^{-\mu t} \qquad (13\text{–}18)$$

where μ is the average rate of transmissions per second from the node. If the switching node is in equilibrium, then the queue contains n packets, and on average, as many packets leave the queue as arrive. The average queue length (buffer size required) is

$$\bar{n} = \frac{r}{\mu - r} \qquad (13\text{–}19)$$

and the average time delay T_q for packets from node entry to exit is given by

$$T_q = \frac{\bar{n}}{\mu} + \frac{1}{\mu} = \frac{1}{\mu - r} \qquad (13\text{–}20)$$

Clearly, from Equations 13–19 and 13–20, the average exit rate μ must equal or exceed the average arrival rate or the buffer will fill up and overflow and the node will be blocked.

■ ■ ■ ■

EXAMPLE 13–4

Packets arrive at a node on the average of 10 per second, whereas node transmissions occur at 12 packets per second. Assume the queuing system is in equilibrium and obeys Poisson statistics.

1. Determine the following probabilities:

   a. Exactly one packet arrives in ½ s.
   b. Two or fewer arrivals in ½ s.
   c. More than three packets are transmitted in ½ s.

2. **a.** What must the minimum size of the node buffer be?
   b. What is the average node delay time?

Solution:

1. a. At the input, $rt = 10 \times 0.5 = 5$. Hence, $\Pr(k, t) = \Pr(1, 0.5 \text{ s}) = e^{-5}(5^1/1!) =$ **0.0337.**

   b. $\Pr(K \leq 2) = e^{-5}[(5^0/0!) + (5^1/1!) + (5^2/2!)] =$ **0.125.** Recall that $0! = 1$.

   c. $\mu t = 12 \times 0.5 = 6$. $\Pr(K > 3) = 1 - \Pr(K \leq 3) = 1 - e^{-6}[(6^0/0!) + (6^1/1!) + (6^2/2!) + (6^3/3!)] = 1 - 0.1512 =$ **0.8488.**

2. a. $\bar{n} = r/(\mu - r) = 10/(12 - 10) =$ **5 packets,** minimum storage size.

   b. $T_q = 1/(\mu - r) = 1/(12 - 10) =$ **0.5 s.**

Synchronous Protocols

A *protocol* is a set of rules to effect the orderly transfer of data from one location to another using common facilities. Synchronous system protocols include provisions for synchronizing the receiver clock, determining which station controls the link (the primary station) and which station is secondary, delineating the control and message groups, detecting errors and providing for retransmission of erroneous messages, establishing message transparency, and data flow maintenance. Protocols are classified according to their framing technique.

1. Character-oriented protocols (like IBM's BSC, or "bisync") use special binary characters to separate different segments of the transmitted information frame.

2. Byte-count protocols (like DEC's Digital Data Communication Message Protocol—DDCMP) use a header followed by a count indicating the number of characters to follow and the number already received error-free. DDCMP can be used on parallel or serial, synchronous or asynchronous, and point-to-point or multipoint systems.

3. Bit-oriented protocols (BOPs) include HDLC, SDLC, and CCITT's X.25; these will be defined later. The BOP frame is made up of well-defined fields between 8-bit start and stop flags. The bits of each field, except the information field, are encoded with address, control, counting, and error checking functions.

These synchronous protocols are used on half- or full-duplex data links allowing for two-way communication and automatic retransmission requests (ARQ) of erroneous frames.

The two most common protocols, bisync and SDLC (considered a subset of HDLC), will be discussed.

Binary Synchronous Communications (Bisync)

Bisync is a synchronous protocol using serial, binary-coded characters comprising text information (message) and/or heading information (message identification and destination). It describes the transmission codes, data-link control and operations, error checking, and message format. The control characters designate and control various portions of the information to be transmitted. Bisync can accommodate three transmission code sets, EBCDIC, US-ASCII, and six-bit Transcode. It can operate in point-to-point or multipoint networks. Stations on the data link idle in a "sync search" mode waiting to identify a transmitted sync character (SYN = 0110100 in ASCII). Most communications systems wait for at least two successive SYN characters before raising a "character available" flag. Once the receiver's clock is synchronized, the remainder of the protocol is followed for receiving the transmitted message and maintaining the circuit. Figure 13–32 shows the bisync format.

In point-to-point operations, a station bids for use of the line using the control character ENQ (inquiry): this character is 1010000 in ASCII and says, "I have a message for you." If both stations bid simultaneously, "with contention," the station that persists gains control of the line unless a priority system is established.

In a multipoint connection, each station is assigned a unique address. Once the station responds to a poll or selection in the affirmative (ACK), message transmission can start. A multipoint link of more than ten stations requires special considerations.

A header for station control and priority usually precedes the message (text) and is started with a SOH (start of header) character; the header is not required. The message consists of one or more blocks and is preceded by the STX (start of text) character. All but the last message block are ended with EOB (end of block); the last block ends with ETX (end of text).

For error control, each transmitted character is checked (vertical redundancy check) and, as each block of the message is completed, the block check character is transmitted. CRC-16 is used for 8-bit codes like EBCDIC (and ASCII with transparency), whereas CRC-12 is used for 6-bit codes. Transmission is terminated with the EOT (end of transmission) character.

Other very important protocol characters in bisync are ACK (positive acknowledgment),* NAK (negative acknowledgment), WACK (wait, or delay, before ACK is acceptable), and DLE (data-link escape). Figure 13–33 shows a couple of examples to illustrate the bisync protocol interaction. Data-link escape is used to provide supplementary control characters. For instance, DLESTX is used to put

---

* ACK0 and ACK1 are positive "go-ahead" responses used alternately during the text block, thus providing a running check to ensure that each reply corresponds to the immediately preceding message block. In ASCII, ACK0 is transmitted as the two-character pair DLE0 and ACK1 is actually DLE1.

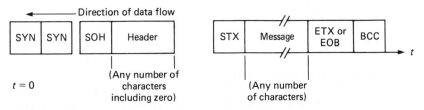

FIGURE 13–32 Bisync protocol format.

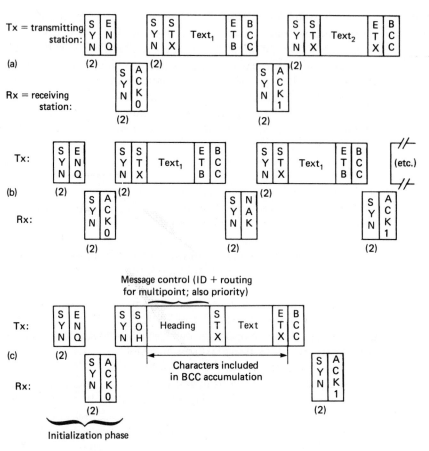

FIGURE 13–33 Bisync protocol controlled data link; a few examples.

the message block into transparency mode. The *transparency* mode allows for the transmission of any binary data, including bit sequences which the receiver would recognize as a control character. This sometimes presents problems because the message might include the bit sequence DLESTX, which, when detected by the receiver, terminates the transparency mode and the message. The solution is for the transmitter to add a DLE character before such recognizable bit sequences and transmit DLEDLESTX. The receiver is required to remove the first DLE character it detects and treat the following bits as pure data.

There are other synchronous data-link control protocols that avoid this and other difficulties and provide for more flexibility and speed. These are bit-oriented protocols.

Bit-Oriented Protocols: SDLC, HDLC, and X.25

One undesirable feature of the character-oriented protocols such as bisync is the need for the DLE character for message transparency. Bit-oriented protocols

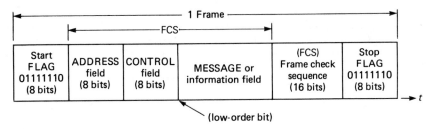

FIGURE 13-34 Bit-oriented protocol (BOP) format; SDLC example.

(BOPs) such as SDLC, HDLC,* and CCITT's X.25 do not use characters of well-defined bits within the address, control, and frame check fields. These fields, as seen in the SDLC format of Figure 13–34, are delineated by the start and stop flags.

Notice that the start (and stop) flag, 01111110, has six consecutive 1s. SDLC protocol requires that, between flags, the transmitter insert an extra zero following any five consecutive 1s it must transmit. The receiver is required to remove the 0 from the data stream anytime it detects five consecutive 1s followed by a 0. In this manner the bit sequence of a flag is obvious, and any bit sequence is allowed, with zero insertion if necessary, during the message. The data terminal and communications equipment synchronize on the flag bit pattern.

Other characteristics of the BOP formatting are as follows:

- The control field defines the function of the frame by one of three 8-bit formats—supervisory, information transfer, and unnumbered.

- A provision is made for transmitting a sequence of information frames—each of which is numbered—and making sure that they are received without error in the proper order.

- Errors are detected by a 16-bit CRC (CCITT V41 with generating polynomial $x^{16} + x^{12} + x^5 + 1$). The "block check character" computed from the address, control, and information fields makes up the frame check sequence. If a frame is in error, the receiver does not advance its received frame count N_r. It is the responsibility of the transmitter to determine that the receiver has not accepted the frame by keeping track of the receivers' frame count.

The address field serves the same purpose as the address (or return address) on a letter mailed through the post office, except that the address is always that of the secondary station or stations. The address field is 8 bits, with the low-order bit following the starting flag. Sixteen bits may be used if more than 256 secondary stations are in the network.

---

* SDLC (synchronous data-link control) is considered a subset of the American National Standards Institute's (ANSI) ADCCP (advanced data communications control procedure), which is essentially a functional equivalent of the International Standards Organization's HDLC (high-level data-link control). The CCITT's X.25 standard also specifies the International Standards Organization's HDLC for the DTE-to-DCE access procedure.

FIGURE 13–35 Eight-bit control-field format for supervisory frame in SDLC protocol.

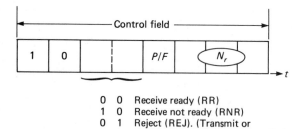

0 0   Receive ready (RR)
1 0   Receive not ready (RNR)
0 1   Reject (REJ). (Transmit or retransmit, starting with frame Nr.)

The function of a transmitted frame is either supervisory or data transfer (or some other network function). Supervisory frames are used to confirm received information frames, convey ready or busy conditions, and to report frame sequence errors. The control field format is shown in Figure 13–35. The P/F bit is used for polling ($P = 1$) by the primary station or is transmitted by the secondary station to indicate the final frame ($F = 1$) it is sending. N_r = number, in binary, of the next frame expected by the receiver and should check with the transmitter's next sent frame N_s. When a supervisory frame is transmitted, the information field is deleted.

When information frames are transmitted, the control field is as shown in Figure 13–36. Frame efficiency is defined by

$$\eta_{fr} = \frac{\text{message bytes}}{(\text{message bytes}) + (\text{overhead bytes})} \tag{13–21}$$

where the frame of Figure 13–34 is seen to contain 6 overhead bytes (48 bits) for flags, address, control, and frame check sequence fields. Clearly, the frame efficiency is improved by long messages. Of course, since this is for synchronous protocols, no parity or stop/start bits are used.

All this inevitably leads to the question of information efficiency. The term most often used for data transmission efficiency is *throughput*. In simplest terms, *throughput* is the rate at which useful information is received per total transmitted bit rate. Another general statement describing throughput is "useful traffic normalized to channel speed."

One of the differences between SDLC and HDLC is in the information field length. With HDLC, the message can have any number of bits, whereas for SDLC the information field must be a multiple of 8 bits. In each 8-bit grouping, the low-order bit is sent first. Another difference is that a control-field option allows up to 7 frames to be transmitted before an ACK is required, for SDLC; for HDLC, 127 frames is the limit. Most packet-switched networks specify use of the X.25 standard. As previously mentioned, the HDLC BOP frame is specified in X.25. The packet is placed in the information field, and the remainder of the frame is,

FIGURE 13–36 Control-field format for information transfer frame.

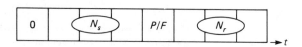

FIGURE 13-37 Control-field format for an unnumbered frame.

1	1	Code	P/F	Code

	Code (2-bit)	Code (3-bit)	
Commands	00	100	Unnumbered poll (UP)
	00	000	Unnumbered information (UI)
	00	001	Set normal response mode (SNRM)
	00	010	Disconnect (DISC)
	10	000	Set initialization mode (SIM)
	11	101	Exchange station identification (SID)
	00	111	Test (TEST)
	10	011	Configure (CFGR)
Responses	00	000	Unnumbered information (UI)
	00	110	Unnumbered ACK (UA)
	10	000	Request initialization mode (RIM)
	11	000	Disconnect mode (DM)
	10	001	Frame reject (FRMR)
	11	101	Exchange station ID (XID)
	00	111	Test (TEST)
	00	010	Request disconnect (RD)
	10	011	Configure (CFGR)

literally, the data-link access and control protocol between the user and the packet-switched network.

Unnumbered frames have a control-field format as illustrated in Figure 13–37. The 2-bit and 3-bit codes shown below the control field are used for such functions as initializing secondary stations (RIM and SIM), controlling the response mode of secondaries (SNRM, DISC, TEST, and XID), and procedural—including errors (UI, UA, DM, FRMR, RD, and CFGR).

When SIM is transmitted, both N_s and N_r are initialized to 0; otherwise the N_s and N_r frame counts of the sender and receiver advance sequentially to keep track of frames.

For a description of these and other details, consult IBM's Synchronous Data Link Control—General Information (GA 27-3093-2, File No. GENL-09).

13-8 ■ OPEN SYSTEM INTERCONNECT AND ISDN

Open System Interconnections (OSI)

Protocols used between user terminals and large, common data transport networks covering a national or international span have been organized into specific categories called layers by the International Standards Organization (ISO). This scheme is called the *open system interconnection* (OSI) reference model and is illustrated in Figure 13–38. The OSI model is a protocol architecture based on layers starting with the *physical* connection, access setup, maintenance, and takedown of layer 1 and progresses up through the data-*link* protocol layer 2, providing control and error handling of the user-to-network access (the HDLC protocol is an example of a standard at this level). Layers 3 and 4 are called the *network* and

FIGURE 13–38 OSI reference model layers.

7	Application
6	Presentation
5	Session
4	Transport
3	Network
2	Data Link
1	Physical

transport layers, respectively, and are concerned with end system–to–end system connectivity, and finally layers 5–7 handle network functions between users that are beyond reliable data transfer in the network; these are the *session*, *presentation*, and *application* protocol layers.

An example of a layer 3 network standard is the CCITT's X.25 recommendation used by packet-switching networks. The X.25 standard specifies a DTE-DCE interface and includes specifications for a *physical layer* called X.21 (or X.21 bis, which is similar to RS-232-C), a *link access* protocol (LAP) layer called LAP-B, which is a subset of the asynchronous balanced mode of the HDLC protocol standard, and a packet (network) layer.

The packet layer of X.25 specifies a *virtual-circuit* service consisting of a *virtual call* (VC) dynamically established virtual circuit using a call setup and takedown procedure similar to X.21, or a permanently assigned virtual circuit. The *permanent virtual circuit* (PVC) service standard establishes protocols for handling data and interrupts, flow control and reset indicators, restart requests/indicators if needed, and diagnostics. The virtual call standard adds call setup and takedown protocols to the functions needed for a permanent virtual circuit. Volumes are published and available on details of the standardized OSI reference model and CCITT recommendations/standards. A few of the CCITT X-series recommenda-

TABLE 13–9 Some CCITT X-Series Recommendations for Packet-Switched Digital Networks (PSDNs)

X.25	Synchronous DTE to DCE interface
X.21/X.21 bis	Physical level DTE to DCE interface
X.3	Start/stop parameters for protocol adapters (PADs)
X.28	Start/stop DTE to terminal PAD protocols
X.29	Host PAD to terminal PAD protocol
X.31	Interface for ISDN
X.32	Switched access interface to the PSDN
X.75	International gateway interface
X.128	International numbering plan for PSDNs
X.200	CCITT OSI reference models for PSDNs

tions pertaining to packet-switched data networks (PSDNs) are listed in Table 13–9.

Integrated Services Digital Network (ISDN)

Digital data transmission systems have evolved and proliferated to the point where a worldwide digital-only network much like the pervasive telephone system is being implemented. This system is called the *integrated services digital network* (ISDN). In the United States, numerous field trials have been completed and some ISDN services are in use by paying customers.

Services offered by ISDN will include all those presently offered or soon to be offered, in addition to new services such as private virtual networks, high-speed digital facsimile, voice/data encryption, voice-annotated text messaging, and more intelligence in the public switched networks. Merging these services along with circuit switched data, packet-switched data, telemetry, telefax, still and slow-motion pictures, with broadband ISDN providing higher transmission rates and picture phone with videoconferencing, will result in a wide range of integrated services.

One of the scenarios that illustrates ISDN capabilities is described as follows: Consider a videoconference among three users. While the visual and audio interaction continues, one party sets up a data link on a second (B) channel by using a D signaling channel. Thus, graphs and other visual data are transmitted between ISDN terminals and can be recorded on peripherals such as printers, plotters, and disk drives while simultaneously being displayed for all three participants in a separate segment of their ISDN video displays.

The ISDN is being designed to give users a uniform view of a wide variety of applications, including low-rate telemetry, interactive data, voice facsimile, high-rate data, and even video with broadband ISDN (B-ISDN). It promises to provide uniform access that can result from the open system interconnection of the ISDN interface protocol specification, and the flexible cost-effective sharing of facilities from an intelligent software controlled network.

While the standards are still evolving, the key hardware architectures at the customers' premises are based on the R, S, T, and U interfaces, illustrated in Figure 13–39, while the two major service capacities are the Basic and the Primary rate lines for small and large users, respectively.

The interfacing between users and the ISDN network is composed of the reference points just mentioned, various network-supplied terminations (NT$_1$, NT$_2$, and a mix of the two, NT$_{1,2}$), and subscriber terminals (TE$_1$ and TE$_2$). The TE$_1$ subscriber equipment is built specifically to be ISDN supporting, while the TE$_2$ is non-ISDN-specific equipment and requires an ISDN terminal adapter, TA; see Figure 13–40.

The basic rate access structure used at the T and S reference points consists of two B channels at 64 kb/s and one D channel at 16 kb/s. This structure is referred to as $2B + D$ service. The B channels carry traffic consisting primarily of circuit-switched, packet-switched, or nonswitched point-to-point facilities. The D channel can be used for circuit setup signaling and some data packet switching and

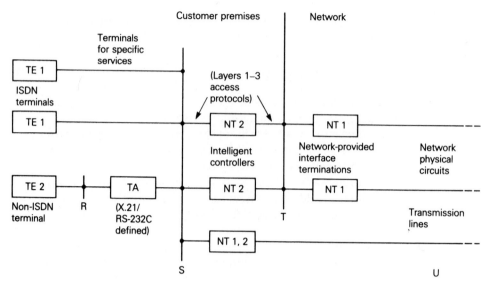

FIGURE 13–39   Some accesses to the ISDN network.

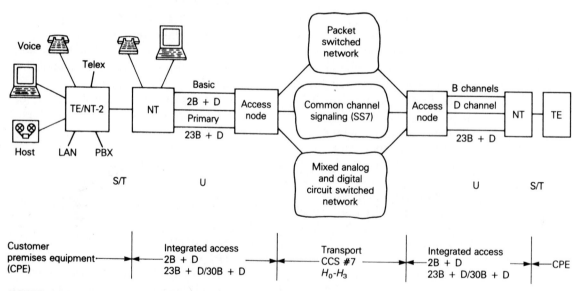

FIGURE 13–40   ISDN service/system block diagram.

telemetry such as house alarm monitoring, as an example. The D channel is primarily a user-to-network control line that is separate from the data-carrying B channels. ISDN will also incorporate an intranetwork, interswitch control line called the *signaling system* number *seven* (SS7).

The primary rate access structure is $23B + D$ in the North American system, or $30B + D$ in the European system. These two service structures correspond to the multiplexed 24-channel T1, 1.544-Mb/s line versus the 30-channel, 2.048-Mb/s European rate. Both systems utilize a common 8 k samples/second, 8-bit PCM standard.

The interface for the primary rate access can be implemented in various H-channel structures. The so-called H_0 channel utilizes 384 kb/s of the available capacity and may have a separate D channel of up to 64 kb/s for signaling. An H_1 structure would use the full 1536 kb/s of the capacity for high-rate data, facsimile or sound. Other H structures—H_2, H_3—are less well defined and allow for various multiplexed low-, medium-, and high-rate configurations.

■ PROBLEMS

1. **a.** How many bits are used for stop (ending each character) in ASCII?
   b. Are they at *mark* or *space* level?
   c. How many characters can be defined by a 7-bit code?

2. A full ASCII (11 bits/character) message is shown in Figure 13–41. This is the NRZ format.
   a. Determine the message.
   b. Is this odd or even parity?
   c. What is the baud rate (bits/second)?
   d. Which character is not printed (nondisplayable)? What happens when this character is decoded?

3. Compute the frame efficiency (information/total bits) for full ASCII.

4. The following CCITT-2 data are received on a 20-mA loop: 11011, 00001.
   a. What should a printer display?
   b. What type of communication system would you expect to make the above transmission?

5. **a.** Write the "ARQ" code equivalent to the transmission of Problem 4.
   b. What advantage does the "ARQ" code have over CCITT-2?

6. What unique feature of the Gray code makes it the choice for encoding slow telemetry?

7. Sketch a cyclic redundancy check generator, including the number of register sections required for a CRC-16 generator polynomial of $G(x) = x^{16} + x^{15} + x^2 + 1$.

8. Input 1011010 to the parity generator of Figure 13–2 with $P_i = 1$ (for even parity). Show the 1 or 0 at each device output. Confirm that $P_o = 0$.

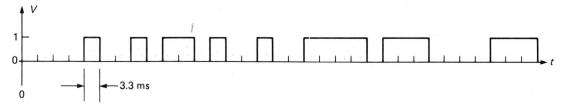

FIGURE 13–41

9. Input 10110100 to Figure 13–2 and confirm a valid character check (P_o = 1, true). Repeat for 10110101, showing all outputs, and confirm an invalid character (P_o = 0, false).

10. Decode the following block of ASCII characters, and then determine even parity for both the vertical (VRC) and the longitudinal (LRC) checks.

   a. 0100000   b. 0001110
   c. 0100101   d. 1011011
   e. 1000101   f. 1010101
   g. 0101010   h. 1001010
   i. 1000010   j. 1100000

   Repeat, using a–e as a block. Repeat, using f–j as a block.

11. a. Determine the probability of error for a channel having Gaussian noise and S/N = 17 dB.
   b. What signal voltage level is required if the rms noise level is 500 mV?

12. A channel with a Gaussian noise distribution has 50 mV rms of noise. What signal level must be detected in order for the channel to achieve a BER of 10^{-8}?

13. How many errors would be expected for a 1-megabyte-long message in a Gaussian noise channel with S/N = 12 dB? (1 byte = 8 bits.)

14. Describe the following types of data transmission:

   a. Simplex   b. Half-duplex
   c. Full-duplex   d. Parallel
   e. Serial   f. Asynchronous
   g. Synchronous

15. a. What is the main function of a UART?
   b. The internal clock of a UART for transmitting 9600 b/s operates at what frequency?
   c. Why is such a high frequency used internally?

16. What is a buffer?

17. An active 20-mA-loop receiver input is shown in Figure 13–42.

FIGURE 13–42

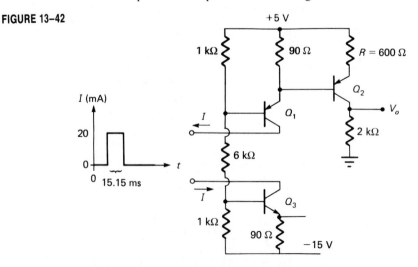

a. What voltage should be measured at the base of Q_1?
b. What value of voltage (referenced to ground) is the emitter of Q_1 capable of reaching when $I = 20$ mA (assume Q_2 removed)?
c. Sketch V_o corresponding to the current waveform. (V_{CE} (sat) $= 0.2$ V.)
d. What value should the 2 kΩ be in order to maximize V_c?
e. If, on the average, there are six characters/word, determine the words/minute, assuming full ASCII is being received.
f. Determine the maximum total transmission line resistance before Q_1 and Q_2 saturate.

18. Determine the cutoff frequency for an RS-232-C terminator circuit if R_L and C_L have maximum allowable values.

19. Sketch the letter "T" in ASCII (7 bits) as it would appear on the BB wire of an RS-232-C interface carrying 100 b/s. Assume ± 15 V.

20. What is the major advantage of the physical characteristics of the EIA's RS-422 interface?

21. If only six interface circuits (wires) are used in the RS-232-C interface of Figure 13–19, what would they be?

22. a. How much time is required to transmit the character NAK in full ASCII on the reverse channel of a system using the AT&T 202 modem?
b. What is the maximum bit rate for this modem (primary channel)?
c. Determine the modulation index at full baud rate for the 202.

23. a. List the originate and answer *mark/space* frequencies for the 103 modem.
b. What would the baud rate have to be to transmit minimum shift key (MSK)?

24. What ASCII character (in binary) is transmitted to synchronize the receiver data clocks when using the BSC (bisync) protocol?

25. Sketch any one of the bisync protocol sequences (frame).

26. The message of Figure 13–43 is received on a data link using Bisync with ASCII plus parity.
a. Using the format of Figure 13–43, sketch an appropriate response to the received signal.
b. Is there an error in the received signal? Where?
c. What is the message? Give the answer as a six-word English expression (not just the character).

FIGURE 13–43

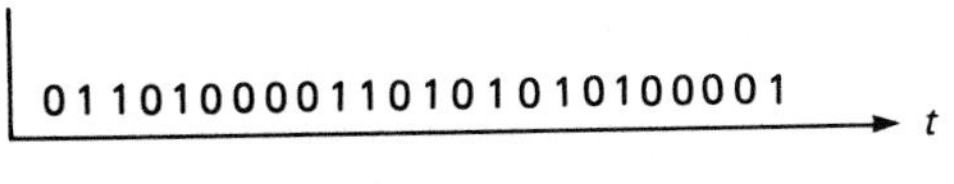

011010000110101010100001 → t

27. Figure 13–44 shows an example of which type of connection: point-to-point, multipoint, multidrop, or circuit switching?

FIGURE 13–44

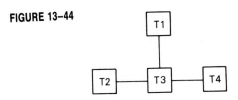

28. **a.** Name a bit-oriented protocol.
    b. Compare the major advantage of this protocol to a character-oriented protocol.
    c. Which protocol requires zero-insertion for certain data conditions?

29. **a.** Sketch a frame for a common bit-oriented protocol.
    b. Where does the packet go in this frame when used for an X.25 packet-switched, data-link control (access) protocol?

30. Automatic request for retransmission is accomplished in the Synchronous Data Link Control protocol by transmission of which: frame counts, repeat, NAK, or bisync?

31. **a.** Assuming no zero-insertions, what is the minimum number of 8-bit bytes that must be transmitted in each BOP frame to achieve a frame efficiency of 63% (comparable to full ASCII)?
    b. Repeat for 50% efficiency.
    c. What is the minimum nonzero efficiency (1-byte message)?

32. Packets arrive at a switching node at a rate of 1000 per minute. Assuming that the random arrivals are independent and equiprobable for any time period (that is, Poisson distributed), compute the probability that:
    a. No packets will arrive during a given 0.2-s interval.
    b. Fewer than three packets arrive in 0.2 s.
    c. Three or fewer arrive in 0.2 s.
    d. More than three arrive in 0.2 s.

33. Describe the S interface in the ISDN architecture.

13-9 ■ COMPUTER ANALYSIS

Using the computer for communications system and circuit analysis has become a virtual necessity in education and industry. What follows are examples of programs and/or readily available simulation software for solving problems of the type covered in this chapter.

All the programs are interactive in the sense that an introductory sentence or short paragraph describes the purpose of the program and the user is then prompted for input data.

The programs for Chapter 13 are as follows:

- **ASCII to FSK conversion:** The program of Figure 13–45 takes keyboard inputs and converts them to the equivalent 4-cycle/bit FSK signal.

- **Bit error rate of a channel with Gaussian noise:** Assuming 200 mV rms of Gaussian noise, the program of Figure 13–46 will calculate the signal level required to achieve a specified bit error rate (probability of error). Note that in all these programs, a few changes in the BASIC program will allow related problems to be solved.

- **Telephone or packet-switch probabilities** for single-server network engineering problems: The program of Figure 13–47 computes the probability of receiving, or transmitting, a specified number of calls or packets in a specified period of time based on the typically assumed Poisson random process.

```
10 REM **** ASCII TO FSK CONVERSION FOR FSK TRANSMISSION ****
20 REM *        EXAMPLE OF STUDENT COMPUTER ASSIGNMENT        *
30 REM ****          FOR PROFESSOR P. YOUNG, AT ASU        ****
40 CLS
50 PRINT "THIS PROGRAM TAKES KEYBOARD INPUTS"
60 PRINT "AND CONVERTS THEM INTO FSK OUTPUT IN REAL TIME.":PRINT
70 PRINT "ENTER A NUMBER OR OTHER KEYBOARD CHARACTER FOR FSK TRANSMISSION."
80 PRINT "AN ASCII-EQUIVALENT 4-CYCLE/BIT FSK SIGNAL WILL BE OUTPUT."
90 PRINT"THE 7-BIT ASCII, 4-CYCLE/BIT SIGNAL CAN BE SCREEN-PRINTED (VERTICALLY)"
100 PRINT "IF YOU REMOVE THE REM STATEMENT ON LINES 190, 260, 270, AND, 300."
110 PRINT "TO QUIT, PRESS THE SPACE BAR.      PRESS ANY KEY, NOW:"
120 A$=INKEY$
130 IF A$=" " GOTO 280
140 IF A$="" GOTO 120
150 PRINT A$;
160 'LPRINT A$
170 'THIS CONVERTS THE INPUT NUMBER TO ASCII FORMAT.
180 A%=ASC(A$)
190 'WE NOW ROLL THE BITS OFF OF THE ASCII WORD 7 TIMES TO CHECK ALL BITS.
200 FOR K%=1 TO 7:B%=INT(A%/2)
210 IF A% = B%*2 THEN SOUND 800,3:PRINT "1010":GOTO 230; ELSE SOUND 2000,3:
    PRINT "1100":'LPRINT "1100";
220 GOTO 240
230 'LPRINT "1010";
240 A%=B%
250 NEXT
260 'LPRINT
270 GOTO 120
280 END
```

FIGURE 13–45

```
10 REM * BIT ERROR RATE OF A CHANNEL WITH GAUSSIAN NOISE *
20 REM *      EXAMPLE OF STUDENT COMPUTER ASSIGNMENT      *
30 REM ****       FOR PROFESSOR P. YOUNG, AT ASU       ****
40 CLS
50      PRINT"THIS PROGRAM WILL FIND THE APPROXIMATE NUMBER"
60      PRINT"OF ERRORS EXPECTED IN A GIVEN N-BIT BINARY MESSAGE"
70      PRINT"OVER A NOISY CHANNEL WITH 200 mV OF GAUSSIAN NOISE ."
80      PRINT
90      PRINT
100 INPUT"The S/N of the system is, (positive integers only) S/N (dB) = ";A
110     IF A=6 OR A<6 THEN 340
120     IF A=7 THEN 340
130     IF A=8 THEN 340
140     IF A=9 THEN 360
150     IF A=10 THEN 380
160     IF A=11 THEN 400
170     IF A=12 THEN 420
180     IF A=13 THEN 440
190     IF A=14 THEN 460
200     IF A=15 THEN 480
210     IF A=16 THEN 500
220     IF A=17 THEN 520
230     IF A=18 OR A>18 THEN 540
240     PRINT
250     PRINT"THE NUMBER OF MESSAGE BITS TRANSMITTED ";
260     INPUT BITS
270     NUMERR=BER*BITS
280     X=A/20
290     VS=(10^X)*(.2)
300     PRINT"THE APPROXIMATE NUMBER OF ERRORS EXPECTED (AVG) = ",NUMERR
310     PRINT
320     PRINT"REQUIRED SIGNAL VOLTAGE LEVEL ="VS" VOLTS PEAK."
330     END
340     LET BER=.5
350     GOTO 240
360     LET BER=5*.01
370     GOTO 240
380     LET BER=.001
390     GOTO 240
400     LET BER=5*.001
410     GOTO 240
420     LET BER=.0001
430     GOTO 240
```

FIGURE 13-46   (page 1 of 2)

```
440    LET BER=.00001
450    GOTO 240
460    LET BER=.000001
470    GOTO 240
480    LET BER=1E-08
490    GOTO 240
500    LET BER=1E-10
510    GOTO 240
520    LET BER=1E-12
530    GOTO 240
540    LET BER=1E-14
550    GOTO 240
560 END
```

FIGURE 13-46   (page 2 of 2)

```
10 REM *   TELEPHONE/PACKET ARRIVAL PROBABILITIES   *
20 REM *   EXAMPLE OF STUDENT COMPUTER ASSIGNMENT   *
30 REM *       FOR PROFESSOR P. YOUNG AT ASU         *
40  CLS
50 PRINT"THIS IS A PROGRAM TO DETERMINE THE PROBABILITY"
60 PRINT"OF RECEIVING A SPECIFIED NUMBER OF CALLS OR PACKETS"
70 PRINT"IN A SPECIFIED PERIOD OF TIME."
80 PRINT
90 PRINT
100 PRINT
110 PRINT"ENTER THE RATE OF CALL ARRIVALS. THAT IS, THE"
120 PRINT"NUMBER OF CALLS PER SECOND:"
130 INPUT R
140 PRINT
150 PRINT"ENTER THE TIME PERIOD THAT THE CALLS WILL BE"
160 PRINT"RECEIVED IN, IN SECONDS:"
170 INPUT T
180 PRINT
190 PRINT"ENTER THE TOTAL NUMBER OF CALLS TO BE RECEIVED"
200 PRINT"IN THE TIME PERIOD:"
210 INPUT N
220 PRINT
230 PRINT"ENTER THE NUMBER FOR THE PROBABILITY TO BE COMPUTED:"
240 PRINT
250 PRINT" 1) THE PROBABILITY OF EXACTLY" N "CALLS IN" T "SEC."
260 PRINT" 2) THE PROBABILITY OF EQUAL OR LESS THAN" N "CALLS IN"
 T "SECONDS."
270 PRINT" 3) THE PROBABILITY OF MORE THAN" N "CALLS IN" T "SEC."
280 INPUT C
290 PRINT:PRINT
```

FIGURE 13-47   (page 1 of 2)

```
300 IF C=1 THEN 400
310 IF C=2 THEN 500
320 IF C=3 THEN 500
330 PRINT
340 PRINT
350 PRINT"WOULD YOU LIKE TO TRY ANOTHER ? Y OR N"
360 INPUT X$
370 CLS
380 IF X$="Y" THEN 50
390 END
400 REM SUBROUTINE TO CALCULATE PROBABILITY OF EXACTLY X CALLS
410 LET F1=1
420 FOR I=2 TO N
430 LET F1= F1*I
440 NEXT I
450 P=2.718^-(R*T)*((R*T)^N)/F1
460 PRINT"THE PROBABILITY OF EXACTLY" N "CALLS IN" T " SECONDS"
470 PRINT"GIVEN THE RATE OF" R "CALLS PER SECOND IS" P "."
480 GOTO 330
490'SUBROUTINE TO CALCULATE THE PROBABILITY OF AT LEAST X CALLS
500 P=0
510 FOR I= 0 TO N
520 F1=1
530 FOR G=2 TO I
540 LET F1=F1*G
550 NEXT G
560 P= P+2.718^-(R*T)*((R*T)^I)/F1
570 NEXT I
580 IF C=3 THEN 620
590 PRINT"THE PROBABILITY OF AT LEAST" N "CALLS IN" T "SECONDS,
    GIVEN"
600 PRINT"THE RATE OF" R "CALLS PER SECOND IS" P "."
610 GOTO 330
620 P=1-P
630 PRINT"THE PROBABILITY OF MORE THAN" N "CALLS IN" T "SECONDS,
    GIVEN"
640 PRINT"THE RATE OF" R "CALLS PER SECOND IS" P "."
650 GOTO 330
660 END
```

FIGURE 13-47   (page 2 of 2)

14

Transmission Lines and Waveguides

■ INTRODUCTION

Radio signals propagating (traveling) through space are, like sunlight, periodic electromagnetic waves and propagate at the speed of light, $c = 3 \times 10^8$ m/s (186,000 m/s). If the period T of the wave is known, then the distance between equivalent points on the wave can be determined by the familiar $d = vt$. Thus the length of one cycle of the wave, called a *wavelength*, is determined from

$$\lambda = vT \qquad \text{(14–1a)}$$

or

$$\lambda = c/f \qquad \text{(14–1b)}$$

where $f = 1/T$ is the frequency of the periodic wave traveling through space at the velocity of light.

Transmission lines and waveguides provide a structure for guiding electromagnetic waves from one place to another. The distinction between "circuit connec-

FIGURE 14–1 Coaxial transmission line and traveling wave.

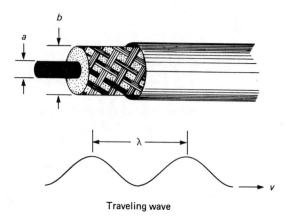

Traveling wave

tion wires," a transmission line, or a waveguide is very broad but relates to the structural dimensions as compared to the wavelength of the signal being propagated.

As an example, the familiar coaxial cable illustrated in Figure 14–1 may be used to connect an audio signal to an oscilloscope. However, at higher frequencies where the signal wavelength is small compared to the cable length, transmission-line behavior affects the "connection" and the signal transfer. At even higher frequencies, where the cross-sectional dimensions of the cable are greater than one-half of the signal wavelength, the signal will propagate primarily by waveguide behavior. The difference between transmission line and waveguide signal propagation behavior will be explained later in the chapter. ■

14–1 ■ TRANSMISSION LINES

Figure 14–2 illustrates a few of the more common transmission-line structures. Transmission lines are characterized by two (or more) conductors that provide a complete current loop. For the coaxial line, the high-potential, sending-end current is conducted by the center wire, and the return current path is provided by the flat conducting braid. Transmission-line propagation of a signal is by means of the variation of the electric and magnetic fields produced by the movement of electric changes. The electromagnetic (EM) wave thus produced propagates primarily in the dielectric (nonconductor) material that keeps the two electrical conductors separated. Dielectric materials include polyethylene, Teflon, air, and pressurized gases (with insulating washers or a helical spacer for support).

For the open-wire and twin-lead transmission lines of Figure 14–2b and c, the EM wave propagates in the space between and around the two conducting wires. The dielectric sheath "ribbon" and spacers maintain a constant separation between the wires to maintain a balanced EM field for best propagation characteristics. Although simpler and less expensive than the coaxial structure, the ribbon results in less structural integrity and more radio frequency interference because

FIGURE 14–2 Transmission lines.

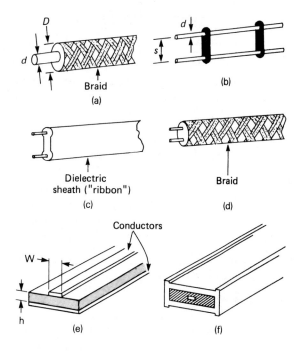

the EM fields are not confined and shielded. The balanced, shielded line of Figure 14–2d with the shield grounded is a vast improvement.

Because of the mirrorlike symmetry of the electric and magnetic fields illustrated in Figure 14–3 for the two-wire line, the fields and propagation characteristics will be unaffected if a large ground plane is slid horizontally between and equidistant from the two wires. If the lower wire is then removed and the ground plane is allowed to carry the return current, a transmission line very much like the *microstrip* line of Figure 14–2e results. The *strip-line* structure of 14–2f is used in place of microstrip for better confinement of EM fields and improved propagation characteristics. Notice from the illustration that the upper and lower ground planes are connected for good balance between the upper and lower fields.

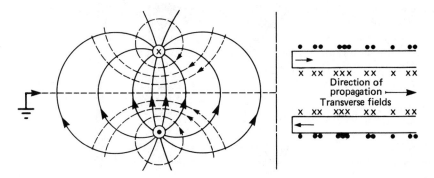

FIGURE 14–3 Electric (solid) and magnetic (dashed) fields in a twin-lead transmission line.

TEM Wave Propagation

The electric and magnetic fields sketched in Figure 14–3 are at all points perpendicular to each other and the signal is propagating into the page. Seen broadside to the two-wire, lossless transmission line, both the electric and magnetic fields are always into or out of the page—that is, transverse to the direction of propagation. The energy in these *transverse electromagnetic* (TEM) waves propagates along the transmission line at a velocity

$$v_p = \frac{c}{\sqrt{\epsilon_r}} \qquad\qquad (14\text{--}2)$$

where ϵ_r is the dielectric constant of the transmission line relative to air (or space).

In free space, RF signals also propagate as TEM waves and the ratio of electric field strength to magnetic field strength of any single wave is a constant with a value approximately equal to 377 V/m per amperes/meter. This value has units of ohms, is called *wave impedance*, and is calculated from

$$Z_w = \sqrt{\frac{\mu_0}{\epsilon_0}} = 377 \ \Omega \qquad\qquad (14\text{--}3)$$

where the permeability of space is $\mu_0 = 4\pi \times 10^{-7}$ H/m and the permittivity (dielectric constant) is $\epsilon_0 = 8.854$ pF/m. Similarly, a uniform, lossless transmission line has a definite ratio of electric-to-magnetic field strength and also voltage-to-current ratio. This ratio is found from the electrical circuit parameters of the transmission line and is called the *characteristic*, or *surge*, *impedance* Z_0 of the line. The electromagnetic properties of the simplest and most common transmission-line structures have been analyzed and the results, assuming no internal power losses, are tabulated in Table 14–1.

Transients and Reflections

Figure 14–4 illustrates the distributed electrical parameters of a model transmission line. The incremental sections show the series inductance/meter L, resistance/meter R, the shunt capacitance/meter C, and conductance (leakage resistance)/meter G. The middle section shows the leakage components distributed along a balanced two-wire line. The transmission line thus reveals its electrical network characteristic, which may be analyzed to yield an expression describing the behavior of voltage and current waves traveling along the line. Such waves are called *traveling waves*.

From Figure 14–4 a voltage wave traveling down the line (from left to right on the x-axis) will experience incremental voltage drops of $(dv/dx)(\Delta x)$, which by Kirchhoff's law yields*

$$(Ri) \ \Delta x + \left(L \frac{di}{dt} \right) \Delta x = -\left(\frac{dv}{dx} \right) \Delta x \qquad\qquad (14\text{--}4)$$

---

* The ac voltage drop across an inductance L is given by Equation 1–1 and for a capacitance C by Equation 1–3.

TABLE 14–1 Transmission-Line (Circuit) Characteristics (Refer to Figure 14–2 for dimensions.)

	$Z_0(\Omega)$	L (H/m)	C (F/m)
Twin-lead	$\dfrac{120}{\sqrt{\epsilon_r}} \ln 2s/d$	$\dfrac{\mu}{\pi} \ln 2s/d$	$\dfrac{\pi\epsilon}{\ln 2s/d}$
Coaxial	$\dfrac{60}{\sqrt{\epsilon_r}} \ln D/d$	$\dfrac{\mu}{2\pi} \ln D/d$	$\dfrac{2\pi\epsilon}{\ln D/d}$
Microstrip (after H. A. Wheeler)	$Z_0 = 377h/\{\sqrt{\epsilon_r}\,W[1 + 1.74(\epsilon_r)^{-0.07}(W/h)^{-0.836}]\}$		

The ϵ_r for typical materials used in transmission lines are (at 10 GHz): polystyrene 2.5, polyethylene 2.3, and teflon 2.1.

Also, the decrease in current due to ac shunting is

$$(Gv)\,\Delta x + \left(C\,\frac{dv}{dt}\right)\Delta x = -\left(\frac{di}{dx}\right)\Delta x \qquad (14\text{--}5)$$

Solving these equations under the generally justified assumption of very low power losses ($R = G = 0$) yields

$$\frac{d^2v}{dx^2} = LC\,\frac{d^2v}{dt^2} \qquad (14\text{--}6a)$$

and

$$\frac{d^2i}{dx^2} = LC\,\frac{d^2i}{dt^2} \qquad (14\text{--}6b)$$

These equations describe traveling waves. Similar equations written in terms of the appropriate physical parameters are found throughout the physical sciences to mathematically describe waves of all sorts found in nature.

The solutions to the wave equations, Equation 14–6, give the voltage across the line and current along the line as a function of time and distance traveled. As an example,

$$v = F(\sqrt{LC}\,x \pm t) \qquad (14\text{--}7)$$

FIGURE 14–4 Distributed circuit model of a transmission line.

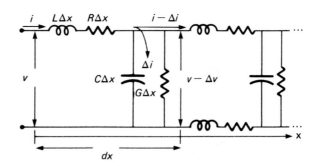

where the + or − gives the direction of travel (+ right, − left). And the velocity at which the voltage wave propagates (travels) is

$$v_p = \frac{1}{\sqrt{LC}} \qquad \text{(14–8)}$$

It also can be shown that the ratio of voltage to current for a single wave on the lossless transmission line network is given approximately by

$$Z_0 = \sqrt{\frac{L}{C}} \qquad \text{(14–9)}$$

and more generally by

$$Z_0 = \sqrt{\frac{Z}{Y}} = \sqrt{\frac{R + sL}{G + sC}} \qquad \text{(14–10)}$$

where s is a frequency operator. For the particular case of sinusoidal waves, $s = j\omega$.

■ ■ ■ ■

EXAMPLE 14–1

A very low loss coaxial transmission line has 30 pF/ft of distributed capacitance and 75 nH/ft of inductance. Determine the following:

1. The capacitance of a 3-ft length of this line used as an oscilloscope probe
2. Z_0
3. The velocity of propagation for a voltage and current transient (velocity relative to a TEM wave in free space)
4. The time required for an input transient to reach the oscilloscope (see part 1)
5. The ratio of shield diameter to center conductor diameter of the coax

Solution:

1. 30 pF/ft × 3 ft = **90 pF.** This can greatly decrease the high-frequency response of a circuit under test.
2. $Z_0 = \sqrt{L/C} = \sqrt{75 \times 10^{-9}/(30 \times 10^{-12})} = $ **50 Ω.**
3. $v_p = 1/\sqrt{LC} = 1/\sqrt{75 \times 30 \times 10^{-21}} = $ **666.7 × 10⁶ ft/s.** $v_p = (666.7 \times 10^6 \text{ ft/s})$ (1 mi/5280 ft) = 126,263 mi/s, so that $v_p/c = 126,263/186,000 = $ **0.679**—a little more than two-thirds the speed of light.
4. $d = v_p t.$ $t = 3 \text{ ft}/(666.7 \times 10^6 \text{ ft/s}) = $ **4.5 ns.**
5. Table 14–1 gives $Z_0 = (60/\sqrt{\epsilon_r})\ln D/d.$ From Equation 14–2, $\sqrt{\epsilon_r} = c/v_p = 1/0.679$, and from part 2, $Z_0 = 50$ Ω. Therefore, $50 \times 1.473/60 = 1.228 = \ln D/d.$ By the definition of logarithms, $D/d = e^{1.228} = $ **3.41.**

To better understand the circuit behavior of transmission lines (TL) with propagating waves and reflections from discontinuities, consider the voltage and current surges along an infinite length TL after closing the switch of Figure 14–5. The

FIGURE 14–5 Infinite-length transmission-line voltage and current transients.

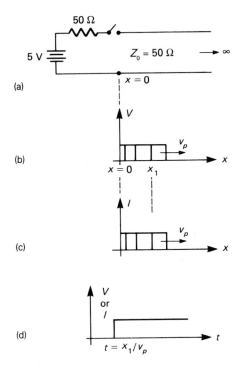

characteristic ac impedance of the TL is 50 Ω; therefore, voltage division with the 50-Ω resistor in series with the 5-V battery will result in a continuous 2.5-V pulse and 2.5 V/50 Ω = 50 mA current pulse propagating along the line, as shown in b and c. An observer making voltage and current measurements at point x_1 somewhere down the line will measure zero until time $t = x_1/v_p$ (see Figure 14–5d) and then measure 2.5 V and 50 mA dc continuously thereafter.

The same result can be achieved for a line of finite length ℓ if a load Z_L of value equal to Z_0 is connected to the TL. This is illustrated in Figure 14–6, where part a shows that the TL to the right of the separation is infinite in length and has a

FIGURE 14–6 Equivalent transmission-line terminations.

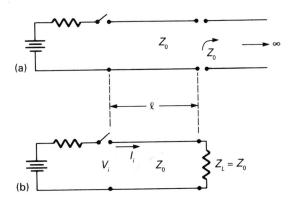

characteristic impedance of Z_0. Hence, as seen in part b, the circuit is identical to Figure 14–6a and $V_i = 2.5$ V, $I_i = 50$ mA. This condition does not change because there are no reflections when the TL is terminated with $Z_L = Z_0$ (impedance-matched). The power $P = 2.5$ V $\times$ 50 mA $= 125$ mW, which would have continued down the TL, is now being consumed in the 50-Ω load Z_L.

Suppose now that *no* load is attached to the end of the finite line. In this case the problem will start out the same as before, but when the voltage and current waves reach the open-circuited end, there will be reflection resulting in voltage and current waves traveling to the left toward the generator. Then we have to analyze whether or not a reflection will occur at the generator end. If a reflection does occur, we will have to follow it back to the "load" and repeat the process until steady state occurs. We would expect that, for an open-circuited lossless TL, a steady state will be reached with V along the line equal to 5 V and $I = 0$.

To analyze reflections, a *reflection coefficient*

$$\Gamma = \rho \angle \theta \qquad (14\text{–}11)$$

is calculated at each end of the line. The reflection coefficient is analyzed as follows: At any point on the line, the resultant voltage and current must be determined by the sum of the incident and reflected waves passing the point. That is,

$$V = V^+ + V^- \qquad (14\text{–}12)$$

and

$$I = I^+ + I^- \qquad (14\text{–}13)$$

where V^+ is the incident voltage wave and V^- is the reflected voltage wave. The reflected wave V^- is found from $V^- = \Gamma V^+$; that is,

$$\Gamma \equiv \frac{V^-}{V^+} \qquad (14\text{–}14)$$

where the voltages and consequently Γ are, in general, complex quantities (phasors) with magnitude and phase as indicated in Equation 14–11. Thus, at a load the percentage of the voltage arriving at the load that is reflected from the load is called the *reflection coefficient at the load*.

Because of current reversal, the voltage reflection coefficient has the opposite sign to that of the current reflection coefficient Γ_I. Consequently,

$$\Gamma_I = -\Gamma \qquad (14\text{–}15)$$

The incident current is $I^+ = V^+/Z_0$ and at the same point the reflected current is $I^- = \Gamma_I I^+ = -I^+$. Also,

$$V = V^+ + V^- = V^+ + \Gamma V^+$$

$$I^+ = \frac{V^+}{Z_0}$$

and

$$I^- = \frac{-V^-}{Z_0}$$

where the minus sign indicates current direction opposite to I^+. The load impedance is

$$Z_L = \frac{V}{I} = \frac{V^+ + V^-}{I^+ + I^-}$$

$$Z_L = \frac{V^+ + V^-}{V^+/Z_0 - V^-/Z_0} \qquad (14\text{–}16)$$

Solving this for V_L^-/V_L^+ at the load produces Γ_L in terms of impedances as

$$\frac{V^-}{V^+} = \Gamma_L = \frac{Z_L - Z_0}{Z_L + Z_0} \qquad (14\text{–}17\text{a})$$

$$= \frac{(Z_L/Z_0) - 1}{(Z_L/Z_0) + 1} \qquad (14\text{–}17\text{b})$$

or
$$\Gamma_L = \frac{Y_0 - Y_L}{Y_0 + Y_L} \qquad (14\text{–}17\text{c})$$

Because Z_L is typically complex, Equation 14–17 reinforces the fact that $\Gamma = \rho \underline{/\theta}$ is a complex quantity with magnitude ρ and phase angle θ.

Turning now to the open-circuited TL of Figure 14–7, we see that $Z_L = \infty$, so that $\Gamma_L = (\infty - 50)/(\infty + 50) = 1$ and $\Gamma_I = -\Gamma = -1$. The problem is now solved as follows: Initially, $V^+ = E[Z_0/(Z_0 + R_g)] = 5 \text{ V}(50/100) = 2.5 \text{ V}$, and

$$I^+ = \frac{2.5 \text{ V}}{50 \text{ }\Omega} = 50 \text{ mA}$$

These voltage and current waves propagate down the transmission line, charging the line capacitance and inductances as they go. At the open-circuited end of the line, the incident voltage will be $V^+ = 2.5 \text{ V}$ and the incident current is $I^+ = 50$ mA. From before, $\Gamma = 1$, so that $V^- = +2.5 \text{ V}$ and $I^- = \Gamma_I I^+ = -\Gamma I^+ = -1 \times 50$ mA $= -50$ mA; that is, the reflected voltage has the same amplitude and phase as the incident voltage, whereas the reflected current has the opposite phase but same amplitude as that of the incident wave.

The voltage at the end of the line will suddenly go from zero to $V = V^+ + V^- = 2.5 + 2.5 = 5 \text{ V}$, and the current will become $I = I^+ + I^- = 50 \text{ mA} + (-50 \text{ mA}) = 0$, as one would expect at an open circuit. The reflected current extinguishes the initial 50 mA on the line as it travels toward the generator, as shown in Figure 14–8. When the reflected wave reaches the generator (sending end of the TL), it is

FIGURE 14–7 Open-circuited transmission line.

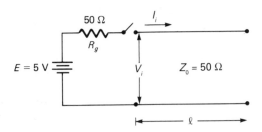

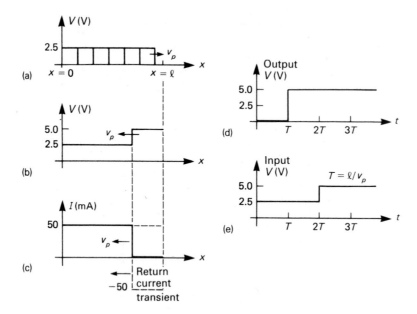

FIGURE 14-8 Transient on open-circuited transmission line. (a) Line voltage just before transient reaches end ($x = \ell$). (b) and (c) show line voltage and current after reflection from open circuit. (d) and (e) show line voltage versus time at output and input, respectively.

a wave traveling on a $Z_0 = 50\ \Omega$ TL that is "terminated" by a "load" $R_g = 50\ \Omega$. The reflection coefficient for this *matched* condition is $\Gamma = (50 - 50)/(50 + 50) = 0$ and no reflections occur. The transmission line hereafter is found to have 5 V at all points along the line and $I = 0$. This is just what you would anticipate very shortly after connecting an open-ended coaxial cable to a 5-V battery. A very long length of (delay) line and an oscilloscope are needed to observe the transients. Indeed, this arrangement forms the basis for time-domain reflectometry (TDR), used for measurements to analyze the conditions of transmission lines.

14–2 ■ SINUSOIDAL SIGNALS AS TRAVELING WAVES

A sinusoidal signal traveling in the positive x direction on a lossless transmission line can be expressed as

$$v = A\ \cos(\omega t - \beta x) \qquad\qquad (14\text{–}18)$$

where β is the transmission-line phase constant (radians/unit length). A stop-action plot of Equation 14–18 is shown in Figure 14–9. An observer traveling on a fixed phase point on the wave moving at a constant velocity v_p has a phase

FIGURE 14-9 Sinusoidal traveling wave traveling in positive x direction.

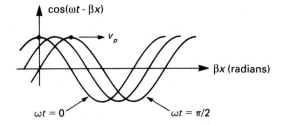

velocity found by differentiating $\omega t - \beta x = $ constant. The result is

$$\frac{dx}{dt} = v_p = \frac{\omega}{\beta} \qquad (14\text{--}19)$$

Also, from Equation 14–1b, $v_p = \lambda f$, so that

$$\beta = \frac{2\pi}{\lambda} \qquad (14\text{--}20)$$

If a constant loss in amplitude of α (nepers/unit length) occurs all along the length of the line, then

$$v \text{ (with losses)} = (Ae^{-\alpha x})\cos(\omega t - \beta x) \qquad (14\text{--}21)$$

Equation 14–21 is the real part of the exponential expression

$$v = (Ae^{-\alpha x})e^{j(\omega t - \beta x)} = Ae^{-\alpha x}e^{-j\beta x}e^{j\omega t}$$

for which the phasor portion is

$$V = Ae^{-(\alpha + j\beta)x} \qquad (14\text{--}22)$$

A general solution of the wave equation (Equation 14–6) for a voltage wave traveling in either direction is

$$V = A_1 e^{\gamma x} + A_2 e^{-\gamma x} \qquad (14\text{--}23)$$

where γ is the *propagation constant* determined from transmission line parameters as $\gamma = \sqrt{ZY}$, just as the characteristic impedance is determined from $Z_0 = \sqrt{Z/Y}$ (Equation 14–10). For sinusoidal signals the calculations are made from

$$Z_0 = \sqrt{\frac{R + j\omega L}{G + j\omega C}} = R_0 + jX_0 \qquad (14\text{--}24)$$

and
$$\gamma = \alpha + j\beta \qquad (14\text{--}25)$$
$$= \sqrt{(R + j\omega L)(G + j\omega C)}$$

To calculate the total signal attenuation over a distance traveled of x, solve $V = Ae^{-\alpha x}$ by taking the natural logarithm of both sides to give $-\alpha x = \ln(V/A)$ nepers. To put the attenuation in decibels, note that $20 \log_{10} e = 8.686$, so that the voltage (or current) attenuation for a total distance traveled x is

$$\text{attenuation (dB)} = 8.686\alpha x \qquad (14\text{--}26)$$

Sinusoidal Signal Reflections and Standing Waves

A transmission line with discontinuities and mismatched impedance conditions ($Z_L \neq Z_0$) will have reflections that produce steady-state *standing waves*. This is the same wave phenomenon as for vibrating strings and pressure waves in musical instruments and other sinusoidally forced, mechanical systems (including quartz crystals).

Equations 14–11 through 14–17 are still valid for sinusoidal signals and reflections. Figure 14–10 illustrates a sinusoidal generator, with $Z_g = Z_0$, driving a long lossless transmission line ($\ell \gg \lambda$ of the signal), which is terminated by a complex load impedance Z_L. By Equation 14–17, if $Z_L = Z_0$, then no reflections occur and no standing waves appear (Figure 14–11a). However, larger values of load impedance will produce reflections from the load, causing the voltage measured along the line to vary, as shown in Figure 14–11b and c; these patterns are called *standing waves*. In part c for no load ($Z_L = \infty$), the voltage measured at the load is exactly twice the voltage along a lossless "flat" or matched line (Figure 14–11a). This voltage value A is also the open-circuited generator voltage. For $Z_L < Z_0$, $V(\ell)$ will be less than A/2 and Figure 14–11d illustrates the lower limit.

The voltage at any point d from the load is, in general,

$$V(d) = v_i \left[\frac{e^{\gamma d} + \Gamma e^{-\gamma d}}{e^{\gamma \ell} + \Gamma e^{-\gamma \ell}} \right] \qquad (14\text{–}27\text{a})$$

$$= v_i \left[\frac{1 + \Gamma e^{-2\gamma d}}{1 + \Gamma e^{-2\gamma \ell}} \right] \frac{e^{\gamma d}}{e^{\gamma \ell}} \qquad (14\text{–}27\text{b})$$

and for a lossless line ($\alpha = 0$, $\gamma = j\beta$),

$$V(d) = v_i \left[\frac{Z_L \cos \beta d + j Z_0 \sin \beta d}{Z_0 \cos \beta \ell + j Z_L \sin \beta \ell} \right] \qquad (14\text{–}28)$$

where Euler's equalities are used to equate sinusoids and complex exponentials. Notice the discontinuity in the standing wave pattern of Figure 14–11c at the generator output terminals (left side). The impedance seen by the generator is clearly not Z_0 as it was for the matched load of part a; it is less than Z_0 and will be complex. On the other hand, if ℓ were equal to $\lambda/2 + \lambda/4 = 3\lambda/4$ (or any integer multiple thereof), then the generator will be driving an apparent short circuit, as seen by the voltage null in part c. The impedance at any point d from the load can

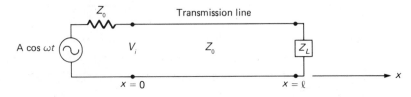

FIGURE 14–10   Generator and load transmission-line terminations.

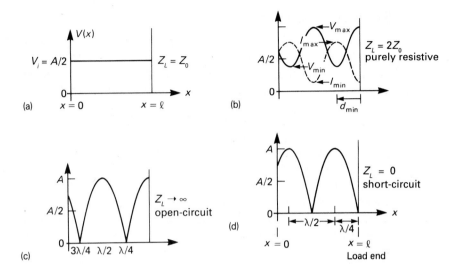

FIGURE 14–11 Voltage measurements on line of length ℓ for various loads. Standing waves create these voltage (and current) patterns.

be calculated from the general expression

$$Z(d) = \frac{V(d)}{I(d)}$$

$$= \frac{V^+(d) + V^-(d)}{I^+(d) + I^-(d)}$$

$$= Z_0 \left[\frac{e^{\gamma d} + \Gamma e^{-\gamma d}}{e^{\gamma d} - \Gamma e^{-\gamma d}} \right] \qquad (14–29)$$

or for the lossless line ($\alpha = 0$),

$$Z(d) = Z_0 \left[\frac{e^{j\beta d} + \Gamma e^{-j\beta d}}{e^{j\beta d} - \Gamma e^{-j\beta d}} \right] \qquad (14–30a)$$

$$= Z_0 \left(\frac{Z_L \cos \beta d + jZ_0 \sin \beta d}{Z_0 \cos \beta d + jZ_L \sin \beta d} \right) \qquad (14–30b)$$

$$= Z_0 \left[\frac{Z_L/Z_0 + j \tan \beta d}{1 + j(Z_L/Z_0)\tan \beta d} \right] \qquad (14–30c)$$

where the identity

$$e^{\pm j\beta d} = \cos \beta d \pm j \sin \beta d$$

is used. These equations are used in engineering analysis, but they are computationally tedious. However, as seen in Section 14–3, graphical techniques along with a few laboratory measurements are faster and more convenient for solving microwave circuit problems.

Slotted Line and VSWR

A *slotted line* is a device with a probe extending into a transmission line through a slot. The probe can be slid along the slot, and by means of a connection to a voltmeter, the voltage may be read at any point. Plots such as those illustrated in Figure 14–11 can be produced from slotted-line measurements. This device also provides a very convenient means for computing the magnitude of the reflection coefficient and for determining the load impedance.

The ratio of maximum voltage to minimum voltage of the standing wave is defined as the *voltage standing-wave ratio*,

$$\text{VSWR} = \frac{V_{max}}{V_{min}} \tag{14--31}$$

V_{max} occurs where the incident and reflected phasors are exactly in phase and add to a maximum. V_{min} occurs where the phasors are exactly out of phase and subtract. The ratio is

$$\text{VSWR} = V^+(1 + \rho)/V^+(1 - \rho) \tag{14--32}$$

$$\text{VSWR} = (1 + \rho)/(1 - \rho)$$

where ρ is the magnitude of the reflection coefficient.

Equation 14–32 can be solved for calculating the *magnitude* of the reflection coefficient as

$$\rho = \frac{\text{VSWR} - 1}{\text{VSWR} + 1} \tag{14--33}$$

To determine the load impedance Z_L, the distance d_{min} from the load to the first voltage *minimum* is used as d in Equation 14–30. Also, the voltage minimum occurs at an impedance minimum; consequently, the minimum impedance on the line can be calculated from

$$Z_{min} = \frac{Z_0}{\text{VSWR}} \tag{14--34}$$

Z_{min} can range in value from Z_0 for a matched load impedance ($\Gamma = 0$, VSWR = 1.0), to zero for a short or open load ($\rho = 1$, VSWR = ∞). Also the maximum impedance on the line can be determined by

$$Z_{max} = Z_0(\text{VSWR}) \tag{14--35}$$

■ ■ ■ ■

EXAMPLE 14–2

A lossless 50-Ω transmission line with $\epsilon_r = 2$ is measured with a slotted line at 1 GHz to have $V_{max} = 10$ mV and $V_{min} = 2$ mV. The first voltage minimum is 2 inches from the load. Determine

1. The operating wavelength λ.
2. VSWR.
3. The magnitude of the reflection coefficient.

4. The distance in wavelengths between the load and first voltage minimum.

5. The value of minimum impedance.

6. The distance in millimeters between the voltage maximum and minimum.

Solution:

1. $\lambda = (c/\sqrt{\epsilon_r})/f = (3 \times 10^8 \text{ m/s})/(\sqrt{2} \times 10^9) = \textbf{212 mm.}$

2. $\text{VSWR} = 10 \text{ mV}/2 \text{ mV} = \textbf{5.}$

3. $\rho = (5 - 1)/(5 + 1) = \textbf{0.67.}$

4. $d_{min} = (2 \text{ in})(25.4 \text{ mm/in}) = 50.8 \text{ mm.}$ $d_{min}/\lambda = 50.8 \text{ mm}/212 \text{ mm} = 0.24.$ Therefore, $d_{min} = \textbf{0.24}\boldsymbol{\lambda}.$

5. $Z_{min} = 50 \text{ }\Omega/5 = \textbf{10 }\boldsymbol{\Omega}.$

6. Voltage maxima and minima are separated by a quarter-wavelength $\lambda/4$. Thus, $d = 212 \text{ mm}/4 = \textbf{53 mm.}$

14–3 ■ THE SMITH CHART

Calculating impedances for different lengths of transmission lines with complex loads can become very tedious. Fortunately a very convenient graphical aid was devised by P. H. Smith in 1938. The improved version, still in use today, was published in the January 1944 issue of *Electronics Magazine*.

The Smith chart* of Figure 14–12 is used to graphically solve Equation 14–30 for the effective impedance at any point on a transmission line when the load impedance or slotted-line measurements are known. Indeed the usefulness of this device is legendary among microwave technologists.

All the impedances on the Smith chart are normalized to Z_0 of the particular transmission line in use; that is, if $Z_L = 25 - j47.5 \text{ }\Omega$ on a 50-Ω line, then the normalized impedance from

$$z = \frac{Z}{Z_0} \qquad \textbf{(14–36)}$$

is

$$z_L = \frac{Z_L}{Z_0} \qquad \textbf{(14–37)}$$

$$= 0.5 - j0.95$$

This impedance point is plotted on Figure 14–12. It may be noticed that solving Equation 14–17b for z_L in terms of Γ yields

$$z_L = \frac{1 + \Gamma}{1 - \Gamma} \qquad \textbf{(14–38)}$$

---

* A very convenient tool is the 9-in. circular slide-rule version, called the *transmission-line calculator* (available from Analog Instruments, P.O. Box 808, New Providence, NJ 07974).

IMPEDANCE OR ADMITTANCE COORDINATES

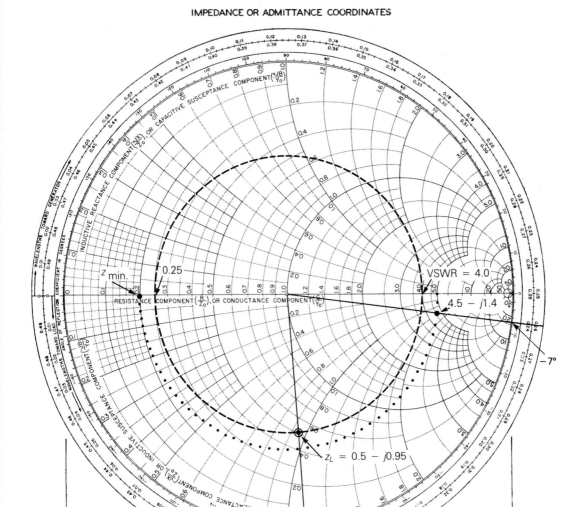

FIGURE 14–12   The Smith chart.

The VSWR created by this load is found for the lossless transmission line by swinging an arc from z_L to the right half of the center line of the chart (VSWR can vary from 1.0 to ∞). The center of the arc is the center of the chart where $z = 1.0$, $Z = Z_0$, and VSWR $= 1.0$. This arc is part of a *constant VSWR circle* that passes through all possible impedance points along the transmission line being considered. As seen on Figure 14–12, VSWR $= 4$ for this mismatched transmission line. The maximum impedance measured along this transmission line will be $Z_{max} = 4 \times Z_0 = 200\ \Omega$, and the minimum will be $Z_{min} = Z_0/4 = 0.25 \times 50\ \Omega = 12.5\ \Omega$. Please note that the constant VSWR circle passes through the purely resistive impedance $z = 0.25 + j0$, where $Z = 12.5\ \Omega$.

The scales at the bottom of Figure 14–12 are very convenient for finding the voltage and power reflection coefficients (REFL. COEF. / VOL. (voltage) and PWR.), return loss (defined in Section 14–4; LOSS IN DB—RETN.), standing-wave ratio in decibels (STANDING WAVE / IN DB), and solving TL problems for lossy transmission lines (TRANSM. LOSS / 1DB STEPS). For the present example, the line from the Smith chart centerpoint through z_L extends out to the inner periphery circle that includes the "ANGLE OF REFLECTION COEFFICIENT IN DEGREES." The value is $\theta = -86°$. If a compass is spread from Smith chart center to z_L and then moved to the REFL. COEF. / VOL. scale where one point is placed on the CENTER line, the other compass point will strike at $\rho = 0.60$. Combining θ and ρ in Figure 14–12, the reflection coefficient at the load is $\Gamma = 0.60 \underline{/-86°}$. This result is confirmed by exact calculations using Equation 14–17a or b; $\Gamma_L = 0.605 \underline{/-85.4°}$.

■ ■ ■ ■

EXAMPLE 14–3

1. Determine the load impedance and component values for Example 14–2.

2. Determine the load admittance.

3. Determine the reflection coefficient at the load.

Solution:

1. From part 4 of Example 14–2, $d_{min} = 0.24\lambda$ from the load. The voltage minimum occurs for the purely resistive impedance $z_{min} = z_0/\text{VSWR} = \frac{1}{5} = 0.2$. ($Z_{min} = 50\ \Omega/5 = 10\ \Omega$, which—when normalized—is $10\ \Omega/50\ \Omega = 0.2$.)

   Start at $z_{min} = 0.2$ and move toward the load a distance of $d_{min} = 0.24\lambda$ (see the dotted arc on Figure 14–12).

   The line drawn through the center of the chart and 0.24 wavelength on the "wavelengths toward the load" scale passes through the constant VSWR circle (dotted arc) at the load.

   Read the normalized impedance—approximately $z_L = 4.5 - j1.4$, so $Z_L = 50(4.5 - j1.4) = \mathbf{225 - j70\ \Omega}$.

   This is a 225-Ω resistor in series with a 70-Ω reactance. At 1 GHz,

$$C = \frac{1}{2\pi f X_c} = \mathbf{2.3\ pF}$$

2. The load admittance is found as follows:

   Plot the normalized load impedance $z_L = 4.5 - j1.4$.

   Draw a line through z_L and center of the chart. Extend this line until it passes through the constant VSWR = 5.0 circle on the opposite side (180° or $\lambda/2$ wavelengths) of the chart.

   The admittance is the inverse of the impedance. Therefore, y_L is exactly 180° on the opposite side of the chart from z_L. The load admittance is approximately $y_L = 0.2 + j0.07$ or $Y_L = Y_0/z_L = (1/50 \ \Omega)(0.2 + j0.07) = \mathbf{0.004 + \textit{j}0.0014 \ S}$.

   Check: $Y_L = 1/Z_L = 1/(225 - j70) = 0.0042\underline{/17.3°} \ S$.

3. From the Smith chart, $\theta = -7°$ and the VOL. REFL. scale below the Smith chart yields $\rho = 0.68$ for $\Gamma_L = 0.68\underline{/-7°}$. An exact calculation using Equations 14–17b and 14–17a is

$$\Gamma_L = \frac{(225 - j70) - 50}{(225 - j70) + 50} = \frac{(4.5 - j1.4) - 1}{(4.5 - j1.4) + 1} = 0.676\underline{/-7.3°}$$

14–4 ■ IMPEDANCE-MATCHING TECHNIQUES

If an antenna or other microwave load produces a mismatch condition on a lossless transmission line, then the transmitter or other microwave source also will probably not be impedance-matched. The result will be inefficient transmission of power and possibly overheating and damage to the transmitter that must dissipate the reflected power.

Incident and reflected power are related by using the magnitude of Equation 14–14 as

$$\rho = \frac{V^-}{V^+}$$

Then

$$\rho^2 = \frac{(V^-)^2}{(V^+)^2} = \frac{P^-}{P^+} \tag{14–39}$$

Reflections and power losses are usually specified in decibels as return loss (RL) and mismatch loss (ML). *Return loss* gives the amount of power reflected from a load and is computed from

$$\text{RL (dB)} = -10 \log \rho^2 = -20 \log \rho \tag{14–40}$$

Or use the scale at the bottom of Figure 14–12 labeled REFLECTION/LOSS IN DB/RETN. The amount of power transmitted to the load, P_L (or P_T), is determined from

$$P_L = P^+ - P^- \tag{14–41a}$$

$$= P^+(1 - \rho^2) \tag{14–41b}$$

The fraction of incident power not reaching the load because of mismatches and reflections is

$$\frac{P_{\text{load}}}{P_{\text{incident}}} = \frac{P_{\text{L}}}{P^+} = 1 - \rho^2 \qquad (14\text{--}42)$$

Hence the *mismatch loss* (or reflection loss) is computed from

$$\text{ML (dB)} = -10 \log(1 - \rho^2) \qquad (14\text{--}43)$$

Or use the Figure 14–12 scale labeled REFLECTION/LOSS IN DB/REFL. The minus signs in Equations 14–40 and 14–43 make the values of RL (dB) and ML (dB) positive, since ρ and $1 - \rho^2$ are fractions less than unity.

The two simplest techniques for impedance matching close to the load so that most of the length of transmission line is "flat" ($\Gamma = 0$) is to use a *stub tuner* or a *quarter-wave transformer.*

The *stub tuner* technique is based on the fact that a short- or an open-circuited length of lossless transmission line will present a pure reactance at the input end of the line. Short-circuited stubs are usually preferred to open circuits because of fringing effects. If $Z_{\text{L}} = 0$ in Equation 14–30c, then a length of line ℓ has input impedance

$$Z_i = jZ_0 \tan \beta\ell \qquad (14\text{--}44)$$

This relationship is seen graphically in Figure 14–12 as the perimeter of the Smith chart where $z = 0 \pm jx$.

Tuning stubs are most conveniently placed in parallel for matching (see Figure 14–13), so we will henceforth work with admittances. A short circuit has an admittance of $y = 1/z = 1/0 \to \infty$. Therefore, to produce a purely *inductive* admittance (susceptance) of $y = 1/jx = -jb$ using the Smith chart, the following procedures apply:

Short-Circuit Stub

1. Start at ∞ (right-hand extreme of chart) where the "wavelengths toward the generator" is marked 0.25λ.

FIGURE 14–13 TL with short-circuited stub tuner.

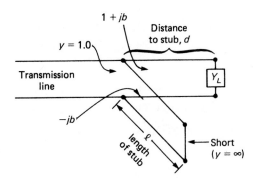

2. Proceed clockwise (toward the generator) on the perimeter to the value *b*.

3. Draw a straight, radial line from the chart center through *b* to the "wavelengths toward the generator" scale and record the value as *A*.

4. The length of short-circuited line needed is $\ell = A - 0.25$ wavelengths.

The distance *d* in wavelengths from the load to the point where the stub is to be connected is determined as follows:

1. Plot the normalized load admittance y_L. Draw a line from the chart center through y_L to the "wavelengths toward the generator" scale and record the value as *B*.

2. Draw a constant VSWR circle clockwise toward the generator until it intersects the constant resistance circle "$y = 1 \pm jb$." Draw a radial line from the chart center through this point, extend it to the "wavelengths toward the generator" scale, and record the value as *C*.

3. The length of transmission line between the load and the stub is $d = B - C$, wavelengths.

An example will illustrate the technique (also see Figure 14–13).

■ ■ ■ ■

EXAMPLE 14–4

1. Determine the position *d* and length ℓ of a short-circuited tuning stub for a match on the transmission line of Examples 14–2 and 14–3.

2. What value of lumped-circuit *L* or *C* could be used? Figure 14–14 illustrates the circuit components.

FIGURE 14–14 Location of quarter-wave transformer for matching purely resistive load.

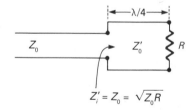

$$Z_i' = Z_0 = \sqrt{Z_0 R}$$

Solution:

1. From the results of Example 14–3, part 2, $y_L = 0.2 + j0.07$ (at 0.01λ).

   Proceed on the constant VSWR circle toward generator to $1 + j1.8$ (at 0.183λ). We need a shunt of $-j1.8$ to produce $y = 1.0$ for a match ($z = 1/y = 1$, $Z = Z_0 z = 50\ \Omega$).

   Mark $-j1.8$ on the Smith chart. We need a circuit component with this admittance placed $0.183\lambda - 0.01\lambda = \mathbf{0.173\lambda}$ from the load. $d = 0.173 \times 212$ mm $= \mathbf{36.7\ mm}$.

   $-j1.8$ is at 0.33λ, which is $\ell = 0.33\lambda - 0.25\lambda = \mathbf{0.08\lambda}$ for the short-circuit ($y = \infty$) stub. 0.08×212 mm $= \mathbf{17\ mm}$ stub.

2. The matching component is $y = -j1.8$. This is an *inductance* of $x = 1/y = 1/-j1.8 = j0.56$ and $X_L = 0.56 \times 50\ \Omega = 28\ \Omega$. $L = 28/2\pi 10^9 = \mathbf{4.5\ nH}$.

Impedance matching with a *quarter-wavelength transformer* is accomplished on the Smith chart by moving from the load on a constant VSWR circle until the (horizontal) real impedance line is intersected. The impedance looking toward the load from this connection is purely resistive, $z = r$ and $R = Z_0 r$.

A quarter-wave transformer transforms any real impedance to any other real impedance by selecting an appropriate characteristic impedance Z_0' for the transformer. To match R to Z_0 of the source impedance, Z_0' must be

$$Z_0' = \sqrt{Z_0 R} \qquad \text{(14–45)}$$

This can be proved by solving Equation 14–28 with $d = \lambda/4$ and $Z_L = R$, but let us do it with the Smith chart as follows (refer to Figure 14–14):

$$z_L' = \frac{R}{Z_0'} \qquad \text{(normalizing } R\text{)} \qquad \text{(14–46)}$$

Moving 0.25λ on the Smith chart inverts the load; therefore,

$$z_i' = \frac{1}{z_L'} = \frac{Z_0'}{R} \qquad \text{(14–47)}$$

Denormalizing, we have

$$Z_i' = Z_0' z_i' = \frac{(Z_0')^2}{R} \qquad \text{(14–48)}$$

If this impedance is to match Z_0 of the source, then $Z_0 = Z_0'^2/R$ and the $\lambda/4$ transformer must have a characteristic impedance given by Equation 14–45.

14–5 ■ S-PARAMETERS

Most students are already familiar with *H*-, *Y*-, and *Z*-parameters for characterizing two- (or more) port networks and devices. However, the most widely used method for characterizing microwave networks and devices is by *S*-parameters. One reason for using *S*-parameters is that their definition is based on driving and terminating the network ports with 50-Ω impedance devices rather than open- or short-circuits (and high VSWRs). Another reason is that microwave circuit analysis and design is based on the reflection (or scattering) of electromagnetic waves. Thus "scattering parameters" are appropriate to work with and, with modern network analyzers, they are easy to measure; see Figure 14–15.

In Figure 14–16, a_1 and a_2 represent *incident voltage waves*; likewise b_1 and b_2 represent *reflected voltage waves* on a two-port network. The four waves are related by Equation 14–49 and written in vector or matrix form in Equation 14–50.

$$b_1 = S_{11}a_1 + S_{12}a_2 \qquad \text{(14–49a)}$$

$$b_2 = S_{21}a_1 + S_{22}a_2 \qquad \text{(14–49b)}$$

$$\begin{bmatrix} b_1 \\ b_2 \end{bmatrix} = [S] \begin{bmatrix} a_1 \\ a_2 \end{bmatrix}$$ **(14–50)**

If the generator and load impedances Z_g and Z_L are matched to Z_0 of the measuring system and connecting transmission lines, then S_{11} and S_{22} are the input and output port reflection coefficients Γ_i and Γ_o, respectively. Also, $|S_{21}|^2$ and $|S_{12}|^2$ are the forward and reverse insertion power gains, respectively, of the two-port network. In terms of power, a_1 and b_2 of the defining matrix are actually $|a_1|^2 =$ available power from the matched generator and $|b_2|^2 =$ power delivered to a matched load.

If a matched load is connected to the output (port 2), there will be no reflection from the load, so that $a_2 = 0$, which, from Equation 14–49a, yields

$$S_{11} = \frac{b_1}{a_1}\bigg|_{a_2=0}$$ **(14–51a)**

FIGURE 14–15   HP 8410S network analyzer system and Smith chart display. (From HP Application Note AN117-1, Microwave Network Analyzer Applications, June 1970.)

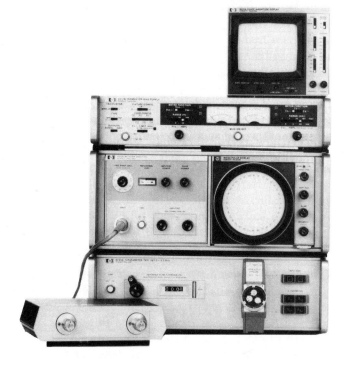

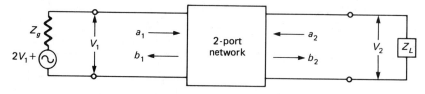

FIGURE 14–16   Two-port network and signals.

Similar measurements yield

$$S_{22} = \frac{b_2}{a_2}\bigg|_{a_1=0} \qquad \text{(14–51b)}$$

$$S_{12} = \frac{b_1}{a_2}\bigg|_{a_1=0} \qquad \text{(14–51c)}$$

$$S_{21} = \frac{b_2}{a_1}\bigg|_{a_2=0} \qquad \text{(14–51d)}$$

Whereas scattering-parameters are, like reflection coefficients, complex quantities having magnitude and phase, a fundamental insight may be gained from the following simple examples.

■ ■ ■ ■

EXAMPLE 14–5

1. Find the S-parameters for the two-port network of Figure 14–17. ($Z_0 = 50\ \Omega$.)
2. Calculate the return loss at the input with $Z_L = Z_0$.
3. Determine the insertion loss for the network when using generator and termination of Z_0.

Solution:

1. S_{11}: Terminate the output in Z_0 and determine Γ at the input; see Figure 14–18a. $25 \parallel 50 = 16.67\ \Omega$ and $Z_1 = 25 + 16.67 = 41.67\ \Omega$. $S_{11} = \Gamma_i = (Z_1 - 50)/(Z_1 + 50) = -0.09$, or $S_{11} = \mathbf{0.09\underline{/180°}}$.

   S_{22}: Terminate the input in Z_0 and find Γ looking back into the output terminals; see Figure 14–18b. $Z_2 = (50 + 25) \parallel 25 = 18.75\ \Omega$. $\Gamma = (18.75 - 50)/(18.75 + 50) = -0.45 = S_{22} = \mathbf{0.45\underline{/180°}}$.

   S_{21}: Drive port 1 with the 50-Ω generator and open-circuit voltage of $2V_1^+$. (The voltage incident on a matched load will be V_1^+.) Measure voltage at port 2 across Z_0; see Figure 14–19a. Then

$$S_{21} = \frac{V_0}{V_1^+} = \frac{V_2}{V^+} \qquad \text{(14–52)}$$

$V_2 = \{(25 \parallel 50)/[(25 \parallel 50) + 75]\}(2V_1^+) = 0.364V_1^+.\ S_{21} = V_2/V_1^+ = \mathbf{0.364.}$

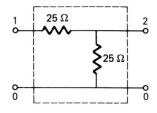

FIGURE 14–17 Two-port network for analysis.

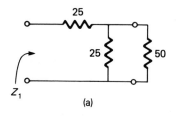

(a)

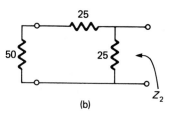

(b)

FIGURE 14–18

FIGURE 14–19

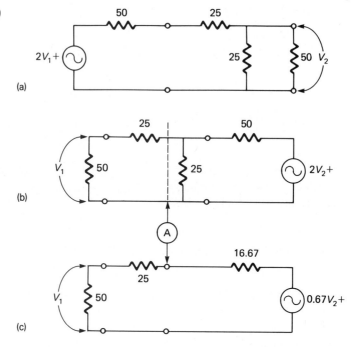

S_{12}: See Figure 14–19b and c. First, thevenize the circuit at A: $R_{TH} = 25 \parallel 50 = 16.67\,\Omega$ and $V_{TH} = [25/(25 + 50)] \times 2V_2^+ = 0.67V_2^+$. From the equivalent circuit (part c), $V_1 = [50/(50 + 25 + 16.67)](0.67V_2^+) = 0.364V_2^+$.

$$S_{12} = \frac{V_1}{V_2^+} \qquad\qquad\qquad \textbf{(14–53)}$$

$$= \mathbf{0.364}$$

2. From part 1, $\Gamma = \rho\underline{/\theta} = 0.09\underline{/180°}$. From Equation 14–39, RL (dB) $= -20 \log(0.09) = \textbf{20 dB}$. The higher this value, the better for efficient coupling into the two-port network.

3. The forward power gain of the network will be $\mid S_{21}\mid^2 = (0.364)^2 = \textbf{0.132.}$ This represents a power loss of $-10 \log 0.132 = \textbf{8.8 dB.}$

■ ■ ■ ■

EXAMPLE 14–6

1. Find the S-parameters for the network of Figure 14–17 if the series 25-Ω resistor is changed to an inductor with an impedance of $+j25\,\Omega$.

2. Confirm the results on Touchstone,* an S-parameter CAD package.

Solution:

1. Following the procedure of Example 14–5,

S_{11}: $Z_1 = j25 + 16.67\,\Omega$

* Touchstone is a registered trademark of EEsof, Westlake Village, Calif.

Therefore, $\quad S_{11} = \Gamma_i = \dfrac{(16.67 + j25) - 50}{(16.67 + j25) + 50} = \dfrac{41.67\underline{/143°}}{71.2\underline{/20.56°}} = \mathbf{0.585\underline{/122.5°}}$

S_{22}: $\quad Z_2 = (50 + j25) \,\|\, 25 = 17.67\underline{/8.16°} = 17.5 + j2.5$

Hence, $\quad S_{22} = \Gamma_2 = \dfrac{Z_2 - 50}{Z_2 + 50} = \dfrac{32.6\underline{/175.6°}}{67.55\underline{/2.12°}} = \mathbf{0.483\underline{/173.5°}}$

S_{21}: $\quad V_2 = \dfrac{16.67(2V_1^+)}{(50 + j25) + 16.67} = 0.468 V_1^+\underline{/-20.56°}$

```
Touchstone (TM) -Ver(1.45-Lot 101) -Ser(17472-3746- 1000)
         RLNET.CKT    01/18/89 - 12:46:06

! Name : RL network/Example  problem (complex)
! Date : 01-18-89
! Purpose : Calculate S-parameters

DIM
  FREQ GHZ
  RES OH
  IND NH
  CAP PF
  LNG MIL
  TIME PS
  COND /OH
  ANG DEG

CKT
  IND 1 2 L=10
  RES 2 0 R=25
  DEF2P 1 2 2PORT

OUT
  2PORT S11
  2PORT S21
  2PORT S12
  2PORT S22

FREQ
  STEP 0.3979
```

(a)

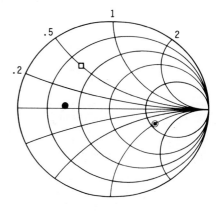

EEsof - 01-18-89 - 12:41:19 - RLNET

□ S11
　2PORT

■ S21
　2PORT

○ S12
　2PORT

● S22
　2PORT

f1: 0.397900
f2: 0.397900

(b)

Touchstone (TM) -Ver(1.45-Lot 101)-Ser(17472-3746- 1000)
         RLNET.CKT   01/18/89 - 12:29:56

FREQ-GHZ	RE(S11) 2PORT	IM(S11) 2PORT	RE(S21) 2PORT	IM(S21) 2PORT	RE(S12) 2PORT	IM(S12) 2PORT	RE(S22) 2PORT	IM(S22) 2PORT
0.397900	-0.315	0.493	0.438	-0.164	0.438	-0.164	-0.479	0.055

(c)

FIGURE 14–20　Results of Touchstone analysis for Example 14–6.

Hence, $$S_{21} = \frac{V_2}{V_1^+} = \mathbf{0.468\underline{/-20.56°}}$$

S_{12}: The thevenized part is identical to Example 14–5. $R_{TH} = 16.67\ \Omega$ and $V_{TH} = 0.67 V_2^+$. Hence,

$$S_{12} = \frac{V_1}{V_2^+} = \frac{50(0.67)}{(50 + j25) + 16.67} = \mathbf{0.468\underline{/-20.56°}}$$

The results may be written in complex form as follows: $S_{11} = -0.3143 + j0.493$, $S_{22} = -0.48 + j0.0547$, $S_{21} = 0.4382 - j0.1643 = S_{12}$.

2. The results of an S-parameter analysis using the software package Touchstone are presented in Figure 14–20. The circuit (CKT) file (Figure 14–20a) lists the circuit components between nodes 1, 2, and 0 (ground) as indicated in Figure 14–17. Notice from the DIM list that L = 10 is in nanohenrys and R = 25 is in ohms. The OUTPUT section lists the parameters to be computed, and the FREQUENCY section lists a single, discrete frequency 0.3979 GHz (at which 10 nH $\sim j25$). Figure 14–20b shows the complex S_{11} through S_{22} quantities on a Smith chart plot. As implied by $f_1 = f_2 = 0.397900$ GHz, these parameters may be swept. Figure 14–20c is a tabular output of the results. Observe that the third-place accuracy differences are due to round-offs in the frequency and calculations.

■ ■ ■ ■

EXAMPLE 14–7

A 50-Ω microwave, integrated-circuit (MIC) amplifier has the following S-parameters: $S_{11} = S_{22} = 0.01\underline{/10°}$, $S_{21} = 10\underline{/180°}$, and $S_{12} = 0.002\underline{/-75°}$. Determine

1. The input VSWR and return loss.
2. Forward and reverse insertion power gains in a 50-Ω system.

Solution:

1.
$$\text{VSWR} = (1 + \rho)/(1 - \rho)$$
$$= \frac{1 + |S_{11}|}{1 - |S_{11}|} \tag{14–54}$$
$$= \frac{1 + 0.01}{1 - 0.01} = \mathbf{1.02}$$

This amplifier is very well matched to Z_0. RL (dB) = $-20 \log 0.01 = \mathbf{40\ dB}$. There is very little reflection at the input (and output) terminals.

2. Forward power gain is $|S_{21}|^2 = 100$, which is $10 \log 100 = \mathbf{20\ dB}$. The reverse power leakage will be $|S_{12}|^2 = (0.002)^2 = 4 \times 10^{-6}$, or -54 dB. The amplifier is virtually unilateral with 34 dB of output-to-input isolation.

14–6 ■ MICROSTRIP AMPLIFIER DESIGN

Microwave devices are characterized by S-parameters at a given frequency. Variations with frequency are usually shown on Smith chart plots in the data sheets for these devices. Transistors are characterized as two-port devices because they are almost always used in common emitter amplifiers so that the emitter lead or ribbons are grounded and are therefore common to the input (base) and the output (collector) ports.

The gain of a single-stage amplifier can be analyzed by considering power gain G_0 of the active device (transistor) connected directly to 50-Ω (Z_0) generator and load; G_1 is the gain (or loss) provided by an output matching network, and G_2 is the gain (or loss) provided by a matching network between the active device output and the load.

If input and output matching networks provide complex conjugate matches between transistor and the generator and load, then the transducer power gain—defined as the ratio of power delivered to the load to source available power—is maximum and for a unilateral transistor ($S_{12} = 0$) is given by

$$G_{max} = G_1(max)G_0G_2(max) \qquad \text{(14–55a)}$$

$$= \frac{1}{1 - |S_{11}|^2} |S_{21}|^2 \frac{1}{1 - |S_{22}|^2} \qquad \text{(14–55b)}$$

If the matching networks at input and output produce reflection coefficients Γ_1 and Γ_2, respectively, and $\Gamma_1 \neq S_{11}^*$ and $\Gamma_2 \neq S_{22}^*$, then less-than-maximum power gain results given by

$$G = \frac{1 - |\Gamma_1|^2}{|1 - S_{11}\Gamma_1|^2} |S_{21}|^2 \frac{1 - |\Gamma_2|^2}{|1 - S_{22}\Gamma_2|^2} \qquad \text{(14–56)}$$

Thus, the gain provided by the ith port matching network is

$$G_i = \frac{1 - |\Gamma_i|^2}{|1 - S_{ii}\Gamma_i|^2} \qquad (i = 1 \text{ or } 2) \qquad \text{(14–57)}$$

The most straightforward matching network design technique is to use a single-stub tuner at the input and at the output.

The input stub is placed at the input connector and, in combination with a length d_i of line to the amplifying device, transforms the source impedance to the complex conjugate of the active device input admittance. Input bias is connected via the ac shorted end of the stub or, if the shortest length of stub is open-circuit, a quarter-wavelength transformer is attached near the active device input lead. The latter technique is illustrated in Figure 14–21.

Also illustrated in Figure 14–21 is the output single-stub tuner with a short-circuited stub and 50-Ω microstrip line (length d_o) used for transforming the output 50-Ω load to the transistor output admittance complex conjugate y_0^*. Collector

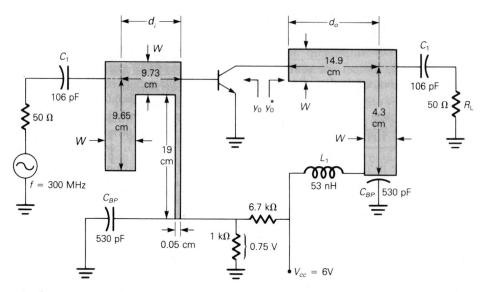

FIGURE 14–21   Microstrip amplifier diagram. Shaded areas represent the copper strip of microstrip transmission lines.

bias is applied at the ac shorted end of the stub, and a choke L_1 improves ac isolation with the power supply. Design of the microstrip transmission lines to complete the amplifier implementation is discussed in the next section.

Microstrip Design

The ease of manufacture of microstrip transmission line has made microstrip line of major importance for microwave circuits, especially at the lower microwave frequencies. The fundamental principle for determining the characteristic impedance of a transmission line is given by

$$Z_0 = \frac{1}{v_p C} \qquad (14\text{–}58)$$

where C is the line capacitance/unit length and v_p, the velocity of propagation, is given by

$$v_p = \frac{c}{\sqrt{\epsilon_{r,\text{eff}}}} \qquad (14\text{–}59)$$

where $\epsilon_{r,\text{eff}}$ is the effective relative dielectric constant, or permittivity, of the dielectric material through which the electromagnetic (EM) field propagates, and $c = 3\text{E}10\text{-}8$ m/s. As illustrated in Figure 14–22, the electric field emanates from the conducting strip of width W, passes through a layer of air ($\epsilon_{r_0} = 1$) and then down through a solid dielectric (ϵ_r) sheet of thickness (or height) h to terminate on the ground-plane conductor.

FIGURE 14–22 Electric and magnetic field lines of propagating signal on microstrip transmission line.

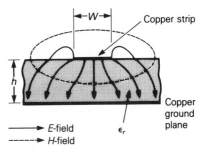

Because of the discontinuity of the EM field at the air-solid dielectric interface, quasi-TEM propagation (and dispersion) results, and a set of largely empirical equations must be solved for calculating the effective dielectric constant as well as the W/h ratio required to achieve a desired characteristic microstrip line impedance Z_0. For best accuracy, two sets of equations are used for both Z_0 and $\epsilon_{r,\text{eff}}$ at frequencies below which dispersion may be neglected. The critical frequency is given by

$$f_{\text{GHz}} = 0.3 \sqrt{\frac{Z_0}{h\sqrt{\epsilon_r - 1}}} \qquad (14\text{–}60)$$

where h is in centimeters.

To compute W/h assuming the strip conductor thickness is negligible ($t/h \leq 0.005$), first estimate W/h using the curves of Figure 14–23, and then use one of the following equations, based on a breakpoint at $W/h = 2$.

FIGURE 14–23 Characteristic impedance versus width-to-height ratio of microstrip line.

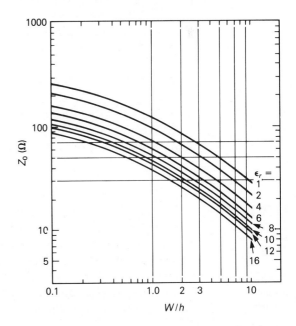

For $W/h \leq 2$:

$$\frac{W}{h} = \frac{8e^A}{e^{2A} - 2} \qquad (14\text{-}61)$$

where

$$A = \frac{Z_0}{60} \sqrt{\frac{\epsilon_r + 1}{2}} + \left[\left(\frac{\epsilon_r - 1}{\epsilon_r + 1} \right) \left(0.23 + \frac{0.11}{\epsilon_r} \right) \right] \qquad (14\text{-}62)$$

Or, for $W/h \geq 2$:

$$\frac{W}{h} = \frac{2}{\pi} \left\{ B - 1 - \ln(2B - 1) + \frac{\epsilon_r - 1}{2\epsilon_r} \left[\ln(B - 1) + 0.39 - \frac{0.61}{\epsilon_r} \right] \right\} \qquad (14\text{-}63)$$

where

$$B = \frac{592.2}{Z_0 \sqrt{\epsilon_r}} \qquad (14\text{-}64)$$

To compute the effective dielectric constant for negligible strip conductor thickness ($t/h < 0.005$), use one of the following equations:

For $W/h \leq 1$:

$$\epsilon_{r,\text{eff}} = \frac{\epsilon_r + 1}{2} + \frac{\epsilon_r - 1}{2} \left[\left(1 + \frac{12h}{W} \right)^{-1/2} + \left(1 - \frac{W}{h} \right)^2 / 25 \right] \qquad (14\text{-}65)$$

Or, for $W/h \geq 1$:

$$\epsilon_{r,\text{eff}} = 0.5 \left[(\epsilon_r + 1) + (\epsilon_r - 1) \left(1 + \frac{12h}{W} \right)^{-1/2} \right] \qquad (14\text{-}66)$$

The value of h is determined by the availability of suitable microwave circuitboard material. Common values of h for RT duroid (Rogers Corp., Chandler, Ariz.) are 31 mils and 62 mils (1.575 mm), and $2 \leq \epsilon_r \leq 2.6$. For most applications, Equations 14–65 and 14–66 are accurate when computing the propagation wavelength from

$$\lambda_g = \frac{c}{f \sqrt{\epsilon_{r,\text{eff}}}} \qquad (14\text{-}67)$$

since quasi-TEM propagation is assumed.

An example microwave amplifier design that includes the microstrip implementation is now given.

■ ■ ■ ■

EXAMPLE 14-8

Microstrip Amplifier Design

1. Design a 50-Ω microstrip amplifier for maximum gain at 1 GHz. Include provisions for coupling and biasing of the BJT transistor that has $S_{11} = 0.5\underline{/108°}$, $S_{12} = 0$, $S_{21} = 3.16\underline{/-88°}$, $S_{22} = 0.7\underline{/-142°}$. The board has $\epsilon_r = 4$ and a dielectric thickness of 1.3 mm. Use the shortest lengths of lossless microstrip line.

   a. Show lengths and widths of all lines, in millimeters.
   b. Sketch the complete amplifier with capacitor, inductor, and bias resistors.

2. a. Compute the gain in decibels for the amplifier.
   b. Confirm the result on the Touchstone CAD package.

Solution:

1. First, calculate the wavelengths in microstrip for 50 Ω and $Z_0 = 120$ Ω. The effective dielectric constant for the 50-Ω lines is computed from Equation 14–66 as

$$\epsilon_{r,\text{eff}} = \frac{\epsilon_r + 1}{2} + \frac{\epsilon_r - 1}{2}\left(1 + 12\,\frac{h}{W}\right)^{-1/2}$$

$$= \frac{4 + 1}{2} + \frac{4 - 1}{2}\,[1 + (12)(0.5)]^{-1/2} = \mathbf{3.07}$$

where $W/h = 2$ from the Z_0 graph estimate and subsequent calculation. Hence,

$$\lambda_g = \frac{c}{\sqrt{\epsilon_{r,\text{eff}}}\,f} = \frac{3 \times 10^{11}\text{ mm}}{\sqrt{3.07}\;10^9} = 171.3\text{ mm}$$

and $W = 2(1.3\text{ mm}) = 2.6$ mm for the 50-Ω matching microstrip. For the quarter-wave bias line, $Z_0 = 120$ Ω, $W/h \sim 0.35$ and, by calculation, $A = 3.3$ and $W/h = 0.292$, which is less than one; therefore,

$$\epsilon_{r,\text{eff}} = \frac{4 + 1}{2} + \frac{4 - 1}{2}\left[\left(1 + 12\,\frac{1}{0.35}\right)^{-1/2} + 0.04(1 - 0.35)\right] = 2.78$$

Consequently, $W = (0.35)1.3 = 0.47$ mm. Finally,

$$\frac{\lambda_g}{4} = \frac{1}{4}\frac{3 \times 10^{11}\text{ mm}}{\sqrt{2.78}\;10^9} = \mathbf{45\text{ mm}}$$

Next, complex-conjugately match the input and output; refer to Figure 14–24.

Input: Plot $S_{11}^* = 0.5\underline{/-108°}$. Move 180° to $y_{11}^* = 0.8 + j1$ (0.15λ); the input network must have this complex admittance. From $y = 1$, we need a *short-circuit* stub to get to $y_1 = 1 - j1.14$ (0.3335λ). The stub is $y_s = j1.14$ (0.365λ), so $\ell_{s_i} = 0.365 - 0.25 = \mathbf{0.115\lambda = 19.7\text{ mm}}$. Then move on TL "toward generator" to y_{11}^* (0.15λ); $d_i = (0.150 + 0.5) - 0.3335 = 0.3165\lambda$, $d_i = \mathbf{54.2\text{ mm}}$.

Output: Plot $S_{22}^* = 0.7\underline{/142°}$. Move 180° to $y_{22}^* = 1.35 - j2.2$ (0.303λ); the output matching network must have this admittance when seen at the collector looking toward the output 50-Ω load. From the output load, $y = 1$, we need an *open-circuited* stub to get to $y_2 = 1 + j1.94$ (0.1865λ). The stub is $y_s = j1.94$ (0.174λ), so $\ell_{so} = 0.174 - 0 = 0.174\lambda = \mathbf{29.8\text{ mm}}$. Now, move on TL "toward generator" to y_{22}^* (0.303λ); $d_o = (0.303 - 0.1865) = 0.1165\lambda$, $d_o = \mathbf{20\text{ mm}}$.

Biasing: The short-circuited input stub allows a direct connection with the (R_1, R_2) resistor divider from V_{cc}. The output open-circuited stub, however, requires the use of a quarter-wave transformer from the Q_1

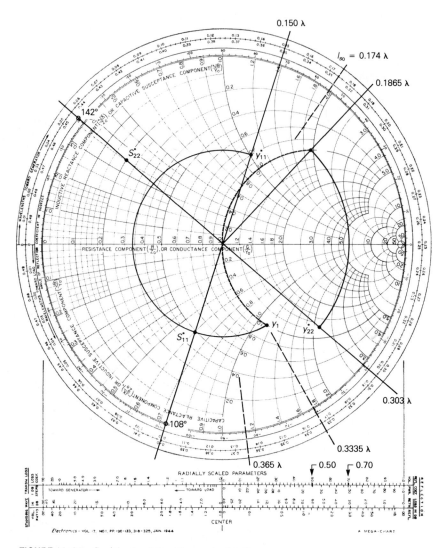

FIGURE 14-24 Smith chart for Example 14-8.

collector to a bypass capacitor at V_{CC}. If we add base and collector coupling capacitors C_c, the final amplifier schematic is illustrated in Figure 14–25, and a photograph of an undergraduate lab rendition using copper tape (Figure 14–26a) and type-N connectors is presented in Figure 14–26. (The input also has a λ/4 stub.)

2. a. $G_{max} = \dfrac{1}{1 - (0.5)^2} \, (3.16)^2 \, \dfrac{1}{1 - (0.7)^2}$

$= 1.25 \text{ dB} + 10 \text{ dB} + 2.92 \text{ dB} = \textbf{14.2 dB}$

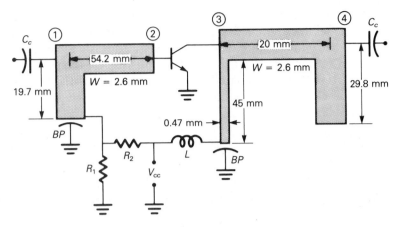

FIGURE 14–25   Microstrip amplifier dimensions for Example 14–8.

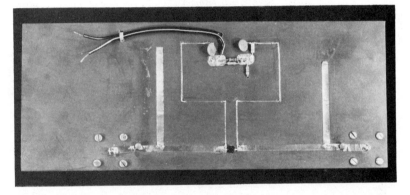

FIGURE 14–26a   A laboratory microstrip amplifier circuit: circuit side.

FIGURE 14–26b   Ground-plane side with type-N connectors.

b. The circuit file for the Touchstone analysis is given in Figure 14–27a. The nodes for the CKT section are shown circled on the circuit diagram of Figure 14–25; note that only the ac transmission line and transistor portion of the amplifier is analyzed. The transistor S-parameter plots defined in the OUT, FREQ, and GRID sections of the circuit file are displayed in Figure 14–27c, where S_{21} is the upper line trace.

FIGURE 14–27a   Touchstone circuit file for Example 14–8 microwave amplifier using S-parameters.

```
Touchstone (TM) -Ver(1.45-Lot 101)-Ser(17472-3746- 1000)
          EET478.CKT    01/03/80 - 02:11:44

!Name: Nick G. Miliotis
!Date  11-17-88
!PURPOSE CONFIRM RESULTS OF PROB 1: SWEEP 0.5GHZ to 1.5GHZ

CKT
     TLIN 1 2 Z=50 E=114 F=1
     TLSC 1 0 Z=50 E=41 F=1
     S2PA 2 3 0 PHY288
     TLIN 3 4 Z=50 E=42 F=1
     TLOC 4 0 Z=50 E=63 F=1
     TLSC 3 0 Z=120 E=90 F=1
     DEF2P 1 4 AMP

OUT
     AMP DB(S21) GR1
     AMP DB(S11) GR1
     AMP DB(S22) GR1

FREQ
     SWEEP 0.5 1.5 0.5

GRID
     RANGE 0.5 1.5 0.5
     GR1 -60 15 5
```

```
Touchstone (TM) -Ver(1.45-Lot 101)-Ser(17472-3746- 1000)
          PHY288.S2p  01/04/80 - 01:16:16

! PHY288.S2P
! VCC = 12V, IC = 5MA
# GHZ S MA R 50
! SCATTERING PARAMETERS:
! F(GHZ)     S11        S21       S12        S22
.500    .65  95    4.5 -70    0 0    .76 -130
1.00    .50 108    3.16 -88   0 0    .70 -142
1.50    .42 120    2.5 -92    0 0    .62 -150
```

FIGURE 14–27b   Transistor S-parameter-versus-frequency file for Touchstone CAD analysis.

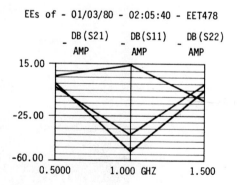

FIGURE 14–27c   Plots of S-parameters (in dB). Gain is top line trace.

14–7 ■ WAVEGUIDES

Electromagnetic energy can be guided as it propagates through space (or other dielectrics) if it is surrounded by a structure that forces the energy to bend or reflect back and forth in a zigzag path down the guide. Figure 14–28 illustrates the most common waveguide configurations. The solid dielectric rod shown is most commonly used for guiding light, which is electromagnetic energy with wavelength near the visible spectrum. This waveguide is described in more detail in Chapter 18, which is entirely devoted to optical fiber technology.

The hollow metal pipes illustrated in Figure 14–28a, b, and c are used at frequencies below that of light and above about 2 GHz. When the term *waveguide* is used, it will be in reference to the hollow metal structure, of which the most common configuration is the rectangular waveguide of Figure 14–28a. The typical metals used in waveguides, copper, aluminum, and brass, must be good electrical conductors to act as a mirror for reflecting the electromagnetic wave back and forth down the guide to its destination.

A good conductor has a very shallow skin depth; therefore, very little current exists in the surface of the interior walls and very little power is dissipated in the waveguide. *Skin depth* δ is a measure of the depth to which an electromagnetic wave can penetrate a conductor. From Equation 1–14 for a sinusoidal wave, $\delta = 1/\sqrt{\sigma\pi\mu_0 f}$ and $\sigma_{cu} = 5.8 \times 10^7$ S/m for copper. At a depth of δ, the electric field has decreased exponentially in strength to 37% of that at the surface. The waveguide wall thickness must be greater than 5δ.

In the discussions to follow, the conductivity of the waveguide wall will be considered high enough to assume zero skin depth and 100% reflection of electromagnetic waves.

Boundary Conditions

The signals that propagate in waveguides are idealized as uniform plane waves with wavefronts perpendicular to the direction of propagation. As indicated previ-

FIGURE 14–28 Waveguides.
Metal: (a) Rectangular,
(b) circular, (c) double-ridged.
(d) Glass optical fiber.

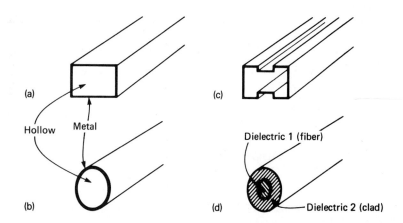

(a)

(c)

Hollow   Metal

Dielectric 1 (fiber)

(b)

(d)   Dielectric 2 (clad)

ously, the electromagnetic waves are guided by reflections of the plane waves from highly conductive metal surfaces. The traveling electric and magnetic field configurations may be very complicated, but they can be predicted mathematically by the equations formulated in 1865 by the Scottish physicist James C. Maxwell.*

The boundary conditions at the reflecting waveguide surfaces are as follows: (1) Any electric field at the (perfect) conductor surface must be perpendicular (normal) to the surface; no parallel (tangential) component of electric field can exist, $E_t = 0$. (2) Any magnetic field at the conductor surface must be parallel to the surface; no component of magnetic field normal to the surface can exist, $H_n = 0$.

Waveguide Modes

The boundary conditions determine the particular electromagnetic field configurations, called *modes*, which can propagate in a waveguide. Unlike the transverse electromagnetic (TEM) mode, which characterizes the field configuration in space and transmission lines, numerous different modes can propagate in waveguides. However, the numerous modes are variations of two basic field configurations, *transverse electric* (TE) modes and *transverse magnetic* (TM) modes.

To get a better idea of a waveguide mode configuration, imagine two identical TEM waves crossing each other in space (Figure 14–29). In Figure 14–29 solid

---

* These are dynamic vector field equations, which are expanded versions of Ampere's law, Coulomb's law, Faraday's law, and a closed magnetic loop law added by Maxwell.

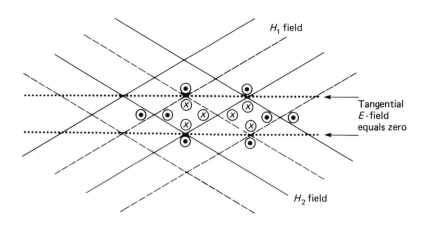

FIGURE 14–29 Two electromagnetic fields passing in space.

lines mark the positive peaks and dashed lines the negative peaks of the sinusoidal magnetic field H. Notice that the electric field, which is everywhere perpendicular to the magnetic field, cancels along the horizontal dotted lines. The electric field vectors reinforce and are at a maximum halfway between the dotted lines. If the conductive walls of a rectangular waveguide are located along these dotted lines (into the page), then the TE mode can propagate.

Figure 14–30a shows a three-dimensional sketch of the fields of Figure 14–29; it is clear that the electric field is transverse to the waveguide axis. Also note that exactly one-half of a wavelength (at the signal frequency) fits between the left and right walls. This is specified as a TE_{10} mode, and the waveguide dimension a is given by

$$a = \frac{\lambda}{2} \qquad (14\text{--}68)$$

The wavelength given by the dimension $2a$ is the longest wavelength that can propagate in a rectangular waveguide. TE_{10}, called the *fundamental mode*, is the lowest-frequency mode that can propagate, and its frequency, called the *cutoff frequency* of the waveguide, is computed from $f = v_p/\lambda$ as

$$f_c = \frac{v_p}{2a} \qquad (14\text{--}69)$$

Figure 14–30b illustrates that if the waveguide is made twice as wide or if the frequency of the propagating signal is doubled, an entirely different field configuration called the TE_{20} mode may propagate. The 2 in TE_{20} indicates that the electric field across the a dimension of the rectangular waveguide has two half-cycle peaks, as shown by field-intensity arrows in Figure 14–30b. The 0 indicates that no half-cycle *magnetic field* peaks occur across the b dimension of the guide.

FIGURE 14–30 Electromagnetic fields in rectangular waveguide. (a) TE_{10} fundamental mode ($a = \lambda/2$). (b) TE_{20} mode ($a = \lambda$).

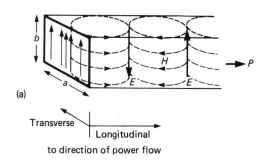

(a)

Transverse
Longitudinal
to direction of power flow

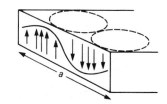

(b)

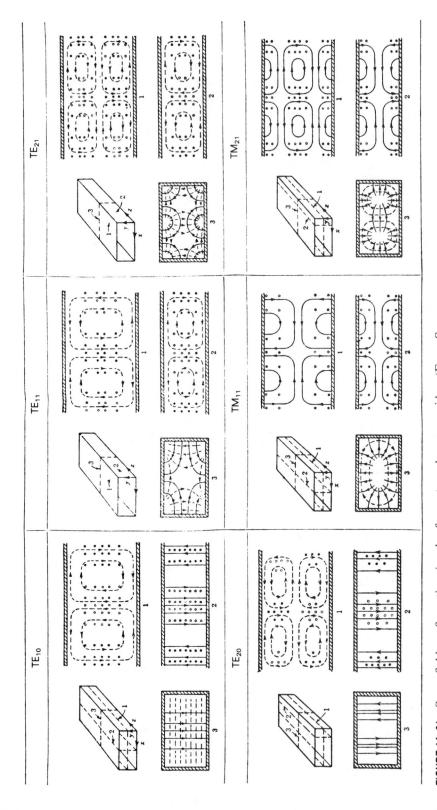

FIGURE 14-31 Some field configurations/modes for rectangular waveguides. (From S. Ramo and J. Whinnery, *Fields and Waves in Modern Radio*, 1956. Reprinted by permission of John Wiley & Sons, Inc.)

As seen in Figure 14–30a, the magnetic field component is entirely longitudinal (along the direction of power flow). Thus the *order* of the mode *mn* for a TE_{mn} or a TM_{mn} mode is equal to the number of maxima across the *a* and *b* dimensions, respectively, of the waveguide cross section.

Figure 14–31 illustrates some of the field configurations (modes) that satisfy Maxwell's equations for a rectangular waveguide. The cutoff frequency for higher-order modes can be calculated for a rectangular waveguide from

$$f_c = \frac{c}{2\sqrt{\mu_r \epsilon_r}} \sqrt{\left(\frac{m}{a}\right)^2 + \left(\frac{n}{b}\right)^2} \qquad \textbf{(14–70)}$$

$v_p = c/\sqrt{\mu_r \epsilon_r}$ is the velocity of propagation of electromagnetic *wavefronts* if the waveguide is filled with a dielectric other than air (or a vacuum). Figure 14–32 shows the relative cutoff frequencies of a few modes for the rectangular waveguide with two different dimension, or *aspect*, ratios, *b/a*.

Coupling to Waveguides

Coupling microwave signals into and between waveguides can be accomplished by appropriately designed slots, windows, and other openings. Coupling to and from a coaxial transmission line is accomplished by probes or loops. Appropriate dimensions are necessary to achieve an impedance match.

Figure 14–33 illustrates a quarter-wave-probe wave launcher in which the coax center conductor extends by a quarter-wavelength ($\lambda/4$) into the waveguide, and the guide end is closed at a quarter-guide-wavelength ($\lambda_g/4$) from the probe (λ_g is defined in the next section.) The position shown will place the *probe* at the peak of the *electric field* for launching a TE_{10} mode. Other TE modes can be excited by

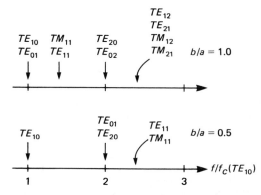

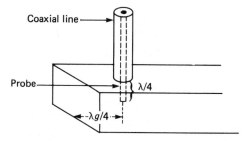

FIGURE 14–32 Frequency of propagation relative to TE_{10}. (From S. Ramo and J. Whinnery, *Fields and Waves in Modern Radio*, 1956. Reprinted by permission of John Wiley & Sons, Inc.)

FIGURE 14–33 Technique for probe coupling to excite TE_{10} propagation in rectangular waveguide from coaxial line.

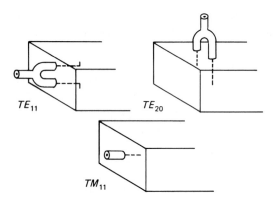

TE_{11}

TE_{20}

TM_{11}

FIGURE 14–34 Coax-to-rectangular waveguide coupling with probes.

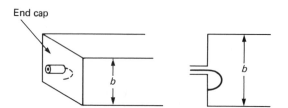

End cap

b

b

FIGURE 14–35 Loop coupling to excite a TE_{10} mode in a rectangular waveguide.

probes as shown in Figure 14–34. A probe in the end cap will launch a TM_{11} or possibly higher-order modes.

The TE_{10} mode can also be launched using a coupling loop in the end cap, as shown in Figure 14–35. The loop produces a longitudinal magnetic wave, and the corresponding electric field will be transverse to the direction of power flow.

Use of Smith Chart with Waveguide Transmissions

The direction of the plane waves that zigzag their way down the waveguide by reflections off the waveguide walls is shown as a "ray" in Figure 14–36. Figure 14–36 shows the top view of a rectangular waveguide with a TE_{10} mode propagating from left to right. From the fields shown in Figure 14–29, we know that the electric field vector of the TE_{10} mode has a single phase reversal ($\lambda/2$) from one side wall of the guide to the other. In time t the wavefront travels a distance ct, whereas the signal energy moves parallel to the guide wall a distance $v_g t$, and the apparent phase moves $v_p t$. This is a relativistic effect: What you observe depends on your observation point. From the point of view of *energy* propagating *along* the *waveguide*, the phase velocity appears to exceed the speed of light, and the guide (or group) wavelength λ_g appears to be longer than the wavelength of the transmitted signal.

FIGURE 14–36 Top view of rectangular waveguide with the TE_{10} mode propagating.

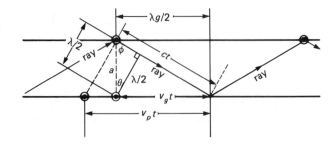

The similar triangles of Figure 14–36 provide a means of solving for the relationships between λ_g, λ, and the waveguide cross-sectional dimension a. From similar triangles,

$$\frac{\lambda_g/2}{ct} = \frac{\lambda/2}{a} \qquad (14\text{--}71)$$

where $ct = \sqrt{a^2 + (\lambda_g/2)^2}$. Cross multiplying and solving for λ_g gives

$$\lambda_g = (\lambda/a)\sqrt{a^2 + (\lambda_g/2)^2} \qquad (14\text{--}72)$$

Squaring both sides and solving for λ_g yields

$$\lambda_g = \frac{\lambda}{\sqrt{1 - (\lambda/2a)^2}} \qquad (14\text{--}73)$$

The waveguide fundamental cutoff wavelength is $\lambda_c = 2a$, so that

$$\lambda_g = \frac{\lambda}{\sqrt{1 - (\lambda/\lambda_c)^2}} \qquad (14\text{--}74)$$

Equation 14–74 gives the (apparent) wavelength along the guide λ_g in terms of the transmission signal wavelength λ and the waveguide cutoff wavelength λ_c.

The velocity at which signal energy propagates along the waveguide axis is

$$v_g = \frac{c\lambda}{\lambda_g} = c\sqrt{1 - (\omega_c/\omega)^2} \qquad (14\text{--}75)$$

and the phase velocity along the waveguide axis is

$$v_p = \frac{c\lambda_g}{\lambda} \qquad (14\text{--}76)$$

The wave impedance of Equation 14–3 for free space is modified for a rectangular waveguide and called the *waveguide characteristic impedance*. For TE mode propagation,

$$Z_g = \frac{377\lambda_g}{\lambda} \qquad (14\text{--}77)$$

and for TM modes,

$$Z_g = \frac{377\lambda}{\lambda_g} \qquad (14\text{--}78)$$

■ ■ ■ ■
EXAMPLE 14–9

A 10-GHz signal is coupled to a 1-m-long copper waveguide that has dimensions of $a = 0.787$ in., $b = 0.394$ in. The output end is loaded in 285 Ω. Determine the following:

1. Skin depth
2. Wavelength and frequency of the lowest-frequency signal that will propagate down the guide

3. The modes, if any, that will propagate

4. The ratio of signal to guide wavelength

5. The VSWR

6. The return loss

7. The impedance seen at the input end

Solution:

1. $\delta = 1/\sqrt{(5.8 \times 10^7)\pi(4\pi \times 10^{-7})(10^{10})} = \mathbf{0.66 \ \mu m}$

2. $\lambda_c = 2a = (2 \times 0.787 \text{ inch}) (2.54 \text{ cm/inch}) = \mathbf{4 \ cm}$
$f = c/\lambda = (300 \text{ Mm/s})/0.04 \text{ m} = \mathbf{7.5 \ GHz}$

3. $b/a = 0.394/0.787 = \frac{1}{2}$. $f/f_c = 10 \text{ GHz}/7.5 \text{ GHz} = 1.33$. By Figure 14–32, the 10-GHz TE_{10} mode will propagate because it is 33% above the guide cutoff frequency. The next possible modes for the guide with $b/a = 0.5$ would be TE_{01} or TE_{20}; however, the signal frequency would have to be increased above $2 \times 7.5 \text{ GHz} = 15 \text{ GHz}$.

4. $\lambda = (300 \text{ Mm/s})/10^{10} \text{ cycles/s} = 3 \text{ cm}$. $\lambda_g = 3 \text{ cm}/\sqrt{1 - (3 \text{ cm}/4 \text{ cm})^2} = 4.54$ cm. $\lambda/\lambda_g = 3/4.54 = \mathbf{0.661.}$

5. $Z_g = (1/0.661)377 \ \Omega = 570 \ \Omega$. For purely real impedances, VSWR $= Z_L/Z_0$ or Z_0/Z_L, whichever is greater than one. VSWR $= 570 \ \Omega/285 \ \Omega = \mathbf{2.0.}$

6. $\rho = (\text{VSWR} - 1)/(\text{VSWR} + 1) = 0.33$. RL(dB) $= -20 \log 0.33 = \mathbf{9.54 \ dB.}$

7. $d = 1 \text{ m}$, which, in wavelengths, is $1 \text{ m}/\lambda_g = 100 \text{ cm}/4.54 \text{ cm} = 22.048\lambda$. Place $z_L = 285/570 = 0.5$ on the Smith chart. Move 0.048λ from $z_L = 0.5$ toward the generator on the constant VSWR circle. This gives $z_i = 0.53 + j0.23$, so $Z_i = 570 \ \Omega(0.53 + j0.23) = \mathbf{302 + j131\Omega.}$

14–8 ■ DIRECTIONAL COUPLERS

A directional coupler is an extremely important device for microwave measurements. It is a passive device that allows for (continuous) signal sampling, signal injection, and the measurement of incident and reflected power to determine VSWR.

Figure 14–37 is a schematic representation of the directional coupler. The input (port 1) is coupled with very low loss to the main output (port 2). The sampled output is port 3, and port 4 must be terminated in 50 Ω. If the coupler is reversed, with port 2 as the input and port 1 as the main output, then port 4 is the sampled output and port 3 must be terminated in 50 Ω.

There are various construction techniques which give the three- (or four-) port device its highly directional characteristics. The *loop*-type couples the electric and

FIGURE 14–37 Directional coupler.

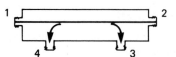

magnetic fields in such a way as to produce cancellation at port 3 for signal inputs at port 2. Signal inputs to port 1, however, will reinforce for outputs at port 3. Another type of construction uses two *coupling holes* between the main line and the sampling (or coupling) port. The holes have λ/4 (90°) spacing, so that a signal traveling in the forward direction is in phase at the coupling holes, but a signal coupling back travels λ/2 (180°) and cancels.

The *coupling factor* expresses the ratio, in decibels of the input port power to the sampled port power; that is,

$$\text{coupling factor (dB)} = 10 \log \frac{P_1}{P_3} \qquad \textbf{(14–79)}$$

The device insertion loss (main line power transfer) is

$$\text{insertion loss (dB)} = -10 \log \frac{P_2}{P_1} \qquad \textbf{(14–80)}$$

The amount of *directivity* is determined as follows (with port 4 terminated in Z_0):

- With input of P_i at port 1 and Z_0 on port 2, measure power at port 3, $P_3(\text{I})$. [$P_i/P_3(\text{I})$ is the coupling factor.]
- Then reverse the coupler and put P_i at port 2 and Z_0 on port 1. Measure the power at port 3, $P_3(\text{II})$. [$P_i/P_3(\text{II})$ is called the reverse coupling factor.]

$$\text{directivity (dB)} = 10 \log \left[\frac{P_3(\text{I})}{P_3(\text{II})} \right] \qquad \textbf{(14–81)}$$

■ ■ ■ ■

EXAMPLE 14–10

A 50-Ω directional coupler is tested with the following results: Port 4 is terminated in 50 Ω. All generators and power meters are 50 Ω. P_1 (in) = 5 mW, P_3 = 0.1 mW; then, with P_2 (in) = 5 mW, P_3 = 50 nW. Determine

1. The coupling factor.
2. The directivity.
3. The amount of power reaching the main output port with 5 mW input to P_1; and the device insertion loss.

Solution:

1. Coupling factor (dB) = 10 log (5 mW/0.1 mW) = **17 dB.** Or, P_1 (in) = 5 mW = +7 dBm and $P_3(\text{I})$ = 0.1 mW = −10 dBm. Then 10 log[P_1 (in)/$P_3(\text{I})$] = +7 dBm − (−10 dBm) = **17 dB.**

2. Directivity (dB) = 10 log[0.1 mW/(50×10^{-6} mW)] = **33 dB.** This is also found from: reverse coupling factor (dB) − coupling factor (dB). Reverse coupling factor (dB) = 10 log[5 mW/(50×10^{-6} mW)] = 50 dB, and 50 dB − 17 dB = **33 dB.**

3. With P_1 (in) = 5 mW, 0.1 mW is lost to the sampling port. Therefore, P_2 = 5 mW − 0.1 mW = **4.9 mW.** Insertion loss (dB) = −10 log 4.9 mW/5 mW = **0.09 dB.**

■ PROBLEMS

1. Determine the wavelength in free space for the following signals (use English and metric units):
   a. 10-kHz audio
   b. 1000-kHz AM carrier
   c. 100-MHz FM carrier

2. A signal with a half-wavelength of 1 cm can propagate by waveguide behavior in a coaxial transmission line filled with Teflon ($\epsilon_r = 2.1$). Determine the signal frequency.

3. An open-wire telephone line has the following parameters: $L = 4$ mH/mi, $C = 8 \times 10^{-3}$ µF/mi, $R = 10$ Ω/mi, and $G = 0.4$ µS/mi. (*Hint:* $\sqrt{A\underline{/\theta}} = \sqrt{A}\ \underline{/\theta/2}$). Determine
   a. Z_0 at 3 kHz.
   b. Frequency at which $R = \omega L$.

4. A lossless, twin-lead transmission line 10 m long has 1-mm lead diameters and lead separation of 1 cm. Assume $\epsilon_r = 1.44$ and $\mu_r = 1$. Determine
   a. Z_0.
   b. Total inductance.
   c. Total capacitance.

5. **a.** Use Equation 14–9 and Table 14–1 to prove that $Z_0 = (120/\sqrt{\epsilon_r})\ln(2s/d)$ for a twin-lead transmission line.
   b. Repeat the problem for $Z_0 = (60/\sqrt{\epsilon_r})\ln(D/d)$ of coax.

6. What must be the ratio of outer-conductor (braid) diameter to center-conductor diameter for a 75-Ω Teflon coaxial transmission line?

7. A short length of alumina ($\epsilon_r = 10$) microstrip has been constructed with a very thin conductor. The conductor width is 3 mm and thickness of dielectric 8 mm. Determine the characteristic impedance.

8. Determine the voltage reflection coefficient for a 75-Ω line with the following terminating impedances:
   a. 25 Ω
   b. 50 Ω
   c. 75 Ω
   d. 112.5 Ω
   e. 225 Ω

9. A lossless 50-Ω line with a 100-Ω terminating load is suddenly connected to a 50-Ω generator with a constant open-circuit voltage of 10 V dc. Determine
   a. Voltage at the line input immediately after the connection to the generator.
   b. Voltage and current at the load after the first reflection.

10. How many reflections will there be from the load end of the line for Problem 9? Make a sketch of v_i (at line input) and v_L versus time.

11. Determine the reflection coefficient for a 50-Ω line with each of the following loads:
    a. 50 Ω   **d.** $50 + j50$ Ω
    b. $-j50$ Ω   **e.** $25 - j100$ Ω
    c. $j100$ Ω   **f.** $Y_L = 0.04 + j0.01$ S

12. Determine γ, α, and β for the telephone line of Problem 3.

13. Determine the attenuation in nepers and in decibels for a 1-km length of RG58A/U coax with a loss of 5 dB/100 ft.

14. Slotted-line ($\epsilon_r = 1$, $Z_0 = 50\ \Omega$) measurements at 2.2 GHz yield the following data: $V_{max} = 50$ V and $V_{min} = 25$ V (occurring 10 mm from the load). Determine
  a. VSWR.
  b. Magnitude of the reflection coefficient.
  c. Maximum value of impedance on the transmission line.
  d. Distance in wavelengths between the load and first maximum impedance point.

15. Find the input impedance, in ohms, for a 50-Ω transmission line, 3.3λ long, if the load is $20 + j75\ \Omega$.

16. Find the input impedance in ohms, VSWR, and reflection coefficient of a transmission line (TL) 4.3λ long when $Z_0 = 100\ \Omega$ and $Z_L = 200 - j150\ \Omega$.

17. The maximum voltage on an *air dielectric* TL is 10 V peak. The closest minimum is 1 m away and measures 4 V pk. Determine each of the following:
  a. VSWR
  b. Frequency of sinusoidal generator driving the line

18. A TL has a VSWR at the load of 1.9. What is the VSWR at the generator if the line has 2 dB of loss? (Use the Smith chart.)

19. Two transmission lines are connected in series. The generator is connected to a 50-Ω line of length $\ell_{02} = 0.159\lambda$. This is connected to a 75-$\Omega = Z_{01}$ line of length $\ell_{01} = 0.116\lambda$, terminated in $Z_L = 25 - j37.5\ \Omega$. Determine the impedance seen by the generator.

20. a. Determine the return loss at the generator in Problem 19.
  b. If 4 mW is incident at the input line from a 50-Ω generator, what will be the reflected power and the power transmitted down the line?

21. A 50-Ω transmission line is connected to an antenna with $Z_{ant} = 100 - j60\ \Omega$. Determine the following for a single-stub tuner:
  a. Shortest distance to the stub (from the antenna load)
  b. Value and type of *reactance* needed to tune (resonate) the line
  c. Length of short-circuited stub

22. A load of $100 - j60\ \Omega$ is to be matched to a 50-Ω line with only a quarter-wave ($\lambda/4$) transformer.
  a. Where should the transformer be located?
  b. Determine Z_0 of the transformer.
  c. If $f = 1$ GHz, determine the length of the transformer in meters if the transmission line dielectric constant is $\epsilon_r = 2.25$.

23. Determine the 50-Ω S-parameters and input return loss for the circuit of Figure 14–38.

FIGURE 14–38

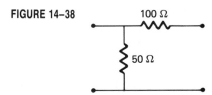

24. Find the S-parameters for a lossless open-circuited 50-Ω transmission line $5\lambda/6$ long (300 electrical degrees).

25. a. Determine the 50-Ω S-parameters for the circuit of Figure 14–39.
   b. Change the 50-Ω resistor to a 50-Ω capacitor and repeat **a**.

FIGURE 14–39

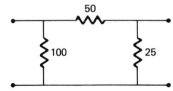

26. A device has $S_{11} = S_{22} = 0.6\underline{/-180°}$, $S_{12} = S_{21} = 0.707\underline{/-90°}$. Find each of the following:
   a. VSWR
   b. Forward insertion gain (loss) when the load is Z_0
   c. Input return loss

27. Design a maximum-gain 963-MHz, 50-Ω microstrip amplifier for a transistor with $S_{11} = 0.42\underline{/80°}$, $S_{22} = 0.60\underline{/-37°}$, $S_{21} = 4\underline{/150°}$, $S_{12} \sim 0$ (unilateral). Use the shortest possible 50-Ω line lengths on an alumina substrate ($\epsilon_{r,\text{eff}} = 9.7$, $h = 0.08$ cm); assume lossless transmission lines.
   a. Show lengths of lines in centimeters.
   b. Compute the line widths in centimeters. (Use $Z_0 = 120\ \Omega$ for any quarter-wave lines.)
   c. Draw the completed amplifier including bias networks.

28. Compute the gain in decibels for the amplifier of Problem 27.

29. A hollow rectangular waveguide with cutoff frequency of 2.66 GHz is used at 3.7 GHz. Determine
   a. The guide wavelength λ_g
   b. The phase velocity
   c. The guide propagation velocity

30. What are the (two) electric and magnetic field conditions at the surface of a perfect conductor that force electromagnetic waves to propagate only by zigzagging (reflecting back and forth) down a metal waveguide?

31. A hollow brass X-band (8.2 − 12.5 GHz) waveguide designed as JAN RG 52/U (or WR 90) has dimensions of 22.860 mm and 10.160 mm.
   a. Determine the wavelength and frequency of the lowest-frequency signal that will propagate.
   b. Determine the cutoff frequency for the TE_{11} mode.
   c. How many modes can propagate if $f = 18$ GHz?

32. The waveguide of Problem 31 is 2 m long and terminated by $Z_L = 377\ \Omega$. For a 10-GHz signal, determine the following:
   a. The ratio of signal-to-waveguide propagation wavelengths
   b. Waveguide impedance (it is not 377 Ω)
   c. VSWR
   d. Impedance at the input end
   e. If 1 kW is incident at the load, how much power is reflected?

33. A 50-Ω three-port directional coupler with 10 dB of forward coupling and essentially infinite directivity has 10 mW at port 1.
   a. Determine the power measured with 50-Ω power meters at ports 3 and 2.
   b. Repeat if forward coupling is 25 dB.

34. An ideal three-port 50-Ω directional coupler is connected to an unknown load Z_L. When connected in the forward direction, $P_3 = 1$ mW; with the coupler reversed, $P_3 = 0.25$ mW. Determine each of the following:
   a. VSWR
   b. RL (dB) and reflection coefficient
   c. Two values of Z_L that could cause the VSWR

14-9 ■ COMPUTER ANALYSIS

Using the computer for communications system and circuit analysis has become a virtual necessity in education and industry. What follows are examples of programs and/or readily available simulation software for solving problems of the type covered in this chapter.

All the programs are interactive in the sense that an introductory sentence or short paragraph describes the purpose of the program and the user is then prompted for input data.

The programs for Chapter 14 are as follows:

- Figure 14–40 is a transmission-line propagation velocity, impedance, and wavelength computation program.

```
10 REM *       TRANSMISSION LINE PROPAGATION VELOCITY,        *
15 REM *            IMPEDANCE AND WAVELENTH                   *
20 REM *      EXAMPLE OF STUDENT COMPUTER ASSIGNMENT          *
30 REM ****        FOR PROFESSOR P. YOUNG, AT ASU          ****
40   CLS
50 PRINT"FOR CALCULATING VELOCITY OF PROPAGATION, CHARACTERISTIC"
60 PRINT"IMPEDANCE AND WAVELENGTH FOR A TRANSMISSION LINE; GIVEN"
70 PRINT"Er, C AND F (for lower microwave frequencies)."
80 CC=3*10^8 : PRINT : PRINT "********************************"
90 INPUT "Enter the dielectric constant, Er =";EFF
100 PRINT
110 PRINT"Enter the capacitance-per-meter of the line,"
115 INPUT" in picofarads. ";PF
120 PRINT:C=PF/1E+12
130 INPUT "Enter the frequency in megahertz, f =";MHZ
135 F=1000000!*MHZ
140 PRINT : PRINT "********************************************"
150 VP=CC/EFF^.5 : PRINT
160 PRINT "Vp=" VP "meters/second"
170 Z=1/(VP*C) : PRINT
180 PRINT"Characteristic impedance of the line is, Zo = "Z "ohms"
190 LAMDA=VP/F : PRINT
200 PRINT "The wavelength, LAMDA =" LAMDA "meters/cycle."
210 PRINT
220 END
```

FIGURE 14-40

- **Transmission-line analysis:** Given R, C, L, and G per mile for a transmission line, the program of Figure 14–41 computes the characteristic impedance, the complex propagation constant parameters alpha and beta, and the propagation velocity, at a given signal frequency.

- **Microstrip characteristic impedance:** The BASIC program of Figure 14–42 calculates Z_0 of microstrip transmission line given the strip width W, dielectric depth h, and relative dielectric constant ϵ_r.

- **PSpice TL transient response:** Figure 14–43 shows the circuit file and results of a transient situation for a mismatched transmission line.

- See Figure 14–27 for Touchstone software files and an analysis plot.

```
10 REM **** TRANSMISSION LINE ANALYSIS PROGRAM ****
20 REM *   EXAMPLE OF STUDENT COMPUTER ASSIGNMENT   *
30 REM ****    FOR PROFESSOR P. YOUNG, AT ASU    ****
40 CLS:PRINT
50 PRINT"THIS PROGRAM CALCULATES Zo, PROPAGATION CONSTANT,ALPHA,
60 PRINT"BETA & Vp, GIVEN R ,F, L, G, AND C (EQUATION 14-10):"
70 PRINT:INPUT "ENTER R, IN OHM/MILE:",R  :PRINT
80 PRINT:INPUT "ENTER FREQUENCY, IN MHZ:",MHZ  :PRINT
90 INPUT "ENTER L, IN uH/MILE:",UH:PRINT
95 INPUT "ENTER G, IN (SIEMENS/MILE):",G
100 PRINT :INPUT "ENTER C, IN PICOFARAD/MILE:",PF
110 PRINT :PRINT
120 F=MHZ*1000000! :L=UH/1000000! :C=PF/1E+12  :PI = 3.14159265#
130 REM CALCULATIONS
140 K=2*PI*F*L :M=(R*R)+(K*K) :M1=SQR(M) :IF R=0 THEN R=1E-16
145 KX=ATN(K/R)
150 K1=KX*180/PI :K2=2*PI*F*C :M2=SQR((G*G)+(K2*K2))
155 IF G=0 THEN G=1E-16
160 K3=(ATN(K2/G))*(180/PI) :W=SQR(M1/M2) :W1=(K1-K3)/2
170 PRINT :PRINT"Zo = ";W;"OHMS AT AN ANGLE OF  ";W1;"DEGREES."
180 Q=SQR(M1*M2):Q1=(K1+K3)/2  :PRINT
190 PRINT"PROPAGATION CONSTANT=";Q;"AT AN ANGLE OF";Q1;"DEG/MILE"
200 Q1=Q1*(PI/180) :S=(COS(Q1))*Q :S1=(SIN(Q1))*Q :PRINT
210 PRINT "ALPHA =";S;"NEPER/MILE.      BETA =";S1;"RADIANS/MILE"
220 VP=(2*PI*F)/S1    :PRINT
230 PRINT "Vp =";VP;" MILES/SECOND" :PRINT
240 PRINT"COMPUTE ANOTHER ?"
250 INPUT "IF YES, ENTER 1. IF NO, ENTER <RETURN>";Y
270 IF Y=1 THEN 40:END
```

FIGURE 14–41

```
10 REM *   MICROSTRIP CHARACTERISTIC IMPEDANCE   *
20 REM * EXAMPLE OF STUDENT COMPUTER ASSIGNMENT *
30 REM *     FOR PROFESSOR P. YOUNG, AT ASU      *
40 CLS
50 PRINT "The following program is used to calculate the"
60 PRINT "characteristic impedance (Zo) of microstrip"
70 PRINT "transmission lines assuming zero or negligible"
75 PRINT "thickness, t, of the strip conductor(i.e., t/h is"
80 PRINT "less than 0.005), given the strip width, W,dielectric"
90 PRINT "depth, h, and relative dielectric constant, Er."
100 PRINT
110 PRINT "**************************************************"
120 PRINT:PRINT "Enter W (width of strip conductor ) in inches"
130 INPUT W
140 PRINT
180 PRINT
190 PRINT"Enter h (thickness of dielectric substrate) in inches"
200 INPUT H
210 PRINT
220 PRINT "Enter Er (relative dielectric constant;  1 in air )"
230 INPUT ER
240 EFF=((ER+1)/2)+(((ER-1)/2)*(((1+((12*H)/W))^-.5)))
250 PRINT "**************************************************"
260 PRINT
270 PRINT "Eff (Effective relative dielectric constant of the "
280 PRINT "substrate) ="EFF
290 RATIO=T/H
300 PRINT
310 PRINT"The ratio of t/h ="RATIO
320 RAT=W/H
330 PRINT
340 PRINT "The ratio of W/h="RAT
350 Z=377*H/(EFF^.5*W*(1+1.74*ER^-.07*(W/H)^-.836))
360 PRINT
370 PRINT "The characteristic impedance is, Zo =" Z "ohms."
380 END
```

FIGURE 14–42

```
    *** 01/31/89 *** Evaluation PSpice (Jan. 1988) *** 02:27:25 ***
    *FILE NAME:  TLTRAN.OUT   (TRANSMISSION LINE TRANSIENT ANALYSIS)
    ****       CIRCUIT DESCRIPTION
    VIN 1 0 PULSE(0 6.66667 0 0 0)
    RIN 1 2 50OHM                        (50 ohm source impedance)
    TL1 2 0 3 0 Z0=150 TD=5NS            ( Zo > Zs << RL )
    *TL1 IS 0.949 METERS LONG
    RL 3 0 50MEG                    (Virtually an open-circuit load)
     .TRAN 1NS 50NS
     .PRINT TRAN V(2) V(3)
     .PLOT TRAN V(2) V(3)
     .PROBE
     .END
    ******************************************************************
      TIME            V(2)          V(3)     [Last of the I/O data file]
      3.000E-09     5.000E+00    0.000E+00
      4.000E-09     5.000E+00    0.000E+00
      5.000E-09     5.000E+00    3.840E-01
      6.000E-09     5.000E+00    9.616E+00
      7.000E-09     5.000E+00    1.000E+01
      8.000E-09     5.000E+00    1.000E+01
      9.000E-09     5.000E+00    1.000E+01
      1.000E-08     5.096E+00    1.000E+01
      1.100E-08     7.404E+00    1.000E+01
      1.200E-08     7.500E+00    1.000E+01
      1.300E-08     7.500E+00    1.000E+01
      1.400E-08     7.500E+00    1.000E+01
      1.500E-08     7.500E+00    9.808E+00
      1.600E-08     7.500E+00    5.192E+00
      1.700E-08     7.500E+00    5.000E+00
      1.800E-08     7.500E+00    5.000E+00
      1.900E-08     7.500E+00    5.000E+00
      2.000E-08     7.452E+00    5.000E+00
      2.100E-08     6.298E+00    5.000E+00
      2.200E-08     6.250E+00    5.000E+00
      2.300E-08     6.250E+00    5.000E+00
      2.400E-08     6.250E+00    5.000E+00
      2.500E-08     6.250E+00    5.096E+00
      2.600E-08     6.250E+00    7.404E+00
      2.700E-08     6.250E+00    7.500E+00
      2.800E-08     6.250E+00    7.500E+00
      2.900E-08     6.250E+00    7.500E+00
      3.000E-08     6.274E+00    7.500E+00
      3.100E-08     6.851E+00    7.500E+00
      3.200E-08     6.875E+00    7.500E+00
      3.300E-08     6.875E+00    7.500E+00
      3.400E-08     6.875E+00    7.500E+00
      3.500E-08     6.875E+00    7.452E+00
      3.600E-08     6.875E+00    6.298E+00
      3.700E-08     6.875E+00    6.250E+00
      3.800E-08     6.875E+00    6.250E+00
      3.900E-08     6.875E+00    6.250E+00
      4.000E-08     6.863E+00    6.250E+00
      4.100E-08     6.575E+00    6.250E+00
      4.200E-08     6.563E+00    6.250E+00
      4.300E-08     6.563E+00    6.250E+00
      4.400E-08     6.563E+00    6.250E+00
      4.500E-08     6.563E+00    6.274E+00
      4.600E-08     6.563E+00    6.851E+00
      4.700E-08     6.563E+00    6.875E+00
```

FIGURE 14–43   (page 1 of 2)

```
*** 01/31/89 *** Evaluation PSpice (Jan. 1988) *** 02:27:25 ***
*FILE NAME:  TLTRAN.OUT   YOUNG   ***   TRANSIENT ANALYSIS ***
LEGEND:
(*): V(2) -  0           2              4              6              8
(+): V(3) - -5           0              5             10             15
 TIME / VOLTS_____V( )_____
0.0E+0/0.0E+0*          +              .              .                  .
1.0E-9/5.0E+0.          +              .       *      .                  .
2.0E-9/5.0E+0.          +              .       *      .                  .
3.0E-9/5.0E+0.          +              .       *      .                  .
4.0E-9/5.0E+0.          +              .       *      .                  .
5.0E-9/5.0E+0.          .+             .       *      .                  .
6.0E-9/5.0E+0.          .              .       *  +.                     .
7.0E-9/5.0E+0.          .              .       *   +                     .
8.0E-9/5.0E+0.          .              .       *   +                     .
9.0E-9/5.0E+0.          .              .       *   +                     .
1.0E-8/5.1E+0.          .              .       *   +                     .
1.1E-8/7.4E+0.          .              .           +              *      .
1.2E-8/7.5E+0.          .              .           +               *     .
1.3E-8/7.5E+0.          .              .           +               *     .
1.4E-8/7.5E+0.          .              .           +               *     .
1.5E-8/7.5E+0.          .              .           +               *     .
1.6E-8/7.5E+0.          .              .+          .               *     .
1.7E-8/7.5E+0.          .              +           .               *     .
1.8E-8/7.5E+0.          .              +           .               *     .
1.9E-8/7.5E+0.          .              +           .               *     .
2.0E-8/7.5E+0.          .              +           .              *      .
2.1E-8/6.25E0.          .              +           .  *                  .
2.2E-8/6.25E0.          .              +           .  *                  .
2.3E-8/6.25E0.          .              +           .  *                  .
2.4E-8/6.25E0.          .              +           .  *                  .
2.5E-8/6.25E0.          .              +           .  *                  .
2.6E-8/6.25E0.          .              .       +   .  *                  .
2.7E-8/6.25E0.          .              .        +  .  *                  .
2.8E-8/6.25E0.          .              .        +  .  *                  .
2.9E-8/6.25E0.          .              .        +  .  *                  .
3.0E-8/6.25E0.          .              .        +  .  *                  .
3.1E-8/6.85E0.          .              .        +  .        *            .
3.2E-8/6.88E0.          .              .        +  .         *           .
3.3E-8/6.88E0.          .              .        +  .         *           .
3.4E-8/6.88E0.          .              .        +  .         *           .
3.5E-8/6.88E0.          .              .       +   .         *           .
3.6E-8/6.88E0.          .              . +         .         *           .
3.7E-8/6.88E0.          .              . +         .         *           .
3.8E-8/6.88E0.          .              . +         .         *           .
3.9E-8/6.88E0.          .              . +         .         *           .
4.0E-8/6.86E0.          .              . +         .        *            .
4.1E-8/6.58E0.          .              . +         .      *              .
4.2E-8/6.56E0.          .              . +         .      *              .
4.3E-8/6.56E0.          .              . +         .     *               .
4.4E-8/6.56E0.          .              . +         .     *               .
4.5E-8/6.56E0.          .              . +         .     *               .
4.6E-8/6.56E0.          .              .   +       .     *               .
4.7E-8/6.56E0.          .              .   +       .     *               .
4.8E-8/6.56E0.          .              .   +       .     *               .
4.9E-8/6.56E0.          .              .   +       .     *               .
5.0E-8/6.57E0.          .              .   +       .     *               .
```

FIGURE 14–43   (page 2 of 2)

15

Antennas and Radiowave Propagation

■ INTRODUCTION: FIELD CONCEPTS

In the study of wave propagation in this and the last chapter, it is assumed that the reader is familiar with elementary stationary electric and magnetic field concepts such as the following.

$$F = \frac{qq_1}{4\pi\epsilon d^2} \qquad \text{(mks units)} \tag{15–1}$$

where F (newtons) is the magnitude of the force on two charges q and q_1 (coulombs), separated by d (meters) in an undisturbed medium of ϵ (farads/meter) permittivity.

$$E = \frac{F}{q} = \frac{q_1}{4\pi\epsilon d^2} \qquad (15\text{--}2)$$

where E (volts/meter) is the magnitude of the *electric field intensity*, or *electric field strength*, a distance d from charge q_1 due to the static electric field produced by q_1.

$$D = \epsilon E = \frac{q_1}{4\pi d^2} \qquad (15\text{--}3)$$

where D (coulombs/meter^2), called *electric flux density*, gives the amount of electric flux due to q_1, which passes through a sphere of *radius d* in an isotropic medium of permittivity ϵ. The surface area of the sphere is $4\pi d^2$. An *isotropic* medium exhibits the same properties in all directions.

$$F = \frac{m_1 m_2}{4\pi\mu d^2} \qquad (15\text{--}4)$$

where F is the magnitude of force between two magnetic poles of pole strength m_1 and m_2 (webers, Wb) separated by d (meters) in an undisturbed medium of μ (henrys/meter) permeability.

$$H = \frac{I}{2\pi d} \qquad (15\text{--}5)$$

where H (amperes/meter) is the magnitude of the *magnetic field strength d* meters from a filament (fine wire) carrying a constant current of moving charge I.

$$B = \mu H \qquad (15\text{--}6a)$$

$$= \frac{\mu I}{2\pi d} \qquad (15\text{--}6b)$$

where B (teslas, T) is the *magnetic flux density* (1 T = 1 Wb/m^2) associated with the field strength H. Equation 15–6b will give the magnetic flux density at a distance d around the current filament in an isotropic medium.

The force on a charged particle q with constant velocity v in a magnetic field is determined from

$$\mathbf{F} = q v \times \mathbf{B} \qquad (15\text{--}7)$$

where the magnitude of the force depends on the angle θ between the velocity vector (particle direction) and the magnetic field direction. Thus, $F = qvB \sin\theta$ and the direction of the force is given by the right-hand curl rule between the directions of $\mathbf{v}$ and $\mathbf{B}$. The boldface indicates a vector quantity with magnitude and direction. ■

15-1 ■ ANTENNA RADIATION

In Chapter 14 we saw that an open-circuited, two-wire transmission line driven by a sinusoid will have a standing wave pattern with a voltage maximum at the open end (Figure 14–11c). At a distance one-quarter wave from the end ($d = \lambda/4$), the voltage standing wave is at a minimum. This transmission line, fortunately, loses very little energy by radiation because the fields of the two wires cancel. However, if the wires are folded out from the $\lambda/4$ points, a *dipole antenna* is formed (see Figure 15–1). Now the electromagnetic fields will be in the same direction and are additive. The electric and magnetic field components far enough away from the immediate vicinity of the antenna cannot return to the antenna and are radiated into space (radiation field). The sinusoidally varying E-field gives rise to a sinusoidally varying H-field, and vice versa. Thus the electromagnetic wave sustains itself as it propagates away from the antenna free of electrical conductors. The ratio of magnitudes E/H is a constant in free space and is called the *wave impedance*. From Equation 14–3, $Z_w = 377\ \Omega$. The velocity of propagation is the speed of light, $c = \sqrt{\mu_0\epsilon_0} = 3 \times 10^8$ m/s.

Polarization

In free space the E- and H-fields are at all times transverse (at right angles) to each other and to the direction of propagation. In fact, the direction of propagation of energy is found with the right-hand rule and the Poynting vector

$$\mathcal{P} = \mathbf{E} \times \mathbf{H} \tag{15–8}$$

the magnitude of which is the power density in watts per square meter. With the right hand open and the fingers pointing in the direction of the E-field vector, curl the fingers toward the direction of the $\mathbf{H}$-vector and the thumb will point in the direction of signal propagation (see Figure 15–2a). The sinusoidal variations of the E- and H-fields propagating in space are seen in Figure 15–2b. With E the peak electric field strength,

$$H = (\beta/\omega\mu_0)E \tag{15–9}$$

is the peak magnetic field strength.

FIGURE 15–1 Half-wave dipole antenna.

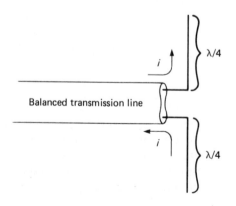

FIGURE 15–2 Transverse electric (E) and magnetic (H) fields. $\mathcal{P}$ is the Poynting vector.
(a) Vector representations and (b) sinusoidal variation of E and H for one cycle of propagation along the z-axis.

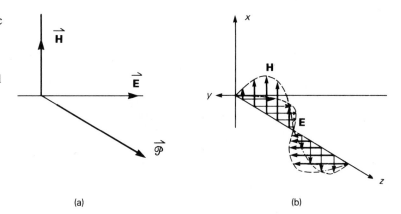

(a)   (b)

$$\beta = \frac{\omega}{v_p} \qquad (15\text{–}10)$$

is the phase propagation constant as in Equation 14–20. Taking the ratio of E/H from Equation 15–9 and substituting Equation 15–10 and $v_p = c = \sqrt{\mu_0 \epsilon_0}$ for free space, it is confirmed that the impedance of space is

$$Z_w = \frac{E}{H} = \sqrt{\frac{\mu_0}{\epsilon_0}} \qquad (15\text{–}11)$$

$$= 376.7 \ \Omega$$

Unless acted on by propagation medium changes in ϵ or μ, the E- and H-field orientations with respect to the earth's surface will remain unchanged. By convention, the orientation of the electric field with respect to the earth's surface is called the *polarization* of the signal wave. If E is vertical, the transmitted signal is called *vertically polarized*; if horizontal, the signal is called *horizontally polarized*. If the **E**-vector is rotating, it is called *circularly polarized*.

For simple antenna structures, the electric field is generally parallel to the plane of the active element; that is, a dipole horizontally oriented with respect to the earth will radiate a horizontally polarized signal. To receive the maximum signal strength, a dipole receiving antenna should be oriented parallel to the electric field. Antennas follow the *reciprocity principle*—the same for transmitting as for receiving. However, this does not mean that a small receiving antenna can necessarily withstand the extreme electric potentials that a transmitting antenna is designed to withstand.

Radiation Pattern and Power Density

The radiation pattern of an antenna is usually compared to that of a theoretical source, a point-source *isotropic radiator*. The point source transmits P_t watts of power that radiates isotropically to form spherical wavefronts. At a distance d from the isotropic radiator, the power has spread out to occupy a spherical surface area of $4\pi d^2$. Hence, the *power density* $\mathcal{P}$ on the wavefront in watts per square

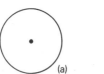

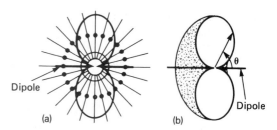

FIGURE 15–3 Radiation pattern for theoretical isotropic radiator.

FIGURE 15–4 Dipole radiation pattern.

meter is

$$\mathscr{P} = \frac{P_t}{4\pi d^2} \qquad (15\text{--}12)$$

The radiation pattern, shown in Figure 15–3a for the theoretical isotropic radiator, is a polar plot of the electric field intensity in the horizontal plane. As seen in three dimensions (Figure 15–3b), the radiation characteristic for the isotropic source is omnidirectional. The two-dimensional plots for various antennas are made by electric field strength measurements far from the antenna. The polar plots of the measurements are usually called the *antenna pattern*. Such a pattern for the horizontal plane is shown in Figure 15–4a, along with the three-dimensional representation in b of an ideal $\lambda/2$ dipole antenna. This pattern is that of the electric field strength E_θ at a great distance from the antenna (far field). The shape is given by

$$E_\theta = \frac{\cos(0.5\pi \cos \theta)}{\sin \theta} \qquad (15\text{--}13)$$

for the half-wavelength dipole. The angle θ is measured from the dipole direction ($\theta = 0$) and has a maximum of unity at $\theta = 90°$.

15–2 ■ RECEIVED POWER AND ELECTRIC FIELD STRENGTH

Equation 15–12 indicates that a transmitted electromagnetic wave spreads out as it travels through space so that a receiving antenna of effective surface area A, parallel to the wavefronts, will only collect P_r watts, given by

$$P_r = \mathscr{P}A_e\eta \qquad (15\text{--}14)$$

where η is the antenna efficiency. A_e is referred to as the effective *aperture* of the antenna.

The electric field strength in volts/meter on the wavefront d meters from a transmitter is determined from $\mathscr{P} = E^2/Z$, which from Equation 15–12 and $Z =$

E/H yields

$$E = \sqrt{\mathscr{P}Z} \tag{15-15a}$$

$$E = \frac{\sqrt{30P_t}}{d} \tag{15-15b}$$

Equations 15–15 indicate that the electric field strength decreases with distance as $1/d$ from the source, whereas, from Equation 15–12, the power density is decreasing as $1/d^2$ with distance.

15-3 ■ DIPOLE (HERTZ) ANTENNA

The simple dipole, Hertz* antenna is shown in Figure 15–5 with the voltage distribution and its variation with time (dashed). The antenna is usually made with two $\lambda/4$ wires or hollow tubes. Additional structural support will, of course, be required. The overall length required to produce a resonant structure is $\lambda/2$ or multiples thereof, where $\lambda = c/f$. The actual $\lambda/2$-dipole antenna is usually cut 5% shorter than the theoretical value to account for (capacitive) fringing effects at the ends. The uncut $\lambda/2$ dipole has a terminal impedance of $Z = 73 + j42.5\ \Omega$.

The current distribution along the dipole is, of course, zero at the ends but rises to a maximum at the center. At the center of the dipole, the voltage is a minimum so the impedance is minimal at the center. For the resonant $\lambda/2$ dipole, the antenna impedance Z_d at the feedpoint (shown in Figure 15–5) is about 72 Ω (resistive), depending somewhat upon the element diameter and I^2R loss component. Most of the 72 Ω (68–70 Ω) represents an equivalent radiation power dissipation and is called *radiation resistance*. It is important to match the transmission line to the antenna impedance Z_d in order to minimize VSWR and power losses on

---

* Named for the German physicist Heinrich Hertz of the late 1800s. He used dipole antennas in proving Maxwell's equations.

FIGURE 15-5   Hertz half-wave dipole antenna.

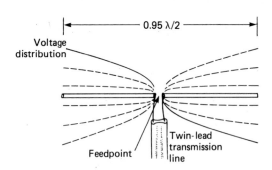

0.95 $\lambda/2$

Voltage distribution

Twin-lead transmission line

Feedpoint

the transmission line. The power input to the antenna is

$$P_i = \frac{Z_d I^2}{2} \tag{15-16}$$

where I is peak sinusoidal current.

Antenna Gain (dB*i*)

The antenna pattern for the ideal half-wave dipole is shown in Figure 15–4. Notice that the radiation field strength is a maximum at the center of the antenna length, and the transmitted power will be concentrated in two loops two-dimensionally and a toroidal or "donut" shape three-dimensionally. The dotted circle on Figure 15–4 is the field intensity for the isotropic radiator; the dipole has minimal transmission or reception off the ends of the antenna, but the maximum radiation exceeds that of the isotropic radiator. The excess is about 2.15 dB and is called the *maximum gain* or *directivity* of the dipole with respect to the theoretical isotropic radiator. Since the gain comparisons for antennas are made with the isotropic radiator pattern as a reference, the gain in decibels is written as dB*i*. Thus, the maximum power gain of an ideal half-wave dipole is 2.15 dB*i*; the power ratio is 1.64.

15–4 ■ DIRECTIVITY, GAIN, AND APERTURE

Antenna structures that concentrate power in one direction at the expense of radiation in other directions have *directional gain* compared to the isotropic radiator. The maximum directive gain is called the directivity of the antenna and is found from theoretical calculations. The *directivity* may be defined as the ratio of the maximum transmitted radiation intensity to the average radiated intensity.

The actual *power gain* of an antenna is determined from measurements and includes internal I^2R and other nondirective power losses. A particular antenna that is inefficient due to such power losses will require more input power to achieve the same power density compared to the theoretical antenna. Hence,

$$G = \eta D \tag{15-17}$$

where η is the efficiency and D is the directivity.

When computing the maximum power intercepted by an antenna, the antenna's effective area, or effective aperture, is used. It can be shown that the *effective aperture* A_e of an antenna is given by

$$A_e = \frac{\lambda^2}{4\pi} D \tag{15-18}$$

As stated at the end of Section 15–3, the directivity of a half-wavelength dipole is 1.64. Hence, the effective aperture (or area) is, for the $\lambda/2$ dipole,

$$A_e = 0.13\lambda^2 \tag{15-19}$$

Also, from Equations 15–17 and 15–18, the gain of an antenna is given by

$$G = \eta \, \frac{4\pi}{\lambda^2} \, A_e \qquad\qquad \textbf{(15–20)}$$

and is therefore directly proportional to the operating frequency squared.

15-5 ■ BEAMWIDTH

Also associated with the directivity of an antenna is the beamwidth. Two lines are drawn from the center of the antenna pattern plot through the points where the radiated power is 3 dB less than at the maximum gain point. The angle formed by the two lines is the beamwidth. Beamwidth and maximum directive gain are both measures of the directivity of an antenna. The beamwidth will be determined for a directional antenna array (see discussion of the Yagi antenna). For the half-wave dipole, beamwidth is 78°. Highly directive, high-gain antennas have much narrower beamwidth.

15-6 ■ THE GROUNDED VERTICAL (MARCONI) ANTENNA

An antenna encountered by most people is the quarter-wave vertical, found on most automobiles. It is named after the Italian engineer Guglielmo Marconi, who invented the grounded quarter-wave antenna and contributed much to transatlantic radio around the turn of the century.

One implementation of this antenna is illustrated in Figure 15–6. A quarter-wavelength in height and mounted on an electrically conductive ground plane, this antenna is one-half of a dipole and is usually fed from a coaxial transmission line with the outer shield connected to the ground plane. The currents and the voltage pattern on this monopole structure are the same as for one-half of a dipole antenna; the voltage value at the input, however, is only one-half of that for the dipole. Consequently, the input impedance of the λ/4 vertical is half that of the

FIGURE 15-6 Marconi quarter-wave vertical antenna.

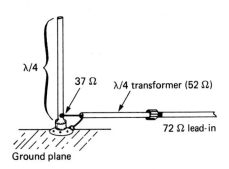

FIGURE 15-7 Idealized radiation pattern for λ/4 Marconi antenna with perfect ground plane.

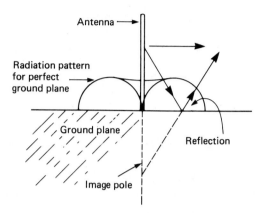

dipole, approximately 37 Ω. Notice also in Figure 15–6 that a 52-Ω quarter-wave transmission-line transformer will match the 37-Ω antenna to a 72-Ω lead-in.

The electrically conductive ground plane is a very important part of the overall antenna and its behavior. Ideally, the ground plane provides mirrorlike reflections of energy radiated from the vertical pole. The vertical element is sometimes called a *whip* due to its motion in the wind.

The radiation pattern, as seen in Figure 15–7 for a Marconi with perfect ground plane, is one-half that of a vertical dipole in free space. The half-loops illustrated form a surface of revolution around the vertical antenna so that it appears from the earth's surface to be omnidirectional, thus the nickname "omni." This pattern depends on the constructive interference of signals from the active pole and the apparent image pole; therefore complete reflection is required.

For many applications the ground plane conductivity is relatively poor, resulting in the radiation pattern (depending on the conductivity) illustrated in Figure 15–8. In addition to the radiation patterns shown in Figures 15–7 and 15–8, a surface wave propagates along the ground plane. The surface wave will account for all the standard broadcast AM daylight propagation. For HF and VHF, however, the surface wave dissipates rapidly.

Counterpoise

The transmission efficiency, radiation pattern, and input impedance can be brought closer to theoretical for vertical antennas on poor conductivity surfaces by burying a large number of heavy copper wires (such as number 8 AWG), called

FIGURE 15-8 Marconi radiation with imperfect ground plane (conductivity not infinite).

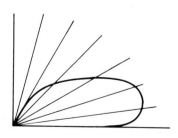

FIGURE 15-9 Example of counterpoise system for Marconi antenna.

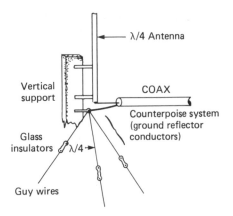

radials, in the ground around the antenna. One hundred or more equally spaced radials, each as long as the antenna (or up to λ/4), and perhaps joined with a ring-shaped ground strap and deep spikes, can provide an excellent *ground* (or *earth*) *mat*.

The same idea is incorporated into the antenna system illustrated in Figure 15-9. This vertical antenna is raised well above the ground, perhaps on a tower, and supported horizontally by metal cable *guy wires*. The guy wires immediately below the antenna are cut to λ/4 in length and connected to the transmission-line ground return or coaxial shield.

Any additional length required for the guy wire supports are electrically isolated by glass insulators. The λ/4 guy wires connected into the transmission-line ground will provide the reflecting surface required for the Marconi antenna. This *ground reflector* technique is called a *counterpoise* system.

The angle of the counterpoise wires can be adjusted to change the input impedance of the antenna to, for example, 50 Ω and also to optimize the angle of the radiation pattern. Quite often guy wires are not necessary, and the counterpoise system is composed of quarter-wavelengths of tubing placed horizontally or on a slight downward-sloping angle from the antenna base.

15-7 ■ TELEVISION ANTENNAS AND ARRAYS

One of the first things noticed when observing TV receiving antennas on house-tops is that the many rods (elements) incorporated have a horizontal orientation with respect to the earth. From the section on polarization earlier in this chapter, the correct conclusion is that these antennas are designed to receive a horizontally polarized signal. By contrast, the vertical antennas on automobiles are designed to receive vertically polarized AM broadcast signals (540–1600 kHz). These automobile verticals can also receive horizontally polarized FM transmissions because signal reflections from automobile and other surfaces will change the polarization angle. Indeed, the "rabbit ears" indoor TV antenna operates best in a V pattern.

15-8 ■ THE FOLDED DIPOLE

The active element in many TV receiving antennas is a folded dipole. It is made from a small-diameter aluminum tube that is folded into a λ/2 flattened loop, as illustrated in Figure 15–10. The folded configuration gives this antenna greater structural integrity and strength than the two-piece dipole, but its electrical properties are a more important consideration.

The folded dipole (FD) voltage pattern is shown with dashed lines in Figure 15–10. Since the current can go around the corners of the FD antenna instead of reflecting from open-circuited ends, there is a full wave (λ) of current on this antenna. Indeed, for the same input power, the folded dipole input current will be one-half that of the half-wave dipole ($I_{FD} = \frac{1}{2}I_D$), with the same input voltages. As a result, the antenna input impedance for the folded dipole is four times that of the dipole; that is,

$$P_{FD} = P_D$$

$$I_{FD}^2 Z_{FD} = I_D^2 Z_D$$

which becomes $(\frac{1}{2}I_D)^2 Z_{FD} = I_D^2 Z_D$, and therefore

$$Z_{FD} = 4Z_D \qquad\qquad\qquad \textbf{(15–21)}$$

Since $Z_D = 72\ \Omega$ ideally, the folded-dipole input impedance is about 288 Ω, so that a 300-Ω balanced, twin-lead transmission line is used for antenna lead-ins.

The most important improvement of a folded dipole over the dipole is in bandwidth. The range of *frequencies* over which the antenna output is within 3 dB of the maximum output is the antenna bandwidth. Bandwidth can also be defined in terms of input VSWR; as an example, VSWR = 2.0 may be used. The bandwidth of resonant antennas such as the dipole ranges from 8% to 16%, depending primarily on the wire (or tube) diameter; the larger the diameter, the lower the reactance and therefore lower Q. The bandwidth of a folded dipole is typically 10% greater than that of the equivalent dipole and exhibits two peaks like a double-tuned circuit. This is an important consideration when wideband signals or a wide range of stations are to be received.

A simple folded dipole antenna can be made for FM reception using 1.5 m of 300-Ω twin-lead. Twist and solder the ends and then cut one wire halfway from one end to form the antenna input terminals.

FIGURE 15–10   Folded dipole.

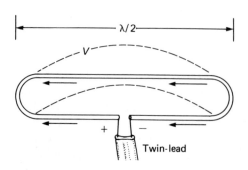

15-9 ■ TURNSTILES, YAGIS, AND OTHER ARRAYS

FM transmitters typically use folded-dipole turnstile antennas in a vertically stacked array or an equivalent vertical traveling-wave transmission-line antenna structure with slots to form a horizontally polarized but vertically expanded transmission pattern. Antenna gains in excess of 10 dB over a single dipole may be realized by this technique. (In commercial practice, antenna gains are usually compared to a reference dipole in free space.)

Dipoles placed at right angles to each other and 90° out of phase form an array called a *turnstile* antenna. Two forms of the turnstile and its radiation pattern are shown in Figure 15–11. The resultant radiation pattern is the sum of the individual dipole patterns, producing a nearly omnidirectional pattern, in the same sense as a Marconi. However, the polarization is horizontal when the crossed dipoles are mounted on a horizontal plane.

Standard broadcast stations use vertical Marconi antennas located in expansive fields whenever possible. For good ground-plane conductivity, the field should be wet (marshlands), or ground mat radials are laid out just under the earth's surface. In order to increase the gain and optimize directivity of the horizontally omnidirectional Marconi antenna, an array of $\lambda/4$ Marconi antennas is constructed. The most common arrangement is the three-Marconi linear array illustrated in Figure 15–12. The phasing of the signals sent to the individual antennas is monitored at the control station in order to optimize the array radiation pattern. Cancellations and reinforcements of the radiated fields will produce an optimized area coverage for programmed material. Typical spacing between vertical radiators is 90 electrical degrees ($\lambda/4$).

Yagi-Uda

The most commonly used antenna for TV reception is the *Yagi-Uda*, named after the two Japanese men responsible for the research and publication work. The

FIGURE 15–11 Turnstile techniques.

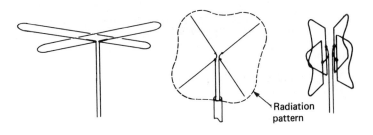

Radiation pattern

FIGURE 15–12 Three-Marconi linear array.

FIGURE 15-13 Basic Yagi with
folded dipole.

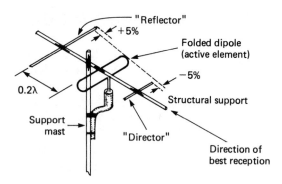

Yagi antenna, as it is called, is a linear array consisting of a dipole and two or more parasitic (nonactive) elements that increase the gain and directivity of the antenna.

The basic three-element Yagi using a folded dipole as the active element is illustrated in Figure 15–13. The folded dipole is one-half wavelength, and a parasitic element is positioned in front of and behind the dipole (with respect to the direction of signal propagation). The reflector is behind the dipole and is a straight aluminum rod cut approximately 5% longer than the dipole. The director is cut approximately 5% shorter than the dipole and placed in front of it. Spacing between the active and both parasitic (no electrical connection) elements is approximately 0.2λ (0.15–0.25λ) for a maximum directivity of about 9 dB.

Signal energy coming from the direction of the director strikes all three conductive elements and excites them into resonance. Reradiation of energy from the parasitic elements *adds constructively* at the dipole for an overall power gain. Additional reflectors produce very little extra gain over the three-element unit; however, more directors increase the gain significantly—2.5 dB with three additional directors, but only 4 dB for eight additional. Five additional directors, for a total of seven elements, provide excellent gain, but practical considerations usually limit Yagis to five total elements.

The antenna signal strength pattern for a Yagi is seen in Figure 15–14. Notice how the parasitic elements have increased the directivity over that of a dipole. Notice also the large *main lobe* (north) and small *minor* lobes behind the main

FIGURE 15-14 Yagi antenna
pattern.

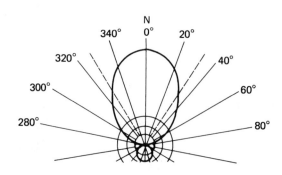

lobe. Actual electric field strength values are not important because the information we want from this pattern, beamwidth and front-to-back ratio, can be derived from relative values.

The *front-to-back ratio* (FBR) is the ratio of maximum signal off the front of the antenna (main lobe peak) to the maximum signal off the back. This ratio is expressed in decibels and is another indicator of the directivity of an antenna. FBRs exceeding 30 dB can be achieved with well-designed and constructed Yagis. An example will help clarify the FBR.

■ ■ ■ ■

EXAMPLE 15–1

Determine the beamwidth and front-to-back ratio (FBR) for the Yagi pattern of Figure 15–14.

Solution:

The radial grid out to the peak of the main lobe can be divided into 10 equal electric field units (volts/meter).

The electric field is down 3 dB at 0.707 of the peak, which is 7.07 radial units. The dashed lines from the pattern-center pass through the main lobe at 7.07, with angles of approximately 33° east of north and 33° west of north. The beamwidth is $\vartheta = 33° + 33° = \mathbf{66°}$.

The electric field is 10 divisions front and about 1.8 divisions back. Therefore, the front-to-back ratio in decibels is

$$\text{FBR (dB)} = 20 \log \frac{10}{1.8} = \mathbf{14.9 \text{ dB}}$$

The VHF TV band covers 54–216 MHz (channels 2–13). Consequently, more than one folded dipole, each cut to a different length, are used to increase the antenna bandwidth. Also, Yagis are often stacked vertically with one wavelength spacing in order to increase the overall array bandwidth, gain, and omnidirectionality for optimum coverage.

Another popular Yagi for television reception is illustrated in Figure 15–15. This antenna has multiple swept-angle (60°) dipoles for good VHF reception and

FIGURE 15–15 Yagi television antenna.

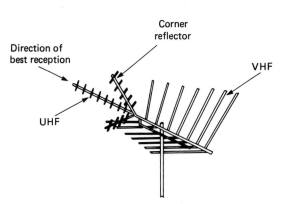

an eight-dipole Yagi with a *corner reflector* for UHF reception. The parasitic elements of a corner-reflector structure perform the same function as the reflector element of the in-line array Yagi. The difference is that the corner-reflector structure places the reflective elements in a vertical plane with respect to the plane of the active Yagi elements. The UHF elements are relatively short and therefore present very little obstruction for VHF reception. The total UHF band covers 470–890 MHz (channels 14–83).

Log-Periodic Dipole Array

Another very important moderate-gain, highly directional, broadband array antenna is the *log-periodic* array illustrated in Figure 15–16. This antenna consists of a horizontal array of half-wavelength dipoles, with the longest, number 1, cut for the lowest frequency (channel 2 for VHF TV) and succeeding dipoles cut shorter and positioned closer to the one before it, starting from number 1. The dipole elements are successively cross-wired, as indicated in Figure 15–16.

The ratio of succeeding dipole lengths and separation distances is a constant given by

$$\tau = \frac{\ell_2}{\ell_1} = \frac{\ell_3}{\ell_2} = \cdots$$

$$= \frac{D_2}{D_1} = \frac{D_3}{D_2} = \cdots \qquad \textbf{(15–22)}$$

where $\ell_1 = \lambda/4$ (lowest frequency desired) and D_1 may be found from the spread angle given by

$$\alpha = \tan^{-1} \frac{\ell_1}{D_1} \qquad \textbf{(15–23)}$$

The characteristic impedance of a two-wire transmission line is given by

$$Z_0 = 276 \log_{10}\left(\frac{2g}{d}\right) \qquad \textbf{(15–24)}$$

FIGURE 15–16   Log-periodic antenna.

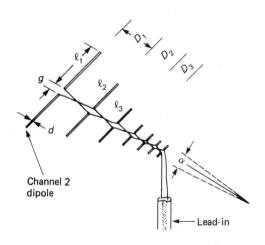

Channel 2
dipole

Lead-in

where d is the diameter of the elements used and g is the gap between the two wires. By defining the antennas for an impedance to match the lead-in Z_0, compute the dipole gap g from Equation 15–24.

Values of the design ratio τ between 0.7 and 0.9 are typical. Impedances Z_0 of 300 Ω or 600 Ω for twin-lead or open-wire transmission-line lead-ins are used. The gap is the same for all dipoles in the log-periodic array. (Student-built lab antennas have return loss measurements reaching −20 dB or better.)

Broad bandwidth is obtained because of the many dipoles resonating at different frequencies. Directivity and gain come from the Yagi principle of using reflectors and directors. For instance, when receiving a station for which dipole number 2 is resonant, the number 1 can act as a reflector while numbers 3 and 4 act as directors.

■ ■ ■ ■

EXAMPLE 15–2

Design a miniature (lab-size) 300-Ω log-periodic antenna for UHF TV (470–890 MHz) using number 8 AWG copper wire (0.128-in. diameter), spread angle 40°, and design ratio of 0.9.

Solution:

$300/276 = 1.09 = \log_{10} 2g/d$. Hence, $g = 10^{1.09} d/2 = 0.78$ in. It is good design practice to slightly overcover the frequency range, so we will make the longest dipole for 470 MHz less 10%, or 423 MHz, and the shortest for 890 MHz plus 10%, or 979 MHz. Then $\ell_1 = \lambda/4 = 6.98$ in., so the longest dipole will be 13.96 in. $\ell_2 = \tau\ell_1 = 0.9 \times 6.98 = 6.28$ in., and so forth, up to $\ell_9 = 3.00$ in. Since $\lambda/4$ (979 MHz) = 3.06 in., we need nine dipoles spaced as follows: Using Equation 15–23 with $\alpha = 40°$ and $\ell_1 = 6.98$ in., $D_1 = 8.3$ in., $D_2 = \tau D_1 = 7.49$ in., and so on, up to $D_8 = 3.98$ in.

15–10 ■ LOOP ANTENNAS

The main advantage of using a simple loop of wire for an antenna is small size and wide bandwidth. As shown in Figure 15–17a and b, the loop can take various shapes. Square or round, the loop shape has little effect on the far-field radiation, but the diameter should be less than about $\lambda/16$. Hence a wide range of frequencies may be received.

FIGURE 15–17 (a), (b) Loop antennas. (c) Ferrite rod with loops.

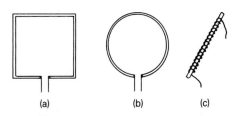

(a)   (b)   (c)

FIGURE 15–18   Radiation pattern of loop antenna.

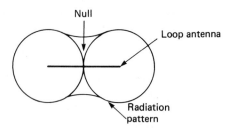

The antenna radiation pattern is illustrated in Figure 15–18 and has the toroidal or donut shape shown (the same as a dipole), with $D = 1.5$ and 90° beamwidth; polarization is linear and in the same plane as the loop. The small size and radiation pattern make the loop antenna ideal for radio direction-finding (RDF) applications. The antenna is turned until a null in the signal strength is observed. The transmitting station is then at right angles to the plane of the loop.

The sensitivity of the antenna can be increased by increasing the number of turns in the loop. AM receivers in the MF broadcast band use a *ferrite rod antenna* with many loops of wire. The ferrite rod helps greatly in concentrating the magnetic flux lines, the same as a transformer core, and the voltage induced in the antenna is directly proportional to the number of turns.

15–11 ■ PARABOLIC DISH ANTENNAS

The most popular antennas up to the microwave range (1 GHz) are the Marconi verticals for omnidirectional medium-frequency applications and the log-periodic and Yagis for broadband broadcast and high-gain, point-to-point radio links. For point-to-point microwave radio transmission, however, horn-fed parabolic-reflector (''dish'') antennas are used. These antennas can provide extremely high gain and directivity and have been used for many years by railroad and utility company radio links, as long distance multiplexed telephone relay links, and are increasingly visible in satellite transmission systems.

The basic component in the microwave dish antenna is the parabolic reflector. This is the same parabolic reflector that is used in flashlights to focus light from a small bulb into a beam of light with high intensity. The ray-collimating property (making the rays parallel to each other) for the parabolic is illustrated in Figure 15–19.

F is the *focal point*—the point at which all incoming parallel rays converge. It is also defined in general for all paths FAA', FBB' such that the distances

$$FA + AA' = FB + BB' = \text{a constant} \qquad \textbf{(15–25)}$$

An equivalent construction is to use the right triangle with hypotenuse E. Then it is recognized that when $\theta = 0$, $E = F$ and otherwise

$$E + E \cos \theta = 2F \qquad \textbf{(15–26a)}$$

FIGURE 15-19 The parabolic reflector.

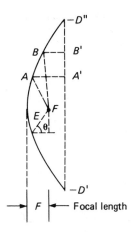

That is, the equation of the parabola is

$$E = \frac{2F}{1 + \cos \theta} \qquad (15\text{--}26b)$$

The mouth diameter $D_r = D'D''$ should be greater than one wavelength at the frequency of operation. The *focal length* is the distance F; the antenna *aperture* is defined by

$$\text{aperture} = \frac{\text{focal length}}{\text{mouth diameter}} = \frac{F}{D_r} \qquad (15\text{--}27)$$

Using Equation 15–18, the directive gain (directivity) for a parabolic reflector (dish) antenna is determined as

$$D = \frac{4\pi}{\lambda^2} (A_e) \qquad (15\text{--}28a)$$

$$= \frac{4\pi}{\lambda^2} \left(\frac{\pi D_r^2}{4} \right) \qquad (15\text{--}28b)$$

$$= \left(\frac{\pi D_r}{\lambda} \right)^2 \qquad (15\text{--}28c)$$

where D_r is the diameter of the dish reflector. The power gain G with respect to isotropic is therefore, from Equation 15–20,

$$G = \eta \left(\frac{\pi D_r}{\lambda} \right)^2 \qquad (15\text{--}29a)$$

$$= \frac{5.4 D_r^2 f^2}{c^2} \qquad (15\text{--}29b)$$

(assuming 55% surface efficiency), which, in decibels, for frequency in megahertz and D_r in feet, is

$$G \text{ (dB}i) = 20 \log f_{\text{MHz}} + 20 \log D_{\text{ft}} - 52.5 \text{ dB} \qquad (15\text{--}30)$$

The half-power beamwidth can be approximated from

$$\vartheta = \frac{70\lambda}{D_r} \text{ degrees} \qquad\qquad (15\text{–}31)$$

■ ■ ■ ■
**EXAMPLE
15–3**

Use Equations 15–30 and 15–31 to estimate the power gain and beamwidth for NASA's 64-m (210-ft) parabolic antenna used in the deep space network for tracking spacecraft at planetary distances. Use 8.4-GHz (*X*-band) frequency.

Solution:

G (dBi) = 20 log 8400 + 20 log 210 − 52.5 dB = 78.5 + 46.4 − 52.5 = **72.4 dBi**

(The actual efficiency of the 64-m antenna is about 42%, so G = 71.3 dBi.)
Beamwidth is $\vartheta = 70c/fD_r = 70(9.836 \times 10^8 \text{ ft/s})/[(8.4 \times 10^9)(210)] = \mathbf{0.039°}$.

Next to the parabolic reflector, the most important feature of a microwave dish antenna is the feed mechanism. The antenna *feed* is the mechanism by which signal energy is radiated toward the parabolic reflector for transmission, or collected at the focal point when receiving. The most common feed mechanisms use a *conical horn* antenna, including the Cassegrain system.

Conical Horn Antennas

The *conical horn* illustrated in Figure 15–20 is a microwave antenna consisting of a truncated cone, and the truncated part of the cone tip is terminated in a circular waveguide that connects the antenna to the transmitter or receiver. When used as an antenna by itself, the cone angle θ, referred to as the *flare angle*, is made at approximtely 50°, and the length of the truncated cone controls the antenna gain.

When used as the primary feed for a parabolic dish, the flare and length can be adjusted for optimum microwave illumination of the reflector. The simplest feed technique is illustrated in Figure 15–21, in which the mouth of the conical horn is located at the focal point of the reflector. In Figure 15–21 the conical horn is supported by four struts, and the waveguide is attached to one of the struts. Often an RF low-noise amplifier (LNA) is located immediately after the feed horn to provide 20–50 dB of gain before the losses incurred in the waveguide or microwave coaxial transmission line. This will improve the receive system noise figure

FIGURE 15–20   Conical horn.

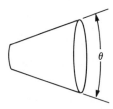

θ

FIGURE 15–21 Microwave antenna with parabolic reflector.

and threshold. Figure 15–22 shows a 5-m-diameter parabolic reflector with a dual-polarization conical feed horn. The gain is 44.1 dBi at 4 GHz with 1.1° beamwidth.

A very important feed mechanism is the *Cassegrain system*. In this system a small conical feedhorn antenna at the center of the parabolic reflector radiates energy toward a convex hyperbolic-shaped reflector, called the *secondary reflector*. One of the foci of the hyperboloid reflector coincides with the focus of the microwave dish so that the dish is illuminated evenly, thereby resulting in a well-focused beam. The Gregorian-fed system is similar except that an elliptical reflector is used instead of the hyperboloid.

For some applications the Cassegrain-fed hyperboloid can become relatively large. In order to reduce the shadow in the primary reflector radiation created by this structure, the feed horn can be extended and the secondary reflector reduced

FIGURE 15–22 Cassegrain feed system.

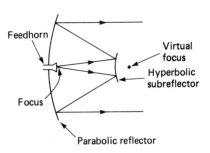

Feedhorn

Virtual focus

Hyperbolic subreflector

Focus

Parabolic reflector

FIGURE 15–23 Cassegrain-fed parabolic antenna.

in diameter. This is shown in the 5-m (16-ft), 44.5-dB gain at 4-GHz Cassegrain-feed antenna of Figure 15–23. The beamwidth is 1.26°, and the first side lobes are down by more than 45 dB.

All feed system structures create some obstruction to the transmitted or received beam, resulting in side lobes in addition to the main beam. Excitation of side lobes necessarily results in power losses and also possible interference to (and from) other systems near the antenna. Assuming that the antenna designer has done the best possible job, it is incumbent upon the user to be aware of this potential problem and minimize it by appropriate site selection and layout.

A variation on the microwave dish antenna is the *hog-horn* antenna illustrated in Figure 15–24. This antenna consists of a horn antenna connected to a section of a parabolic reflector. This technique eliminates some of the extraneous side lobes due to obstructions such as the Cassegrain reflector, struts, waveguide, and the horn antenna itself; the result will be lower noise pickup. Another important feature is that the point of energy concentration (the focus) does not move as the antenna is rotated about the vertical axis.

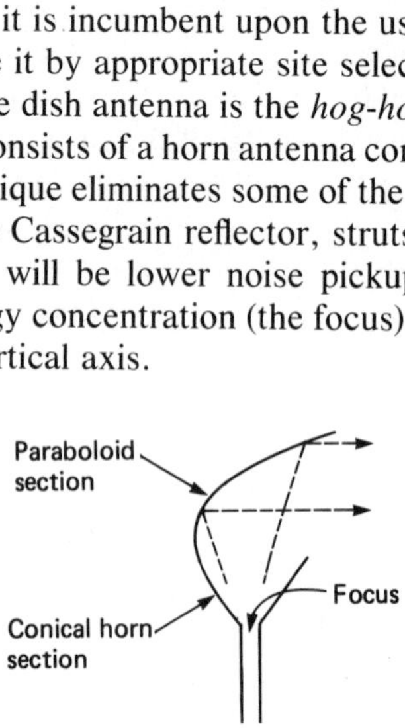

FIGURE 15–24 Hog-horn microwave antenna.

Paraboloid section

Conical horn section

Focus

Space and time do not allow for a comprehensive coverage of antenna design theory. The interested reader can find such coverage in *Antenna Theory and Design* by Warren L. Stutzman and Gary A. Thiele.

15-12 ■ WAVE PROPAGATION THROUGH SPACE

RF signal transmission from one antenna to another can be divided into two major categories, *line-of-sight* and *HF radio*.* Line-of-sight (LOS) propagation is by *direct* electromagnetic wave expansion from the source antenna. Waves produced by dropping an object in a pond provide an analogy. Propagation in the HF region of the spectrum (3–30 MHz) consists of some combination of a *groundwave* component and a *skywave* component, where the groundwave is made up of *direct* and *surface* waves.

Microwave radio links use point-to-point, line-of-sight transmission. These TEM waves are nearly planar and propagate in straight lines unless they encounter changes in the dielectric constant or the conductivity of the medium. Reflections due to conductivity changes in the propagation medium were considered in Chapter 14 for waveguides. A change in the dielectric constant of the transmission medium can result in reflection and refraction. Direction changes can also result from diffraction.

Refraction

The amount of refraction of an electromagnetic wave due to changes in dielectric constant can be predicted by the use of *Snell's law*, expressed by

$$\frac{\sin \theta_1}{\sin \theta_2} = \frac{\sqrt{\epsilon_2}}{\sqrt{\epsilon_1}} \qquad \textbf{(15-32)}$$

Snell's law can be derived from the geometry of Figure 15–25. Wavefront AA' is completely inside of medium 1 and has just reached the interface at point A. In T seconds, point A' travels a distance $v_1 T$ to point B'; in the same amount of time, point A travels a distance $v_2 T$ to point B. BB' is wavefront AA', T seconds later and is completely inside medium 2. The distance AB' is the common boundary during time T and forms the hypotenuse for two right triangles, $AA'B'$ and ABB'. Since $\sin \theta_1 = v_1 T / AB'$ and $\sin \theta_2 = v_2 T / AB'$, the ratio is

$$\frac{\sin \theta_1}{\sin \theta_2} = \frac{v_1}{v_2} \qquad \textbf{(15-33)}$$

Equation 15–32 follows, since $v = c/\sqrt{\epsilon}$ for each medium. Also,

$$n_1 \sin \theta_1 = n_2 \sin \theta_2 \qquad \textbf{(15-34)}$$

---

* *Microwave tropospheric scatter* techniques in which line-of-sight transmissions are scattered by refraction and reflection from the troposphere (sea level to about 35,000 ft or 11 km) fit somewhere between LOS and HF.

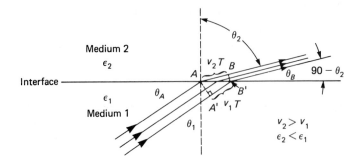

FIGURE 15–25 Refraction at interface of two dielectrics of differing permittivity.

where *n* is the *index of refraction* of the medium. Index of refraction is the ratio of free-space velocity to the medium propagation velocity:

$$n \equiv \frac{c}{v_p} \tag{15–35a}$$

$$n = \sqrt{\epsilon_r} \tag{15–35b}$$

Diffraction

Diffraction is the phenomenon that accounts for the fact that some energy can be received by antennas that are just inside the shadow created by an obstacle. This phenomenon is explained by *Huygen's principle*, which states that each point on a given wavefront can be considered to be a point radiation source (isotropic radiator). Thus, when the wavefront strikes an opaque obstacle, points of the wave near the edge of the obstacle radiate some energy into the shadow region.

HF Radio Waves

HF radio and standard AM broadcast antennas are very large and necessarily built close to the earth; for example, a 1000-kHz, λ/4 Marconi antenna for AM broadcast is a 72- to 75-m (236- to 246-ft) tower. An idealized radiation pattern for this antenna over a perfectly conducting surface would approximate the pattern of Figure 15–7. Thus energy would be radiated into the sky and along the ground; that is, the radiation pattern consists of a skywave and a groundwave. The *groundwave* consists of a vector sum of direct radiation from the antenna plus reflections from the earth. For the nonidealized, real case in which the closeness of the earth to the antenna precludes the reflection of plane waves (they are still spherical), a *surface wave* is produced that contributes to the groundwave. As seen in Figure 15–8 for the case of poor ground conductivity, the surface wave and the groundwave are attenuated, and most of the signal takes off into the sky. The *takeoff angle* (TOA) is defined as the angle above the horizon at which the transmitting antenna has maximum radiation. For vertically polarized antennas, the TOA depends primarily on ground conductivity; for horizontally polarized antennas, frequency and antenna heights *h* above ground are important. The relationship is approximately

$$\text{TOA} = \sin^{-1}\left(\frac{\lambda}{4h}\right) \qquad (15\text{--}36)$$

Also, as the signal frequency increases, the surface waves and groundwaves weaken.

For these reasons, transmissions at HF and above use elevated antennas and rely on skywave (also called *ionospheric*) propagation.

Long-Distance HF Radio Transmission

Highly reliable (90% or better) long-distance communications links can be operated at HF. These communication links take advantage of the fact that HF signal waves can be reliably reflected by the ionosphere and returned to the earth at great distances from the transmitter. By using multiple reflections (*multihop operation*) between the earth and ionosphere, reliable HF radio links in excess of 6000 km (3730 mi) are being operated.

The maximum usable frequency (MUF) for such a transmission may be determined by solving Snell's law with $\theta_2 = 90°$ (see Figure 15–25). As seen in the figure, 90° is the *critical angle* because, for smaller θ_2 and θ_1, the signal is not reflected back to earth and escapes into space unless other layers exist to provide reflection.

Thus the maximum usable frequency can be found from

$$f_{\text{MUF}} = \frac{f_v}{\cos \theta_1} \qquad (15\text{--}37\text{a})$$

or
$$f_{\text{MUF}} = f_v \sec \theta_1 \qquad (15\text{--}37\text{b})$$

where f_v is the maximum frequency for which reflection of a vertically transmitted signal will occur. Equation 15–37b is the *secant* equation in which θ_1 is the angle between the radiated ray and the normal to the surface of the earth.

The Ionosphere

Various layers of gases in the earth's atmosphere become ionized due to bombardment by particle radiation from cosmic rays, meteor activity, and especially sunspot activity. The four major layers illustrated in Figure 15–26 are collectively called the ionosphere.

Layer D, 50–90 km above the earth, exists only during the daylight hours. *Layer E*, 90–140 km above the earth, is also a daylight phenomenon. This layer depends on the sun's ultraviolet radiation and also exhibits sporadic reflectivity irregularities. The f_v is approximately 4 MHz.

The F_1 layer exists 140–250 km above the earth, has f_v of about 5 MHz and tends to rise at night, merging with the F_2 layer. The F_2 layer is present day and night but at varying altitudes. During the day it resides at about 250–300 km, and at night it merges with F_1 and covers 150–350 km above the earth's surface.

FIGURE 15-26 The four major layers of the ionosphere.

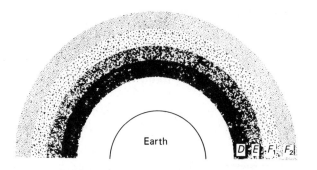

Operating reliable communication links that take advantage of refractions and reflections between the ionosphere and earth's surface takes a good deal of planning. The operating frequency must be changed a few times over a 24 hour period to coincide with ionospheric changes. The optimum operating frequencies are predicted and published monthly by the U.S. National Bureau of Standards. Also, the TOA and various earth and human-made obstacles (including RF interference) must be included in the planning.

Other considerations include *skip* or *quiet zones, multipath, fades,* and normal fixed attenuations over a given path, called *space loss* or *space attenuation.*

The Skip Zone

Figure 15–27 illustrates the skip zone; two hops of a *multihop transmission* are shown. The transmitting antenna is located at *A* and radiates a groundwave, a direct wave, and a skywave. Dissipation of the groundwave and loss of direct radiation leave an electromagnetic shadow between points *C* and *E*, where the skywave returns to earth after ionospheric scattering. The region between points *C* and *E* is a zone of no reception, or *skip zone.*

Of course, an antenna beyond the groundwave and over the horizon at point *D* may still be able to pull in the signal if it is tall enough. The required height is found from the approximation

$$h_r = \tfrac{1}{2}(d - \sqrt{2h_t})^2 \qquad\qquad (15\text{--}38)$$

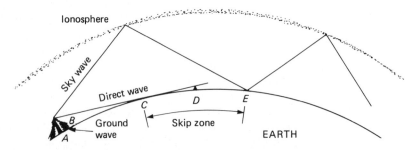

FIGURE 15-27 Multihop long-distance HF transmission.

FIGURE 15–28 Multipath reception.

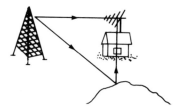

where h_r and h_t for the receiving and transmitting antennas are in feet, and d is the distance in miles between over-the-horizon antennas.

Multipath Distortion

Objects with conducting surface dimensions on the order of one-half wavelength or more will reflect radio signals. This phenomenon was used to advantage for parasitic array antennas. However, when the reflections are from buildings and other electrically conducting objects within a few miles of the receiving antenna, distortion may result.

A well-known example of multipath distortion in television reception results in "ghosts." As illustrated in Figure 15–28, a TV transmission is received via two paths. The direct signal is received first and is relatively strong; the same but weaker signal reflected from an object such as a building or mountain arrives later at the receiving antenna. The later signal will be recorded on the TV screen as a ghost. The source of such multipath distortion problems can usually be located by a quick calculation and a search. The calculation is demonstrated in Example 15–4.

Multipath problems in TV systems have prompted recommendations for changing TV transmission to *circular polarization*. The reason is that the phase reversal in a reflected horizontally polarized wave (as in present TV transmission) is received by the TV antenna just as well as the unreflected signal; however, a signal with right-circular polarization will have left-circular polarization when reflected, and a receiving antenna built for right-circular polarization will not pick it up. An antenna with an active element wound in a *helix* (cockscrew shape) can be wound with a right spiral or a left spiral. If everyone could then agree to buy a new *helix antenna*, we could say good-bye to TV ghosts.

■ ■ ■ ■

EXAMPLE 15–4

Determine the difference in path lengths between the direct and multipath signal that produces a TV ghost displaced by 2 cm from the direct image on a 19-in. screen.

Solution:

A 19-in. diagonal screen with 4 : 3 aspect ratio (standard width-to-height dimensioning) has a width of $19 \times 4/5 = 15.2$ in. (38.6 cm). Horizontal sweep time is $1/15.75$ kHz $= 63.5$ μs, of which about 53 μs is visible. Thus, the 2-cm delay amounts to $(2/38.6)53$ μs $= 2.75$ μs difference in signal arrival times. The distance $d = v_p t = 3 \times 10^8$ m/s $\times 2.75$ μs $= $ **825 m** (about a half mile).

15-13 ■ POWER LOSS IN SPACE

The natural loss of signal power between a transmitter and receiver has three major components: power density decrease with distance, called *path-loss* or *free-space attenuation*, *absorption* due to molecules in the earth's atmosphere, and signal *fading* phenomena that are terrain and weather dependent.

Free-Space Attenuation and Atmospheric Absorption

Equation 15–12 shows that if the power radiated from the transmitter is P_t and the distance to the receiving antenna is d, then the power density at the receiving antenna is

$$\mathscr{P}_r = \frac{P_t}{4\pi d^2} \qquad (15\text{–}39)$$

The power collected by the receiving antenna is

$$P_{A_r} = \mathscr{P}_r A_{e_r} \qquad (15\text{–}40)$$

where A_{e_r} is the receiving antenna effective area. As seen in Equation 15–20, the effective area is proportional to the antenna gain (including efficiency) and $\lambda^2/4\pi$. Leaving the antenna gain (directivity and efficiency) for computing the power reaching the actual receiver, and recalling that $\lambda = c/f$, then $A_e/G = c^2/(4\pi f^2)$. Hence,

$$P_{A_r} = \frac{P_t}{4\pi d^2}\left(\frac{c^2}{4\pi f^2}\right) = \left(\frac{c}{4\pi f d}\right)^2 P_t \qquad (15\text{–}41)$$

The power lost between that transmitted from the transmitting antenna and the amount *collected* by the receiving antenna is called the *space attenuation*.

The gain from transmitter to receiver is written as

$$\frac{P_{A_r}}{P_t} = \left(\frac{c}{4\pi}\right)^2 \frac{1}{d^2}\frac{1}{f^2} \qquad (15\text{–}42)$$

or, in decibels using d in kilometers and f in megahertz,

$$G_s\ (\text{dB}) = 20\log\left(\frac{c}{4\pi 10^3 10^6}\right) + 20\log\left(\frac{1}{d_{\text{km}}}\right) + 20\log\left(\frac{1}{f_{\text{MHz}}}\right) \qquad (15\text{–}43a)$$

$$= -32.44 - 20\log(d_{\text{km}}) - 20\log(f_{\text{MHz}}) \qquad (15\text{–}43b)$$

The space loss L_s is the inverse of Equation 15–42, or the negative of Equation 15–43; that is, using Equation 15–43b, the space loss is computed with

$$L_s\ (\text{dB}) = 32.4\ \text{dB} + 20\log d_{\text{km}} + 20\log f_{\text{MHz}} \qquad (15\text{–}44)$$

Converting kilometers to miles will change the value of the proportionality constant by $-20\log(0.62\ \text{mi/km}) = 4.15\ \text{dB}$, to $36.6 \sim 37\ \text{dB}$; therefore, the space loss in decibels is approximately

$$L_s \text{ (dB)} = 37 \text{ dB} + 20 \log f_{\text{MHz}} + 20 \log d_{\text{mi}} \qquad \textbf{(15–45)}$$

The earth's atmosphere accounts for power loss due to *absorption* by electrons, uncondensed water vapor, and molecules of various gases, most noticeably oxygen. Absorption has a sharp peak around 60 GHz for oxygen and around 21 GHz for water vapors. Free electrons in the atmosphere cause increasing losses for lower frequencies. Graphs such as Figure 15–29 allow the system designer to choose transmission frequencies in the "windows" between high-absorption regions.

Fading and Fade Margin

All radio links in general and terrestrial (earth) links in particular require the inclusion of some *gain margin* (excess gain) in the system gain budget to account for path losses that vary over time. These gain variations, called *fades*, are caused by rainfall and other short-term as well as long-term propagation disturbances.

For satellite systems, fading is due principally to heavy rainfall, and in some regions 6 dB of excess system gain is included as a *fade margin* for systems below 10 GHz. In the case of terrestrial links, it has been found that certain types of terrain and climatic conditions such as dry, windy, mountainous areas are less

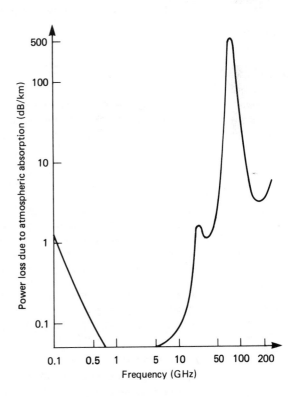

FIGURE 15–29 Atmospheric power loss.

TABLE 15-1 Fade Margin for Various Reliability Objectives for Terrestrial Links. (Interpolate for Other Values. 6 dB Maximum Will Cover Most Satellite Links Under 10 GHz.)

Single-Hop Propagation Reliability (%)	Fade Margin (dB)
90	8
99	18
99.9	28
99.99	38
99.999	48

subject to deep fades than are hot, still, or humid, flat terrain conditions. Since the primary mechanism in fading seems to be multipath, the preceding suggests that the route to choose is one that generally mixes up or randomizes the physiographic conditions on a long-term basis.

For terrestrial links, the actual amount of fade margin to include in the system gain calculation for a *propagation reliability* specified by the user can be determined by a number of different methods without making actual field measurements over a long period of time. The simplest method is to assume that the random fading over time follows a *Rayleigh* type of distribution. Then Table 15–1 may be used for determining an appropriate fade margin.

■ PROBLEMS

1. The force between two electrically charged particles in empty space will double if the distance between the particles is _____ .

2. The force between two ideal magnetic poles is inversely proportional to distance between them; true or false? Why?

3. Determine the magnetic field strength 2 m from a very long current filament carrying 100 mA dc.

4. Under ideal conditions a radio signal is bounced off our moon. If the round trip time from the earth antenna is measured at 17 s, what would be the distance between the earth and moon?

5. A dipole antenna is used to receive a horizontally polarized transmission. Should the antenna be mounted vertically or horizontally for best reception?

6. Determine the exact length of a typical dipole antenna to receive a 108-MHz FM transmission.

7. **a.** If the distance between a transmitter and receiver is doubled, by how much will the electric field strength change at the receiver?
   b. What is the effect on received power?

8. If a receiver could use an ideal isotropic antenna, it would receive −40 dBm. How much could be received if an ideal half-wave dipole were used?

9. Sketch radiation patterns for Hertz and Marconi antennas.

10. Write a computer program to compute points for, or plot, the radiation pattern of Equation 15–13.

11. The input current to a well-designed Marconi antenna is 10 A rms.
    a. Determine the input power.
    b. What is the theoretical current at the upper tip of the antenna?

12. Why is a counterpoise system often used with a Marconi antenna?

13. As an FM antenna, what is the main advantage of a folded-dipole over an ordinary dipole?

14. Sketch a turnstile antenna and its radiation pattern.

15. An antenna pattern is shown in Figure 15–30. Determine
    a. Beamwidth.
    b. Front-to-back ratio (in dB).
    c. Is this pattern most likely for a dipole, omni, or Yagi?

FIGURE 15–30

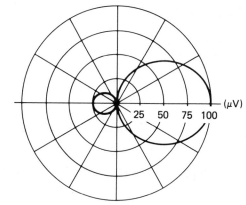

16. Determine the lengths for the reflector and director elements used in a 60-MHz Yagi television antenna if the active element is exactly a half-wavelength long.

17. **a.** Why would a log-periodic antenna be a much better choice for TV reception than a six-element Yagi?
    b. Where would the Yagi have the advantage?

18. **a.** Determine the diameter required for a parabolic reflector if the directive power gain of a 2-GHz antenna is to be 30 dB.
    b. What will be the half-power beamwidth?

19. An X-band (10-GHz) dish antenna must have a 1° beamwidth.
    a. What must be the parabolic diameter?
    b. If 55% efficient, what will be the antenna gain?

20. A 55% efficient parabolic reflector "dish" antenna measures 0.5 m from the focal point to the back of the dish. Also, the dish physically ends at the focal plane ($\theta = 90°$).

    a. What is the distance from the surface of the dish at $\theta = 45°$ to the focal plane, $D'\text{-}D''$?

    b. Compute both the *directivity* and *gain*, both in dBi, when transmitting a 6-GHz signal.

21. The 3-dB *beam*width of a 70% efficient microwave dish antenna is 0.1° at 4 GHz. Compute the gain in dBi.

22. If the dish of Problem 21 has the feed horn at the center of the dish surface and has a small reflector at the focal point, what is the name of the feed system?

23. What will be the propagation velocity of an electromagnetic wave in a dielectric with relative dielectric constant of 1.22?

24. An RF signal in space enters a dielectric material with an index of refraction of 1.5. The entrance angle is 30° with respect to the interface. Determine the angle of refraction with respect to the normal to the interface.

25. **a.** What phenomenon makes it possible for RF energy to be received even behind obstacles?

    b. What basic principle is involved?

26. How high must a receiving antenna be to receive a signal directly from an over-the-horizon transmitter whose antenna is 30 ft high? The distance between antennas is 15 mi.

27. **a.** What causes ghosts on a television screen?

    b. A ghost is displaced 1 in. on a 13-in.-diagonal TV screen. Determine

      (1) The time between signal receptions, and

      (2) The difference in path lengths.

28. A satellite has a 4-GHz, 10-W, solid-state transmitter and a microwave dish antenna with 30 dB of gain. The earth receiving antenna is 25,000 mi away and has a diameter 56 times greater than the satellite antenna. The earth receiver has a noise figure of 6 dB and bandwidth of 20 MHz. Determine

    a. The satellite effective radiated power (EIRP) in dBm.

    b. Space attenuation.

    c. Gain of receiving antenna.

    d. Signal power into the earth receiver.

    e. Signal-to-noise power ratio, S/N (dB), at the receiver IF output.

15–14 ■ COMPUTER ANALYSIS

Using the computer for communications system and circuit analysis has become a virtual necessity in education and industry. What follows are examples of programs and/or readily available simulation software for solving problems of the type covered in this chapter.

All the programs are interactive in the sense that an introductory sentence or short paragraph describes the purpose of the program and the user is then prompted for input data.

The program for Chapter 15 is as follows:

- **Transmission link analysis:** The program of Figure 15–31 computes the power and signal-to-noise ratio for a receiver over a free-space propagation path of given length. Also, receiver and transmitter parabolic-reflector antenna gains are computed and provided as output.

  Data prompted for will be the transmitter frequency and power, antenna gain or dish diameter, transmission path length in miles or kilometers, receiver antenna gain or dish diameter, noise figure, and bandwidth if given.

```
10  REM *            TRANSMISSION LINK         *
20  REM *               ANALYSIS               *
30  REM *           By P. Young, ASU           *
40  CLS: PRINT
50  PRINT"  This program computes EIRP, antenna gains, space loss,"
60  PRINT"  received power and S/N for a typical satellite link. "
70  PRINT"  Inputs will be the transmitter frequency and power "
80  PRINT"  (in watts or dBm), the sum of branching and feed-cable"
90  PRINT"  losses, antenna gain or parabolic reflector diameter"
100 PRINT"  transmission path length (miles or kilometers), link"
110 PRINT"  power fade margin, receiver antenna gain or diameter"
120 PRINT"  and, if S/N is desired, receiver noise figure and"
130 PRINT"  bandwidth are requested."
140 PRINT: PRINT: PI=3.14159
150   INPUT"The signal carrier frequency is (in MHz): ",FO:PRINT
160   PRINT"The transmitter output power level is"
170   INPUT"(either dBm or watts): " ,PO :PRINT
180   INPUT"This is in  1) dBm, or  2) watts. (Which?): " ,POX
190   PRINT: IF POX=2 THEN 200 ELSE GOTO 210
200   PO=4.343*LOG(PO*1000) :PRINT :REM  Change watts to dBm.
210   PRINT"The total transmit branching and feed-cable losses"
220   INPUT"are (in dB): " ,LBLF :PRINT
230   PRINT"Enter the antenna gain in dB, or enter the diameter"
240  INPUT"of the Xmtr parabolic reflector in meters: ",GAT:PRINT
250  INPUT"This is  1) gain, or  2) diameter. (Which?): " ,GATX
260   PRINT: IF GATX=2 THEN 270 ELSE GOTO 280
270   GT = .55*(PI*GAT/(300/FO))^2 :REM  Calculate ant. gain
273   GAT = 4.343*LOG(GT) :REM 10/LN(10) = 4.343
275 PRINT:PRINT"  The Xmtr antenna gain is, G ="GAT"dB."
280 PRINT: EIRP = PO - LBLF + GAT
290 PRINT"  The EIRP from the antenna is, EIRP ="EIRP"dBm."
300   PRINT: REM * Compute space path attenuation *
310  INPUT"Enter the transmission path length in miles or km: ",L
320  PRINT: INPUT"This is in  1) miles, or  2) kilometers" ;LX
330  PRINT: IF LX=2 THEN 340 ELSE 350
340  L = L*.62
350  LS = 36.6 + 8.6859*LOG(L) + 8.6859*LOG(FO)
360  PRINT"  The free-space attenuation is"LS"dB." :PRINT
364  INPUT"Enter the link margin (fade, etc.), in dB: ",FM :PRINT
370  PRINT"Enter the receive antenna gain in dB, or the diameter"
380  INPUT"of the parabolic reflector in meters: ",GAR:PRINT
390  INPUT"This is  1) gain, or  2) diameter. (Which?): " ,GARX
400  IF GARX=2 THEN 410 ELSE GOTO 440
410  GR = .55*(PI*GAR/(300/FO))^2 :REM  Calculate ant. gain
423  GAR = 4.343*LOG(GR) :REM 10/LN(10) = 4.343
430  PRINT:PRINT"  The receiver antenna gain is, G ="GAR"dB."
440    PRINT: REM * Compute the receiver input power.
```

FIGURE 15–31 (page 1 of 2)

```
445   S = EIRP - LS + FM + GAR
450   PRINT"  The receiver input power is S(dBm) = "S"dBm.":PRINT
460      INPUT"Input the receiver noise figure, in dB: ",NF:PRINT
470      INPUT"What is the system bandwidth, in hertz ";BW:PRINT
480      SNR = S - (-174 + NF + 4.343*LOG(BW))
490   PRINT"The predetection signal-to-noise ratio is "SNR"dB."
500   PRINT"**************************************************"
510   END
```

FIGURE 15–31   (page 2 of 2)

16

Basic Television

■ INTRODUCTION

Many of the fundamentals of information processing and transmission using electronic circuits have been presented in previous chapters. One communications system to which we have all been exposed—television—incorporates elements from all we have studied. Although a one-chapter discussion of television is necessarily an incomplete coverage of a vast subject, we can present some of the unique features of TV systems and circuits.

Many exciting developments are occurring in the television industry today. Stereo sound is being added, but probably the most notable development is high-definition TV (HDTV), which promises to double the picture resolution, to provide a cinemascope-type wider picture with a 16:9 (versus the present 4:3) aspect ratio, and to reduce vertical line crawl by changing from interlaced to progressive scanning. Clarifying these and other parameters is the objective of this chapter. ■

16–1 ■ OVERVIEW: STUDIO TO VIEWER

An overall view of a typical television link is illustrated in Figure 16–1. A studio site is usually located in an easily accessible location for programming and production. Here the visible scene is transformed into a video signal, and the audible voice and other sound is transformed into an audio signal.

A microwave studio-to-transmitter link (STL) relays the information signals to a transmitter site, where they are demodulated and prepared for broadcast. The broadcast site is usually in a remote elevated location for direct high-power VHF or UHF broadcast transmission to viewers. The TV receiving antenna, transmission line, and receiver circuitry provide the audio and video electrical signals for reproducing the sound and picture information by means of transducers, a speaker for audio and a picture tube or flat-screen monitor for video.

Television is in an exciting evolutionary stage. Direct satellite microwave broadcast is already a reality as are stereo sound and more than one picture displayed on the screen. TV receivers with digital video processing provide the storage and signal processing needed for simultaneous multiple picture display and other features. Indeed, digital processing of sound, video, and special visibly displayed characters will make it possible to receive and decode direct satellite transmissions even from other countries with incompatible scanning techniques* and languages. The next stage of development is high-definition (high-resolution) television (HDTV) using bandwidth compression techniques to make all-digitally encoded TV transmission commonplace.

* The three major TV system standards are the NTSC (National Television System Committee) in North America and Japan; SECAM in France, the Middle East, the former Soviet Union, and Eastern Europe; and PAL in the rest of the world.

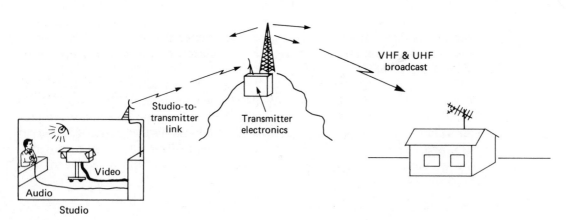

FIGURE 16–1   Television link.

16–2 ■ FREQUENCY ALLOCATIONS

The NTSC standard approved by the FCC in the early 1940s included the basic formats for picture scanning, black-and-white (luminance) AM video, FM sound, channel frequency allocations, and other details of transmission.

The approved channel frequency assignments are outlined in Table 16–1. All channels (stations) are allocated 6 MHz of bandwidth. The channels in the VHF band start at 54 MHz with channel 2 and end at 216 MHz in channel 13 (well within the 30- to 300-MHz spectrum segment defining VHF). The break in the VHF channel assignments between channels 6 and 7 is the segment given to FM and some other users while the television industry was trying to agree on what they wanted. The UHF segment of the frequency spectrum is an uninterrupted band from 470 MHz (low end of channel 14) to 890 MHz (high end of channel 83).

The NTSC standard allows for a video carrier placed 1.25 MHz above the low end of the 6-MHz allocated bandwidth and an audio carrier placed 4.5 MHz above the video carrier. The video signal amplitude-modulates the picture carrier and is filtered to eliminate most of the lower sideband. Only a 0.75-MHz "vestige" of the lower sideband remains. This vestigial sideband (VSB) AM technique is used to conserve bandwidth and minimize the low-frequency phase distortion and receiver complexity that would result if true single-sideband suppressed-carrier were used. The 4.2-MHz video bandwidth required to provide sharp pictures would otherwise have required 8.4 MHz of transmission bandwidth even before the sound signal was included. Figure 16–2 shows the transmission spectrum as defined in the FCC standard and the specific frequencies used by channel 2.

Broadcast standards in the rest of the world are either *phase-alternating line* (PAL) or *système électronique couleur avec mémoire* (SECAM—electronic color

TABLE 16–1 NTSC Frequency Assignments

VHF		*UHF*	
Channel Number	**Assigned Bandwidth**	**Channel Number**	**Assigned Bandwidth**
2	54–60 (MHz)	14–83	470–890 MHz in 6-MHz bandwidth increments
3	60–66		
4	66–72	(Channels 70–83 are now allocated for land mobile service. Old licenses are renewable.)	
5	76–82		
6	82–88		
FM and others	88–174		
7	174–180		
8–13	180–216 MHz in 6-MHz bandwidth increments		

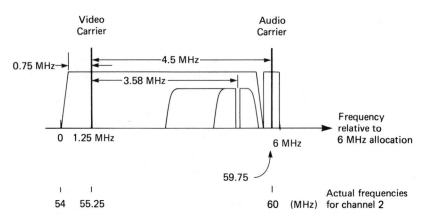

FIGURE 16–2 Frequency allocation for each NTSC station.

system with memory). The most widespread of these is PAL. The International Radio Consultative Committee, abbreviated CCIR, lists nine variants of PAL. One of these, PAL-M, is virtually identical to NTSC, with 525 lines and 60 fields per second, 6-MHz RF channel bandwidth and spacing, 4.2-MHz video bandwidth, 0.75-MHz vestigial width, negative video modulation, and 4.5 MHz for the sound-carrier position above the picture carrier. Many of these details are illustrated in Figure 16–2. All other PAL variants use 625 lines per frame and 50 fields per second.

PAL-B uses a 7-MHz RF channel bandwidth and 5-MHz video bandwidth, and the sound carrier is placed 5.5 MHz above the picture carrier. The rest of the PAL variants use 8-MHz channel spacing, but the video bandwidth is either 5, 5.5, or 6 MHz. Some countries even use combinations of the variants in their cable television systems, such as 7-MHz channel spacing for the first 15 channels and 8-MHz spacing for the others.

16–3 ■ TV CAMERAS

A TV camera is a transducer for changing the light energy from a visible image (scene) to a continuous electrical signal. There are many names for these cameras: Vidicon, Image Orthicon, Plumbicon, Saticon, Vistacon, Leddicon, and so on. Most are proprietary improvements on the Vidicon and the old Image Orthicon vacuum tubes. Both of these camera tubes include an electron gun that produces a steady, well-focused stream (beam) of electrons that strike a target where the scene image has been projected by the camera lens system. The electron beam is electromagnetically deflected from left to right and top to bottom as it scans, in a sequence of horizontal lines, the image on the target. One complete scanning (or reading) of the image from top to bottom provides a *field*. (Because of the image inversion through the lens, the actual scanning in the camera is left to right, bottom to top.) More details on the scanning process will follow.

The light-to-electrical image process for the *Vidicon* tube is provided by a photoconductive (or photoresistive) material applied to one side of the optically transparent metal target. The metal target is attached to the inside of the Vidicon glass tube faceplate with the photoconductive material on the electrically scanned side of the target. Light from the scene is projected onto the photoconductive material in proportion to the light intensity at each point of the projected image. The complete current path is shown in Figure 16–3. As the electron beam scans the target image, the total current i varies with the resistance of the photoconductive material, and the video signal voltage is produced across the fixed resistor R.

The Vidicon is used in applications where high light levels can be maintained because it is relatively insensitive. However, the simplicity and low cost of the Vidicon make it attractive for many applications.

The *Image Orthicon* uses a photoemissive material that reproduces the image in an electrical charge on the target. The scanning electron beam impinges on the target, and the electrons that are not absorbed by the target at a particular point are accelerated back to the rear of the tube into a multistage electron multiplier-amplifier. The instantaneous output current is directly proportional to the light intensity at the particular point in the scene. Each horizontal scanning sweep produces the video signal for one of the 525 lines that make up each frame.

Three tubes are usually used in a color camera. A red, green, or blue lens is placed in front of each tube, and the electrical output signals provide the color information by the relative proportions of these three basic colors. Color and the color signal are discussed in a later section.

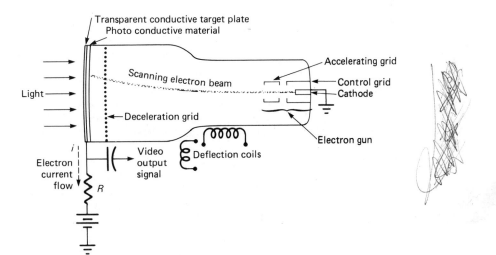

FIGURE 16–3 Vidicon camera tube.

16–4 ■ THE SCAN RASTER AND SYNCHRONIZING SIGNALS

The relative intensity of the visual image is electrically recorded line by line in the scanning process much the same as reading this page. As the continuous electrical signal representing one horizontal line of the scene is completed, the signal is blanked (darkened) during the quick retrace to begin the next line.

In order to avoid the "flicker" (fast-flashing) effect reminiscent of old-time movies, an interlace technique is used along with a scanning rate that is fast enough for our persistence of vision to integrate the frames into a smoothly changing picture. Figure 16–4 illustrates the interlaced scanning technique. The entire scene is scanned once from top to bottom using 262½ horizontal lines (the solid lines in the figure); this produces one picture *field*. After the vertical retrace, blanked so that no streaks are seen by the viewer, the second field is started at the upper center of the picture. The interlaced lines of the second field (dashed) fall between those of the first field. The two fields constitute one complete frame with 525 lines. (In most of Europe, 625 lines per frame and 25 frames per second are used.) The picture sweep pattern of Figure 16–4 is referred to as the *scan raster*. The individual lines of the raster are easily seen on a large-screen home TV set, especially when no picture is being received.

In the United States, each field takes approximately ⅟₆₀ s, and the two fields produce one frame (a frame rate of ~30 frames/s). Since there are 525 horizontal lines/frame, the horizontal line sweep rate is (525 lines/frame) × (29.97 frames/s) = 15,734.25 lines/s. Of the 63.5 μs/line, only 53 μs are visible because 10.5 μs are taken for the (blanked) retrace (also called *flyback*). During the time between the end of one line and the beginning of the next, the scanning system generates a short *horizontal sync pulse* that is transmitted so that the TV receiver can synchronize with the transmitting station. When the transmitted signal includes color (chrominance), an additional synchronizing signal called the *color burst* is transmitted during the horizontal retrace time. (Color transmissions are covered in sections to follow.)

Figure 16–5 shows the last lines of the *composite video signal* for a black-and-white transmission. Note that a high voltage corresponds to a dark portion of the scene. The last few horizontal lines of the picture are followed by the beginning of the vertical retrace period, called the *vertical blanking interval*. The vertical retrace takes about 1⅓ ms, meaning that 21 horizontal lines cannot be displayed during the vertical blanking interval. This time is now being used for transmitting reference (VIR) and test (VITS) signals, as well as special displayed information such as teletext and closed captions for speech text.

FIGURE 16–4 Interlaced scan raster. Line 1 of the first field starts at upper left corner of picture tube.

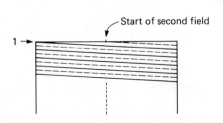

FIGURE 16–5 Bottom line of picture and composite video signal for black-and-white transmission.

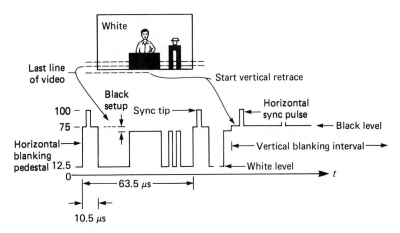

The voltage level, called *black setup*, established for a black part of the scene is set for 65–70% of the sync tip level (maximum allowable voltage, see Figure 16-5). Hence the blanking level *pedestal*, at 75%, will most definitely cut off the electron beam of the TV receiver during horizontal and vertical retrace. The vertical (and horizontal) synchronizing pulses would be blacker than black and can only be seen on a home TV by adjusting the vertical sync control to roll the vertical blanked interval down on-screen after increasing the brightness and decreasing the contrast (that is, wash out the picture). The white reference level is 10–15% of maximum and is shown in Figure 16–5 as 12.5%.

16–5 ■ VIDEO BANDWIDTH AND RESOLUTION

The two narrow table legs at the right of the scene in Figure 16–5 will be clearly distinguishable only if the signal reaching the TV receiver picture tube (cathode-ray tube, CRT) can change fast enough. This means that the video signal and system bandwidth, especially the receiver video amplifier bandwidth, limits the picture resolution.

The NTSC limit on video bandwidth is 4.2 MHz, meaning that the fastest on-off cycle variation in the picture cannot exceed about 0.24 μs. This limits the picture resolution to a little over 100,000 resolvable light cycles or 200,000 discrete black-and-white picture elements. Example 16–1 illustrates how this is determined.

■ ■ ■ ■

EXAMPLE 16–1

Determine the theoretical number of distinguishable picture elements for an NTSC standard TV transmission.

Solution:

(53 μs/picture line)/(0.24 μs/black-white cycle) = 221 light cycles/line, or 442 individual picture elements/line. Each frame (two fields interlaced) contains 525

lines, of which $2 \times 21 = 42$ are blanked during the two vertical retraces. Hence, 442 elements/line $\times$ (525 $-$ 42) lines = **213,486 picture elements.** So that there are not any dark edges or corner areas on the viewing screen, some of the picture elements occur off-screen.

16-6 ■ TRANSMITTED VIDEO AND AUDIO SIGNALS

The black-and-white (B&W) video signal of Figure 16–5 is amplitude-modulated on the picture carrier and filtered for VSB transmission. Transmitter power is saved by using the *negative transmission* format (white = low voltage) because most scenes are light and relatively few are dark (night scenes). The maximum EIRP licensed by the FCC is 500 kW for UHF and 100 kW for channels at the low end of the VHF band. Typical VHF transmitters operate in the 15–25 kW peak envelope power (PEP) range, and the antennas have 5–15 dB of gain over isotropic.

The audio bandwidth is 50 Hz–15 kHz. The audio is frequency-modulated on a completely separate sound carrier using 75 μs preemphasis. Full deviation for the FM sound is ±25 kHz, making it one-third that of FM radio. The FM sound signal is added to the VSB-AM video signal before the antenna by the use of a diplexer, a combination of separate filters and quarter-wave transformer transmission lines. The radiated audio FM signal is supposed to be maintained between 10% minimum and 20% maximum of the peak video transmitter power. Carrier frequencies are to be within ±1 kHz of the nominal values (see Section 16–2).

16-7 ■ BLACK-AND-WHITE TV RECEIVERS

Tuner, IFs, and Detectors

Figure 16–6 shows a basic receiver block diagram. The VHF tuner consists of input filters, an RF amplifier, an active mixer, and a local oscillator. Varactor-diode tuners are common, and many tuners use phase-locked synthesized LOs similar to that diagrammed in Figure 16–7.

The received signal is down-converted to an IF, called the *video IF*, centered near 44 MHz. The UHF tuner consists only of a two-pole cavity filter, a diode mixer, and an LO, also housed in a tuned cavity. When switched to a UHF channel, the VHF LO is disabled and the RF amp and the input to the transistor mixer stage are tuned to provide additional IF amplification.

Like most AM superheterodyne receivers, the IF provides most of the gain, selectivity, and gain control for linearity. Figure 16–8 shows a typical video IF bandwidth. The low gain for the sound carrier is made up in the sound IF. The 50% amplitude points coincide with the color and picture carriers to compensate for vestigial sideband transmission. Video signals within ±0.75 MHz of the pic-

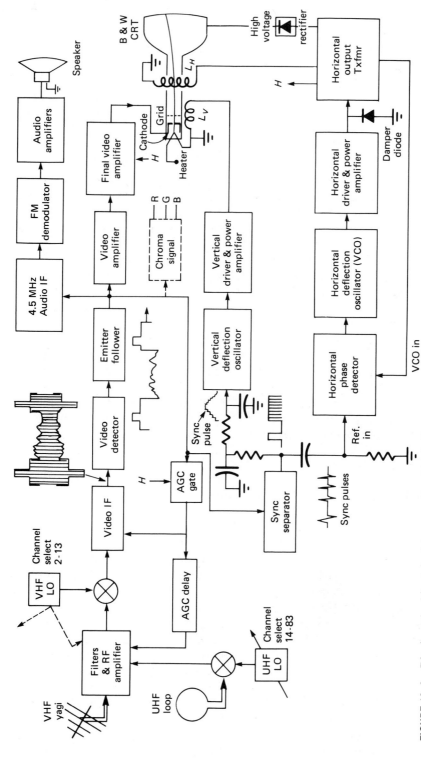

FIGURE 16–6 Black-and-white television receiver block diagram.

FIGURE 16-7 Phase-locked loop synthesized channel selection system.

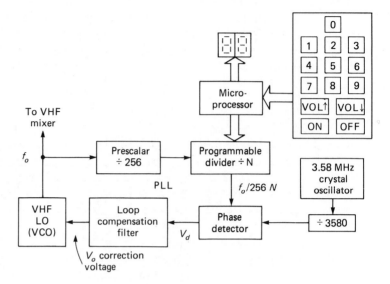

ture carrier have a full AM modulation index because both sidebands are transmitted; however, the higher-frequency video signals have only one-half the index. The IF bandpass compensates by deemphasizing the low frequencies (those close to the picture carrier).

To achieve the desired bandwidth and selectivity, either stagger tuning and high-Q wavetraps are used or the IF is made broadband and a surface acoustic wave (SAW) filter is used.

The video detector is a peak AM detector for demodulating the VSB AM signal. This AM detector does not produce the AGC voltage because most TV receivers use a gated-AGC technique. However, in B&W receivers, this detector is used as a second mixer for down-converting the sound carrier from its first IF frequency (41.25 MHz) to a 4.5-MHz sound IF. The LO signal for this second conversion is the video IF carrier of 45.75 MHz that is, by specification, precisely 4.5 MHz from the sound carrier. This arrangement is referred to as *intercarrier sound*. The sound IF, detector, and audio preamp are usually contained in a single IC chip. The detector is typically either a quadrature FM detector (see Chapter 9) or a phase-locked loop.

FIGURE 16-8 Video IF passband.

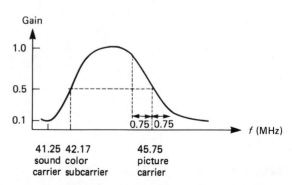

Many color sets use a separate wideband detector for the sound IF conversion so that wavetraps (high-Q resonant circuits) can eliminate the sound carrier from the video detector. This helps to eliminate a 920-kHz beat frequency interference between the 4.5-MHz signal and 3.58-MHz color subcarrier. This interference results in a herringbone pattern or voice-related jitter in the picture.

Gated AGC

The video IF signal will have the format shown in the schematic of Figure 16–6 for B&W transmission. Since the video information varies from line to line, the average value of the signal varies and an averaging detector is inappropriate for TV AGC. A peak detector would, however, have too long a time constant. This is especially troublesome when rapidly changing multipath effects from passing aircraft cause the picture to flutter.

Gated or *keyed* AGC is able to overcome these problems and provide a relatively fast-acting low-noise AGC response. This is accomplished by sampling the demodulated composite video signal only when the horizontal sync pulse tips are present. The sync tips are transmitted at a constant 100% video level and are used to establish an appropriate IF gain. Also, the maximum peak envelope power is transmitted during sync pulses, and therefore the optimum S/N occurs during the AGC sampling instants.

Figure 16–9 shows a typical gated-AGC circuit and controlled IF amplifier. The composite video input is shown with a noise impulse during a video line. Pulses from the receiver's horizontal sweep oscillator supply the PNP transistor with an

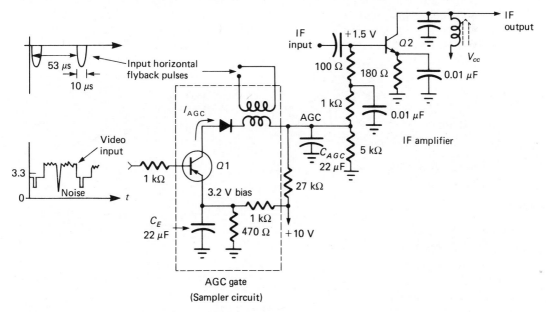

FIGURE 16–9 Video IF amplifier bias and gated AGC.

appropriate negative collector bias voltage, but only during the time when video sync pulses are present. Consequently the AGC gate is cut off except during sync pulse tips, and the noise pulse shown has no effect on the AGC line. The AGC capacitor C_{AGC} smooths the gate current I_{AGC} and provides the automatic gain control of IF amplifier Q_2.

Video Amplifiers and CRT Bias

Figure 16–10 shows typical TV video amplifier and picture tube bias circuitry. Q_1 is part of an emitter-follower current (and power) amplifier whose main purpose is to provide high impedance to the video detector and have low enough output impedance to drive four circuits—the sync separator, AGC system, color system (or sound IF in B&W sets), and, of course, the video system voltage amplifier.

The common-emitter amplifier with Q_2 provides up to 35 dB of voltage gain. The gain control potentiometer R_1 sets the peak-to-peak video signal at the cathode of the CRT to 50 V or more. Very low peak-to-peak signal voltage gives *low contrast* in the picture (washed-out picture), whereas large peak-to-peak signals can result in too much contrast, extreme black and white areas but no grays. The CRT average *brightness* level is set by the potentiometer R_2 that varies the CRT beam bias current.

The B&W CRT is a vacuum tube with $B^+ = 9$–15 kV bias on an anode consisting of a conductive layer of graphite material coating the inside of the tube near the flat-front picture screen. The inside surface of the screen is coated with a special mixture of phosphors that emit white light after electrons with high kinetic energy from the beam have raised electrons in the phosphor atoms to high potential-energy states. A photon of light is released during the excited-state–to–normal-state transition.

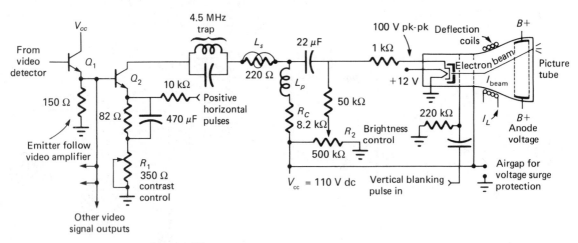

Video amplifier

FIGURE 16–10 Typical video amplifier section.

As shown in Figure 16–10, the cathode has a much lower positive bias voltage, about 25 V, so that the electron beam accelerating potential is essentially full B^+.

Two input points are shown for beam blanking. Positive pulses from the horizontal output transformer are fed to the emitter of Q_2. When present, a pulse will reverse-bias Q_2, thereby allowing the collector voltage to rise to full V_{cc}. This positive pulse, coupled to the cathode of the CRT, will completely cut off the electron beam during horizontal retrace. Vertical retrace blanking is accomplished by coupling a negative pulse to the CRT control grid from the vertical sweep circuit output.

If the full 4.2-MHz video bandwidth is to be amplified, the shunt circuit capacitance in the collector-to-CRT path must be compensated at high frequencies. Series inductance L_s is chosen to resonate with the shunt capacitance of the CRT to peak up the amplifier near 3 MHz. This is referred to as *series peaking*. The 1-kΩ resistor in series with the CRT cathode ensures that the resonant Q is low enough to extend the bandwidth almost an octave, from about 2.5 MHz to 4 MHz. Typically L_s is wound on a resistor such as the 220 Ω shown in Figure 16–10. This swamps (de-Qs) the inductance to prevent ringing.

A similar peaking technique can be provided by L_p in series with the collector resistor R_c. This inductor is chosen to parallel-resonate the circuit shunt capacitance at about 3 MHz. *Shunt peaking*, as it is called, can increase the upper end of the frequency response by almost an octave.

Sync Separator

The sync separator circuit receives the composite video signal and must strip away the video information, then separate horizontal from vertical sync pulses.

Figure 16–11 shows the basic sync separator circuit. With no base bias resistor, Q_1 is off except during the most positive peaks of the composite video signal. During these horizontal and vertical sync pulses, the base-emitter junction is forward-biased, coupling-capacitor C charges with the polarity shown, and the

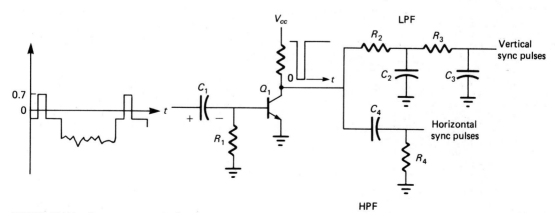

FIGURE 16–11 Sync separator circuit.

collector voltage drops to nearly zero. Between the pulse peaks, C_1 remains negatively charged to cut off Q_1, although there is a slight discharge through R_1.

The high-frequency horizontal and vertical sync pulses are separated by the RC high-pass (HPF) and low-pass filter (LPF) networks, as shown in Figure 16–11. The LPF *integrates* the pulses, and only the long vertical pulses add up to enough charge on C_3 to produce a vertical sync signal. The HPF *differentiates* the pulses and only the pulse edges produce output for horizontal synchronization.

Vertical Deflection and Sync

The picture screen on the TV receiver lights up when the electron beam is swept horizontally at 15,734 lines per second and vertically downward at nearly 30 fields (60 frames) per second. The sweep system that produces this raster, described in Section 16–3, consists of the vertical and horizontal oscillators, driver amplifiers, and output power amplifiers that provide the current for the electromagnetic coils in the yoke assembly around the neck of the CRT. The electromagnetic fields deflect the electron beam of current I_{beam} that averages approximately 1 mA. The deflection coil currents are a couple orders of magnitude greater.

The vertical deflection (sweep) oscillator is typically an astable multivibrator (see Chapter 2). The sync pulses are added to the capacitor-charging voltages such that the onset of conduction is altered, thereby forcing the oscillator period to synchronize with the TV station's vertical scan rate.

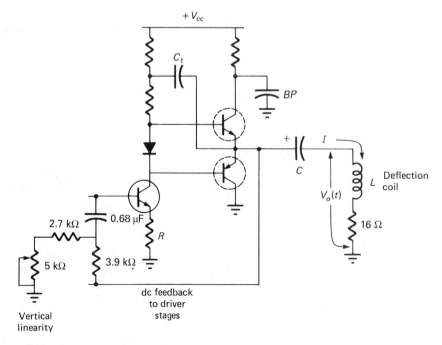

FIGURE 16–12   Vertical power amplifier.

A transformerless, push-pull, vertical deflection, power amplifier is shown in Figure 16–12. This circuit, including the bootstrapping capacitor C_1, was discussed in Chapter 7. Here it must supply deflection coil L with a linear current ramp reaching nearly 1 A. Since the voltage across an inductor is directly proportional to the derivative of current (Equation 1–1), in order for the current to be a ramp, the applied voltage waveform $V_o(t)$ would have to be rectangular. Furthermore, any circuit resistance, such as the 16 Ω in Figure 16–12, whose voltage is directly proportional to current, will require an additional modification in $V_o(t)$. These fundamental design relationships are illustrated with an example.

■ ■ ■ ■

**EXAMPLE
16–2**

Figure 16–13a shows an ideal deflection current waveform that rises from 0 to 600 mA in approximately 15 ms and returns to zero in approximately 2 ms.

1. Determine the value of L for the deflection coil if the voltage drop across the ideal inductor is a constant 2.4 V during the 15-ms trace period.
2. Determine the voltage drop across L during the 2-ms retrace.
3. Sketch accurately the applied $V_o(t)$ for the circuit that includes 16 Ω of equivalent series resistance.

Solution:

1. $L = (V)(dt/di)$. Since the current slope is a constant,

$$\frac{\Delta i}{\Delta t} = \frac{600 \text{ mA} - 0}{15 \text{ ms}} = 40 \text{ A/s},$$

then $L = 2.4/40 = \textbf{60 mH.}$

FIGURE 16–13 Vertical deflection coil waveforms.

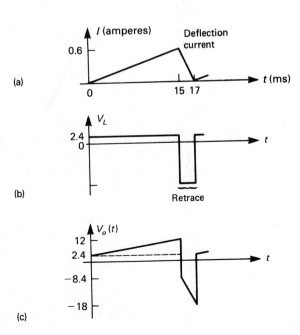

2. Assuming the current slope is constant as shown,

$$\frac{\Delta i}{\Delta t} = \frac{0 - 600 \text{ mA}}{2 \text{ ms}} = -300 \frac{\text{A}}{\text{s}}.$$

Then the voltage drop across L during retrace will be a constant

$$V_L = L \frac{\Delta i}{\Delta t} = (60 \text{ mH}) \times \left(-300 \frac{\text{A}}{\text{s}}\right) = -18 \text{ V}.$$

3. The voltage drop across the 16 Ω is $V_R = iR$. Hence, from the current waveform, $V_R(t = 0) = 0 \cdot 16 \Omega = 0$, $V_R(t = 15 \text{ ms}) = 600 \text{ mA} \cdot 16 \Omega = 9.6 \text{ V}$. Also $V_R(t = 17 \text{ ms}) = 0$. Since $V_o(t) = V_L(t) + V_R(t)$, then the result as seen in Figure 16–13 will be linear between the following four points: $V_o(t = 0) = 2.4 + 0 = 2.4 \text{ V}$, $V_o(t = 15 \text{ ms}^-) = 2.4 + 9.6 = 12 \text{ V}$, $V_o(t = 15 \text{ ms}^+) = -18 + 9.6 = -8.4 \text{ V}$, and $V_o(t = 17 \text{ ms}) = -18 - 0 = -18 \text{ V}$. The result is the trapezoidal deflection voltage sketched in Figure 16–13c.

As indicated in Figure 16–12, the output voltage $V_o(t)$ is fed back to the driver amplifier. This stage usually includes a relatively low-value coupling capacitor in the feedback path and a potentiometer that allows for vertical linearity adjustments. Also, the vertical picture height can be adjusted by varying the gain of the oscillator or driver.

Horizontal Deflection and CRT High Voltage

The horizontal deflection (sweep) signal is derived from the horizontal oscillator and driver. The deflection current and anode voltage for the picture CRT are taken from the *flyback transformer* of a transformer-coupled power amplifier. A simplified version of the horizontal output stage is shown in Figure 16–14.

Transistor Q_1 is a power transistor with a heat sink indicated by the dashed line. Q_1 performs the function of a switch, and is on for less than one-half of the 63.5-μs horizontal sweep period. R_1 and C_1 provide for class C bias to help hold Q_1 off. D_1 is the *damper diode* that dampens ringing that would be visible as vertical bars of intensity on the picture screen when no picture is present. D_1 conducts and provides a current path for I_L during this first half of the horizontal trace. Inductance L_1 represents the CRT horizontal deflection coil. Transformer T_2 is the horizontal output transformer (HOT), also called the *flyback*, and C resonates the entire output circuit at 50–70 kHz to provide for a half-cycle high-voltage pulse for retrace. This retrace pulse is stepped up to around 10 kV, rectified and filtered for the picture tube anode B^+ voltage.

Derivation of the horizontal sweep current I_L is explained with the help of Figure 16–15. Switch S, the collector-emitter of Q_1, shorts due to a positive base voltage, and I_L builds up from zero to a maximum, at which point S opens to begin the retrace. The CRT electron beam is then on the extreme right side of the picture tube. With an open-circuited S, the peak current of I_L flows into C, forming an underdamped 50-kHz LC-tuned circuit, so that approximately one-half cycle of

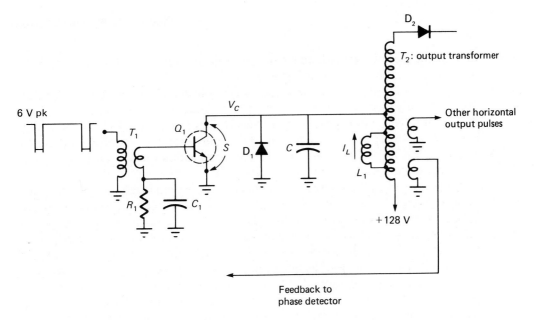

FIGURE 16-14 Horizontal power output amplifier and transformer.

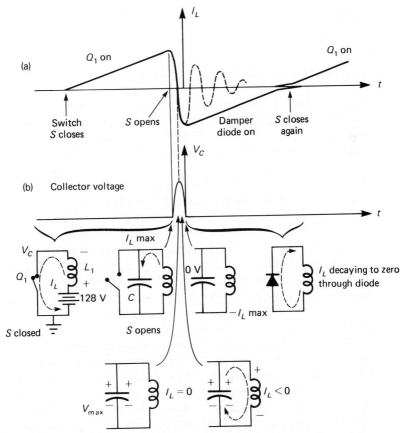

FIGURE 16-15 Horizontal sweep waveforms and circuit configuration.

this ringing current brings I_L to a negative peak as shown; and the scanning beam has retraced to the extreme left side of the CRT.

Figure 16–15b shows that the voltage from collector to ground across Q_1 makes approximately a one-half cycle positive pulse during retrace when the transistor is not a short. As V_c goes negative, however, D_1 forward-biases and becomes a short across L so that I_L can discharge "slowly" without ringing, thereby providing the ramp current for the left-hand half of the horizontal sweep. Transistor Q_1 then conducts to complete the horizontal line trace.

If the peak of the collector voltage reaches, for example, 667 V, then the high-voltage flyback transformer winding needs only a 15:1 turns ratio to provide 10 kV for the CRT anode. Note that the CRT bias power is not as awesome as might be anticipated because the average beam current is less than 1 mA, so that the power requirement is less than 10 W for black-and-white TV. Also, although not shown in Figure 16–14, the horizontal output transformer has other secondary windings to provide horizontal pulses for video blanking, AGC gating, color TV burst gating, low-voltage rectifiers for receiver power, and feedback (as shown) to provide a signal for horizontal oscillator synchronization by means of the phase-locked loop AFC circuit.

Phase-Locked Horizontal Sync

The type of system used for frequency control of the horizontal deflection oscillator is a phase-locked loop (PLL). Although the phase detector circuit is commonly labeled AFC, the horizontal synchronizing system is not the Crosby AFC system studied in Chapter 9. The Crosby AFC technique is a frequency-sensed negative-feedback system, and a small steady-state frequency error always exists to produce the correction signal. The PLL, on the other hand, is a phase-sampled negative-feedback system and, while a small static phase error exists to produce phase corrections, the steady-state frequency is *exactly* that of the reference horizontal sync signal from the video sync pulses.

Another reason for using this system rather than the bias-point sync technique of the vertical system is that the input filter is a sync separator high-pass network, which passes much more noise than the vertical system low-pass input. The PLL loop bandwidth is made narrow, however, so that the noise ultimately has little effect. The result is often likened to a flywheel effect.

A typical horizontal phase-frequency control circuit diagram for television is shown in Figure 16–16, and the block diagram is shown in Figure 16–17. R_0, C_1, and the phase detector input resistance form the high-pass filter for the sync separator. The horizontal sync pulses are very narrow and negative-going in order to switch diodes D_1 and D_2 of the phase detector circuit.

The VCO input from the horizontal output transformer (HOT, or flyback) are pulses that are integrated by $R_3 C_3$ to produce the (dual-slope) ramp seen in Figure 16–17a and b. The steep part of the ramp is sampled by the switching of D_1 and D_2. If the VCO ramp and the horizontal sync pulses occur as shown in Figure 16–17a, then the average (dc) value of V_d will be zero. If, however, the VCO phase is delayed in time as illustrated in Figure 16–17b, then the average value of the

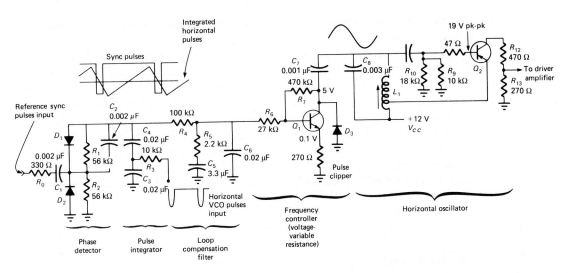

FIGURE 16-16 Horizontal phase-locked loop circuitry.

phase-detector output voltage V_d will be positive. After filtering out all of the switching signals by the loop filter and PLL compensation network (R_4, R_5, and C_5), the positive dc voltage is used by the frequency- and phase-controller circuit to correct the phase of the VCO. The amount of residual phase error remaining depends on loop gain k_L, and the difference between the video sync pulse frequency and the *free-running* (unlocked-loop) frequency of the horizontal voltage-controlled oscillator. The relationship, from Chapter 10, is $\theta_e = \Delta f / k_L$. The loop

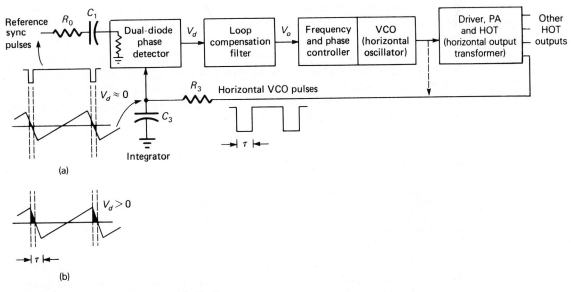

FIGURE 16-17 Horizontal PLL block diagram.

compensation network prevents overshoot and ringing that results in a symptom called *piecrusting*.

Techniques used in frequency- and phase-controller circuits vary among manufacturers. The most popular circuit in vacuum-tube sets was the reactance tube. A solid-state version, the reactance modulator, is discussed in Chapter 9. A different technique, shown in Figure 16–16, is from an RCA color TV receiver. Q_1 acts as a voltage-variable resistance to control the effective amount by which C_7 shunts the VCO resonant tank (C_8 and L_1). D_3 clips the negative-going part of the VCO sinusoid and provides a rectified 5 V dc for Q_1 collector bias. A couple of kHz of frequency range may be achieved by this circuit. If the VCO drifts outside this range before the PLL acquires horizontal sync, then the horizontal hold control L_1 must be retuned.

16–8 ■ COLOR TELEVISION

The color that is seen on the screen of typical TV sets is emitted from special phosphors. Most of the phosphor compounds used are the same as those mixed together for black-and-white cathode-ray picture tubes. However, for the color CRT, three separate phosphors are placed on the screen—one emits red, one emits green, and the third emits blue light when struck by electrons. The three phosphors are kept separate but are arranged very close to each other either in alternating vertical stripes or as tiny dots in a tight matrix of delta (Δ) patterns. These arrangements, along with the three separate electron beams and a shadow mask, are illustrated in Figure 16–18.

The shadow mask is actually very close to the screen and is used to ensure that only the three color phosphors that form a single picture element are struck by electrons at a given instant. This is important because the relative amounts of red, green, and blue light emitted determine the hue (color) perceived by the viewer.

The Venn diagram of Figure 16–19 gives a gross breakdown of some of the colors produced by different combinations of red, green, and blue light. For example, the center of the diagram marked W indicates that all three of the primary colors—red, green, and blue—must be emitted in order for white to be perceived.

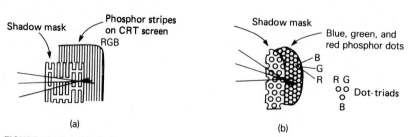

FIGURE 16–18 (a) Striped phosphors and slotted shadow mask. (b) Dot-triad phosphors and holed shadow mask.

FIGURE 16-19   Color Venn diagram.

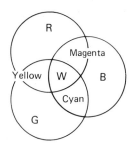

If the blue gun* fails, then a true white cannot be produced, and the normally white parts of the scene will be yellow. Hence, yellow is said to be the complement of blue (see Figure 16–19).

Red is the longest-wavelength electromagnetic wave to which the human eye is sensitive; blue is a short wavelength (only violet is shorter); green is approximately halfway in between. Hence all the colors of the rainbow can be produced by the various mixtures of R, G, and B. Unlike the subtractive system whereby mixing red, green, and blue paint (or overlapping red, green, and blue lenses) produces black (or very dark), adding R, G, and B light emissions produces white. In fact, the correct mix of R, G, and B to produce the white perceived on a TV screen is 0.11 B + 0.59 G + 0.30 R; that is, 11% blue, 59% green, and 30% red camera-tube output signals. These values correspond to our eye sensitivities, and our eyes are most sensitive to green.

Chroma Signals

The transmitted signals required to produce the color on an otherwise black-and-white picture signal (called *luminance*) are referred to as the *chrominance*, or just *chroma*. The chroma signal is a 3.579545 MHz ± 10 Hz phasor (or vector). The phase angle indicates the hue (*tint* or type of color), and the magnitude indicates the *saturation* level or color intensity. A red fire truck is an example of a very saturated red object. Pink is desaturated red.

The reference phase for proper demodulation of the chroma signal at the receiver is transmitted in short bursts during each horizontal retrace period. The phase-color diagram of Figure 16–20 shows the chroma signal phase for the primary colors (R, G, and B) and their complements (180° away) relative to the color-burst phase reference. The phase orientation of Figure 16–20 corresponds to that of a vectorscope that is often used in color troubleshooting.

The chroma phasor is produced by summing two quadrature 3.58-MHz subcarrier signals, modulated by I and Q signals developed from the red, green, and blue camera-tube outputs at the studio. A more detailed description of these color signals is now given and diagrammed in Figure 16–21.

---

* Color CRTs either have a single electron gun and a three-way beam splitter or they have three separate guns.

FIGURE 16–20 Phase-color diagram related to vectorscope display.

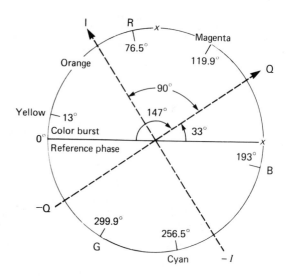

Chroma I and Q Signals

As indicated in Figure 16–21, the input light image is focused and reflected through three colored lenses. The three lenses filter the light into red, green, and blue components of the scene, and the corresponding camera tubes produce the electrical signals R, G, and B. These time-varying signals are processed in a matrix of resistors and inverting amplifiers to develop the Y (luminance), I (in-phase), and Q (quadrature) video signals. Filters allow the Y signal to be approximately 4.1 MHz for sharp black and white illumination, but I and Q are limited to 1.5 MHz and 0.5 MHz, respectively. The human eye has higher resolution to orange than to reddish-blue colors, and since the I signal phase is toward orange, it is allowed the wider bandwidth.

The I and Q signals are translated up to the high-frequency portion of the video bandwidth in balanced AM modulators. The 3.579545-MHz subsidiary carrier was chosen to be exactly one-half of an odd multiple of the 15.73426-kHz horizontal pulse rate (see the frequency multiplier and divider in Figure 16–21). Since all the signals modulating the video transmitter carrier are chopped up into 15.734-kHz horizontal line segments, this choice of chroma subcarrier frequency allows for the *interleaving* of luminance and chrominance harmonics such that the *interference* from alternate lines on the picture screen will *cancel* in the viewer's eye.

The output of each balanced AM modulator is double-sideband/suppressed-carrier (DSB-SC). The I modulator output is vestigial-sideband-filtered to remove two-thirds of the upper sideband, so that all chroma signal energy falls within the 4.2-MHz video bandwidth. The phase-quadrature DSB-SC I and Q signals are linearly added to form the chroma phasor. The result is an analog version of the QPSK modulator circuit of Figure 17–14. Notice from Figure 16–20 that by varying the amplitude of the ±I and ±Q signals, any color in the visible spectrum may be transmitted. A mathematical description of this *quadrature-multiplexed* signal is included in Figure 8–20; also, a minimum required receiver block diagram is shown in Figure 8–21.

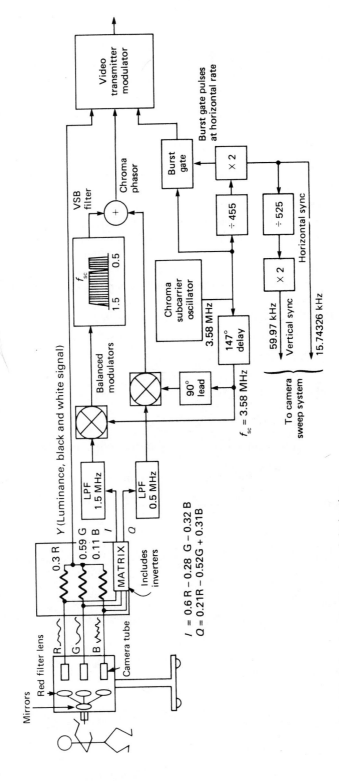

FIGURE 16–21 Colorplexed video signal derivation.

As illustrated in Figure 16–21, the baseband Y signal, the 3.58-MHz subcarrier quadrature-multiplexed chroma signal, and the reference-phase burst all modulate the video transmitter. The resultant signal modulating the transmitter is called the *colorplexed video* signal.

Color TV Receiver Circuits

A color TV receiver has a three-beam CRT, as previously described. These three electron beams must *converge* on the holes or slots in the shadow mask and strike the designated red, green, or blue phosphor on the screen. Convergence in the center area of the screen is achieved by turning two permanent-magnet rings located on the neck of the picture tube. This alignment is called *static convergence*. Convergence is a much more complicated alignment challenge when the beams are scanning the outer perimeter of the screen. Circuits on a *dynamic convergence* board and associated electromagnets in an assembly called the CRT *yoke* are used for achieving full-screen convergence. In addition to the dynamic convergence board and its parabolic waveform generators for ''pincushion'' control, a high-voltage regulator circuit is used for precisely setting the CRT-anode accelerating voltage.

The receiver circuitry for demodulating and processing the luminance (Y) signal is the same as that described for the black-and-white TV system. Indeed this compatibility was designed into the NTSC's 1953 FCC-approved requirements for television broadcast in North America. Consequently, the luminance signal is demodulated with the rest of the colorplexed video signal of Figure 16–22, and the black-and-white Y signal is amplified in the video amplifier.

Chroma Circuits. The additional circuitry required for demodulating and processing the chrominance (phasor) signal is shown in the block diagram of Figure 16–23. The primary function of chroma demodulation is provided by the two DSB-SC *product detectors* in phase-quadrature to each other. Actually, the 90° phase difference between the 3.58-MHz reference inputs to these detectors is not an absolute requirement. Two detectors operating with some large phase-reference spread are required, however, in order to demodulate the chroma phasor that can have a phase in any phase quadrant. One scheme, called the XZ system, uses approximately 57°. The most commonly used scheme is the R-Y/B-Y, or U/V, system where U = 0.493(B-Y) and V = 0.877(R-Y).

FIGURE 16–22 Video signals.
(a) Black and white composite.
(b) Colorplexed video signal.

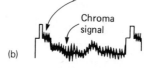

(a)

≥ 8 cycles of color burst on back porch

Chroma signal

(b)

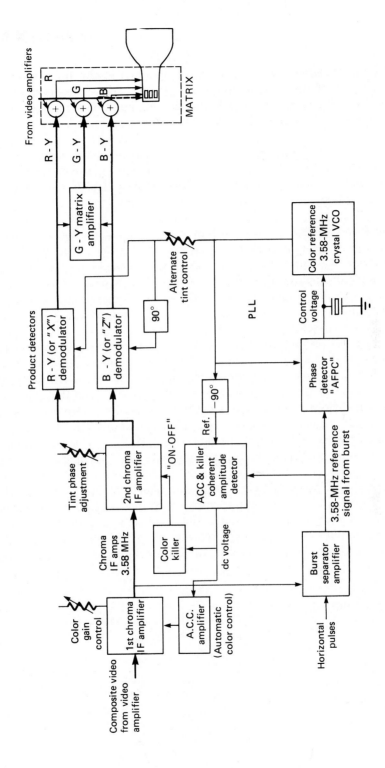

FIGURE 16–23 Chroma circuitry block diagram.

R-Y/B-Y Demodulators. The phase-color diagram of Figure 16–20 has two small x's, one at 90° and the other at 180° with respect to the color-burst phase. The R-Y detector operates in phase quadrature (90°) to the color-burst reference phase, and the B-Y detector operates at phase quadrature to the R-Y detector.

Typical demodulator circuits used are the integrated-circuit balanced detector of Figure 10–8 and the more conventional discrete component balanced demodulator of Figure 16–24. These are the same kind of phase-coherent demodulators discussed in Chapter 8 for DSB-SC and SSB-SC and in Chapters 9 and 10 for phase detectors.

Color Reference Signal. The 3.58-MHz reference signal for the R-Y and B-Y product detectors is generated by a crystal-stabilized VCO called the *color reference oscillator*. Various tuned-circuit oscillators are used by different manufacturers. A simple example is the Colpitts of Figure 2–5a with the crystal from base to ground. A small amount of VCO tuning can be achieved with a varicap (varactor) tuning diode in parallel with the crystal or by other reactance control circuits (see VCOs in Chapter 9).

The color oscillator is phase-locked to the transmitted color reference burst. The phase detector is virtually identical to the product detector of Figure 16–24 and is labeled "AFPC"—*automatic frequency and phase control*. The reference input to this phase detector is from a high-Q tuned circuit that rings continuously from the 3.58-MHz burst that is separated from the colorplexed video signal.

As seen in the chroma circuitry block diagram of Figure 16–23, the composite video signal is amplified in the first 3.58-MHz IF chroma amplifier. If a color transmission is being received, then the colorplexed composite video signal from the receiver video detector has an eight-to-ten cycle burst of color reference signal on the "back porch" of each horizontal blanking pulse. This eight-to-ten cycle burst is passed by the *burst separator circuit* that is gated from horizontal flyback pulses. The eight-cycle reference signal burst is sustained between horizontal blanking intervals by a high-Q circuit tuned to 3.58 MHz. This signal is used for three functions: phase reference for demodulating the chroma phasor via the

FIGURE 16–24   Balanced product-detector color demodulator.

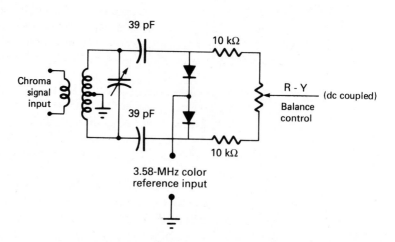

color-reference PLL, automatic color-level control (a chroma-IF AGC), and the color-killer system.

Color and Tint Control.   The chroma IF amplifiers are tuned to 3.58 MHz and provide bandpass amplification for the 2 MHz of quadrature-multiplexed DSB-SC modulated color phasor. The exact shape of the chroma IF response depends on the rolloff characteristics of the video IF. The overall result, however, must cover the 3.58 ± 0.5 MHz of the Q-phasor and compensate for the 3.58 $(+0.5, -1.5)$-MHz VSB I-phasor.

AGC for this linear-AM IF amplifier is derived from the sustained reference-burst signal previously described and is called *automatic color control* (ACC). The detector can be a noncoherent AM-type peak detector but is most often a quadrature phase detector. The ACC and color-killer detector of the block diagram (Figure 16–23) is a phase detector like that of Figure 16–24. When the phase reference to this detector is in quadrature to that of the PLL reference, then the ACC/killer phase detector acts as a coherent peak-amplitude detector. The principle behind coherent peak detectors can be observed from the phase-detector waveforms of Figure 10–4c. It is, incidentally, the same principle used in telephone touch-tone decoder circuits (Figures 12–31 and 12–32) and coherent satellite transponders. The virtue of coherent AGC (or ACC, in this case) is that, despite the presence of noise, there is no output voltage from this peak detector until the PLL locks up to the color burst.

The color-killer circuit disables the chroma IF unless the presence of the color burst indicates that a color transmission is being received. The objective is to prevent B&W video signal noise from passing through the chroma system and causing color noise flecks, called *confetti*, on the screen during B&W signal reception.

ACC holds the amplitude of the chroma signal constant with respect to the amplitude of the color burst so that the color saturation (intensity of color) is correct. Fine control of this parameter is provided externally to the viewer by means of an IF gain adjustment potentiometer on the set, called the *color control*. The actual hue (red, yellow, orange, and so forth) is defined by the phase of the chroma signal relative to the phase-locked, color-reference oscillator and can be fine-tuned by a phase-control potentiometer available to the viewer, called *tint control*. The phase-shifting varicap tuning diode is located either in the second chroma IF tuned circuit or the demodulator color-reference signals.

Color Matrixing.   Two product detectors producing R-Y and B-Y signals are enough to yield R, G, and B signals for controlling the three electron beams in the color CRT. The R-Y and B-Y outputs shown in the block diagram of Figure 16–23 are combined in a matrix amplifier to produce G-Y. Then these three signals must be added to the Y luminance output of the video amplifier system so that the sums yield R, G, and B.

Adding Y to the R-Y, G-Y, and B-Y signals is done either in the chroma demodulator circuits or in the CRT. The latter technique is referred to as *matrixing in the CRT*, where the R-Y, G-Y, and B-Y signals go to the three control grids, and the Y signal goes to the cathodes. The former technique adds Y right at the chroma demodulators. The disadvantage here is that the circuitry following the demodulators must have 4 MHz of bandwidth instead of the narrower chroma bandwidth because the Y signal is full-video. The full-video RGB signals are typically applied to the color CRT screen grids. Whatever the technique, color television is a complex system to which many people devote extraordinary energy, but the results are often enjoyable.

16–9 ■ HIGH-DEFINITION TELEVISION

In late 1992, the FCC released a channel plan, whereby existing broadcasters of the standard NTSC television format will be assigned a second channel for high-definition TV (HDTV) broadcasting. Thus TV broadcast stations that want to transmit HDTV in the next 15 years will be required to broadcast simultaneously using the NTSC standard and the HDTV standard on a separate channel. However, near the year 2007, the NTSC standard will be phased out, and the channels allocated to NTSC broadcast will have to be given back to the FCC for HDTV allocations. During 1992, six proposed HDTV systems were tested under the auspices of the FCC. In early 1993 the FCC decided which would be accepted as the standard format for high definition television in the United States.

Although the HDTV broadcast signal will have some analog characteristics, all the proposals featured a compressed digital information format. The advent of digital television technology in North America became a reality in 1993.

All-digital HDTV will offer certain specific features and characteristics. The most obvious to the user will be the wide-screen aspect ratio of 16 : 9, compared to NTSC's 4 : 3. The horizontal and the vertical resolution will be effectively doubled, with 1050 or 787.5 lines per frame with progressive scanning, compared to NTSCs 525 lines with interlaced scanning. Also, there is five-channel* compact disk-quality stereo sound.

Because of the simulcast requirement mentioned at the beginning of this section, the frequency spectrum of the transmitted HDTV signal must not produce interference in existing NTSC-format receivers. Likewise, a transmitted NTSC signal must not produce interference in HDTV receivers. The use of binary signaling can allow the elimination of sync, subcarriers, and even high-power carriers, because this system information can be digitized and transmitted by suppressed-carrier digital modulation techniques. Fortunately, and by design, digital video signals can be detected very accurately with 20- to 30-dB-lower signal-to-noise (S/N) ratios than analog. This means that HDTV will be transmitted with much lower average power than present NTSC signals.† It also means that the digital

---

* Arthur Cole, "HD Proponents Outline Changes," *TV Technology Magazine* (December 1992): 13.

† For NTSC, low VHF = 100 kW (20 dBk, or decibels above 1 kW), high VHF = 316 kW (25 dBk), and UHF = 5000 kW (37 dBk) of transmitted power. Average power requirements for HDTV transmissions are 12 to 18 dB less than NTSC.

signals will be less interference-prone in a high-power NTSC environment. When error-correction coding is added to the HDTV digital signals prior to transmission, the interference susceptibility will be even less.

A sampled and digitized (see Chapter 11) video signal will possess a random noiselike character. Extensive processing for compression and error-correction coding (see Chapter 13) will further randomize the data. Time-dispersion filters are used to spread out certain coherent pulses, thereby reducing their TV screen visibility. If necessary, scrambling (see Chapter 17) techniques may be employed. Low-level, random noiselike signals intruding into an NTSC receiver will produce much less observable and objectionable video interference than a coherent signal would. Hence, interference avoidance will be designed in from the start.

Although the proposed approaches featured digital information formats provided by advanced digital signal processing (DSP) techniques, both digital modulation and analog modulation for transmission of the DSP compressed HDTV picture signal were proposed. An example of each of these approaches will give a flavor of the techniques involved. Then a discussion of some of the digital compression techniques for video will allow an insight into modern information processing and transmission.

One transmission approach, referred to by the acronym of DSC-HDTV, uses a multilevel vestigial sideband amplitude-modulation technique. As an example for 4-VSB, 2 bits at a time of the digital information to be transmitted are used to set the amplitude of an RF carrier in a balanced modulator. The amplitude is determined by the information content of the 2 bits—that is, by the bit pattern (00, 01, 10, or 11). Of course, four levels can be uniquely defined by 2 bits. Hence, a two-level to four-level converter as represented in Figures 17–13 and 17–25 might be used.

As the data continue, the output of the balanced modulator is an AM double-sideband, suppressed-carrier (DSB/SC) RF signal that can be transmitted or up-converted to a higher center frequency. To make the receiver demodulation simpler and more accurate, a low-power carrier-related pilot is inserted into the transmitted spectrum, and the lower sideband of the DSB/SC signal is filtered out to halve the required RF transmission bandwidth, as illustrated in Figure 16–25. This approach is called *digital spectrum compatible* (DSC) high-definition television—that is, DSC-HDTV. Here, 4-VSB DSC-HDTV will require 16.5 dB S/N at the receiver in order to limit the bit error rate (BER) to fewer than 1 error per million bits transmitted when forward error correction (FEC; Chapters 13 and 17) is applied. Without FEC, the minimum SNR will be 21 dB. Since fiber optic and coaxial signal environments are less noisy than terrestrial broadcast, 16-VSB can be used for CATV transmission. For 16-VSB, 28 dB minimum S/N is required to maintain at least 10^{-6} BER; 33 dB are needed if uncorrected.

The DSC-HDTV system proposed included a 16:9 aspect ratio, 787.5 lines/frame progressively scanned, and a 59.94 Hz frame rate. Luminance bandwidth of 34 MHz and two chrominance bandwidths of 17 MHz each were chosen. The sampling frequency is 75.3 MHz, and there are 720 vertical and 1280 horizontal picture elements (pixels) on the screen. There are four channels of audio at 125.87 kb/s each, for a total audio data rate of 503.48 kb/s. All the sampled luminance

FIGURE 16–25 HDTV signal spectra. (a) DSC-HDTV. (b) DigiCipher. (c) AD-HDTV. (d) Standard NTSC.

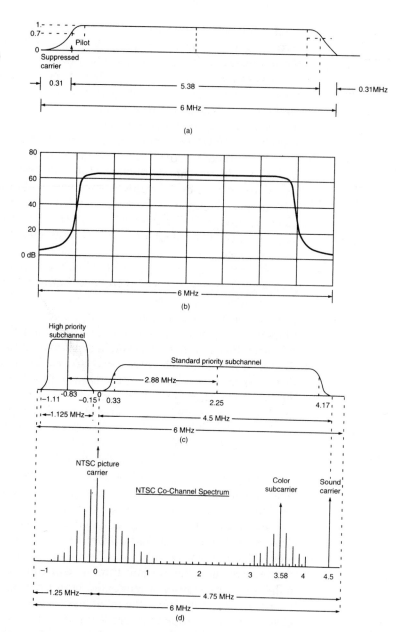

and chrominance video data are compressed to just 17 Mb/s and, after the digital audio and ancillary data streams are multiplexed together, the four-level VSB modulator input is 21.52 Mb/s. The 2 bits per symbol 4-VSB low-power transmission is at a rate of 10.76 million symbols per second (10.8 megabaud).

Channel Compatible DigiCipher uses a 32- or 16-QAM digital modulation technique (Chapter 17). The total data rates are about 25 Mb/s for the 32-QAM and 20 Mb/s for the 16-QAM. These *M*-ary digital modulation techniques provide a bandwidth compression–like spectrum efficiency. As an example, since $2^4 = 16$, 16-QAM theoretically provides 4 bits of information per transmitted symbol. Hence the transmitted baud rate is 5 megasymbols per second (4.878 M symbols/s, actually) for either scheme. The frequency of the suppressed carrier is in the center of the television channel bandwidth, and the transmission spectrum is nearly flat across the full 6 MHz. With its higher information rate capability, 32-QAM provides higher picture performance but requires at least 16.5 dB carrier-to-noise ratio. However, 16-QAM has a lower reception threshold (12.5 dB) and, hence, wider area coverage for a given transmitted power level. The choice is determined by the circumstances of the station operator. In order to minimize interference during the simulcast years, HDTV receivers will use comb filters to reduce energy in the regions where the high-power NTSC video, chrominance, and audio carriers are transmitted. NTSC receivers have protection from HDTV transmissions by virtue of the lower power requirements for digital transmission.

Yet another digital transmission approach (designated AD-HDTV) utilizes two in-band 32-QAM transmitters with a combined 24-Mb/s information rate. A 1125-MHz-bandwidth 32-QAM signal is centered 830 kHz below where an NTSC video carrier would be located, and a lower-power 4.5-MHz-bandwidth 32-QAM signal is centered 2.25 MHz above the frequency at which an NTSC video carrier would be transmitted. The lower-frequency, narrowband, higher-power signal delivers a ''high-priority channel'' that produces an acceptable picture and CD-quality sound using a total 4.8-Mb/s information rate. The low-power, wideband signal delivers 19.2 Mb/s of ''low-priority'' data to produce the full HDTV image. This two-tier system offers a 1050-line, interlaced scan with the standard U.S. 59.94-Hz field rate. All these schemes feature or can adapt to the 1050/59.94, 1125/60, or Europe's 1250/50 HDTV formats.

Objectives of the advanced digital (AD) HDTV system are to avoid interference from the simulcast high-power NTSC video carrier while also providing a digital transmission scheme that degrades gracefully at receivers far from the transmitter, where signal strength becomes marginal. Observers of analog amplitude-modulated TV or radio will sometimes accept some amount of spotty noise-like degradation in order to receive some important event. However, a digital receiver with marginal signal strength loses synchronization timing and blocks of data, which causes chunks of the picture and/or sound to be lost and then requires time (up to many milliseconds) to recover. The AD-HDTV proposal was the first to incorporate the internationally touted compression standard produced by the Motion Picture Experts Group (MPEG).

Video Compression Techniques

Without compression and M-ary digital modulation techniques (see Chapter 17), the information content of an HDTV signal would require data rates of about 1 Gb/s. A television channel transmission bandwidth can support approximately 20 Mb/s reliably. Hence, the information content of the video, audio, and other data (synchronization, closed caption, and so forth) must be minimized to achieve the 50-to-1 data reduction. Since the bulk of the information is in the luminance and two chrominance signals, video compression is a must.

The principal objective of the video compression process is to eliminate as much superfluous and redundant color and brightness information as possible without it being noticed by the viewer. Compression techniques exploit image correlations and human image-formation limitations in both space and time. Compression within a given frame (intraframe) is also called *spatial compression*. Frame-to-frame (interframe) compression is also called *temporal* (time-related) *compression*. Video compression is the end result of the electronic manipulation of video information in preparing images for a minimum storage space or a minimum transmission bandwidth while providing acceptable picture quality.

The single most important means of reducing superfluous video information is the removal of frame-to-frame (temporal) redundancy. Most program material is of relatively stationary scenes; little motion is occurring. Consequently, most of the image repeats without change over many frames at 30 frames/s (fps) in the United States and 25 fps in Europe. So, an electronic representation of the image can be stored at the receiver and the first frame can be repeated, except for alteration of the moving parts. Of course, changing only a few areas of an image implies a finely segmented screen and fast, sophisticated signal processing. This is a scenario ready-made for digital television, in which the electronic representation of an image is in a binary form.

The luminance (Y) and two chrominance (U and V) signal streams of each frame are sampled at very high rates as discrete picture elements called *pixels* (sometimes abbreviated *pels*), with the luminance being sampled faster than the U and V signals for high spatial-frequency resolution.* The samples that form each pixel are quantized into a number, M, of amplitude levels determined by the required resolution, where, for binary, M equals 2 raised to the number of equivalent bits of information. Horizontal and vertical intensity information requires more dynamic range than image color, so the Y samples are assigned more bits. Each pixel then has three sets of bits, representing the Y, U, and V samples. An idea of the amount of video storage capacity and output data rate necessary may be found by examining the European format: If 1150 (or, perhaps, 1152) lines per frame are projected on the screen using 1440 pixels with each pixel using 16 bits (4 : 2 : 2), then the dynamic random access memory (DRAM) required will be approximately 26.5 Mb/frame. Further, at a 50-Hz refresh rate, the 20 ms per frame will produce a data rate of approximately 1.3 Gb/s. The 20 ms/frame assumes two interlaced fields per frame and 100 Hz per field (the present slight flicker is to be eliminated by doubling the field rate). Clearly, HDTV presents challenges to digital hardware designers. To reduce the data rate to allow transmission over standard 4-MHz (NTSC), 5- and 5.5-MHz (PAL and variants), and 6-MHz (SECAM) bandwidths, a major reduction in the amount of data to be transmitted is called for.

Various algorithms, both standardized and proprietary, are used to process pixel samples for compression. Standards include those by the Joint Photographic

---

* For example, 75.3 million samples per second for the DSC-HDTV system.

Experts Group (JPEG) for still-picture compression, CCITT's H.261 for video teleconferencing, and the Moving Pictures Expert Group (MPEG) for full-motion video compression. For motion video, the most widely used algorithms are based on *predictive coding*, in which each pixel is compressed by quantizing the difference between its present value and a predicted value so that the difference can be used, and *transform coding*, in which the time-domain pixel samples are mathematically transformed into frequency-domain sample values. (For example, see Fourier series and transforms in Chapter 3.)

For typical compression methods, the image pixels are separated into nonoverlapping 8 by 8 square blocks (arrays). Each block of 16 pixels of (Y, U, V) samples is then transformed using a two-dimensional discrete cosine transform (DCT) with an 8 by 8 kernal.* The resulting 64 coefficients represent the frequency contents of the given block. Most of the DCT coefficients compute to a value of zero and will not need to be transmitted, thus achieving data compression. The DCT coefficient value in the upper left-hand corner of the 8 by 8 array is a measure of the 0-Hz (dc, or average) value of the block. The other 63 entries are ac coefficients representing the relative strengths of signal terms with increasing horizontal frequency from left to right and for terms with increasing vertical frequency from top to bottom. If there is a large amount of spatial redundancy in the picture, many of the medium- and high-frequency coefficients are small or zero. Techniques such as this that produce high compression ratios ($10:1$ to $200:1$) are referred to as *lossy* in that the reconstructed picture at the receiver is not identical to the original. Lower compression methods ($3:1$) are used in sensitive applications such as medical imaging in order to eliminate possible false artifacts in the image that could result in an incorrect diagnosis.

A simplified HDTV video compression scheme is illustrated in Figure 16–26. A predicted picture is produced in the main feedback loop by applying motion compensation to a previous picture stored in frame memory. (A discussion of motion estimation and compensation is deferred for now.) The difference between the new input picture and the predicted picture is DCT transformed and quantized with a variable step size determined by the buffer. The result is to down-scale a large-value data point to a lower-precision value requiring fewer bits—that is, rounding off—and discarding coefficients below a value determined by the output buffer. This is a lossy (lost information) algorithm. If done correctly, the observer (viewer) will not notice a change.

The quantized DCT coefficients to the variable-length coder (VLC) are reordered into a one-dimensional array (a serial data stream) by reading out the entries of the 8 by 8 array along a zigzag path diagonally from the upper left corner component c_{11} to c_{21} to c_{12}, etc., to the 64th component, c_{88}. The coefficients are losslessly encoded, typically using Huffman coding to reduce the number of bits actually used, and then run-length coded to take advantage of repeating values, especially zeros. Run-length coding reduces the coefficients to a number of events, each of which represents a nonzero coefficient and the number of preced-

* Peng Ang, Peter Ruetz, and David Auld, "Video Compression Makes Big Gains," *IEEE Spectrum Magazine* (October 1991): 16–19.

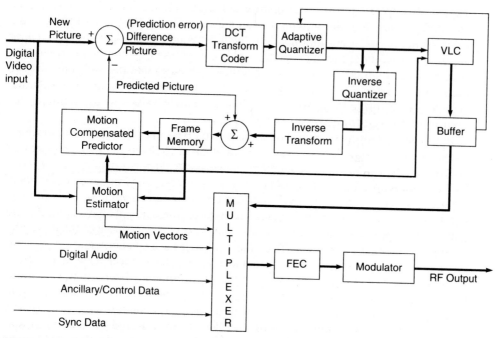

FIGURE 16–26 HDTV compressed digital video transmitter.

ing zero coefficients. Hence another level of compression is achieved. The buffer accepts data at varying rates and sends constant-rate picture-difference video data to the output multiplexer, where digital audio, sync data, ancillary control data, and video motion vectors are combined. After forward error control such as convolutional coding is applied, the data stream is sent to one of the modulators mentioned in the preceding section and discussed in Chapter 17.

To address motion prediction and compensation, return to the top of Figure 16–26. The predicted picture is subtracted from the actual incoming new picture. This picture (or frame) difference information is, after further processing to remove visual redundancies, sent to the output multiplexer for transmission to the receiver decoder for correcting its prediction. At the same time, the DCT and quantized difference data are taken from the adaptive quantizer, as seen in Figure 16–26 at the start of the main feedback path, and used to construct the next predicted and motion-compensated picture.

To improve motion estimation in the predicted current picture, the preceding picture data, stored in the frame memory block, are compared to the current picture data on a pixel-block by pixel-block basis in the motion estimation functional block in order to find matching areas in the preceding frame. When a nearly matching area is found, a motion vector is generated by the motion estimator and sent to the motion (compensated) predictor. Both interframe (temporal, frame-to-frame) and intraframe (spatial, within a frame) information are processed. A motion vector must indicate what part of the preceding frame should be used to predict a specific area of the current frame. Using motion prediction and compensation, the magnitude and direction of picture changes are determined. The motion vectors are sent to the motion predictor, the VLC, and the multiplexer for transmission to the TV receiver.

Motion prediction and compensation by digital processing provides an additional feature for video transmission.* With well-established standardized definitions, including transmission of headers and descriptors with the data, long-term compatible upgrading of image quality provided to the receiver decoder can be accomplished without any need for hardware upgrades in decoders. The ability to extend the performance, features, or technology of the system in the future has been considered in terms of extensibility, scalability, and interoperability. These are terms taken from the computer industry that are now criteria applied to HDTV standardization.

Extensibility is the ability to extend operations to diverse display technologies that will be adaptable for use with new, higher-resolution displays developed in the future. *Scalability* provides the capability to scale various video parameters upward or downward. Such scaling will allow the user to choose different aspect ratios, number of frames per second (frame rate), or the number of scan lines at the point of purchase or while viewing. Thus the number of lines on the screen will be measured in lines per inch as is done for computers, rather than television's present lines per frame (525/1050, for example). Also, multirate encoding, multirate displays, and variable complexity decoders will be available for a variety of resolutions and costs. *Interoperability* is already well established in the computer industry. It is the part of standardization that allows the use of a variety of receivers and display devices. The thrust in all of this is to design more flexibility and compatibility with computers and other media devices into digital HDTV systems, allow consumers various hardware options, and allow for future upgrades without hardware changes.

■ PROBLEMS

1. Determine the frequencies of the video and audio carriers for channel 10.

2. Explain the differences in how light is changed to electrical signals in the Vidicon and Image Orthicon–type camera tubes.

3. The PAL system in Europe uses 625 lines per frame and 25 frames per second. Assuming a video bandwidth of 5 MHz, determine:
   a. The horizontal sweep rate.
   b. The number of distinguishable picture elements, assuming 84% of each horizontal line and 90% of the 625 lines are visible.

4. Refer to the PLL channel synthesizer of Figure 16–7. Determine the following:
   a. The phase detector reference input frequency.

* S. Merrill Weiss, ''Examining Compression Techniques,'' part of a multipart series in *TV Technology Magazine* (October 1992): 35, 38, and 39.

b. The value of N for the programmable divider when tuned to channel 5. (*Hint:* Calculate the LO frequency f_0 from the video carrier RF and IF frequencies.)

5. What are the advantages of the sampled AGC system used in TV receivers?

6. Refer to Figure 16–10.
   a. How much bias current must Q_1 have in order for Q_2 to have a 2-mA collector-bias current? (Assume very high beta transistors and $R_1 = 350\ \Omega$.)
   b. Determine the maximum voltage gain of the video amplifier (maximum contrast), assuming R_2 is in the center of its range, $I_{C2} = 10$ mA, and the ac cathode impedance is 224 kΩ. (Inductors and bypass capacitors are shorts.)
   c. If the electron beam is 1 mA average, what is the maximum possible dc voltage on the cathode? (Set the brightness control for maximum brightness.)
   d. If the CRT anode voltage is 10 kV, find the apparent dc impedance of the vacuum tube. (Include part **c** conditions.)

7. **a.** What component is included in Figure 16–10 for shunt peaking?
   b. What component is included for series peaking?
   c. Explain the purpose of peaking for a video amplifier.

8. What value should L_s be to resonate with 30 pF at 3 MHz?

9. In Figure 16–11, the base-emitter junction of Q_1 is used as a rectifier. What component is primarily responsible for keeping the collector voltage at V_{cc} between sync pulses? Why?

10. A 5-V battery is suddenly applied to an ideal 40-mH vertical deflection coil.
    a. After 15 ms, what will be the peak current?
    b. What will be the maximum negative voltage across the inductor during the flyback period if the vertical sweep frequency is exactly 60 cycles/s and the coil is discharged through a 1-Ω resistor?
    c. Sketch the vertical sweep current and voltage waveforms.
    d. Repeat the above if the coil has 4 Ω of series resistance.

11. Give two specific functions of the damper diode.

12. If the peak collector voltage of Q_1 in Figure 16–14 reaches 120 V and there are 20 turns on T_2 from the collector tap point to the 128-V power supply, what must be the total number of turns in order to produce a 9-kV pk pulse for CRT anode voltage?

13. Why is a phase-locked loop used to synchronize the horizontal sweep?

14. R_4, R_5, and C_5 of Figure 16–16 form a lead-lag, frequency-compensation network for the horizontal sync system.
    a. Determine the lead and lag corner frequencies, assuming the phase detector output impedance is 110 kΩ and R_6 is open.
    b. Determine the corner frequency due to C_6.
    c. Determine the maximum and minimum horizontal frequencies for $L_1 = 30$ mH, assuming the voltage-variable resistance circuit can vary from 0 to infinite ohms.

15. A color TV screen is white until the red gun fails. What is then seen on the screen?

16. Give three reasons for transmitting color burst.

17. **a.** The color phasor has what phase for the I signal?

    b. If green is to be transmitted, what are the approximate relative magnitudes of the I and Q vectors; use Figure 16–20 for a graphical approximation.

18. a. Why does the I signal have a wider bandwidth than the Q signal?

    b. What is the minimum bandwidth of the chroma amplifier system in the receiver?

19. Why is a color oscillator necessary in a color TV? (Specify modulation types, and so on.)

20. The tint control actually controls the _____ of the chroma signal, and the color control actually controls the _____ of the chroma signal.

17

Digital Radio and Space Communication

■ INTRODUCTION

In the past, terrestrial line-of-sight and satellite microwave systems have used analog FM with its excellent noise-suppression qualities. Many of these systems still use FM, but the baseband (modulation) signals are increasingly digital to allow real-time transmission, as well as the simple data storage and information rate-change techniques offered by digital signal processors.

Since digital processing of data, voice, and video has become so widespread and efficient bandwidth utilization so crucial, higher-level (*M*-ary, or multistate) signaling techniques are being used. These techniques take advantage of digital phase modulation, called *phase-shift key* (PSK), and a combination of amplitude and phase-shift keying designated APK or, more commonly, *quadrature-amplitude modulation* (QAM).

Also, relatively new techniques have been developed for effectively utilizing the bandwidth, power, and broadcast capability of repeaters. The multiple-access techniques used today in ground-satellite networks and even local and wide area networks are *frequency-division multiple access* (FDMA), *time-division multiple access* (TDMA), and *code-division multiple access* (CDMA). And these can be employed with either fixed assignment (FDMA) or demand assignment (DAMA).

Transmission over terrestrial, global, and extraterrestrial distances requires the use of modulated carriers and, for modern applications, FSK, *M*-ary PSK, and QAM are used. Digital radio incorporates all of the digital modulation techniques discussed in this chapter. However, the constant-envelope signals provided by *M*-ary PSK and QAM are still receiving most of the attention for digital telecommunication satellite applications.

A representative, terrestrial digital radio system is the AT&T DR 6–30. The repeater (actually consisting of two receivers and two transmitters) is illustrated in the block diagram of Figure 17–1. The DR 6–30 operates within the 6-GHz common-carrier frequency allocation and conforms to the required 30-MHz maximum of half-power bandwidth. The received signal is 16-QAM-modulated, with a 90.524-Mb/s data stream (equivalent to 1344 voice channels) consisting of two 44.736-Mb/s DS-3 data streams that can be made up from seven DS-2 lines (which is four T1 lines; see Table 17–1).

The receiver RF system has a two-stage GaAs FET preamplifier with a nominal gain of 15 dB and a noise figure of 2.8 dB (the system NF is 4 dB). The LO is derived from a 1-GHz transistor oscillator phase-locked to a 4-MHz TCXO through a ÷256 prescaler and is stable to approximately ±2 ppm over a 0–50°C temperature range. The receiver threshold is −78 dBm for a bit error rate (BER) of 10^{-3} (1 error per 1000 bits), but the system threshold is maintained to better than 10^{-6} BER.

The 16-QAM signal is demodulated to baseband and then regenerated for minimum noise accumulation. The parity is checked and corrected, and service channel bits are extracted and inserted. Nonregenerative-type repeaters merely downconvert to IF for filtering and then up-convert and boost the power for retransmission.

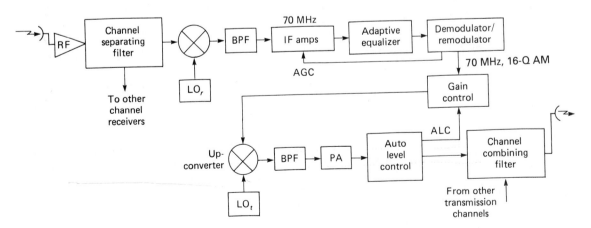

FIGURE 17–1 Digital radio block diagram—AT&T's DR 6–30.

TABLE 17–1 U.S. Hierarchy for Telephone Multiplexing

Designation	Rate (Mb/s)	Number of Telephone Voice Channels
DS-1 (T1)	1.544	24
DS-2	6.312	96
DS-3	44.736	672
DS-4	274.176	4,032

The transmitter amplifier is a 30-dB gain, four-stage GaAs PA with 2.5 W of power when backed off 2.5 dB from saturation. The linearity of the overall system is very important because of the QAM modulation. Therefore, amplitude is carefully controlled in the receiver and the transmitter. In addition, an adaptive equalizer allows for correction of amplitude to within 1 dB and delay distortion to within ± 1 ns over ± 12 MHz of the RF bandwidth.*

The preceding circuit performance parameters have all been introduced in previous chapters in the context of analog radio and baseband digital systems, so now the digital modulation and transmission techniques are discussed. Then the chapter concludes with satellite multiple-access techniques. ■

17–1 ■ MODEMS AND DIGITAL MODULATION TECHNIQUES

As noted in Chapter 13, M-ary digital modulation techniques are used in high-data-rate synchronous modems for digital communication over telephone circuits. These modems can transmit end-to-end data rates of up to 9600 b/s (information rate). How it is possible to transmit 9.6 kb/s over a 3-kHz bandwidth telephone line is explained in this section. The digital modulation schemes listed for standard telephone system modems in Table 17–2 are explained in subsequent sections of this chapter.

Phase-Shift Key (PSK)

Phase-shift keying is a carrier system in which only discrete phase states are allowed. Usually 2^n phase states are used, and $n = 1$ gives 2-phase (binary, BPSK), $n = 2$ gives 4-phase (quadriphase, QPSK), $n = 3$ gives 8-phase, $n = 4$ gives 16-phase, and so forth.

Binary PSK is a two-phase modulation scheme in which a carrier is transmitted (0° phase, reference) to indicate a MARK (or digital 1) condition, or the phase is reversed (180°) to indicate a SPACE (or digital 0). The phase shift does not have to

---

* I am indebted to J. J. Kenney of Bell Laboratories for his excellent article, "Digital Radio for 90 Mb/s, 16-QAM Transmission at 6 and 11 GHz," in the August 1982 issue of the *Microwave Journal*.

TABLE 17-2 Standard Telephone System Modem Characteristics

CCITT (V.XX), AT&T (Bell) Specification	Bit Rate (bps)	Modulation Scheme	Wires/ Duplex	Sync/ Async	Remarks Dial-up (D) Leased (L)
Bell 103	0–300	FSK	2/full	async	D, L
Bell 202	1200	FSK	4/half	async	D, L
V.22, Bell 212	1200/600	QPSK/FSK	2/full	either	D, L
V.22bis	2400	16-QAM	2/full	either	(fallback: 212/103); D, L
V.26, Bell 201	2400	QPSK	2/half, 4/full	sync	D
V.26bis	2400	QPSK	4/full	sync	L
V.26ter	2400	QDPSK	2/full	sync	L
Bell 208	4800	8-PSK	2/half, 4/full	sync	L
V.27	4800/2400	8-DPSK/QDPSK	4/full	sync	L
V.27ter	4800/1200	8-DPSK/QDPSK	2/half	sync	L
Bell 209	9600/2400	16-QAM	4/half	sync	L
V.29	9600	16-QAM (modified)	4/half	sync	L
V.32	9600	32-QAM*	2/full	sync	L
V.33	14400	32-QAM*	4/full	sync	L
V.35	56000		4/full	sync	L

* With trellis-coded modulation (TCM) and echo cancellation.

be 180°, but this allows for the maximum separation of the digital states. Maximizing the phase-state separation is important because the receiver has to distinguish one phase from the other even in a noisy environment. A simple block diagram for BPSK is shown in Figure 17–2.

The switches shown in Figure 17–2 can be implemented with numerous devices and circuits, but two of the more popular circuits useful over a wide range of frequencies—up to 4 GHz—are the balanced modulator using Schottky barrier (hot-carrier) diodes, as seen in Figure 17–3, and dual-gate FET switches. This diode modulator is obviously an important circuit because you have seen it before as a balanced mixer and a balanced AM modulator (double-sideband/suppressed-carrier) and demodulator. It is also the basis for higher-level PSK and QAM modulators and demodulators.

When the digital voltage is high, D_1 and D_3 conduct with ground G being the data current return path. A high-frequency carrier coupled across T_1 will then be connected through closed switches D_1 and D_3 to the primary of T_2. Notice the instantaneous phase of the carrier vector at T_2 given by the ± signs. If the data polarity reverses, D_1 and D_3 reverse-bias (open), whereas D_2 and D_4 conduct as short circuits. But notice that, for the same instantaneous carrier phase at T_1, the phase at T_2 will be reversed from that of the previous data condition—the carrier

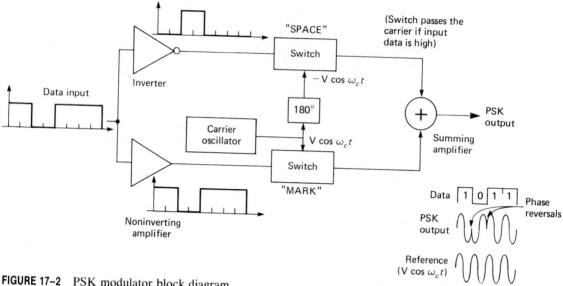

FIGURE 17-2   PSK modulator block diagram.

phase has been shifted by 180°. This is a very simple phase-reversal modulator. So the block diagram of Figure 17–2 can be implemented as shown in Figure 17–4.

Demodulator

Like any other PM receiver, the PSK demodulator must have an internal signal whose frequency is exactly equal to the incoming carrier in order to demodulate the received signal.

With analog PM, a phase-locked loop (PLL) is used that locks up to the received carrier. However, BPSK with its sudden phase reversals is equivalent to

FIGURE 17-3   Balanced PSK modulator circuit. If T_1 and T_2 are wound so that they do not invert, the carrier phase will be 180° (referenced to the input carrier) when the data is a digital 0. A digital 1 will produce no phase shift.

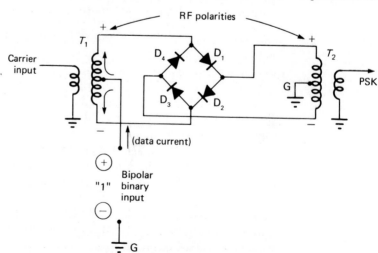

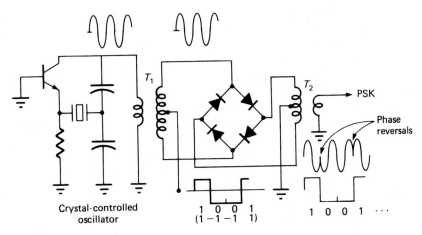

FIGURE 17-4 BPSK modulation.

constant-amplitude DSB-SC, and no discrete carrier component is present for directly locking the PLL. The BPSK signal is written

$$s(t) = A \cos[\omega_c t + \theta_d(t)] \tag{17-1}$$

where $\theta_d = 0$ or π rad depending on whether the digital input is $d = 1$ or -1. Substituting for the two possible phase states, we see that Equation 17–1 can be equivalently written as

$$s(t) = A \, d(t)\cos \omega_c t \tag{17-2}$$

As with DSB-SC demodulation, a Costas loop (see Section 8–8) can be used. The Costas loop is a PLL that combines quadrature- ("at right angles") detector outputs to derive the PLL error voltage. The error voltage then controls the VCO frequency that becomes the *recovered carrier* signal needed for demodulation. If the recovered carrier (VCO output) signal is

$$vco(t) = 2 \cos \omega_c t \tag{17-3}$$

then the appropriate demodulator is a phase detector, input with the BPSK signal as seen illustrated in the lower branch of Figure 17–5. As discussed in Chapter 10, the phase detector is a dc-coupled mixer that performs a circuit function mathematically equivalent to multiplication. Thus, the phase-detector product produces an output given by

$$v_{PD}(t) = (2 \cos \omega_c t) \times (A \, d(t)\cos \omega_c t) \tag{17-4a}$$

$$= 2A \, d(t)\cos^2 \omega_c t \tag{17-4b}$$

$$= 2A \, d(t)[0.5 + 0.5 \cos 2\omega_c t] \tag{17-4c}$$

$$= A \, d(t) + A \, d(t)\cos 2\omega_c t \tag{17-4d}$$

Equation 17–4d consists of the demodulated baseband information (data) signal $d(t)$ scaled by a constant factor A, and the right-hand part is a BPSK signal at the

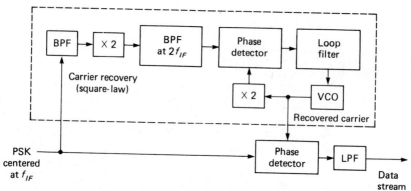

FIGURE 17-5 PSK carrier recovery and demodulation.

second harmonic of the carrier frequency that can be filtered out in a low-pass filter.

Another technique for carrier recovery is the square-law or doubler method. This is similar to the data-clock synchronizing described later in this chapter and is illustrated in Figure 17–5. The square-law device is a frequency doubler that produces a discrete spectral component at twice the carrier frequency. The other components are filtered out, and the carrier second-harmonic component, along with its noise, is used to phase-lock a frequency-doubled VCO. When the noise is averaged out, the VCO frequency is precisely equal to the received carrier. A generalized block diagram of the PSK demodulator with carrier recovery for phase reference is shown in Figure 17–6.

Differential PSK

Differential phase-shift keying (DPSK) means that the data is transmitted in the form of discrete phase shifts Δθ, where the phase reference is the previously

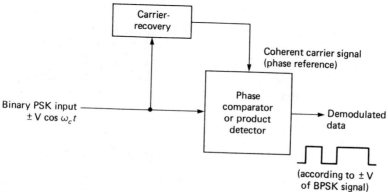

FIGURE 17-6 PSK demodulator block diagram.

transmitted signal phase. The obvious advantage of this technique is that the receiver, as well as the transmitter, does not have to maintain an *absolute* phase reference against which the received signal is compared.

In DPSK, the data stream is initially processed in an exclusive-NOR gate and then made bipolar as shown in Figure 17–7 before phase modulation of the carrier. Figure 17–7 also shows the computer output data, the bipolar voltages applied to the BPSK modulator, and the final transmitted DPSK phase shift.

The demodulator of Figure 17–8 shows the same DPSK signal transmitted by the modulator of Figure 17–7. The inputs to the phase detector are the received signal and the same signal delayed by an amount equal to one bit period. The phase detector produces a positive voltage when the input phases are the same. The comparator is used to regenerate the data pulses and make them TTL computer compatible.

As is usually the case, we still don't get something for nothing. DPSK does not require a phase-locked loop demodulator; however, the bit error rate (BER) will be worse by a couple decibels of S/N than coherent demodulation of PSK because both inputs, signal and reference, to the DPSK phase detector are noisy.

Despite all this discussion, binary PSK offers less than 10 dB improvement over FSK when coherent detection is used, and considerable complexity is added to the modem hardware.

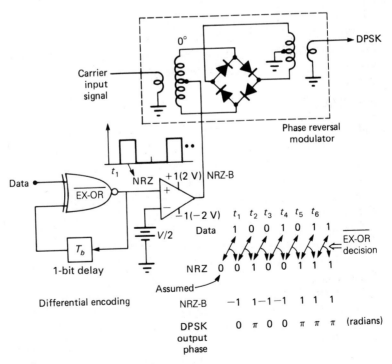

FIGURE 17–7 Differential PSK (DPSK) modulator.

FIGURE 17-8 Differential PSK
(DPSK) demodulator.

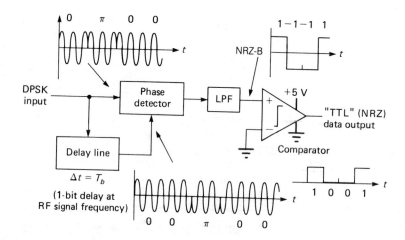

Beyond Binary

Digital data systems for high-speed transmission over bandwidth-limited media, such as telephone lines and microwave links, use higher- (than binary) level modulation techniques; that is, they send more information per transmitted signal transition (symbol) than binary systems. Instead of *bin*ary, then, this is *M*-ary encoding.

The most widely used multilevel modulation schemes today are *M*-ary PSK and QAM (quadrature AM, a combination of PSK and ASK). Quaternary (four-level) PSK allows twice the information density of binary PSK and forms the basis for understanding all sorts of quadrature-carrier modems and digital microwave systems.

QPSK

Quaternary or quadriphase PSK offers twice as many data bits per carrier phase change than does BPSK. Hence, QPSK (4-PSK) finds wide application in high-speed carrier-modulated data transmission systems. Most data modems for synchronous transmission at data rates of 2400 b/s on voice-grade telephone lines, and at higher rates on broadband circuits and microwave digital radio links, use differentially encoded QPSK.

The system data transfer rate is 2400 b/s, but the transmission line signaling rate is 1200 symbols per second (baud)—a "symbol" in this case being a phase shift.

A typical choice for the four relative phase shifts in a differentially encoded QPSK system is shown in Figure 17–9. Here the phases $\pm\pi/4$ and $\pm3\pi/4$ are seen relative to the horizontal axis that represents the phase of the previous baud (transmitted phase shift).

This figure is referred to as a *constellation*, as if the Xs represent stars on an astronomy chart. Since there are four possible phase changes for each data transition, each change in phase, $\Delta\theta$, represents one of four points in the constellation.

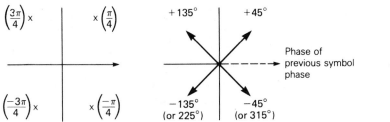

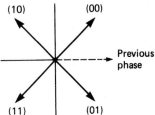

FIGURE 17-9 Phase constellation for differential QPSK.

FIGURE 17-10 One possible dibit constellation for differential QPSK.

Recall that the word *bit* is a contraction of *binary* and *digit*. Thus, compared to a two-level system, four levels define twice as many binary states, so we call each transmitted phase change a *dibit* or di (= double) binary digit. An example of how the dibits might be defined for the differential QPSK constellation above is shown in Figure 17–10. Notice how successive phase shifts correspond to dibits that change by only 1 bit; this is a Gray code arrangement.

The circuitry for generating QPSK is forthcoming, but to get the idea behind how a four-level modulation might be generated, let us look at the block diagram of Figure 17–11. Note that this block diagram is general, and the voltage generated (V_{pm}) could be used to produce various four-level signals such as four-level intensity modulation of a laser.

When they arrive at the transmitter modulator, the incoming serial binary data bits are temporarily stored in pairs. Each binary digit pair (dibit) is then given a voltage level that, when applied to the phase modulator, produces a phase shift corresponding to that of the Figure 17–9 constellation.

When the assigned voltage level causes a differential phase shift relative to the previous phase, this is called *differential* QPSK, or DQPSK. An example of a differential four-level voltage for DQPSK is shown in Figure 17–12.

Compare the data stream to the transmitted phase and notice that, if the data bits are coming in at 2400 b/s, the transmitted phase shifts are at a rate of 1200 symbols per second—the baud rate. The receiver reverses the process and produces 2400 b/s. Thus, by using QPSK, the transmission line bandwidth can be

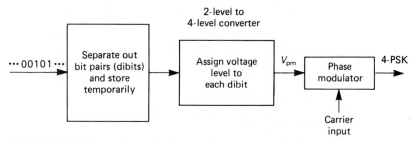

FIGURE 17-11 Generalized 4-PSK modulator block diagram.

FIGURE 17–12 Four-level voltage and DQPSK output, Δθ.

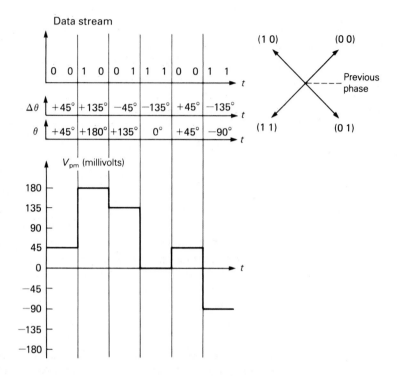

one-half that of BPSK or FSK transmitting the same amount of information (bits per second).

In order to realize this two-to-one bandwidth savings, however, a lot of thought and design effort goes into optimizing the transmitted QPSK spectrum—specifically, spectrum symmetry and bandwidth minimization.

Spectrum symmetry primarily reduces signal propagation distortions caused by multipath and other transmission-media interferences. Symmetry is achieved by scrambling. *Scrambling* is a process of pseudorandomizing the input data to eliminate fixed spectral components that occur during long strings of 1s or 0s and helps maintain bit synchronization. Fixed spectral components can produce greater interference to adjacent channel signals than randomly occurring components. Scrambling can also be used to provide some transmission security. *Bit stuffing* is used when different bit-rate data streams are combined into the synchronous data stream. This is usually required when multiplexing data from different sources.

Bandwidth minimization is accomplished by transmitter output filtering as usual and, more importantly, by premodulation shaping of the data pulses. Raised-cosine pulse shaping was discussed in Chapter 12.

In addition to producing a four-level signal and QPSK from a single data stream, some systems multiplex two separate data streams to give two-channel operation on a time-division multiplex basis. A logic (with analog output) circuit capable of producing the four-level signal V_{pm} from two TTL data bits (A and B) is shown in Figure 17–13.

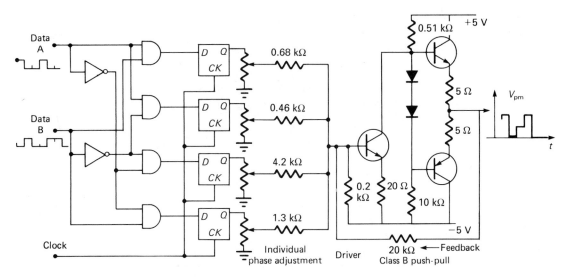

FIGURE 17-13 Two-level to four-level converter, which produces the modulation voltage for QPSK from two separate binary data streams or the two bits forming each dibit.

For generating QPSK, we could apply the two bits forming each dibit to two BPSK modulators (like Figure 17–3) that are in phase quadrature (90°) to each other. This circuit is shown in Figure 17–14.

This is a very important circuit to conceptualize because it forms the basis for deriving higher-level (8, 16, 32, . . .) PSK, QAM, and other quadrature-carrier modulation schemes. Incidentally, there is nothing new about this quadrature-carrier modulator; a continuous-phase version is used to produce color on your TV set.

To understand how the output constellation is produced by this circuit, let a_1 and b_1 assume the values for each dibit pair and follow the phase of the oscillator signal through the circuit. If no phase inversion occurs across transformers (primary-to-secondary), and a_1, the first of two bipolar (± 1) data bits, causes two of the four diodes in the upper phase-reversal modulator to conduct (and two to be reverse-biased), then the constellation shown in Figure 17–14a will result. Examples for 00 and 01 are shown in Figure 17–15. The input data register might be implemented as in Figure 17–14b.

Offset QPSK (OQPSK) or staggered QPSK (S-QPSK) is produced in a similar circuit to QPSK except that the second of the bits, b_2, input to the serial shift register is delayed one-half symbol period before being applied to its phase-reversal modulator. This results in instantaneous carrier phase transitions of only 0° or $\pm 90°$ from the previous phase state, as compared to QPSK's 0°, $\pm 90°$, and 180°. Hence, OQPSK does not have the complete phase-reversal transients (and deep envelope notches) of QPSK when bandlimited. The primary circuit difference is that the diode quads are replaced with a true multiplier circuit.

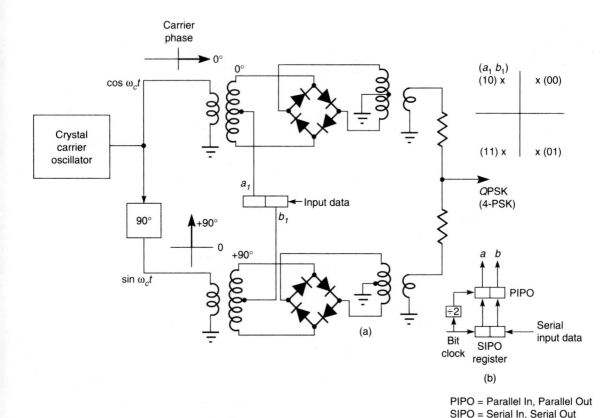

FIGURE 17–14   Quadrature-carrier QPSK modulator.

QPSK Demodulation

A block diagram showing QPSK demodulation and data recovery processing is given in Figure 17–16. This system includes circuitry for recovering a coherent carrier for the phase detector. Various techniques are used for carrier recovery, including the Costas loop and multiply/divide schemes.

Following demodulation and filtering of the in-phase and quadrature signals, the data pulses are regenerated (squared up) and processed to remove the receiver phase ambiguities, followed by parallel-to-serial conversion of the I and Q data.

The recovery of a clock signal from a synchronous data transmission is examined following quadrature demodulation and carrier recovery.

Quadrature Demodulation

The question that arises with the QPSK demodulator is why two data output lines, I and Q, are shown. After all, a single phase representing a 10, 01, 11, or 00 is transmitted, and it would be a simple matter to have the demodulated voltage level

FIGURE 17-15 Generating the QPSK constellation for the quadrature-carrier circuit of Figure 17-14. Each phase-reversal modulator reverses the carrier phase for a binary 1.
(a) Dibit 00 produces $\theta_0 = \pi/4$.
(b) Dibit 01 produces $\theta_0 = 3\pi/4 = -\pi/4$.

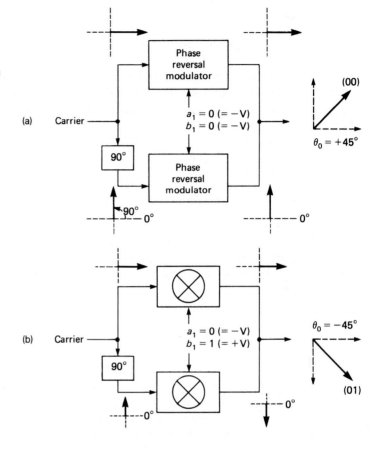

FIGURE 17-16 QPSK receiver block diagram.

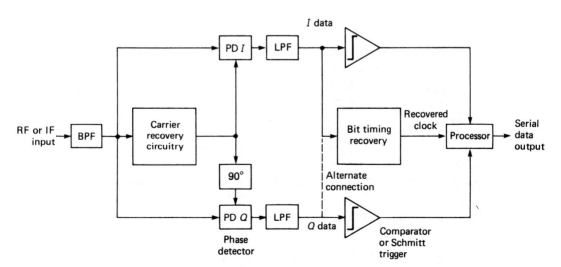

trigger the appropriate two bits. The problem is ambiguity. The output of a phase detector, like PDI of Figure 17–16, is ambiguous regarding which phase was transmitted. A review of phase detector basics will explain this (also see Section 10–1).

Recall that the phase detector is a mixer or product detector with a dc-coupled output. Like all mixers, the function performed mathematically is multiplication. In the case of a phase, or product detector, the two input signals have the same carrier frequency; therefore the difference or IF frequency is *dc*—or in the case where one of the signals is phase-modulated, the dc voltage varies with the phase variation. If the phase detector inputs are $\cos(\omega_c t + \theta_d)$ and $\cos \omega_c t$, as shown in Figure 17–17, then the output will be

$$\cos(\omega_c t + \theta_d) \times \cos \omega_c t = \tfrac{1}{2} \cos[(\omega_c t + \theta_d) + \omega_c t]$$
$$+ \tfrac{1}{2} \cos[(\omega_c t + \theta_d) - \omega_c t]$$
$$= \tfrac{1}{2} \cos(2\omega_c t + \theta_d) + \tfrac{1}{2} \cos \theta_d \qquad \textbf{(17–5)}$$

where θ_d is the transmitted data phase.

A low-pass filter is used to attenuate the second harmonic term so that the result is

$$V_o = \tfrac{1}{2} \cos \theta_d \qquad \textbf{(17–6)}$$

θ_d is the modulated phase ($\pm\pi/4$, $\pm3\pi/4$), and $V_o = \tfrac{1}{2} \cos \theta_d$ is the output dc voltage representing the appropriate dibit. Figure 17–18 shows the four possible outputs of the *I* detector. The ambiguity is that if $\tfrac{1}{2} \cos(\pm45°) = +0.35$ is the detector, it is not known if $+45°$ or $-45°$ was received; the same ambiguity exists for -0.35 V and $\pm135°$.

In order to resolve the ambiguities, we must use a second phase detector operating in quadrature (90°), called PDQ in Figure 17–16, and a two-level decision circuit to establish the final output full-binary data stream. Figure 17–19 shows the PDQ outputs for QPSK. Notice that the voltage for $-\pi/4$ and $+3\pi/4$

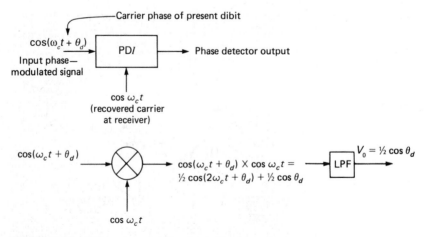

FIGURE 17–17 Phase detector for detecting data phase θ_d.

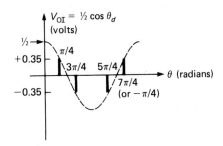

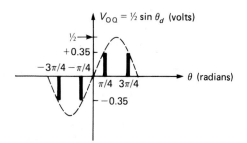

FIGURE 17–18 In-phase phase detector (PDI) output showing the output voltages corresponding to the four discrete phases of the QPSK input.

FIGURE 17–19 Quadrature phase detector (PDQ) outputs for QPSK.

have the opposite polarity to those at the output of PDI. Hence a logic decision circuit can examine the two phase detector outputs and determine the transmitted phase. Having correctly determined the received dibit phase, the logic circuitry produces the corresponding two bits of serial data.

Carrier Recovery

Because of transmission problems, especially in digital radio systems with multipath, fades, and other phase nonlinearities, the data bits are usually scrambled in a way that will make the transmitted QPSK spectrum quasi-random. This makes for a very complex receiver.

The problem is finding a carrier signal to lock to in the midst of a quasi-random keyed, phase-reversal modulated signal spectrum (see the spectrum of Figure 17–20). One type of carrier-recovery circuit for QPSK is the ×4 multiplier tech-

FIGURE 17–20 Spectrum of PSK signal modulated with quasi-random data.

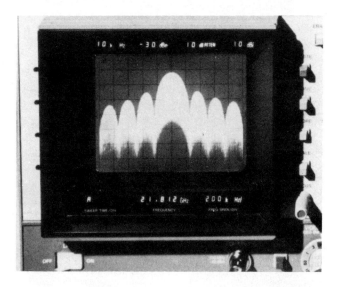

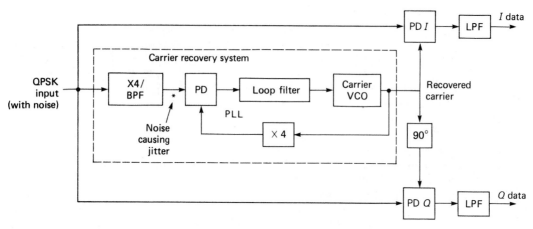

FIGURE 17–21   QPSK demodulator, including the carrier recovery system.

nique shown in Figure 17–21. It can be shown mathematically that spectral lines at the carrier frequency can be produced by frequency-multiplying the random QPSK. After filtering, we divide back down using a PLL to recover the carrier. Clearly, this is the same technique as used for BPSK carrier recovery (where ×2 was used).

In the QPSK demodulator using the ×4 method of carrier recovery shown in Figure 17–21, phase-reversal transitions as well as other noise will cause the recovered coherent carrier to jitter with phase and frequency noise. *Phase jitter* of the receiver phase reference must be minimized in order to minimize *inter-symbol interference* in multiplexed data systems and to minimize the bit error rate.

Bit Timing Recovery

Once we get the demodulated *I* and *Q* data from the in-phase and quadrature detectors, it is necessary to produce a clock from the synchronous (and scrambled) data bit stream. Figure 17–22 is an example of a recovery scheme that produces pulses at amplitude transitions and uses a narrowband filter (a PLL) to

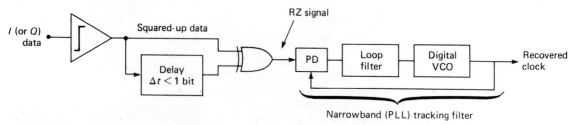

FIGURE 17–22   Clock recovery circuit for bit timing recovery.

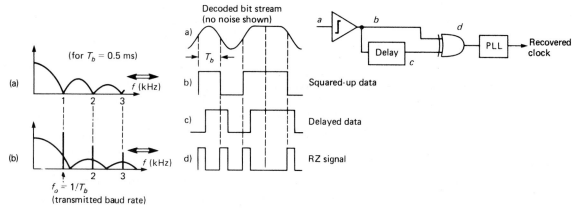

FIGURE 17-23 Bit timing (clock) recovery for quasi-random synchronous data.

establish a "clean" clock signal; that is, we synchronize a VCO with one of the spectral lines created in the process of producing an RZ (return-to-zero) signal from the I (or the Q) data stream. The explanation of this is illustrated in Figure 17–23. The data is quasi-random, having been bit-stuffed and scrambled at the transmitter. The frequency spectrum of such a random bit stream is shown in Figure 17–23b. The 274-Mb/s (AT&T's T4M) microwave PCM system uses scrambling.

The spectrum of the squared-up data pulses has two notable qualities: (1) this is a *continuous spectrum* because of the assumption of random data, equally probable 1s and 0s; and (2) there are *no* spectral lines to which a clock could be synchronized.

When the RZ signal d is produced by the exclusive-OR, there is a change in the continuous spectrum (Figure 17–23d). The continuous part we don't care about; what is important is the fact that discrete spectral lines are produced at frequencies that are harmonics of the transmitted symbol rate, $nf_b = 1$ kHz, 2 kHz, 3 kHz, and so on. If this spectrum is passed through a narrow bandpass filter so that all but one of the spectral lines are attenuated, we could then use this signal to synchronize the clock oscillator.

In particular, a phase-locked loop will both synchronize and filter simultaneously, so that the recovered (sync'ed) clock is the VCO output. This is the narrowband tracking filter of Figure 17–22. The 274-Mb/s T4M system uses the narrowband PLL; T1 systems recover timing with tuned circuits set for the 1.544-Mb/s data rate.

Modems with Higher Than Four-Level PSK and QAM

Multilevel signaling schemes are often compared on the basis of bandwidth efficiency by a parameter called *information density* D_i. Information density is defined by

$$D_i = \frac{f_i}{B_i} \qquad\qquad (17\text{--}7)$$

where f_i = information transfer rate (bits/s) and B_i = information bandwidth (Hz). Thus information density is given in b/s/Hz. Much disagreement exists over how to define the bandwidth of digital signals. For the purpose of comparing multilevel transmission systems, the absolute minimum bandwidth of an ideal (Nyquist) channel is used. Both f_i and B_i were defined in Chapter 12.

As an example, the theoretical maximum for binary (baseband) signaling is $D_i = 2$ b/s/Hz, since two bits can be transmitted per cycle in an ideal noiseless channel. When these signals are modulated on a carrier, both the upper and lower sidebands are usually transmitted. Consequently, the bandwidth doubles and D_i is cut in half.

Modems transmitting more than 2400 bits/s use synchronous formats and higher than four-level encoding techniques to improve the information density and transmit more information in a given bandwidth. These M-ary techniques encode n bits per symbol, where $n = \log_2 M$, which comes from solving the basic relationship $M = 2^n$. As an example, $n = 2$ for four-level QPSK and 2 bits (one dibit) are transmitted per symbol.

8-PSK transmits $n = 3$ bits (called *tribits*) per transmission line change (symbol). Consequently, 4800 b/s can be sent at 1600 baud (4800/3 = 1600) using 8-PSK. Examples of such modems are the AT&T types 208A and 208B. The 208A operates full-duplex on four-wire private (leased) lines, whereas the 208B operates half-duplex on two-wire switched (telephone) lines; both transmit synchronous data; refer to Table 17–2. The four-wire system incorporates two complete transmission lines: one for full-bandwidth transmitting and the other for full-bandwidth receiving.

Note that these are modulated-carrier systems that produce upper and lower sidebands; consequently the maximum theoretical information density D_i is determined from $D_i = n$ (b/s/Hz). As examples, BPSK can theoretically achieve 1 b/s/Hz, QPSK promises 2 b/s/Hz, and 8-PSK, 3 b/s/Hz. Operating QPSK systems are achieving slightly over 1.9 b/s/Hz.* This is limited by system distortion, timing jitter, and other degradations.

Higher-level PSK signals can be generated by combining more BPSK modulators, as was done to produce the QPSK circuit of Figure 17–14. Of course, the fixed 90° phase shift between the BPSK sections must be reduced (compared to QPSK) to accommodate the new vectors. More often, however, the QPSK scheme is maintained, and additional vectors are produced by varying the amplitudes of the separate quadrature carriers with attenuators in each BPSK section. Not only can M-ary PSK be implemented with such a scheme, but QAM can as well.

* R. L. Freeman, *Telecommunication Transmission Handbook*, 2nd ed. (New York: Wiley Interscience, 1981), p. 606.

Quadrature Amplitude Modulation (QAM)

Modems transmitting 9600 b/s on narrowband telephone circuits use 16-level PSK or QAM. The AT&T 209A modem transmits synchronous data at 9600 b/s full-duplex using 16-QAM on four-wire equalized (phase- and frequency-compensated) leased lines. Microwave radio systems using 16-QAM are Bell Lab's DR 6–30 (Figure 17–1) and the DR 11–40. A Rockwell-Collins MRD-5102 uses 256-QAM.

QAM is a multilevel quadrature-carrier system similar to that described for QPSK, except that different amplitudes are included in the signal-state space (constellation). A comparison of 8-PSK and two possible 8-QAM constellations is shown in Figure 17–24.

The constellations of Figure 17–24 can be implemented with the QPSK circuit of Figure 17–14 using the PIN diodes and multilevel converter scheme such as shown in Figure 17–13. Thus the amplitudes of the quadrature carriers are adjusted to produce the various constellations. An extension of this concept is illustrated in the linear combining of two QPSK signals to produce the 16-QAM signal of Figure 17–25. The modulator and demodulator block diagrams are also shown in this figure. As an example for Figure 17–25, consider the phase of the carrier oscillator to be along the horizontal axis and the 90°-shifted carrier to be along the vertical axis, as in Figure 17–14. Now suppose for the 16-QAM modulator of Figure 17–25 that two serial bits (0 followed by 0, for example) have entered the *a* port and are mapped into a voltage level of −3 V. For the same period of time (1 baud), suppose that two serial bits (0 followed by 1, for example) in the *b* port are mapped into +1 V. The transmitted phasor would be the sum of a long vector (3 V) pointing to the left (negative sign) and a short vector (1 V) pointing upward (positive sign). Thus, the resultant would have an amplitude of $\sqrt{7}$ ($\sqrt{(-3)^2 + 1^2}$) and a phase of approximately 161.6° ($180° + \tan^{-1}[1/(-3)]$). This phasor would transmit the 4 bits of binary information $a = 00_2$ and $b = 01_2$ or, if serially combined, 0001_2; the subscript 2 denotes a binary number.

If *M*-ary PSK and QAM can be generated by essentially equivalent circuits, what are the trade-offs? One obvious consideration is that QAM, because it is an AM signal, requires linear amplifiers at the transmitter output and receiver, whereas PSK does not. However, QAM systems can outperform PSK in transmitted power and noise immunity trade-offs.

FIGURE 17–24 Comparison of 8-PSK and two 8-QAM constellations.

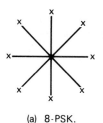

(a) 8-PSK.

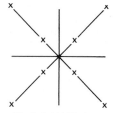

(b) 8-QAM (4 phase, 2 amplitude).

(c) 8-QAM (8 phase, 2 amplitude).

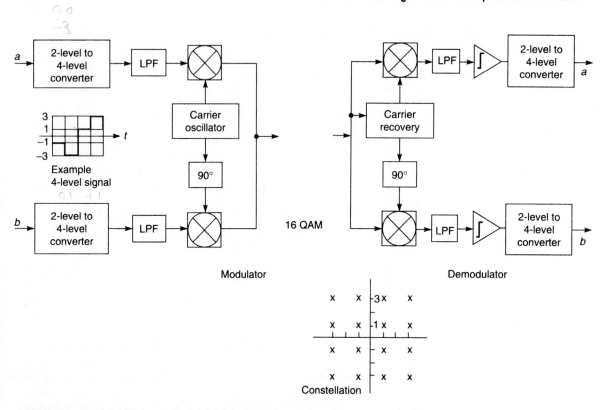

FIGURE 17–25   16-QAM modulator, demodulator, and constellation.

17–2 ■ NOISE AND ERROR RATE PERFORMANCE

If the longest vectors in the constellations of Figure 17–26 have equal peak power, then clearly the average signal power for an 8-QAM constellation is less than that of the 8-PSK, with its equal-length vectors. On the other hand, if the QAM power is increased until it is equal to the average PSK power, then the distance between

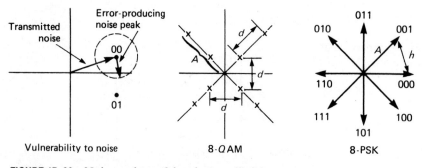

FIGURE 17–26   Noise and two 8-level constellations.

FIGURE 17–27 Data error rate versus carrier-to-noise power ratio.

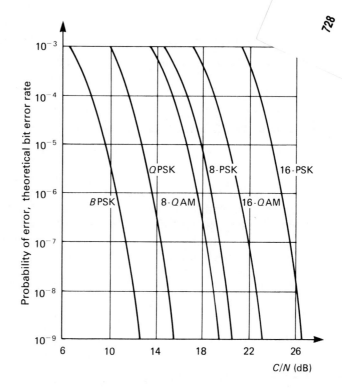

any two points in the constellation will be greater for QAM. Hence, QAM has better noise performance than PSK. As seen in Figure 17–26, the distance between points in the constellation is proportional to noise immunity, because when peak noise vectors exceed one-half the distance between points, the probability of error is great.

Figure 17–27 compares the probability of error (bit error rate, BER) for various digital-modulation types, given an rms carrier-to-noise ratio, C/N. C/N is calculated in the absolute minimum RF bandwidth (two-sided Nyquist) for the transmitted symbol (baud) rate.

■ ■ ■ ■

EXAMPLE 17–1

1. Compare the magnitude of allowable peak noise for 8-PSK and the particular 8-QAM constellation of Figure 17–26. The maximum peak signal amplitudes for both systems are equal ($V_{max} = A$).

2. Determine the power savings for the QAM, assuming an equal probability of transmission for each of the eight states.

3. Use the curves in Figure 17–27 to compare the required C/N for 8-PSK and 8-QAM to achieve a system BER of 1 error in 10^8 bits.

Solution:

1. From the geometry of the 8-QAM, $A = d + (d/2)\sqrt{2} = 1.707d$, so a noise peak of $A_n = d/2 = 0.293A$, occurring at a receiver sample time, can cause an error.

For the 8-PSK, $A_n = h/2 = A \sin(45°/2) = 0.383A$. Comparing noise peaks, PSK/QAM $= 0.383A/0.293A = 1.307$ (**30.7%**). This is not an S/N or C/N ratio.

2. Power savings for the QAM of question 1: $P_{avg} \propto A^2 \times$ (number of states with amplitude A). $P_{8\text{-PSK}}/P_{8\text{-QAM}} = (8 \times A^2)/[(4 \times A^2) + 4 \times (d\sqrt{2}/2)^2]$, where $d = A/1.707$. Hence, $P_{8\text{-PSK}}/P_{8\text{-QAM}} = 8/[4 + 4(0.414)^2] = $ **1.707** (2.32 dB).

3. 8-QAM requires about $C/N = 19$ dB for BER $= 10^{-8}$, and 8-PSK requires about **20 dB**. The QAM can tolerate more noise.

E_b/N_0

Noise performance curves such as those of Figure 17–27 are plots of the complementary error function for Gaussian noise channels (see Chapter 13). In addition to C/N, these curves are also plotted against another very important noise performance parameter for digital transmission systems, E_b/N_0 (dB). E_b is the detected *energy per bit*, and N_0 is the *noise spectral density*.

If the transmitted information rate is f_b, then the time per bit is

$$T_b = \frac{1}{f_b} \tag{17–8}$$

and the energy per bit is

$$E_b = ST \tag{17–9}$$

where S is the signal power. Hence

$$S = E_b f_b \tag{17–10}$$

and the total noise $[kT \text{ (dBm/Hz)} + NF \text{ (dB)}]$ in system bandwidth B is

$$N = N_0 B \tag{17–11}$$

where N_0 is the total noise spectral density in W/Hz.

The limiting factor to the number of levels M that can reasonably be used in an M-ary modulation scheme is noise. The Shannon limit for information rate capacity is Equation 12–11b, $C = B \log_2(S/N + 1)$, where B is the system bandwidth. Equation 12–11b can then be expressed for digital data systems as

$$C = B \log_2\left[\left(\frac{E_b f_b}{N_0 B}\right) + 1\right] \tag{17–12}$$

Figure 17–28 compares the data error probability rates for FSK and a few PSK modulation schemes, including the noncoherent differential two-level (DPSK). Notice that coherent 2-PSK and 4-PSK (BPSK and QPSK) have the same probability of error. Also, the noncoherent FSK and DPSK follow exponential curves of

$$P_{e,\text{FSK}} = 0.5 \exp\left(\frac{-0.5E_b}{N_0}\right) \tag{17–13}$$

and

$$P_{e,\text{DPSK}} = 0.5 \exp\left(\frac{-E_b}{N_0}\right) \tag{17–14}$$

FIGURE 17–28 Data error rate comparisons for various binary signaling schemes.

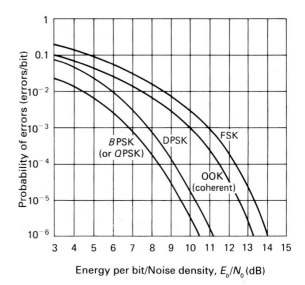

Energy per bit/Noise density, E_b/N_0 (dB)

respectively, whereas the coherent schemes, including on-off key, follow the error-function relationship.

Bandwidth Requirements for Digital Radio

The bandwidth efficiency of *M*-ary data modulation techniques makes the added circuit complexity worthwhile when operating in bandwidth-limited environments. The absolute minimum bandwidth for a modulated signal is twice the modulating signal rate.

For NRZ pulse formats, the worst-case carrier switching rate is from consecutive high and low symbols that produce a fundamental rate of $f_b/2$. The modulated signal bandwidth must then be at least the transmission-line baud rate f_B in order to include the minimum upper and lower sidebands. Hence, for binary PSK the absolute minimum system bandwidth is equal to the information rate f_i because $f_B = f_i$ for binary. BPSK has a bandwidth efficiency of 1 b/s/Hz.

For QPSK with its four-level encoding, 2 bits of information are transmitted for each carrier shift, so the transmission system bandwidth need be only one-half the information (computer data) rate. For an *M*-ary PSK or QAM system, the minimum *transmission system* bandwidth is

$$B = \frac{f_i}{\log_2 M} \qquad (17–15)$$

For 8-PSK, then, the minimum bandwidth is $B = f_i/\log_2 8 = f_i/3$. The bandwidth efficiency is 3 b/s/Hz, also called the *information density*.

Digital radios do not transmit only one set of spectral lines. A typical common-carrier microwave link multiplexes numerous synchronous and asynchronous digitized voice and data signals into a single baseband signal for transmission. The

FIGURE 17–29   FCC mask.

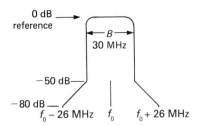

microwave signal is very sensitive to nonlinear circuit phase slopes, multipath, and other distortions. Also, the FCC forbids the transmission of constant discrete spectral lines; consequently, data scramblers are used for quasi-randomizing the baseband signal. This produces a symmetrical RF spectrum (see Figure 17–20). The FCC has defined a bandwidth "mask" within which the microwave signal must be confined.

The *FCC mask* shown in Figure 17–29 is specifically for the 5.925–6.425 GHz band, in which 30 MHz of bandwidth is allowed. Other heavily used frequency bands and their allowable bandwidths are 2.11–2.13 GHz (3.5 MHz bandwidth), 2.16–2.18 GHz (3.5 MHz), 3.7–4.2 GHz (20 MHz), and 10.7–11.7 GHz (40 MHz).

When high data rates and rectangular pulses are modulated on the microwave carrier, the spectrum (see Figure 17–20) usually exceeds the FCC mask. Consequently, bandwidth-efficient modulation schemes are used. The typical TTL (NRZ) or RZ data pulses are reshaped, usually into raised-cosine pulses that have a fast spectral rolloff outside of the first lobe. Also, raised-cosine pulses produce low intersymbol interference (ISI).

In addition to the premodulation pulse shaping, multiple Nyquist filtering is done in the IF's after modulation, after the up-converter, and after the final power amplifier that has traditionally been a slightly saturated (backed off from saturation) traveling-wave tube (TWT) for high power and efficiency.

17–3 ■ SPACE AND FREQUENCY DIVERSITY

The propagation phenomena that cause signal fading were discussed in Chapter 15. The problem of transmission loss during fades can be overcome by overpowering the system signal budget, but diversity techniques are used where deep fades must be reliably accommodated.

Studies show that most fading phenomena are frequency-dependent, and the probability of a deep fade at two different frequencies separated by roughly 2% is quite low.

Frequency diversity can be designed into a microwave link by providing two or more transmitter and matching receiver pairs that operate at different frequencies. In addition to the improved reliability with respect to fades, system redundancy is available during a circuit failure.

FIGURE 17-30 Microwave link with space diversity.

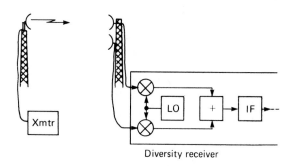

Diversity receiver

Postdetection receiver logic circuitry can monitor the BER performance of the link and switch channels if the error rate threshold is exceeded.

Space diversity reception is illustrated in Figure 17–30. The transmitter radiates simultaneously to two antennas separated vertically by at least six wavelengths. The probability of a deep fade over both paths simultaneously is small.

The received signals are down-converted and combined at the IF. This system can be used to improve C/N when both received signals are of comparable levels. This is possible if signal delays in both paths have been equalized at the carrier frequency so that the carrier peak *voltages* add constructively, giving as much as a 6-dB increase. The random noise in the two paths will not be correlated, so the noise *power* adds (rms) for an increase of as much as 3 dB. The improvement in C/N can then be as much as 3 dB.

17-4 ■ SYSTEM GAIN, POWER, AND NOISE BUDGETS

In designing a UHF or microwave communications link, especially via a satellite, one of the first steps is to determine the system power gains and losses. With this information the trade-offs between transmitter power, antenna gains, and receiver sensitivity (noise figure and gain) may be determined.

Usually the fixed constraint for the system is the receiver signal-to-noise ratio or, for digital systems, the error rate. The link designer uses the curves of Figures 17–27 and 17–28 to convert BER to a system signal-to-noise quantity such as C/N or E_b/N_0.

The system power gain is calculated by considering all of the losses and gains that the transmitted signal experiences as it propagates from the transmitter output to the receiver input. As an example, consider the line-of-sight transmission link of Figure 17–31.

The transmitter (Xmtr) develops power P_t measured in dBm or dBW. This power is coupled through circulators used to couple other transmitters (for other channels) to the antenna feed system while minimizing VSWR and power "blasts" from other transmitters. Some power loss L_{B_t} is experienced because of this coupling and *branching* in the transmitter output system.

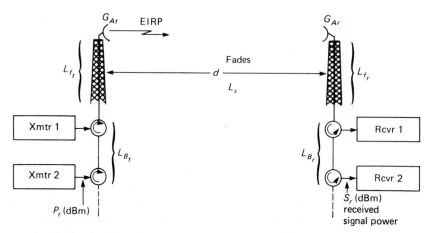

FIGURE 17–31   Components in system gain and power budget.

The remainder of the transmitter output power is fed to the antenna, typically some distance from the transmitter. This distance is kept to an absolute minimum in satellites. The feed system—including losses due to cables or waveguides, flanges, and associated VSWRs—is included in the system gain budgets as *feeder losses* L_f. If exact losses for a system are unknown, then a reasonable estimate is 10 dB per 100 m of feeder.

Microwave dish antennas have energy-concentrating properties due to their geometries that give a considerable gain over a theoretical isotropic radiator. Equation 15–29, $G_A = \eta(\pi D/\lambda)^2$, is used to compute the antenna gain when the diameter and frequency are known. At this point we could calculate the radiated power or EIRP—the effective isotropic radiated power. The EIRP in decibels with respect to 1 W is found from

$$\text{EIRP (dBW)} = P_t \text{ (dBW)} - L_{B_t} \text{ (dB)} - L_{f_t} \text{ (dB)} + G_{A_t} \text{ (dB)} \qquad \textbf{(17–16)}$$

The major loss of power in the system is the space, or path, attenuation L_s. The power or energy radiated from an antenna is not guided by waveguides or cables; thus it will spread out in all directions as it propagates through space. In addition to power density loss, there are numerous atmospheric, ionospheric, and other anomalous propagation disturbances that result in loss of signal power. Some of the propagation disturbances are fixed; others vary over time. The variable losses will be combined into the category called *fades*, and the system gain margin, called *fade margin* (F.M.), is determined from Table 15–1 for a given system reliability objective. The fixed-path loss or space attenuation (Equation 15–42) has been found to be proportional to the square of the distance d traveled by the signal and also to the square of the signal frequency, $L_s \propto d^2 f^2$. With the constants of proportionality being 92.4 dB for kilometer-GHz and 36.6 dB for miles-MHz, the space attenuation power loss will be

$$L_s \text{ (dB)} = 92.4 \text{ dB} + 20 \log d_{\text{km}} + 20 \log f_{\text{GHz}} \qquad \textbf{(17–17)}$$

or $L_s \text{ (dB)} = 36.6 \text{ dB} + 20 \log d_{\text{mi}} + 20 \log f_{\text{MHz}}$, which is Equation 15–45.

As an example, a terrestrial microwave radio 2-GHz transmission over 50 km (31 mi) will result in 132.4 dB of power loss. Clearly, a satellite in a geosynchronous orbit 35,880 ± 113 km (22,300 ± 70 mi) above the equator will have a path loss considerably greater. The geosynchronous orbit, however, has the special significance of allowing the satellite to orbit at the same rate as the earth's spin (24-h period). Thus, the satellite appears over the same point on the earth's equator continuously.

Received Signal Power and System Gain

The information of the last few paragraphs gives a basis for computing the system gain by which the signal power received over a communications link may be determined. The system gain is defined by

$$G_{\text{sys}} \text{ (dB)} = P_t \text{ (dBm)} - S_r \text{ (dBm)} \qquad \textbf{(17–18)}$$

The signal power S_r reaching the receiver is proportional to the power transmitted and all the gains and losses between the transmitter output and receiver input. Expressed mathematically,

$$S_r \propto \frac{P_t G_{A_t} G_{A_r}}{L_B L_f (\text{F.M.}) L_P} \qquad \textbf{(17–19)}$$

The received signal power calculated in decibels will be

$$S_r \text{ (dBm)} = [P_t \text{ (dBm)} - L_B \text{ (dB)} + G_{A_t} \text{ (dB)}]^* - [37 \text{ dB} + 20 \log f + 20 \log d]\dagger$$
$$- \text{ F.M. (dB)} + G_{A_r} \text{ (dB)} - L_f(\text{dB}) \qquad \textbf{(17–20)}$$

where P_t (dBm) = transmitter output power, in dBm (Remember, dBm = dBW + 30 dB.)

L_B (dB) = total branching losses, from transmitter output to the antenna-feed system, including losses due to circulators; and receiver antenna-feed system to the receiver input, including power splits if more than one receiver is to process the same signal (Use 3 dB unless more specific information is known.)

L_f (dB) = total transmit and receive antenna-feed losses (Use 10 dB/100 m (328 ft) for typical air-filled coax or waveguide, unless more specific information is known.)

G_{A_t} (dB) = transmit antenna gain, in dB (This is actually in dB*i*, since antenna gains are referenced to the gain of an omnidirectional "isotropic" point source.)

[ ]* = effective isotropically radiated power from the transmit antenna (EIRP)

[ ]† = L_s (dB), the space path loss of a signal at frequency f (MHz) over a distance d (mi) of empty space

F.M. (dB) = fade margin to account for variable path losses (Use 6 dB for satellite communication links below 10 GHz and Table 15–1 for land links.)

G_{A_r} (dB) = receiver antenna gain, in dB. (See G_{A_t} above.)

To illustrate, the following example is given for a line-of-sight land link.

Determine the signal power received 40 mi from a 2.11-GHz, 1-W transmitter. Both receiver and transmitter are 164 ft away from their 3-m antennas. The worst-case propagation reliability is to be 99.99%; that is, outages will occur for less than 0.01% of the time over a year (52 min during the year).

Solution:

From Equation 17–20 and $\lambda = c/f = 3 \times 10^8/2.11 \times 10^9 = 0.142$ m:

$$S_r \text{ (dBm)} = \left[10 \log \frac{1 \text{ W}}{1 \text{ mW}} - 3 \text{ dB} + 10 \log 0.55 \left(\frac{\pi \cdot 3}{0.142}\right)^2\right]$$

$$- [37 \text{ dB} + 20 \log 2110 + 20 \log 40]$$

$$- 38 \text{ dB (see Table 15–1)} + 34 \text{ dB} - \left(\frac{10 \text{ dB}}{328 \text{ ft}}\right) \times (2 \times 164 \text{ ft})$$

$$= [61 \text{ dBm EIRP}] - [135.5 \text{ dB path loss}] - 14 \text{ dB}$$

$$= \mathbf{-88.5 \text{ dBm}}$$

The power is, from -88.5 dBm $= 10 \log S_r/1$ mW, $S_r = $ **1.4 pW.** This is the minimum power received for the system because we have included worst-case fading.

System Threshold

System threshold is given by a minimum performance criterion for a particular application. This must be translated into a minimum signal level or, more often, a minimal signal-to-noise power ratio.

Having determined the amount of signal power at the input to the receiver, we can determine the carrier-to-noise ratio C/N or the E_b/N_0 ratio in the IF system at the detector. These are extremely important system parameters because the quality of the demodulated information depends on the predetection signal-to-noise ratio. These ratios not only determine the noisiness of received voice/video transmissions and error rate of data transmissions but also will determine in an FM* system whether or not the receiver will be captured by the signal or by the noise (see *capture effect* in Section 9–7).

All we need to know is the receiver noise figure (NF) and the net predetection bandwidth (*B*). *Net bandwidth* refers to the overall bandwidth of the receiver, including IF narrowing (see Chapter 5). Actually, for the type of high-performance receivers we are considering here, the IF amplifiers themselves are made wideband (low-circuit *Q*), and a multipole bandpass filter is placed between the mixer

---

* Most microwave line-of-sight (LOS) links, both terrestrial and satellite, are FM systems. This is giving way to purely digital-modulated carriers, but for now we find many common carriers taking information at baseband or already modulated in different formats such as SSB-SC, digital PCM, VSB (vestigial sideband, like TV video), and FM'ing it onto the microwave carrier.

and first IF amplifier. This filter determines the predetection bandwidth and must be wide enough to pass the modulated signal with very low amplitude and phase distortion.

The thermal noise power at the receiver input is $N = kTB$ and can be expressed in dBm by Equation 5–24:

$$N_{th} \text{ (dBm)} = -174 \text{ dBm} + 10 \log B$$

The total noise power will include the system noise figure NF.

Using the results of Example 17–2, we can calculate the C/N at the detector for a 36-MHz bandwidth and 3.94-dB receiver noise figure as

$$\frac{S_r}{N} \text{ (dB)} = S_r \text{ (dBm)} - N \text{ (dBm)}$$

$$\frac{S_r}{N} \text{ (dB)} = -88.5 + 174 - 10 \log B - NF$$

$$= +6 \text{ dB}$$

This is a very poor predetection C/N. Granted, a 36-MHz bandwidth seems terribly wide, but this is the bandwidth allowed for satellite transponders of common carriers (AT&T, GTE, MCI, and so forth).

At the transmitter, there are ways to improve the *postdetection S/N*, such as *encoding techniques** and, for FM, preemphasis. However, some types of demodulators, in particular an FM discriminator (a noncoherent detector), have a "threshold" that we must exceed or pay a heavy penalty—as much as 20 dB of degradation in an already poor S/N (see Figure 9–37).

Assuming the ideal, low modulation index, FM threshold characteristic of Figure 9–37, the received signal level required to barely exceed FM threshold is

$$S_r \text{ (threshold)} = -161 \text{ dBm} + 10 \log B + NF \text{ (dB)} \qquad \textbf{(17–21)}$$

where $-174 \text{ dBm} + 13 \text{ dB} = -161 \text{ dBm}$ puts the predetection signal just above the knee at $C/N = +13$ dB. Equation 17–21 also gives the received signal level required to achieve a demodulator output S/N of $+33$ dB. For signals stronger than this FM threshold, the output S/N increases linearly with the increase in S_r.

G/T

A very useful receiver system noise performance parameter is the G/T ratio. This parameter is often specified for satellite earth stations instead of signal-to-noise ratio because it is independent of the receiver bandwidth. G is the receiver antenna power gain, and T is the system noise temperature. The ratio is expressed in decibels by

$$\frac{G}{T} \text{ (dB/K)} = 10 \log \frac{G}{T} \qquad \textbf{(17–22)}$$

---

* Up to 6 dB coding gain for convolutional coding ($K = 24$, $R = \frac{1}{2}$) when sequentially decoded.

From Equation 17–19 we see that the received signal strength is directly proportional to the receiver antenna gain G_{A_r}. The noise contribution of the receiver is given by the noise figure ratio NR or, equivalently, by the system noise temperature T_{sys}. The receiver noise temperature is $T_e = T_0(\text{NR} - 1)$, where $T_0 = 290$ K by convention. Indeed, one reason that noise temperature is used instead of noise figure is to avoid the arbitrary 290 K reference temperature that often is invalid in satellite-earth applications.

When noise temperatures are specified for receiver components, the system noise temperature is found from

$$T_{\text{sys}} = T_{e1} + \frac{T_{e2}}{G_1} + \frac{T_{e3}}{G_1 G_2} + \cdots + \frac{T_{en}}{G_1 G_2 \cdots G_{n-1}} \qquad \textbf{(17–23)}$$

where T_{e1} includes the sum of antenna noise temperature (including space noise and any feed-in device noise) and receiver RF amplifier noise temperature. G_1 is the power gain of the receiver's RF amplifier, less feed-in and input filter losses. Notice the simplicity of Equation 17–23 compared to the noise figure formula (Equation 4–15).

When the system noise temperature is known, the *noise spectral density* needed to calculate the E_b/N_0 ratio is found from

$$N_0 = k T_{\text{sys}} B \qquad \textbf{(17–24)}$$

Then from Equation 17–10, $S = E_b f_b$, and $N = N_0 B$,

$$\frac{S}{N} = \frac{E_b f_b}{N_0 B} \qquad \textbf{(17–25)}$$

Combining Equation 17–25 with the link budget equation (Equation 17–20), we may write the link budget in terms of E_b/N_0 as

$$\frac{E_b}{N_0} \text{ dB} = \text{EIRP (dBm)} - L_s \text{ (dB)} + \frac{G}{T} \text{ (dB/K)} + k_{\text{dB}} + f_b \text{ (dB)} \qquad \textbf{(17–26)}$$

where the EIRP is computed with Equation 17–16, L_s is the space loss from Equation 17–17, $k_{\text{dB}} = -174$ dBm/Hz is Boltzmann's constant, and f_b (dB) $= 10 \log f_b$ is the NRZ data bit rate. Equation 17–26 leaves out receive system feed losses, branching losses, and fade margin, and assumes the receiver uses an absolute minimum Nyquist data filter (Equation 11–7 with a rolloff factor of zero).

17–5 ■ SATELLITE ACCESS FOR GLOBAL NETWORKING

There has been, and will increasingly be, demand for communications services on a national and global basis. The range of services and connectivity between suppliers and users is limited to a certain degree by available technology, but future services are limited only by the imagination.

Presently most users of voice (plain old telephone service, POTS) and data services within large metropolitan areas and, to a lesser degree, surrounding rural areas have interconnectivity via the switched public telephone system. The service providers are, primarily, the 7 Bell regional holding companies (BRHCs) through the 22 Bell Operating Companies (BOCs) in North America and the Public Telephone and Telegraph systems (PTTs) in Europe.

National networks have connected the large metropolitan areas together by various combinations of common carriers using long lines (transmission lines) and microwave radio, and now fiber-optic and satellite links. Presently, fiber-optic systems have the advantage of relatively low cost and access serviceability, whereas satellite service is quicker to install and offers large-area (continental) coverage capability not possible with fiber, microwave radio, or long lines.

Satellite Systems

Telecommunication services being provided—including telephone, television, facsimile, Telex, electronic mail, teleconferencing, and others—are increasingly digital. Most domestic satellites (Domsats), which include COMSTAR, WEST-STAR, TELSTAR, GALAXY, RCA's SATCOM, and Canada's ANIK, are largely single-sideband frequency-division multiplex (SSB/FDM), and FDM/FM analog, with frequency-division multiple access (FDMA). However, new systems will incorporate time-division multiple access (TDMA), modern digital modulation techniques, on-board switching (satellite switched, SS-TDMA), baseband signal processing, and multiple-beam antennas with frequency reuse. An example of a transponder scheme incorporating TDMA, FDMA, and satellite-switched TDMA (SS-TDMA) is NASA's Advanced Communications Technology Satellite (ACTS) system deployed in 1992; see Figures 17–32 and 17–33.

NASA's ACTS program is aimed at developing the high-risk advanced satellite technologies required to greatly increase the capacity of future satellite communications systems. For the 1992 experimental flight system, three fixed-beam antennas are used on the satellite to cover the east and west coasts of the United States using Ka-band frequencies. In addition, two narrow scanning beam antennas are used to scan other areas of the United States. The spatial separation of beams—which allows for frequency reuse—and the use of the Ka-band with its wide bandwidth potential permit very large data capacities per satellite.

At the satellite, received signals in the Ka-band (27.5–30 GHz) are converted to IF and then processed in one of two ways. Either they are switched to one of the transmitters and up-converted for relay to an earth station—this method is used for high-data-rate (220 megasymbols/s) fixed-beam trunk traffic—or, the received signals are switched to one of two demodulators for processing, buffering, and switching at baseband, followed by remodulation (perhaps with FEC encoding) on a down-link carrier for transmission to earth.

Two scanning antennas receive low-data-rate (110/55 megasymbols/s) signals on a Ka-band carrier. The 55-megasymbols-per-second rate is used when forward error control (FEC) must be incorporated to overcome signal attenuation due to rain (rain fade). The FEC scheme chosen is rate $1/2$, constraint-length 5, convolutional encoding.

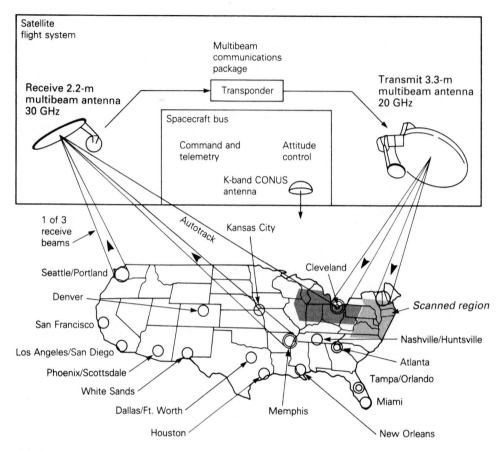

FIGURE 17-32   Advanced Communications Technology Satellite system of NASA.

As indicated in the block diagram of Figure 17–33, a separate low-noise receiver is used with each of the antennas and the received signals are down-converted to IF (3.3 GHz) and routed by a 3-input, 2-output IF switch. The third IF input to this switch is from one of three fixed-beam antennas. The low-data-rate signals are demodulated, and the data is sequentially stored during a time frame (for example, TDMA mode) in 64-bit-wide CMOS RAMs.

The 64-bit words are FEC decoded (if necessary), cross-connect-routed under master earth terminal control in a 3×3 digital switch ($N = 3$) and buffered for FEC encoding (if used), modulated, and then transmitted on one of two down-link scanning beams or one of the three fixed-beam antennas. Modulation is serial minimum-shift-key (SMSK; Chapter 13) and the down-link carrier frequency is in the 19- to 20.2-GHz range.

Multiple-Access Techniques

The most common satellite system today consists of a satellite, rocket launched and placed in a geosynchronous orbit over the equator (typically above the Atlan-

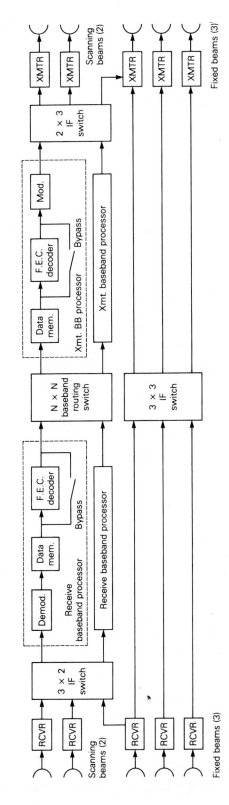

FIGURE 17-33 Advanced Communications Technology Satellite transponder.

tic Ocean). Thus, the "bird"—as the satellite is often referred to—is orbiting the earth at an altitude of 22,300 ± 70 mi, and moving in the same direction and speed as the earth's natural rotation. Hence, the satellite appears to be stationary over a fixed point above the earth. Geosynchronously orbiting satellites with their microwave dish(es) are positioned side by side, separated by as little as 2° of arc for use as microwave relay stations.

The most common microwave carrier frequencies in use are near 6 GHz for the uplink, and near 4 GHz for the downlink. The 6/4-GHz band is referred to as *C-band*. More recently launched satellites are most likely operating in 14 GHz (up) and 12 GHz (down) band called *Ku-* ("Kay-you") band. Clearly, the required antenna size for *Ku*-band is four times smaller than for *C*-band—an obvious hardware advantage for newer systems. The most recently utilized commercial satellite carrier frequency band is *Ka* (30/20 GHz).

C-band satellites are permitted to use a total bandwidth of 500 MHz, centered at its assigned carrier frequency. A satellite typically carries 12 transponders (automatic receiver/transmitter), each using 36 MHz of the available 500 MHz allowed. The 12 transponders are frequency-division-multiplexed (FDM) with 4-MHz guardbands between each transponder's spectral allocation, and the 20 MHz remaining of the 500 MHz is split into two 10-MHz guardbands at the low and high ends of the 500-MHz satellite allocation.

The number of transponders in a satellite can be doubled to 24 if every other transponder in the 500-MHz allocation is fed from a quadrature-polarized antenna feed mechanism. That is, as illustrated in Figure 17–34, 12 odd-numbered transponders receive and transmit via a vertically polarized antenna feed, whereas the 12 even-numbered transponders receive and transmit on frequencies offset by 20 MHz from the carrier frequencies of the odd-numbered transponders but are on the horizontally polarized antenna feed. (The frequency offset helps in the problem of minimizing cross talk between transponders.) This technique is called *frequency reuse* by means of dual-quadrature antenna polarization.

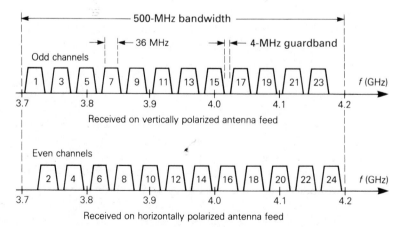

FIGURE 17–34 Typical *C*-band satellite transponder channel frequency plan.

Typically, the use of a given transponder is leased by a company or common carrier to provide data transmission services to multiple users. The way in which a transponder's 36-MHz bandwidth is used or accessed is by one of a few common *multiple-access* techniques. The three most commonly employed multiple-access methods are

- Frequency-division multiple access (FDMA)
- Time-division multiple access (TDMA)
- Code-division multiple access (CDMA)

The major distinction between the three multiple-access methods listed are that in *FDMA* multiple users share a portion of the available transponder bandwidth and power on a continuous basis, whereas in *TDMA* the entire transponder bandwidth and power are used by each of the multiple users but only for a prescribed time interval (slot). For *CDMA* the entire transponder bandwidth is typically used on a continuous basis by all users, but the available transponder power is shared between the users.

Frequency-Division Multiple Access (FDMA)

Frequency-division multiple access is employed as *multiple channel per carrier* (MCPC), in which a few channels of information are frequency-division multiplexed at baseband and the resulting baseband signal modulates an RF carrier, or as a *single channel per carrier* (SCPC), in which each channel modulates a separate RF carrier and the modulated carriers are summed. An example of the multiple-channel-per-carrier FDMA format is illustrated in Figure 17–35. In this example, seven earth stations are continuously, simultaneously using one entire transponder of a 24-transponder satellite. Each earth station is assigned a carrier frequency and allocated a fixed amount of bandwidth: In this example, seven different ground-station bandwidth capacities used in recent INTELSAT satellites illustrate one possible arrangement for utilizing the entire 36-MHz transponder bandwidth.

In Figure 17–35, observe that station F has been assigned a carrier frequency of 6420 MHz and a 5-MHz bandwidth (which includes 1-MHz guardbands) for transmitting 60 voice frequency (VF) channels of telephone conversation. Each VF channel has 4 kHz bandwidth and is initially single-sideband/suppressed-carrier (SSB-SC) modulated in a scheme that allows each to be frequency-division multiplexed (FDM) to form *groups* of 12 VF channels; refer to Section 8–7. Then five groups, one or more of which are destined for other countries (say, stations D, E, and G), are similarly FDM'ed to form a 60-VF group referred to as a *supergroup*. The preceding supergrouping is usually done at a national switching center where the supergroups are serially transmitted on a radiolink carrier to the earth station, demodulated, and then frequency-modulated on the ground station *uplink* carrier; thus the overall modulation scheme is referred to as FDM on FM, denoted FDM/ FM. A quick calculation indicates that the station F baseband signal would require 240 kHz of bandwidth. However, a system maintenance and routing facility con-

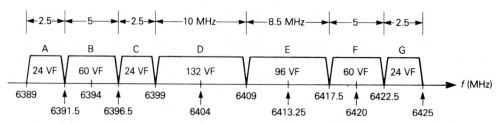

FIGURE 17–35a　Uplink spectrum, *C*-band INTELSAT IV–V series.

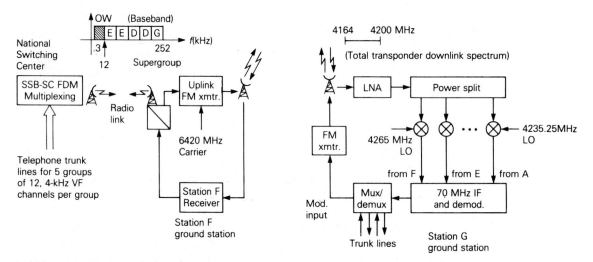

FIGURE 17–35b　Transmission/reception example for ground stations. Downlink frequencies are 2225 MHz below the uplink frequencies.

sisting of teleprinter, tone-telegraph, and voice, all in a 300 Hz to 12 kHz bandwidth and called the *orderwire* (OW), is included to make the FM baseband extend to 252 kHz.

The 6420-MHz uplink carrier and 5-MHz bandwidth allocated to station F includes information destined for any of the other six stations, some of which are located in different subscriber countries. The 6389–6425-MHz *C*-band transponder uplink spectrum illustrated in Figure 17–35a is the top of the usable 6-GHz uplink band. The transponder must receive and translate the entire spectrum down to a 3.7–4.2 GHz *C*-band downlink spectrum for transmission back to earth. Thus, to receive all the other stations, each ground station must use six receivers and cover the particular 36-MHz portion of the total 500 MHz of the lower *C*-band spectrum.

Figure 17–35b includes a possible arrangement for the station G ground station that includes the two-group transmitter multiplexer and two-group receiver demultiplexer. The receive system consists of a low-noise amplifier (LNA) located as close as possible to the antenna (typically, a 30-m-diameter dish) feed horn to minimize signal losses and maximize system noise figure. The *C*-band downlink spectrum is then coupled to six mixers for down-conversion of each of the six modulated carriers to a combined 70 MHz (center frequency) IF; note that this IF needs to be only $2 \times (36 - 2.5 \text{ MHz}) = 67$ MHz in total bandwidth.

After FM demodulation, two groups, either or both of which may be received from any of the other six stations, will be sent to the switching center for filtering and demultiplexing for further routing to individual telephone subscribers.

The earth-station carrier frequency and bandwidth allocations in this FDM/FM/FDMA example are usually assigned on a fixed, long-term basis referred to as *fixed-assignment, multiple-access* (FAMA), usually pronounced "fame-ah." Another allocation scheme, called *demand-assignment multiple access* (DAMA), usually pronounced "dame-ah," allows for continuous use of the entire transponder bandwidth by all the traffic presented to the various ground stations in the system by assigning any available carrier frequency on a statistically multiplexed basis. Of course, availability of an unused carrier is limited by the number of modulators available at a given station.

FDMA with DAMA

As described in the previous paragraph, DAMA can be used in an FDMA system. Perhaps the best example of an actual FDMA system with DAMA goes by the acronym SPADE, for *s*ingle-channel per carrier, *P*CM, multiple-*a*ccess *d*emand-assignment *e*quipment. This system was developed for and used on the INTELSAT series IVA and V satellites.

The main distinguishing feature of a SPADE system is that the 36-MHz transponder bandwidth is divided into 794 separate 45-kHz bandwidth signal paths plus 160 kHz used for continuous system control via a common signaling channel (CSC). A ground-station transmitter incorporates many frequency synthesizers in the 70-MHz IF band, which, on command from the demand assignment and switching unit controlled from CSC, produces one of 397 possible frequencies that becomes the carrier for a QPSK modulator. The digital bit stream input to this modulator comes from a codec that digitizes a telephone channel into the 64-kb/s PCM format discussed in Sections 11–4 and 12–7. The 397 possible QPSK channels are up-converted and transmitted in the transponder frequency spectrum below a CW pilot signal supplied by the CSC reference station with a frequency exactly centered in the transponder frequency spectrum. The 397 QPSK channels (BW = 45 kHz each) located above the CW pilot signal provide the return voice or data channel for full-duplex operation over the satellite link. That is, the information between communicating parties is transmitted on 45-kHz-bandwidth QPSK carriers exactly 18.045 MHz apart in the transponder bandwidth.

For inherently bursty traffic applications, the single-channel-per-carrier (SCPC) scheme with DAMA just described has an overall throughput efficiency (net information transferred over the system per unit time) advantage over fixed-assignment schemes for the same reason that packet switching has higher throughput, in general, over plain old telephone-like space switching: When a preassigned circuit is set up and is not being used, its bandwidth is unavailable to anyone else demanding service.

TDMA

The access technique employed in most modern digital telecommunication satellite systems is time-division multiple access (TDMA). TDMA is based on dividing access time among all stations on the network and allowing each station to use the full system bandwidth and power to transmit a burst of binary data on command by the master station. In a generic sense, then, TDMA is somewhat similar to multipoint operation on a complex computer network.

A satellite TDMA system is controlled by the master control ground station (MCS), and the full bandwidth and transmission power of the satellite transponder are used in precisely synchronized time slots by each of the network ground stations, all operating at the same uplink frequency; refer to Figure 17–36. The uplink bursts from active ground stations must be synchronized by the MCS so as to arrive at the satellite transponder in N time slots following an initializing reference burst.

As illustrated in Figure 17–37, N station bursts following the reference burst constitute a *frame*. Every station in the network can transmit at least one burst per frame and burst lengths generally may be varied in duration to meet changing traffic patterns.

In order to synchronize the transmission process, avoid burst overlaps, and provide other network signaling, each burst begins with a preamble similar to that illustrated in Figure 17–37. The *carrier-recovery and bit timing* pattern is typically 60 bits long and is used in the ground-station receiver for coherent carrier phase lock and data clock synchronization. This is followed by a 20-bit *unique word* provided by the reference ground station for establishing an accurate time reference or, in some systems, for station identification. In other systems, such as the International Telecommunication Satellite Corporation INTELSAT, station identification is provided by a 6-bit *station ID* code. The *control-signal channel* is used for supervisory and configuration control, while the teletype (TTY) and voice

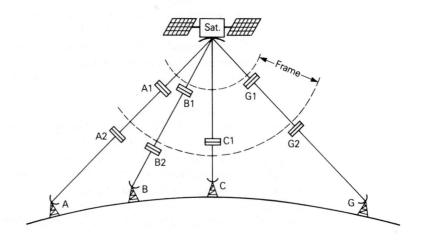

FIGURE 17–36 A time-division multiple-access satellite link.

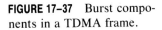

FIGURE 17–37 Burst components in a TDMA frame.

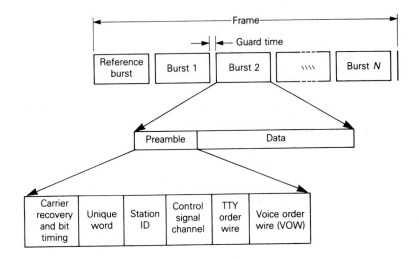

order wire (VOW) bits provide for ground-station system status reporting and maintenance activity coordination.

The INTELSAT VI series of satellites have 12 transponders that are dedicated exclusively to TDMA operation; other transponders provide FDMA operation. The modulation employed is QPSK with coherent, as opposed to differential, demodulation. The use of four-state PSK effectively doubles the bandwidth efficiency so that the nominal 120-Mb/s data rate is transmitted as 60 megasymbols/s. The TDMA frame length is nominally 2 ms and forward error control (FEC) is applied to selected traffic bursts.

Because of the bursty nature of voice (on average only 40% of the call duration is active with speech), the INTELSAT TDMA system will incorporate *digital speech interpolation* (DSI) to improve the voice channel utilization. DSI is an activity-dependent transmission technique in which the speech activity of a channel is sensed and, during speech pauses, the channel can be switched to pick up active voice bursts. This technique, which is the digital equivalent of *time-assignment speech interpolation* (TASI), also has application to packet-switched networks.

A major advantage of TDMA over FDMA is that full transponder bandwidth and power can be used for each channel (station) during its allocated time slot, whereas, to minimize intermodulation distortion (IMD) and cross talk, FDMA is operated at reduced power levels—"backed off" power amplifier (typically a traveling-wave tube, TWT) drive levels—and consequently lower efficiency. That is, if the satellite is transmitting a total of 10 W (+10 dBW) and 100 equal-power FDMA carriers, then each carrier will use no more than 100 mW (−10 dBW). On the other hand, TDMA has very stringent network timing and is much more organizationally complex than FDMA. This is being found to be more than offset by the magnitude of digital data traffic, the store and forward capability, forward error control (FEC) encoding capability (especially during rain-caused signal-fading periods), and emerging digital signal-processing-enhancement techniques.

CDMA/Spread Spectrum

A relatively new multiple-access technique is code-division multiple access (CDMA), also called spread spectrum multiple access. (In this book, SSMA will not be used as an acronym for the spread spectrum technique in order to reserve SS for denoting satellite-switched multiple-access techniques.) As implied, the bandwidth of a CDMA signal is greatly increased by the spreading process. However, the price paid in bandwidth expansion is rewarded by virtual elimination of CW interference, allowing all users to occupy the entire system bandwidth continuously (no TDMA-type bursts), and by enhancing signal-to-noise.

CDMA can be implemented in direct-sequence or frequency-hopped techniques of spread spectrum; these are described in the following section. The direct-sequence (DS-SS) technique is a good candidate for local area networks where power levels among stations are relatively equal, and clock synchronism can be maintained. In this arena, the CDMA process is cost effective for low-duty cycle, low-average data rate implementations. Cost efficiency exists because all signals occupy the same frequency band (hence require no networkwide burst-level timing), do not suffer from collision-caused retransmission delays, and are initiated without slot reservations.

Use of CDMA/spread-spectrum techniques in a satellite system requires that the power received from each ground-station transmitter must, at the satellite transponder, be equal. This is because efficient satellite systems typically employ IF limiting or a saturated traveling-wave tube amplifier (TWTA) in the transmitter, and the weaker signals will be suppressed; see *capture effect* in Section 9–7. A partial exception to the equal-power rule may be applied for linear, high-dynamic-range transponders.

In *code-division multiple access* for satellites, each of the transmitting ground stations is assigned a separate and unique binary code sequence $c_i(t)$. This unique code, in effect, becomes a key with which a receiving station can recover the transmitted information. As illustrated in Figure 17–38, the binary data $m_i(t)$ of a transmitting station is first modulated either by BPSK, QPSK, or perhaps FSK, followed by a spectrum multiplication by the spreading code $c_i(t)$ and then amplified and coupled to the station antenna. The pseudorandom nature of a particular code sequence $c_1(t)$ is often referred to as pseudorandom noise (PRN) and must have high autocorrelation properties with itself, $\overline{c_1(t) \times c_1(t)}$, and low cross-correlation properties with the other station codes, as expressed by

$$\overline{c_1(t) \times c_2(t)} \ll \overline{c_1(t) \times c_1(t)} \qquad \textbf{(17–27)}$$

where the overbar represents an averaging (integrating) operation. Correlation was discussed in Section 4–7. In general, the spectrum multiplication process may be frequency hopping (FH) or direct sequence and uniquely correlates the modulated signal with the code sequence. At the same time, this process spreads the modulated spectrum over the full satellite transponder bandwidth and all stations transmit simultaneously.

As mentioned previously, at the satellite transponder all signals must have approximately equal signal strength if IF limiting or a saturated TWTA is em-

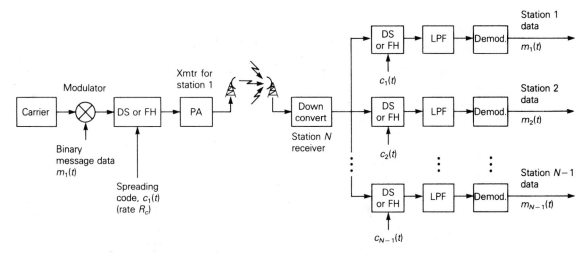

FIGURE 17-38 Generic CDMA transmitter/receiver block diagram; spread spectrum technique.

ployed in order to avoid suppression of the weaker signals; see *capture effect* in Section 9–7. As illustrated for a system of N ground stations (Figure 17–36), a ground station designed to receive all of the other stations must power-split the downlink signal into $N - 1$ correlators, each with the $N - 1$ unique codes of the other stations. Each decorrelator compares its unique code with the entire received signal spectrum and correlation occurs when the unique code sequence exactly matches one of the received CDMA spreading signals. The output will consist of a single, despread but still modulated BPSK, QPSK, or FSK station signal, and the other $N - 2$ stations look like a very wideband noise. Filtering greatly reduces the noise, and demodulation recovers the message. The codes used in CDMA systems must possess high auto- (self-) correlation and uniformly low cross-correlation properties. Spread spectrum techniques, of which CDMA is one particular application, are covered in the next section.

17-6 ■ SPREAD SPECTRUM SYSTEMS

Spread spectrum techniques have been employed in military transmission systems for many years for providing a low probability of intercept (LPI) or detection, for interference rejection, and for information security in general. Military and civilian organizations have utilized spread spectrum techniques for high-accuracy range measurements and positioning, even on a global scale. These systems, such as the Global Positioning System (GPS), use several satellites in synchronous and asynchronous orbits.

The transmitter block diagram in Figure 17–38 illustrates a generic spread spectrum transmitter. The essential difference between a spread spectrum transmitter and that for any other modulated carrier communication system is an additional modulation by a high-rate (R_c), discrete pseudorandom code. The pseudorandom code sequence $c(t)$ must have a rate R_c much greater than the binary message rate R_m. That is,

$$R_c \gg R_m \tag{17–28}$$

by two or three orders of magnitude ($\times 100$–$\times 1000$). Also, $c(t)$ must be *statistically independent* of the message signal $m(t)$. If these two criteria are met, then the modulated message signal spectrum will be increased (spread) by an amount given by

$$G = \frac{R_c}{R_m} \tag{17–29}$$

where G is called the *processing gain*, and the resulting spread spectrum signal can be accurately demodulated only if the receiver possesses a matched despreading circuit using $c(t)$. By far the most common despreading technique employs simple correlation of the received spread spectrum signal with the stored reference code.

Some of the most common techniques for achieving the spreading function are as follows:

1. *Direct-sequence* (DS): A sequence $\{c(t)\}$ of binary polar (± 1) pulses linearly multiplies the modulation $m(t)$ or modulated carrier, $\cos \omega_0 t$, producing a double-sideband/suppressed-carrier signal $s_{DS}(t)$, where

$$s_{DS}(t) = c(t)m(t)\cos(\omega_0 t) \tag{17–30}$$

2. *Frequency hopping* (FH): Here $\{c(t)\}$ is demultiplexed to control a k-bit frequency synthesizer used as a frequency-hopping local oscillator (LO). The resulting 2^k pseudorandomly generated LO frequencies f_{LO} are then used in a mixer to sequentially up-convert the modulated carrier to hopped frequencies given by $\{f_h\}$ on each FH chip of duration T_c, where T_c is the inverse of R_{FH} and

$$R_{FH} = \frac{R_{c(t)}}{2^k} \tag{17–31}$$

3. *Hybrid-DS/FH* techniques that include both DS and FH characteristics to mitigate against the susceptibility of either taken alone.

Direct-Sequence

As defined previously, direct-sequence spread spectrum (DS-SS) results when a message-modulated signal is multiplied by spreading waveform $c(t)$ in a balanced mixer called the spreading correlator. The spreading code rate is $R_c = 1/T_c$, where T_c is the time duration of a single plus-or-minus (bipolar) pulse, called a *chip*, and chips have a 100–1000 times shorter duration than a data message bit ($T_c \ll T_m$). Consequently, the transmitted spectrum will be 100–1000 times greater than the bandwidth of the initial BPSK, QPSK, or FSK data-modulated carrier, having been finely chopped up by a wideband, unique spreading code sequence. The resulting spectrum is highly correlated with the spreading code; correlation was discussed in Section 4–7.

The spreading codes used in spread spectrum systems are primarily *maximal-length sequences*, called *m-sequence*, and *Gold* codes. Gold codes are combinations of maximal-length codes and were invented in 1967 at Magnavox Corporation specifically for multiple-access CDMA applications. Relatively large sets of Gold codes exist that have well-controlled cross-correlation properties.[*]

M-sequence codes are by far the most widely used in spread spectrum systems. They can be generated as illustrated in Figure 17–39 using a modulo-2 adder (ex-OR gate) and an *r*-stage serial shift register. Table 17–3 lists the shift-register tap points that produce the maximal length given by

$$N = 2^r - 1 \qquad\qquad (17\text{--}32)$$

where N is the number of chips of duration T_c before the random sequence repeats. Hence, the *m*-sequence code length period in seconds is given by

$$T_{\text{DS}} = NT_c \qquad\qquad (17\text{--}33)$$

for a direct-sequence spread spectrum system.

For a DS-SS system, the processing gain given by Equation 17–29, $G = R_c/R_m$, and the desirable parameters of this technique, such as interference rejection, low probability of intercept, and antijam characteristics, are increased by 20–30 dB over those of systems without spreading. Figure 17–40 illustrates the major components of a DS-SS system. In this implementation, the data clock is derived from and is therefore coherent with the spreading code, and the data modulator is a balanced mixer that produces BPSK; the correlator is also a balanced mixer (single-sideband/suppressed-carrier modulator). If the transmitter power amplifiers have a net voltage gain of A, then the transmitted BPSK/DS-SS signal can be expressed as

$$s_{\text{DS}}(t) = Ac(t)m(t)\cos \omega_0 t \qquad\qquad (17\text{--}34)$$

and has a $(\sin x)/x$ spectrum, as illustrated in the figure, centered at f_0 with a first-lobe null bandwidth of

$$\text{BW}_{\text{DS-SS}} = 2R_c \qquad\qquad (17\text{--}35)$$

in hertz, and the spectrum nulls occur at

$$f_{\text{null}} = \pm nR_c \qquad (n = 1, 2, \dots) \qquad\qquad (17\text{--}36)$$

in hertz.

The pseudorandom spectral lines in the spectrum of $s_{\text{DS}}(t)$, labeled $S_{\text{DS}}(f)$, are harmonics of R_c/N. That is,

$$f_{\text{line}} = \pm f_0 \pm \frac{nR_c}{N} \qquad\qquad (17\text{--}37)$$

---

[*] R. E. Ziemer and R. L. Peterson, *Digital Communications and Spread Spectrum Systems* (New York: Macmillan, 1985), p. 404.

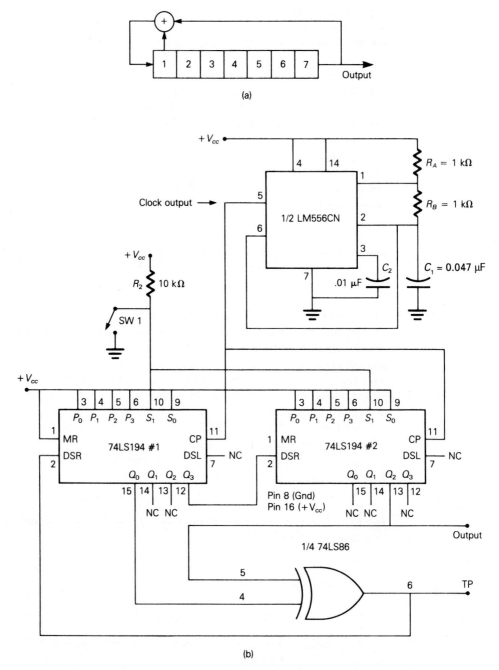

FIGURE 17-39 (a) Block and (b) schematic for a seven-stage serial, right-shift register and ex-OR gate implementation of a [7,1] *m*-sequence PRN code generator.

TABLE 17–3 Maximal-length PRN Code-Generator Configurations

Number of Stages	Code Length N	Tap Points for Maximal Length (Fewest-Tap Configurations, Only)
2	3	[2, 1]
3	7	[3, 1]
4	15	[4, 1]
5	31	[5, 2]
6	63	[6, 1]
7	127	[7, 1]
8	255	[8, 5, 3, 1]
9	511	[9, 4]
10	1,023	[10, 3]
11	2,047	[11, 1]
12	4,095	[12, 6, 4, 1]
13	8,191	[13, 4, 3, 1]
14	16,383	[14, 12, 2, 1]
15	32,767	[15, 1], [15, 4]

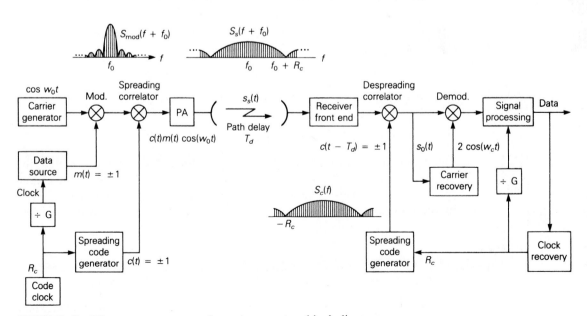

FIGURE 17–40 Direct-sequence spread spectrum system block diagram.

where, from Equation 17–33,

$$\frac{R_c}{N} = R_{DS} \qquad\qquad\qquad (17\text{--}38)$$

is the rate at which the m-sequence code repeats. Clearly, the transmit spectrum for DS-SS with high processing gain is determined primarily by the spreading characteristics.

■ ■ ■ ■

EXAMPLE 17–3

A DS-SS system requires 33 dB of CW interference rejection, uses a 10-stage shift register m-sequence generator, and transmits 500-b/s binary message data on a 1-GHz carrier. Determine the following:

1. Required code generator clock rate.
2. First-null bandwidth of the data.
3. Spectral line spacing in the transmitted signal.
4. First-lobe bandwidth and sketch of the (positive-only) frequency spectrum.

Solution:

1. The processing gain is $G = 10^{G(dB)/10} = 10^{3.3} = 2000$. The required code clock rate is $R_c = R_m G = 1$ *megachips* per second **(1 Mc/s).**

2. The first null in a 50% duty cycle squarewave (average shape of pseudorandom binary data) is "every even harmonic"; that is, $f_{null} = 1/T_m = f_m = $ **500 Hz.** (The first data spectral line is $1/(2T_m) = 250$ Hz because of the usual 2 bits/cycle of NRZ formats.)

3. Assuming the appropriate taps are used in the code generator to provide a maximal-length code, then

$$N = 2^{10} - 1 = 1023 \text{ chips} \qquad \text{(per cycle)}$$

Then from Equation 17–38, the code period, in hertz, is

$$R_{DS} = \frac{10^6 \text{ chips/s}}{1023 \text{ chips/cycle}} = \textbf{977.5 Hz}$$

4. The input data spectrum first null is 500 Hz, from part 2. When modulated, the full first-null RF bandwidth will be $2 \times 500 = 1000$ Hz. The point of the spread spectrum technique is to increase the bandwidth of the data-modulated carrier by the processing (gain); hence, the transmitted first-lobe bandwidth will be

$$BW_{DS\text{-}SS} = 2R_m G = (1000 \text{ Hz}) \times 2000 = 2 \times 10^6 \text{ Hz}$$

Otherwise, use Equation 17–35:

$$BW_{DS\text{-}SS} = 2R_c = 2(10^6) = \textbf{2 MHz}$$

Individual spectral lines will appear in the spectrum pseudorandomly at every harmonic of 977.5 Hz above and below the transmission carrier frequency of

FIGURE 17-41 Example problem frequency spectrum.

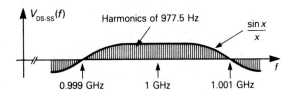

1 GHz. The amplitude of the spectral lines above and below the carrier will decrease in a $(\sin x)/x$ fashion with nulls every 977.5 Hz $\times$ 1023 = 1 MHz. Refer to Figure 12–5 with $\tau = 1/R_c$ and $T = 1/R_{DS}$). Of course, we could have used Equation 17–37. The results are sketched in Figure 17–41.

Frequency-Hopping Spread Spectrum

As illustrated in Figure 17–42, the message data input to a frequency-hopping spread spectrum transmitter is first modulated and then up-converted using a local oscillator whose frequency is switched by a k-bit PN code taken from a spreading code generator with serial output, $c(t)$.

The result at the output, $s_{FH}(t)$, is that at any time instant, $t = iT_H$, the information-modulated carrier is transmitted at frequency $f_h(iT_H)$ with an information bandwidth that fills a frequency slot B_H in the overall system bandwidth B_s (see Figure 17–43).

With k bits addressing the frequency synthesizer, there can be n LO frequencies, where

$$n = 2^k \qquad (17\text{–}39)$$

If the spreading spectrum code rate is $R_c = 1/T_c$, then it takes kT_c code clock periods to fill up the k-stage serial-in, parallel-out demultiplexer before each hop

FIGURE 17-42 Frequency-hopped spread spectrum system block diagram.

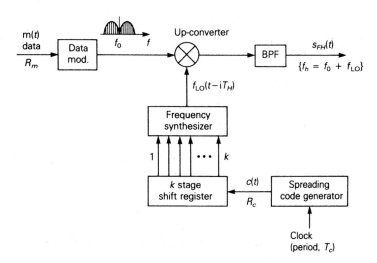

FIGURE 17–43 FSK spectrum hops for FH-SS system.

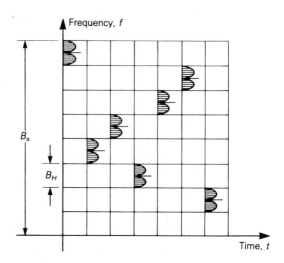

can occur; thus the hop period is

$$T_H = kT_c = \frac{\log_2 n}{R_c} \tag{17-40}$$

and the hopping rate is computed with

$$R_H = \frac{1}{T_H} = \frac{R_c}{\log_2 n} = \frac{R_c}{k} \tag{17-41}$$

or
$$R_c = kR_H \tag{17-42}$$

It is very common to choose binary FSK (BFSK) for the data modulator and to transmit no more than one of the two FSK frequencies, f_M and f_S (mark and space), during each hop. Each hopped frequency slot must have a bandwidth B_H capable not only of encompassing f_M or f_S, but also of providing for nonoverlapping switching spectra, that is, orthogonality. The frequency difference between the FSK mark/stop frequencies should be

$$f_M - f_S \geq \frac{2}{T_H} = 2R_H \tag{17-43}$$

and in order to guarantee sufficient separation during decoding, the transmit frequency slot B_H should be twice the tone separation for spectral orthogonality. That is,

$$B_H = 4R_H \tag{17-44}$$

Such a system operating at one information bit per hop is referred to as *slow-FH*.

For binary slow-FH at 1 bit/hop, the system bandwidth B_s would contain n frequency slots such that

$$n = \frac{B_s T_H}{4} \tag{17-45}$$

Fast-FH will result if the data symbol frequency (1 bit) is transmitted over several hops. That is,

$$R_H = qR_d \qquad \qquad (17\text{--}46)$$

where q is always an integer greater than zero.

The system gain with respect to hostile intrusions or signal-path interference is in the decreased probability of transmitting the data in a given frequency band. Hence,

$$G = \frac{R_H}{R_d} \qquad \qquad (17\text{--}47)$$

Since $R_H = R_c/2^k$, the code generator clock must exceed that used for DS-SS.

Direct sequence spread spectrum systems are the most straightforward conceptually and analytically. However, as frequency synthesizer technology improves, especially in speed, frequency-hopped spread spectrum systems are being implemented more frequently than DS-SS.

■ PROBLEMS

1. **a.** Sketch the BPSK modulator of Figure 17–4 and show the complete data current path when a digital 1 is applied to the input pin.
   b. Using dashed lines, show that a +RF phase on the secondary of T_1 will be connected to the top lead of T_2.

2. Repeat Problem 1 for a digital 0 input and show the phase reversal.

3. Starting with "1" as the first transmitted bit, 101001011110 is to be encoded for differential PSK transmission.
   a. Assuming that the initialization bit to the encoder is a 0, determine the NRZ encoded data sequence (refer to Figure 17–7).
   b. Determine the transmitted DPSK phase sequence from the Figure 17–7 transmitter.

4. Determine the output TTL data sequence for the DPSK demodulator of Figure 17–8 if the input phase sequence is $0\pi\pi0\pi\pi00000\pi$. (The zero is the first received phase, and the initialization phase at the delayed input to the phase detector is also zero degrees.)

5. **a.** Sketch the QPSK constellation defined by $(00) = -45°$, $(01) = +45°$, $(11) = +135°$, and $(10) = -135°$.
   b. For the data sequence of Problem 3 and constellation of part **a**, write the output phase sequence.
   c. If the carrier at the output of each phase reversal modulator (Figure 17–14) is 2 V rms at 1592 Hz and the QPSK modulator serial input data rate is 300 b/s,
   (1) Write the output QPSK signal $V_{QPSK}(t)$ with $a_1(t)$ and $b_1(t)$ as the binary modulation of the $\cos \omega_c t$ and $\sin \omega_c t$ carriers, respectively.
   (2) Show the calculation for the required absolute minimum QPSK output bandwidth.

6. If QPSK modulation is used, a transmission line capable of 1800 phase-states per second can transfer a computer-to-computer information rate of _____.

7. One carrier recovery technique for QPSK demodulation uses a ×4/filter/ ÷4 scheme. The same technique for BPSK uses ×2. What does the (nonlinear) multiplication produce?

8. What advantage does differential-QPSK have over QPSK?

9. **a.** Reverse the diodes in the schematic of Figure 17–14 and determine the new QPSK constellation.
   b. Change the 90° phase shifter in Figure 17–14 to −90° and determine the new QPSK constellation.

10. Calculate the theoretical information density for the following:
    a. QPSK
    b. 8-PSK
    c. 4-QAM
    d. 16-QAM
    e. 64-QAM

11. Identify, with a few letters and numbers, the constellations of Figure 17–44.

12. **a.** Show the "tribits" for the data sequence of Problem 3.
    b. Write the corresponding transmitted phase sequence for the 8-PSK constellation of Figure 17–26.

13. Referring to the distances d and h of Figure 17–26, compare the 16-QAM of Figure 17–25 to 16-PSK. Make the amplitude of the maximum transmitted signal voltage A the same for both systems. Put the ratio of h/d in decibels for comparison.

14. **a.** If the maximum transmitted vector in the 16-QAM constellation of Figure 17–25 is 100 V rms, determine the long-term average power transmitted from a 50-Ω antenna if each point in the constellation is equally likely to be transmitted. (*Note:* The answer must be less than $100^2/50 \ \Omega = 200$ W.)
    b. Show the amount (in dB) by which the QAM power can be increased to make the average transmitted power equal to PSK.

15. A 16-QAM system must maintain less (better) than a 10^{-7} BER. What is the minimum allowable receiver carrier-to-noise ratio?

16. **a.** Compare the error rate performance for 8-QAM and 8-PSK with $C/N = 18$ dB.
    b. How many errors are expected to occur for each if 10 Mb of data are transmitted?

17. PCM is modulated on a microwave carrier as 8-PSK.
    a. What is the theoretical C/N (in dB) necessary at the receiver in order for the system to make only one error in 100,000 bits?
    b. How many more dB of system gain is required of 16-PSK for the same BER?

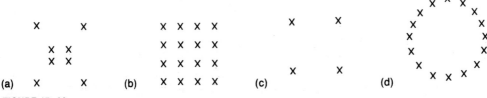

(a) (b) (c) (d)

FIGURE 17–44

18. An FSK system must maintain less than a 10^{-3} BER. What is the minimum allowable received E_b/N_0?

19. **a.** Compare the error rate performance for coherent and differential PSK (binary) for a system energy/bit-per-noise-density ratio of 7 dB.
    b. How many errors are expected for each if 10 Mb of data are transmitted?

20. Draw the FCC mask using actual frequencies for a 6-GHz microwave transmitter output spectrum.

21. **a.** Explain frequency diversity.
    b. Explain space diversity.

22. What is the minimum recommended distance in meters for antenna separation in a 2-GHz space-diversity receiving system?

23. An SBS satellite is 22,300 mi from an earth station. The earth station transmitter power output is 1 kW at 6 GHz. Fifty meters of EWP52 waveguide (4 dB/100 m) is used for the antenna feed, and the branching losses are 0.7 dB (circulators and VSWRs). Determine:
    a. The effective radiated power (with respect to isotropic) from the 5-m-diameter microwave dish.
    b. The power at the satellite transponder input if the receiving antenna is 1.2 m in diameter, branching loss is 3 dB, and EWP52 waveguide is 1 m.

24. A receiving system with noise temperature of 200°C and 22-MHz IF bandwidth receives an FM transmission of −100 dBm.
    a. Determine the C/N (dB).
    b. Would a discriminator be operating above threshold?

25. What is the G/T rating for a 4-GHz receiving system with a 2.3-dB system noise figure and a 30-m-diameter, 55% efficient microwave dish antenna.

26. A *Ku*-band, 14-GHz uplink, 12-GHz downlink satellite system employs an 11.22-W TWTA power amplifier and a 1.92-m "dish" antenna. The line-of-sight distance from the satellite to the ground-station receiver is 24,000 mi (38.7 Mms). The ground-station receiving antenna has a diameter of 10 m, a low-noise amplifier (LNA) with NF = 4 dB (including received space noise, antenna, and LNA noise), and a noise bandwidth of 50 MHz. Determine the following:
    a. Satellite EIRP in dBW and dBm.
    b. Satellite antenna beamwidth in degrees.
    c. Ground-station antenna gain–to–effective system noise temperature ratio, G/T, in decibels.
    d. $S/N \ (= C/N)$ at the LNA output, in decibels.
    e. Determine the maximum receiver bandwidth if the demodulator must operate above FM threshold.

27. **a.** Figure 17–45 shows an example of: (A) TDMA, (B) DDMA, (C) FDMA, (D) TAMA, (E) none of the above.

FIGURE 17–45

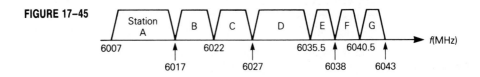

   b. In which access method will all stations transmit simultaneously? (A) TDMA, (B) DDMA, (C) FDMA, (D) TAMA, (E) none of the above.

   c. For which access method would the total bandwidth of a satellite be available to each station? (A) TDMA, (B) DDMA, (C) FDMA, (D) TAMA, (E) none of the above.

28. A receiver has 36-MHz bandwidth and 10-dB noise figure. For a -90-dBm carrier and 100 kbps data, compute the following:

   a. Magnitude of the power spectral density (dBm/Hz of RF bandwidth) when the carrier is BPSK modulated with a serial random, bipolar data stream if we assume that all of the spectral components are uniformly distributed over the main-lobe bandwidth. Answer in dBm/Hz and W/Hz.

   b. What is the maximum possible processing gain G for a direct-sequence spread spectrum system using this receiver?

29. a. Draw a four-stage m-sequence (maximum-length) code generator.

   b. Show the register contents over 20 clockings. (Initialize with all 1s.)

   c. Show by inspection the number of chips/period.

   d. Compute the theoretical number of chips/period, N.

   e. Repeat **a**, **b**, and **c** for a [4,2] generator.

30. An FH-BPSK signal is modulated with 10 kb/s data and the system bandwidth is 10 MHz. The BPSK signal is filtered, allowing only two spectral components on either side of the carrier. Find

   a. The maximum number of hopping frequencies.

   b. The number of bits (chips) to be used in the spreading generator.

   c. The hopping rate.

17-7 ■ COMPUTER ANALYSIS

Using the computer for communications system and circuit analysis has become a virtual necessity in education and industry. What follows are examples of programs and/or readily available simulation software for solving problems of the type covered in this chapter.

All the programs are interactive in the sense that an introductory sentence or short paragraph describes the purpose of the program and the user is then prompted for input data.

The programs for Chapter 17 are as follows:

■ **BPSK modulator:** The PSpice analysis of Figure 17–46 was performed on a BPSK modulator circuit (Figure 17–3). The circuit file demonstrates the modeling of a center-tapped transformer on SPICE, and the output file shows the time-domain results of the digital data-modulated PSK waveform.

■ **BIT error rate:** The program of Figure 17–47 computes the probability of error for noncoherent FSK and differentially encoded binary PSK (DPSK), both as a function of the receiver energy/bit per noise density ratio (E_b/N_0).

■ **Transmission link analysis:** See the Computer Analysis section of Chapter 15.

```
******** 5/26/88******** Demo PSpice (May 1986) *******14:42:18********
BPSK MODULATOR
.WIDTH IN=80 OUT=80
IFR 1 0 SIN(0 5M 150MEG)
R1 1 0 50
L1 1 0 100N
L2 3 4 100N
L3 4 5 100N
K12 L1 L2 .96
K23 L2 L3 .96
K13 L1 L3 .96
RDC 5 0 10MEG
D1 3 6 DIODE
D2 6 5 DIODE
D3 8 3 DIODE
D4 5 8 DIODE
.MODEL DIODE D
L4 6 7 100N
L5 7 8 100N
L6 9 0 100N
K46 L4 L6 .96
K56 L5 L6 .96
K45 L4 L5 .96
RLOAD 9 0 50
L7 4 7 100N
L8 11 0 100N
K78 L7 L8 0.96
RLO 11 0 50
ILO 11 0 PULSE(.2 -.2 0 .0 .0 6NS 12NS)
.TRAN .2NS 20NS
.PLOT TRAN V(9) V(1)
.END
```

FIGURE 17-46   (page 1 of 3)

```
TRANSIENT ANALYSIS    *: V(9)
                      +: V(1)
(*+)----------- -2.000D-01   -1.000D-01    .000D+00    1.000D-01    2.000D-01
    TIME      V(9)    - - - - - - - - - - - - - - - - - - - - - - - -
 2.800D-09 -8.208D-04 .           .            *   +           .            .
 3.000D-09  1.735D-02 .           .            . *    +        .            .
 3.200D-09  3.515D-02 .           .            .   *    +      .            .
 3.400D-09  5.223D-02 .           .            .      *   +    .            .
 3.600D-09  6.808D-02 .           .            .        *   +. .            .
 3.800D-09  8.150D-02 .           .            .          *  .+ .           .
 4.000D-09  9.299D-02 .           .            .           *.  +            .
 4.200D-09  1.007D-01 .           .            .            *   +           .
 4.400D-09  1.061D-01 .           .            .            .*   +          .
 4.600D-09  1.069D-01 .           .            .            .*   +          .
 4.800D-09  1.054D-01 .           .            .            .*+             .
 5.000D-09  9.908D-02 .           .            .            X               .
 5.200D-09  9.066D-02 .           .            .            X.              .
 5.400D-09  7.807D-02 .           .            .          +*                .
 5.600D-09  6.393D-02 .           .            .         +*                 .
 5.800D-09  4.673D-02 .           .            .      +  *                   .
 6.000D-09  2.880D-02 .           .            .   .+   *                    .
 6.200D-09  9.426D-03 .           .            . + .*                        .
 6.400D-09 -1.029D-02 .           .          +   *.                          .
 6.600D-09 -2.949D-02 .           .        +   *  .                          .
 6.800D-09 -2.601D-02 .           .        +    *  .                         .
 7.000D-09  1.744D-02 .           .        +     . *                         .
 7.200D-09  3.815D-02 .           .     +        .    *                      .
 7.400D-09  5.349D-02 .           .   +          .      *                    .
 7.600D-09  6.420D-02 .           . +            .      *                    .
 7.800D-09  7.200D-02 .          .+             .       *                    .
 8.000D-09  7.367D-02 .          . +            .       *                    .
 8.200D-09  7.482D-02 .          . +            .       *                    .
 8.400D-09  6.902D-02 .          .  +           .      *                     .
 8.600D-09  6.278D-02 .           .   +         .     *                      .
 8.800D-09  5.067D-02 .           .      +      .    *                       .
 9.000D-09  3.826D-02 .           .        +    .  *                         .
 9.200D-09  2.190D-02 .           .          +. *                            .
 9.400D-09  5.424D-03 .           .           .*+                            .
 9.600D-09 -1.261D-02 .           .         *  .    +                        .
 9.800D-09 -3.057D-02 .           .       *    .       +                     .
 1.000D-08 -4.751D-02 .           .     *      .          +                  .
 1.020D-08 -6.419D-02 .           .   *        .           +.                .
 1.040D-08 -7.742D-02 .          . *           .           +                 .
 1.060D-08 -9.026D-02 .        .*             .             . +              .
 1.080D-08 -9.775D-02 .        *             .              .  +             .
 1.100D-08 -1.047D-01 .      *.              .              .  +             .
```

FIGURE 17-46   (page 2 of 3)

```
1.120D-08 -1.052D-01 .         *.            .            . +      .
1.140D-08 -1.052D-01 .         *.            .            . +      .
1.160D-08 -9.831D-02 .          *            .          .+        .
1.180D-08 -9.106D-02 .         .*            .          +         .
1.200D-08 -7.873D-02 .          . *          .       + .          .
1.220D-08  3.939D-02 .            .          . *   +   .          .
1.240D-08  3.903D-02 .            .          .  +*     .          .
1.260D-08  2.515D-02 .            .          . + *     .          .
1.280D-08  7.465D-03 .            .        +  .*       .          .
1.300D-08 -1.135D-02 .            .       +  *.        .          .
1.320D-08 -2.963D-02 .            .     +   *.         .          .
1.340D-08 -4.767D-02 .            .    +   * .         .          .
1.360D-08 -6.319D-02 .           .+    *     .         .          .
1.380D-08 -7.812D-02 .         +  . *        .         .          .
1.400D-08 -8.870D-02 .        + .*           .         .          .
1.420D-08 -9.828D-02 .       +   *           .         .          .
1.440D-08 -1.024D-01 .       +   *           .         .          .
1.460D-08 -1.052D-01 .       +*.             .         .          .
1.480D-08 -1.021D-01 .       +*              .         .          .
1.500D-08 -9.775D-02 .       X               .         .          .
1.520D-08 -8.798D-02 .       . X             .         .          .
1.540D-08 -7.695D-02 .        .*+            .         .          .
1.560D-08 -6.178D-02 .         . *  +        .         .          .
1.580D-08 -4.564D-02 .         .   *   +     .         .          .
1.600D-08 -2.717D-02 .         .      *   +. .         .          .
1.620D-08 -8.161D-03 .         .       *.  + .         .          .
1.640D-08  1.103D-02 .            .*    .  + .         .          .
1.660D-08  3.027D-02 .            .   *     .  +       .          .
1.680D-08  4.751D-02 .            .      *   .  +      .          .
1.700D-08  6.429D-02 .            .       *  .  +      .          .
1.720D-08  7.717D-02 .            .          . *   +   .          .
1.740D-08  8.915D-02 .            .          .   *   + .          .
1.760D-08  9.588D-02 .            .          .    * . + .         .
1.780D-08  1.014D-01 .            .          .     * +  .         .
1.800D-08  1.013D-01 .            .          .     * +  .         .
1.820D-08  9.989D-02 .            .          .    *+    .         .
1.840D-08  9.334D-02 .            .          .       X. .         .
1.860D-08  8.262D-02 .            .          .   +*.    .         .
1.880D-08 -4.382D-02 .            .  *       .      + . .         .
1.900D-08 -4.890D-02 .            .  *       .    +     .         .
1.920D-08 -3.560D-02 .            . *       .   +      .          .
1.940D-08 -1.785D-02 .           .  * +     .         .          .
1.960D-08  1.213D-03 .            .     + * .         .          .
1.980D-08  1.999D-02 .           +   . *    .         .          .
2.000D-08  3.870D-02 .       +        .   * .         .          .
```

FIGURE 17-46 (page 3 of 3)

```
10 REM * ERROR PROBABILITY RATE OF A SIGNAL WITH NOISE *
20 REM ***   By P. Young   ***  5-22-88  ***   ASU   ***
30 CLS
40 PRINT "This program computes error probability rates"
50 PRINT "for noncoherent FSK and DPSK (Figure 17-28).":PRINT
60 PRINT "You must enter the energy/bit, Eb, and the noise"
70 PRINT "spectral density, No. Note that, if you do not know"
80 PRINT "these specifically, make up the ratio from Eb/No(dB)."
90 PRINT "----------------------------------------------------"
100 INPUT "FSK: Please enter the value for Eb";EB:PRINT " "
110 INPUT "Please enter the value for No";NO:PRINT
120 LET ANS1=.5*(EXP(-.5*EB/NO))
130 PRINT "The probability of error for noncoherent FSK is";ANS1;"."
140 PRINT " " :PRINT
150 INPUT "DPSK: Please enter the value for Eb";EE:PRINT
160 INPUT "Please enter the value for No";NN:PRINT
170 LET ANSW2=.5*(EXP(-EE/NN))
180 PRINT "The probability of error for DPSK is";ANSW2;"."
190 END
```

FIGURE 17-47

18

Fiber-Optic Communication

■ INTRODUCTION

Communicating by light goes back as far as eyesight. Indeed, modulation of a light source in a classical communication link, including the use of a code, has been practiced by the military for centuries. This system, called "flashing light signaling," is still used today. The modulation consists of a high-intensity lamp with a parabolic reflector for focusing and a shutter system that is opened and closed according to the Morse code. The receiving operator reads the light flashes to decode the message. The information rate of this system is, however, mechanically and physiologically limited and the decoding not easily automated.

The invention of lasers in the late 1950s stimulated interest in communication by light, and the development in 1970 of optical fibers with only 20 dB of loss per kilometer (compared to more than 1000 dB/km previously) spurred the proliferation of fiber-optic guided, light communications technology. Now, optical fiber links span both the Atlantic and Pacific oceans as well as Europe and the North American continent. These developments have spurred network protocols and standards.

This chapter reviews the fundamentals needed for understanding typical fiber-optic concepts, link design parameters, applications, and standards. The overall communication link parameters are common to those already studied—the link power budget, bandwidth, distortion, noise, modulation—but devices and other components have unique characteristics. The discussion begins with an overall link description. ■

18–1 ■ THE OPTICAL FIBER COMMUNICATION LINK

Like all communication links, the fiber-optic link consists of a transmitter, transmission medium (optical fiber), and a receiver. Figure 18–1 shows the basic block diagram of the system. Systems such as local area networks (LANs) have a huge number of terminals (transmitters and receivers) and operate over relatively short distances. These systems use highly reliable, inexpensive light-emitting diodes (LEDs) as the light source with simple, inexpensive coupling mechanisms. Wideband, long-haul links take advantage of high-power coherent laser light sources. These links transport cable television, telephone, and high-speed data.

The advantages of optical fiber as a transmission medium are due to the fact that the fiber is made of a dielectric (glass or plastic) and the signal it carries is light. There is neither a conductive path nor metallic connection between the two ends of the link. Glass and plastic fibers are very lightweight, flexible, and resilient.

Wide bandwidth over long distances is possible because the attenuation of optical fiber is not frequency dependent, as is electrical coaxial cable, and high data rates can be transported over long distances without repeaters or equalizers.

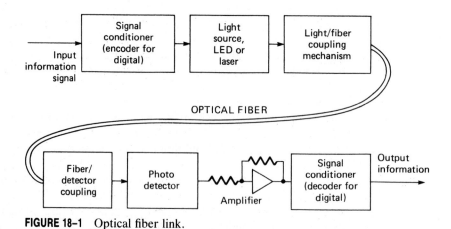

FIGURE 18-1 Optical fiber link.

Since the carrier of signal information is optical and the waveguide is a dielectric, optical-fiber transmission is not affected by conventional electrical interference. The signal will not be affected by stray RF or other electromagnetic energy, high voltage, or lightning.

Since glass is unaffected by most chemicals and solvents, optical fiber can be used in chemical plants as well as in oil and gas refineries. A broken fiber will not cause a spark, leading to an explosion. Short circuits cannot occur, thus alleviating concern over power supply–overload problems. Also, since there is no electrical connection between the two ends of the transmission system, there will not be ground loops with signals other than those specifically desired. All these technological and other advantages make the fiber-optic medium a good choice for many applications.

18–2 ■ LIGHT

Light is one of nature's most important information carriers. The human eye can detect electromagnetic energy in the frequency range of 4.3×10^{14} Hz to about 7.5×10^{14} Hz, wherein the lower frequency is identified as red and the higher as violet. It is the potential bandwidth and electromagnetic integrity of this information carrier that makes optical communications so important. As an example, a 430,000-GHz (4.3×10^{14}-Hz) carrier in a 10% bandwidth system could carry 43,000 Gb/s of data—the equivalent of 3.6 million TV programs or more than 4 billion analog telephone conversations.

One of the technological constraints in dealing with light is our ability to fabricate small enough circuits such as integrated optical circuits. From $\lambda = v/f$ we compute that the wavelength of light identified as red is about 0.7 μm (micrometers, or microns). Obviously, we can't use photolithography (light) to make sharp printed-circuit tracks with submicron dimensions. However, shorter-wavelength sources such as X rays are being used (X-ray lithography).

The light wavelengths being used in fiber-optic systems include 0.77–0.86 μm and 1.1–1.6 μm, with the most popular being 0.82, 1.3, and 1.55 μm. Since 0.82 μm can also be written as 820 nm (nanometers) or 8200 Å (angstroms, 10^{-10} meters), you can see that dimensions are approaching atomic levels.

The amount of *energy W* in light is proportional to its frequency.

$$W = hf \tag{18–1a}$$

$$W = \frac{hv}{\lambda} \tag{18–1b}$$

where $h = 6.626 \times 10^{-34}$ J · s (joule-seconds) is Planck's constant and v is the velocity of propagation.

One joule is 1 W · s of energy, and 1.6×10^{-19} J is 1 electron-volt (1 eV) of energy. There still exists the unresolved duality of light; that is, some light behavior is best described by wave phenomena, whereas particle behavior, due to a packet of energy called a *photon*, best describes other light behavior.

■ ■ ■ ■

EXAMPLE 18–1

1. Determine the energy in a photon of red (0.82-μm) light in free space.
2. Determine the power for a 0.1-μs burst of this red light.

Solution:

1. $W = 6.626 \times 10^{-34}$ (J · s) $\times 3 \times 10^8$ (m/s)/0.82 μm = **2.42 × 10⁻¹⁹ J/photon** = 1.5 eV/photon.

2. $P = W/t = 2.42 \times 10^{-19}$ W · s/10^{-7} s = **2.42 pW.**

18–3 ■ LIGHT PROPAGATION IN GLASS FIBERS

Critical Angle and Reflections

In the dielectric medium called *free space*, light propagates at a speed of about 3×10^8 m/s. In water this velocity is reduced by about 25%, and in various types of glass light is about 33% to 47% slower. The index of refraction is, from Equation 15–35, $n = c/v_p = \sqrt{\epsilon_r}$. Thus n is seen to be inversely proportional to the propagation velocity in the medium; Table 18-1 gives values for various materials.

A waveguide for propagating light can be made from a strand of glass the thickness of a human hair. When light is coupled into the fiber end, it will propagate by waveguide action, reflecting back and forth off the sides of the waveguide. However, the waveguide we are considering is not surrounded by a mirrorlike conductor, but rather by a dielectric with a different refractive index from that of the fiber core.

When a single electromagnetic wave is incident on a smooth interface separating the fiber and surrounding medium of different refractive index, the wave will either be totally reflected back into the fiber, or partially reflected and partially refracted, with part escaping from the fiber core. Figure 18–2 illustrates these two possibilities. Snell's law is used to determine the *critical angle of incidence* $\theta_{1_{cr}}$. This is the smallest angle from the normal for which total reflection occurs. The critical angle occurs when $\theta_2 = 90°$. At this angle, the refracted wave travels parallel to the interface. Snell's law as derived in Chapter 15 is expressed by

TABLE 18–1 Index of Refraction for Various Materials

Vacuum	1.00
Water	1.33
Glass (approximately)	1.5
Fused quartz	1.46
Diamond	2.0
Silicon	3.4
Gallium arsenide (GaAs)	3.6

FIGURE 18-2 Angles for refraction or total reflection.

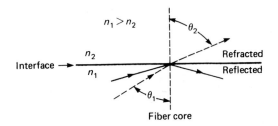

Equation 15–34, $n_1 \sin \theta_1 = n_2 \sin \theta_2$. Since $\sin 90° = 1$, the critical angle of incidence is

$$\theta_{1_{cr}} = \sin^{-1}\left(\frac{n_2}{n_1}\right) \qquad (18\text{-}2)$$

Please note that, for total reflection of the incident light, n_2 must be less than n_1. This is because $\sin \theta_1 = n_2/n_1 < 1$ for total reflection. Note also that Snell's law shows that when total reflection occurs and the light stays in medium 1, the angle of reflection is equal to the angle of incidence because "n_2" $= n_1$ in this case.

■ ■ ■ ■

EXAMPLE 18-2

What is the critical angle beyond which an ideal underwater light source will not shine into the air above?

Solution:

$$\theta_{1_{cr}} = \sin^{-1}(1.0/1.33) = \mathbf{48.6°}$$

Optical fibers for communication applications are made typically with a glass core of $n_1 = 1.5$ and surrounded ("coated," if you will) with glass or plastic of slightly lower refractive index—$n_2 = 1.485$ is common. This gives the fiber a large value for the critical angle of incidence (81.9°).

Numerical Aperture and Reflectance

The three most important factors that limit the usefulness of fiber-optic communication systems are *input coupling* of the light, *power losses* in the fiber, and waveguide *dispersion* that limits system bandwidth. Input coupling and dispersion are affected by the numerical aperture of the fiber.

The amount of useful light that can be coupled into a fiber is limited by the numerical aperture and reflectance. Of the light arriving at right angles to the face of the fiber, about 96% enters the glass fiber and 4% is reflected. Hence, for two fibers coupled across an airgap, there will be about 8% of power loss due to the *reflectance* or reflectivity of the glass-air interface. The reflectivity of the glass surface is found from

$$\text{reflectivity} = \left(\frac{n_1 - n_a}{n_1 + n_a}\right)^2 \tag{18–3}$$

where $n_a = 1$ for air.

Numerical aperture (NA) is a measure of the maximum angle of acceptance at the input for light rays that can be totally reflected inside the fiber. This does *not* mean that all rays within this angle will propagate down the waveguide, however. The acceptance angle θ_i is illustrated in Figure 18–3. Snell's law and some geometry will show that the numerical aperture can be determined for input from air ($n = 1$) by

$$\text{NA} \equiv \sin \theta_i(\text{max}) = \sqrt{n_1^2 - n_2^2} \tag{18–4}$$

If the input medium is not air, then the result of Equation 18–4 must be divided by n of the input medium. Note from Figure 18–3 that θ_i defines a solid cone of accepted light rays.

When devices with different NAs are coupled closely, a power loss proportional to the square of the NA ratio is incurred if the receiving NA is smaller. This is given by

$$\text{NA loss} = -20 \log \left(\frac{\text{NA}_R}{\text{NA}_s}\right) \tag{18–5a}$$

where $\text{NA}_s > \text{NA}_R$. When coupling from a small-diameter LED to a larger-diameter, step-index fiber with NA_f, the loss will be

$$\text{NA loss} = -20 \log \text{NA}_f \tag{18–5b}$$

Equations 18–5a and b are the same except that an LED has NA = 1, approximately. A laser cone of radiation is usually so small that virtually no NA loss is incurred when coupled closely to a receiving device.

Another factor that determines the amount of power accepted by the fiber is the relative cross-sectional area of the fiber compared to the source device. If the fiber

FIGURE 18–3 Light entering optical fiber inside the cone of acceptance defined by $\theta_i(\text{max})$ will be totally reflected at the core-cladding interface and propagate in the fiber core.

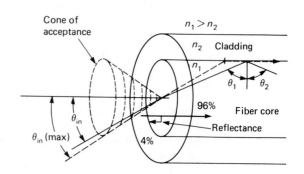

cross-sectional area A_f is smaller than that of the light source* cross section A_s, then the lost power is proportional to the ratio of areas:

$$P_f = P_s\left(\frac{A_f}{A_s}\right) \qquad (A_f < A_s) \tag{18–6a}$$

or

$$P_f = P_s\left(\frac{D_f}{D_s}\right)^2 \tag{18–6b}$$

or

$$\text{loss} = -20 \log\left(\frac{D_f}{D_s}\right) \tag{18–6c}$$

This loss is unity (0 dB) if the receiving fiber (or other device) is larger than the source. These equations show that large fibers with large NAs are efficient. However, large NA and diameter can result in large mode-dispersion in the fiber waveguide.

Optical Fiber Waveguides

The optical fiber waveguide discussed thus far is assumed to have a core with constant index n_1 surrounded by a cladding material of lower index n_2. This type of fiber waveguide is called *step index*, and the index profile is shown in Figure 18–4. Notice the characteristic step change in refractive index from core to cladding.

Recall from Chapter 15 that not all rays within a waveguide will propagate. Only those at certain discrete angles, for which the phase of the twice-reflected wave is the same as the incident wave, will propagate. The particular electric and magnetic field configurations which satisfy all the necessary conditions are called *modes* of propagation.

If the diameter D of the core is small enough—on the order of one or two wavelengths—then only a single mode will propagate in the waveguide. This

---

* Equations 18–5 and 18–6 assume a Lambertian type of radiation pattern. A light source is Lambertian if the power-versus-radiation angle follows a cosine relationship, $P(\theta) = P_0\cos \theta$. This type of power decrease with angle from the normal is typical of light reflected from the surface of this page with a small light source directly overhead.

FIGURE 18–4 Index profile for step-index fiber.

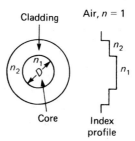

Cladding   Air, $n = 1$

n_2

n_1

Core

Index profile

dominant mode is an HE_{11} mode; not a purely TM or TE mode, it is rather a hybrid mode in which neither the longitudinal electric nor magnetic fields are zero. The hybrid modes are designated as HE or EH, depending on whether the longitudinal component of H or E makes a larger contribution to the transverse field.

The number of modes that can exist in a step-index fiber waveguide of core diameter D has been determined by numerical methods and can be found from a plot by Gloge in Figure 18–5. The horizontal axis is written in terms of a *normalized frequency* determined from fiber parameters as

$$V = \frac{D\pi(\text{NA})}{\lambda} \tag{18–7}$$

where λ is the free-space wavelength of the light. For small values of V use Figure 18–5, but for $V \gg 1$, the number of modes propagating in a fiber can be estimated with

$$N(\text{modes}) = \frac{V^2}{2} \tag{18–8}$$

Figure 18–5 assumes the typical fiber waveguide that has a small percentage of difference in n_2 with respect to n_1. This core-cladding index difference is usually written as

$$\Delta = \frac{n_1 - n_2}{n_1} \tag{18–9}$$

and for the typical optical fiber

$$\text{NA} \approx n_1 \sqrt{2\Delta} \tag{18–10}$$

Notice in Figure 18–5 that for single-mode propagation $V \leq 2.4$; otherwise multimode operation exists. You may remember the importance of the value 2.4 in regard to the Bessel functions (Figure 9–5) used for FM. The connection here is

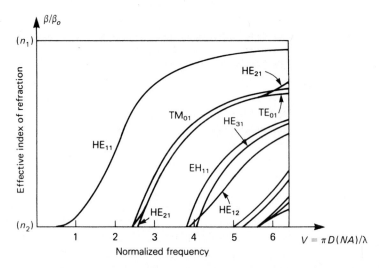

FIGURE 18–5 Optical fiber waveguide propagation modes for a step-index fiber. (From D. B. Keck, in *Fundamentals of Optical Fiber Communication*, ed. M. K. Barnoski, Academic Press, 2nd ed., 1981, p. 18.)

that Bessel functions are needed to solve the field equations for cylindrical wave-guides.

■ ■ ■ ■

EXAMPLE 18-3

A fiber-optic waveguide will have a core of index 1.5 and cladding 1.485.

1. How many modes will propagate if the core diameter is 4 μm and an 820-nm light source is used?

2. What is the maximum core diameter for single-mode propagation?

Solution:

$$NA = \sqrt{(1.5)^2 - (1.485)^2} = 0.21$$

1. $V = 4\pi(0.21)/0.82 = 3.2$. From Figure 18–5, **4 modes** will propagate (HE_{11}, TE_{01}, TM_{01}, HE_{21}).

2. $D(max) = 2.4(0.82)/(0.21\ \pi) = $ **2.98 μm.**

Multimode operation has the advantage of propagating greater power than does single-mode. However, pulse spreading due to modal dispersion limits the pulse and information rate.

Bandwidth-Distance Product

Fibers with large numerical aperture and large diameter will allow more light power to propagate. Both factors allow more useful light to be coupled into the fiber, and a large diameter allows more modes to be excited and carry energy. However, step-index fibers with large diameters are very limited with respect to a very important information capacity parameter in communication applications—the *bandwidth distance (cable length) product, B·L,* in Hz·km. Various distortion effects in step-index fibers, primarily intramodal dispersion, limit the bandwidth-distance product to about 20 MHz·km.

A 20-MHz·km limitation means that a 10-km (6.2-mi) fiber-optic link with no repeaters will have only 2 MHz of bandwidth. Small-diameter fibers with single-mode propagation can have a bandwidth-distance product orders of magnitude greater so that only power and laser-noise limit the repeaterless-link distance. In 1992, 800 Gb/s · km was achieved in the laboratory.

A special fabrication process that produces a *graded-index core* can yield a bandwidth-distance product in excess of 2 GHz·km (two orders of magnitude over step-index multimode) and yet allow the higher power-carrying capability of large-diameter multimode operation. Graded-index fibers are discussed in Section 18–6.

Dispersion and Pulse Spreading

With power losses in optical fibers now below 1 dB/km, the principal limitation to the use of guided light for high-speed data communication is the distortion of light pulses due to dispersion in the fiber waveguide.

Just as in the case of a metal waveguide, dispersive mechanisms in glass fibers cause distortion of the transmitted signal due to the variation of propagation velocity for different wavelengths in the transmitted light. This is called *intramodal* or *chromatic dispersion* and has the same result as *group delay* distortion in metal waveguides. The source of this dispersion is a nonlinear ϵ_r and propagation velocity versus wavelength of the fiber material; thus the name *material dispersion*. This dispersion tends to be worse in fiber-optic systems because light sources like LEDs and lasers rarely produce only one output wavelength.

While better for lasers than for LEDs, the different light components on both sides of the main signal are much stronger with light sources than for their microwave counterparts. Thus, the source *linewidth* or *spectral width*, illustrated in Figure 18–6, is not just random noise, as is typical of microwave oscillators. Rather it consists of spurious wavelengths that remain with the transmitted pulse and, because of the wavelength-dependent dispersion of optical waveguide materials, will result in a received light pulse that is wider (pulse spread) than the one transmitted. Communication systems are often operated at 1.3 μm because most optical fibers have minimum chromatic dispersion at this wavelength.

With LEDs, the spurious light around the center wavelength is due to random noncoherent light emission. However, the resonant cavity that results in coherent light for laser diodes (LD) is much longer than the desired wavelength, λ. That is, a very long wavelength and all its shorter harmonics, called longitudinal modes, can exist in the physically long cavity, L. The Gaussian-shaped amplitude envelope exhibited by the spectral lines of Figure 18–6 is the shape of the gain versus wavelength of the lasing material in the semiconductor diode. The particular mix of elements, such as aluminum, phosphorus, indium, gallium, and arsenic, determines the energy and, consequently, the center wavelength, λ, of the laser emission spectrum. The spacing between the longitudinal modes in the spectrum, $\Delta\lambda_s$, can be determined from

$$\Delta\lambda_s = \frac{\lambda^2}{2n\text{L}} \qquad\qquad (18\text{–}11)$$

where n is the lasing material index of refraction.

A more serious dispersive mechanism for large-diameter, step-index fibers is intermodal or simply *modal dispersion*. Each mode excited by a spectrally pure light source will have the same frequency and free-space wavelength. However, as illustrated in Figure 18–7, each mode has its own distinct propagation angle in

FIGURE 18–6 Spectrum for LEDs.

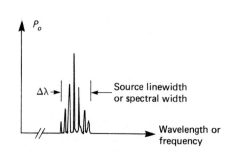

FIGURE 18-7 Higher-order modes (shown as steeper rays) take longer to propagate through the fiber.

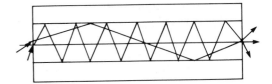

order to satisfy boundary conditions. Consequently, higher-order modes with their steeper angles must travel a greater distance over a given length L of fiber.

For a given digital data pulse, the pulse of light at the receiver, due to low-order modes, extinguishes at the end of the pulse period T_b, but the light from the highest-order mode is still arriving at the detector and finally extinguishes τ_{ps} seconds later. The amount of pulse spread can be determined by tracing the light rays for each mode and calculating the propagation time over that path. The analysis is as follows: Consider the steepest-angle ray of Figure 18-7 compared to the straight-through ray. Let the time for the highest-order mode to travel once from the core center, reflect off the core-cladding interface, and return to the core center be t_{max}.

The base of the triangle formed by the highest-order mode ray is the straight, lowest-order mode ray that requires t_1 to traverse the base of distance ℓ. The time to traverse distance ℓ in Figure 18-8 is

$$t_1 = \frac{\ell n_1}{c} \qquad (18\text{--}12)$$

where the light velocity in the core is

$$v = \frac{c}{n_1} \qquad (18\text{--}13)$$

Also,

$$t_{max} = \frac{\ell n_1}{c \sin \theta} \qquad (18\text{--}14)$$

where θ is the angle of the highest-order mode ray with the normal to the core-cladding interface. The difference between t_1 and t_{max} gives the pulse spread; that is,

$$\tau_{ps} = t_{max} - t_1 \qquad (18\text{--}15)$$

FIGURE 18-8 Highest- and lowest-order (ℓ) ray traces for single reflection.

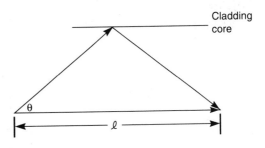

When Equations 18–12 and 18–14 are substituted and total fiber length L is considered, the total pulse spread will be

$$\tau_{ps} = \frac{Ln_1}{c}\left(\frac{n_1}{n_2} - 1\right) = \frac{Ln_1^2\,\Delta}{cn_2} \qquad\qquad \textbf{(18–16a)}$$

$$\approx \frac{Ln_1\,\Delta}{c} \qquad\qquad \textbf{(18–16b)}$$

Thus, for a given fiber waveguide, pulse spread is directly proportional to the fiber's length.

Maximum Allowable Data Rate

The maximum data rate $f_B(\text{max})$ for a digital fiber-optic link may be determined from the dispersive pulse spread per unit length σ, where

$$\sigma = \frac{\tau_{ps}}{L} \qquad\qquad \textbf{(18–17)}$$

with units of nanoseconds/kilometer. If very narrow return-to-zero (RZ) data is transmitted, as shown in Figure 18–9, then the allowable pulse spread is one period; that is, if the pulse spread is T, the "zero" will be received as a "one." The maximum bit (baud) rate can be analyzed for RZ data pulses of duty cycle $d =$

FIGURE 18–9 Pulse spread causes errors in digital data transmission.

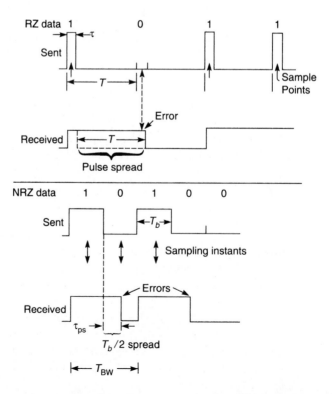

τ/T with the aid of Figure 18–9. Left as an exercise for the reader, the result is that the maximum (noiseless) baud rate for RZ light pulses is

$$f_B(\text{max})_{\text{RZ}} = \frac{2 - d}{2\tau_{\text{ps}}} \qquad (18\text{–}18)$$

For NRZ pulses, a different analysis is required but the result is the same if $d = 1$ is used in Equation 18–18. That is, the maximum baud rate is found from

$$\frac{T_B}{2} = \tau_{\text{ps}} \qquad (18\text{–}19)$$

$$f_B(\text{max}) = \frac{1}{2\tau_{\text{ps}}} \qquad (18\text{–}20)$$

As an example, if the total dispersion of a 10-km fiber is known to be 100 ns/km at a given operating wavelength, then the maximum NRZ baud rate will be 1/2(100 ns/km × 10 km) = 0.5 Mb/s. Note that noise jitter will increase the probability of error (bit error rate) over and above these considerations.

■ ■ ■ ■

EXAMPLE 18–4

A 10-km-long, 50-μm-diameter glass fiber with $n_1 = 1.5$ and $n_2 = 1.485$ is to transmit NRZ data pulses. Determine the pulse spreading per unit length and the maximum allowable data rate.

Solution:

From Example 18–3 we know that this wide-diameter fiber will have many more than four modes propagating. Therefore, multimode operation exists on this step-index fiber. From Equation 18–16, the modal dispersion can be expressed in terms of $\tau_{\text{ps}}/L = n_1 \, \Delta/c = (n_1 - n_2)/c = 0.015/3 \times 10^5$ km/s = **50 ns/km.** This will be orders of magnitude more than other dispersion effects.

The maximum data rate for a 10-km link with the NRZ pulses can be $f_B = 1/2(50$ ns/km × 10 km) = **1 Mb/s.**

Fiber Power Losses

Power loss or signal attenuation in an optical fiber is caused by *impurity absorption*, *scattering*, and *radiation loss*.

Absorption is primarily due to the presence of water ions (OH⁻) and transition metal impurities in the glass material. The peak at around 0.88 μm (or microns), and the large ones at 0.95 and 1.4 μm in Figure 18–10 are due to OH ions. Figure 18–10 shows an older curve for attenuation versus light wavelength in low-loss optical fiber. However, it illustrates some of the inherent loss mechanisms that have been almost completely removed in recent years. Only the OH⁻ absorption between 1.2 and 1.5 μm exhibits excess loss over the Rayleigh scattering lower limit for modern fibers. Notice that the loss trend at shorter wavelengths follows a $1/\lambda^4$ relationship. This type of loss is caused by light *scattering* in the material due

TABLE 18-2 Corning Corguide Optical Fiber*

Optical Parameters		50/125 LDF™ Fiber	62.5/125 LNF™ Fiber	85/125 dBF™ Fiber	100/140 SDF™ Fiber
Attenuation range	@ 850 nm	2.4–3.0 dB/km	3.0–3.2 dB/km	2.8–3.5 dB/km	3.5–4.5 dB/km
	1300 nm	0.6–1.2 dB/km	0.7–0.9 dB/km	0.7–1.5 dB/km	1.5–2.0 dB/km
Temperature dependence	@ 850 nm				
	CPC (250 μm):	≤0.2 dB/km	≤0.2 dB/km	≤0.3 dB/km	≤0.2 dB/km
	CSB (500 μm):	≤0.3 dB/km	≤0.3 dB/km	≤0.3 dB/km	≤0.3 dB/km
	@ 1300 nm:				
	CPC (250 μm):	≤0.2 dB/km	≤0.2 dB/km	≤0.3 dB/km	≤0.2 dB/km
	CSB (500 μm):	≤0.3 dB/km	≤0.3 dB/km	≤0.3 dB/km	≤0.3 dB/km
Temp. dependence range		−60°C to +85°C	−60°C to +85°C	−60°C to +85°C	−60°C to +85°C
Numerical aperture		0.200 ± 0.015	0.275 ± 0.015	0.260 ± 0.015	0.290 ± 0.015
Core diameter		50.0 ± 3.0 μm	62.5 ± 3.0 μm	85.0 ± 3.0 μm	100.0 ± 4.0 μm
Bandwidth range	@ 850 nm	400–1000 MHz·km	160–200 MHz·km	100–200 MHz·km	100–300 MHz·km
	1300 nm	400–1500 MHz·km	200–600 MHz·km	200–400 MHz·km	100–500 MHz·km
Min. zero disp. wavelength		1297 nm	1332 nm	1322 nm	—
Max. zero disp. wavelength		1316 nm	1354 nm	1342 nm	—
Max. zero disp. slope		0.101 ps/nm²·km	0.097 ps/nm²·km	0.100 ps/nm²·km	—
Attenuation difference 1380 nm − 1300 nm		≤3.0 dB/km	≤1.0 dB/km	≤3.0 dB/km	≤3.0 dB/km
Physical Parameters					
Clad diameter		125.0 ± 2.0 μm	125.0 ± 2.0 μm	125.0 ± 2.0 μm	140.0 ± 6.0 μm
Core noncircularity		≤6.0%	≤6.0%	—	≤6.0%
Clad noncircularity		≤2.0%	≤2.0%	—	≤4.0%
Core-clad offset		≤3.0 μm	≤3.0 μm	≤3.0 μm	≤3.0 μm
Standard length range		1.1, 1.7, 2.2 km	1.1, 1.7, 2.2 km	1.1, 1.7, 2.2 km	1.1, 1.7, 2.2 km
Standard coatings		CPC3 (250 μm) CSB3 (500 μm)	CPC3 (250 μm) CSB3 (500 μm)	CPC3 (250 μm) CSB3 (500 μm)	CPC2 (250 μm) CSB2 (500 μm)
Standard	CPC (250 μm):	50 kpsi	50 kpsi	50 kpsi	50 kpsi
Proof test	CSB (500 μm):	100 kpsi	100 kpsi	100 kpsi	100 kpsi

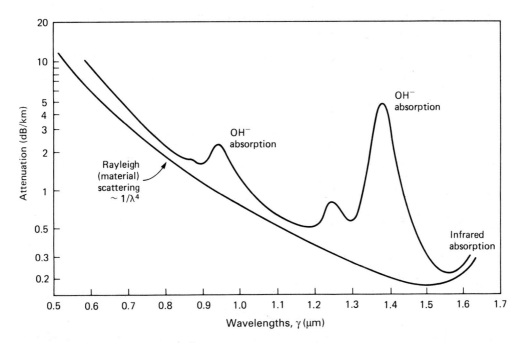

FIGURE 18–10 Power loss mechanisms in optical fibers. Add about 0.3 dB to the Rayleigh curve to estimate attenuation for modern fibers.

to tiny inhomogeneities frozen into the glass during its manufacture. Another scattering mechanism, called *waveguide scattering*, is due to irregularities at the core-cladding interface. For modern fibers, typical loss values may be estimated by adding about 0.3 dB to the Rayleigh loss curve of Figure 18–10. Also see Table 18–2 for Corning Corguide optical fiber parameters and data. This table is for multimode fiber. For example, column 1 is for a fiber with a 50-μm core diameter and 125-μm cladding. Single-mode fiber will have a core diameter of less than 10 μm.

Fiber bends of less than about 10 cm (radius of curvature) will cause light rays to leave the core, pass through the cladding, and be lost due to *radiation*. Another source of radiation loss is caused by *microbends* that develop when the fiber is packed into reinforcing cables. Small axial distortions cause coupling from one mode to another. Coupling into too high a mode (ray angle) will also result in loss due to radiation or, at the least, propagation in the cladding.

Other losses at cable interface connections are discussed later for specific connections.

18–4 ■ TRANSMITTER DEVICES AND CIRCUITS

The most commonly used light sources for fiber-optic communication are solid-state devices, as opposed to very high-power gas lasers. The three most important devices are the light-emitting diode (LED), the injection laser diode (LD or ILD), and the neodymium: yttrium aluminum garnet (Nd : YAG) solid-state laser.

All three of these light sources can be intensity (power level)-modulated by varying the input drive current. The LED produces noncoherent emissions; laser output waves are coherent. To produce the coherent light, laser structures include an optical resonant cavity, one end of which is highly reflective, the other partially reflective and partially transmissive. The resonant cavity provides for wavelength selectivity and higher transmitted power. Consequently lasers have good power in a narrow spectral width and can be used for single-mode operation with small-diameter, step-index fibers, whereas LEDs must be used with graded-index fibers for high data-rate systems.

Communication LEDs

Light-emitting diodes (LEDs) made especially for optical communication systems usually have one of two structures similar to those illustrated in Figures 18–11 and 18–12. Figure 18–11 is a schematic representation of a Burrus *surface-emitting* LED, named after its pioneer, C. A. Burrus at Bell Labs. The doped gallium arsenide (GaAs) pn junction is forward-biased by applying a dc potential difference across the metal contacts. When the potential is high enough, the electrons and holes will have enough added energy to enter the depletion region, where some recombine in the usual way to produce heat, and some recombine radiatively to produce light. The fraction of the recombination process that produces light is called the *internal quantum efficiency*.

The wavelength of the light produced is determined from Equation 18–1b, where the energy W is approximately that of the material bandgap energy that for GaAs is 1.43 eV. Since this produces $\lambda = 0.905$ μm, which falls in a high-attenuation region for most fibers, dopants such as aluminum, indium, and phosphorus

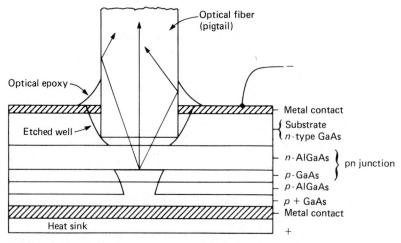

FIGURE 18–11   Burrus etched-well laser diode.

FIGURE 18-12 Stripe contact, edge-emitting structure for fiber-optic LED or laser.

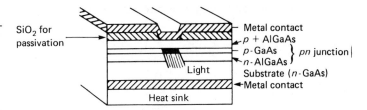

By etching a well through the top substrate and attaching a microlens, or by bonding a fiber directly as shown in Figure 18–11, the light may be guided to an output connector. A short length of fiber called a *pigtail* is provided for this purpose. Units with output power exceeding 1 mW are available.

The LED illustrated in Figure 18–12 is a *stripe contact edge-emitting* structure. The metalization for the upper contact is applied after a stripe has been etched in the silicon dioxide passivation (isolation) layer. The active recombination region will now be a waveguide channel confined between two different alloy layers above and below. This is referred to as a *double heterostructure* (DH). A 50-μm stripe width will provide an appropriate channel dimension for coupling from the cleaved edge to typical 50- to 100-μm, graded-index fibers.

Injection Lasers

Like LEDs, lasers depend on the *spontaneous emission* of light when high-energy (excited) electrons in the conduction band drop to the lower-energy valence band. In semiconductor laser diodes, enough electrons are injected into the depletion region by a high forward-bias that most of the lowest conduction states are filled, thereby producing a *population inversion* of excited states. At this point, if high-energy photons impinge on excited-state electrons and stimulate them to drop to the ground state (valence band) and give off a photon in the process, then the emitted photon is in phase with the incident photon and the resultant emission is known as *stimulated emission.*

The active area of a laser diode is confined in a cavity of optically reflective surfaces so that the coherent in-phase light energy can build up by resonant feedback—that is, light amplification of stimulated emissions. Such a structure is called a Fabry-Perot cavity. One end of the cavity has a partially transmissive (and partially reflective) surface to allow coupling to a fiber. A few milliwatts CW of light with very narrow spectral width (about 2 nm) and beam width (5–10°) are typical for Fabry-Perot semiconductor laser diodes.

Figure 18–13 illustrates what is called the *L-I* curve for a semiconducting diode laser. It shows the light output power for a typical laser diode (LD) as a function of the input bias current. Below the lasing threshold current I_{th}, the device behaves essentially as an incoherent light emitter (an LED), with low spatial and temporal

FIGURE 18-13 Double-hetero-structure diode power versus bias.

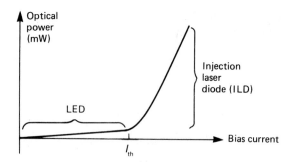

coherence. Above the threshold, laser action occurs; that is, the number of participating spatial modes collapses and the emitted optical spectrum narrows significantly, which is characteristic of a Fabry-Perot laser.

The double-heterostructure of Figure 18–12 is used for *injection laser diodes* (ILD). The stripe contact, edge-emitting DH structure of Figure 18–12 can be used noncoherently as an LED for biases up to the lasing threshold I_{th}, whereupon stimulated emissions produce coherent laser action.

The *Neodymium : YAG* laser is produced with a solid-state rod, rather than the semiconductor diode of the injection laser. The Nd : YAG rod is cut to an appropriate length for in-phase feedback from the mirrored and semi-mirrored end surfaces. A flashlamp pump is included in the housing and either parallels the Nd : YAG rod or is wrapped in a helix around it. The flashlamp emits light into the rod to excite electrons and produce an excited-state population inversion necessary for lasing. The Nd : YAG device is the most widely used of the solid-state family of lasers for communications applications. (Solid-state and semiconductor lasers are in separate families.)

Direct Modulation Transmitters

The linearity of the LED and ILD sections of the light power curve (*L-I* curve, Figure 18–13) allows for power- (amplitude-) modulation of these devices. The bias current level can be pulsed for digital amplitude shift-key (ASK) or on-off key (OOK). For analog AM, the current level is varied from a fixed bias level.

An LED transmitter showing the diode and current drive circuitry is shown in Figure 18–14. This is a 20-megabaud (NRZ data) fiber-optic transmitter from Motorola's book, *Optoelectronic Device Data*. Input data is shaped in two high-speed NAND gates, and the output of U_2 controls the base current to the MPS6515. When the U_2 output is low, D_1 conducts and base current is shunted through U_2 so the LED will be off. When U_2 is high, D_1 is back-biased and full base current allows the LED to conduct. C_1 improves the rise time of the drive circuitry while the differential amplifier scheme is nonsaturating and fast like ECL logic. The 2N4400 is a constant current source that is partially temperature-compensated by the 1N914 diode in its base circuit.

For laser modulators with digital OOK modulation, the "off" current is usually set just below lasing threshold (as opposed to zero) in order to reduce the switch-

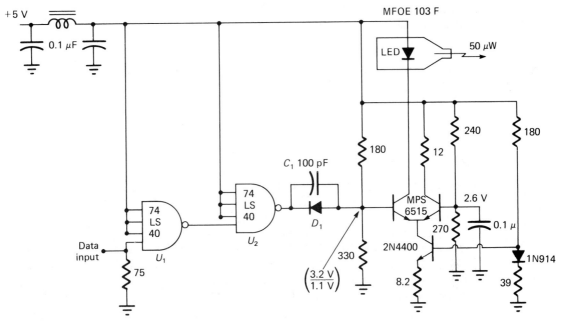

FIGURE 18–14   20-megabaud fiber-optic transmitter.

ing current range. The drive circuitry is much more complicated for lasers than for LEDs because of the temperature sensitivity of the threshold current and power linearity due to the changes in internal quantum efficiency with temperature. In fact, the ultra linear distributed feedback (DFB) laser used for multichannel CATV and other applications typically requires a thermal electric cooler (TEC) to maintain constant temperature for constant wavelength and optical power. A basic thermal electric couple consists of p- and n-type semiconductor material that can act to pump heat in either of two directions to cool or to increase temperature. When a positive dc voltage is applied to the n-type element, electrons pass from the p- to the n-type element and the cold side temperature will decrease as the electrons pass from a low energy level in the p element to a higher energy level in the n element while heat is being absorbed. Reversing the dc polarity reverses the current, and heating will occur. Temperature can be controlled to within 0.01°C. A packaged laser with DFB laser chip and TEC is illustrated in Figure 18–15.

The single longitudinal mode DFB laser diode chip included in Figure 18–15 is characterized by a Fabry-Perot cavity with a built-in corrugated structure called a periodic Bragg reflector. As the photonic fields build up in the cavity, only the mode with the Bragg period (and Bragg wavelength) is reflected constructively, and this mode will lase, whereas others will be suppressed from oscillation. As a consequence, the output light spectrum is virtually a single wavelength with all the optical power that manages to get coupled to the fiber through lenses, optical

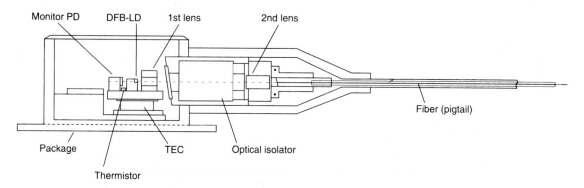

FIGURE 18–15   Cross-section sketch of a distributed-feedback (DFB) laser with TEC and optical isolator.

isolator, and connectors. Ultra linear lasers with second- and third-order modulation distortions better than 70 dB and SNRs better than 57 dB are producing 15 mW or more available with modulation bandwidth in excess of 1 GHz. DFB lasers are produced for both 1310-nm and 1550-nm applications.

External Modulators

Another technique for producing intensity modulation that can provide more optical power than a DFB laser is to use high-power lasers like the Nd : YAG laser with an external modulator. The external modulator usually employed is a Mach-Zehnder interferometer. The Mach-Zehnder modulator consists of an optical waveguide Y-branch power splitter, two separated optical waveguide strips covered with electrode material, and a combiner at the device output. The optical waveguides are made by diffusing titanium into a lithium-niobate (LiNbO$_3$) substrate. Between the input splitter and output combiner, the electrode-covered waveguides are separated so they do not cross-couple. The center electrodes deposited over each optical waveguide path can be electrically biased to change the refractive index of the titanium-diffused waveguide. When the bias is set to a few volts, the difference in propagation velocity results in the polarized light being phase delayed in one waveguide relative to the other. With no voltage applied, the optical path lengths are equal and the light is coupled together in phase with constructive interference to produce maximum output intensity. Approximately 15 V difference between the electrodes results in 180° phase delay difference so that the combined light intensity is a minimum—20 dB or so below the maximum transmission. Between the two extreme voltages, output light intensity is a raised cosine function of the applied voltage. For example, if V_{max} is required to extinguish the light, then the light power output-to-input transfer function of the modulator as a function of signal voltage v is

$$\frac{P_o}{P_i} = \frac{1}{2} + \frac{1}{2}\cos\left(\frac{\pi}{V_{max}}v\right)$$

(18–21a)

$$= \cos^2\left(\frac{\pi}{2V_{max}}\, v\right) \qquad\qquad \textbf{(18–21b)}$$

When biased to $V_{max}/2 \equiv V_{\pi/2}$, the transfer function for ac-coupled modulating signals is a raised sinusoid:

$$\frac{P_o}{P_i} = \frac{1}{2}\left[1 - \sin\left(\frac{\pi}{V_{\pi/2}}\, v_{mod}\right)\right] \qquad\qquad \textbf{(18–22)}$$

When coupling a laser to the Mach-Zehnder intensity modulator, single-polarization light must be used. If the modulator is separate from the laser, the connecting fiber must be polarization-maintaining. Polarization-maintaining fiber (PMF) has a very complex refractive index profile and, as is implied, the polarization of the input light is maintained as the light propagates through the fiber. Besides coupling from a high-power Nd:YAG laser to a Mach-Zehnder AM modulator, PMF is used when coherent optical detection receivers are employed.

Fiber Amplifiers and Dispersion-Shifted Fiber

Most of the long-distance land-based fiber installations of the late 1980s and early 1990s used 1.3-μm-wavelength optics. However, more and more new installations are using dispersion-shifted fiber with 1.55-μm optics for low fiber loss and to take advantage of erbium-doped fiber amplifiers.

An erbium-doped fiber amplifier (EDFA) consists of approximately 20 m of optical fiber doped with erbium ions (Er^{3+}), a 980-nm or 1480-nm pump laser, and a wavelength-division multiplexing (WDM) coupler. Amplification is achieved by coupling the 1550-nm signal-bearing light and the pump laser through the WDM into the 20 m of erbium-doped fiber. The pump can be coupled into the fiber input or output, or both.

Erbium absorbs the 980 nm or 1480 nm pump light, thereby raising electrons to higher energy levels. The input 1550 light stimulates the higher level–to–lower energy electron transitions that produce more coherent photons at 1550 nm. Up to 30 dB of gain can be achieved. The input 1550 light power keeps the excess photon generating process saturated for high efficiency, but little distortion of the intensity modulated signal occurs. When fully saturated, the EDFA can achieve 15 dBm or more of output power. Finding fiber doping materials capable of producing suitable gain at the shorter wavelength of 1300 nm is still a subject of intense research. At present, the performance of 1550-nm EDFAs is far superior to what has been achieved at 1300 nm.

Dispersion-shifted fiber employs different dopants, such as germanium, and/or a different refractive index profile than step-index single-mode fiber to shift the zero-dispersion wavelength from 1310 nm to 1550 nm, where fiber exhibits minimum loss. A good rule of thumb is that single-mode dispersion-shifted fiber (SMF/DS) exhibits attenuation of 0.2 dB/km at 1550 nm and 0.4 dB/km at 1310 nm. The single-mode field diameter of the propagating light is 8 μm at 1550 nm and 6.5 μm at 1310 nm. Dispersion is approximately 2.5 ps/nm · km and the dispersion slope is 0.08 ps/nm² · km.

18–5 ■ RECEIVER DEVICES AND CIRCUITS

The most important devices used for detecting light in fiber-optic systems are the PIN (positive-intrinsic-negative doped) diode, the avalanche photodiode (APD), and the phototransistor (including photo-darlingtons). For small size, high sensitivity, fast response time, and low noise, silicon photodiodes are used.

Photodiode Light Detectors

A simple photodiode structure is illustrated in Figure 18–16. The received light passes through an antireflective, transparent glass covering and into the depletion region of a reverse-biased *pn* junction. If the light has more energy than the material bandgap, then the energy can excite electrons to jump from the valence band of the semiconductor material into the conduction band, thus creating hole/electron pairs. Photodiodes are operated with reverse bias so that when conduction electrons and holes appear, they are quickly attracted to the external circuit as current.

The photodiode characteristics before and after light is received are illustrated in Figure 18–17, where I_{SC} is the value of current (called *photocurrent*) that can flow into a very low-impedance load. The ratio of current to incident light power is called *responsivity*. The diode leakage current at the operating reverse-bias voltage is called the *dark current*.

PIN Photodiodes

A *PIN diode* is like the normal *pn* junction just described except that the p and n regions are separated by a very lightly n-doped region that can be considered intrinsic (undoped). This structure allows for very low reverse-bias leakage current and relatively wide junction width to reduce junction capacitance, thereby improving response speed.

Silicon PIN diodes are used extensively in the 800–900-nm light range. They exhibit quantum efficiencies from 30% to 95%, responsivity of about 0.65 $\mu A/\mu W$,

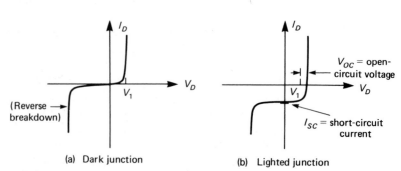

(a) Dark junction   (b) Lighted junction

FIGURE 18–17 Photodiode I-V curve.

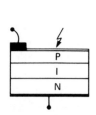

FIGURE 18–16
PIN diode structure.

and low noise generation. Above 900 nm, InGaAs and germanium PIN diodes are used because silicon has very low responsivity for long wavelengths. The responsivity of InGaAs at 1300 nm is about 0.6, InGaAsP (indium gallium arsenide phosphide) is 0.75 to 0.85, and Ge is 0.45 μA/μW. Also, germanium has notoriously high leakage (dark) current.

Avalanche Photodiodes (APD)

The avalanche mechanism of diodes can be utilized in a photodiode to greatly increase the number of conduction electrons and thereby improve the responsivity. Such a diode has heavily doped *pn* regions and a narrow avalanche region. When reverse-biased, the avalanche region in the *pn* junction has a very high electric field strength. A photogenerated electron can gain enough energy to knock valence electrons from their atoms (ionization) upon collision. This process can of course cascade, thereby multiplying the available current. As long as the multiplication process can be limited, the device can be used as a photodiode.

The major limitations to the use of APDs are the temperature sensitivity of the current multiplication and the electrical noise inherent in the avalanche process. However, APDs are often used in long-distance networks due to their high detection sensitivity.

Receiver Circuits

Fiber-optic receivers consist of the photodiode or phototransistor and operational amplifiers to increase the current gain and provide a compatible circuit output level. The basic configurations (simplified) are shown in Figure 18–18. Figure 18–18a shows a transimpedance amplifier with very low input impedance. The photodiode remains reverse-biased by a positive bias voltage (not shown) applied to the plus input terminal. Any current from the diode I_d flows through R_f to produce the output voltage

$$V_o = I_d R_f \qquad (18\text{--}23)$$

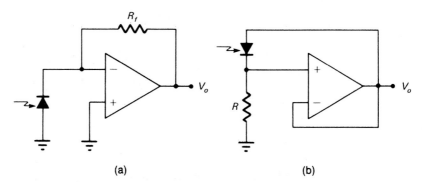

(a) (b)

FIGURE 18–18 Basic receiver configurations.

The gain of the transimpedance amplifier is output divided by input, or

$$G = \frac{V_o}{I_d} = R_f \qquad (18\text{--}24)$$

with units of ohms, but is usually given in decibels from

$$G \text{ (dB)} = 20 \log R \qquad (18\text{--}25)$$

where R is the equivalent of V_o/I_i.

Any current generated by the photodiode in Figure 18–18b causes the voltage at the plus terminal of the IC to go negative. Because of the feedback from output to the inverting terminal, the op-amp has unity voltage gain (but high current gain), so that

$$V_o = -I_d R \qquad (18\text{--}26)$$

for Figure 18–18b.

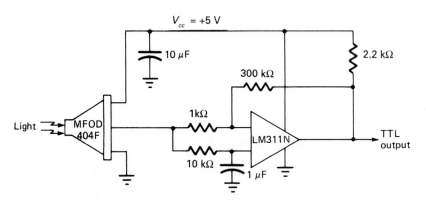

FIGURE 18–19 Fiber-optic receiver for digital data.

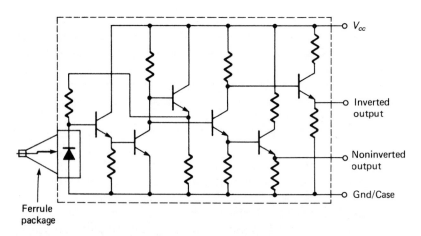

FIGURE 18–20 Integrated detector/preamplifier, Motorola MFOD404F.

Fairly simple digital data receivers can be made using integrated detector pre-amplifiers. The schematic for one such receiver using the Motorola MFOD404F and the LM311N voltage comparator is shown in Figure 18–19. This design from Motorola's *Optoelectronic Device Data* is suggested for two-megabaud data rates. The integrated detector preamplifier, shown in Figure 18–20, has a 10-Mb/s capability.

■ ■ ■ ■

EXAMPLE 18–5

The PIN photodiode of Figure 18–18a has a responsivity of 0.66 $\mu A/\mu W$ and a dark current of 130 nA.

1. Determine the resistor value that would provide 1 V for a received signal of -20 dBm.

2. What will be the minimum output voltage?

Solution:

1. -20 dBm $= 10\ \mu W$ so that $I_d = 0.66 \times 10\ \mu m = 6.6\ \mu A$.
   $R_f = 1\ V/6.6\ \mu A = $ **150 kΩ.**

2. The minimum diode current is the dark current. Thus,
   $V_o(min) = (130\ nA)(150\ k\Omega) = $ **0.02 V.**

18–6 ■ TRANSMISSION TECHNOLOGY

The components needed between transmitters and receivers in a fiber-optic link are fibers and/or cables consisting of numerous fibers, connectors and splices, power splitters, and directional couplers.

Fiber Technology

A very common method for fabricating an optical communication fiber is to begin with a 1-m-long, 10-mm-diameter tube of very pure glass (silica or quartz) and dope the interior of the tube with precisely regulated quantities of silica and desired dopants. This is the vapor-phase oxidation method. Common dopants for the high-index core are oxides of germanium and phosphorus. After the core is formed, fluorine or boron oxide is used to form the lower-index cladding. The doped tube is then collapsed by heating, thus forming what is called the *preform*.

Figure 18–21 shows the system whereby the solid preform is slowly fed into a high-temperature furnace. The tip of the melt is fed to a take-up spool that thereafter draws the fiber through the system. The fiber diameter is precisely monitored by light beams and a feedback signal to control the pull rate. A silicon or Kynar coating is applied to the fiber to protect the outer surface. The final length of the monofilament fiber is about 1 km. Single lengths of fiber 20 km in length are common for CATV applications.

FIGURE 18–21 Optical fiber manufacturing.

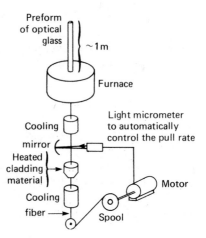

If a core with a single refractive index is formed, a step-index fiber is produced. Graded-index fibers are made by varying the dopants during the preform deposition stage to form a continuous gradation of index from low to highest at the very center of the core.

Refractive index versus radius for the step- and graded-index fibers are shown in Figure 18–22, along with a two-step index fiber called the W-*index guide*. The W-index profile guide allows single-polarization propagation instead of the usual randomly polarized light transmissions.

After the fiber is formed and coated it is usually covered with a polyethylene jacket and then encased in a Kevlar yarn braid. This amazing material has a tensile strength approximately equal to steel. This single-fiber cable is covered with a protective jacket made of polyethylene, polyurethane, or polyvinyl chloride (PVC).

Multifiber cables have various configurations. Some are made by slipping individual fibers with their first polyethylene jacket loosely into plastic buffer tubes that are either (1) connected laterally in ribbons (ribbon cable), or (2) grouped in bundles of two to more than a dozen, or (3) aligned in a circle around a single Kevlar or other tensile strength member, with the entire package encased in a PVC jacket.

Another cabling package used by Western Electric is illustrated in Figure 18–23. This package has 12 ribbons of 12 close-packed coated fibers. These 144

FIGURE 18–22 Index of refraction profiles. (a) Step-index. (b) Graded-index. (c) W-index.

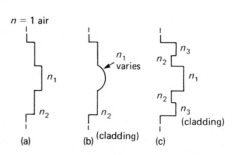

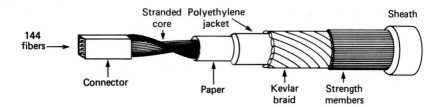

FIGURE 18-23 Western Electric fiber cable.

fibers are then wrapped in paper, encased in a polyethylene jacket and a Kevlar-type braid that is surrounded by steel wires embedded in a plastic protective sheath. There even exists a mass splice for aligning all 144 individual fibers when cables longer than 1 km are needed.

Connectors, Splices, and Couplers

It may seem that coupling light from one piece of glass to another would be simple enough. However, the power losses incurred when coupling 1-km lengths of fiber together usually exceed the attenuation over the 1-km length of glass unless great care or an expensive connector is used.

The sources of optical signal loss are listed here and illustrated in Figure 18–24.

• *Air gap* Air gap losses include *reflectance* and loss due to diverging light rays. Reflectance has been discussed, and Equation 18–3 is used to calculate this loss. As mentioned, each glass-to-air interface can result in 4% (0.18-dB) loss. The loss between air and GaAs sources and detectors, however, is 32% (1.7 dB).

Diverging light rays can miss the core altogether or fall outside of the cone of acceptance of the receiving fiber. Such a loss of power, often referred to as *end-separation* loss, can be calculated from

$$L_{es} = \frac{D/2}{D/2 + \ell \tan[\sin^{-1}(NA/n)]} \qquad (18\text{–}27)$$

where D are the fiber diameters, ℓ is the separation, and n is the media index of refraction. (Sometimes the gap is filled with epoxy and other index-matching materials.)

FIGURE 18-24 Sources of optical signal loss.

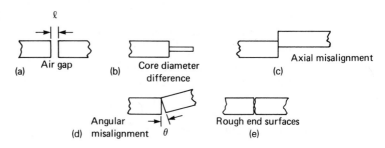

- *Core diameter difference*   This difference was discussed in Section 18–3. Equations 18–5 and 18–6 are used to calculate this loss.

- *Axial misalignment*   Also called *lateral* or *radial displacement*. If the displacement is as much as 15% of the smaller core diameter, the loss will be about 1 dB. Displacement of 5% will keep the loss below 0.3 dB.

- *Angular misalignment*   The formula is very complex and is a function of numerical aperture NA, refractive index of the separating medium, and of course the angle θ. For typical fibers with NA = 0.5 and $\theta = 2°$, the loss will be about 0.5 dB.

- *End-surface roughness*   The ends of connecting fibers must be highly polished (less than 0.5 μm grit) if scattering losses are to be avoided. Also, the polishing must leave the end faces as flat and parallel as possible. The best way is to use a sapphire or diamond scribing tool to nick the fibers and then pull until the fiber snaps.

Even *splicing* two fibers by heating and fusing the glass can result in losses of around 0.1 dB; the waveguides simply do not maintain a perfect size, shape, and angle. Splicing by the application of special epoxies results in more loss but it is a relatively simple process.

Various alignment techniques are used when splicing. Mechanical alignment to various degrees of precision can be accomplished by wedging the fiber ends between three slightly deformable rods that are then secured with shrink tubing or a similar technique using precision spheres. The fiber ends can also be wedged into a V-groove slot and secured with a V-shaped top plate cover and shrink tubing. Again, the material used for the V-groove and cover should be soft enough to deform slightly for accommodating fibers of various diameters but not so soft as to allow the fiber ends to bend and crack each other or allow cold flow.

Splicing techniques are numerous, ingenious, and important for communication links in excess of 10 km. But for maximum system flexibility over short distances, such as local area network (LAN) applications, rematable connectors are used.

Optical fiber *connectors*, like their RF-cable connector counterparts, allow for system flexibility. However, the mechanical tolerances are tighter than even precision microwave connectors, yet their losses can be much greater. Basic connector types are either *butt-coupled* or *lens-coupled*.

Spherical and hemispherical lenses are used to collimate the light rays to minimize power losses. This technique is used more for coupling LEDs, lasers, and detectors to fibers than for fiber-to-fiber connections.

The most commonly used fiber-connection technique is to butt the two fiber ends together with minimum gap and maximum alignment, while still allowing for remating when required. Tubes, sleeves, resilient ferrules, and the previously mentioned alignment techniques are used. Threaded connectors such as the tapered ferrule/bushing type of Figure 18–25 are common. Part (a) shows the ferrule-mounted light source or detector, and part (b) shows the fiber-to-fiber connector in which the slightly deformable ferrule slips very snugly into the receiving

FIGURE 18-25 Ferrule mounting and connectors.

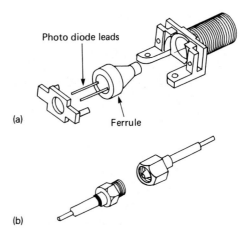

(a)

(b)

bushing. Standard connectors are the FC, including the FC-APC (angled physical contact), ST, SC, SMA, mini-BNC, D4, Biconic, and even an FDDI/MIC.

The coupling of light from one or a few fibers into many (multiplexing) is accomplished with (among other techniques) the *star coupler* illustrated in Figure 18-26a and b. The star-coupler circuit symbol is illustrated in Figure 18-27.

The *reflective star coupler* is an enlarged section of fiber called a *rod* with numerous connectors on one end and a mirrored surface or other reflecting mechanism on the other. Light coupled into this "mixing" rod will be reflected to all of

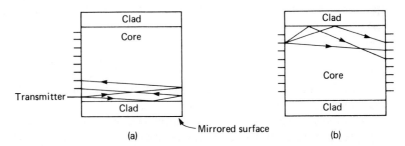

FIGURE 18-26 (a) Reflective and (b) transmissive star couplers.

FIGURE 18-27 Star-coupler circuit symbol.

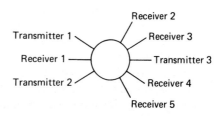

the connections, including the transmitters. Clearly, only one transmitter should be on at a given time and all fibers will receive this signal.

The *transmissive star coupler* offers isolation between ports on opposite sides of the device. In Figure 18–26b fibers on the left can transmit to those on the right, and vice versa, but with good design very little light is reflected back. This directionality is advantageous in some systems; it is also more efficient than the reflective star because not all connections to the device receive power. The transmissive star has another advantage in allowing full-duplex operation (between two stations only), whereas the reflective star offers half-duplex operation.

A simple optical *directional coupler* can be made by allowing two fibers placed side by side to touch over a very short length while applying heat to the connection. If the fibers are allowed to fuse while pulling them into a biconical taper, good directionality can be achieved. In fact, this technique applied to more than two fibers twisted together, heated, and pulled will form a star coupler. Again the fused portion of the connection should have the biconical taper.

Couplers are available in multiple port configurations, including $1 \times N$ (one by N, where N equals the number of desired output ports; $N = 3, 8, 16$, for example), called *tree couplers*, and $M \times N$ *star couplers* (where M equals the number of desired input ports and N equals the number of desired output ports). A simple split may be specified for equal (50%/50%) or unequal (20%/80%, 30%/70%, etc.) splitting ratios.

Generally, there are three wavelength ranges, or windows, of interest in fiber-optic communication systems: 850-nm, 1310-nm and 1550-nm wavelengths. Wavelengths can be multiplexed for propagation down one fiber, greatly increasing the amount of data that can be transported. A device called a wavelength division multiplexer (WDM) allows the coupling of light of the various wavelengths from the single fiber into separate fibers. The coupling efficiencies for these devices are wavelength-dependent. However, a WDM can be designed for wavelength independence over a broad wavelength range; an example is the single-mode broadband coupler (SMBC).

Some of the most important performance parameters for fiber-optic couplers and wavelength-division multiplexers are insertion loss, wavelength dependence, directivity, polarization stability, thermal stability, mechanical considerations, and isolation.

Insertion loss (IL) is the combination of excess loss (EL) plus coupling loss (CL); see Equation 18–28 for a coupler with two outputs, O_1 and O_2, and one (or two) inputs, I_1 (and I_2). *Excess loss* is considered the amount of loss attributable to the manufacturing process. On any given device, excess loss is the sum of the output powers with respect to the input power. *Coupling loss*, or splitting ratio, the amount of light split or coupled, is the power of a given output port with respect to the sum of all output powers. A total worst-case insertion-loss value may also include effects due to wavelength dependence, polarization, and thermal considerations.

Wavelength dependence is defined as the variation in insertion loss as a function of wavelength. The wavelength dependence can be device-specific, with certain devices offering minimum wavelength dependence. The wavelength division multiplexer/demultiplexer is the other extreme, where considerable wavelength dependence is desirable.

$$IL \text{ (dB)} = CL \text{ (dB)} + EL \text{ (dB)} \tag{18–28a}$$

$$CL \text{ (dB)} = -10 \log(CL) \tag{18–28b}$$

$$CL = \frac{O_2}{O_1 + O_2} \tag{18–28c}$$

$$EL \text{ (dB)} = -10 \log\left(\frac{O_1 + O_2}{I_1}\right) \tag{18–28d}$$

$$Dir \text{ (dB)} = 10 \log\left(\frac{I_2}{I_1}\right) \tag{18–29}$$

If there are two or more input ports, the *directivity* (often referred to as near-end cross talk) is the amount of power observed at a given input port with respect to another input port; see Equation 18–29. If bidirectional, directivity can be measured for the output ports as well.

Polarization stability is specified only for single-mode couplers. It is an indication of insertion-loss variation as the polarization state of the input light is varied. Polarization stability values of ±0.1 dB are readily achieved with the fused biconical taper technology.

Thermal stability is a measure of insertion-loss variation as the device undergoes various temperature extremes. Each application can possess unique environmental requirements. The thermal stability of a device is often limited by the user's choice of cabling and coupler packaging materials.

Mechanical considerations include the package materials and dimensions, cable retention (the change in insertion loss as a tensile pull force is applied to the fiber leads), flex (insertion-loss variation as the device is bent), shock, and vibration (insertion-loss variation as a function of shock or vibration).

Isolation (often referred to as far-end cross talk) is primarily used in reference to WDM products. Isolation is a measure of light at an undesired wavelength at any given port. For example, if a given wavelength is desired at output port 1, then the amount of this wavelength seen at output port 2 with respect to the amount of the same wavelength at port 1 is the far-end cross talk for that wavelength. Isolation can be reduced through the use of fiber-optic in-line filters placed on the output ports.

18–7 ■ FIBER-OPTIC DATA LINK ANALYSIS

Most of the basic concepts necessary to determine the suitability of the myriad of fiber-optic components for a particular communication link have been discussed. The objective in this section is to summarize by examining some example design problems.

Analog Link Analysis

Signal.   A laser or LED is modulated by coupling signal current to the laser bias current. The result is light intensity modulation (IM). This modulation is a power modulation, not a voltage vector phenomenon. The optical modulation index, m, is defined by the fractional difference between the peak launched power, P_{pk}, and the average (unmodulated) launched light power, P_L, and is given by

$$m = \frac{P_{pk} - P_L}{P_L} \qquad (18\text{–}30)$$

For a fiber link loss of L_f and a photodetector responsivity of $\mathcal{R}$ amperes current per watt of incident light power (A/W), the average output current from the photodiode will be

$$I_o = P_L L_f \mathcal{R} \qquad (18\text{–}31)$$

and the rms information signal current will be

$$I_s = \frac{m}{\sqrt{2}} \mathcal{R} P_L L_f \qquad (18\text{–}32)$$

Signal power out of the photodetector is proportional to current squared, as given by

$$I_s^2 = \tfrac{1}{2} m^2 (\mathcal{R} P_L L_f)^2 \qquad (18\text{–}33)$$

From Chapter 5, on amplitude modulation, the $(\tfrac{1}{2}) \cdot m^2$ in Equation 18–33 looks familiar (see Equation 5–8, for example). However, electrical signal power being proportional to the *square* of optical power seems strange. The results here are a consequence of defining the optical modulation index (OMI) in terms of a power ratio and also the fact that a photodetector is a power detector. Perhaps the most difficult part of all is that a 10-dB optical loss results in 20-dB loss of electrical signal power.

Noise.   LED and laser diode sources are typically noisier than microwave generators. Therefore, to determine signal-to-noise ratio for a fiber-optic link, the light-source noise must be considered. A number of noise phenomena in laser diodes result in noise with a spectrum that is quite flat, with frequency up to nearly the relaxation resonance* of the laser. The total mean-squared noise spectral intensity relative to the coherent, average output light power squared is referred to as the laser's relative intensity noise (RIN). RIN is normally given in decibels per hertz of bandwidth.

Like any diode, a photodetector diode produces an output *shot* noise, given by Equation 4–3 as

$$I_{sn} = \sqrt{2qBI_o} \qquad (18\text{–}34)$$

where $q = 1.6 \times 10^{-19}$ C (charge per electron).

---

* Basically, there is an oscillation between the spontaneous recombination lifetime of injected electrons and the photon lifetime in the LD cavity. Also, see Amnon Yariv, *Optical Electronics*, 3rd ed. (New York: Holt, Rinehart and Winston, 1985), pp 188–193.

Beyond the photodetector, receiver electronics produces noise, primarily thermal. Thermal noise voltage is given by Equation 4–2 as

$$E_n = \sqrt{4kTBR} \qquad (18–35)$$

where R is the equivalent noise-producing resistance. As discussed in "Receiver Circuits" in Section 18–5, photodetector current is usually converted to voltage in a transimpedance amplifier. The predominant noise in a transimpedance amplifier such as the simple, low-frequency version depicted in Figure 18–18 is due to the gain-determining feedback resistor; call it R. The noise power density can be computed using Equations 4–2, 4–5, and 4–10 combined as

$$\frac{I_n^2}{B} = \frac{4kT}{R} \qquad (18–36)$$

Hence, the transimpedance amplifier noise current, referenced to a bandwidth of B hertz, is

$$I_{tzn} = \sqrt{4kTB/R} \qquad (18–37)$$

with units of amperes per square root hertz ($A/\sqrt{Hz}$). When the noise figure (ratio), NR, of the electronics following the transimpedance amplifier is included, the noise current from all postphotodiode amplifiers is

$$I_{an} = \sqrt{4kTB(NR)/R} \qquad (18–38)$$

Total receiver noise referenced to the transimpedance amplifier input is a sum of the laser RIN contribution I_{RIN}, photodetector shot noise I_{sn}, and receiver amplifier noise which is dominated by transimpedance amplifier noise, I_{an}. Adding on a power basis, the total receiver noise is proportional to

$$I_{total}^2 = I_{RIN}^2 + I_{sn}^2 + I_{an}^2 \qquad (18–39)$$

where

$$I_{RIN}^2 = (RIN)B(P_L L_f \mathcal{R})^2 \qquad (18–40)$$

The optical power received at the photodetector is the amount launched, P_L, reduced by all link signal losses L_f, including connectors, splices, and other coupling-loss mechanisms; that is,

$$P_r = P_L L_f \qquad (18–41)$$

For an intensity-modulated laser with an OMI of m, the signal-to-noise power ratio in a bandwidth of B hertz can be calculated using

$$S/N = \frac{I_s^2}{I_{RIN}^2 + I_{sn}^2 + I_{an}^2} \qquad (18–42)$$

where I_s = signal current
I_{RIN} = laser noise at receiver
I_{sn} = photodiode shot noise
I_{an} = receiver amplifier noise in picoamperes/square root Hz

The link signal-to-noise ratio is calculated as

$$S/N = \frac{0.5(m\mathscr{R}Pr)^2}{[(\text{RIN})B(\mathscr{R}Pr)^2] + [2qB\mathscr{R}Pr] + [4kTB(\text{NR})/R]} \quad (18\text{–}43)$$

where $\mathscr{R}$ = photodetector responsivity
Pr = received optical power
B = video bandwidth
q = electron charge
m = RMS modulation index
RIN = laser relative intensity noise ratio
k = Boltzmann's constant (see Equation 4–1)
T = receiver temperature in kelvins
NR = receiver noise ratio (from NF)
R = receiver transimpedance in ohms

A statistical analysis* for modulating a DFB laser to the onset of clipping with N channels of uncorrelated composite TV video gives a maximum optical modulation index *per channel* of

$$m = \frac{0.348}{\sqrt{N}} \quad (18\text{–}44)$$

■ ■ ■ ■

**EXAMPLE
18–6**

Sixty NTSC (4-MHz video bandwidth) video carriers are to be transmitted on a 20-km fiber-optic link using a 3.16-mW, 1300-nm DFB laser diode. Total RIN, including other fiber-related noise, is known to be −153 dB/Hz. Total link losses may be determined at 0.4 dB/km, a good rule of thumb these days for 1300-nm wavelength. To keep second- and third-order distortions below 65 dB, an OMI of 3.5% (per carrier) is used. If the receiver photodetector has 0.85 responsivity and the receiver transimpedance amplifier plus electronics produces receiver amplifier noise of 7.5 pA/$\sqrt{\text{Hz}}$, determine the following:

1. Average optical power, in dBm, at the receiver (output of the fiber).
2. The system carrier-to-noise ratio (CNR) in decibels.

Solution:

1. Launched power is $10 \cdot \log(3.12 \text{ mW}/1 \text{ mW}) = 5$ dBm and $L_f = 20$ km × 0.4 dB/km = 8 dB (a gain of −8 dB, or 1/6.31 reduction), so received light power is $P_r = 5$ dBm − 8 dB = **−3 dBm** (0.5 mW).
2. $\mathscr{R} = 0.85$ A/W, $m = 0.035$: $I_s^2 = \frac{1}{2}m^2(\mathscr{R}P_L L_f)^2 = 0.5 \times 0.035^2(0.85 \times 0.5 \times 10^{-3})^2 = \mathbf{0.11}$ **nA².** Received laser noise power is proportional to $(\text{RIN})B(\mathscr{R}Pr)^2 = 5 \times 10^{-16} \times 4 \times 10^6(0.85 \times 0.5 \times 10^{-3})^2 = \mathbf{3.6 \times 10^{-16}}$ **A².** PD shot noise is proportional to $2qB\mathscr{R}Pr = 3.2 \times 10^{-19} \times 4 \times 10^6 \times 0.85 \times$

---

* David Grubb and Yudhi Trisno, "AM Fiber Optic Trunks—A Noise and Distortion Analysis," *National Cable Television Association, NTCA Technical Papers* (1989): 225–230.

$0.5 \times 10^{-3} = \mathbf{5.44 \times 10^{-16}\ A^2}$. Receiver amplifier noise is proportional to $I_{rn}^2 = (7.5 \times 10^{-12}\ A/\sqrt{Hz})^2 \times 4 \times 10^6\ Hz = \mathbf{2.25 \times 10^{-16}\ A^2}$. Hence,

$$C/N = \frac{0.11 \times 10^{-9}}{(3.6 + 5.44 + 2.25) \times 10^{-16}} = 97.4 \times 10^3 = \mathbf{49.9\ dB}$$

A MathCAD program with a solution to this problem is given in Figure 18–34 of the computer analysis section.

Bandwidth/Data Rate

One design consideration faced when using LEDs: LEDs have very poor spectral purity, meaning that the power they produce is spread across a wide frequency range (wide linewidth). For reasonably high data rates (low dispersion requirement) over long distances (high power requirement), graded-index fibers are used because they allow large-diameter fibers with multimode operation. The dispersion (and pulse spreading) is low because the graded-index core (see Section 18–6) allows the various modes to propagate at different speeds, yet arrive at the receiver at approximately the same time. Figure 18–28 illustrates that the higher-order propagation modes have steep angles but travel faster because they spend more time in low refractive index regions of the fiber (and $v_p \propto 1/n$). In the past AT&T has used relatively inexpensive and noncritical LEDs with graded-index fibers for their short-run "lightphone" links. However, long-distance networks use single-mode fibers at 1.3 μm and 1.55 μm.

■ ■ ■ ■

**EXAMPLE
18–7**

Assuming a spectrally pure light source (therefore no chromatic dispersion), what is the maximum allowable fiber dispersion to handle a 200-Mb/s, 20% duty cycle, RZ data format on a 10-km link?

Solution:

From Figure 18–9, the pulse spread can reach $\tau_{ps} = T_b - \tau/2$, where $T_b = 1/200$ Mb/s = 5 ns and $\tau = 0.2 T_b = 1$ ns. Hence, $\tau_{ps} \leq 4.5$ ns and $\tau_{ps}/L \leq 4.5$ ns/10 km = **0.45 ns/km**. Although theoretically possible for graded-index fiber, this

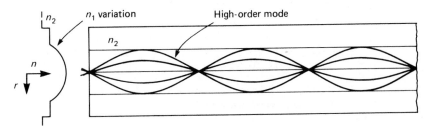

FIGURE 18–28 Multimode propagation with low pulse spread due to index of refraction variation in a graded-index fiber.

requirement really calls for a laser with its low linewidth and a small-diameter, step-index fiber (single-mode operation).

Using Equations 18–17 and 18–18, the same result is achieved: From Equation 18–18, $\tau_{ps} = (2 - 0.2)/2(200\mu) = 4.5$ ns and from Equation 18–17, $D = \tau_{ps}/L = 4.5$ ns/10 km = **0.45 ns/km.**

The bandwidth/data rate can also be limited by the amount of power received over the link. The reason is that a single photon has a finite amount of energy, and the energy in a single pulse (1 transmitted bit of information) has an *energy per bit* of

$$E_b = P_b \tau_b \qquad\qquad (18–45)$$

where P_b is the power per pulse and τ_b is the effective pulse (bit) duration. Since the power is most conveniently measured with an average-reading photometer, we determine the power per bit from

$$P_b = \frac{T_b}{\tau_b} P \qquad\qquad (18–46)$$

where P is the average power and τ_b/T_b is the duty cycle of the pulse bit stream.

Power Budget

The previous example illustrates the need for an adequate power budget for a fiber-optic link. The various power-loss mechanisms encountered have been discussed previously. Figure 18–29 illustrates the points where power is lost. Power losses at the transmitter-to-fiber L_t and fiber-to-receiver L_r interfaces consist of reflectance (also known as *Fresnel loss*, see Equation 18–3), NA loss (Equation 18–5), area difference loss (Equation 18–6), air-gap loss (Equation 18–25), and misalignment losses. Fiber-to-fiber coupling loss L_c results from reflectance, air gaps, and misalignment. These are usually specified by the connector manufacturer, but if not, 2 dB is a worse case assumption. Of course, the fiber loss L_f that

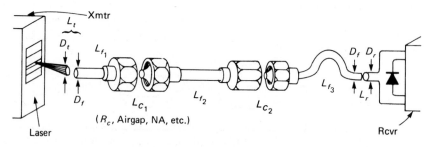

FIGURE 18–29   Power loss points in fiber-optic links.

is highly wavelength dependent (see Figure 18–9) produces the greatest loss. A fiber-optic link power budget example is now given.

■ ■ ■ ■

EXAMPLE 18–8

Received data pulses have a 20% average duty cycle and 1-ns pulse width. If −71.1 dBm is measured with an average-reading photometer:

1. Find the energy per bit.
2. How many photons are in one received light pulse if the transmitter is operating at 1.55 μm?
3. What is the shortest-wavelength detector that would respond to the received pulses?

Solution:

1. −71.1 dBm = 0.776×10^{-7} mW. Each pulse has power of

$$P_b = \frac{0.776 \times 10^{-10} \text{ W}}{0.2} = 0.388 \text{ nW}$$

The energy/bit is $E_b = 0.388$ nW $\times$ 1 ns = **0.388×10^{-18} J** (or W·s).

2. One photon at 1.55 μm has energy of

$$W = \frac{hc}{\lambda} = \frac{1.9875 \times 10^{-25} \text{ J·m}}{1.55 \text{ μm}} = 1.28 \times 10^{-19} \text{ J}$$

There must be 0.388×10^{-18} J$/(1.28 \times 10^{-19}$ J per photon) = **3 photons/bit.**

3. One to three photons per pulse is a theoretical but not a practical limit. A detector capable of responding to $\lambda = hc/(0.388 \times 10^{-18}$ J) = **512 nm** could theoretically respond to the received signal. Ten photons per bit represents a practical lower limit.

■ ■ ■ ■

EXAMPLE 18–9

A repeaterless fiber-optic communication link is to be 10 km long. The LED produces 8 mW at 0.82 μm, has an emitting area with a diameter of 80 μm, and is closely coupled to a short fiber pigtail of 80 μm diameter. Since the diode radiation pattern is nearly Lambertian, assume NA = 1, while the pigtail has NA = 0.30.

Step-index fibers of 1 km length with NA = 0.3, core diameter 50 μm, and n = 1.5 are used. Connectors, each with 2 dB of loss from all mechanisms, are to be used. The receiver uses a ferrule connector with a very short fiber, and diode with an antireflecting surface and an input diameter of 100 μm and NA = 0.5. The air gap is essentially zero. Determine the detected light power.

Solution:

L_t From Equations 18–5 and 18–6, $P_f = P_s[NA_f^2 \times (D_f/D_{LED})^2]$. The loss is $L_t(dB) = -20 \log(0.3 \times 50$ μm$/80$ μm) = **14.5 dB**. Power in the pigtail is −5.5 dBm (+9 dBm − 14.5 dB).

L_c and L_f There will be 10 fibers and 11 connectors for L_c (dB) = 22 dB. One long spliced cable would be no more than L_c (dB) = 11 × 0.5 dB = 5.5 dB. L_f = 10 × 2 dB/km = 20 dB, where 2 dB/km is from Figure 18–9.

L_r Since the diode area and NA is larger than the fiber, only reflectance loss is included. From Equation 18–3, the loss is 10 log{1 − [(1.5 − 1)/(1.5 + 1)]²} = 0.18 dB. Clearly, reflectance loss is negligible in most applications.

The total link loss is L (dB) = 14.5 + 22 + 20 + 0.18 = 56.68 dB. P_t (dBm) = 10 log 8 mW/1 mW = +9 dBm. Therefore, P_d (dBm) = +9 dBm − 56.68 dBm = **−47.7 dBm—about 17 nW.**

18–8 ■ FIBER-OPTIC COMMUNICATION APPLICATIONS

International Networking

Perhaps the most remarkable commercial application of optical fiber technology for communications is the undersea transatlantic (TAT) and the transpacific (HAW/TPC) fiber telecommunication links. If all undersea FO links are considered, nearly 100,000 km of undersea cable with optical fiber and repeaters will exist in the mid-1990s.

TAT-8 was installed in 1988 with a system cost of one-third of a billion U.S. dollars by AT&T (USA), Standard Telephones and Cables (England), and Submarcom (France). The system consists of 125 repeaters spaced approximately every 50 km with cable made up of the fiber covered by an elastomer, surrounded by steel and copper sheathing, and encased in a jacket of polyethylene. The overall cable diameter is approximately 0.8 in. (2 cm), becoming slightly thicker near shallow water.

The 1300-nm wavelength light is generated by InGaAsP lasers modulated with up to 280 Mb/s binary data and voice, capable of providing the equivalent of 4000 voice channel telephone calls. The total fiber length underwater is nearly 5800 km (3600 mi), with the longest stretch (5080 km) between Tuckerton, N.J., and a point between England and France. From this point, there is a 268-km branch to Penmarch, France, and a 450-km branch to Widemouth, England. The receiver photodetectors in the repeaters are InGaAs PIN diodes. Using two fibers per cable, the system handles 4000 two-way voice or equivalent full duplex data transactions for a total system capacity of 560 Mb/s. The most recent TAT system, TAT-9, will have double this capacity and include Canada and Spain with the United States, France, and the United Kingdom.

Meanwhile, in the Pacific Ocean a fiber-optic link (HAW-4) will connect California and Hawaii with a 4200-km undersea cable; the cities are San Francisco and Honolulu, respectively. From Oahu, the remainder of the Pacific Ocean is covered by a 9100-km transpacific (TPC-3) cable to Guam and Boso, Japan. The 13,000-km (8060-mi) total fiber link is designed for 50-km repeater spacing similar to the TAT-8 system. A feeder cable will also connect Japan with Hong Kong and Korea. Also, a GP-2 system will extend the TPC-3 system to the Philippines and Taiwan.

Local Area Networking

On a less grandiose scale, high-speed local and metropolitan area networking (LAN and MAN) with fiber optics are emerging with standards such as the *High Speed Channel Standard* (HSC) and the *Fiberoptic Distributed Data Interface* (FDDI). While both standards are still evolving, the data-to-fiber interface, FDDI standard, is aimed at digital data transmission at 100 Mb/s, whereas the HSC is for point-to-point links of 800–1600 Mb/s with 32-bit data words plus 4 parity and 8 control bits for a total of 44. In general, a LAN is considered to be a link up to 3 km in length, a MAN is somewhere between 3 km and 100 km, and a wide-area network (WAN) is an extended MAN.

The FDDI standard of 2 km (1.24 mi) maximum node-to-node length and 100-Mb/s data rate will accommodate most digital fiber-optic links using multimode, graded-index 62.5-μm-diameter fiber achieving 200–500 MHz·km bandwidth-distance product at 1300 nm. (Table 18–2 has representative fiber data.) In the 1300-nm wavelength range, the zero dispersion window makes single-mode fiber very attractive. For single-mode propagation, the fiber core diameter must be approximately 10% of what a graded-index fiber is typically allowed, and therefore fiber connector alignment is much more critical in order to avoid excessive power loss.

The X3T9.5 American National Standards Institute (ANSI) committee has developed the FDDI specification based on a backbone of dual counterrotating fiber-optic rings (a primary ring and a secondary ring) using a timed-token protocol capable of transmitting data at 100 Mb/s in each ring; see Figure 18–30. FDDI backbone links can provide a superhighway for interconnecting LANs and wide-area networks (WANs). Between stations, an individual multimode fiber length may be as long as 2 km; however, networking can be used to transfer data over distances up to 200 km to nearly 1000 stations.

The X3T9.5 working group has specified four functional layers that make up the FDDI standard. They are the media access control (MAC) document, physical layer protocol (PHY) document, physical medium dependent (PMD) document, and the station management (SMT) document. The MAC is responsible for interpreting packets, token passing, and packet framing. The PHY standard is responsible for clock recovery and encoding and decoding the input signals from the PMD layer. The PMD defines and characterizes the fiber-optic drivers and receivers, cables, connectors, optical bypass provisions, and physical hardware–related characteristics. SMT provides overall control of the FDDI ring by monitoring, managing, and configuring the ring. SMT is also responsible for connection management—the logical connection between stations.

FDDI specifies three station types: dual-attachment station, single-attachment station, and concentrator. The dual-attachment station has two PMD entities to accommodate the dual counterrotating ring. A single-attachment station has one PMD entity and is connected to a concentrator, which is then connected to the ring. Dual-attachment stations connect to both rings so damage to the cable plant will allow the ring to operate in a wraparound configuration from the primary to the secondary portion of the ring. SMT logic identifies the problem and makes the

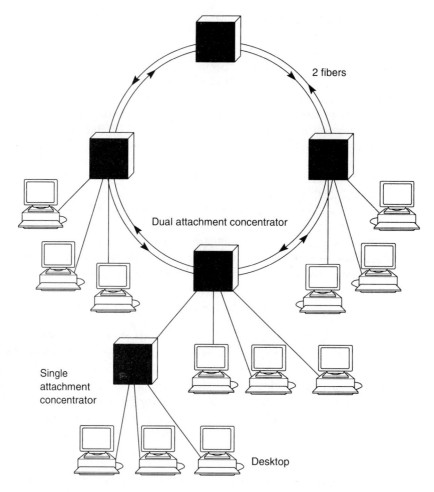

2 fibers

Dual attachment concentrator

Single
attachment
concentrator

Desktop

FIGURE 18–30 Typical FDDI network with dual counterrotating fiber-optic rings.

appropriate changes in the signal path. Failure within a dual-attachment station, if identified by that station, will result in the station switching to the optical bypass mode. Optical switches within the station direct the light to the next station.

The FDDI concentrator may be the single most important station type. The concentrator has many functions that permit connectivity to single-attachment stations and bridges to slower-speed networks, such as Ethernet (IEEE 802.3 standard) and Token Ring (IEEE 802.5). A dual-attachment concentrator is a dual station that offers additional attachment for stations that are not part of the dual ring. The concentrator connected to a bridge allows slower stations or other LANs access to the FDDI dual ring without increasing the protocol overhead of the backbone ring.

The PMD specification defines the dimensions and tolerances of the media interface connector (MIC) receptacle at the backplane of FDDI stations. This (duplex) receptacle has been agreed upon as the boundary between the station and cable plant. Some MICs connect only to dual-attachment stations on the backbone ring, and others connect only to single-attachment stations on the back of the concentrators. The distinction is important because station management dictates that stations on the ring must be connected in the correct order. Stations that are improperly connected on the ring might result in a primary ring being connected to a secondary ring, thereby preventing the network from wrapping back on itself in the event a station shuts down or the cable plant malfunctions. Even the initial ring configuration would be impossible, since peer-to-peer station identification would be out of order. The duplex receptacle is keyed in order to solve these problems, thus allowing stations to be connected to the correct fiber channel.

The media specified in the PMD document is optical fiber. However, for connecting workstations and high-performance personal computers at the desktop that are 100 m or less from room wiring closets, shielded twisted-pair wire may be used for the connection to the concentrator. Optical fiber core/cladding dimensions are not limited by the FDDI standard but 62.5–125-μm multimode fiber is mentioned. The operating wavelength for the 100-Mb/s data rate is 1300 nm, and the end-to-end fiber-link budget should be limited to 11 dB. This figure includes fiber, connectors, splices, and optical switches. Modal dispersion is limited in the PMD specification to allow a bandwidth distance product of at least 500 MHz·km.

For fiber rates up to 2000 Mb/s, the limiting factor for a fiber system is dispersion. Therefore, single-mode operation is used, especially for long-haul applications. Then the spectral linewidth of the light source becomes critical in minimizing dispersion. Since a laser is the source of choice for single-mode long-haul applications, the transmitter will require dynamic power level control with power foldback, and active cooling techniques.

SONET.   An optical network transmission standard that has international connectivity is the *Synchronous Optical Network* standard. Originally developed for the United States by Bell Communications Research (Bellcore) of New Jersey, SONET has been modified and adopted by the United States's T1X1 (digital hierarchy and synchronization) subcommittee of the American National Standards Institute (ANSI) and the International Telegraph and Telephone Consultative Committee (CCITT). SONET defines standard optical signals, a synchronous data frame structure for multiplexed digital traffic, and network operations procedures.

The basic building block for data transport interfacing and multiplexing is called the Synchronous Transport Signal, level 1 (STS-1). The STS-1 bit rate is 51.84 Mb/s, which accommodates up to the DS3 (28 T1 lines), 672 multiplexed telephone voice channels, all operating in a synchronous-like mode. Unlike the digital data frames of Chapter 13, the STS-1 frame is two-dimensional with 90 columns and 9 rows of 8-bit bytes (B in Figure 18–31). Flag framing, sender ID, and destination address functions are included in the section and line overhead bytes indicated in the figure. The STS-1 frame is transmitted byte by byte, beginning with column 1 of row 1, followed by a second framing byte at column 2, row 1, and

FIGURE 18–31 The STS-1 frame.

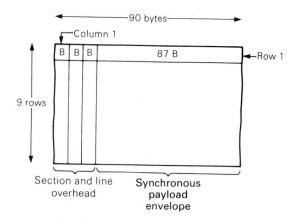

then an ID byte that precedes the start of the transmitted message in the synchronous payload envelope (SPE).

Figure 18–32 shows more details of the functions of the section and line overhead bytes. The section layer consists of overhead bytes that are processed by all equipment in the synchronous optical network, including repeaters. As indicated in Figure 18–32, the two framing bytes that show the start of each STS-1 frame are followed by an STS-1 identification byte. These three bytes are the only bytes that

FIGURE 18–32 Section and line overhead bytes.

1 byte

Section overhead	Framing A1	Framing A2	STS-1 ID C1
	BIP-8 B1	Orderwire E1	User F1
	Data com D1	Data com D2	Data com D3
Line overhead	Pointer H1	Pointer H2	Pointer action H3
	BIP-8 B2	APS K1	APS K2
	Data com D4	Data com D5	Data com D6
	Data com D7	Data com D8	Data com D9
	Data com D10	Data com D11	Data com D12
	Growth Z1	Growth Z2	Orderwire E2

First 3 columns of each STS-1

Path layer
Trace J1
BIP-8 B3
Signal label C2
Path status G1
User channel F2
Multiframe H4
Growth Z3
Growth Z4
Growth Z5

Path layer overhead (carried as part of SONET payloads)

are not subjected to scrambling before transmission. The first three bytes shown in the second row of the STS-1 frame are section overhead bytes that provide for section error monitoring (8-bit bit-interleaved parity; BIP-8), an orderwire channel byte for network maintenance personnel communications, and a channel for unspecified user applications. The last three section overhead bytes provide a channel to carry network maintenance and provisioning data. A second layer of the section overhead is an LAPD link protocol layer. Higher-level layers will be added in future updates to the SONET standard.

Line overhead bytes are processed by all SONET equipment except network repeaters. These bytes provide information for *pointing* to the start of a synchronous payload envelope in a multiplexed, high-data-rate, byte-interleaved STS-*N* frame; a line error monitoring BIP-8; a 2-byte automatic protection switching (APS) message channel; a 9-byte data communications channel; and an orderwire channel. Two additional bytes are included for future network growth. In addition to section and line overhead information, there is a single columnar-oriented *path layer* overhead of 9 bytes that is part of the SONET STS-1 payload and is processed only by payload terminating equipment. Multiple STS-1s may be multiplexed together to form higher-rate STS-*N*s, where $N = 3, 9, 12, 18, 24, 36$, and 48. An STS-48, for example, would have a rate of 48×51.84 Mb/s $= 2488.32$ Mb/s.

The basic building block for CCITT-specified international links consists of three STS-1 multiplexed frames called the synchronous transport module—level 1 (STM-1). The STM-1 operates at 155.52 Mb/s, exactly three times the STS-1 rate, conveniently accommodating the European rate of 139.264 Mb/s derived from the 2.048-Mb/s primary rate signal, as compared to the North American 1.544-Mb/s T1 line rate. An *optical carrier*—level 1 (OC-1) is obtained after scrambling an STS-1 frame (to eliminate long strings of 1s and 0s) and then intensity modulated on the optical carrier. Higher-rate multiplexing will yield OC-*N* SONET signals.

■ PROBLEMS

1. What and when was the breakthrough in optical fiber technology for communication applications?

2. Light at 1.55 μm in air has what (a) frequency, (b) energy (in electron volts), and (c) power in a 50-ns burst?

3. Repeat Problem 2 for a 0.2-μs pulse of light with an 820-nm wavelength.

4. a. Determine the critical angle (of incidence) at a glass/fused-quartz interface.
   b. Calculate the numerical aperture for an optical fiber made of these materials.
   c. For total reflection, should the fiber core material be the glass or the fused-quartz? Why?

5. A fiber core is glass with $n = 1.5$.
   a. Calculate the reflectivity (in percent) from air.
   b. What value must the cladding refractive index n be in order to produce an optical fiber with NA = 0.2?

6. An optical fiber is made with a core of flintglass ($n = 1.62$) and a cladding of crown glass ($n = 1.51$).
   a. What is the critical angle at the core/cladding interface?
   b. Determine the numerical aperture for light input to this fiber.
   c. What is the maximum angle for light entering the fiber from air? Make a sketch to illustrate, using the results of parts **a** and **c**.

7. An 820-nm wavelength light source is used with the fiber of Problem 6.
   a. Determine the maximum core diameter for single-mode propagation.
   b. How much power loss will result due to NA difference if the light from the fiber is coupled into one with $NA = 0.20$?

8. How many modes will propagate in a 3-μm diameter, step-index, glass/fused-quartz FO waveguide for 1300-nm light?

9. a. Determine the maximum core diameter for single-mode propagation of 1.3-μm light in a glass/fused-quartz FO waveguide.
   b. Approximately how many modes will propagate if the diameter is made 50% larger than for part **a**?

10. A Fabry-Perot laser diode cavity 0.395 μm long is made with InGaAs. The index of refraction of InGaAs is 3.0 and the laser gain peaks at 1310 nm.
    a. Determine the frequency difference between output longitudinal modes.
    b. Find the wavelength separation between longitudinal modes.
    c. If the full-width, half-of-maximum (FWHM) spectral width of the transmitted spectrum is approximately 2.9 μm, show with a sketch (and a calculation) that five longitudinal modes will be included in the FWHM.

11. A 30-km FO link is to handle 200-Mb/s NRZ data. What is the maximum dispersion in ns/km for the multimode fiber, assuming a spectrally pure light source?

12. a. What will be the maximum allowable 25%–RZ data rate for the 20-km-long fiber of Problem 6 with multimode operation?
    b. How much attenuation can be expected at 820 nm?

13. Prove Equation 18–18 using Figure 18–9.

14. A 2-mW light source has an effective diameter of 4 μm and radiates light in a Lambertian pattern. Determine the coupling power loss (in dB) and the power coupled into a glass/fused-quartz fiber of each core diameter.
    a. 2 μm
    b. 20 μm
    c. 200 μm

15. a. Explain laser action.
    b. What are the dopants used for making GaAs LEDs and lasers produce light at (1) 820 nm and (2) 1550 nm?

16. The diodes of Figure 18–13 drop 0.7 V when conducting.
    a. Determine the normal bias voltages on both bases of the MPS 6515 differential amplifier.
    b. What will be the voltage on the input base of this differential amplifier when U_2 is saturated ($V_0 \simeq 0.2$ V).
    c. Will the LED be on or off when U_2 is saturated?
    d. Is the data input high or low when the LED is on? Show why.

17. The average power received in a photodiode detector is -63 dBm, and the 1.3-μm light is on/off in equal 300-ns time periods. Determine (a) the received energy/pulse (joules and eV) and (b) the number of photons/pulse.

18. The PIN diode of Figure 18–16b has a responsivity of 0.5 μA/μW and a dark current of 50 nA. Assuming a low-gain, extremely high-Z_i IC,
  a. What value of R would yield 2 V at the output with a -30-dBm light input?
  b. What will be the minimum V_o?

19. A Mach-Zehnder modulator has minimum loss for 0 V bias and maximum loss at 14 V. With 7 V bias, a 0.5-V peak-to-peak modulation voltage is applied.
  a. Show that the average launched light attenuation is 6 dB.
  b. Prove that the optical modulation index is 5.6%.

20. How much power loss (in dB) results due to an air gap of 10 μm between 25-μm diameter fibers with NA = 0.4? (Ignore reflectance.)

21. Describe two types of star couplers.

22. A 1 by 2 passive optical splitter has an input of $+3.5$ dBm and outputs of 0.4 mW and 1.6 mW. If directivity is infinite (no reflections), determine the following:
  a. The coupling loss in percent and in decibels for each of the two outputs
  b. The excess loss of the coupler

23. a. Determine the energy/bit for an optical receiver receiving -60 dBm average power if 50-ns-wide light pulses at 33.3% duty cycle are received.
  b. How many photons/pulse are received if $\lambda = 1300$ nm?

24. An 8-km repeaterless FO link has the following component specifications: connectors, 2 dB each (all loss mechanisms); 1-km sections of 52-μm-diameter graded-index fiber (NA = 0.4); an 820-nm, 50-mW laser transmitter; and a pigtailed 3-μm-diameter photodetector with numerical aperture of 0.30. The laser has an output diameter and beamwidth much less than 52 μm and is connected through a microlens to the first fiber.
  a. Determine the power level just after each of the nine connectors.
  b. How much current will the photodetector produce if its responsivity is 0.60 A/W?

25. A 1.3-μm wavelength single-mode laser emits 0.5 mW into a short pigtail (NA = 0.6, diameter 20 μm). The link is 4 km in length, with plastic connectors good for about 2 dB of loss each. The 1-km lengths of step-index fiber each have a diameter of 62.5 μm and 0.3 numerical aperture. The receiver pigtail has a diameter of 30 μm and NA = 0.2; the photodiode detector is closely coupled with a lens and has a responsivity of 0.6 A/W. Determine
  a. The power in dBm after each of the five connectors.
  b. The photodiode output current.
  c. Confirm your answers using the FO link, BASIC computer analysis program.
  d. What value of photodetector dark-current would be required to ensure a 38-dB dynamic range?

26. Write a computer program to analyze dispersion and pulse spreading problems for RZ and NRZ light pulses.

27. A Fabry-Perot laser launches 1 mW of optical power and has a RIN of -138 dB/Hz. It is modulated with a single sinusoid at an OMI of 0.2. The receiver photodetector has a responsivity of 0.86, a postdetector noise current (density) of $7 \text{ pA}/\sqrt{\text{Hz}}$, and a bandwidth of 5 MHz (PAL video bandwidth).

a. Determine the detected signal-to-noise ratio for a 5-dB and a 10-dB link.

b. Write your own program or use MathCAD to prove your answer to part **a.**

18–9 ■ COMPUTER ANALYSIS

Using the computer for communications system and circuit analysis has become a virtual necessity in education and industry. What follows are examples of programs and/or readily available simulation software for solving problems of the type covered in this chapter.

All the programs are interactive in the sense that an introductory sentence or short paragraph describes the purpose of the program and the user is then prompted for input data.

The program for Chapter 18 is as follows:

- **Fiber-optic link analysis:** The program of Figure 18–33 computes the light power at the output of each connector in the link and also the power input to the photodiode. It will also compute the detector output current if the detector responsivity is given.

   Data prompted for will be similar to that required in Example 18–9: source power and pigtail diameter; NA and wavelength; total system fiber length; connector losses; fiber NA and diameter; and photodetector NA, diameter, and responsivity.

- **Analog signal link analyis:** The MathCAD program of Figure 18–34 solves for carrier-to-noise ratio of analog signal transmission over a fiber-optic link. The particular data shown in Figure 18–34 shows Example 18–6, part 2.

```
10 REM *   FIBER OPTIC  LINK  *
20 REM *    DESIGN/ANALYSIS    *
30 REM *  BY P. YOUNG,   ASU  *
40 CLS : PRINT"          FIBER OPTIC LINK DESIGN/ANALYSIS"
50   PRINT "    This program will solve for the light power"
60   PRINT " at the output of each connector in the link and"
70   PRINT " also the power input to the detector chip. It"
80   PRINT " will also compute the detector output current if"
90   PRINT " detector responsivity is known."
100 PRINT "    Data prompted-for will be similar to that "
110 PRINT " required in Example 18-8: Source and pigtail "
120 PRINT " power, diameter, NA and wavelength; total system"
130 PRINT " fiber length; connector losses; fiber NA and "
140 PRINT " diameter; and photodetector NA, diameter, and "
150 PRINT " responsivity."
160 PRINT"*****************************************************":PRINT
170  INPUT "Enter the laser or LED source power, in milliwatts:";SP
180 SPDBM = 10*LOG(SP)/LOG(10)
```

FIGURE 18–33   (page 1 of 2)

```
190   PRINT
200   PRINT "Enter the source wavelength in microns (.82, 1.3, or"
210   INPUT "1.55):";LMDA :PRINT
220   PRINT "Enter the source diameter. If unknown, or microlens"
230   INPUT "coupled, enter the pigtail diameter:";SD
240 PRINT
250   PRINT "Enter the source numerical aperture. Use 1 if"
260   INPUT "unknown:";SNA :PRINT
270   INPUT "Enter the pigtail NA:";PIGNA  :PRINT
280 IF SNA < PIGNA THEN 300
290 SPNADB = -20*LOG(PIGNA/SNA)/LOG(10)
300   PRINT "Enter the link (total fiber) length, in kilometers. Use
310   INPUT "integer values only:";TD
320 NCON = TD + 1 :PRINT
330   INPUT "Enter the fiber NA:";FIBNA  :PRINT
340 IF PIGNA < FIBNA THEN 360
350 LTNADB = -20*LOG(FIBNA/PIGNA)/LOG(10)
360   INPUT "Enter the fiber diameter in microns:";FD
370   PRINT
380 IF FD > SD THEN 400
390 LTDB = -20*LOG(FD/SD)/LOG(10)
400   INPUT "Enter the total loss per connector, in decibels:";CON
410   PRINT
420 LCDB = NCON * CON
430 IF LMDA = .82 THEN LF = 2
440 IF LMDA = 1.3 THEN LF = .6
450 IF LMDA = 1.55 THEN LF = .2
460 LFDB =LF * TD
470   PRINT "Enter the diameter of the photodetector or the pigtail,"
480   INPUT "which ever is the smallest, in microns:";DD :PRINT
490   INPUT "Enter the detector NA:";DNA :PRINT
500   INPUT "Enter the detector responsivity:";R
510 PRINT"**********************************************************"
520 IF FIBNA < DNA THEN 540
530 LRNADB = -20*LOG(DNA/FIBNA)/LOG(10)
540 IF DD > FD THEN 560
550 LRDB = -20*LOG(DD/FD)/LOG(10)
560     REM   COMPUTE POWER INTO EACH 1km LENGTH OF FIBER
570 K=2 :PRINT
580 P1 = SPDBM -SPNADB - LTNADB - LTDB - LTDDB - CON
590 PRINT"The power into the first of the"TD", 1km lengths of fiber is"
600 PRINT"P 1 = "P1", dBm."  :PRINT
610    FOR K = 2 TO TD
620 PK = P1 - (K-1)*CON - (K-1)*LF
630 PRINT"The power into fiber"K"of the "TD" fibers is "
640 PRINT"P"K" = "PK", dBm."
650 PRINT
660    NEXT K
670 PPIG = PK - CON - LF
680   PRINT "The power into the receiver pigtail is PR = "PPIG", dBm."
690 PDBM = PPIG - LRNADB - LRDB
700 PD = 10^(PDBM/10)/.000001
710   PRINT
720 PRINT"The power into the photodetector diode is PD =";PD"nW."
730   PRINT
740 IF R = 0 THEN END
750 I = R*PD
760   PRINT "The photodetector output current is I ="I"nanoamps."
770 END
```

FIGURE 18-33   (page 2 of 2)

System CNR Model: By Mark Tucker

This Mathcad program predicts the system CNR for an intensity modulated fiber optic CATV link. The input variables include: photodiode responsivity, laser RIN, modulation index per channel, launched laser power, link budget, receiver noise current and channel bandwidth. The model does not predict the effects of IIN (Interferometric Intensity Noise) on system CNR. Derating the value of laser RIN about 2-3 dB will compensate for this omission.

$R := .85$ Photodiode reponsivity in A/W
$RIN := -153$ Effective Laser RIN in dB/Hz
$Pl := 5$ Launched laser power in dBm
$ll := 0 .. 10$ Link budget (loss) in dB

$Ircvr := 7.5 \cdot 10^{-12}$ Receiver noise current in Amps per Root Hz

$BW := 4 \cdot 10^{6}$ Channel bandwidth: 4 MHz for NTSC, 5 MHz for PAL

$M := .035$ (appr 60 chan)   Optical Modulation Index per channel
            (Low distortion)
$Pr_{ll} := Pl - ll$ (dBm)

$Pw_{ll} := .001 \cdot 10^{Pr_{ll} \cdot 0.1}$ (watts)

$Isn_{ll} := 3.2 \cdot 10^{-19} \cdot R \cdot Pw_{ll} \cdot BW$ (Amps squared)

$Irin_{ll} := \left[R \cdot Pw_{ll} \right]^2 \cdot BW \cdot 10^{0.1 \cdot RIN}$ (Amps squared)

$Ircvr_{ll} := BW \cdot Ircvr^2$ $Int_{ll} := Isn_{ll} + Irin_{ll} + Ircvr_{ll}$ (Amps squared)

$Isig_{ll} := 0.5 \cdot \left[M \cdot R \cdot Pw_{ll} \right]^2$ $CNR_{ll} := 10 \cdot \log \left[\dfrac{Isig_{ll}}{Int_{ll}} \right]$ $CNRsn_{ll} := 10 \cdot \log \left[\dfrac{Isig_{ll}}{Isn_{ll}} \right]$
      (amps squared)   (dB)   (dB)

FIGURE 18-34   (page 1 of 2)

System CNR (dB)

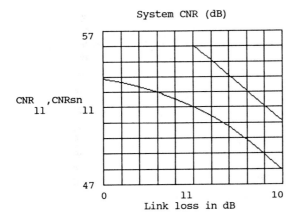

CNR_{11} ,$CNRsn_{11}$

Red (top) curve: shot noise
limited CNR (quantum limit)

Green (bottom) curve:
system CNR

Link loss in dB

(Amps squared) System Noise Contributions

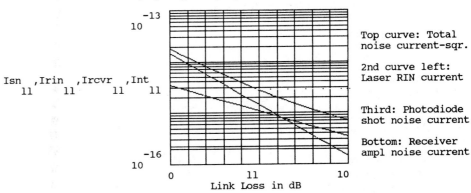

Isn_{11} ,$Irin_{11}$,$Ircvr_{11}$,Int_{11}

Top curve: Total
noise current-sqr.

2nd curve left:
Laser RIN current

Third: Photodiode
shot noise current

Bottom: Receiver
ampl noise current

Link Loss in dB

CNR_{11} (dB)	
53.8707	0 dB link
53.6327	
53.3414	
52.9871	
52.5596	
52.0481	
51.4424	
50.7331	
49.9128	
48.9768	
47.9231	10 dB link

FIGURE 18-34   (page 2 of 2)

Appendices

A

Introducing the Impulse Function

A very useful function for describing frequency spectra is the Dirac *delta* function $\delta(t)$. As illustrated in Figure A–1, the delta function exists only at the time instant at which the argument, in parentheses, goes to zero. Figure A–1b illustrates that $\delta(t - t_0)$ is a zero-width impulse that occurs when $t - t_0 = 0$—that is, at $t = t_0$. Also called the *unit impulse*, $\delta(t)$ has unit area as illustrated in Figure A–2. This figure also best describes how $\delta(t)$ is defined mathematically, namely,

$$\int_{t=-\infty}^{\infty} g(t)\delta(t)\, dt = g(0) \qquad \text{(A–1)}$$

where $g(t)$ is any function that is continuous at $t = 0$. Notice that the product of $g(t)$ and $\delta(t)$ integrated over all time results in producing the value of $g(t)$ evaluated for the time instant at which the delta function occurs. An example is

$$\int_{-\infty}^{\infty} g(t)\delta(t - t_0)\, dt = g(t_0) \qquad \text{(A–2)}$$

It is observed in Equation A–2 and Figure A–2 that we could have defined $\delta(t)$ using the two equations

$$\int_{-\infty}^{\infty} \delta(t)\, dt = 1 \qquad \text{(A–3a)}$$

$$\delta(t) = \begin{cases} \infty & t = 0 \\ 0 & t \neq 0 \end{cases} \qquad \text{(A–3b)}$$

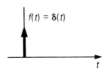

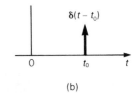

FIGURE A–1 The unit impulse Dirac delta function.

(a)   (b)

FIGURE A-2

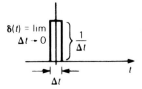

where Equation A–3a indicates that the area enclosed by $\delta(t)$ is unity. Also, observe that

$$\int_{-\infty}^{\infty} \delta(t)e^{-j2\pi ft}\, dt = e^0 = 1 \qquad\qquad \textbf{(A–4a)}$$

Multiplying both sides of Equation A–4a by $e^{j2\pi ft}$ and integrating over all f yields

$$\int_{-\infty}^{\infty} \int_{-\infty}^{\infty} \delta(t)e^{-j2\pi ft}\, dt\; e^{j2\pi ft}\, df = \int_{-\infty}^{\infty} 1e^{j2\pi ft}\, df \qquad\qquad \textbf{(A–4b)}$$

Then, using Equation A–3a,

$$\int_{-\infty}^{\infty} \int_{-\infty}^{\infty} \delta(t)\, dt\, df = \int_{-\infty}^{\infty} e^{j2\pi ft}\, df$$

from which it is observed that

$$\delta(t) = \int_{-\infty}^{\infty} e^{j2\pi ft}\, df \qquad\qquad \textbf{(A–5)}$$

Since the right side of Equation A–5 is the inverse Fourier transform of unity, we see that the Fourier transform of $\delta(t)$ is unity. This means that the frequency spectrum of the impulse function has unity magnitude and is uniform from dc to infinite frequency.

Another important property of the impulse function is its relationship to the unit step function. The unit step function previously defined is related to the unit delta function by

$$\int_{-\infty}^{t} \delta(\tau)\, d\tau = u(t) \qquad\qquad \textbf{(A–6)}$$

where, again, the time variable τ is used in the integrand; τ is a dummy time variable. Because of Equation A–6 we can write

$$\frac{du(t)}{dt} = \delta(t) \qquad\qquad \textbf{(A–7)}$$

That is, an ideally differentiated step function produces an impulse.

B

Filter Attenuation Curves*

* Howard W. Sams & Co., Editorial Staff. *ITT Reference Data for Radio Engineers*. Reproduced with permission of the publisher, Howard W. Sams & Co., Indianapolis, 1975.

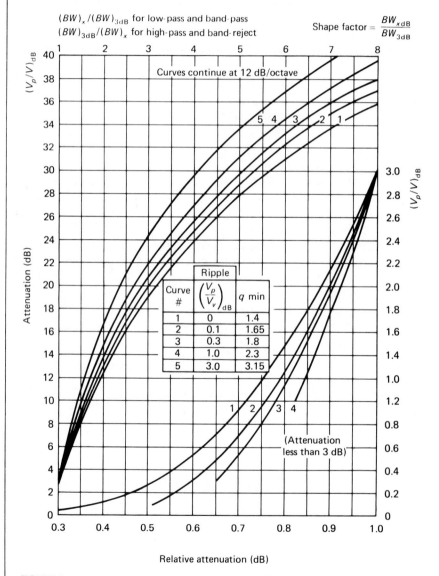

FIGURE B-1   Relative attenuation for a 2-pole network

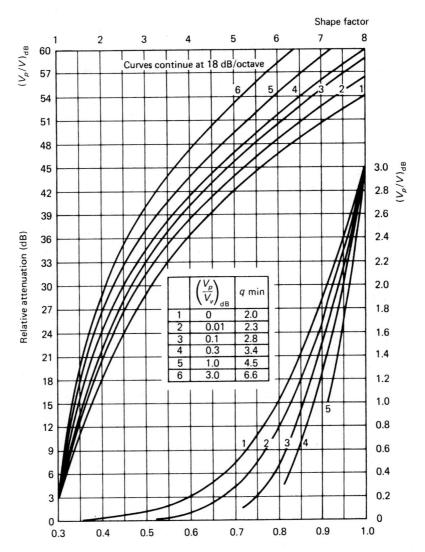

FIGURE B–2 Relative attenuation for a 3-pole network.

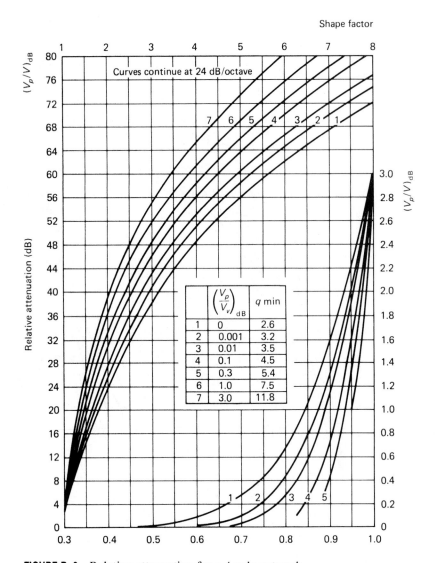

FIGURE B–3 Relative attenuation for a 4-pole network.

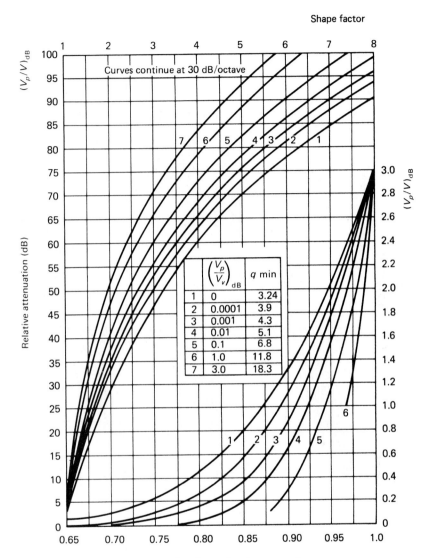

FIGURE B–4 Relative attenuation for a 5-pole network.

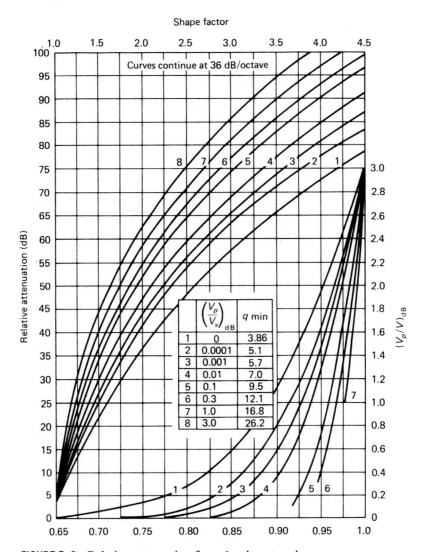

FIGURE B–5 Relative attenuation for a 6-pole network.

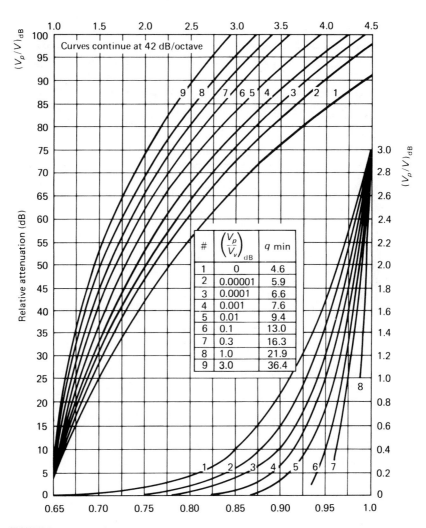

FIGURE B-6 Relative attenuation for a 7-pole network.

C

Feedback Analysis for Transconductance Amplifier

Figure C–1 is a common-emitter amplifier with feedback resistor R_F. The circuit configuration is voltage-sampled/shunt-feedback. Since the input is a parallel circuit, it is convenient to use the Norton equivalent as seen in Figure C–2.

The basic amplifier ($i_F = 0$) has

$$v_i = \frac{R_1 R_F}{R_1 + R_F} i_s \qquad \text{(C–1)}$$

where $R_1 = R_s \parallel R_i$, and

$$v_o = -\frac{R_2}{R_2 + R_C} A_v v_i$$

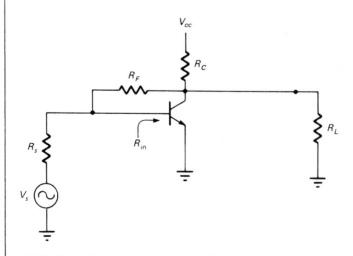

FIGURE C–1 Transconductance amplifier.

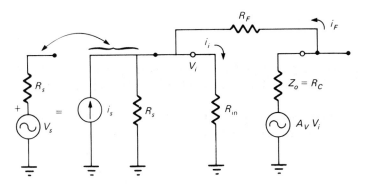

FIGURE C–2 Equivalent circuit.

where $R_2 = R_L \parallel R_F$ for the amplifier loaded by the next stage. For our analysis

$$v_o = \frac{R_F}{R_F + R_C} A_v v_i \qquad \text{(C-2)}$$

For shunt feedback analysis it is most convenient to characterize the circuit gain as $v_o/i_s = A_C$, which is actually a transconductance gain. The overall circuit with feedback is called a *transconductance amplifier*.

Using Equations C–1 and C–2 yields

$$A_C \equiv \frac{v_o}{i_s} = \frac{R_F}{R_F + R_C} A_v \frac{R_1 R_F}{R_1 + R_F} \qquad \text{(C-3)}$$

and the feedback improvement factor is $T = 1 + A_C B$, where

$$B = \frac{i_F}{v_o} = \frac{-1}{R_F} \qquad \text{(C-4)}$$

Consequently,

$$T = 1 + \left[\frac{A_v R_F R_1}{(R_F + R_C)(R_F + R_1)} \right] \qquad \text{(C-5)}$$

For our circuit $R_F \gg R_C$ and $R_F \gg R_1$, so

$$T = 1 + A_v \left(\frac{R_1}{R_F} \right) \qquad \text{(C-6)}$$

Notice that for the case in which the amplifier is a very-high-gain IC so that $A_v B \gg 1$, the closed-loop circuit gain will be $A_{v_f} = -A_v/(A_v R_1/R_F) = -R_F/R_1$, which you recognize for the inverting op-amp.

D

Derivation of Frequency Response for PLLs

The PLL frequency and step responses are determined as follows. From Figure D–1, we can see that, for an input of θ_i and VCO output phase θ_o, the phase difference $\theta_e = \theta_i - \theta_o$ is the input to the phase detector. From this, $\theta_i = \theta_e + \theta_o$. Now, by working backward around the loop from θ_o, we see that $\theta_o = (2\pi k_o/s)V_o{}^*$ and $V_o = k_A k_\phi \theta_e$. Thus $\theta_o = (2\pi k_o/s)k_A k_\phi \theta_e$. Substituting this into $\theta_i = \theta_e + \theta_o$, we get $\theta_i = \theta_e + [(2\pi k_o k_A k_\phi)/s]\theta_e$ or $\theta_i = (1 + [2\pi k_o k_A k_\phi/s])\theta_e$.

To determine how the output voltage V_o varies with input-phase changes, use $V_o = k_A k_\phi \theta_e$ to get $\theta_i = \{1 + [(2\pi k_o k_A k_\phi)/s]\}[V_o/(k_A k_\phi)]$. The output/input transfer function is V_o/θ_i which, from our last expression, will be

$$\frac{V_o}{\theta_i} = \frac{s k_\phi k_A}{s + 2\pi k_o k_A k_\phi} \qquad\text{(D–1)}$$

for an uncompensated PLL.

The frequency response and transient behavior of the PLL are determined from the poles of the transfer function—that is, where the denominator of Equation

---

* s is the frequency variable and, for sinusoidal VCO modulating signals, $s = j\omega$. k_o/s shows both the integrating effect of the VCO and the decrease of modulating sensitivity (k_o) with increasing frequency.

FIGURE D–1 Uncompensated PLL.

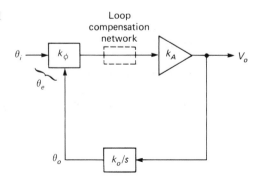

D–1 goes to zero. Clearly, the denominator approaches 0 when $s = -2\pi k_o k_A k_\phi = -k_v$.

$$\frac{V_o}{\theta_i} = \frac{s k_\phi k_A}{s + k_v} \qquad (\text{D–2})$$

shows the same kind of transient response as an RC high-pass filter: $V_o/V_i = (sRC)/(sRC + 1) = s/(s + \omega_c)$, where $\omega_c = 1/RC$ is the cutoff frequency in rad/s.

An input step-phase change will track-out to zero in the steady state just like an input voltage-step settles out to zero after the initial transient for the high-pass RC filter of Figure D–2. However, for input *frequency* changes, the PLL responds like a low-pass filter because, from $\omega = d\theta/dt$, $\omega_i = s\theta_i$ and $f_i = (1/2\pi)s\theta_i$, so that $V_o/\theta_i = V_o/[(2\pi f_i)/s] = (sV_o)/(2\pi f_i)$. Consequently,

$$\frac{V_o}{f_i} = \frac{2\pi k_\phi k_A}{s + k_v} \qquad (\text{D–3})$$

which shows that V_o changes with f_i similarly to a low-pass RC filter, where $V_o/V_i = 1/(sRC + 1) = \omega_c/(s + \omega_c)$, with $\omega_c = 1/RC$ is the corner frequency in rad/s.

The bandwidth of the uncompensated PLL is $\omega = k_v$ and the Bode plot is shown in Figure 10–21. The transient response for this simple PLL will, like the LPF, simply be that of a charging capacitor when the input to the PLL is a step change of frequency. This is shown mathematically by the use of tables of Laplace transforms as follows: If f_i takes a step change of Δf at $t = 0$, then $F_i(s) = \Delta f/s$. Solving for the output voltage (VCO input signal) with Equation D–3 and $2\pi k_\phi k_A = k_v/k_o$ yields

$$V_o(s) = \frac{\Delta f k_v/k_o}{s(s + k_v)} \qquad (\text{D–4})$$

A table of inverse transforms will show that the time response of V_o for the Δf input step is

$$v_o(t) = \left(\frac{\Delta f}{k_o}\right)(1 - e^{-k_v t}) \qquad (\text{D–5})$$

This is a rising exponential with rise time $1/k_v$ and steady-state value of $V_o(\text{SS}) = \Delta f/k_o$.

FIGURE D–2 (a) Response of PLL output V_o to input-phase step change $\Delta\theta_i$. Notice the resemblance of this response to that expected for the high-pass RC filter of (b) with an input voltage step.

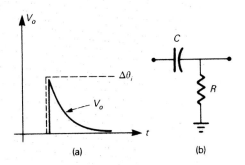

(a)

(b)

The slow-rising output voltage from an uncompensated PLL may be unacceptable, especially in FSK demodulator applications. The solution is to compensate the loop and speed up the response. This is called *loop compensation.*

Loop Compensation

It may seem contradictory, but the response of the uncompensated first-order loop can be speeded up by inserting a phase-lag network in the loop between the phase detector and amplifier. Such a network is the RC integrator shown in Figure D–3. The result will be to produce a second-order system, the phase of which can be controlled, thereby controlling the rise time and regenerative nature of the loop.

The frequency response is derived in exactly the same way as above except that, every place $k_\phi k_A$ appears, the lag network transfer function $F(s)$ is included—that is, $k_\phi k_A \to k_\phi F(s) k_A$.

Equation D–1 becomes

$$\frac{V_o}{\theta_i} = \frac{sk_\phi k_A F(s)}{s + 2\pi k_o k_\phi k_A F(s)} \tag{D-6}$$

and Equation D–3 becomes

$$\frac{V_o}{f_i} = \frac{2\pi k_\phi k_A F(s)}{s + k_v F(s)} \tag{D-7}$$

Since $F(s) = \omega_c/(s + \omega_c)$ for lag compensation then, after some algebra, we find that the equivalent to Equation D–4 for a step change of input frequency is

$$V_o(s) = \frac{\Delta f \, k_v \omega_c / k_o}{s(s^2 + \omega_c s + k_v \omega_c)} \tag{D-8}$$

Compare this equation to the frequency response of a general second-order system,

$$H(s) = \frac{\omega_n^2}{s(s^2 + 2\delta\omega_n s + \omega_n^2)} \tag{D-9}$$

whose inverse Laplace transform and time response is

$$h(t) = 1 - \frac{e^{-\delta\omega_n t}}{\sqrt{1 - \delta^2}} \sin[\omega_n t \sqrt{1 - \delta^2} + \theta] \tag{D-10}$$

where $\theta = \tan^{-1}\sqrt{(1/\delta^2) - 1}$.

FIGURE D-3 Phase-lag network.

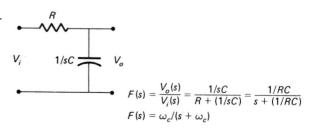

$$F(s) = \frac{V_o(s)}{V_i(s)} = \frac{1/sC}{R + (1/sC)} = \frac{1/RC}{s + (1/RC)}$$

$$F(s) = \omega_c/(s + \omega_c)$$

The comparisons yield the following relationships for *lag compensation*:

$$\omega_n^2 = k_v \omega_c \qquad (10\text{–}21)$$

and

$$\delta = \frac{\omega_c}{2\omega_n} = \frac{\omega_c}{2\sqrt{k_v\omega_c}} = \frac{1}{2}\sqrt{\frac{\omega_c}{k_v}} \qquad (10\text{–}20)$$

Equation D–8 may now be rewritten as

$$V_o(s) = \frac{(\Delta f/k_o)\omega_n^2}{s(s^2 + 2\delta\omega_n s + \omega_n^2)} \qquad (\text{D–11})$$

so the time response with compensation is

$$v_o(t) = \frac{\Delta f}{k_o}\left\{1 - \frac{e^{-\delta\omega_n t}}{\sqrt{1 - \delta^2}}\sin[\omega_n t\sqrt{1 - \delta^2} + \theta]\right\} \qquad (\text{D–12})$$

where

$$\frac{\Delta f}{k_o} = V_o(\text{SS}) \qquad (\text{D–13})$$

For $0 < \delta < 1$, the PLL with lag compensation has a response to a frequency-step input much the same as an *underdamped* second-order low-pass* filter. The response is plotted in Figure 10–27. For lead-lag compensation, it is easy to show by simple algebra, for the network of Figure 10–33, that

$$F(s) = \frac{R_2}{R_1 + R_2}\frac{s + \omega_2}{s + \omega_1} \qquad (\text{D–14})$$

The transfer function, $F(s) = V_o/V_i(s)$, is found by a simple voltage-divider derivation where the complex reactance of the capacitor C is $1/sC$. As an example, the bandpass filter (BPF) of Figure D–4 can be shown to have a voltage-divider

---

* Here is a listing for identifying second-order filter types from their Laplace frequency-response expression:

General 2nd-order circuits

$$\frac{As^2 + Bs + C}{s^2 + \alpha s + \beta}$$

The numerator varies according to filter type as follows:

Low-pass: A & $B = 0$ LPF: $\dfrac{C}{s^2 + \alpha s + \beta}$

Bandpass: A & $C = 0$ BPF: $\dfrac{Bs}{s^2 + \alpha s + \beta}$

High-pass: B & $C = 0$ HPF: $\dfrac{As^2}{s^2 + \alpha s + \beta}$

In some cases, the constant term C remains in the numerator for the high-pass circuit.

FIGURE D–4

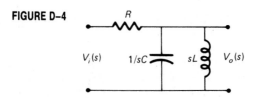

transfer function expressed as

$$\frac{V_o(s)}{V_i(s)} = \frac{sL}{s^2RLC + sL + R} \qquad \textbf{(D–15)}$$

The only math required is algebra.

E

Communications Software

Perhaps the most notable change in technology development in the late 1980s was the emphasis on computer-aided design/analysis and simulation. The use of the computer as an engineering tool has proliferated rapidly in the workplace, driven by the high cost of equipment and laboratory development time. Developing familiarity with the tools of the trade is an absolute must for all students of engineering and technology.

Computer-aided techniques for analysis and design fall into two categories: *formula-based* approaches, in which the computer is programmed with complex formulas to be evaluated; and *simulation-based* approaches, in which the computer is programmed to simulate actual voltage and current signal waveforms that propagate through the circuit or system. The formula-based approach is characterized by programs that solve for one variable from a multivariable equation. The simulation-based approach is characterized by integration subroutines and transform-domain techniques, and the output is usually either a time-domain waveform or a frequency-domain spectrum plot. Both approaches are found in this text.

A brief description will be presented of four of the most widely known software packages for communication circuit simulation and analysis: SPICE and PSpice, Micro-CAP II, Touchstone, and, for general formula solutions (and technical word processing), MathCAD. Three other packages are presented as well: Compact, Eagleware, and Hewlett Packard. System simulators tend to be very specialized, with digital signal processing software packages such as DSPlay, and even more specialized communications system (link) analysis/design packages such as TESS, or even SYSTID, TOPSIM, CMSP, CHAMP, LINK, ICSSM, ICS, BOSS, and MODEM which are typically not even available in the public domain. One of the first circuit simulation software packages accessible to students was PSpice. PSpice is the personal computer (PC) version of the Simulation Program, Integrated Circuit Emphasis (SPICE) package, originally developed at the University of California at Berkeley.

PSpice will calculate the magnitude and phase of voltages and currents in the time and frequency domains for analog circuits consisting of passive and active devices. The circuit to be analyzed is drawn with numbered nodes so that a structured file can be written to define each node-to-node circuit component type and value. Then the signal source transient and/or sinusoidal waveform is written into the input section of the file and the output section defines the type of analysis to be performed as transient, swept sinusoidal frequency, or noise response; dc bias levels at all nodes are automatically included. Other capabilities include Fourier analysis, component tolerance effects (Monte Carlo statistical yield prediction, temperature and dc variation effects), high-resolution node waveform probing, and even switches and digital logic simulations. Typical analyses and their I/O files for common communication RF circuits are provided throughout this book. PSpice now comes with a digital simulation extension allowing analog/digital simulation. Also, the 1988 version includes an interactive shell and internal word processor.

MicroSim Corporation has added a package called **Genesis.** Genesis includes a schematic capture program called **Schematics,** which provides a unified system for creating, editing, simulating, and viewing schematics, with a direct interface to PSpice and Probe. A fully integrated symbol editor allows the user to create and edit parts symbols for use in the schematic editor. Version 5.0 of the Circuit Analysis package is required, as is Windows 3.0.

Schematics provides pull-down menus and dialog boxes for specifying analysis parameters and running simulations directly from the circuit schematic. There is no need to exit the system and invoke another software package to perform a circuit analysis. If device-simulation parameters need adjustment after running a simulation, they can be modified and the simulation can be rerun. Netlists for PSpice are generated automatically and can be examined on the screen. Schematic pins and net names are used rather than arbitrary node numbers. The electrical rule checker inspects the connections on the schematic before the simulation is run. An important convenience is that Probe may be activated without leaving the Schematics environment. The Microsoft Windows environment allows for the simultaneous display of dc, ac, and transient analysis using three plot windows on the computer screen. The schematic or graphical waveforms can be saved to the Windows Clipboard for use in a document.

Micro-CAP II has similarities to PSpice as a circuit simulator. The most notable difference to the user is that, for Micro-CAP, the circuit is formed on the screen by user entry of three-letter identifiers for library-defined generators and circuit components. The software then draws the device on the screen complete with its circuit parameter value. A standardized spacing between nodes is built into the cursor arrow keys, and straight-line segments are easily entered for connecting the ground ends of shunt components. The netlist (circuit file) is automatically created from the circuit diagram screen, called the *Designer*, and up to four screens (pages) can be connected for large circuits. Micro-CAP II also has a digital circuit analysis program called Micro-Logic II.

Spectrum Software's **Micro-CAP III** and **IV** are larger and faster on 386/486 computers. Simulating circuits as large as 10,000 nodes, the new Micro-CAP products feature an interactive graphical interface with windows, pull-down menus, and on-line Help support. High-speed matrix solvers, Monte Carlo statistical yield analysis and a parameter-stepping capability allow for quick determination of any component parameter change. Also, results are plotted during the analysis to provide immediate graphical feedback, including direct schematic waveform probing.

Touchstone is primarily a linear microwave circuit analysis tool. It is based on a nodal input file similar to the Berkeley SPICE format but requires input circuit, output frequency, plotting grid, and other blocks in the input file.

Touchstone runs on PCs and/or mainframes and includes linear analysis, circuit tuning, optimization, Monte Carlo parameter-variation yield prediction of RF and microwave circuits, and even waveguides. As a microwave tool, Touchstone is an S-parameter analyzer and provides Smith chart plots in addition to multivariable rectangular plots. A listing of all the device and microwave/waveguide circuits that Touchstone is capable of analyzing, tuning, and optimizing requires a broader background in microwave circuits than has been presented in this book.

Actually, Touchstone is only one of a growing line of file-compatible CAD/ CAE tools produced by EEsof. Another program in the Touchstone compatible family is Microwave SPICE. If an excessively large signal such as 1-V peak-to-peak is input to a common-emitter amplifier circuit in a linear analysis tool such as PSpice, Micro-CAP, or Touchstone, the output signal for a frequency sweep analysis can be 100 or more volts—even assuming a 6-V power supply voltage! The reason is that these programs initially analyze the transistor dc bias values (bias point) and then the ac analysis assumes a linear transistor model. For analysis beyond the linear limits of a microwave circuit, the nonlinear analysis capabilities of Microwave SPICE or Libra should be used.

Microwave SPICE analyzes both lumped element and microwave circuits providing for ac, dc, and transient power analysis as well as distortion, noise, and ac sensitivity, in both the time and frequency domains. Display results for up to four circuit nodes on a single plot for power compression in multistage amplifiers and oscillators, plot conversion loss versus power level in mixers, and measure output power versus input power and output power versus frequency. The results of nonlinear analyses on Microwave SPICE can be loaded into Touchstone for combined linear and nonlinear analyses.

EEsof's *Libra* combines in one package the nonlinear device models from Microwave SPICE with the complete linear microwave element model library of Touchstone. It offers a fast, efficient method for designing networks where nonlinear devices are interspersed among large numbers of linear elements. Libra performs harmonic-balance simulation by combining frequency-domain and nonlinear time-domain analysis.

Other members of the Touchstone family of integrated software are *MiCAD* and *MICmask*. MiCAD is used for producing, editing, and optimizing production drawings and physical circuit layouts of the microstrip and stripline circuits developed using the other Touchstone programs. It includes an interactive graphics editor for modifying the physical layouts and an interactive optimizer used in conjunction with Touchstone for changing the circuit dimensions to compensate for changes resulting from layout constraints. The artwork and other documentation are created, including layered displays of the physical circuits. Up to 300 individual layers are allowed in a hierarchical database with up to 8 colors per drawing.

When satisfied that the layout is ready, MICmask postprocessing programs interface with various desktop, photo, and drafting plotters to produce the paper drawing. Then the plotter pin can be replaced with MICknife and the plotter paper replaced with rubylith so that the camera-ready rubylith can be cut right at the workstation.

Other members of the Touchstone software family are *LineCalc* for electrical/ physical transmission-line calculations from VHF through millimeter waves, *E-Syn* for performing lumped element synthesis of interstage and filter network impedance and noise matching and translating the results into a Touchstone circuit file, *Touchstone Sr.* for modeling custom or company proprietary elements for adding to the Touchstone library, *Microwave filter design* programs for wideband and narrowband filter circuits up to *Ku*-band, and *Touchstone/RF* for radio frequency applications. Also, Touchstone is capable of importing *S*-parameter and other circuit data from network analyzers for comparison with manufacturer's data or for other analyses.

EEsof has released a design software package called *jOMEGA* specifically for frequencies below 3 GHz. It features schematic entry, linear and nonlinear circuit simulation with bias-dependant small- and large-signal BJT models, as well as nonlinear diode and JFET model libraries containing voltage and current sources plus correlated and uncorrelated noise sources. Lumped-element device models include resistors, capacitors, inductors, coupled coils, and transformers. Distributed-element models, such as transmission lines (coax, microstrip, and stripline), coupled-line elements, dispersive microstrip, and discontinuities such as junctions, curves, bends, and open and short circuits, are also included.

Based on harmonic-balance analysis, *jOMEGA* combines the flexibility of time-domain simulation with the speed of frequency-domain simulation for fast nonlinear analyses. This feature allows the user to analyze and optimize such circuit functions as mixer conversion loss, power amplifier efficiency and I/O, and oscillator-output harmonic content in a mouse-driven graphical environment. Also, a major feature of *jOMEGA* involves circuit realization and manufacturability. A layout option produces a circuit-board layout and allows the designer to switch back and forth between schematic entry and a physical layout. The user then can key the software to simulate the circuit response of the circuit board to show layout parasitics and space-limitation problems. Working back and forth, the designer can optimize the circuit, trading off schematic values and layout changes. Finally, circuit yields for variations in part tolerances can be analyzed using a statistical simulation algorithm called Shadow model that is faster than the classical Monte Carlo analysis.

OmniSys is EEsof's system simulator. It addresses three important simulation domains: frequency, time, and modulation. The modulation analysis allows carrier signals with complex envelopes, including analog and digital AM, FM, pulse, and others in linear and nonlinear systems. Digital modulations encompass MSK and QPSK, including differential QPSK with analyses that have phase noise and bit-error rate in real-world environments.

Compact Software of Paterson, New Jersey, offers a wide range of powerful linear and nonlinear modules that can be tied together in their Microwave Design Workstation. Some of the capability and features are as follows: *Schematic Capture* consists of a library of symbols with menus for schematic drawing of circuits. All information for analysis and optimization can be placed on the schematic, completely eliminating the need for entering or editing a circuit file by hand.

The **Super-Compact** and **LINMIC+** modules provide linear and electromagnetic field simulation. Features include yield optimization and unrestricted nodal analysis as well as an extensive library of the latest MMIC and finline transmission-line elements. Powerful new algorithms make analysis and optimization fast and accurate.

Nonlinear analysis and optimization is provided by **Microwave Harmonica** and **Microwave Harmonica PC,** which give voltage, current, and power waveforms in both the time and frequency domains. Power, spectral purity, gain, conversion gain, return loss, DC or RF efficiencies, and distortion can be optimized by varying the nonlinear and linear circuit elements and drive and bias levels.

A *layout module* provides automatic layout of multilayer, cell-based and fully hierarchical designs. The library of devices is accessed by the layout module, which places and connects them according to their location in the schematic. Complete text- and polygon-editing capabilities are also supplied. An optional design-rule checker can be added to complete the design process.

The user can select from a wide variety of modules for the *synthesis* of lumped or distributed filter circuits, phase-locked-loop circuits, complex matching networks, oscillators, and transmission lines. Among these modules are **Filter Design Kit, Microwave Filter Design Kit, PLL Design Kit, RF Design Kit, Complex Match, Sonata,** and the **Communications Design Kit.**

Eagleware of Stone Mountain, Georgia, has a circuit simulator in a windows environment. Their **SuperStar** version 4.0 offers simulation and synthesis of narrow- or broadband amplifiers with arbitrary terminations, lumped and distributed elements. Narrow-band, controlled-delay, symmetric and zigzag LC filters with all-pole and elliptic responses are realizable, as are group delay equalization of measured or simulated filters using one to ten second-order all-pass sections. LC, SAW, crystal, and transmission-line oscillators can be designed with low SSB phase noise and low residual FM and PM. The high-speed SuperStar provides for physical and electrical simulation of microstrip, stripline, coax, and coplanar waveguides.

Hewlett Packard's Microwave Design System 85150B provides modeling, analysis, and layout for RF and microwave circuit development. The 85150B integrates a design-capture system, microwave linear and nonlinear simulators, and a microwave artwork generator with a documentation facility. Circuit elements are placed on a gridded circuit page using a mouse, and transmission lines or wires are added to make connections. The circuit symbols appear with any required element parameters to be chosen. Besides the linear simulations, including all sorts of transmission lines and waveguides, the nonlinear simulator allows for analysis and optimization of mixers, oscillators, and amplifiers in limiting. Intermodulation distortion can be modeled with up to six independent inputs, including AM and FM sources. Monte Carlo and sensitivity analysis calculates statistical data for analyzing and optimizing circuit yields for production.

MathCAD is not a circuit simulator. Its application capabilities greatly exceed those of programs dedicated entirely to circuit simulation. With MathCAD, circuit and system equations can be solved and plotted and, like text, can be placed anywhere on a given page. In addition to solving the sums and products of functions, fast Fourier transforms, complex number functions and contour integrals, vectors and matrices, and trigonometric, hyperbolic, log, exponential, and Bessel functions, MathCAD can also solve the integrodifferential equations of electrical circuits.

Statistical problems such as mean, standard deviation, variance, correlation and other linear regression functions, and cumulative normal distribution (Gaussian) and gamma functions are handled by MathCAD. Also, random numbers can be generated and other problems solved such as conditionals and solving for roots or zeros of a function; and these are just some of the built-in functions!

Circuits and Signals Software: Getting Started

A very brief description has been given of the commercially available software demonstrated in various chapters throughout the text. Since the most difficult aspect of using this software is the initial apprehension of getting started, some of the start-up steps will be presented for each package. But first, a few general, initializing comments.

With any of the software packages the very first step is to power up the computer with its operating system in the hard-disk drive or a floppy diskette in drive A. It is usually a good idea to get into the habit of entering the date and time at the operating system prompt. While seemingly unimportant at start-up, this will maintain a chronology in the files created and help you to sort and to appreciate the evolution of your skill and sophistication with the software.

Computer commands used in the rest of this section are based on a DOS-type computer system. The program is installed or loaded into the computer from floppy diskettes, or loaded from the hard-disk drive. If installation is required, a directory listing (type "DIR") will show an "INSTALL.EXE" file (type "INSTALL"); the quotation marks are not to be included when typing the commands. When installed, or when the program only requires loading into the computer memory, a directory listing will show a "⟨program name or initials⟩.EXE" where "⟨program name or initials⟩" is "MC2" for Micro-CAP-II, "PSPICE" (or "PSPICE1") for PSpice, "MCAD" for MathCAD, or "TSTONE" (or "TSTONE1") for Touchstone. After a short loading period, the program will be up and running, awaiting specific user inputs or other commands. If this does not happen, consult the manual or, better, an experienced friend. Program specifics are now presented.

PSpice

PSpice is the personal computer version of SPICE. The user produces a computer file of any analog circuit (limited in size compared to SPICE) consisting of resistors, capacitors, inductors, mutual inductors, voltage and current sources (independent and dependent), and semiconductor devices (standard or user-created). The user begins the process by drawing and numbering the circuit nodes; see Figure 1-32b for example. Then a circuit file list is written describing each component (element) type, the node numbers it is between, and its electrical parameter value; see Figure 1-32a, for example. A node is a connection of two or more circuit element leads, and node 0 is always the circuit ground or common return line. A list of metric suffixes used in PSpice is provided in Table E-1. Note that "milli" is M and "Mega" is MEG (only uppercase letters are used). When the circuit element is a semiconductor, a ".MODEL" statement is included for specifying either a standard device such as a 2N3904 or any desired device parameters.

Time-domain and Fourier transform analyses require a ".TRAN" statement (see the bottom of Figure 2-29a) and a frequency response Bode plot of a circuit requires a ".AC" statement (see bottom of Figure 1-32a). For PSpice versions prior to 1988, the circuit file just described would have required typing by the user into a computer file using a word-processor program such as DOS's EDLIN or other standard wordprocessor program such as Wordstar, WordPerfect, and so forth. The 1988 version of PSpice includes an internal word processor for creating the .CIR file. This package is called *Interactive PSpice*.

Once the circuit file is completed and copied to the PSpice drive or diskette, then the analysis is run by typing "PSpice ⟨filename⟩" where ⟨filename⟩.CIR is the circuit file name with the "circuit" extension. PSpice will load and run the analysis. When completed, a new file "⟨filename⟩.OUT" will have been created during

TABLE E-1 Standard SPICE/PSpice Unit Notation	**PSpice Abbreviation**		
	F	femto	10^{-15}
	P	pico	10^{-12}
	N	nano	10^{-9}
	U	micro	10^{-6}
	M	milli	10^{-3}
	K	kilo	10^{+3}
	MEG	mega	10^{+6}
	G	giga	10^{+9}
	T	tera	10^{+12}

plus:
MIL 25.4 · 10^{-6} m = 1 milli-inch

the run and is available on the drive or diskette containing the PSpice program. This file is an ASCII file and therefore the analysis results may be displayed by typing "TYPE ⟨filename⟩.OUT" and entered.

Micro-CAP II

Like Spice, Micro-CAP II is an analog circuit simulation and analysis program. However, Micro-CAP II (MC2) is more interactive, starts from a circuit diagram instead of a circuit file listing, and, most important, is entirely self-contained rather than requiring a word processor or other computer file writing program as did the old PSpice. On the other hand, it is more demanding in setting up the graphics and, except for simple bode plots, will take at least as much time achieving acceptable results as SPICE, and rankles teachers by using "db" internally and printing out "Db" instead of "dB" for decibels; that is, it does not conform to IEEE standard unit notation. Also, the Fourier analysis results leave much to be desired.

Following the initial INSTALL procedure, Micro-CAP II is started by typing "MC2", or "START", or "START H", depending on your computer graphics card ("H" for Hercules).

When up and running, the first and predominant screen is the designer module, shown in Figure E–1, called the "*Designer*." Other modules, accessed from the designer, are the *Analyzer* (select Transient, AC, DC, or Fourier), *Library Files* (retrieve for editing, delete, or save a file; display file directory), and *Utilities* (change default disk drive, print netlist, print circuit drawing). The circuit components and the symbols displayed on the designer screen when they are entered ("E") are shown in Figure E–2. Table E–2 lists the designer module commands and cursor arrow movements.

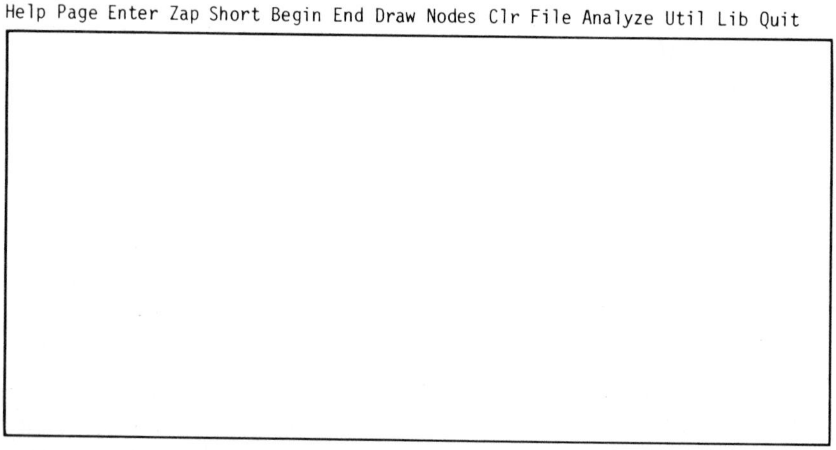

Help Page Enter Zap Short Begin End Draw Nodes Clr File Analyze Util Lib Quit

FIGURE E-1 Micro-CAP II designer screen.

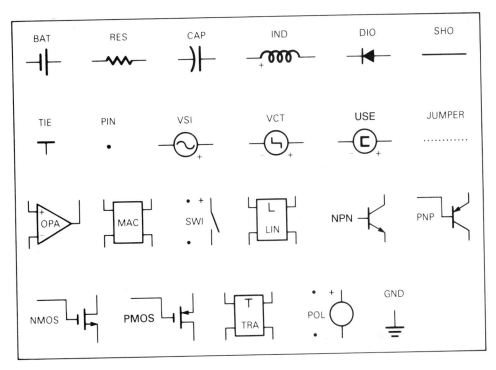

FIGURE E-2 Micro-CAP II components.

TABLE E-2 Micro-CAP II Command Menu (page 1 of 2)

Command	Function	Command	Function
Up arrow	Moves cursor up one grid in normal mode, up a ½ grid in the Jumper mode. When Scroll Lock is on, this key scrolls the window up.	Enter	Initiates component entry routine. Requires user to specify component type, direction, reflection, and parameter. Draws the selected component and adds it to the circuit netlist. Pressing Esc aborts the command.
Down arrow	Moves cursor down one grid in normal mode, down a ½ grid in the Jumper mode. When Scroll Lock is on, this key scrolls the window down.	Zap	Initiates component removal/ change routine. The user places the cursor on one of the nodes of the component and presses "Z". The system lists each component and asks the user to press either "Z" to zap or remove it or "C" to change its parameter. If "Z" is pressed the component is removed. If "C" is pressed the system prompts for a new parameter, the user types the new parameter, and the system redraws the component, showing the new parameter.
Left arrow	Moves cursor left one grid in normal mode, left a ½ grid in the Jumper mode. When Scroll Lock is on, this key scrolls the window left.		
Right arrow	Moves cursor right one grid in normal mode, right a ½ grid in the Jumper mode. When Scroll Lock is on, this key scrolls the window right.		

TABLE E-2 Micro-CAP II Command Menu (page 2 of 2)

Command	Function	Command	Function
Short	Requires the user to specify the direction of the short and then draws a shorting bar between the node the cursor was initially on and the node immediately adjacent and in the direction specified. Positions the cursor at the output node (other end of the short).	File	Allows the user to save, retrieve, or delete a circuit file from the diskette. Allows display of circuit file directory. Esc aborts a save or retrieve operation.
Begin	Puts the system in the Jumper mode. A Jumper is an extended short between two nodes. While in the Jumper mode the system recognizes only cursor commands and the E: End command. This command is used to connect together two widely separated nodes. The jumper will connect the node the cursor was initially on and the node at the cursor when the END command is given. During the Jumper mode the cursor plots a trail and moves in ½-grid increments, allowing the user to draw between components and around nodes.	Analyze	Allows selection of one of four analysis types (transient, AC, DC, or Fourier) on the circuit currently in memory.
		Utility	Displays the Utility menu. Allows printing a netlist, changing the default data diskette drive, color selection for EGA-equipped systems, graphics dump of the drawing, and display of DC voltages from a transient analysis operating point calculation.
		Library	Displays the Standard Components Library. Allows user to save, retrieve, review, edit, or obtain a printout of the libraries.
		Help	Displays a short-form explanation of each command.
End	Terminates the Jumper mode. Creates a short between the initial node and the node the cursor is currently at. Cursor must be on a node (not between) to terminate.	Page	Allows user to switch to one of the four pages of the circuit diagram. The four pages are referred to as 0, 1, 2, and 3. The default page is 0.
Draw	Draws the circuit currently in memory.	Quit	Terminates the Micro-CAP II program. Returns control to the Disk Operating System.
Nodes	Assigns node numbers to the circuit and draws them on the diagram.	Title	Allows user to enter a title. The title then is displayed in lieu of the Command menu until the "T" key is again pressed.
Clear	Clears or erases the circuit currently in memory. Does not erase any circuit saved on the diskette.	Width	Causes the system to draw all graphic figures with a double thickness.

The circuit diagram is started by moving the cursor a little to the left of the blank designer screen using the computer keyboard movement keys and then typing "E". "E" is the command for *entering* a circuit component. When this is entered, the user must respond to four more queries as follows: "Enter type of component" (RES, IND, CAP, etc.; for resistor, inductor, capacitor, and so on); "Direction (R, L, U, D)" (draw the component to the right, left, up, or down); "Reflection (X, Y, none)" (the component or source can be mirror reflected in the *x*- or *y*-axis direction, or None for no reflection); "Parameter" (component value or source waveform shape). The cursor will automatically move to the node at the opposite end of the component entered. The circuit is not complete until return leads are *shorted* together (type "S" followed by "L", left, or "R" for a short to the right) and a circuit ground (type "E", "GND", "D" for the usual downwardly drawn ground symbol).

After the circuit has been created on the designer screen, check the node numbering by typing "N" for node. Sometimes when the circuit is being altered, node numbers get switched. Use "Z" to remove ("zap") a circuit component at a node: Hit the return [Enter] key if the wrong component at a node flashes. The next step before defining and running an analysis is to specify the parameters for sources and special devices such as transformers. For example, a transformer requires specification of the primary winding self-inductance, turns ratio, coefficient of coupling, and the secondary-winding resistance (do not use zero!). Entering these specifications requires editing the Component Library. Go to the Component Library by typing L and selecting the component type from a displayed list. When you have selected "transformer," for example, a number of transformer modes are available by name "alias" or number. If none of these meets your specs, change one of the models and any of the specs individually or one at a time from top to bottom by pressing [Enter] and then "E", for edit, before typing the new parameter value.

Now go to the analysis module by typing "A". Once in the Analyzer module, you choose from a menu of TRANSIENT, AC, DC, or Fourier. Except for the Fourier analyzer, a choice of an analysis type brings to the screen a Prologue for selecting the analysis limits, accuracy, plot grid values, circuit component initial values, DC operating point (active devices), temperature (27°C), number of runs, output device for results, whether or not to save every value computed during the run, and Waveform Dump, which can be used for saving the signal voltage values at a particular node for a later Fourier analysis, or for creating a unique waveform to be used with a special waveform generator routine called *Wavegen*.

Each prologue comes complete with values entered from a previous run; think of these as default values—see Table E-3. You can change one or all of the default values by responding to "Are these correct (. . .)?" by typing "N". The cursor moves to the value at the top of the list and, if you don't want to change it, press [Enter, or Return] and the cursor moves to the next listed value, and so forth. CAUTION: Do not use zero for any of these values! An "N" response to *Default plotting parameters* will bring up the plotting parameter menu such as that of Table E-4 for the AC (frequency domain) analysis in which "tokens" are

TABLE E–3a AC Analysis
Prologue

Lowest frequency	1E6
Highest frequency	1E8
Lowest gain (db)	−80
Highest gain (db)	20
Lowest phase shift	−360
Highest phase shift	90
Lowest group delay	1E-9
Highest group delay	1E-4
Input node number	1
Output node number	2
Minimum accuracy (%)	5
Auto or Fixed frequency step (A,F)	A
Temperature (Low/High/Step)	27
Number of cases	1
Output: Disk, Printer, None (D,P,N)	N
Save, Retrieve, or Normal run (S,R,N)	N
Default plotting parameters (Y,N)	Y
Are these correct (Y,N,ESC=Abort)?	

TABLE E–3b Transient Analysis

Simulation time (Tmax/Tmin/Timestep)	1E-6
Display window (Dmax/Dmin/Printstep)	1E-6
Minimum accuracy (%)	5
Upper trace A waveform spec	1
Upper trace B waveform spec	1
Upper trace range (High,Low)	5
Lower trace A waveform spec	2
Lower trace B waveform spec	2
Lower trace range (High/Low)	6/-1
Zero, Read, Leave initial values (Z,R,L)	L
Edit/review initial values (Y,N)	N
Dump node waveform to user file (Y,N)	N
Calculate DC operating point (Y,N)	N
Temperature (Low/High/Step)	27
Number of cases	1
Output: Disk, Printer, None (D,P,N)	N
Save, Retrieve, or Normal run (S,R,N)	N
Default plotting parameters	Y
Are these correct Y,N,I=Iterate flag, ESC=Abort)?	Y

TABLE E–3c DC Analysis Prologue

As in the other analysis modes, the user selects
the analysis limits during the Prologue menu.

Input voltage range (High/Low)	10	Relaxation factor	2
Output voltage range (High/Low)	10	Temperature (Low/High/Step)	27
Maximum input voltage step	.5	Number of cases	1
Input node number	1	Output: Disk, Printer, None (D,P,N)	N
Output node number	2	Save, Retrieve, or Normal run (S,R,N)	N
Number of iterations	50	Default plotting parameters (Y,N)	Y
Error criteria	.001	Are these correct (Y,N,ESC=Abort)?	

rectangles, squares, or circles delineating each of the three possible curves plotted on the same graph, and "double width" doubles the thickness of all the graphic figures and lines. Finally, since there is a Prologue there must be an Epilog; this is shown in Table E–5 as copied directly from "Micro-CAP II Electronic Circuit Analysis Program, Rev. B," published by Spectrum Software, 1021 S. Wolfe Rd., Sunnyvale, CA 94086, p. 6–12; it comes with the Student Version program.

TABLE E–4 Plotting Parameters

Default plotting parameters	
Plot gain (Y,N)	**Y**
Plot phase (Y,N)	**Y**
Plot group delay (Y,N)	**Y**
Plot tokens (Y,N)	**Y**
Plot Y-axis minor grids (Y,N)	**Y**
Plot frequency minor grids (Y,N)	**Y**
Print numbers on screen (Y,N)	**Y**
Double width plot (Y,N)	**N**

There are 6 options:

1. Whether or not to plot the gain.

2. Whether or not to plot the phase shift.

3. Whether or not to plot the group delay.

4. Whether or not to include plot tokens (an open rectangular box for phase shift and a closed rectangular box for group delay). Plot tokens help to distinguish the 3 potential plots.

5. Whether or not to include minor grids on the Y-axis.

6. Whether or not to include minor grids on the frequency axis.

TABLE E–5   Epilog Options

Following completion of the simulation (or pressing the "Q" key or ESC key), the system displays the Epilog Menu:

1: Another run.

2: Analysis limits.

3: Quit.

4: Dump.

Option 1 returns to the prologue for another analysis.

Option 2 prints the current set of analysis limits out on your printer.

Option 3 returns to the **Designer** module.

Option 4 lets you dump the screen to an EPSON compatible printer. It also lets you type in a title for the graph. Pressing any key restores the menu.

Touchstone

As discussed previously in the general remarks section, Touchstone is a linear microwave analysis tool based on a nodal input file format similar to SPICE. An important difference, however, is the requirement for specifically segmenting the circuit file into sections, called *blocks*, labeled "CKT" (circuit nodal description), "OUT" (results to be computed), "FREQ" (sweep frequency range and step size), and "GRID" (horizontal and vertical axes dimensioning). Except for GRID, which is rarely omitted, these are the minimum required input file blocks. Other blocks include "DIM" for specifying parameter dimensions other than the default values [GHz (frequency), OH (ohms), NH (nanohenrys), PF (picofarads), MIL (milli-inch lengths), PS (picoseconds), /OH (inverse of ohms for conductances), and DEG (electrical lengths or angles in degrees)], "OPT" for circuit element design value or error function optimization, "EQN" for as many as 25 user-defined equations with as many as 50 variables that may be combined to model nonlinear or unique circuit functions, "TOL" for Monte Carlo analysis tolerances, and other less often used control block options such as variables (VAR), processor (PROC), and termination (TERM).

When Touchstone is loaded by entering "TSTONE" or "TSTONE1" at the DOS prompt, the Touchstone screen and main (command) menu comes up. The Touchstone screen is primarily empty (providing the editor/graphics window) with a *status* panel displayed across the upper edge of the screen and a "ready for command" prompt at the left. The command menu is a two-row list of commands with corresponding keyboard *Function Key* numbers across the lower edge of the screen (Figure E–3). The commands in the upper of these two rows are executed by holding down the *shift key* while depressing the desired function key. A list of the commands and their descriptions are found in the Touchstone reference manual. The good news is that the reference manual that accompanies the program

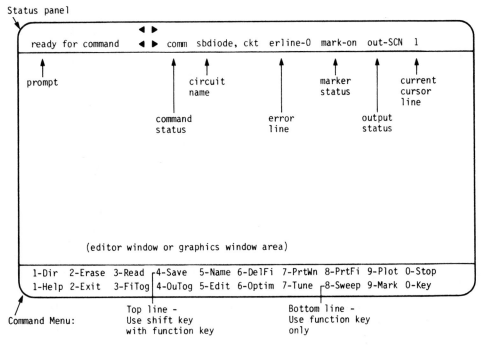

FIGURE E–3 The Touchstone screen.

diskettes is the best and most complete of all the software on the market. The bad news is that the Touchstone package is very expensive; it is intended for corporate purchase, although it may be possible for a university program to get a discount.

To start a new circuit file, press the [F5] function key for "Edit". Then respond to "Name of File" by entering a name for the circuit file to be created. "***Bottom of File***" will appear at the top of the editor window; begin typing your circuit file from the top line down. When the circuit file is completed (see example file, Figure 14–27), press "Exit" [F2] to return to the *command* status. The program is run by pressing the "sweep" command key [F8]. This initiates simulation and loads the output data array.

The File Toggle key, FiTog [F3], is used for changing the screen display between the circuit file and the analysis output results. The Output Toggle key, OuTog [F4], selects an output result for display. The status panel on the command line indicates which output result is being displayed and, normally, "out=SCN" is active indicating the display of tabular results. When the analysis "sweep" is complete, the File Toggle key may be used to toggle back and forth between the circuit file and the analysis output results. Use the Output Toggle key to toggle between results requested in the OUTPUT block of the circuit file: SCN will display tabular listings, GR1 through GR3 will display rectangular-grid plots, and SC1 through SC3 will display Smith chart plots. After selecting the requested grid (GR1–GR3) or (SC1–SC3) and noticing the "out=GR1", for example, in the status line, then press the sweep key [F8] to display the requested parameter versus frequency. To make a hard copy, press the computer [Shift] key along with [F9] for a printout of the GR1 plot. The [F4] File Toggle key is then used for repeating this procedure for other requested analysis results.

Pressing the Save key, [Shift][F4], saves the current circuit or output file, whichever is the active file. The circuit file will be saved under the name shown in the status panel when the Save key is pressed. The output file will be saved under the same name except that the extension will be ".OUT" instead of the ".CIR" for the circuit file. Pressing the Stop key, [Shift][F10], will exit Touchstone and return the computer to the system prompt.

MathCAD

Type "MCAD" to load MathCAD. Or, type "MCAD ⟨filename⟩" and MathCAD will load and the already created file named "filename" will be automatically loaded into MathCAD and appear on the screen. To create a new file once MathCAD has been loaded, begin by typing quotation marks, ", and start typing text; at least type a title. When this is completed, use an arrow key [↓] on the keyboard to move the cursor out of the text region—you will notice that the quotation marks disappear. Provide at least two blank lines before beginning an equation because, as you will observe from the MathCAD-produced figures of Chapter 9, divisions, integrals, and other functions can consume up to four vertical lines.

As an example, the equations of Figure 9–47a were created with the following keystrokes:

t : 0,1/144;1 followed by four right-arrow strokes (two will get the cursor out of the equation region) and then type ko : 4 $\rightarrow$$\rightarrow$$\rightarrow$$\rightarrow$$\rightarrow$ Vpk : 3 $\rightarrow$$\rightarrow$$\rightarrow$$\rightarrow$$\rightarrow$ fm : 2 $\rightarrow$$\rightarrow$$\rightarrow$$\rightarrow$$\rightarrow$ fc : 20

Now, use the Return key [Return, or Enter] twice to get to the left of the screen. (Any notation inside [ ] is either one of the computer control keys such as [Ctrl] or [Alt], or English-language directions to save printing space in this book.)

Continue with

A : 2 $\rightarrow$ [press the right arrow ten times] Vm(t) : Vpk*cos(2*[Alt]p*fm*t) $\rightarrow$ [five times] "Mod. index is:" [Return] ↓ Mf : ko*Vpk/fm

Finally,

s(t) : A*cos(2*[Alt]p*fc*t + 2*[Alt]p*ko*&Vm(t)) $\rightarrow$$\rightarrow$$\rightarrow$ t $\leftarrow$ [nine times] ↑t↓↓0↑ $\rightarrow$ [nine times] $\rightarrow$$\rightarrow$ [Return]

The last two arrows take the cursor out of the equation region and the key-strokes following the ampersand (for integral) and $Vm(t)$ fill in the integral limits.

To produce the plot, move the cursor down a couple of lines using the arrow key, type "Amplitude (volts)", arrow down one line, and type @ and fill in the placeholders starting with the t below the lower horizontal line. Move to the other placeholders using the arrow keys. Move to where the backward-L cursor is blinking to the right and bottom of the lighted placeholder square. When all are filled in, press the F9 function key and wait until all the calculations are completed. A "WAIT" signal flashes on the right side of the command line.

To change the plot size, move the cursor inside of the plot outline and type f for *format*, and then edit the command line just like the text. To get the size shown in Figure 9–47, use logs = 0,0 (to have both *y*- and *x*-axes linear), subdivs = 4,0 (for 4 evenly spaced vertical (*y*-axis) subdivisions—4 grid lines, and no horizontal subdivisions), size 15,55 (for 15 lines by 55 columns), and type = 1 (the default lowercase L that means that calculated points are connected by straight lines. About a dozen options for trace markers are available). Now press the [Return] key. Changes may be made in the plot limits by simply editing any of the values shown. After the change, move inside of the plot outline and press [F9] for a replot. Finally, fill in the time-axis label below the line holding the variable *t* using the text entry "Time (seconds)" including the quotes, and moving the cursor away using the arrow key.

Print the results for a hard copy by pressing the escape [Esc] key and typing the command "selectprinter", then press [Esc] again and type "print" followed by two returns. Finally, save the file by using [Esc] and printing "save" followed by responding to "save as:" by typing a filename and [Return]. Exit MathCAD using the [Esc] key and typing "quit."

Answers to Selected Problems

Chapter 1

1. **(a)** See Figure ANS 1–1a.
   (b) 9.2 MHz
3. $11.3 \ \Omega\underline{/63.7°}$
5. **(a)** 500 Ω, 9.4, 1.06 MHz
   (b) 20 mA, j188.7 mA, −j188.7 mA
   (c) 9.4 = Q
6. **(a)** $70\underline{/45.8°}$ **(b)** $46.6\underline{/-62.2°}$
8. r = 500 Ω, 796 μH, 2.5 kΩ, 995 μH

9. **(a)** 15.9 MHz **(b)** 836 mV
11. **(a)** 51.3 Mrad/s **(b)** 49.3 dB
    (c) 35.3 dB **(d)** 78.5 kHz
    (e) −25.1 dB, 18 times
13. For V_E = 0.1 V_{cc} and V_{BE} = 0.7 V,
    R_E = 729 Ω (use 680), R_1 = 13 kΩ,
    R_2 = 56 kΩ, R_{dn} = 100, C_{dn} = 3800 pF,
    C_E = 3800 pF (use 4700), C_c = 2300 pF.
    X_L = 250 Ω and C(total) = 31.7 pF,
    $C_t \approx$ 25 pF and n_p/n_s = 8.2.

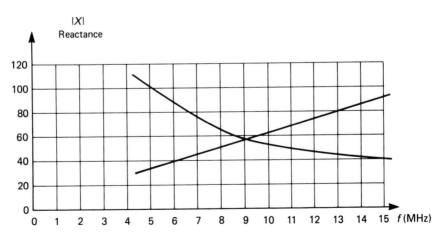

FIGURE ANS 1–1a

16. 2 mH

18. (a) 0.13   **(b)** 5.6 mA $\underline{/141°}$
  (c) j350 mA

21. (b) $10 + j10 \ \Omega$ **(c)** 1.5 μH
  (d) $11.25 - j11.25 \ \Omega$ **(e)** 1126 pF

22. (a) 0.071
  (b) Between critical and optimally coupled
  pled

25. See Figures ANS 1–25a and ANS 1–25b
  for plots.

29. (a) $Z_i(s) = \dfrac{s^2RLC_1 + sL + R}{sRC_1 + 1}$

  (b) $H(s) = \dfrac{V_o(s)}{V_i(s)} = \dfrac{R}{s^2RLC_1 + sL + R}$

30. (a) $Z_i(j\omega) = \dfrac{R - \omega^2RLC_1 + j\omega L}{1 + j\omega RC_1}$

  (b) $H(j\omega) = \dfrac{R}{R - \omega^2LRC_1 + j\omega L}$

Chapter 2

1. B with A at ac ground
2. (b) 0 dB
3. (c) 44.4 pF
4. (b) 40   **(c)** About -15.5 dB
5. 1.89 MHz
6. (a) 507 pF   **(b)** 6.39   **(c)** 0.64   **(e)** No
8. -3.1 dB

9. (a) 515.8 kHz   **(b)** 100 kΩ
  (c) 50 dB   **(f)** -12.9 kHz
11. (a) 581.15 kHz   **(b)** 581.295 kHz

Chapter 3

1. 50 Hz, 5
3. 2.5 V at $f = 0$, 3.18 Vpk at f_0, 1.06 Vpk
  at $3f_0$, 0.64 at $5f_0$, 0 at $2f_0$ and $4f_0$
6. (a) 1.91 Vdc   **(b)** 3 Vpk   **(c)** 1.27 Vpk
  (d) 7.3 mW + 9 mW + 1.6 mW =
  17.9 mW
8. (a) $v(t) \approx 1.91 + 3 \sin 2\pi 60t - 1.27 \times$
  $\cos 2\pi 120t - 0.255 \cos 2\pi 240t$, volts
  (b) $v(t) \approx 3.82 + 2.55 \cos 377t - 0.51 \times$
  $\cos 754t + 0.22 \cos 1131t$, volts
  (c) $v(t) \approx 2.43 \cos 377t + 0.27 \times$
  $\cos 1131t + 0.10 \cos 1885t$, volts
11. (a) 0.75 V, 1.35 V pk, 0.955 V pk
  (b) See Figure ANS 3–11b.
13. (a) See Figure ANS 3–13a.
  (b) Same as part a except the 3 kHz
  component is -1.05 V pk, and 5 kHz is
  zero
  (c) See Figure ANS 3–13c.
  (d) From part c: $v_o(t) = 5 + 6.4 \times$
  $\cos 2\pi(1 \text{ kHz})t - 1.05 \cos 2\pi(3 \text{ kHz})t$
  (e) 16.6%
14. (a) 40%   **(b)** 20%   **(c)** 44.7%
  (d) 3.75 W

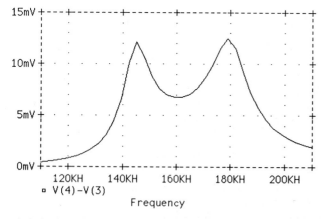

FIGURE ANS 1–25a

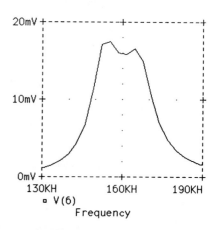

FIGURE ANS 1–25b

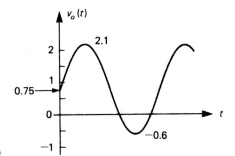

FIGURE ANS 3–11b

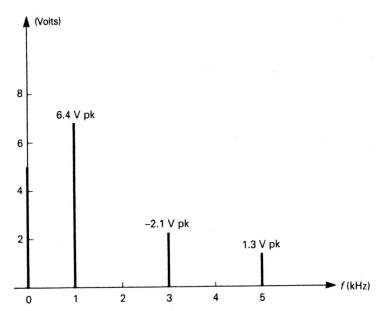

FIGURE ANS 3–13a

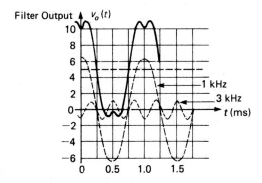

FIGURE ANS 3–13c

16. The complex Fourier series is

$$v(t) = \sum_{n=-\infty}^{\infty} j\frac{A}{2} e^{-j\pi/2}\left(\frac{\sin n\pi/2}{n\pi/2}\right) e^{j2\pi nf_o t}$$

Hint: (1) Use property 1 of Table 3–3 with the positive and the negative pulses, (2) $1 = e^{j\pi/2}e^{-j\pi/2}$, (3) $e^{-j\pi} = e^{-j\pi/2}e^{-j\pi/2}$, and (4) Euler's sine-complex exponential equivalent.

Chapter 4

1. 4×10^{-21} W; Thermal versus shot
2. (a) 4 nV **(b)** 645 pV **(c)** 465 pV
   (d) 116 pV
4. 125 Hz
5. (a) 1.83 mV
   (b) 11.5 µV (14.4 µV for actual circuit bandwidth)
   (c) 229.5 µV (all voltages rms)
   (d) 0.115 divisions
6. (a) 1.56×10^{-14} V²/Hz
   (b) 3.23×10^{-22} A²/Hz
   (c) 88.9 µV rms
7. (a) 5.96 dB **(b)** 855.5 K
8. 4.2 dB
9. (a) 0.4 pW **(b)** 44 dB

Chapter 5

1. (a) 0.56 **(c)** 5.4 kW **(d)** 0.42 kW
   (e) 6.2 kW **(f)** 87.1% **(g)** 9 kHz
   (h) 250 V pk, 49.5 kHz sinusoid
2. (a) 33.3% **(b)** 2.25 W **(c)** 62.5 mW
   (d) 2.63% **(f)** 57.7%
3. (a) See Figure ANS 5–3a.

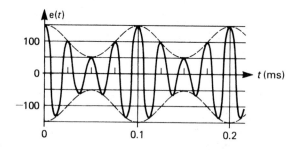

FIGURE ANS 5–3a

(c) $e(t) = 100 \cos(251.3 \times 10^3 t) + 25 \times \cos(188.5 \times 10^3 t) + 25 \cos(314.2 \times 10^3 t)$ volts **(d)** 20 kHz
   (e) 100 W, 6.25 W, 112.5 W **(f)** 94.4%
5. (c) 0.81 mW and 0.93 mW
   (e) −700 mV **(f)** 0.05 µF
6. The average value is −99.8 V.
8. (a) +3 dBm **(b)** 2 V pk **(c)** 3.6 V pk
   (d) −1.8 V dc **(e)** (T = 0.2 ms)
   (f) (dc = 0) **(g)** 0.01 µF
10. 70 dB
12. (a) −97.8 dBm **(b)** 106.7 MHz **(c)** 4
   (d) −97.8, −85.8, −91.8, −92.8, −67.8, −42.8, −17.8, +7.2 dBm or 3 dBm for one 20.8-dB amplifier
   (e) 117.4 MHz
14. −130.8 dBm
15. (a) 6 dB **(b)** +15 dB **(c)** 25 mW

Chapter 6

2. (a) 150 W **(d)** 8.8 W
4. Distortion—elaborate
7. (a) 23.5 kW **(b)** 40 kW
9. 1.008 kW
12. (a) ≈ 120° **(b)** $\theta_g \approx 70°$ **(c)** 312.5 µA
   (d) 580 V pk **(e)** 841 Ω
   (f) 58 W, 77.5%
13. (b) 111 mA **(c)** 11.35 k **(d)** −100 V
   (e) 950 V pk **(f)** 1.5 k **(g)** 2.38 µH
   (h) 5.5 **(i)** 130 W, 69.8%
16. (a) 28 V **(b)** 3.92 Ω
18. 1.33 µH and 0.0068 µF
19. 55 nH and 210 pF
20. (a) 200 − j400 Ω **(b)** 400 mH **(c)** 1.6
   (d) No, 200 Ω ≠ 50 Ω
   (e) See Figure ANS 6–20e.
21. See Figure ANS 6–20e for a time domain plot. A frequency domain plot will also yield the answer to 20(e).
22. (a) $35.355\underline{/+45°}$
   (b) 25 Ω by reducing L to one-half of its value
27. (a) 200 V pk-pk to reduce the plate voltage to zero on peaks
   (b) 6 kHz **(c)** 563 pF **(d)** 18

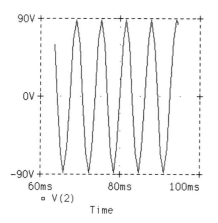

FIGURE ANS 6–20e

28. (a) 3.3 kΩ   **(b)** 6.38 kW
    (c) 2.58 kW   **(d)** 2.72 kW   **(e)** 3.13 kW

Chapter 7

3. (a) 17 dB   **(b)** 67 dB   **(c)** 3 dB
4. (b) 93 mW approx.   **(c)** 9 V pk-pk
6. 1 kΩ
7. (a) $V_g(RF) = 1.7$ V, $V_g(LO) = 6.9$ V
    (b) 1.56 μH   **(c)** 540 mV, 15 V
    (d) 28.9 mW
9. (a) 1.55 kΩ   **(b)** 979 μH   **(c)** 211 μH
    (d) 520 Ω   **(e)** 603 Ω
10. (1) $R_1 = 8.1$ kΩ, $R_2 = 42$ kΩ, $R_3 = 1.1$
    kΩ, $C_1 = 0.9$ pF, $C_2 = 146$ pF, $C_3 =$
    6500 pF, $C_t = 11$ pF
    (2) $n_s/n_p = 2$ for T1, $n_p/n_s = 24$ for T2
    (3) 139 mV (enough) is the LO signal on
    the base of Q_1
11. (a) 1/3   **(b)** 8.94   **(c)** 162 mV pk
13. Chebyshev: (a) 3 poles
    (b) 0.1 dB   **(c)** 127.4   **(d)** 11.4 dB
    Butterworth: (a) 4 poles
    (b) 0 dB by definition   **(c)** 118
    (d) 16.8 dB (Do not use resonators with
    minimum required Q when building this
    filter; too much signal loss results. If
    $Q = 1000$, the Butterworth insertion loss
    is reduced to 1.5 dB.)
14. (b) $Q_{min} = 68$

(c) For infinite Q_u, $L = 2.41$ μH, $C_1 =$
2538 pF, $C_{12} = 88.3$ pF, $C_2 = 2450$ pF
(d) 5.7 dB
15. (a) 6 poles   **(b)** 0.01 dB   **(c)** 3.9 dB
17. (a) 0.6 mW   **(b)** −56.2 dBm
    (c) 1.3 V, 1.08 V
    (d) See Figure ANS 7–17d

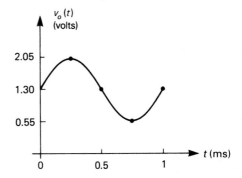

FIGURE ANS 7–17d

18. 5 μW
20. (b) (The Q-point is 14 V, 71 mA)
    (c) 198 Ω
    (d) 6.8   **(e)** 0.462 W, 43.4%
    (f) 191 mW, 17.9%
21. (a) $V_{B2} = 6.70$ V, $I_{C1} = 20$ mA, $V_{C1} =$
    $V_{B3} = 5.3$ V, $V_{E1} = 20$ mV, $V_{B1} = 0.67$
    V, $V_{E3} = 5.95$ V, $I_{C2} = I_{C3} = 16.7$ mA,
    $V_{E2} = 6.05$ V, $I_{B2} = I_{B3} = 166.7$ μA
    (b) $i_o(sat) = 438.5$ mA pk, $v_o(sat) =$
    4.385 V pk, $P_o = 961.4$ mW, $P_{AVG} =$
    1.675 W, eff = 57.4% (3Ω dissipates
    288.4 mW)
22. (b) −80, 20, −6, 1, 0.94, 79 dB
    (c) 51.7 dB   **(d)** 135 kΩ, 21 mΩ
23. (b) 800 kHz, 100 kHz, 1.3 kHz
24. (a) 39.6°, 300 kHz   **(b)** 23.6°, 400 kHz
    (c) 8°, 530 kHz, very marginal
    (d) −24.9°, unstable

Chapter 8

2. (a) $66\frac{2}{3}$%   **(b)** 75.8%
5. $v_o(t) = 2.5 \sin 144513t + 2.5 \sin 333009t$
    $+ 0.83 \sin 622035t + \cdots$
6. 125 mW

7. About 0.56 mV, rms

9. (0.64 V dc), 424.4 mV pk at 10 kHz; 84.9 mV at 20 kHz and 36.4 mV pk at 30 kHz

12. **(a)** 83.3% **(b)** 87.88%
(c) DSB-SC has only 66.7% savings.

14. 1.67

16. Hints: Output up-converter mixer inputs are 2.005 MHz and 32 MHz. Transmitted signal is 34.005 MHz or 29.995 MHz and a balanced mixer is used.

18. **(a)** (SF = 1.69)
(b) For R = 1 kΩ, L = 103 μH, C_1 = 22.705 nF, C_{12} = 933 pF, C_2 = 22.241 nF, C_{23} = 464 pF, C_3 = 22.808 nF, C_{34} = 366 pF, C_4 = 22.906 nF

(c) See Figure ANS 8–18c, part 1 for the 34-dB bandwidth (99.5 kHz–104.5 kHz) and Figure ANS 8–18c, part 2 for the 3-dB bandwidth. Observe that there is zero ripple.

20. 200 parts/billion

22. The first three spectral components (excluding dc) are 990 Hz, 2990 Hz, and 4990 Hz.

24. **(a)** 64–84 kHz **(b)** 24 kHz

Chapter 9

3. **(a)** 20 kHz pk, m_f = 4
(b) 10 kHz pk, m_f = 1 **(c)** No change

5. $2 \cos(6.28 \times 10^8 t) - \{0.4[\sin(31.4 \times$

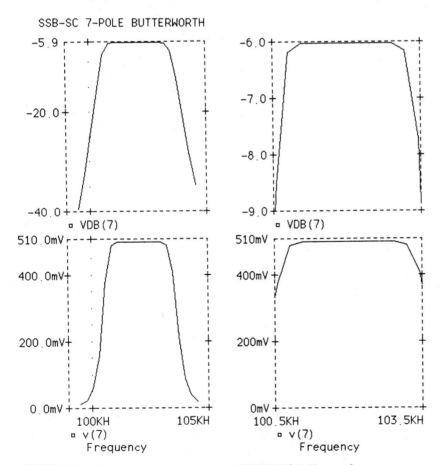

FIGURE ANS 8–18c, part 1

FIGURE ANS 8–18c, part 2

$10^3 t)] \times [\sin(6.28 \times 10^8 t)]\} = 2 \times \cos 628318531 t - 0.2 \cos 628287115 t + 0.2 \cos 628349947 t$. See Figure ANS 9–5.

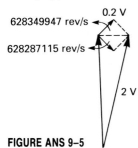

0.2 V

628349947 rev/s

628287115 rev/s

2 V

FIGURE ANS 9–5

7. (a) 1.29 kW **(b)** 1.29 mW **(c)** 24 W
9. (a) 1.4 **(b)** 0.467 V pk @ 10 kHz
   (c) 3.025 kW **(d)** 0.8 V pk
11. (b) 0.34 rads, approx. **(c)** 0.26 rad/V
13. (a) 1.524 MHz **(b)** 1.278 MHz
   (c) 1.726 MHz
14. (a) 68 kHz/V
   (b) m_f = 5.67, or 5.1 by exact slope method.
17. (a) 392 kHz
18. (a) 44 μH **(b)** −25%
   (c) 0.84 MHz/V approx. and 1 MHz/V slope
20. 10 V pk, 5 kHz
21. (a) 11 MHz, 33 MHz, 1 MHz
   (b) No, 29.7 kHz **(c)** Yes, 1856 Hz
   (d) 588 Hz
23. (a) $v_1 = v_2$ = 13.3 V rms @ −90°, V_A = 13.9 V rms, V_0 = 0
   (c) V_0 = −4.1 V dc; V_A = 11.3 V$/\underline{-68°}$, see Figure ANS 9–23c.

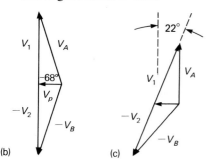

(b)

(c)

FIGURE ANS 9–23c

24. (a) −6.7 V dc **(b)** 20.5 V/MHz
26. 1.5 V
28. (a) Approx. +10 dB **(b)** 63.2 mV rms
31. (a) 12 dB **(b)** 17.4 dB

Chapter 10

3. 2.55 V/rad and 0.044 V/degree
5. (a) 60 kHz/rad **(b)** 377×10^3 s^{-1}
   (c) 111.5 dB
7. 785.4 mV
9. (a) 250 kHz and 30 kHz **(b)** 30 kHz
10. (a) 110 kHz **(b)** +4.5 V dc
   (c) −0.75 V dc **(d)** −1.18 rad
11. 80 kHz
13. (c) The loop gain would have to be increased by 7.16 times.
14. (b) 15.9 μs
15. (a) k_v(dB) = 96 dB at ω = 1 rad/s; k_v = 0 dB at ω = 62.832 k rad/s
   (b) k_v = 0 dB and is down 3 dB at 62.832 k rad/s
   (c) 10 kHz
16. (a) 0.35 **(b)** Approx 30% (Why?)
18. (a) 0.5 V dc **(b)** δ = 0.16
   (c) 0.002 μF
   (d) See Figure 10–32c, and V_0(SS) = 0.5 ± 0.6 V with a period of 2 ms (1 ms/bit).
19. R_1 = 24.5 kΩ, C = 2 μF
21. (a) R_1 = 162 kΩ, R_2 = 5 kΩ
   (b) Use voltage divider rule and remember that s already includes jω for sinusoids.
23. (a) 60 kHz **(b)** 70 kHz **(c)** 10 kHz
26. (b) N = 225 **(c)** 26.975 MHz
   (d) 455 kHz
27. (a) 50 **(b)** 200 kHz **(c)** 0.2
   (d) About 2

Chapter 11

3. (a) 6 kHz
   (b) Frequency foldover occurs (Why?)
6. Expand on Figure 11–15 and include its mirror image. **(a)** 32 k pulses/s
   (b) 15.625 μs

8. See Chapter 3
10. **(b)** Figure ANS 11–10b. Also, for **(b)** the phase-locked loop

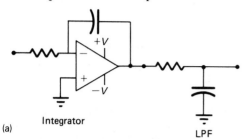

(a)   Integrator

(b)   Integrator

FIGURE ANS 11–10b

15. **(a)** PDM **(b)** 50 kHz **(c)** 2.5 V
16. **(a)** $V_0(t) = -V_{cap}$. Solve Equation 1–3 for V across capacitor where $i(t) = V/R$ $(0 \le t \le W, 0$ otherwise)

   (b) $V_{avg} \equiv (1/T) \int_0^T v(t)\, dt = \dfrac{-VW}{2RC}$

18. $V_0(avg) = (A/T)(t_{ppm} - t_{ck})$
21. 7.966 theoretically, but 8 are used
23. 4.5 kHz (Why?)
24. 6.25 μs
26. **(b)** 1110111011101 . . .
27. **(a)** See Figure ANS 11–27a.
    (b) 110111011101 . . .

28. **(a)** 15.1 MHz **(b)** $(f_c \propto 1/\text{step-size})$
    (c) (clock rate)

Chapter 12

1. At least 5.75, so 6 are used.
3. **(a)** 50 kb/s **(b)** 25000 Hz
5. 16297 b/s
6. 4.1 minimum, so use 5 levels.
7. **(a)** 8.3 μW (−20.8 dBm)
   (b) 3.3 μW (−24.8 dBm)
8. **(a)** 30.27 mV pk **(b)** 1.8 dB
9. **(a)** 4.5 MHz **(b)** 32
10. **(a)** 386 **(c)** 3.088 Mb/s **(d)** Time
    (e) 1.544 MHz
13. 30 dB
15. **(a)** The ratio of peak signal to peak noise voltage is 36.1 dB. However, the signal-to-noise power ratio is 37.9 dB.
    (b) 82 levels require 7 bits.
16. **(a)** 317.7 mV **(b)** 7 **(c)** 0.78
    (e) 1.055 MHz
19. 25 kΩ
20. 640 kΩ
21. **(b):(A)** 2.5 V **(B)** I(LSB) = 0.25 mA
22. **(a)** See Figure ANS 12–22a.
    (b) 46.9 μs
    (c) 11001100 (the LSB could also be a 1)
23. 1011_2
25. **(a)** 90 mV **(b)** 282.8 mV, ideally
26. **(a)** 16.33 Hz **(b)** 0.04 rads **(c)** 8 mV

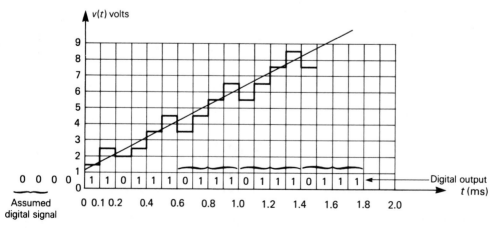

FIGURE ANS 11–27a

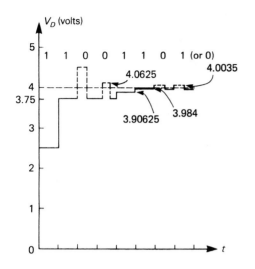

FIGURE ANS 12-22a

Chapter 13

2. (a) "HI" **(b)** Even **(c)** 303 b/s
   (d) "ring"
3. 63.6%
4. 5
7. Referring to Figure 13-8, there are only 2 sections in the rightmost register and 1 in the leftmost.
9. See Figure ANS 13-9.
10. (a) Examples: 8, Q, *, ETX. Total LRC = 0000010; total VRC is 1111101100.
   (c) u, *,), !, ETX; 01100; 0010111 LRC
11. (a) 10^{-12}
   (b) 3.5 V pk (Approx. answers)
12. 281 mV pk
13. 400

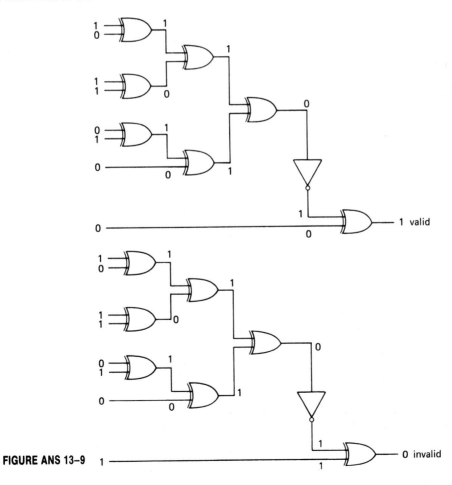

FIGURE ANS 13-9

17. (a) +2.5 V (b) +3.2 V (c) +3.7 V/0
 (d) 60 WPM
18. 9.1 kHz
22. (a) 2.2 s (c) 0.83
23. (b) 400 b/s
25. Example: SYN, SYN, STX, text, ETX,
 BCC
26. (a) Yes
 (c) "I have a message for you."
 (d) 011010000110100010101000 (explain)
28. (c) SDLC
31. (a) 11 bytes
32. (a) 0.03567 (b) 0.3527 (c) 0.5729
 (d) 0.4271

Chapter 14

1. (a) 30,000 m, 98425 ft, 18.64 mi
 (b) 300 m, 984 ft, 0.186 mi
 (c) 3 m, 9.84 ft
3. (a) $708.8 - j47.1 \ \Omega$ (b) 398 Hz
6. 6.12
8. (a) −0.5 (b) −0.2 (c) 0
 (d) +0.2 (e) +0.5
9. (a) 5 V (b) 6⅔ V, 66.7 mA
11. (a) 0 (b) $-1/90°$ (d) $0.447/63.5°$
 (f) $0.368/-163°$
13. 18.88 Np, 164 dB
15. $6.25 - j12.5 \ \Omega$
16. $40 + j57 \ \Omega$, 3.3, 0.538
17. (a) 2.5 (b) 75 MHz
18. About 1.5
19. $107.5 + j50 \ \Omega$
20. (a) 6.8 dB (b) 0.84 mW, 3.16 mW
21. (a) 0.126λ
 (b) Inductive reactance of $j44.25 \ \Omega$
 (c) 0.115λ
22. (a) 0.21λ from load (b) $29.6 \ \Omega$
 (c) 0.05 m
23. $S_{11} = -0.143$, $S_{22} = 0.429$, $S_{21} = 0.286$
 $= S_{12}$; R.L. $= 16.9$ dB at input
25. (a) $S_{11} = -0.111$, $S_{22} = -0.444$, $S_{21} =$
 $0.222 = S_{12}$
26. (a) 4 (b) −3 dB (c) 4.44 dB
27. (a) $d_{in} = 2.7$ cm, $l_s = 1.33$ cm, $d_o =$
 1.24 cm, $l_{so} = 0.96$ cm

 (b) $W_{50} = 0.08$ cm
28. 14.8 dB
31. (a) 4.572 cm, 6.562 GHz
 (b) 16.16 GHz
 (c) 5; TE_{10}, TE_{01}, TE_{20}, TE_{11}, TM_{11}
32. (a) 0.755 (b) $500 \ \Omega$ (c) 1.325
 (d) $600 - j118 \ \Omega$ (e) 19 W
33. (a) 1 mW and 9 mW (b) 0.0316 mW
34. (a) 3 (b) 6 dB and $\rho = 0.5$
 (c) $3Z_0$ or $Z_0/3$

Chapter 15

3. 8×10^{-3} A/m
4. 1.58 million miles
6. 1.32 m
8. 164 nW
11. About 3.7 kW
15. (a) About 70° (b) 12 dB
18. (a) 2 m (b) 5.2°
20. (a) 0.414 m
 (b) $G = 39.38$ dBi, $D = 42$ dBi
21. 65.3 dBi
23. 2.716×10^8 m/s
24. 35.3°
26. 26.3 ft
27. (b2) 1.53 km
28. (a) +70 dBm (b) 197 dB (c) 65 dB
 (d) −62 dBm (e) 33 dB

Chapter 16

1. 193.25 MHz, 197.75 MHz
3. (a) 15.625 kHz (b) 302,400
4. (b) 480.47
6. (a) 10.4 mA ($V_{BE} = 0.7$ V) (b) 2965
 (c) 51 V dc (d) 9.95 MΩ
8. 94 μH
10. (a) 1.875 A (b) −1.875 V
 (c) Almost rectangular waveform
 (d) (a) 0.971 A, (b) −4.855 V, (c) See
 Figure ANS 16–10d.
12. 1500
14. (a) $\omega_1 = 1.43$ rad/s and $\omega_2 = 137.7$ rad/s
 (b) 22.9 k rad/s
 (c) 14.529 kHz to 16.776 kHz

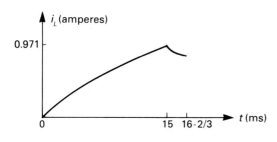

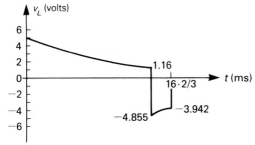

FIGURE ANS 16-10d

17. **(a)** 57°
(b) Approximately $I = -0.44$, $Q = -0.87$

Chapter 17

3. **(a)** 011011000001 **(b)** 0ππ0ππ · · ·
4. 101001011110
5. **(b)** −135, −135, 45, . . . degrees.

(c1) $2.83 \cos\left[10^4 t + \dfrac{\pi}{2}(1 + d_1(t))\right]$

is one-half of it.
(c2) 150 Hz
9. **(a)** (00) = −135°, (01) = 135°, (10) = −45°, (11) = 45°
(b) (00) = −45°, (01) = 45°, (10) = −135°, (11) = 135°. The only sure way is to sketch the circuit and trace through.

10. **(a)** 2 **(c)** 2 **(e)** 6
12. **(a)** 101, 001, 011, 110
(b) −90, 45, 90, 180°
14. **(a)** 111 W **(b)** 2.56 dB
16. **(a)** 8-QAM is 16 times better
(b) 5 for 8-QAM and 80 for 8-PSK
18. 12.59
19. Coherent PSK is about 5 times better
22. 0.9 m
23. **(a)** 28.8 MW **(b)** −62.6 dBm
25. 36.3 dB/K
26. **(a)** 55.5 dBW **(b)** 0.9° **(c)** 33 dB/K
(d) 31.7 dB **(e)** 3.7 GHz
28. **(a)** −146 dBm/Hz **(b)** 22.55 dB
29. **(b)** 11110101100100001110 **(d)** 15
(e,b) 111100111100111110011
30. **(a)** 250 **(b)** 8 **(c)** Assume slow-FH

Chapter 18

2. **(a)** 1.935×10^{14} Hz **(b)** 0.8 eV
(c) 2.56 pW
4. **(a)** 76.7° **(b)** 0.34
5. **(a)** 4% **(b)** 1.487
6. **(a)** 68.8° **(b)** 0.587 **(c)** 35.9°
8. 3; name them.
9. **(a)** 2.92 μm
11. 0.08 ns/km
12. **(a)** 111.2 kb/s
14. **(a)** 15.4 dB, −12.4 dBm
(b) 9.3 dB, −6.3 dBm
17. **(a)** 1875 eV **(b)** 1962
18. **(a)** $R = 4$ MΩ **(b)** 0.2 V
20. 0.74 (1.3 dB)
23. **(a)** 150×10^{-18} J **(b)** 981 photons
24. **(a)** Last, 17 dBm **(b)** 22.3 nA
25. **(a)** Last, −25.6 dBm **(b)** 159.1 nA

Index

ISBN 0-02-431201-0

9 780024 312013

90000>